AF410823

Bioactive Molecules from Extreme Environments II

Bioactive Molecules from Extreme Environments II

Editor

Daniela Giordano

MDPI • Basel • Beijing • Wuhan • Barcelona • Belgrade • Manchester • Tokyo • Cluj • Tianjin

Editor
Daniela Giordano
Consiglio Nazionale delle Ricerche (CNR)
Italy

Editorial Office
MDPI
St. Alban-Anlage 66
4052 Basel, Switzerland

This is a reprint of articles from the Special Issue published online in the open access journal *Marine Drugs* (ISSN 1660-3397) (available at: http://www.mdpi.com).

For citation purposes, cite each article independently as indicated on the article page online and as indicated below:

LastName, A.A.; LastName, B.B.; LastName, C.C. Article Title. *Journal Name* **Year**, *Volume Number*, Page Range.

ISBN 978-3-0365-2718-5 (Hbk)
ISBN 978-3-0365-2719-2 (PDF)

Contents

About the Editor

Daniela Giordano, Ph.D., is a CNR (National Research Council) Scientist in Biochemistry. She obtained a permanent position at CNR in 2011, working first at the Institute of Protein Biochemistry (IBP-CNR) and, since 2014, at the Institute of Biosciences and BioResources (IBBR-CNR) in Naples. She graduated in Pharmaceutical Chemistry at the University of Naples Federico II in 2002 and obtained a Ph.D. at the University of Sacro Cuore in Rome in 2007, spending a period of six months at Northeastern University in Boston. After receiving her Ph.D., she began her research experience with fellowships at the Institute of Protein Biochemistry (IBP-CNR) in Naples, focusing her research on the molecular basis of cold adaptation of oxygen-binding proteins in polar bacteria and fish. She spent many periods as a visiting scientist at the National Institute of Health and Medical Research (INSERM), Paris, the University of Antwerp, and the University of Buenos Aires, studying the recombinant expression of globins of Antarctic bacteria and fish and their kinetic and thermodynamic properties. In 2010, she participated in an Arctic cruise, the TUNU-IV expedition (TUNU is East Greenland in Inuit language), for the collection of Arctic samples. Her research interest is focused on Antarctic and Arctic marine organisms because they are amongst the most vulnerable species to climate change and are a valuable source of natural products that can function as start structures of new molecules for drug discovery. Indeed, she is now leading a new research line on the identification of bioactive compounds as promising industrial products (pharmacy, nutraceutics, cosmeceuticals) from marine polar organisms. The results of her research are summarized in over 60 publications on highly qualified international journals and book chapters.

 *marine drugs*

Editorial

Bioactive Molecules from Extreme Environments II

Daniela Giordano [1,2]

1   Institute of Biosciences and BioResources (IBBR), CNR, Via Pietro Castellino 111, 80131 Napoli, Italy; daniela.giordano@ibbr.cnr.it
2   Department of Marine Biotechnology, Stazione Zoologica Anton Dohrn (SZN), Villa Comunale, 80121 Napoli, Italy

Citation: Giordano, D. Bioactive Molecules from Extreme Environments II. *Mar. Drugs* **2021**, *19*, 642. https://doi.org/10.3390/md19110642

Received: 4 November 2021
Accepted: 10 November 2021
Published: 17 November 2021

Marine organisms are known to produce a wide variety of natural products that are unique in terms of diversity, structural, and functional properties [1].

Marine organisms living in extreme environments experience conditions close to the limit of life, e.g., polar and hot regions, deep sea, hydrothermal vents, marine areas of high pressure or high salinity. They have evolved unique strategies for surviving in these harsh conditions, as well as biosynthesizing novel bioactive compounds, which are potentially useful for pharmaceutical, cosmeceutical, nutraceutical, and biotechnological applications [2]. Research on extreme environments often requires complex and expensive infrastructure, as well as access. Recent technological developments have made these areas more accessible and, currently, research on extreme life is growing fast. However, the biodiversity in these hostile environments is still largely unknown, and it is expected that in the near future, further research will be dedicated to this field.

This Special Issue, as a continuation of the previous Special Issue, "Bioactive Molecules from Extreme Environments" (https://www.mdpi.com/journal/marinedrugs/special_issues/Extreme_Environments accessed on 4 November 2021), includes 10 research articles and 2 reviews, providing a wide overview of the chemical biodiversity offered by different marine organisms inhabiting extreme environments to be used for biotechnological and pharmaceutical applications. The six articles in this Special Issue are focused on the polar regions, which represent an untapped source of marine natural products and are still largely unexplored compared to more accessible sites. Many of these articles refer to Antarctica, which is the coldest and most inaccessible continent on the Earth, where extreme temperatures, light and ice have selected biological communities with a unique suite of bioactive metabolites. The marine organisms of Arctic and Antarctic environments are a reservoir of natural compounds, exhibiting huge structural diversity and significant bioactivities that could be used in human applications.

In Núñez-Pons et al. [3] authors firstly described the regulations on access and benefit sharing requirements for research in polar environments and then provided an overview of the molecules from Antarctic and Arctic marine organisms with promising biological activities. The main target of this review was to investigate bacteria and fungi as microbes and macroorganisms, such as cnidaria, bryozoa, mollusca, echinodermata, sponges, tunicates and macroalgae. The attention was focused on purified marine natural products with elucidated structures showing a biological activity against human pathogens. Terpenes, terpenoids and derivatives are the most commonly found compounds, whereas antimicrobial properties are the most widely reported activities from these environments. This survey was aimed at highlighting the chemical diversity of marine polar life and the versatility of a particular group of biomolecules with interesting biological activities.

Avila and Angulo-Preckler [4] wrote a comprehensive survey on bioactive compounds from heterobranch molluscs, which was very well documented by 876 references. In this paper, the authors analyzed the bioactivity of more than 450 compounds from ca. 400 species of heterobranch molluscs, many of which are from Antarctica. Molecules produced by molluscs, display ecological activities such as predator avoidance, toxicity, antifouling, trail-following and alarm pheromones, sunscreens, UV protection and tissue regeneration, whereas many other compounds display pharmacological activities. The most studied activities were cytotoxicity and anticancer and antibiotic activities.

As molluscs, sponges are known to produce a series of marine natural products with interesting biological activities that could be applied to biomedical applications. Riccio et al. [5] reported a bioassay-guided fractionation of four Antarctic sponges, *Mycale (Oxymycale) acerata*, *Haliclona (Rhizoniera) dancoi*, *Hemimycale topsenti* and *Hemigellius pilosus* that led to the identification of two different chemical classes of molecules, suberitenone A and B and mycalols. The fraction containing the marine mycalol and its analogues, which have already been reported to possess anticancer activity on anaplastic thyroid carcinoma cells [6,7], was identified as the most promising bioactive product. In fact, investigation at the gene and protein levels demonstrated that it may trigger ferroptosis in HepG2 cells.

The majority of marine bioactive compounds derive from microorganisms as an invaluable source for novel chemistry and sustainable production of bioactive compounds, bypassing the limit for the re-collection of marine resources using destructive practices [8]. However, many bioactive molecules obtained from host invertebrates are instead produced by their bacterial symbionts in the marine environment, highlighting their crucial roles in host-associated chemical defense [9]. In Sun et al. [10], spectroscopic and chemical analyses elucidated the structure of three new aspochracin-type cyclic tripeptides, sclerotiotides M–O, together with three known analogues, sclerotiotide L, sclerotiotide F and sclerotiotide B, obtained from the ethyl acetate extract of the fungus *Aspergillus insulicola* HDN151418, which was isolated from an unidentified Antarctic sponge. Sclerotiotides M and N showed antimicrobial activity against a panel of pathogenic strains.

Di Lorenzo et al. [11] reported the structural characterization of the lipopolysaccharide's glycolipid moiety, the lipid A, from three different psychrophilic bacteria belonging to the phylum of Proteobacteria isolated from Terra Nova Bay, Antarctica. Lipopolysaccharides are amphiphilic molecules exposed on the outer membrane of the Gram-negative bacteria that are essential for viability and survival. Under hostile conditions, lipopolysaccharides can undergo desaturation of the acyl chains, a reduction in length and an increase in their branching to provide further protection and facilitate adaptation. In cold-adapted bacteria, lipopolysaccharides display several uncommon structural features. Since lipid A is also involved in the innate immune response in mammals, the study of lipid A from cold-adapted bacteria is of great interest to understand the mechanism of cold adaptation for drug synthesis.

John et al. [12] described the production of copper nanoparticles using the green reduction of $CuSO_4$ at low temperatures (22 °C) by using five Antarctic bacterial strains isolated from a consortium associated with the Antarctic ciliate *Euplotes focardii*. All copper nanoparticles display antimicrobial activity against various types of Gram-negative and Gram-positive bacteria and fungi pathogen microorganisms, including *Escherichia coli*, *Staphylococcus aureus*, and *Candida albicans*. The ability of these bacteria to synthesize copper nanoparticles may represent a mechanism of defense against this heavy metal. In fact, Antarctica is not completely free from contaminants that could reach the Southern Ocean via long-range atmospheric transport from other continents. Metal nanoparticle synthesis using microorganisms is an eco-friendly and sustainable strategy alternative to chemical and physical approaches, which is particularly promising in the bioremediation of the contaminated environment and in the production of antibiotics against various types of pathogenic microorganisms. Most research work in this field has been carried out on silver and gold nanoparticles. Another work on silver nanoparticles from Antarctic bacteria is given in the first book [13].

Du et al. [14] described the discovery and structural elucidation using spectroscopic analyses of two new secondary metabolites pyrrolidinone-bearing lipodipeptides, svalbamides A and B, identified in *Paenibacillus* sp. SVB7, which was isolated from the Svalbard archipelago in the Arctic Ocean. Svalbamides A and B are structurally unique as they contain 3-amino-2- pyrrolidinone amino acid, which is rarely reported in natural products, and 3-hydroxy-8-methyldecanoic acid was occasionally found in natural products from *Paenibacillus* and related bacteria. Svalbamides A and B may function as potential chemo-preventive agents, as they induced quinone reductase activity in murine hepatoma cells.

In addition to the polar regions, extreme are also environments of deep sea, hydrothermal vents, hot/arid regions, etc and other articles review these environments in this Special Issue.

Juhasz et al. [15] reported the structural elucidation of three dermacozines, dermacozines N–P, isolated from the piezotolerant actinomycete strain *Dermacoccus abyssi* MT 1.1^T, from a Mariana Trench sediment in 2006, which was collected at a depth of 10,898 m from the Challenger Deep by the remotely operated submersible Kaiko in 1998. In the past, seven highly colored dermacozines A–G and 4 derivatives dermacozines H–J were isolated from this promising strain [16–18]. Dermacozine N is unique among phenoxazines because it bears a novel linear pentacyclic phenoxazine framework, is never reported as a natural product, and displays near-infrared absorption maxima, making it an excellent candidate for research in biosensing chemistry, photodynamic therapy, and opto-electronic applications. Moreover, dermacozine N possesses weak cytotoxic activity against melanoma (A2058) and hepatocellular carcinoma cells (HepG2), with IC50 values of 51 and 38 mM, respectively.

Singh and et al. [19] described *Bacillus amyloliquefaciens*, a thermotolerant marine strain S185 isolated from offshore the South Sea, China. This marine strain displayed a strong antifungal activity against *Fusarium oxysporum* f. sp. cubense (Foc), which is responsible for a severe fungal disease in banana plants (Panama disease), due to its capacity to produce the antifungal compound iturin A5. This strain is able to grow between 20–50 °C, adapting to variable conditions of pH, salts and temperature. Due to these features, it is a promising candidate for a cost-effective and sustainable biocontrol application for Panama disease in the future.

Xing et al. [20] reported the isolation, structure elucidation, and biological activity of 10 new and 26 known compounds isolated from *Penicillium griseofulvum* MCCC 3A00225, a deep sea-derived fungus isolated from the Indian Ocean sediment. The *Penicillium* species is recognized as the richest source for biologically important and structurally unique secondary metabolites. All isolates were tested for in vitro anti-food allergic bioactivities in immunoglobulin (Ig) E-mediated rat basophilic leukemia (RBL)-2H3 cells. One of these compounds, (-)-cyclopenol, significantly decreased the degranulation release with an IC50 value of 60.3 μM, compared to that of 91.6 μM of the positive control, loratadine.

Ruiz-Domínguez et al. [21] studied the effect of two different drying methods for the recovery of lutein extracted by the supercritical fluid extraction process from the microalgae *Muriellopsis* sp (MCH35), which is isolated from an arid region of the north of Chile, Antofagasta, where it is exposed to high solar radiation. Among microorganisms as a prolific source of bioactive molecules, microalgae are the most diversified photosynthetic organisms with high adaptability to various environmental conditions [22]. Microalgae are rich with bioactive molecules with healthy benefits and have potential applications in pharmacological, nutraceutical, cosmeceutical, and biotechnological sectors. Lutein is a carotenoid belonging to the class of terpenoids, and due to its antioxidant potential, it displays a beneficial role to human health in ameliorate cardiovascular diseases [23], various types of cancer [24], and age-related macular degeneration [25]. Since several strategies have been studied to enhance microalgal carotenoid production, *Muriellopsis* sp. (MCH35) represents a potential candidate for lutein production under intense UV irradiation for biotechnological applications.

Wang et al. [26] reported the identification and characterization of a crustin from the shrimp *Rimicaris* sp. inhabiting the deep-sea hydrothermal vent in Manus Basin (Papua New Guinea). Shrimps belonging to the family Alvinocarididae are particularly abundant in the deep waters of Atlantic, Pacific, and Indian Oceans, especially in hydrothermal vents and cold seeps. Crustin, a cationic peptide of 7–22 kDa, belongs to the class of antimicrobial peptides (AMPs), a class of evolutionarily conserved molecules that exist in almost all organisms and play an important role in the innate immunity of organisms. They directly kill bacteria, fungi, viruses, and parasites, targeting the inner and/or outer membranes of microorganisms in a non-receptor-specific manner, with a rate of resistance several orders of magnitude lower than that of conventional antibiotics.

As Guest Editor of the second edition of this Special Issue, I dedicate this Issue to the memory of Prof. Guido di Prisco, my mentor, who passed away in September 2019. He spent his life studying the structure, function, and evolution of the hemoglobins of polar fishes and inspired my interest in life in Antarctic and Arctic environments. I am also grateful to all the authors who contributed to making this an exceptional Special Issue with new discoveries/research on bioactive molecules from extreme conditions. The papers included in this second Special Issue, as well as in the first Special Issue, highlight the increasing interest in these extreme habitats, which are perceived as important sources for drug discovery. I want to thank all the reviewers and *Marine Drugs* for their support and kind help.

I hope that this collection will provide a unique and valuable reference source for researchers interested in extreme environments and help the scientific community inspire the next generation to further research dedicated to this field.

Funding: This study was funded by the Italian National Programme for Antarctic Research (PNRA) (2016/AZ1.06-Project PNRA16_00043 and 2016/AZ1.20-Project PNRA16_00128).

Institutional Review Board Statement: Not applicable.

Informed Consent Statement: Not applicable.

Data Availability Statement: Not applicable.

Conflicts of Interest: The author declares no conflict of interest.

References

1. Carroll, A.R.; Copp, B.R.; Davis, R.A.; Keyzers, R.A.; Prinsep, M.R. Marine natural products. *Nat. Prod. Rep.* **2021**, *38*, 362–413. [CrossRef] [PubMed]
2. Jaspars, M.; De Pascale, D.; Andersen, J.H.; Reyes, F.; Crawford, A.D.; Ianora, A. The marine biodiscovery pipeline and ocean medicines of tomorrow. *J. Mar. Biol. Assoc. UK* **2016**, *96*, 151–158. [CrossRef]
3. Núñez-Pons, L.; Shilling, A.; Verde, C.; Baker, B.; Giordano, D. Marine Terpenoids from Polar Latitudes and Their Potential Applications in Biotechnology. *Mar. Drugs* **2020**, *18*, 401. [CrossRef] [PubMed]
4. Avila, C.; Angulo-Preckler, C. Bioactive Compounds from Marine Heterobranchs. *Mar. Drugs* **2020**, *18*, 657. [CrossRef]
5. Riccio, G.; Nuzzo, G.; Zazo, G.; Coppola, D.; Senese, G.; Romano, L.; Costantini, M.; Ruocco, N.; Bertolino, M.; Fontana, A.; et al. Bioactivity Screening of Antarctic Sponges Reveals Anticancer Activity and Potential Cell Death via Ferroptosis by Mycalols. *Mar. Drugs* **2021**, *19*, 459. [CrossRef]
6. Cutignano, A.; Nuzzo, G.; D'Angelo, D.; Borbone, E.; Fusco, A.; Fontana, A. Mycalol: A natural lipid with promising cytotoxic properties against human anaplastic thyroid carcinoma cells. *Angew. Chem. Int. Ed.* **2013**, *52*, 9256–9260. [CrossRef]
7. Cutignano, A.; Seetharamsingh, B.; D'Angelo, D.; Nuzzo, G.; Khairnar, P.V.; Fusco, A.; Reddy, D.S.; Fontana, A. Identification and Synthesis of Mycalol Analogues with Improved Potency against Anaplastic Thyroid Carcinoma Cell Lines. *J. Nat. Prod.* **2017**, *80*, 1125–1133. [CrossRef] [PubMed]
8. Soldatou, S.; Baker, B.J. Cold-water marine natural products, 2006 to 2016. *Nat. Prod. Rep.* **2017**, *34*, 585–626. [CrossRef] [PubMed]
9. Blockley, A.; Elliott, D.; Roberts, A.P.; Sweet, M. Symbiotic Microbes from Marine Invertebrates: Driving a New Era of Natural Product Drug Discovery, Divers Distrib. *Diversity* **2017**, *9*, 49. [CrossRef]
10. Sun, C.; Zhang, Z.; Ren, Z.; Yu, L.; Zhou, H.; Han, Y.; Shah, M.; Che, Q.; Zhang, G.; Li, D.; et al. Antibacterial Cyclic Tripeptides from Antarctica-Sponge-Derived Fungus Aspergillus insulicola HDN151418. *Mar. Drugs* **2020**, *18*, 532. [CrossRef]
11. Di Lorenzo, F.; Crisafi, F.; La Cono, V.; Yakimov, M.; Molinaro, A.; Silipo, A. The Structure of the Lipid A of Gram-Negative Cold-Adapted Bacteria Isolated from Antarctic Environments. *Mar. Drugs* **2020**, *18*, 592. [CrossRef]
12. John, M.; Nagoth, J.; Zannotti, M.; Giovannetti, R.; Mancini, A.; Ramasamy, K.; Miceli, C.; Pucciarelli, S. Biogenic Synthesis of Copper Nanoparticles Using Bacterial Strains Isolated from an Antarctic Consortium Associated to a Psychrophilic Marine Ciliate: Characterization and Potential Application as Antimicrobial Agents. *Mar. Drugs* **2021**, *19*, 263. [CrossRef] [PubMed]
13. John, M.S.; Nagoth, J.A.; Ramasamy, K.P.; Mancini, A.; Giuli, G.; Natalello, A.; Ballarini, P.; Miceli, C.; Pucciarelli, S. Synthesis of Bioactive Silver Nanoparticles by a *Pseudomonas* Strain Associated with the Antarctic Psychrophilic Protozoon *Euplotes focardii*. *Mar. Drugs* **2020**, *18*, 38. [CrossRef] [PubMed]
14. Du, Y.; Bae, E.; Lim, Y.; Cho, J.; Nam, S.; Shin, J.; Lee, S.; Nam, S.; Oh, D. Svalbamides A and B, Pyrrolidinone-Bearing Lipodipeptides from *Arctic Paenibacillus* sp. *Mar. Drugs* **2021**, *19*, 229. [CrossRef] [PubMed]

15. Juhasz, B.; Pech-Puch, D.; Tabudravu, J.; Cautain, B.; Reyes, F.; Jiménez, C.; Kyeremeh, K.; Jaspars, M. Dermacozine N, the First Natural Linear Pentacyclic Oxazinophenazine with UV–Vis Absorption Maxima in the Near Infrared Region, along with Dermacozines O and P Isolated from the Mariana Trench Sediment Strain Dermacoccus abyssi MT 1.1T. *Mar. Drugs* **2021**, *19*, 325. [CrossRef] [PubMed]

16. Abdel-Mageed, W.M.; Milne, B.F.; Wagner, M.; Schumacher, M.; Sandor, P.; Pathom-Aree, W.; Goodfellow, M.; Bull, A.T.; Horikoshi, K.; Ebel, R.; et al. Dermacozines, a new phenazine family from deep-sea dermacocci isolated from a Mariana Trench sediment. *Org. Biomol. Chem.* **2010**, *8*, 2352–2362. [CrossRef]

17. Wagner, M.; Abdel-Mageed, W.M.; Ebel, R.; Bull, A.T.; Goodfellow, M.; Fiedler, H.-P.; Jaspars, M. Dermacozines H–J Isolated from a Deep-Sea Strain of Dermacoccus abyssi from Mariana Trench Sediments. *J. Nat. Prod.* **2014**, *77*, 416–420. [CrossRef] [PubMed]

18. Abdel-Mageed, W.M.; Juhasz, B.; Lehri, B.; Alqahtani, A.S.; Nouioui, I.; Pech-Puch, D.; Tabudravu, J.N.; Goodfellow, M.; Rodríguez, J.; Jaspars, M.; et al. Whole Genome Sequence of Dermacoccus abyssi MT1.1 Isolated from the Challenger Deep of the Mariana Trench Reveals Phenazine. *Mar. Drugs* **2020**, *18*, 131. [CrossRef]

19. Singh, P.; Xie, J.; Qi, Y.; Qin, Q.; Jin, C.; Wang, B.; Fang, W. A Thermotolerant Marine Bacillus amyloliquefaciens S185 Producing Iturin A5 for Antifungal Activity against Fusarium oxysporum f. sp. cubense. *Mar. Drugs* **2021**, *19*, 516. [CrossRef]

20. Xing, C.; Chen, D.; Xie, C.; Liu, Q.; Zhong, T.; Shao, Z.; Liu, G.; Luo, L.; Yang, X. Anti-Food Allergic Compounds from Penicillium griseofulvum MCCC 3A00225, a Deep-Sea-Derived Fungus. *Mar. Drugs* **2021**, *19*, 224. [CrossRef]

21. Ruiz-Domínguez, M.; Marticorena, P.; Sepúlveda, C.; Salinas, F.; Cerezal, P.; Riquelme, C. Effect of Drying Methods on Lutein Content and Recovery by Supercritical Extraction from the Microalga Muriellopsis sp. (MCH35) Cultivated in the Arid North of Chile. *Mar. Drugs* **2020**, *18*, 528. [CrossRef] [PubMed]

22. Forján, E.; Navarro, F.; Cuaresma, M.; Vaquero, I.; Ruíz-Domínguez, M.C.; Gojkovic, Ž.; Vázquez, M.; Márquez, M.; Mogedas, B.; Bermejo, E. Microalgae: Fast-growth sustainable green factories. *Crit. Rev. Environ. Sci. Technol.* **2015**, *45*, 1705–1755. [CrossRef]

23. Dwyer, J.H.; Navab, M.; Dwyer, K.M.; Hassan, K.; Sun, P.; Shircore, A.; Hama-Levy, S.; Hough, G.; Wang, X.; Drake, T. Oxygenated carotenoid lutein and progression of early atherosclerosis: The Los Angeles atherosclerosis study. *Circulation* **2001**, *103*, 2922–2927. [CrossRef] [PubMed]

24. Heber, D.; Lu, Q.-Y. Overview of mechanisms of action of lycopene. *Exp. Biol. Med.* **2002**, *227*, 920–923. [CrossRef]

25. Landrum, J.T.; Bone, R.A. Lutein, zeaxanthin, and the macular pigment. *Arch. Biochem. Biophys.* **2001**, *385*, 28–40. [CrossRef]

26. Wang, Y.; Zhang, J.; Sun, Y.; Sun, L. A Crustin from Hydrothermal Vent Shrimp: Antimicrobial Activity and Mechanism. *Mar. Drugs* **2021**, *19*, 176. [CrossRef]

Review

Marine Terpenoids from Polar Latitudes and Their Potential Applications in Biotechnology

Laura Núñez-Pons [1], Andrew Shilling [2], Cinzia Verde [3,4], Bill J. Baker [2,*] and Daniela Giordano [3,4,*]

[1] Department of Integrated Marine Ecology (EMI), Stazione Zoologica Anton Dohrn (SZN), Villa Comunale, 80121 Napoli, Italy; laura.nunezpons@szn.it
[2] Department of Chemistry, University of South Florida, Tampa, FL 33620, USA; ashillin@mail.usf.edu
[3] Institute of Biosciences and BioResources (IBBR), CNR, Via Pietro Castellino 111, 80131 Napoli, Italy; cinzia.verde@ibbr.cnr.it
[4] Department of Marine Biotechnology, Stazione Zoologica Anton Dohrn (SZN), Villa Comunale, 80121 Napoli, Italy
* Correspondence: bjbaker@usf.edu (B.J.B.); daniela.giordano@ibbr.cnr.it (D.G.)

Received: 25 June 2020; Accepted: 25 July 2020; Published: 29 July 2020

Abstract: Polar marine biota have adapted to thrive under one of the ocean's most inhospitable scenarios, where extremes of temperature, light photoperiod and ice disturbance, along with ecological interactions, have selected species with a unique suite of secondary metabolites. Organisms of Arctic and Antarctic oceans are prolific sources of natural products, exhibiting wide structural diversity and remarkable bioactivities for human applications. Chemical skeletons belonging to terpene families are the most commonly found compounds, whereas cytotoxic antimicrobial properties, the capacity to prevent infections, are the most widely reported activities from these environments. This review firstly summarizes the regulations on access and benefit sharing requirements for research in polar environments. Then it provides an overview of the natural product arsenal from Antarctic and Arctic marine organisms that displays promising uses for fighting human disease. Microbes, such as bacteria and fungi, and macroorganisms, such as sponges, macroalgae, ascidians, corals, bryozoans, echinoderms and mollusks, are the main focus of this review. The biological origin, the structure of terpenes and terpenoids, derivatives and their biotechnological potential are described. This survey aims to highlight the chemical diversity of marine polar life and the versatility of this group of biomolecules, in an effort to encourage further research in drug discovery.

Keywords: Arctic/Antarctic; marine bioprospecting; marine natural product; terpene; terpenoid; biotechnological application; drug discovery

---

1. Foreword

The global market for marine biotechnology is expected to reach US\$ \$6.4 billion by 2025 [1]. However, in regard to the polar regions, marine biotechnology has lagged and the potential for many sectors is not yet fully realized. These remote and underexploited habitats are promising sources of environmental and biomedical applications and may provide significant opportunities for compound discovery and bioprospecting [2–6].

The Arctic Ocean (~4.3% of Earth's ocean area) and the meridional part of the Southern Ocean (south of the Antarctic Convergence, ~6.1% of Earth's ocean area) comprise the Polar Seas [7]. In the coldest years, the sea ice cover in these areas can reach up to ~13% of Earth's total surface. The Arctic Ocean is delineated by the International Hydrographic Organization (https://iho.int) as the body of water north of 57° from Earth's equatorial plane. It is mainly characterized by light seasonality and cold temperatures with winter extremes, and comprises a vast ocean accessible to an influx of warm water from the Atlantic and the Pacific Oceans [8]. The Antarctic Ocean encompasses the body of water between 60° South latitude and the continent of Antarctica (https://iho.int). Known as being the driest, windiest and coldest place on Earth, the white continent is isolated from the other continents, geographically and thermally by the Antarctic Circumpolar Current. Its northern border, the Antarctic Polar Front, represents a barrier for latitudinal migration of marine organisms, which renders the local biota as highly endemic [9–11].

Some evidence indicates that the events that generated the confinement of Antarctica occurred about 34 million years ago around the Eocene/Oligocene limit [12]. Polar marine life has evolved to thrive at one of life's extreme limits, with liquid water that cannot get any colder and long periods of darkness, which has selected for species with unique adaptations, including strategies based on often unique secondary metabolites [13,14].

Indeed, the distinctive environment and ecological pressures faced by marine wildlife in polar regions are major drivers of secondary metabolite production and the emergence of compounds with diverse biological activities [15]. Competition for space and food, predator avoidance, pathogen threat and, in general, intra-species chemical communication, promote chemical diversity in marine environments [16]. In addition, marine polar organisms require sophisticated biochemical and physiological adaptations to cope with low temperatures, strong winds, arrested nutrient availability and high UV radiation [6]. Therefore, under extreme polar scenarios, bioactive compounds in the form of secondary metabolites are of major relevance for the survival of marine organisms [13].

Although compounds with varied biotechnological properties have been identified in polar organisms, the most commonly reported activity is toward pathogenic infections. Discovering novel and efficient antimicrobial metabolites is imperative, due to the increasing emergence of antibiotic-resistant or even multidrug-resistant strains of microorganisms [17].

In this review, we first expose the framework regarding the regulations that govern access and benefit sharing requirements in polar environments, and secondly, we report bioactive compounds isolated from marine bacteria, fungi and macroorganisms inhabiting Arctic and Antarctic environments. In particular, the present survey focuses on marine terpenes and their derivatives harboring properties with potential impact on pharmacological remedies towards human diseases.

2. Marine Biotechnology: Governance, Access and Benefit Sharing in the Antarctic and Arctic Environments

Marine natural products have attracted growing commercial interest from the biotech sector, supporting a rich marine biodiscovery pipeline [18]. The global market for marine-derived pharmaceuticals is around US$5 billion [19]. However, despite successful discoveries, few drugs isolated from marine organisms have been approved by the US Food and Drug Administration (FDA) [19–21], e.g., anticancer drugs such as cytarabine (Cytosar-U®®) from the Caribbean sponge *Tectitethya crypta* (=*Cryptotethya crypta*) [22], trabectedin (Yondelis®®) from the ascidian *Ecteinascidia turbinata* [23], eribulin (Halaven®®) from the sponge *Halichondria okadai*, the antiviral vidarabine from the sponge *Tethya crypta* and the analgesic ziconotide (Prialt®®) from the sea snail *Conus magus* [22].

Technological advances are swiftly reducing sequencing costs as demonstrated by the exponential growth of data freely available of public repositories (INSDC, comprising GenBank, EMBL-EBI/ENA and DDBJ) [24]. Nevertheless, there is an urgent need for services able to handle and harmonize the information obtained by heterogeneous standards. In addition, stakeholders, academia, industry, funding agencies, infrastructure providers and scholarly publishers should promote the FAIR (Findable, Accessible, Interoperable and Reusable) concepts. As yet, the Nagoya Protocol on Access to Genetic Resources and the Fair and Equitable Sharing of Benefits Arising from their Utilization of 2010 ("Nagoya Protocol") does not apply to some areas of the ocean [25].

The Convention on Biological Diversity (CBD) [19] defines bioprospecting as the exploration and development of knowledge of genetic and biochemical resources for commercial purposes.

From the perspective of protecting biodiversity, the CBD and the Nagoya Protocol are the most important agreements. The Nagoya Protocol aims to support and supplement domestic legislation by defining the genetic resources and the contractual obligations to be written in mutually agreed terms. The CBD introduced novel legal concepts such as biodiversity, ecosystems, genetic resources, biotechnology, benefit sharing and traditional knowledge. Applying the CBD in Antarctica is no simple matter and still requires more explicit regulation [26–28].

Antarctica represents a space exclusively dedicated to international cooperation where states are not able to exercise sovereign rights. Antarctica is governed by its own regional regime, the Antarctic Treaty System (ATS). Essential to the ATS is the Antarctic Treaty (AT, 1959), which applies to the area south of 60° South latitude. The ATS is composed of five main treaties: the AT, the Convention for the Conservation of Antarctic Seals (1971), the Convention on the Conservation of Antarctic Marine Living Resources (CCAMLR 1980), the Convention on the Regulation of Antarctic Mineral Resource Activities (1988) and the Madrid Protocol amendment to the AT (1991). Article III of the AT requires that parties cooperate by sharing information on research results. Article II of the AT acknowledges the importance of the freedom of scientific investigation [29,30].

Bioprospecting is a rapidly accelerating activity in Antarctica with new players such as Malaysia with strong commercial interests in Antarctic organisms [31,32]. Terrestrial and marine biological material from Antarctica can be accessed either by collecting specimens from the field or from ex situ collections of Antarctic Biological Material (ABM) held in institutions around the world [26]. Bioprospecting was first discussed within the ATS in 1999. Since then it has received regular attention at meetings of the Scientific Committee on Antarctic Research (SCAR) [33], Committee for Environmental Protection (CEP) and the Antarctic Treaty Consultative Meeting (ATCM). Despite being a consistent topic at every ATCM since 2002, there are still many gaps to manage bioprospecting expectations and activities in Antarctica. CCAMLR may regulate travel to Antarctica but cannot legislate commercial profit from bioprospecting.

Antarctica is a hot spot for bioprospecting. A recent study reported patents for 439 Antarctic species that illustrate the great diversity of potential applications from marine cold-adapted organisms [34].

Bioprospecting is also developing in the Arctic, linked to regional features and the role of enterprises both large and small. Five nations, Canada, Denmark, Norway, Russia and the United States, are currently seeking to increase their territory in the Arctic. Companies from North America, Norway, Iceland, Finland, Sweden, Denmark and the United Kingdom are actively developing new biotechnologies based on genetic resources found in the Arctic [35]. Biotechnology research in the Arctic is mainly focused on industrial processes, food technology, pollution control technologies, pharmaceutical and medical products and health related advancements. More than 40 companies are now involved in bioprospecting in the Arctic largely attracted by the unexplored space and unusual biology of Arctic organisms that promise to open new opportunities in applied research [36]. In 2008, Leary [37] noted the recent rise in the number of patents derived from Arctic genetic resources, though no new reviews have appeared in the intervening 10+ years. Norway is leading projects in marine Arctic bioprospecting mostly focused on the Arctic and sub-Arctic waters. In March 2007, the Research Council of Norway launched MabCent-SFI, a marine bioprospecting project to discover and develop high-value, marine-based products for antibacterial, anticancer, anti-inflammatory products, antioxidants and diabetes medicine, as well as novel cold-adapted enzymes.

3. Marine Compounds from Polar Regions

In 2018, 1554 new natural products from polar marine sources were reported, revealing a number of pharmacologically interesting activities [38]. Of the more than 30,000 publications on marine natural products that have been produced, less than 3% were focused on studies from polar environments, largely due to the difficult accessibility and logistics and associated costs [5,16]. The majority of these cold-water natural products up to 2005 came from marine macroorganisms (e.g., sponges, ascidians, mollusks and corals). After 2005, thanks to advances in sampling methodologies and increased access to previously inaccessible areas, natural products isolated from marine polar microorganisms increased from 22% to 71%. Indeed, microbes have turned into a prolific resource for novel chemistry and the sustainable production of bioactive metabolites, overcoming the problem of recollection of samples from the field, which is required for macroorganisms [5].

Polar microorganisms have been used in many biotechnological applications for the production of bioactive molecules, including enzymes [4] for industrial processes such as food technology and natural compounds with antimicrobial, anti-inflammatory and anticancer activities [5,16]. Moreover, cold-adapted fungi and bacteria are considered a source of (i) antifreeze proteins [39] to be used in the food industry [40–42], and for biomedical purposes [43,44]; (ii) extracellular polymeric substances (EPSs), already used in the textile, cosmetic and food industry [45,46], in the remediation of heavy-metal contaminated environments, including low temperature applications [47], and in the pharmaceutical industry [48]; (iii) polyunsaturated fatty acids (PUFA) as promising alternatives to fish oils in the food industry [39,49,50].

Natural compounds with antimicrobial, anti-inflammatory and anticancer activities have been isolated from marine invertebrates in the Arctic region [3,51–53] and in the marine Antarctic environment [54–57].

Bioactive compounds recovered from host invertebrates may be produced by their microbial symbionts in the marine environment [58,59]. Although the ecological function of bacteria-invertebrate interactions in polar areas is only poorly understood, it seems that microbial metabolites may have crucial roles in host-associated chemical defenses and in shaping the symbiotic community structure [60]. Fungi and bacteria isolated from polar sponges and algae, for instance, show wide spectrum antibacterial activity against many microorganisms [61–69].

Bacteria and fungi, as representative of microbial biodiversity, and macroorganisms, such as sponges, macroalgae, ascidians, corals, bryozoans, echinoderms and mollusks, are the target biological taxa of our review. In Table 1 we list purified marine natural products with elucidated structures isolated from organisms inhabiting the polar environments, focusing mostly on those evaluated for biological activity against human pathogens. We excluded studies evaluating crude extracts or impure fractions for bioactivity, while some relevant ecological activities have been highlighted in the table.

Table 1. Bioactive compounds from Polar marine bacteria, fungi, cnidaria, bryozoa, mollusca, echinodermata, sponges, tunicates and macroalgae.

Species	Collection Site (Distribution)	Compound	Molecule Type	Bioactivity	Reference
		BACTERIA **Plylum Actinobacteria** **Class Actinobacteria**			
Arthrobacter sp.	Marine sediment, Terranova Bay, Ross Sea (Antarctica);	monoramnholipids	rhamnolipid	Antimicrobial activity against *Burkholderia cepacia* complex	[70]
	Surface water, sea ice, zooplankton, the deep sea, and meltwater (Arctic Ocean)	arthrobacilins A–C	cyclic glycolipids	Antimicrobial activity against *Vibrio anguillarum* and *Staphylococcus aureus*	[71]
Nocardia dassonvillei	Marine sediment (Arctic Ocean)	N-(2-hydroxyphenyl)-2-phenazinamine, 1,6-dihydroxyl-phenazine, 6-hydroxyl phenazine-1-carboxylic acid, 6-methoxy-1-phenazinol, 2-amino-1, 4-naphthoquinone, 2-amino-phenoxazin-3-one and 2-(N-methylamino)-3-phenoxazone	phenazine, phenoxazine, naphthoquinone	Antifungal activity against *Candida albicans* and high cancer cell cytotoxicity against HepG2, A549, HCT-116 and COC1 cells	[72]
Nocardiopsis SCSIO KS107	Seashore sediment sample, China Great Wall station (Antarctica)	7-hydroxymucidone, 4-hydroxymucidone, germicidin H	α-pyrones	Antibacterial activity against *Micrococcus luteus* and *Bacillus subtilis*	[73]
Rhodococcus sp. B7740	Deep seawater (Arctic Ocean)	isorenieratene (**116**)	carotenoid	Antioxidant activity	[74,75]
		menaquinone $MK_8(H_2)$ (**115**)	isoprenoid quinone	Antioxidant and antiglycation activities	[76]
Streptomyces sp.	Marine surface sediment of the East Siberian continental margin (Arctic Ocean)	arcticoside, C-1027 chromophore-V, III, fijiolides A and B	benzoxazines, glycosylated paracyclophane	Cytotoxic activity against breast carcinoma MDA-MB231 cells and colorectal carcinoma cells (line HCT-116)	[77]
Streptomyces strain 1010	Shallow sea sediment from the region of Livingston Island (Antarctica)	phthalic acid diethyl ester, 1,3-bis(3-phenoxyphenoxy)benzene, exanedioic acid dioctyl ester, 2-amino-9,13-dimethyl heptadecanoic acid	aromatic compounds	No activity tested *	[78]
Streptomyces sp. SCO736	Marine sediment (Antarctica)	antartin (**10**)	zizaane-type sesquiterpene	Cytotoxic activity against A549, H1299 and U87 cancer cell lines by causing cell cycle arrest at the G1 phase	[79]
Streptomyces sp. NPS008187	Alaskan marine sediment (Arctic Ocean)	glyciapyrroles A (**11**), B (**12**) and C (**13**), cyclo(leucyl-prolyl), cyclo(isoleucyl-prolyl), cyclo(phenylalanyl-prolyl)	pyrrolosesquiterpenes, diketopiperazines	Cytotoxic activity against colorectal adenocarcinoma HT-29 and melanoma B16-F10	[80]

Table 1. *Cont.*

Species	Collection Site (Distribution)	Compound	Molecule Type	Bioactivity	Reference
		Plylum Proteobacteria **Class γ-Proteobacteria**			
Pseudoalteromonas haloplanktis TAC125	French Antarctic station Dumont d'Urville, Terre Adélie (Antarctica)	methylamine	Volatile Organic Compounds (VOCs)	Antimicrobial activity against *B. cepacia* complex	[81]
		4-hydroxybenzoic acid	benzoic acid derivative	Antitumor activity against human A459 lung adenocarcinoma cells	[82]
		cyclo-(D-pipecolinyl-L-isoleucine), cyclo-(L-prolyl-L-histidine), cyclo-(L-prolyl-L-alanine), cyclo-(L-prolyl-L-tyrosine), cyclo-(L-prolyl-L-proline), cyclo-(L-alanyl-L-isoleucine), cyclo-(D-pipecolinyl-L-leucine), cyclo-(L-pipecolinyl-L-phenylalanine), L-valyl-L-leucyl-Lprolyl-L-valyl-L-prolyl-L-glutamine and L-tyrosyl-L-valyl-L-prolyl-L-leucine	diketopiperazines	Antioxidant activity	[83]
		pentadecanal	long-chain fatty aldehyde	Anti-biofilm activity against *Staphylococcus epidermidis*	[84]
Pseudomonas sp.	Terranova Bay, Ross Sea (Antarctica)	monoramnholipids	rhamnolipid	Antimicrobial activity against *B. cepacia* complex	[70]
Psychrobacter	Marine sediment from Terranova Bay, Ross Sea (Antarctica)	monoramnholipids	rhamnolipid	Antimicrobial activity against *B. cepacia* complex	[70]
		Plylum Bacteroidetes **Class Flavobacteriia**			
Aequorivita	Marine sediments from Edmonson Point (Antarctica)	R-(+)-N-[15-methyl-3-(12-methyltridecanoyloxy)-hexadecanoyl]glycine and methyl ester derivatives;	aminolipids	Antimicrobial activity against *S. aureus*	[85]
Salegentibacter strain T436	Bottom section of a sea ice floe (Arctic Ocean)	4-Hydroxy-3-nitrobenzoic acid, 4,6-Dinitroguiacol, 4,5-Dinitro-3-methoxyphenol, (4-Hydroxy-3-nitrophenyl)-acetic acid methyl ester, (4-Hydroxy-3,5-dinitrophenyl)-acetic acid methyl ester, (4-Hydroxy-3-nitrophenyl)-acetic acid, (4-Hydroxy-3,5-dinitrophenyl)-acetic acid, (4-Hydroxy-3,5-dinitrophenyl)-propionic acid methyl ester, (4-Hydroxy-3-nitrophenyl)-propionic acid, (4-Hydroxy-3,5-dinitrophenyl)-propionic acid, 2-Chloro-3-(4-hydroxy-3,5-dinitrophenyl)-propionic acid methyl ester, 2-Hydroxy-3-(4-hydroxy-3-nitrophenyl)-propionic acid methyl ester, 2-(4-Hydroxy-3-nitrophenyl)-ethanol, 2-(4-Hydroxy-3,5-dinitrophenyl)-ethyl chloride, 2-(4-Hydroxy-3,5-dinitrophenyl)-ethanol, 2-Nitro-4-(2-nitroethenyl)-phenol, 3,5-Dinitrogenistein, 3-Nitrogenistein, 3-Nitro-1H-indole	aromatic nitro compounds	Antimicrobial activity against *C. albicans, Paecilomyces variotii, Penicillium notatumb, Mucor miehei* Tü 284, *Magnaporthe grisea, Nematospora coryli, Ustilago nuda, Bacillus brevis, B. subtilis, M. luteus, Escherichia coli* K12b, *Proteus vulgaris* and cytotoxic activities	[86,87]

Table 1. *Cont.*

Species	Collection Site (Distribution)	Compound	Molecule Type	Bioactivity	Reference
		Plylum Firmicutes **Class Bacilli**			
Bacillus sp.	Sea mud (Arctic Ocean)	mixirins A, B and C	cyclopeptides	Cytotoxic activity against human colon tumor cells (HCT-116)	[88]
		FUNGI **Plylum Ascomycota** **Class Eurotiomycetes**			
Aspergillus protuberus MUT 3638	Sub-sea sediments, Barents Sea (Arctic Ocean)	bisvertinolone	sorbicillonoid	Antimicrobial activity against *S. aureus*	[89]
Penicillium sp. PR19N-1	Deep-sea sediment, Prydz Bay (Antarctica)	chlorinated eremophilane sesquiterpenes (**14–17**), eremofortine C (**18**)	chloro-eremophilane sesquiterpenes	Cytotoxic activity against HL-60 and A549 cancer cell lines	[90]
		eremophilane-type sesquiterpenes (**19–23, 25, 26**), eremophilane-type lactam (**24, 27**)	eremophilane-type sesquiterpenes	Cytotoxic activity against HL-60 and A549 cancer cell lines	[91]
Penicillium granulatum MCCC 3A00475	Deep-sea sediment, Prydz Bay (Antarctica)	spirograterpene A (**37**), conidiogenone I (**38**) and conidiogenone C (**39**)	tetracyclic spiro-diterpene, cyclopiane diterpenes	Spirograterpene A: antiallergic effect on immunoglobulin E (IgE)-mediated rat mast RBL-2H3 cells	[92]
Penicillium sp. S-1-18	Sea-bed sediments (Antarctica)	butanolide, guignarderemophilane F, xylarenone A (**28**)	furanone derivative, sesquiterpene	Butanolide: inhibitory activity of butanolide against tyrosine phosphatase 1B; xylarenone A: antitumor activity against HeLa and HepG2 cells and growth-inhibitory effects against pathogenic microbes	[93,94]
Penicillium crustosum PRB-2	Deep-sea sediment, Prydz Bay (Antarctica)	penilactones A and B	oxygenated polyketides	Cytotoxic activity against HCT-8, Bel-7402, BGC-823, A549 and A2780 tumor cell lines	[95]
		Class Dothideomycetes			
Strain KF970 (Lindgomycetaceae family)	Sea-water (Arctic Ocean)	lindgomycin, ascosetin	polyketides	Antimicrobial activity against methicillin-resistant *S. aureus*	[96]
		ALGAE-ASSOCIATED MICROBES **Bacteria-Plylum Actinobacteria-Class Actinobacteria**			
Nocardiopsis sp. 03N67	Arctic seaweed (*Undaria pinnatifida*)	cyclo-(L-Pro-L-Met)	diketopiperazine	Anti-angiogenesis activity against human umbilical vein endothelial cells (HUVECs)	[97]

Table 1. *Cont.*

Species	Collection Site (Distribution)	Compound	Molecule Type	Bioactivity	Reference
SPONGE-ASSOCIATED MICROBES					
Bacteria-Plylum Bacteroidetes-Class Flavobacteriia					
Cellulophaga fucicola	Antarctic sea sponge	zeaxanthin (**110**), β-cryptoxanthin (**111**), β-carotene (**112**)	carotenoids	Antioxidant activity	[98]
Zobellia laminarie	Antarctic sea sponge	zeaxanthin (**110**), β-cryptoxanthin (**111**), β-carotene (**112**), phytoene (**113**)	carotenoids	Anti-UV and antioxidant activity and phototoxicity profile in murine fibroblasts	[99]
Bacteria-Plylum Proteobacteria-Class γ-Proteobacteria					
Pseudomonas aeruginosa	*Isodictya setifera*, Ross Island (Antarctica)	cyclo-(L-Pro-L-Val)/cyclo-(L-Pro-L-Leu)/cyclo-(L-Pro-L-Ile)/cyclo-(L-Pro-L-Phe)/cyclo-(L-Pro-L-Tyr)/cyclo-(L-Pro-L-Met)/diketopiperazines, phenazine-1-carboxylic acid, phenazine-1-carboxamide	diketopiperazines, phenazine alkaloids	Antimicrobial activity against *B. subtilis*, *S. aureus* and *M. luteus*	[65]
Fungi-Plylum Ascomycota-Class Leotiomycetes					
Pseudogymnoascus sp. F09-T18-1	Antarctic sponge genus *Hymeniacidon*, Fildes Bay, King George Island (Antarctica)	pseudogymnoascin A, B, C, 3-nitroasterric acid	nitroasterric acid derivatives	Inactive in antimicrobial activity against *P. aeruginosa*, *Acinetobacter baumannii*, *E. coli*, *S. aureus*, a methicillin-sensitive *S. aureus* and methicillin-resistant *S. aureus*, *C. albicans*, *Aspergillus fumigatu*	[100]
CORAL-ASSOCIATED MICROBES					
Fungi-Plylum Ascomycota-Class Eurotiomycetes					
Penicillium sp. SF-5995	Unidentified soft coral, Terra Nova Bay (Antarctica)	methylpenicinoline	pyrrolyl 4-quinoline alkaloid	Anti-inflammatory effect inhibiting NF-κB and MAPK pathways in lipopolysaccharide-induced RAW264.7 macrophages and BV2 microglia	[101]
CNIDARIA					
Plylum Cnidaria-Class Anthozoa					
Alcyonium antarcticum	Terra Nova Bay (Antarctica)	bulgarane sesquiterpene	sesquiterpene	No bioactivity tested (antipredation activity and ichthyotoxicity) *	[102]
Alcyonium antarcticum	Weddell Sea (Antarctica)	alcyopterosins	illudalane sesquiterpenoids	No bioactivity tested (antipredation activity) *	[103]
Alcyonium paessleri	South Georgia Islands (Antarctica)	alcyopterosin A (**29**), C (**30**), E (**31**), H (**32**)	illudalane sesquiterpenoids	Cytotoxic activity against Hep-2 (human larynx carcinoma) and HT-29 (human colon carcinoma) cell lines	[104]
Alcyonium paessleri	South Georgia Islands (Antarctica)	paesslerins A (**33**), B (**34**)	esquiterpenoids	Cytotoxic activity against human tumor cell lines	[105]
Anthomastus bathyproctus	Deception Island, South Shetland Islands (Antarctica)	conjugated cholestane, ergostane and 24-norcholestane steroids (**79–82**)	steroids	Cytotoxic activity against three human tumor cell lines.	[106]

Table 1. *Cont.*

Species	Collection Site (Distribution)	Compound	Molecule Type	Bioactivity	Reference
Anthoptilum grandiflorum	Burdwood Bank, Scotia Arc (Antarctica)	bathyptilone A (**48**), B, C, enbepeanone A	briarane diterpenes, trinorditerpene	Bathyptilone A: cytotoxic activity against the neurogenic mammalian cell line Ntera-2	[107]
Dasystenella acanthina	Kapp Norvegia, Eastern Weddell Sea (Antarctica)	7 polyoxygenated steroids (**83–89**)	steroids	Growth inhibition of several human tumor cell lines LN-caP and K-562	[108]
Gersemia fruticosa	White Sea (circumpolar Arctic)	6 polyoxygenated sterols (**90–95**)	sterols	Cytotoxic activity against human erythroleukemia K-562 cells, HL-60 and P388	[109]
Gersemia fruticosa	White Sea (circumpolar Arctic)	9,11-secosterol (**96**)	sterol	Cytotoxic activity against human leukemia K562, cervical cancer HeLa and Ehrlich ascites tumor cells	[110]
Gersemia fruticosa	Alaskan Beaufort Sea (Arctic Ocean)	gersemiols A–C, eunicellol A (**40**)	diterpenoids	Eunicellol A: antimicrobial activity against MRSA—methicillin resistant *S. aureus*	[111]
Plumarella delicatissima	Plateau of Fascination, Falkland Islands (Antarctica)	keikipukalides A–E (**41–45**), pukalide aldehyde (**46**), norditerpenoid ineleganolide (**47**)	diterpenes, diterpenoid	Cytotoxic activity against leishmaniasis causing a parasite, *Leishmania donovani*, with no cytotoxicity against the mammalian host	[112]
Undescribed soft coral	Scotia Arc (Antarctica)	shagenes A (**35**), B (**36**)	sesquiterpenoids	Cytotoxic activity against leishmaniasis causing a parasite, *L. donovani*, with no cytotoxicity against the mammalian host	[113]
BRYOZOA Plylum Bryozoa-Class Gymnolaemata					
Tegella cf. *spitzbergensis*	Bear Island (Arctic Ocean)	ent-eusynstyelamide B, eusynstyelamides D–F	brominated tryptophan-derived	Antimicrobial activity against bacteria; weak cytotoxicity against the human melanoma A2058 cell line	[114]
Dendrobeania murrayana	Vesterålsfjorden, Northern Norway (Arctic Ocean)	dendrobeaniamine A	guanidine alkaloid	Tested but inactive for cytotoxic, antimicrobial, anti-inflammatory or antioxidant activities	[115]
Alcyonidium gelatinosum	Hopenbanken, Svalbard (Arctic Ocean)	ponasterone A and F	ecdysteroids	Tested but inactive for cytotoxic, antimicrobial, estrogen receptor agonist activities	[52]

Table 1. *Cont.*

Species	Collection Site (Distribution)	Compound	Molecule Type	Bioactivity	Reference
		MOLLUSCA **Plylum Mollusca-Class Gasteropoda**			
Austrodoris kerguelenensis	Anvers Island (Circumpolar Antarctica)	palmadorin A (**49**), B (**50**), D (**51**), M (**52**), N (**53**), O (**54**)	diterpenoid glyceride esters	Inhibition of human erythroleukemia (HEL) cells; Palmadorin M inhibits Jak2, STAT5, and Erk1/2 activation in HEL cells	[116]
		ECHINODERMATA **Plylum Echinodermata-** **Class Holothuroidea**			
Kolga hyalina	Deep sea, Amundsen Basin (Arctic Ocean)	holothurinoside B, kolgaosides A (**120**), B (**121**)	triterpene holostane nonsulfated pentaosides	Kolgaosides A–B: hemolytic activity against mouse erythrocytes and inhibition against Ehrlich ascite carcinoma cells	[117]
Staurocucumis liouvillei	South Georgia Islands (Antarctica)	liouvillosides A (**118**), B (**119**)	trisulfated triterpene glycosides	Activity against herpes simplex virus type 1 (HSV-1)	[118]
		Class Astheroidea			
Asterias microdiscus	Chukchi Sea (Arctic Ocean)	polyhydroxylated steroids A–F	steroids	No activity tested *	[119]
		Class Ophiuroidea			
Ophiosparte gigas	Ross Sea (Antarctica)	cholest-5-ene-2α,3α,4β,21-tetraol-3,21-disulphate (**102**), cholest-5-ene-2β,3α, 21-triol-2,21-disulphate (**103**)	disulfated polyhydroxysteroids	cholest-5-ene-2α,3α,4β,21-tetraol-3, 21-disulphate: cytotoxic activity; cholest-5-ene-2β,3α, 21-triol-2,21-disulphate: cytoprotective activity against HIV-1	[120]
Astrotoma agassizii	Antarctica	disulfated polyhydroxysteroids (**104–106**)	disulfated polyhydroxysteroids	Activity against one DNA (HSV-2) and two RNA (PV-3, JV) viruses	[121]
		SPONGES **Plylum Porifera** **Class Demospongiae**			
Crella sp.	Norsel Point, Amsler Island (Antarctica)	norselic acid A (**97**), B (**98**), C (**99**), D (**100**), E (**101**)	oxidized steroids	Norselic acid A: activity against MRSA, methicillin-sensitive *S. aureus* (MSSA) and vancomycin-resistant *Enterococci faecium* (VREF) and *C. albicans*. All norselic acids were active against leishmaniasis	[122]

Table 1. *Cont.*

Species	Collection Site (Distribution)	Compound	Molecule Type	Bioactivity	Reference
Dendrilla antarctica	Anvers Island (Antarctica)	aplysulphurin (**55**), membranoid A (**56**), B (**57**), C (**58**), D (**59**), E (**60**), G (**61**),H (**62**)	oxidized diterpenoids	Aplysulphurin: activity against *C. albicans*, and Gram-negative antibiotic activity against *S. aureus* and *E. coli*; membranoids: activity against the leishmaniasis	[123,124]
		darwinolide (**63**), tetrahydroaplysulphurin-1 (**65**), membranolide (**66**), glaciolides (**67–68**), cadlinolide C (**69**), dendrillin A (**70**), B (**71**), C (**72**), D (**73**) and semisynthetic derivatives (**74–76**)	oxidized diterpenoids	Darwinolide: selectivity against the biofilm phase of MRSA compared to the planktonic phase; membranolide: activity against MRSA; dendrillin B: activity against *L. donovani*; 76, activity against *Plasmodium falciparum*	[125,126]
		9-11-dihydrogracilin A (**64**)	oxidized diterpenoid	Immuno-modulatory and anti-inflammatory activity in human cell lines	[127]
Geodia baretti	North Sea off the coast of Sweden and the northern coast of Iceland (Arctic Ocean)	barettin and the geobarrettins	diketopiperazine (likely produced by a symbiont)	Moderate antioxidant and anti-inflammatory activities	[53] [128] [129]
Haliclona viscosa	Svalbard Archipelago (Arctic Ocean)	viscosamine viscosaline	3-alkyl pyridinium alkaloids	Antibiotic activity against four separate sympatric bacterial strains	[130] [131]
Kirkpatrickia variolosa	Antarctica	variolins A-D (B most active)	pyridopyrrolopyrimidine	Cytotoxic activity against P388 murine leukemia cell line	[132] [133]
Latrunculia sp.	Aleutian Islands, Alaska (also found in Antarctic specimens)	discorhabdins A, C, R, dihydrodiscorhabdin B	spirocyclic imino-quinones	Anti-HCV (Hepatitis C virus) activity, antimalarial activity and selective antimicrobial activity against MRSA, *Mycobacterium intracellulare* and *M. tuberculosis*.	[134]
				Antiprotozoal activity in vitro (*P. falciparum*)	
	Weddell Sea (Antarctica)	tsitsikammamines	pyrroloiminoquinones	Anticancer and cytotoxic activities	[57]
Lyssodendoryx flabellata	Terra Nova Bay (Antarctica)	terpioside	glycosphingolipid	Inhibition effect in mixed lymphocyte reactions on human cells	[135]
Mycale acerata	Terra Nova Bay (Antarctica)	mycalol	alkyl glyceryl ether lipid	Activity against human thyroid carcinoma cells	[136]

Table 1. *Cont.*

Species	Collection Site (Distribution)	Compound	Molecule Type	Bioactivity	Reference
Plakortis simplex	Sula Ridge Reef, Norwegian Shelf (sub Arctic)	methyl 2-((3R,6S)-4,6-diethyl-6-hexyl-3,6-dihydro-1,2-dioxin-3-yl)acetate	cyclic peroxide (fatty acids)	Selectively inhibited proliferation in gastric cancer (GXF 251L), non-small cell lung cancer (LXFL 529L) and melanoma (MEXF 462NL) cell lines.	[137]
Polymastia boletiformis	Norwegian coast, Western Irish Coast (sub Arctic)	polymastiamide A (**117**), B, C, D, E, F	sulfated steroid-amino acid conjugates	Polymastiamide A: antifungal activity against plant pathogens *Cladosporium cucumerinum* and *Pythium ultimum* and human yeast pathogen *C. albicans*	[138]
				Polymastiamide A: antibacterial activity against *S. aureus*	[139] [140]
Stryphus fortis	Spitsbergen, Svalbard, (Arctic Ocean)	ianthelline	bromotyrosine derivative	Antitumor properties against several malignant cell lines and inhibition of PK activity	[141]
Suberites sp.	King George Isalnd, McMurdo Sound (Antarctica)	suberitenones A (**77**), B (**78**)	oxidized sesterterpenoids	Inhibition of the cholesteryl ester transfer protein (CETP)	[142]

TUNICATES
Plylum Chordata
Class Ascidiacea

Species	Collection Site (Distribution)	Compound	Molecule Type	Bioactivity	Reference
Aplidium meridianum	South Georgia Islands (Antarctica)	meridianins	brominated 3-(2-aminopyrimidine) indoles	Prevention of cell proliferation and induction of cell apoptosis. Inhibition of CDKs, GSK-3, PKA and other kinases in the low micromolar range	[143]
Aplidium sp.	Ross Sea (Antarctica)	rossinones A (**122**), B (**123**)	meroterpenoids	Antiproliferative activity against several cell lines. Antiviral activity against the DNA virus HSV-1 as well as antibacterial and antifungal activity against *B. subtilis* and *Trichophyton mentagrophytes*	[144]
Clavelina lepadiformis	Bergen, Norway	lepadins	decahydroquinoline alkaloid	Lepadin A: anti-cancer activity against leukemia P388, breast cancer (MCF7), glioblastoma/astrocytoma (U373), ovarian (HEY), colon (LoVo) and lung (A549)	[145] [146]
Synoicum adareanum	Anvers Island (Antarctica)	palmerolides A–G	enamide-bearing macrolides	Palmerolide A: Activity against melanoma (UACC-62 LC50), by inhibition of vacuolar ATPase-	[147] [148]
		hyousterones A (**107**), B, C (**108**), D, abeohyousterone (**109**)	ecdysteroids	Activity against colon cancer cells	[149]

Table 1. *Cont.*

Species	Collection Site (Distribution)	Compound	Molecule Type	Bioactivity	Reference
Synoicum pulmonaria	Tromso, Northern Norway (Arctic Ocean)	synoxazolidinones A-C	brominated guanidinium oxazolidinones	Antibacterial activity against MSSA, MRSA and *Corynebacterium glutamicum* as well as antifungal properties against *Saccharomyces cerevisiae*. Active against human melanoma (A2058), breast adenocarcinoma (MCF-7) and colon carcinoma (HT-29) cell line, with noted cytoxicity	[150] [151]
		pulmonarins A, B	brominated methoxybenzoylesters bearing quaternary ammonium mioeties	Acetylcholinesterase inhibitory activity and weak antibacterial activity against *C. glutamicum*	[152]
MACROALGAE **Plylum Rhodophyta** **Class Florideophyceae**					
Delisea fimbriata and *pulchra*	Anvers Island (Antarctica)	fimbrolides and analogues	polyhalogenated furanones	Antimicrobial activity against *S. aureus*, *E. coli*, *C. albicans* and *Streptococcus* sp.	[153]
Plocamium cartilagineum	Anvers Island (Antarctica)	oregonene (1) and similar compounds (2–4), anverenes A (5), B (6), C (7), D (8), E (9)	polyhalogenated monoterpenes	Cytotoxic activity against cervical cancer cells	[154]
Plylum Ochrophyta Class Phaeophyceae					
Desmarestia menziesii	Anvers Island (Antarctica)	Menzoquinone (124)	terpenoid-quinone	Antimicrobial activity against MRSA, MSSA, VREF	[155]

* Natural compounds with no bioactivity tested.

3.1. Marine Chemical Diversity in Polar Regions

In the Antarctic and Arctic marine environment, a huge reservoir of microbial biodiversity has been recognized as promising for the isolation of new antimicrobial metabolites [85,156–159]. Among marine bacteria, *Pseudoalteromonas haloplanktis* TAC125 is a potential untapped source of biologically active natural products [160]. This cold-adapted bacterium produces valuable bioactive secondary metabolites, such as anti-biofilm metabolites [160], antibiotics such as methylamine, a volatile compound active against *Burkholderia cepacia* complex strains [81] and anticancer compounds [82]. The biotechnological potential of *P. haloplanktis* TAC125 is related to its genus *Pseudoalteromonas*, ubiquitously distributed in almost all marine habitats. The recent reconstruction of the largest *Pseudoalteromonas* pangenome allowed the identification of *Pseudoalteromonas* genes for cold adaptation and the production of secondary metabolites, among others [161]. Other Antarctic strains belonging to the same genus have proven useful for cosmetic application, including Antarcticine®® launched by LIPOTEC as an anti-aging product obtained from extracts of marine *Pseudoalteromonas antarctica* [162], and SeaCode®®, a mixture of EPS and other glucidic exopolymers produced by biotechnological fermentation of a *Pseudoalteromonas* sp. isolated in Antarctic waters [163].

On the macroscopic realm, several new natural products have been isolated from marine polar biota, some yielding encouraging properties for biotech discovery [6]. The Arctic bryozoan *Dendrobeania murrayana* revealed the presence of the guanidine alkaloid, dendrobeaniamine A, which was inactive in cytotoxicity, antimicrobial, anti-inflammatory and antioxidant assays [115]. The permethylated hexapeptide friomaramide is a potent inhibitor of liver-stage infection by *Plasmodium falciparum* [164]. Among all marine natural products coming from the polar seas, we will focus in this review on terpenoids and their derivatives (included in Table 1), because they represent one of the most important classes of bioactive molecules commonly found in polar organisms.

3.2. Marine Polar Terpenoids

In marine environments, terpenes, terpenoids and their derivatives display an array of diverse chemical structures, with promising biological activities [165–168]. Terpenes are hydrocarbons that represent a large family of natural compounds, which include primary and secondary metabolites metaphorically biosynthesized from five carbon isoprene units.

Terpenoids, which are sometimes called isoprenoids, are derivatives or modified terpenes. Although sometimes used interchangeably with "terpenes," terpenoids are often polycyclic structures where methyl groups have been moved or removed, and oxygen-containing functional groups have been added. About 60% of known natural products are terpenoids [169].

Modification of the isoprene unit structures leads to a wide structural diversity of derivatives showing diverse bioactivities. Terpenes and terpenoids are often classified by the number of isoprene units added to the parent terpene, and divided into biogenetic subclasses, e.g., monoterpenes, sesquiterpenes, diterpenes, sesterterpenes, triterpenes and tetraterpenes, containing two, three, four, five, six and seven isoprene units, respectively. The triterpenes include sterols and steroids and their derivatives conjugated to other functional group while the tetraterpenes include carotenoids. There are also compounds that contain terpene fragments, such as prenylated polyketides (meroterpenes, indole alkaloids).

Their biosynthesis was described in 1953, by Leopold Ružička as the C_5 rule or biogenetic isoprene rule, reporting the linking of isoprene units "head to tail" to form chains [170]. There are two metabolic pathways that create terpenoids (Figure 1): (a) the mevalonic acid or mevalonate pathway (MVA pathway), which also produces cholesterol and occurs in the cytoplasm of most organisms (e.g., bacteria, archaea, fungi, plants, animals), except green algae; and (b) the 2-C-methyl-D-erythritol 4-phosphate/1-deoxy-D-xylulose 5-phosphate pathway (MEP/DOXP pathway), also known as non-mevalonate pathway, takes place in plastids of plants and green algae, in apicomplexan protozoa and in many bacteria [171]. From these two pathways isopentenyl diphosphate (IPP) and its isomer dimethylallyl diphosphate (DMAPP) are derived, which are sequentially elongated by prenyltransferases to geranyl diphosphate (C_{10}), farnesyl diphosphate (C_{15}) and geranylgeranyl diphosphate (C_{20}) (Figure 1). These acyclic intermediates are then transformed by terpenoid synthases into monoterpenes, sesquiterpenes and diterpenes and further modifications produce triterpenes and tetraterpenes [171].

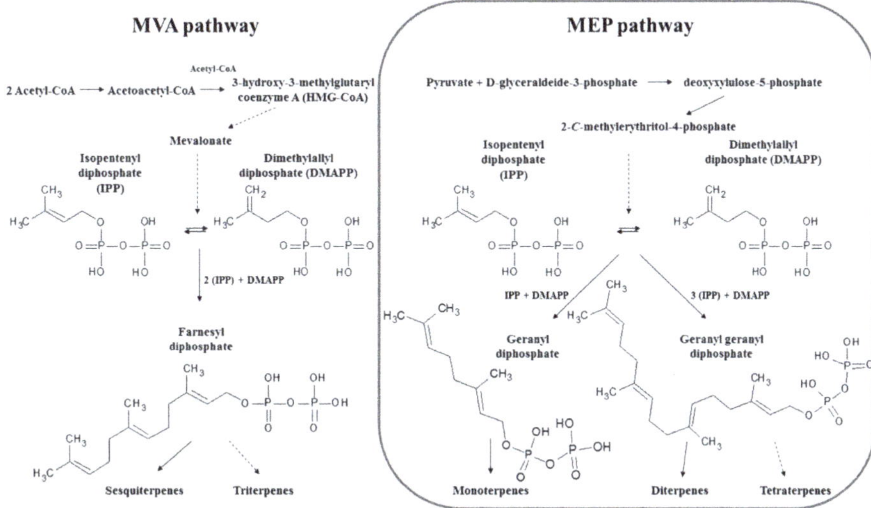

Figure 1. Schematic representation of isoprenoid biosynthesis by mevalonate pathway (MVA pathway) starting from two molecules of Acetyl-CoA and non-mevalonate pathway (MEP pathway) starting from pyruvate and glyceraldeide-3-phosphate. Adapted from [171].

Marine organisms are prolific producers of terpenes and terpenoids. Particularly in cnidarians (corals), mollusks and sponges, these types of compounds are well represented among reported natural products. In ascidians and echinoderms, the proportion of terpene and terpenoid molecules reported to date is lower but they are still relatively common. In total, terpenoids are the most frequently reported natural product class from the Antarctic, displaying a large degree of structural diversity and bioactivity [172]. In the following paragraphs, we will describe some aspects of the biotechnological potential of terpene and terpenoid compounds from polar latitudes.

3.2.1. Monoterpenes and Monoterpenoids

Monoterpenes are comprised of two isoprene units that result from a single condensation between DMAPP and IPP to yield a 10-carbon molecule, which is often further modified (Figure 1). Polyhalogenated monoterpenes are commonly produced by red algae inhabiting the shallow waters of the polar regions. These metabolites are often found in high abundance and can be either linear or cyclic. Like many natural products found in other phyla, the ecological function of these halogenated secondary metabolites is thought to be multi-faceted. The most commonly assumed role is that of defense, conveying deterrence against herbivory and resistance to detrimental fouling species [173]. These compounds may also be used offensively, possibly providing allelopathic competitive advantages to the producers, while deleteriously affecting potential competitors for space and resources [155,174–176].

Likely owing to their defensive and offensive ecological functions in nature, previous investigations into the therapeutic potential of halogenated metabolites in red seaweeds from other regions have yielded a plethora of relevant biological activities, ranging from antimicrobial to perhaps most notably their antitumor properties [155,174–177]. Similarly, a recent chemical investigation of what is generally considered to be an Antarctic subspecies of the red alga *Plocamium cartilagineum* collected from around

Anvers Island along the Western Antarctic Peninsula yielded several polyhalogenated monoterpenes (**1–9**) including anverenes A-E (Figure 2), which displayed cytotoxicity in the low micromolar range (1–13 μM) against a cervical cancer cell line (HeLa) [154].

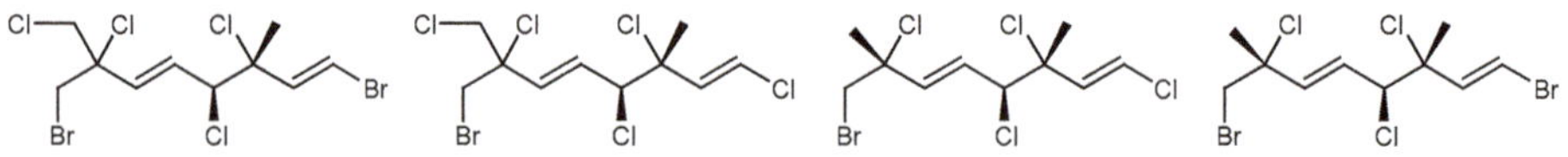

Oregonene A (1)   2   3   4

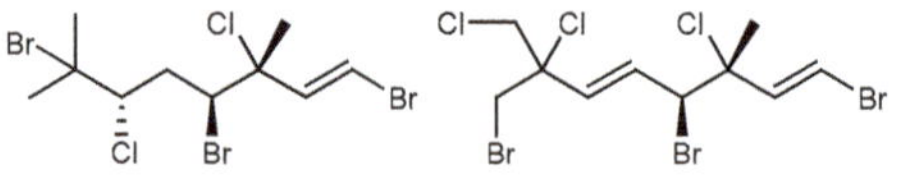

Anverene A (5)   Anverene B (6)

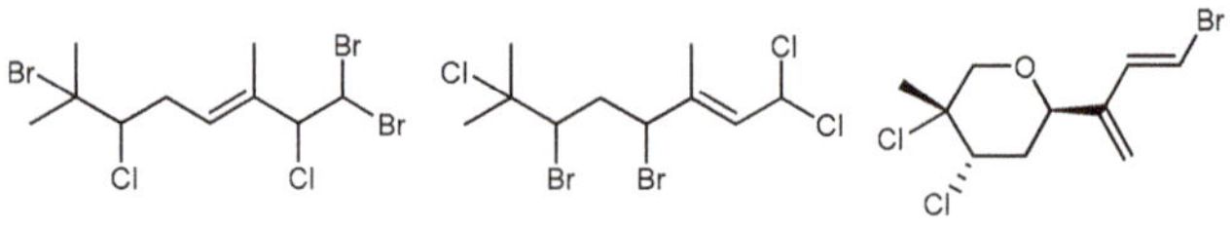

Anverene C (7)   Anverene D (8)   Anverene E (9)

Figure 2. Structures of monoterpenoid compounds, (**1–9**) oregonene A and anverenes A–E from *Plocamium cartilagineum*.

3.2.2. Sesquiterpenes and Sesquiterpenoids

Sesquiterpenes comprise "one and a half monoterpenes" (*sesqui-* prefix means one and a half), thus, three isoprene units. They may be acyclic or contain rings, including many unique combinations. As in other terpenes, biochemical modifications such as oxidation or rearrangement produce the related sesquiterpenoids [169,171]. Sesquiterpenes and sesquiterpenoids are an important class of natural products with a wide range of biological activities as semiochemicals, e.g., defensive agents or pheromones, and commonly found in terrestrial organisms (e.g., plants, fungi and insects). Their occurrence in the marine environment is quite remarkable, with several ecological and pharmacological bioactivities described [178]. Therefore, marine sesquiterpenes and derivatives represent important candidates for natural products in drug discovery [166,179–181].

Streptomyces sp. SCO-736, isolated from an Antarctic marine sediment, was found to produce antartin (**10**) a tricyclic zizaane-type sesquiterpene with an unusual phenyl group (Figure 3) [79]. Tricyclic sesquiterpenes are an important class of marine natural products characterized by structural diversity often unprecedented when compared to terrestrial metabolites. They often display interesting biological activity, such as antifouling, cytotoxic, antibiotic, antifungal, antiparasitic and anti-inflammatory activities [182]. Antartin displays moderate cytotoxicity against a wide range of cancer cell lines, A549, H1299 and U87 cell lines, by causing cell cycle arrest at the G1 phase [79].

Streptomyces sp. NPS008187, isolated from a marine sediment in Alaska, produces three pyrrolosesquiterpenes, glyciapyrroles A (**11**), B (**12**) and C (**13**), along with the known diketopiperazines cyclo(leucyl-prolyl), cyclo(isoleucyl-prolyl) and cyclo(phenylalanyl-prolyl) (Figure 3). Glaciapyrrole A displays moderate antitumor activity against both colorectal adenocarcinoma HT-29 and melanoma B16-F10 tumor cell growth [80].

Although terpenoid substructures are often mixed with aromatic ring systems of various types (e.g., phenols, quinones, coumarines and flavonoids), they are rarely found in combination with pyrroles; there are only a few cases: pyrrolostatin, isolated from a Brazilian soil Actinobacteria *Streptomyces chrestomyceticus* EC40 [183], and two pyrrolosesquiterpenes isolated from soil actinomycete *Streptomyces* sp. Hd7–21 [184].

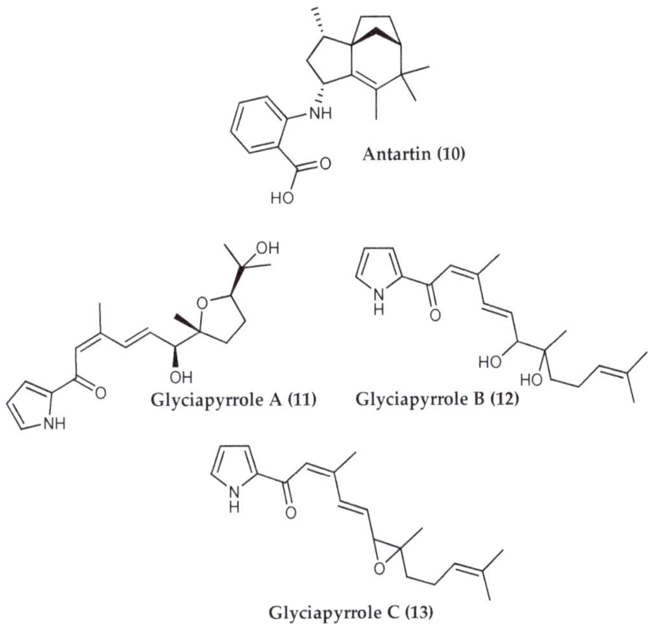

Antartin (10)

Glyciapyrrole A (11)   **Glyciapyrrole B (12)**

Glyciapyrrole C (13)

Figure 3. Structures of the sesquiterpenoid compounds (**10**) antartin A and (**11–13**) glyciapyrroles A–C from *Streptomyces* spp.

Penicillium sp. PR19N-1, a prolific producer of sesquiterpenes, is a fungus isolated from an Antarctic deep-sea sediment in Prydz Bay (−1000 m). It was found to contain four chlorinated eremophilane sesquiterpenes (**14–17**), along with a known sesquiterpenoid eremofortine C (**18**) [90] and five eremophilane-type sesquiterpenes (**19–23**), a new rare lactam-type eremophilane (**24**), together with three previously reported compounds (**25–27**) (Figure 4) [91]. Compound **14** displayed moderate cytotoxic activity against HL-60 and A549 cancer cell lines [90], whereas compound **23** was the most active against the A-549 cells [91]. Eremophilane-type sesquiterpenes, a subclass of sesquiterpenoids, firstly identified in plants [185] and only decades later in fungi [186], are characterized by complex and unique structure and diverse biological properties, such as phytotoxic, antifungal, antibiotic and anticancer bioactivity [179]. They are composed of a terpene skeleton, where the isoprene units are connected head-to-head [187]. They have also been found from a deep marine-derived fungus, *Aspergillus* sp. SCSIOW2, grown after chemical epigenetic manipulation to be induced to produce these secondary metabolites [188] in marine-derived *Xylariaceous* fungus LL-07H239 [189] and in the mangrove endophytic fungus *Xylaria* sp. BL321 [190]. However, these chloro-eremophilane sesquiterpenes are the first examples found in microorganisms [90], and compound **23** represents a rare example of eremophilane-type lactam found in microorganisms [91]. To date, only four lactam-type compounds were reported [191–193].

Penicillium sp. S-1-18, a fungus isolated from Antarctic marine sediments, was reported to produce a new sesquiterpene guignarderemophilane F, with no detectable activity, together with six known compounds, among them the sesquiterpenoid xylarenone A (**28**) (Figure 5) [93]. Compound **28**, previously isolated from the endophytic fungal strain *Xylaria* sp. NCY2, exhibited moderate antitumor activities against HeLa and HepG2 (human liver carcinoma) cells and displayed growth-inhibitory effects against pathogenic microbes [94].

Illudalane sesquiterpenes are a group of potent bioactive products, with antimicrobial and cytotoxic properties, that are primarily found in terrestrial ferns (family Pteridaceae) (e.g., [194]) and fungi (phylum Basidiomycota) (e.g., [195]). In the marine realm, 15 illudalane sesquiterpenoids, alcyopterosins (indicated with letter from A to O) were isolated from the sub-Antarctic soft coral *Alcyonium paessleri*, which was collected at a depth of 200 m near the South Georgia Islands. Alcyopterosins A (**29**), C (**30**), E (**31**) and H (**32**) (Figure 6) demonstrated mild cytotoxicity toward the Hep-2 (human larynx carcinoma) cell line with an IC_{50} of 13.5 µM, while alcyopterosins **29**, **30** and **32** displayed cytotoxicity toward HT-29 (human colon carcinoma) at an IC_{50} of 10 µg mL^{-1} [104].

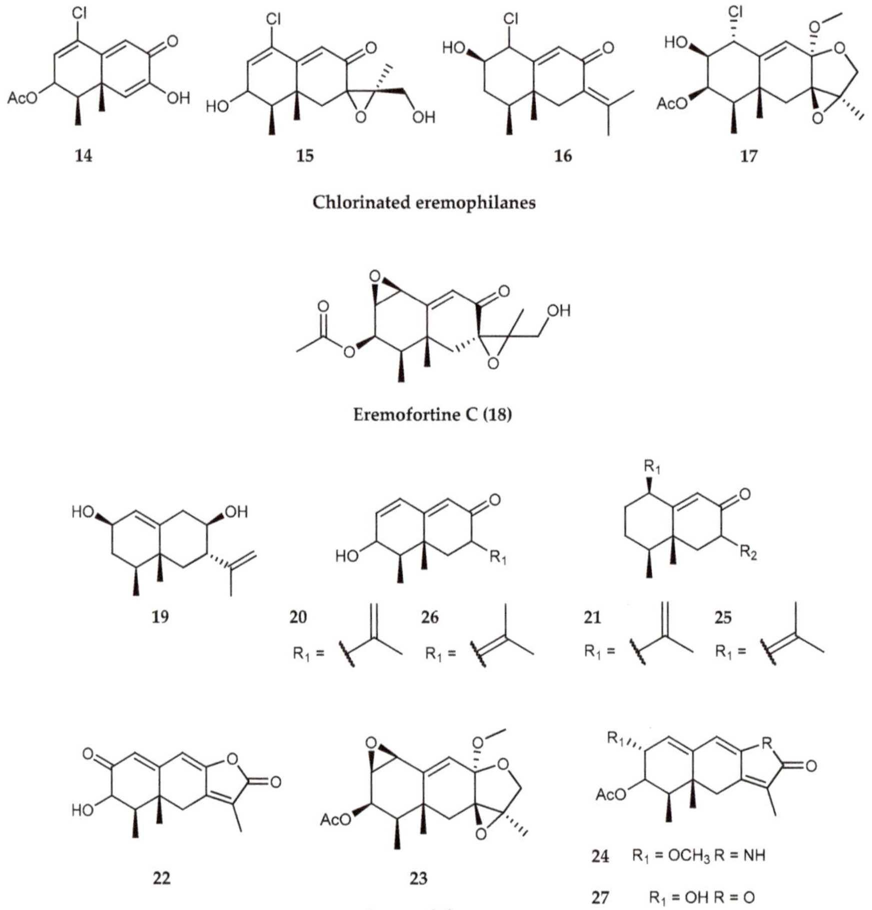

Figure 4. Structures of several sesquiterpenoid compounds from *Penicillium* sp. PR19N-1: (**14–17**) chlorinated eremophilane sesquiterpenes, (**18**) eremofortine C and (**19–27**) several eremophilane-type sesquiterpenes.

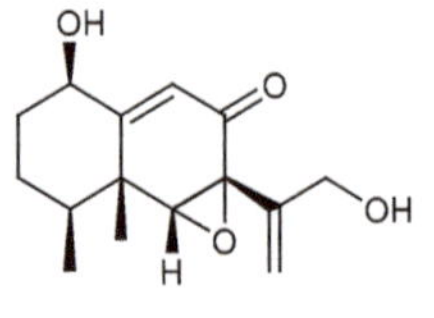

Figure 5. Structure of the sesquiterpenoid compound (**28**) xylarenone A from *Penicillium* sp. S-1-18.

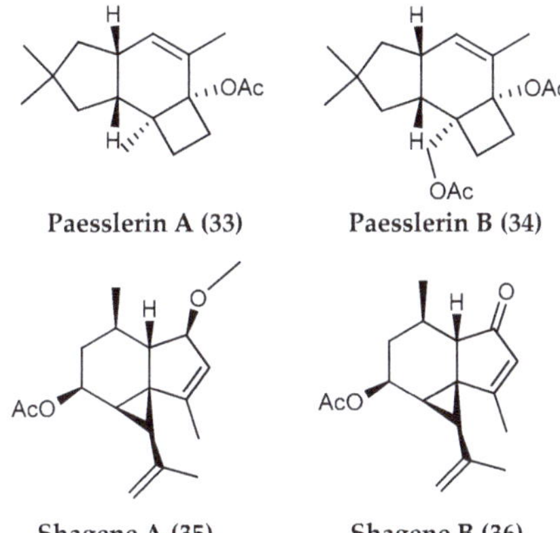

Alcyopterosin A (29)   Alcyopterosin C (30)

Alcyopterosin E (31)   Alcyopterosin H (32)

Figure 6. Structures of the sesquiterpenoid compounds (**29–32**) alcyopterosins A, C, E and H from *Alcyonium paessleri*.

Alcyonium paessleri from South Georgia Island further yielded two new sesquiterpenoids, paesslerins A and B (**33** and **34**) (Figure 7), with moderate cytotoxicity against human tumor cell lines [105]. The structures shown are revised based on the results of total synthesis of the originally proposed structures [196].

Paesslerin A (33)   Paesslerin B (34)

Shagene A (35)   Shagene B (36)

Figure 7. Structures of the sesquiterpenoid compounds (**33, 34**) paesslerins A–B from *Alcyonium paessleri* and (**35, 36**) shagenes A–B from an undescribed soft coral.

Chemical examination of an undescribed soft coral collected from the Scotia Arc in the Southern Ocean resulted in the isolation and characterization of two new tricyclic sesquiterpenoids, shagenes A (**35**) and B (**36**) (Figure 7). Both compounds displayed significant inhibition against the visceral leishmaniasis, causing a parasite, *Leishmania donovani*, with no cytotoxicity against the mammalian host [113].

Octocorals or "horny corals" are conspicuous components on Antarctic sea bottoms, which extensively rely on natural products for protection. Eudesmane sesquiterpenes, ainigmaptilones A and B, seem to be taking part of the chemical arsenal of gorgonians *Ainigmaptilon antarcticus* from the Weddell Sea. Ainigmaptilone A was found to display ecologically relevant activities, in repelling putative sea star predators, and block fouling or microbial infections caused by sympatric bacteria strains and diatoms [197]. Antifouling properties have also been reported in azulenoid sesquiterpenes (i.e., linderazulene, ketolactone 2, and a new brominated C-16 linderazulene derivative) isolated from another Antarctic gorgonian, *Acanthogorgia laxa* [198]. The nudibranch gastropod, *Bathydoris hodgsoni*, from the Weddell Sea revealed high concentrations of a new drimane sesquiterpene, hodgsonal, exclusively allocated in the mantle. Hodgsonal has been hypothesized to be de novo synthesized and to play a defensive role, by analogy with drimane sesquiterpenes in other dorid nudibranchs [199,200].

3.2.3. Diterpenes and Diterpenoids

Diterpenes are composed of "two monoterpene units," or four isoprene subunits [171]. They are pharmacologically interesting compounds for possessing antimicrobial, antiviral, antiparasite, anticancer and anti-inflammatory properties [201].

A rare spirocyclic diterpene, named spirograterpene A (**37**), was isolated from the deep-sea fungus *Penicillium granulatum* MCCC 3A00475 [92] from Prydz Bay, Antarctica, together with two known biosynthetically-related cyclopianes, conidiogenone I (**38**) [202] and conidiogenone C (**39**) (Figure 8) [203]. Spirograterpene A displays an anti-allergic effect on immunoglobulin E (IgE)-mediated rat mast RBL-2H3 cells, displaying 18% inhibition compared with the positive control, loratadine, with 35% inhibition at the same concentration of 20 µg/mL [92]. Cyclopianes, belonging to a rarely reported diterpenoid family, are tetracyclic diterpenes characterized by a highly fused and rigid ring system of 6/5/5/5 skeleton. They were first identified in the fungus *Penicillium cyclopium* in 2002 [204], followed by related compounds, all of which have been obtained only in the *Penicillium* species [202–205]. Spirograterpene A (**37**) is the second example of a diterpene spiro-tetracyclic skeleton with a 5/5/5/5 ring system [92], demonstrating that marine fungi represent a unique source of structurally novel compounds.

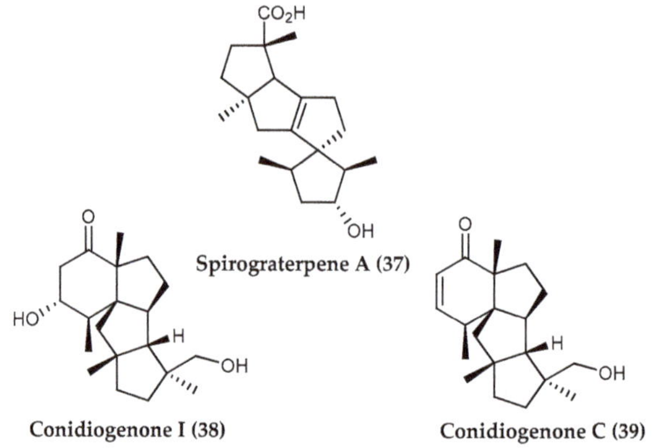

Spirograterpene A (37)

Conidiogenone I (38)

Conidiogenone C (39)

Figure 8. Structures of the diterpenoid compounds (**37**) spirograterpene A and (**38, 39**) conidiogenone I and C from *Penicillium granulatum*.

Among marine invertebrates, cnidarians, and in particular corals, have been found to possess a wide variety of diterpene and diterpenoid products, likely mediating allelochemical interactions [206]. Cold water polar ecosystems are devoid of scleractinian coral reef formations, such as those in the tropics, but instead harbor rich communities of soft-bodied octocorals that are well-known for their chemical diversity [16,207,208]. The Arctic soft coral *Gersemia fruticosa*, collected in the Bering sea, revealed the presence of three diterpenes named gersemiols A−C and another eunicellane diterpene, eunicellol A, which were purified together with the known sesquiterpene (+)-α-muurolene. Eunicellol A (**40**) (Figure 9) was found to exhibit moderate and selective antibacterial activity against methicillin-resistant *Staphylococcus aureus* (MRSA) [111].

Five new furanocembranoid diterpenes, keikipukalides A−E (**41–45**), the known diterpene pukalide aldehyde (**46**) and the known norditerpenoid ineleganolide (**47**) (Figure 10) were isolated from the octocoral *Plumarella delicatissima* collected between 800 and 950 m depth, and demonstrated inhibitory activity against *L. donovani* [112].

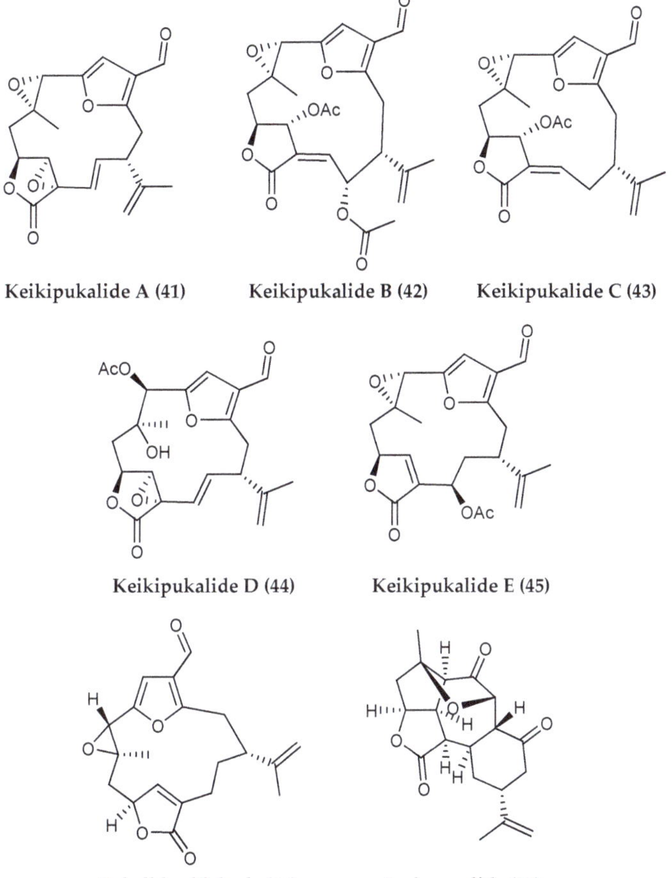

Eunicellol A (40)

Figure 9. Structure of the sesquiterpene compound (**40**) eunicellol A from *Gersemia fruticosa*.

Keikipukalide A (41)   **Keikipukalide B (42)**   **Keikipukalide C (43)**

Keikipukalide D (44)   **Keikipukalide E (45)**

Pukalide aldehyde (46)   **Ineleganolide (47)**

Figure 10. Structures of the diterpenoid compounds (**41–45**) keikipukalides A–E, (**46**) pukalide aldehyde and (**47**) ineleganolide from *Plumarella delicatissima*.

The feather boa sea pen, *Anthoptilum grandiflorum*, is a cosmopolitan pennatulacean octocoral. Specimens collected at 662 and 944 m depth north of Burdwood Bank (Scotia Arc) yielded three new briarane diterpenes, bathyptilone A (**48**) (Figure 11), B and C together with a trinorditerpene, enbepeanone A. Nanomolar cytotoxicity against the neurogenic mammalian cell line Ntera-2 was detected only for bathyptilone A [107].

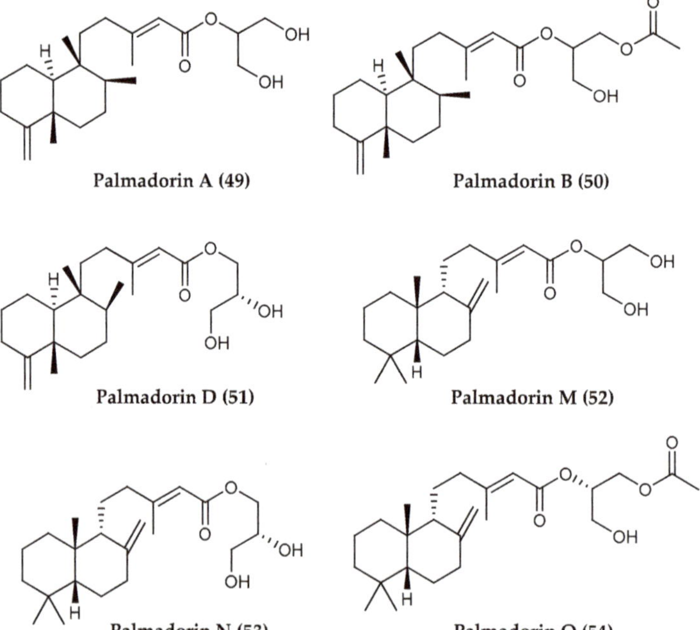

Bathyptilone A (48)

Figure 11. Structure of the diterpene compound (**48**) bathyptilone A from *Anthoptilum grandiflorum*.

The gastropod *Austrodoris kerguelenensis* is considered the most common and conspicuous Antarctic nudibranch, and among the most studied polar species for chemical ecology [13,14,209]. Detailed investigation of specimens from the vicinities of Palmer Station (Western Antarctic Peninsula), McMurdo Sound and the Weddell Sea has resulted in the isolation of a suite of tricyclic diterpenoid 2′-monoglyceryl esters (i.e., austrodorin A–B) [210] and diterpenoic acid glycerides [i.e 2′-acetoxyglyceryl (5R,10R,13R)-labda-8-en-15-oate, 3′-acetoxyglyceryl (5R,10R,13R)-labda-8-en-15-oate] [211], which were hypothesized to be produced by the nudibranch cells as opposed to being accumulated from its sponge diet [200]. Out of the variety of diterpene and diterpenoid products known from this mollusk, palmadorins A (**49**), B (**50**), D (**51**), M (**52**), N (**53**) and O (**54**) (Figure 12) proved to inhibit human erythroleukemia (HEL) cells and palmadorin M blocks Jak2, STAT5 and Erk1/2 activation in HEL cells, in addition to causing apoptosis [116].

Palmadorin A (49)

Palmadorin B (50)

Palmadorin D (51)

Palmadorin M (52)

Palmadorin N (53)

Palmadorin O (54)

Figure 12. Structure of the diterpenoid compounds (**49–54**) palmadorins A, B, D, M, N and O from *Austrodoris kerguelenensis*.

Sponges are known to be a particularly rich source of defensive diterpenoids, and among polar species, Antarctic *Dendrilla* sponges, typically reported as *D. membranosa* (recently revised to *D. antarctica*), stand out as the most prolific producers of bioactive diterpenes. While the chemical ecology of this sponge has been well-studied, investigations into the bioactive potential of Antarctic *Dendrilla* sponges was not carried out until 2004, with aplysulphurin (**55**) isolated from methanolic extracts along with three methyl acetals (**56, 58, 60**) (Figure 13), which displayed moderate antifungal activity against *Candida albicans*, and antibiotic activity against *S. aureus* and *Escherichia coli* [123]. Termed "membranolides B–D" at the time of original publication, more recent investigations have shown these acetals to be artifacts from the methanolysis of aplysulphurin and have yielded several additional semisynthetic methyl acetal variations of the scaffold, now known collectively as membranoids A-H (**56–62**) (Figure 13). As a whole, the membranoids show potent bioactivity against the leishmaniasis causing parasite *L. donovani*, with membranoids B (**57**), D (**59**) and G (**61**) most notably displaying IC_{50} values of 0.8 µM, 1.4 µM and 1.9 µM, respectively, against *L. donovani* infected J774A.1 macrophages, with no discernable cytotoxicity observed towards the healthy variant of human cells [124].

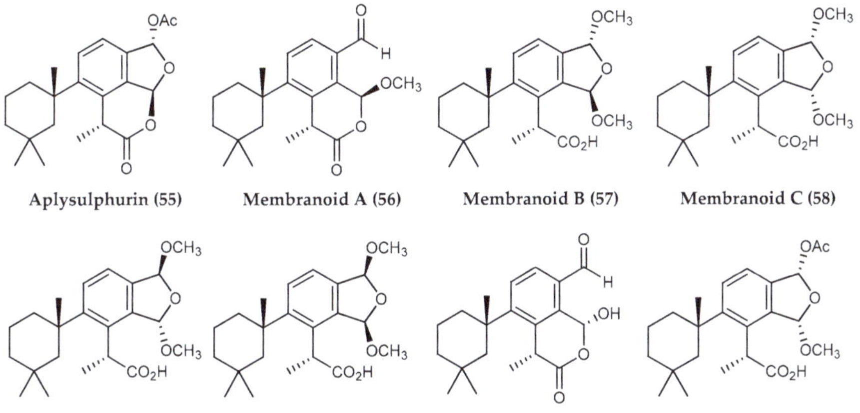

Aplysulphurin (**55**)   Membranoid A (**56**)   Membranoid B (**57**)   Membranoid C (**58**)

Membranoid D (**59**)   Membranoid E (**60**)   Membranoid G (**61**)   Membranoid H (**62**)

Figure 13. Structures of the diterpenoid compounds (**55**) aplysulphurin and (**56–62**) membranolids A, B, C, D, E, G and H from *Dendrilla membranosa*.

Dendrilla sponges from around Anvers Island, Antarctica, have yielded several further diterpenoid natural products, including darwinolide (**63**), which was tested against MRSA and was found to be four times more potent against the biofilm (33.2 µM) than the planktonic form of MRSA (132.9 µM). This type of selective toxicity towards biofilms is rare and a promising lead in the search for antibiofilm-specific antibiotics [125]. A recent continuation of that study revealed a library of diterpenoids containing both known (**64–70**) (Figure 14) and new (**71–73**) natural products along with several additional semisynthetic derivatives (**74–76**) (Figure 15). This small but diverse collection of diterpenoids showed remarkable antibiotic properties against a range of infectious disease models. The most prominent bioactivity including membranolide (**66**) showed >90% eradication of MRSA biofilm at or below concentrations of 25 µg/mL, dendrillin B (**71**), active against *L. donovani* infected J774A.1 at macrophages at an IC_{50} of 3.5 µM, and **76** with 100% inhibition of *P. falciparum* at 5 µg/mL [126]. 9,11-dihydrogracilin A (**64**) isolated from *Dendrilla* sponges collected around the same area, has also recently been shown to display immuno-modulatory and anti-inflammatory properties in human cell lines [127].

3.2.4. Sesterterpenes and Sesterterpenoids

Sesterterpenes are composed of "two and a half monoterpene units" and are typically C_{25}, resulting from an initial condensation between DMAPP and isopentenyl IPP pyrophosphate followed by three additional and consecutive condensations of IPP to add to the growing chain. Sesterterpenes are the longest of the terpenes to be formed in this fashion, as subsequently longer terpenes with 30+ carbons are formed by additional condensation of two preformed phosphorylated isoprene precursors [212].

Sesterterpenoids are not as commonly found in the marine polar environment compared to diterpenes or triterpenes, however, sponges of the *Suberites* genus collected at several spots around Antarctica including King George Island and McMurdo Sound have yielded the polycyclic suberitenones A and B (**77**, **78**) (Figure 16), the latter of which has been shown to inhibit the cholesteryl ester transfer protein (CETP). This CETP protein mediates the transfer of cholesterol ester and triglyceride between high-density lipoproteins (HDL) low-density lipoproteins (LDL) and is a major target for the development of atherosclerotic disease treatments [142].

Darwinolide (63)   9,11-dihydrogracilin A (64)   Tetrahydroaplysulphurin-1 (65)   Membranolide (66)

Glaciolide (67)   68   Cadlinolide C (69)   Dendrillin A (70)

Figure 14. Structures of further diterpenoid compounds from *Dendrilla membranosa*: (**63**) darwinolide, (**64**) 9,11-dihydrogracilin A, (**65**) tetrahydroaplysulphurin-1, (**66**) membranolide, (**67**) glaciolide, (**68**), (**69**) cadlinolide C and (**70**) dendrillin A.

Dendrillin B (71)   Dendrillin C (72)   Dendrillin D (73)

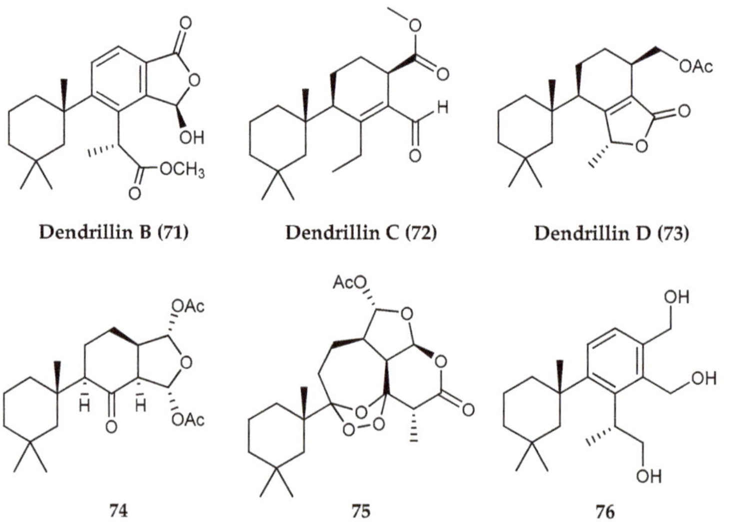

74   75   76

Figure 15. Structures of diterpenoid compounds from *Dendrilla membranosa*, (**71–73**) dendrillins B, C and D and semisynthetic derivatives (**74–76**).

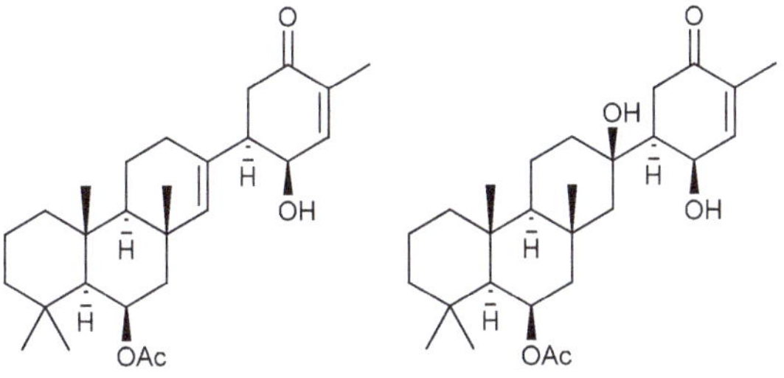

Suberitenone A (77)   **Suberitenone B (78)**

Figure 16. Structures of the sesterterpenoid compounds (**77, 78**) suberitenones A-B from *Suberites* spp.

3.2.5. Triterpenes and Triterpenoids

Triterpenes are formed by six isoprene units, conceptualized as three monoterpene units. Functionalized triterpenes (containing heteroatoms substitutions) should instead be referred to as triterpenoids. Animals, plants and fungi all produce triterpenes and triterpenoids, which exist in ~200 different skeletons and a great variety of structures (e.g., cholesterol) [213].

Steroids

Animals, plants and fungi all produce triterpenes, among which is squalene, the precursor to all steroids. These contain a core moiety of the triterpene cucurbitane, and in practice are biosynthesized from either lanosterol (animals and fungi) or cycloartenol (plants) via the cyclization of squalene. Steroids are further metabolized from squalene via subsequent demethylation to a tri-nor C_{27} skeleton, or further still to even smaller steroids. Steroids have two principal biological functions, being either key components of cell membranes or signaling molecules. Some examples of steroids are vitamin D3, the lipid cholesterol, the sex hormones estradiol and testosterone and the anti-inflammatory drug dexamethasone [214].

The Alcyonacean octocoral *Anthomastus bathyproctus* Bayer 1993 collected in the South Shetland Islands (Antarctica) afforded seven steroids, all of them displaying a cross-conjugated ketone system in the A ring of the tetracarbocyclic nucleus, while their side chains belong to the cholestane, ergostane and 24-norcholestane types. Compounds **79** to **82** (Figure 17) showed diverse in vitro cytotoxicity against the human tumor cell lines MDA-MB-231 (breast adenocarcinoma), A-549 (lung carcinoma) and HT-29 (colon adenocarcinoma) [106].

Dasystenella acanthina collected from Kapp Norvegia (Eastern Weddell Sea, Antarctica) was found to possess seven polyoxygenated steroids (**83–89**) (Figure 18). All compounds displayed some sort of growth inhibitory activity on tumor cell lines, including DU-145 (prostate carcinoma), LN-caP (prostate carcinoma), IGROV (ovarian adenocarcinoma), SK-BR3 (breast adenocarcinoma), SK-MEL-28 (melanoma), A549 (lung adenocarcinoma), K-562 (chronic myelogenous leukemia), PANC1 (pancreas carcinoma), HT29 (colon adenocarcinoma), LOVO (colon adenocarcinoma), LOVO-DOX (colon adenocarcinoma resistant to doxorubicin) and HeLa (cervix epithelial adenocarcinoma). The most affected cell lines were LN-caP and K-562. Compounds **88** and **89** presented broader cytostatic effects, and compound **89** was active against all tested cell lines [108].

Six polyoxygenated sterols (**90–95**) (Figure 19) were isolated from the soft coral *Gersemia fruticosa*, exhibiting a moderate cytotoxic activity against human erythroleukemia K-562 cells and other leukemia cell lines [109].

Gersemia fruticosa was also found to contain further a bioactive 9,11-secosterol steroid, named 24-nor-9,11-seco-11-acetoxy-3,6-dihydroxycholest-7,22(*E*)-dien-9-one (**96**) (Figure 20). This compound was shown to yield growth inhibition (IC_{50} below 10 μM) and cytotoxicity against human leukemia K562, human cervical cancer HeLa and Ehrlich ascites tumor cells in vitro [110].

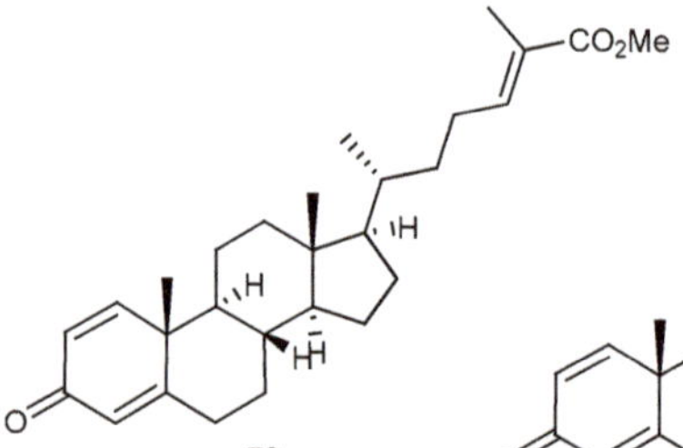

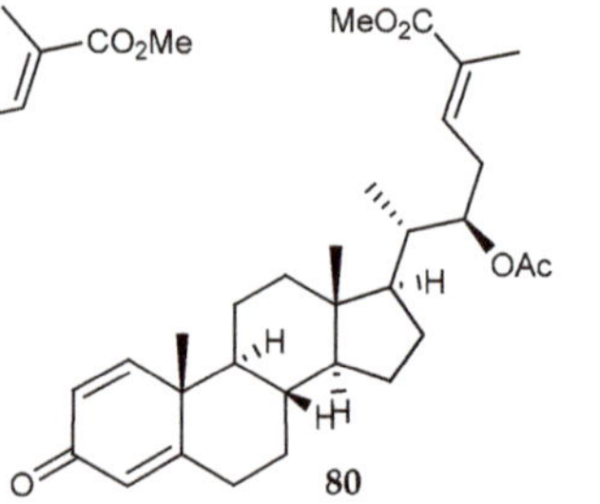

79 80

Conjugated cholestane, ergostane and 24-norcholestane steroids

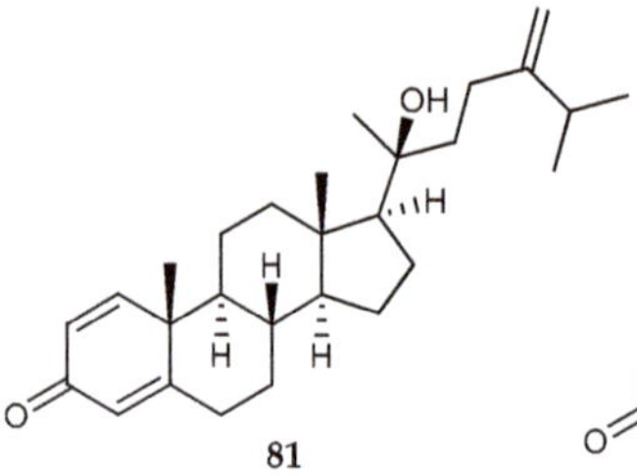

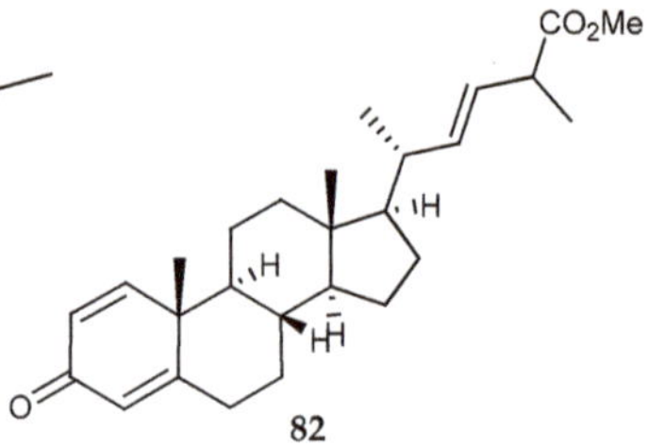

81 82

Figure 17. Structures of several steroid compounds (**79–82**) from *Anthomastus bathyproctus*.

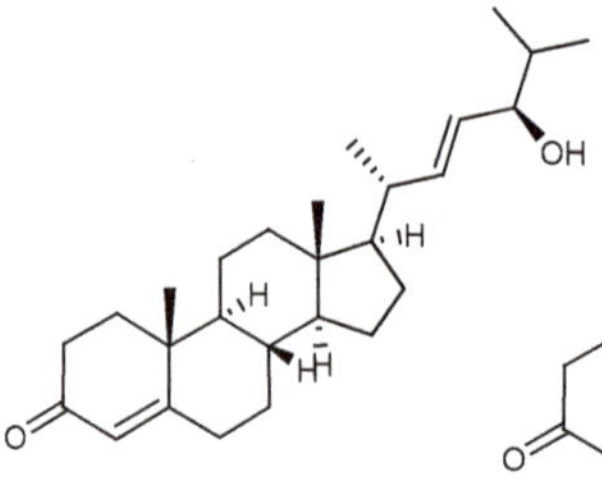

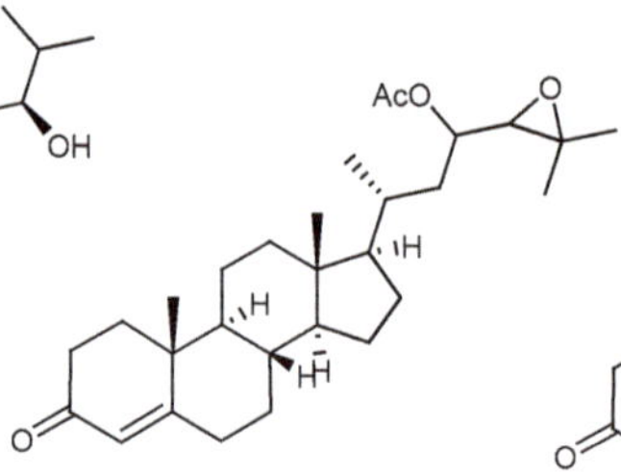

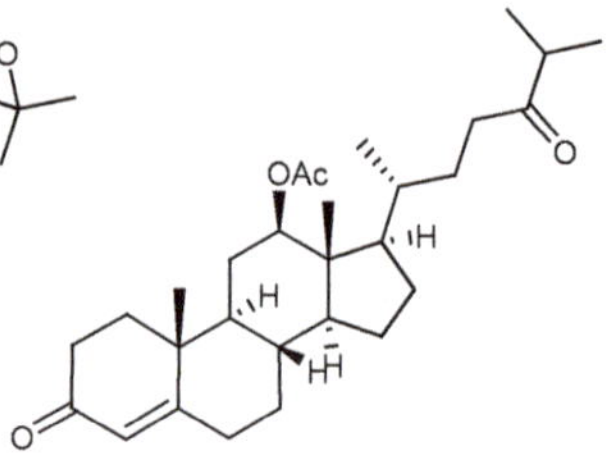

83 84 85

Polyoxygenated steroids from *D. acanthina*

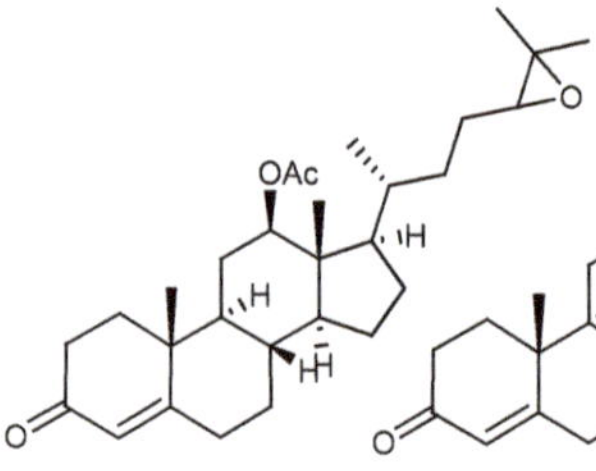

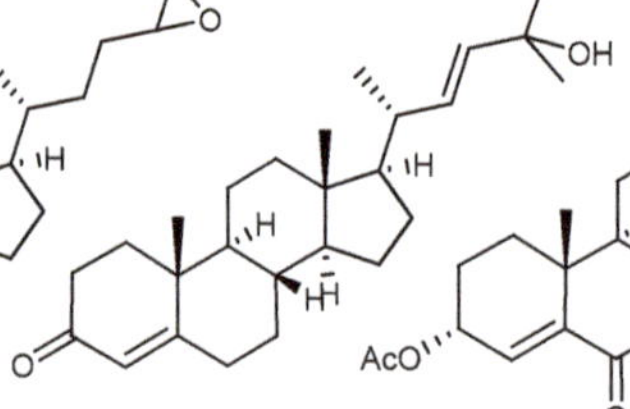

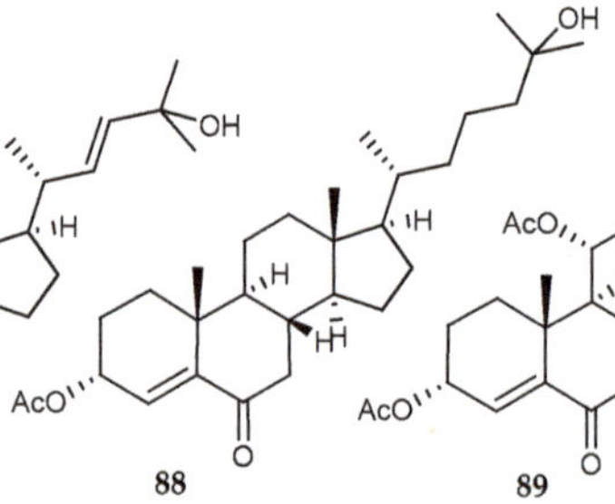

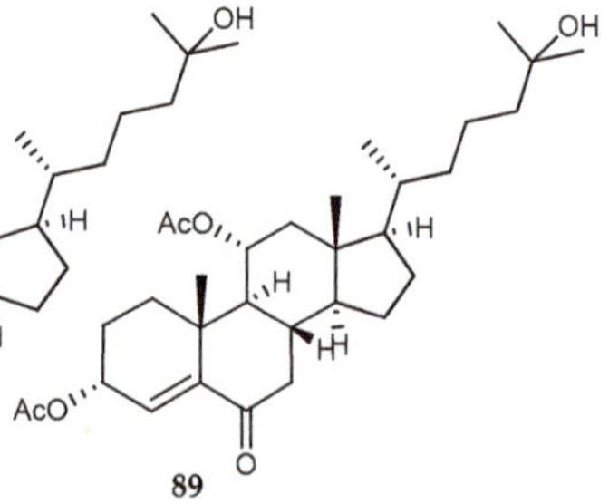

86 87 88 89

Figure 18. Structures of polyoxygenated steroid compounds (**83–89**) from *Dasystenella acanthina*.

Sponges are known to produce a wide variety of bioactive steroids and steroid derivatives, and the species found in polar waters are no exception. A red encrusting sponge within the genus *Crella* is commonly found on the sheer walls in the shallow waters along the Western Antarctic Peninsula, and a group of specimens collected around Norsel Point near Anvers Island yielded norselic acids A-E (**97–101**) (Figure 21). Among these oxidized steroids, the most abundant, norselic acid A, showed inhibitory activity against MRSA and vancomycin-resistant *Enterococci faecium* (VREF) in addition to antifungal activity against *C. albicans*. All of the norselic acids showed low micromolar activity against the leishmaniasis causing protozoan parasite *L. donovani* with potencies ranging from 2.0–3.6 µM [122].

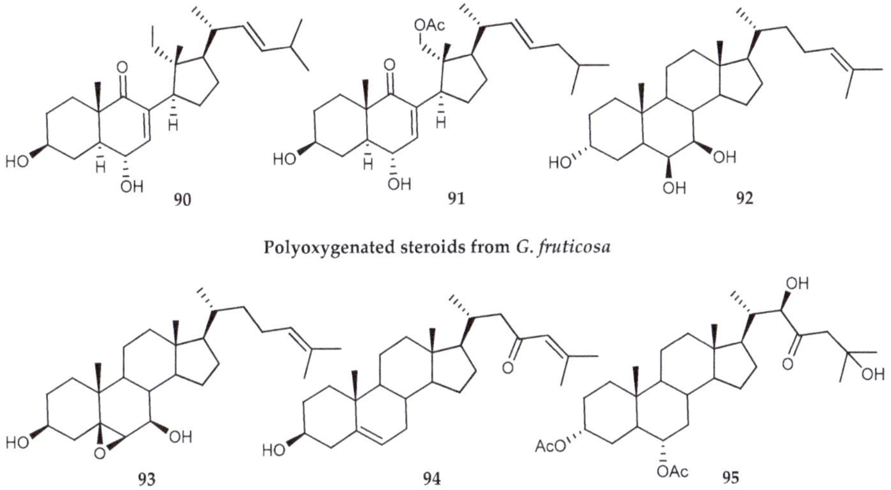

Polyoxygenated steroids from *G. fruticosa*

Figure 19. Structures of polyoxygenated steroid compounds (**90–95**) from *Gersemia fruticosa*.

Secosterol (96)

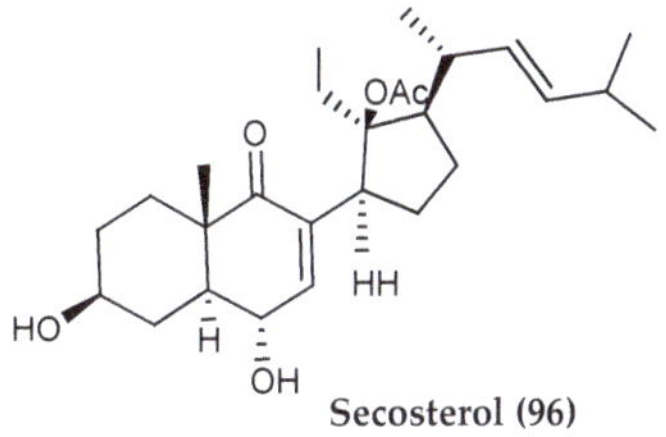

Figure 20. Structure of a further steroid compound (**96**) from *Gersemia fruticosa*.

Norselic acid A (97)

Norselic acid B (98)

Norselic acid C (99)

Norselic acid D (100)

Norselic acid E (101)

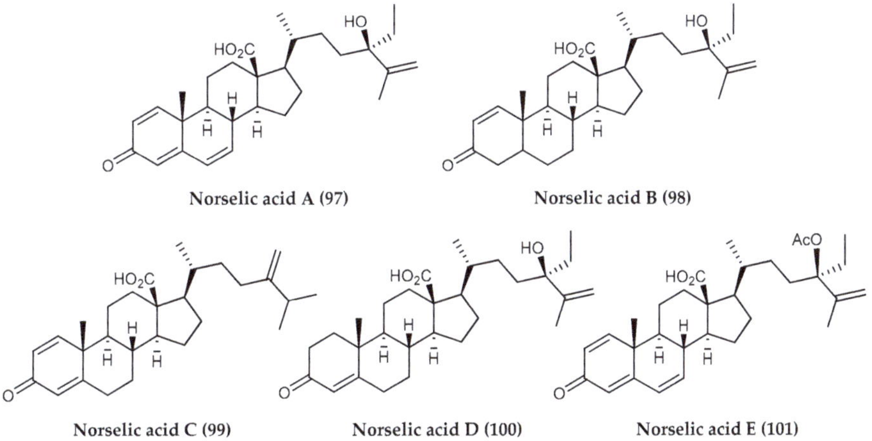

Figure 21. Structures of steroid compounds (**97–101**) from *Crella* sp.

Sulphated polyhydroxysteroids, obtained from Antarctic brittle star species, have been providing promising antiviral properties. Ophiuroid *Ophiosparte gigas*, coming from 70 m depth in the Ross Sea area revealed the presence of cholest-5-ene-2α,3α,4β,21-tetraol-3,21-disulphate (**102**), which was remarkably cytotoxic, along with cholest-5-ene-2β,3α, 21-triol-2,21-disulphate (**103**) with cytoprotective activity against HIV-1 (Figure 22) [120]. Further disulfated polyhydroxysteroids (**104–106**) from another Antarctic ophiuroid, *Astrotoma agassizii* (Figure 22), as well as their synthetic derivatives displayed antiviral activities against one DNA (HSV-2) and two RNA (PV-3, JV) viruses [121].

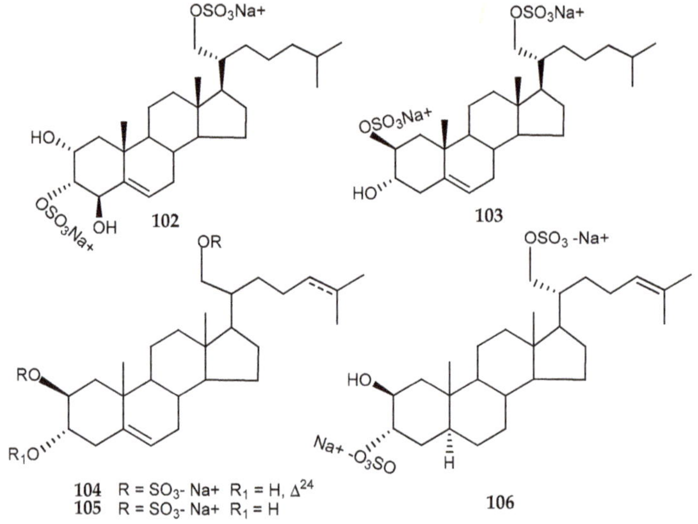

Figure 22. Structures of sulphated polyhydroxysteroids compounds (**102, 103**) from *Ophiosparte gigas* and (**104–106**) from *Astrotoma agassizii*.

Despite being most well-known for their role as hormones regulating molting in arthropods, a surprising array of unique ecdysteroids have been found in the Antarctic ascidian *Synoicum adareanum*, some of which display promising therapeutic activities. Among these are hyousterones A and C (**107, 108**) (Figure 23) which displayed IC_{50} values of 10.7 µM and 3.7 µM, respectively, against the HCT-116 colon cancer cell line, while abeohyousterone (**109**) (Figure 23) was active at 3.0 µM in the same biological assay, all of which were isolated from tunicates around Anvers Island [149].

3.2.6. Tetraterpenes and Tetraterpenoids

Carotenoids

Tetraterpenes are terpenes built from eight isoprene units (four monoterpene units). Carotenoids belong to the category of tetraterpenoids. They are natural isoprenoid pigments derived from head-to-tail condensation of two C_{15} or C_{20} isoprenoid precursors to form a C_{30} or C_{40} backbone, respectively, which are then modified to obtain different carotenoid structures [215]. They are divided into two main classes: carotenes, which are hydrocarbons, and xanthophylls, oxygenated derivatives of carotenes. Due to the long system of conjugated double bonds, they are able to capture and absorb light in the 400–500 nm range, displaying a peculiar strong coloration [216]. For this feature, carotenoids play a critical role in the photosynthesis process and provide photo-oxidative protection to the cells acting as strong antioxidant compounds. Being chemical quenchers of singlet oxygen, they function as potent scavengers of reactive oxygen species (ROS) [216]. They are essential constituents of photosynthetic organisms (e.g., plants, algae and cyanobacteria), but have been found also in fungi and bacteria [217]. Not synthesized by humans or animals, they are present in their blood and tissues, deriving from dietary sources. Being precursors of retinol (vitamin A), they perform a role of particular significance to human health [216].

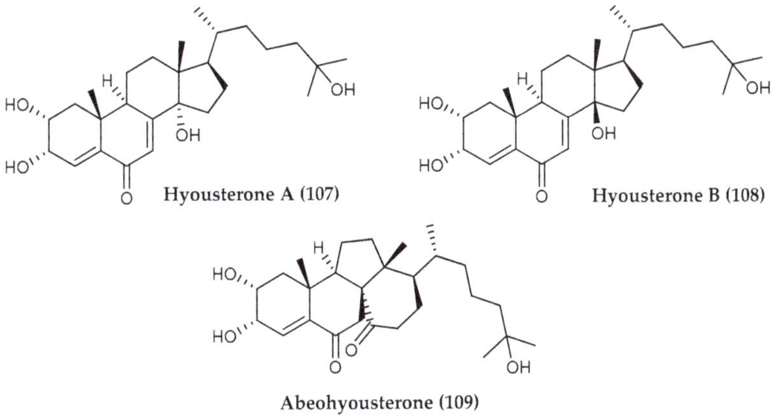

Hyousterone A (107)

Hyousterone B (108)

Abeohyousterone (109)

Figure 23. Structures of ecdysteroid compounds (**107**, **108**) hyousterones A-B and (**109**) abeohyousterone from *Synoicum adareanum*.

In fact, carotenoids from marine environments are strong antioxidants used as nutraceutical ingredients in the food industry and cosmeceutical molecules for the photoprotection against UV radiation [218,219].

Although marine animals do not synthesize carotenoids de novo, they contain significant amounts of carotenoids derived from dietary sources. More than 100 carotenoids have been isolated from sponges, cnidarians, mollusks, crustaceans, echinoderms, tunicates and fishes [220].

Marine organisms inhabiting polar environments have developed a variety of adaptive strategies to cope with UV radiation, including light avoidance mechanisms, synthesis of UV-sunscreens, enzymatic and non-enzymatic quenching of ROS and DNA repair mechanisms [2]. The synthesis of carotenoid pigments in the Antarctic marine organisms belongs to the antioxidant defense mechanisms able to counteract ROS damage [2].

Indeed, some pigmented bacteria owe their colors to the presence of carotenoids. The genome mining of *Marisediminicola antarctica* ZS314T, isolated from intertidal sediments of the cost near the Chinese Antarctic Zhongshan Station in East Antarctica, demonstrated the biosynthetic potential of this orange Actinobacterium in producing carotenoids and their derivatives [221]. *Cellulophaga fucicola* strain 416 and *Zobellia laminarie* 465, yellow and orange pigmented bacteria, respectively [98,99], isolated from Antarctic sea sponges, were found to be resistant to UV-B and UV-C radiation, thanks to the expression of carotenoids isolated and chemically identified by ultra-high-performance liquid chromatography with photodiode array and mass spectrometry detectors. Zeaxanthin (**110**), β-cryptoxanthin (**111**) and β-carotene (**112**) were identified in both strains, whereas two isomers of zeaxanthin was identified only in *C. fucicola* [98] and phytoene (**113**) only in *Z. laminarie* (Figure 24) [99]. These pigments displayed a very high antioxidant activity, although they were shown to be phototoxic in murine fibroblast lines [98,99].

The red-orange strain *Rhodococcus* sp. B7740, isolated from 25 m deep-sea water in the Arctic Ocean, is a promising source of natural carotenoids and isoprenoid quinones, interesting both in amounts and varieties for the application in the food industry [74]. Among them, synechoxanthin (χ,χ-caroten-18,18′-dioic acid) (**114**), a unique aromatic dicarboxylate carotenoid, recently discovered only in some cyanobacteria [222,223], dehydrogenated menaquinones with eight isoprene units [MK-$_8$(H$_2$)] (**115**), produced in higher concentration than that reported in other bacteria [224], and isorenieratene (**116**), an aromatic carotenoid used in smear cheese industry [225], have been identified (Figure 25). The latter is a promising metabolite in future food and medicine applications for its higher stability than β-carotene and lutein in model gastric conditions and for its high retention rate in the gastrointestinal tract after ingestion [74]. Moreover, it has been demonstrated that isorenieratene, able to prevent UV-induced DNA damage in human skin fibroblasts [226], displays a photoprotective effect against UV-B radiation compared with two macular xanthophylls, lutein and zeaxanthin, in the multilamellar vesicles model and human retina cell model [75]. Additionally, MK-$_8$(H$_2$), the main menaquinone from *Rhodococcus* sp. B7740, has a potential application in the field of medicine for its higher antioxidant effect and antiglycation capacity compared with ubiquinone Q10 and MK$_4$ [76].

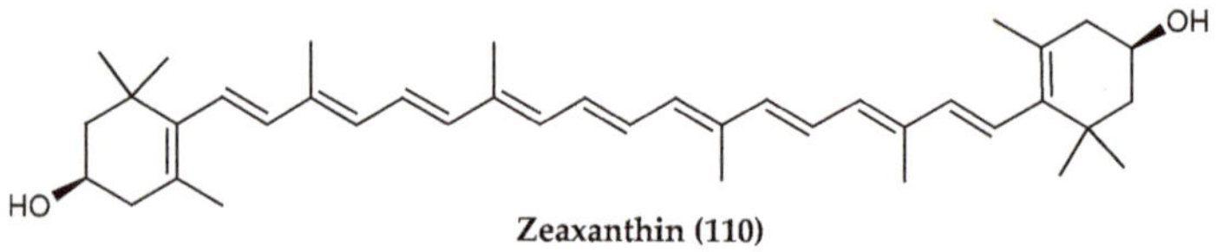

Zeaxanthin (110)

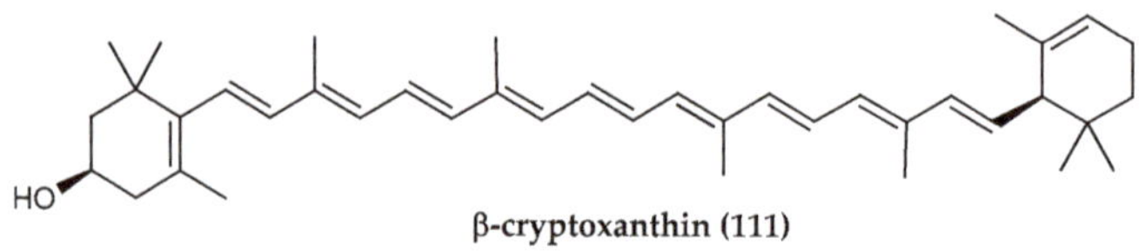

β-cryptoxanthin (111)

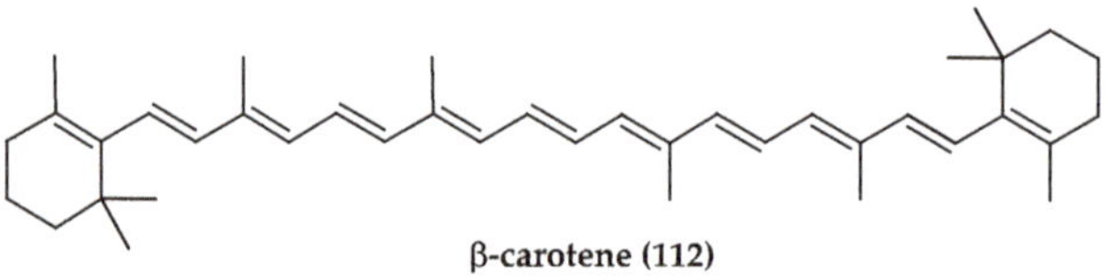

β-carotene (112)

Phytoene (113)

Figure 24. Structures of carotenoids found in bacteria *Cellulophaga fucicola* 416 and *Zobellia laminarie* 465: (**110**) zeaxanthin, (**111**) β-cryptoxanthin, (**112**) β-carotene and (**113**) phytoene.

3.2.7. Triterpene and Triterpenoid Derivatives

Triterpenoid Conjugates

While steroids come in a variety of forms, they can also be found conjugated to other functional groups. One such example of these remarkable molecules are the sulfated steroid-amino acid conjugates known as polymastiamides A–F, isolated from the cold-water sponge *Polymastia boletiformis* collected at various locations along the Norwegian coast. Of these compounds, polymastiamide A (**117**) (Figure 26) has shown activity against plant pathogens *Cladosporium cucumerinum* and *Pythium ultimum* as well as human yeast pathogen *C. albicans* and antibacterial activity against *S. aureus* [138,139].

Triterpenoid Saponins

Triterpenoid saponins are amphipathic glycosides that have one or more hydrophilic glycoside moieties combined with a lipophilic triterpene or steroid derivative, thus, their chemical class can be identified as triterpenoid glycosides. They are well-known as plant-derived allelochemicals, but they have also been obtained from marine organisms, in particular sea cucumbers, where they have been proposed as chemotaxonomic proxies [227,228]. Several bioactive properties have been reported from these compounds, including anti-feedant (repellents), antimicrobial and ichthyotoxic [229,230]. The term saponin derives from the soapwort plant (genus *Saponaria* family Caryophyllaceae), the root from which is used as a soap. Additionally, these triterpene conjugates produce a soap-like foam when shaken in aqueous solutions [230]. The amphipathic properties of saponins make them efficient surfactants, due to their capacity to interact with cell membrane components, e.g., cholesterol and phospholipids, and therefore, they are interesting for the development of cosmetics, drugs and nutraceuticals (nutrient absorption enhancers) [231]. Saponins are also readily soluble in water, and have been proposed as adjuvants, to dissolve active principles in the development of vaccines [232].

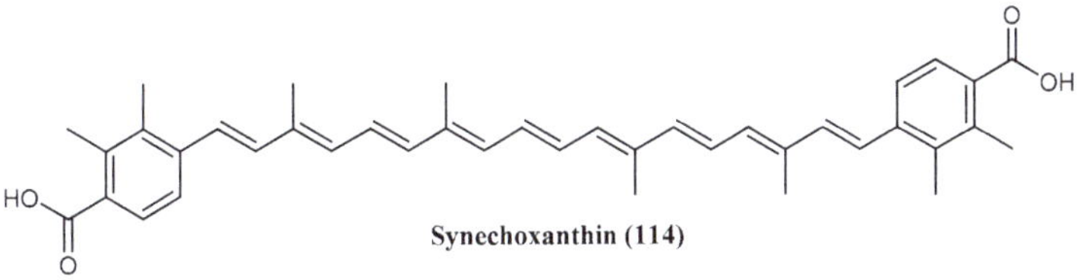

Synechoxanthin (114)

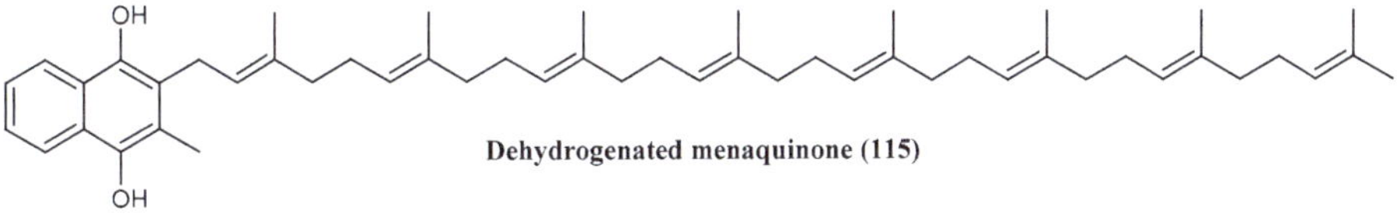

Dehydrogenated menaquinone (115)

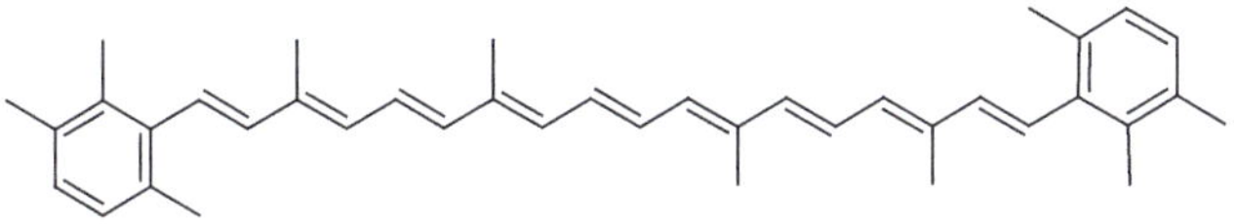

Isorenieratene (116)

Figure 25. Structures of several carotenoids obtained from *Rhodococcus* sp. B7740: (**114**) synechoxanthin, (**115**) dehydrogenated menaquinone and (**116**) isorenieratene.

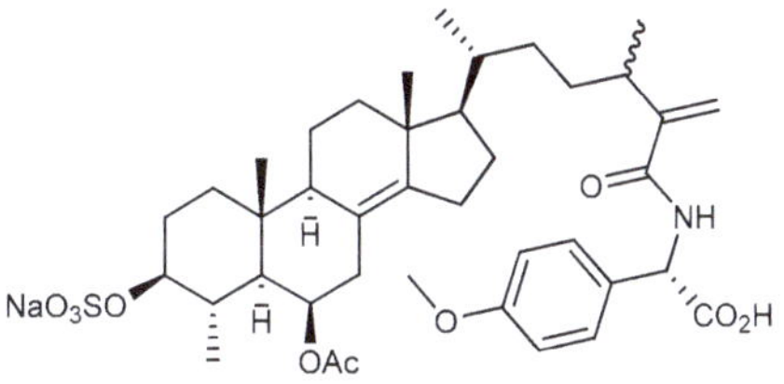

Polymastiamide A (117)

Figure 26. Structure of the triterpenoid conjugate compound (**117**) polymastiamide A from *Polymastia boletiformis*.

Two trisulfated triterpene glycosides, liouvillosides A (**118**) and B (**119**) (Figure 27), both virucidal against herpes simplex virus type 1 (HSV-1), were isolated from the Antarctic cucumarid sea cucumber *Staurocucumis liouvillei,* collected in South Georgia Islands [118].

The deep sea Arctic holothurian echinoderm *Kolga hyalina* collected at Amundsen Basin at 4352–4354 m depth was found to contain holothurinoside B (known from temperate holothurid species [233], as well as two novel triterpene holostane nonsulfated pentaosides, kolgaosides A (**120**) and B (**121**) (Figure 28), both possessing hemolytic activity against mouse erythrocytes, and mild inhibitory action against Ehrlich ascite carcinoma cells. All these triterpene glycosides are structurally close to achlioniceosides A1–A3 from the Antarctic sea cucumber *Rhipidothuria racovitzai* Hèrouard, 1901 (=*Achlionice violaescupidata* [234]), supporting a potential chemotaxonomic value [235].

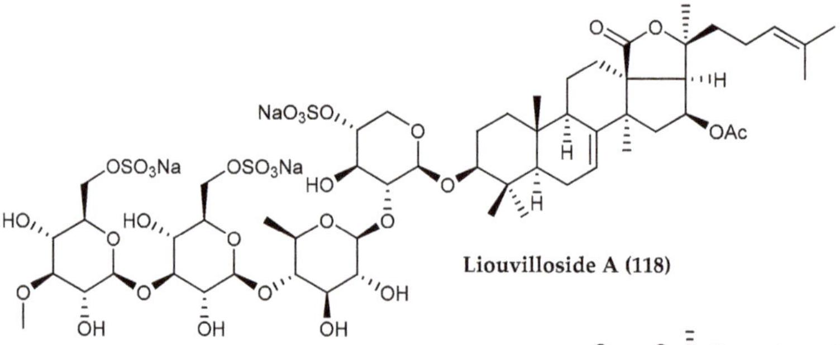

Liouvilloside A (118)

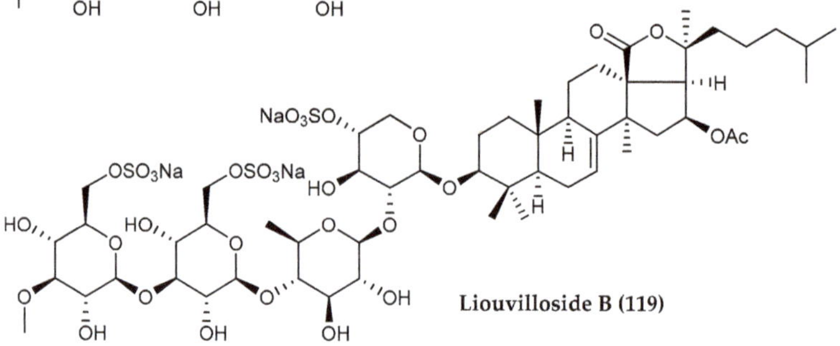

Liouvilloside B (119)

Figure 27. Structures of two triterpenoid saponin compounds (**118, 119**) liouvillosides A–B from *Staurocucumis liouvillei*.

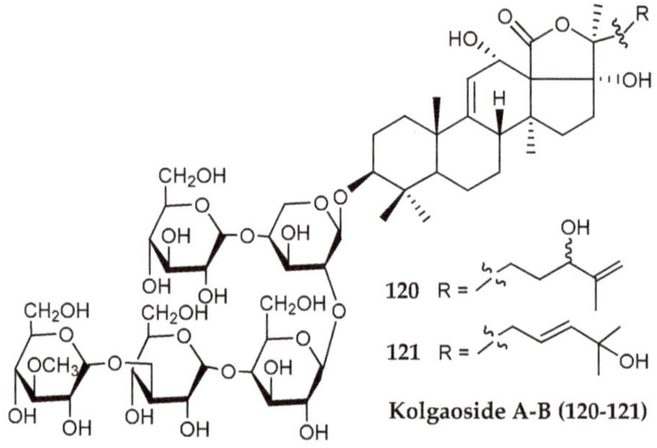

Kolgaoside A-B (120-121)

Figure 28. Structures of two triterpenoid glycoside compounds (**120, 121**) kolgaosides A–B from *Kolga hyalina*.

Meroterpenes

Meroterpenes are molecules with a partial terpenoid structure attached to a shikimate-derived aromatic, usually a phenol, and are commonly found in marine ascidians, sponges and to a lesser extent in soft corals [236]. Bioactive meroterpenes featuring sesquiterpene moieties can also be found in organisms inhabiting the harsh polar regions. Tunicates in the genus *Aplidium* collected in the Ross Sea yielded rossinones A and B (**122, 123**) (Figure 29), which displayed antiproliferative activity against several cell lines with IC_{50} values ranging from 0.084 to 30 μM as well as selective antiviral activity against the DNA virus HSV-1 in addition to antibacterial and antifungal activity towards *Bacillus subtilis* and *Trichophyton mentagrophytes* [144].

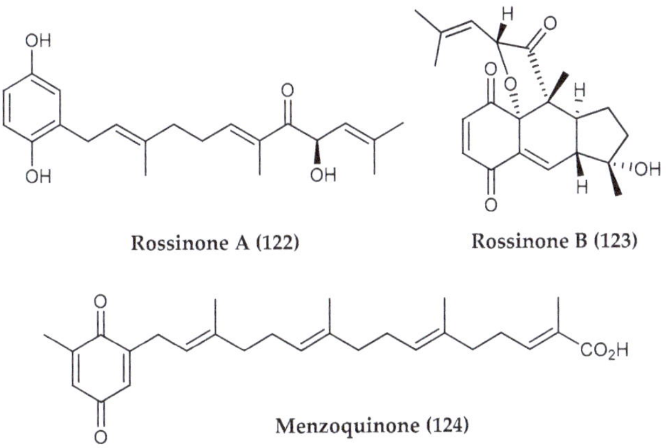

Rossinone A (122)   **Rossinone B (123)**

Menzoquinone (124)

Figure 29. Structures of two meroterepenoid compounds, (**122, 123**) rossinones A-B from *Aplidium* spp., and (**124**) menzoquinone from *Desmarestia menziesii*.

While the majority of compounds reported from polar marine algae are monoterpenes, bioactive diterpenoids in the form of meroterpenes such as menzoquinone (**124**) (Figure 29) have been isolated from *Desmarestia menziesii*, a commonly occurring brown algae that plays a major role in structuring the benthic ecosystems along the northern latitudes of the Western Antarctic Peninsula. This methylated diterpene-quinone bearing a carboxylic acid has shown antimicrobial activity against MRSA and VREF [155].

4. Perspectives and Conclusions

Bioprospecting is a complex issue as it embraces many fields, such as intellectual property rights, scientific research and exploitation of resources in an eco-friendly ethical manner [237–240]. Cooperation is needed to develop sustainable, respectful and appropriate access and benefit-sharing mechanisms for marine resources as well as the promotion of the participation by all states in international negotiations for encouraging innovation and greater equity [241–243].

Bioactive natural products from the sea are in this particular era are timely to develop drugs to fight against ever more frequent and contagious emerging pathogenic agents. One actual example is a potent antiviral molecule obtained from the ascidian *Aplidium albicans*, which is under clinical trials on infected humans with Corona Virus, similar to the 2020 pandemic SARS CoV-19 [244].

Marine organisms from the polar regions could greatly contribute to this growing repertoire of promising bioactive compounds. Indeed, extreme environments are important hot spots of microbial, metazoan and symbiotic cluster diversities, where selective forces have promoted the evolution of unique biosynthetic pathways for secondary metabolite production [14]. In the light of the publication record on molecules with pharmacological potential isolated up to date from Arctic and Antarctic marine taxa, terpene and terpenoid derivatives seem to be the most frequently reported [5]. Furthermore, these compound types often yield remarkable antimicrobial properties, including anti-viral and antitumoral activities. Such cytotoxic actions likely respond to detrimental effects driven by terpenoid products on the structure and function of microbial membranes and cell walls [245]. Therefore, natural bioactive terpenes and terpenoids, in these times of increasing incidence of emerging infectious diseases, antibiotic-resistance pathogenesis and cancer, represent a precious glimmer of hope for drug discovery. Bioprospecting of organisms inhabiting the polar environments has already led to the discovery of new bioactive molecules, mainly enzymes with potential commercial use for food, paper and textile industries [4]. It is thus expected that, in the near future, natural products from polar latitudes with an untapped biotechnological potential will also be included in health products to address upcoming epidemics, and disorders related to emerging and resistant infective vectors.

Author Contributions: All authors contributed to the bibliography of the review and preparation of the final version of the manuscript. D.G. developed the original idea. L.N.-P., B.J.B. and D.G. organized all the materials. All authors have read and agreed to the published version of the manuscript

Funding: This study was funded by the Italian National Programme for Antarctic Research (PNRA) (2016/AZ1.06-Project PNRA16_00043 and 2016/AZ1.20-Project PNRA16_00128) and by CNR project Green&CircularEconomy (FOE-2019, DBA.AD003.139). It was carried out in the framework of the SCAR Programme "Antarctic Thresholds–Ecosystem Resilience and Adaptation" (AnT-ERA).

Acknowledgments: D.G. wishes to thank Valentina Brasiello, Chiara Nobile and Giovanni Del Monaco for technical support and assistance.

Conflicts of Interest: The authors declare no conflict of interest.

References

1. Hurst, D.; Børresen, T.; Almesjö, L.; De Raedemaecker, F.; Bergseth, S. *Marine Biotechnology Strategic Research and Innovation Roadmap: Insights to the Future Direction of European Marine Biotechnology*; Marine Biotechnology ERA-NET: Oostende, Belgium, 2016.

2. Nunez-Pons, L.; Avila, C.; Romano, G.; Verde, C.; Giordano, D. UV-protective compounds in marine organisms from the Southern Ocean. *Mar. Drugs* **2018**, *16*, 336. [CrossRef] [PubMed]

3. Angulo-Preckler, C.; Spurkland, T.; Avila, C.; Iken, K. Antimicrobial activity of selected benthic Arctic invertebrates. *Polar Biol.* **2015**, *38*, 1941–1948. [CrossRef]

4. Bruno, S.; Coppola, D.; di Prisco, G.; Giordano, D.; Verde, C. Enzymes from marine polar regions and their biotechnological applications. *Mar. Drugs* **2019**, *17*, 544. [CrossRef] [PubMed]

5. Soldatou, S.; Baker, B.J. Cold-water marine natural products, 2006 to 2016. *Nat. Prod. Rep.* **2017**, *34*, 585–626. [CrossRef]

6. Tian, Y.; Li, Y.L.; Zhao, F.C. Secondary metabolites from polar organisms. *Mar. Drugs* **2017**, *15*, 28. [CrossRef]

7. Eakins, B.W.; Sharman, G.F. *Volumes of the World's Oceans from ETOPO1*; NOAA National Geophysical Data Center: Boulder, CO, USA, 2010.

8. Polyak, L.; Alley, R.B.; Andrews, J.T.; Brigham-Grette, J.; Cronin, T.M.; Darby, D.A.; Dyke, A.S.; Fitzpatrick, J.J.; Funder, S.; Holland, M.; et al. History of sea ice in the Arctic. *Quat. Sci. Rev.* **2010**, *29*, 1757–1778. [CrossRef]

9. Convey, P.; Bindschadler, R.; di Prisco, G.; Fahrbach, E.; Gutt, J.; Hodgson, D.A.; Mayewski, P.A.; Summerhayes, C.P.; Turner, J. Antarctic climate change and the environment. *Antarct. Sci.* **2009**, *21*, 541–563. [CrossRef]

10. Jayatilake, G.S.; Baker, B.J.; McClintock, J.B. Isolation and identification of a stilbene derivative from the Antarctic sponge *Kirkpatrickia variolosa*. *J. Nat. Prod.* **1995**, *58*, 1958–1960. [CrossRef]

11. Gordon, A. Oceanography of Antarctic Waters. In *Antarctic Oceanology I*; Reid, J.L., Ed.; AGU: Hoboken, NJ, USA, 1971; pp. 169–203.

12. Barker, P.F.; Filippelli, G.M.; Florindo, F.; Martin, E.E.; Scher, H.D. Onset and role of the Antarctic Circumpolar Current. *Deep Sea Res. II Top. Stud. Ocean.* **2007**, *54*, 2388–2398. [CrossRef]

13. Avila, C.; Taboada, S.; Núñez-Pons, L. Antarctic marine chemical ecology: What is next? *Mar. Ecol.* **2008**, *29*, 1–71. [CrossRef]

14. Núñez-Pons, L.; Avila, C. Natural products mediating ecological interactions in Antarctic benthic communities: A mini-review of the known molecules. *Nat. Prod. Rep.* **2015**, *32*, 1114–1130. [CrossRef] [PubMed]

15. Liu, J.T.; Lu, X.L.; Liu, X.Y.; Gao, Y.; Hu, B.; Jiao, B.H.; Zheng, H. Bioactive natural products from the antarctic and arctic organisms. *Mini Rev. Med. Chem.* **2013**, *13*, 617–626. [CrossRef] [PubMed]

16. Lebar, M.D.; Heimbegner, J.L.; Baker, B.J. Cold-water marine natural products. *Nat. Prod. Rep.* **2007**, *24*, 774–797. [CrossRef] [PubMed]

17. Klemm, E.J.; Wong, V.K.; Dougan, G. Emergence of dominant multidrug-resistant bacterial clades: Lessons from history and whole-genome sequencing. *Proc. Nat. Acad. Sci. USA* **2018**, *115*, 12872–12877. [CrossRef] [PubMed]

18. Martins, A.; Vieira, H.; Gaspar, H.; Santos, S. Marketed marine natural products in the pharmaceutical and cosmeceutical industries: Tips for success. *Mar. Drugs* **2014**, *12*, 1066–1101. [CrossRef] [PubMed]

19. Market Research Engine. Available online: https://www.marketresearchengine.com/marine-derived-drugs-market (accessed on 15 May 2020).

20. Mayer, A.S.M. Marine Pharmaceuticals: The Clinical Pipeline. Available online: https://www.midwestern.edu/departments/marinepharmacology.xml (accessed on 27 July 2020).

21. Blunt, J.W.; Carroll, A.R.; Copp, B.R.; Davis, R.A.; Keyzers, R.A.; Prinsep, M.R. Marine natural products. *Nat. Prod. Rep.* **2018**, *35*, 8–53. [CrossRef] [PubMed]

22. Jaspars, M.; De Pascale, D.; Andersen, J.H.; Reyes, F.; Crawford, A.D.; Ianora, A. The marine biodiscovery pipeline and ocean medicines of tomorrow. *J. Mar. Biol. Assoc. U. K.* **2016**, *96*, 151–158. [CrossRef]

23. Schoffski, P.; Dumez, H.; Wolter, P.; Stefan, C.; Wozniak, A.; Jimeno, J.; Van Oosterom, A.T. Clinical impact of trabectedin (ecteinascidin-743) in advanced/metastatic soft tissue sarcoma. *Expert Opin. Pharm.* **2008**, *9*, 1609–1618. [CrossRef]

24. Karsch-Mizrachi, I.; Takagi, T.; Cochrane, G.; International Nucleotide Sequence Database Collaboration. The international nucleotide sequence database collaboration. *Nucleic Acids Res.* **2018**, *46*, D48–D51. [CrossRef]

25. Costello, M.J.; Chaudhary, C. Marine biodiversity, biogeography, deep-sea gradients, and conservation. *Curr. Biol.* **2017**, *27*, R511–R527. [CrossRef]

26. Puig-Marcó, R. Access and benefit sharing of Antarctica's biological material. *Mar. Genom.* **2014**, *17*, 73–78. [CrossRef] [PubMed]

27. Dodds, K.J. Sovereignty watch: Claimant states, resources, and territory in contemporary Antarctica. *Polar Rec.* **2010**, *47*, 231–243. [CrossRef]

28. Petra, D.; Alex, G.O.E.; Bert, V.; Tamara, T. Marine genetic resources in areas beyond national jurisdiction: Access and benefit-sharing. *Int. J. Mar. Coast. Law* **2012**, *27*, 375–433. [CrossRef]

29. Dodds, K. Governing Antarctica: Contemporary challenges and the enduring legacy of the 1959 Antarctic Treaty. *Glob. Policy* **2010**, *1*, 108–115. [CrossRef]

30. Joyner, C.C. Bioprospecting as a challenge to the Antarctic Treaty. In *Antarctic Security in the Twenty-First Century*; Hemmings, A.D., Rothwell, D.R., Scott, K.N., Eds.; Routledge: Abingdon, UK, 2012; pp. 197–214.

31. Tvedt, M.W. Patent law and bioprospecting in Antarctica. *Polar Rec.* **2010**, *47*, 46–55. [CrossRef]

32. Rogan-Finnemore, M. What bioprospecting means for Antarctica and the Southern Ocean. In *International Law Issues in the South Pacific*; von Tigerstrom, B., Ed.; Ashgate: Hampshire, UK, 2005; pp. 199–228.

33. Summerhayes, C.P. International collaboration in Antarctica: The International Polar Years, the International Geophysical Year, and the Scientific Committee on Antarctic Research. *Polar Rec.* **2008**, *44*, 321–334. [CrossRef]

34. Oldham, P.; Hall, S.; Forero, O. Biological diversity in the patent system. *PLoS ONE* **2013**, *8*, e78737. [CrossRef]

35. Eritja, M.C. Bio-prospecting in the Arctic: An overview of the interaction between the rights of indigenous peoples and access and benefit sharing. *Boston Coll. Environ. Aff. Law Rev.* **2017**, *44*, 223.

36. Hoag, H. The cold rush. *Nat. Biotech.* **2009**, *27*, 690–692. [CrossRef]

37. Leary, D. *UNU-IAS Report: Bioprospecting in the Arctic 8, 2008*; UNU-IAS: Minato, Japan, 2013.

38. Carroll, A.R.; Copp, B.R.; Davis, R.A.; Keyzers, R.A.; Prinsep, M.R. Marine natural products. *Nat. Prod. Rep.* **2020**, *37*, 175–223. [CrossRef]

39. Alcaino, J.; Cifuentes, V.; Baeza, M. Physiological adaptations of yeasts living in cold environments and their potential applications. *World J. Microbiol. Biotechnol.* **2015**, *31*, 1467–1473. [CrossRef] [PubMed]

40. Munoz, P.A.; Marquez, S.L.; Gonzalez-Nilo, F.D.; Marquez-Miranda, V.; Blamey, J.M. Structure and application of antifreeze proteins from Antarctic bacteria. *Microb. Cell Fact.* **2017**, *16*, 138. [CrossRef] [PubMed]

41. Bang, J.K.; Lee, J.H.; Murugan, R.N.; Lee, S.G.; Do, H.; Koh, H.Y.; Shim, H.E.; Kim, H.C.; Kim, H.J. Antifreeze peptides and glycopeptides, and their derivatives: Potential uses in biotechnology. *Mar. Drugs* **2013**, *11*, 2013–2041. [CrossRef] [PubMed]

42. Venketesh, S.; Dayananda, C. Properties, potentials, and prospects of antifreeze proteins. *Crit. Rev. Biotechnol.* **2008**, *28*, 57–82. [CrossRef] [PubMed]

43. Qadeer, S.; Khan, M.A.; Ansari, M.S.; Rakha, B.A.; Ejaz, R.; Iqbal, R.; Younis, M.; Ullah, N.; DeVries, A.L.; Akhter, S. Efficiency of antifreeze glycoproteins for cryopreservation of Nili-Ravi (*Bubalus bubalis*) buffalo bull sperm. *Anim. Reprod. Sci.* **2015**, *157*, 56–62. [CrossRef] [PubMed]

44. Lee, S.G.; Koh, H.Y.; Lee, J.H.; Kang, S.H.; Kim, H.J. Cryopreservative effects of the recombinant ice-binding protein from the arctic yeast *Leucosporidium* sp. on red blood cells. *Appl. Biochem. Biotechnol.* **2012**, *167*, 824–834. [CrossRef]

45. Kumar, A.S.; Mody, K.; Jha, B. Bacterial exopolysaccharides-A perception. *J. Basic Microbiol.* **2007**, *47*, 103–117. [CrossRef]

46. Freitas, F.; Alves, V.D.; Reis, M.A. Advances in bacterial exopolysaccharides: From production to biotechnological applications. *Trends Biotechnol.* **2011**, *29*, 388–398. [CrossRef]

47. Caruso, C.; Rizzo, C.; Mangano, S.; Poli, A.; Di Donato, P.; Nicolaus, B.; Di Marco, G.; Michaud, L.; Lo Giudice, A. Extracellular polymeric substances with metal adsorption capacity produced by *Pseudoalteromonas* sp. MER144 from Antarctic seawater. *Environ. Sci. Pollut. Res. Int.* **2018**, *25*, 4667–4677. [CrossRef]

48. Poli, A.; Anzelmo, G.; Nicolaus, B. Bacterial exopolysaccharides from extreme marine habitats: Production, characterization and biological activities. *Mar. Drugs* **2010**, *8*, 1779–1802. [CrossRef]

49. Bianchi, A.C.; Olazabal, L.; Torre, A.; Loperena, L. Antarctic microorganisms as source of the omega-3 polyunsaturated fatty acids. *World J. Microbiol. Biotechnol.* **2014**, *30*, 1869–1878. [CrossRef] [PubMed]

50. Xue, Z.; Sharpe, P.L.; Hong, S.P.; Yadav, N.S.; Xie, D.; Short, D.R.; Damude, H.G.; Rupert, R.A.; Seip, J.E.; Wang, J.; et al. Production of omega-3 eicosapentaenoic acid by metabolic engineering of *Yarrowia lipolytica*. *Nat. Biotechnol.* **2013**, *31*, 734–740. [CrossRef]

51. Abbas, S.; Kelly, M.; Bowling, J.; Sims, J.; Waters, A.; Hamann, M. Advancement into the Arctic region for bioactive sponge secondary metabolites. *Mar. Drugs* **2011**, *9*, 2423–2437. [CrossRef] [PubMed]

52. Hansen, Ø.K.; Isaksson, J.; Glomsaker, E.; Andersen, H.J.; Hansen, E. Ponasterone A and F, ecdysteroids from the Arctic bryozoan *Alcyonidium gelatinosum*. *Molecules* **2018**, *23*, 1481. [CrossRef]

53. Di, X.; Rouger, C.; Hardardottir, I.; Freysdottir, J.; Molinski, T.F.; Tasdemir, D.; Omarsdottir, S. 6-Bromoindole derivatives from the Icelandic marine sponge *Geodia barretti*: Isolation and anti-inflammatory activity. *Mar. Drugs* **2018**, *16*, 437. [CrossRef]

54. Figuerola, B.; Sala-Comorera, L.; Angulo-Preckler, C.; Vazquez, J.; Jesus Montes, M.; Garcia-Aljaro, C.; Mercade, E.; Blanch, A.R.; Avila, C. Antimicrobial activity of Antarctic bryozoans: An ecological perspective with potential for clinical applications. *Mar. Environ. Res.* **2014**, *101*, 52–59. [CrossRef] [PubMed]

55. Berne, S.; Kalauz, M.; Lapat, M.; Savin, L.; Janussen, D.; Kersken, D.; Avguštin, J.A.; Jokhadar, Š.Z.; Jaklič, D.; Gunde-Cimerman, N.; et al. Screening of the Antarctic marine sponges (Porifera) as a source of bioactive compounds. *Polar Biol.* **2016**, *39*, 947–959. [CrossRef]

56. Moles, J.; Nunez-Pons, L.; Taboada, S.; Figueroa, B.; Cristobo, J.; Avila, C. Anti-predatory chemical defences in Antarctic benthic fauna. *Mar. Biol.* **2015**, *162*, 1813–1821. [CrossRef]

57. Li, F.; Janussen, D.; Peifer, C.; Perez-Victoria, I.; Tasdemir, D. Targeted isolation of tsitsikammamines from the Antarctic deep-sea sponge *Latrunculia biformis* by molecular networking and anticancer activity. *Mar. Drugs* **2018**, *16*, 268. [CrossRef]

58. Thomas, T.R.; Kavlekar, D.P.; LokaBharathi, P.A. Marine drugs from sponge-microbe association—A review. *Mar. Drugs* **2010**, *8*, 1417–1468. [CrossRef]

59. Blockley, A.; Elliott, D.; Roberts, A.P.; Sweet, M. Symbiotic microbes from marine invertebrates: Driving a new era of natural product drug discovery. *Div. Distrib.* **2017**, *9*, 49. [CrossRef]

60. Lo Giudice, A.; Rizzo, C. Bacteria associated with marine benthic invertebrates from polar environments: Unexplored frontiers for biodiscovery? *Diversity* **2018**, *10*, 80. [CrossRef]

61. Furbino, L.E.; Godinho, V.M.; Santiago, I.F.; Pellizari, F.M.; Alves, T.M.; Zani, C.L.; Junior, P.A.; Romanha, A.J.; Carvalho, A.G.; Gil, L.H.; et al. Diversity patterns, ecology and biological activities of fungal communities associated with the endemic macroalgae across the Antarctic peninsula. *Microb. Ecol.* **2014**, *67*, 775–787. [CrossRef]

62. Henríquez, M.; Vergara, K.; Norambuena, J.; Beiza, A.; Maza, F.; Ubilla, P.; Araya, I.; Chávez, R.; San-Martín, A.; Darias, J.; et al. Diversity of cultivable fungi associated with Antarctic marine sponges and screening for their antimicrobial, antitumoral and antioxidant potential. *World J. Microbiol. Biotech.* **2014**, *30*, 65–76. [CrossRef] [PubMed]

63. Godinho, V.M.; Furbino, L.E.; Santiago, I.F.; Pellizzari, F.M.; Yokoya, N.S.; Pupo, D.; Alves, T.M.; Junior, P.A.; Romanha, A.J.; Zani, C.L.; et al. Diversity and bioprospecting of fungal communities associated with endemic and cold-adapted macroalgae in Antarctica. *ISME J.* **2013**, *7*, 1434–1451. [CrossRef]

64. Maida, I.; Bosi, E.; Fondi, M.; Perrin, E.; Orlandini, V.; Papaleo, M.; Mengoni, A.; de Pascale, D.; Tutino, M.L.; Michaud, L.; et al. Antimicrobial activity of *Pseudoalteromonas* strains isolated from the Ross Sea (Antarctica) versus Cystic Fibrosis opportunistic pathogens. *Hydrobiologia* **2015**, *761*, 443–457. [CrossRef]

65. Jayatilake, G.S.; Thornton, M.P.; Leonard, A.C.; Grimwade, J.E.; Baker, B.J. Metabolites from an Antarctic sponge-associated bacterium, *Pseudomonas aeruginosa*. *J. Nat. Prod.* **1996**, *59*, 293–296. [CrossRef]

66. Mangano, S.; Michaud, L.; Caruso, C.; Brilli, M.; Bruni, V.; Fani, R.; Lo Giudice, A. Antagonistic interactions between psychrotrophic cultivable bacteria isolated from Antarctic sponges: A preliminary analysis. *Res. Microbiol.* **2009**, *160*, 27–37. [CrossRef]

67. Orlandini, V.; Maida, I.; Fondi, M.; Perrin, E.; Papaleo, M.C.; Bosi, E.; de Pascale, D.; Tutino, M.L.; Michaud, L.; Lo Giudice, A.; et al. Genomic analysis of three sponge-associated *Arthrobacter* Antarctic strains, inhibiting the growth of *Burkholderia cepacia* complex bacteria by synthesizing volatile organic compounds. *Microbiol. Res.* **2014**, *169*, 593–601. [CrossRef]

68. Papaleo, M.C.; Fondi, M.; Maida, I.; Perrin, E.; Lo Giudice, A.; Michaud, L.; Mangano, S.; Bartolucci, G.; Romoli, R.; Fani, R. Sponge-associated microbial Antarctic communities exhibiting antimicrobial activity against *Burkholderia cepacia* complex bacteria. *Biotechnol. Adv.* **2012**, *30*, 272–293. [CrossRef]

69. Papaleo, M.C.; Romoli, R.; Bartolucci, G.; Maida, I.; Perrin, E.; Fondi, M.; Orlandini, V.; Mengoni, A.; Emiliani, G.; Tutino, M.L.; et al. Bioactive volatile organic compounds from Antarctic (sponges) bacteria. *New Biotechnol.* **2013**, *30*, 824–838. [CrossRef] [PubMed]

70. Tedesco, P.; Maida, I.; Palma Esposito, F.; Tortorella, E.; Subko, K.; Ezeofor, C.C.; Zhang, Y.; Tabudravu, J.; Jaspars, M.; Fani, R.; et al. Antimicrobial activity of monoramnholipids produced by bacterial strains isolated from the Ross Sea (Antarctica). *Mar. Drugs* **2016**, *14*, 83. [CrossRef] [PubMed]

71. Wietz, M.; Månsson, M.; Bowman, J.S.; Blom, N.; Ng, Y.; Gram, L. Wide distribution of closely related, antibiotic-producing *Arthrobacter* strains throughout the Arctic Ocean. *Appl. Environ. Microbiol.* **2012**, *78*, 2039–2042. [CrossRef] [PubMed]

72. Gao, X.; Lu, Y.; Xing, Y.; Ma, Y.; Lu, J.; Bao, W.; Wang, Y.; Xi, T. A novel anticancer and antifungus phenazine derivative from a marine actinomycete BM-17. *Microbiol. Res.* **2012**, *167*, 616–622. [CrossRef] [PubMed]

73. Zhang, H.; Saurav, K.; Yu, Z.; Mándi, A.; Kurtán, T.; Li, J.; Tian, X.; Zhang, Q.; Zhang, W.; Zhang, C. α-Pyrones with diverse hydroxy substitutions from three marine-derived *Nocardiopsis* strains. *J. Nat. Prod.* **2016**, *79*, 1610–1618. [CrossRef] [PubMed]

74. Chen, Y.; Xie, B.; Yang, J.; Chen, J.; Sun, Z. Identification of microbial carotenoids and isoprenoid quinones from *Rhodococcus* sp. B7740 and its stability in the presence of iron in model gastric conditions. *Food Chem.* **2018**, *240*, 204–211. [CrossRef]

75. Chen, Y.; Guo, M.; Yang, J.; Chen, J.; Xie, B.; Sun, Z. Potential TSPO ligand and photooxidation quencher isorenieratene from Arctic Ocean *Rhodococcus* sp. B7740. *Mar. Drugs* **2019**, *17*, 316. [CrossRef]

76. Chen, Y.; Mu, Q.; Hu, K.; Chen, M.; Yang, J.; Chen, J.; Xie, B.; Sun, Z. Characterization of MK(8)(H(2)) from *Rhodococcus* sp. B7740 and its potential antiglycation capacity measurements. *Mar. Drugs* **2018**, *16*, 391. [CrossRef]

77. Moon, K.; Ahn, C.-H.; Shin, Y.; Won, T.H.; Ko, K.; Lee, S.K.; Oh, K.-B.; Shin, J.; Nam, S.-I.; Oh, D.-C. New benzoxazine secondary metabolites from an Arctic Actinomycete. *Mar. Drugs* **2014**, *12*, 2526–2538. [CrossRef]

78. Ivanova, V.; Oriol, M.; Montes, M.-J.; Garcia, A.; Guinea, J. Secondary metabolites from a Streptomyces strain isolated from Livingston Island, Antarctica. *Z. Naturforsch. C Biosci.* **2001**, *56*, 1–5. [CrossRef]

79. Kim, D.; Lee, E.J.; Lee, J.; Leutou, A.S.; Shin, Y.H.; Choi, B.; Hwang, J.S.; Hahn, D.; Choi, H.; Chin, J.; et al. Antartin, a cytotoxic zizaane-type sesquiterpenoid from a *Streptomyces* sp. isolated from an Antarctic marine sediment. *Mar. Drugs* **2018**, *16*, 130. [CrossRef] [PubMed]

80. Macherla, V.R.; Liu, J.; Bellows, C.; Teisan, S.; Nicholson, B.; Lam, K.S.; Potts, B.C.M. Glaciapyrroles A, B, and C, pyrrolosesquiterpenes from a Streptomyces sp. isolated from an Alaskan marine sediment. *J. Nat. Prod.* **2005**, *68*, 780–783. [CrossRef] [PubMed]

81. Sannino, F.; Parrilli, E.; Apuzzo, G.A.; de Pascale, D.; Tedesco, P.; Maida, I.; Perrin, E.; Fondi, M.; Fani, R.; Marino, G.; et al. *Pseudoalteromonas haloplanktis* produces methylamine, a volatile compound active against *Burkholderia cepacia* complex strains. *New Biotechnol.* **2017**, *35*, 13–18. [CrossRef]

82. Sannino, F.; Sansone, C.; Galasso, C.; Kildgaard, S.; Tedesco, P.; Fani, R.; Marino, G.; de Pascale, D.; Ianora, A.; Parrilli, E.; et al. *Pseudoalteromonas haloplanktis* TAC125 produces 4-hydroxybenzoic acid that induces pyroptosis in human A459 lung adenocarcinoma cells. *Sci. Rep.* **2018**, *8*, 1190. [CrossRef]

83. Mitova, M.; Tutino, M.L.; Infusini, G.; Marino, G.; De Rosa, S. Exocellular peptides from Antarctic psychrophile Pseudoalteromonas Haloplanktis. *Mar. Biotechnol.* **2005**, *7*, 523–531. [CrossRef] [PubMed]

84. Casillo, A.; Papa, R.; Ricciardelli, A.; Sannino, F.; Ziaco, M.; Tilotta, M.; Selan, L.; Marino, G.; Corsaro, M.M.; Tutino, M.L.; et al. Anti-biofilm activity of a long-chain fatty aldehyde from Antarctic *Pseudoalteromonas haloplanktis* TAC125 against *Staphylococcus epidermidis* biofilm. *Front. Cell. Infect. Microb.* **2017**, *7*, 46. [CrossRef] [PubMed]

85. Chianese, G.; Esposito, F.P.; Parrot, D.; Ingham, C.; de Pascale, D.; Tasdemir, D. Linear aminolipids with moderate antimicrobial activity from the Antarctic gram-negative bacterium *Aequorivita* sp. *Mar. Drugs* **2018**, *16*, 187. [CrossRef]

86. Al-Zereini, W.; Schuhmann, I.; Laatsch, H.; Helmke, E.; Anke, H. New aromatic nitro compounds from *Salegentibacter* sp. T436, an Arctic sea ice bacterium: Taxonomy, fermentation, isolation and biological activities. *J. Antibiot.* **2007**, *60*, 301–308. [CrossRef]

87. Schuhmann, I.; Yao, C.B.F.-F.; Al-Zereini, W.; Anke, H.; Helmke, E.; Laatsch, H. Nitro derivatives from the Arctic ice bacterium *Salegentibacter* sp. isolate T436. *J. Antibiot.* **2009**, *62*, 453–460. [CrossRef]

88. Zhang, H.L.; Hua, H.M.; Pei, Y.H.; Yao, X.S. Three new cytotoxic cyclic acylpeptides from marine Bacillus sp. *Chem. Pharm. Bull.* **2004**, *52*, 1029–1030. [CrossRef]

89. Corral, P.; Esposito, F.P.; Tedesco, P.; Falco, A.; Tortorella, E.; Tartaglione, L.; Festa, C.; D'Auria, M.V.; Gnavi, G.; Varese, G.C.; et al. Identification of a sorbicillinoid-producing *Aspergillus* strain with antimicrobial activity against *Staphylococcus aureus*: A new polyextremophilic marine fungus from Barents Sea. *Mar. Biotechnol.* **2018**, *20*, 502–511. [CrossRef] [PubMed]

90. Wu, G.W.; Lin, A.Q.; Gu, Q.Q.; Zhu, T.J.; Li, D.H. Four new chloro-eremophilane sesquiterpenes from an Antarctic deep-sea derived fungus, *Penicillium* sp PR19N-1. *Mar. Drugs* **2013**, *11*, 1399–1408. [CrossRef] [PubMed]

91. Lin, A.; Wu, G.; Gu, Q.; Zhu, T.; Li, D. New eremophilane-type sesquiterpenes from an Antarctic deepsea derived fungus, *Penicillium* sp. PR19 N-1. *Arch. Pharm. Res.* **2014**, *37*, 839–844. [CrossRef]

92. Niu, S.; Fan, Z.-W.; Xie, C.-L.; Liu, Q.; Luo, Z.-H.; Liu, G.; Yang, X.-W. Spirograterpene A, a tetracyclic spiro-diterpene with a fused 5/5/5/5 ring system from the deep-sea-derived fungus *Penicillium granulatum* MCCC 3A00475. *J. Nat. Prod.* **2017**, *80*, 2174–2177. [CrossRef] [PubMed]

93. Zhou, Y.; Li, Y.H.; Yu, H.B.; Liu, X.Y.; Lu, X.L.; Jiao, B.H. Furanone derivative and sesquiterpene from Antarctic marine-derived fungus *Penicillium* sp. S-1–18. *J. Asian Nat. Prod. Res.* **2018**, *20*, 1108–1115. [CrossRef]

94. Hu, Z.-Y.; Li, Y.-Y.; Huang, Y.-J.; Su, W.-J.; Shen, Y.-M. Three new sesquiterpenoids from *Xylaria* sp. NCY2. *Helv. Chim. Acta* **2008**, *91*, 46–52. [CrossRef]

95. Wu, G.W.; Ma, H.Y.; Zhu, T.J.; Li, J.; Gu, Q.Q.; Li, D.H. Penilactones A and B, two novel polyketides from Antarctic deep-sea derived fungus *Penicillium crustosum* PRB-2. *Tetrahedron* **2012**, *68*, 9745–9749. [CrossRef]

96. Wu, B.; Wiese, J.; Labes, A.; Kramer, A.; Schmaljohann, R.; Imhoff, J.F. Lindgomycin, an unusual antibiotic polyketide from a marine fungus of the Lindgomycetaceae. *Mar. Drugs* **2015**, *13*, 4617–4632. [CrossRef]

97. Shin, H.J.; Mondol, M.A.M.; Yu, T.K.; Lee, H.-S.; Lee, Y.-J.; Jung, H.J.; Kim, J.H.; Kwon, H.J. An angiogenesis inhibitor isolated from a marine-derived actinomycete, *Nocardiopsis* sp. 03N67. *Phytochem. Lett.* **2010**, *3*, 194–197. [CrossRef]

98. Silva, T.R.; Canela-Garayoa, R.; Eras, J.; Rodrigues, M.V.N.; Dos Santos, F.N.; Eberlin, M.N.; Neri-Numa, I.A.; Pastore, G.M.; Tavares, R.S.N.; Debonsi, H.M.; et al. Pigments in an iridescent bacterium, *Cellulophaga fucicola*, isolated from Antarctica. *Antonie Leeuwenhoek* **2019**, *112*, 479–490. [CrossRef]

99. Silva, T.R.; Tavares, R.S.N.; Canela-Garayoa, R.; Eras, J.; Rodrigues, M.V.N.; Neri-Numa, I.A.; Pastore, G.M.; Rosa, L.H.; Schultz, J.A.A.; Debonsi, H.M.; et al. Chemical characterization and biotechnological applicability of pigments i solated from Antarctic bacteria. *Mar. Biotechnol.* **2019**, *21*, 416–429. [CrossRef] [PubMed]

100. Figueroa, L.; Jiménez, C.; Rodríguez, J.; Areche, C.; Chávez, R.; Henríquez, M.; de la Cruz, M.; Díaz, C.; Segade, Y.; Vaca, I. 3-Nitroasterric acid derivatives from an Antarctic sponge-derived *Pseudogymnoascus* sp. fungus. *J. Nat. Prod.* **2015**, *78*, 919–923. [CrossRef] [PubMed]

101. Kim, D.-C.; Lee, H.-S.; Ko, W.; Lee, D.-S.; Sohn, J.H.; Yim, J.H.; Kim, Y.-C.; Oh, H. Anti-inflammatory effect of methylpenicinoline from a marine isolate of *Penicillium* sp. (SF-5995): Inhibition of NF-κB and MAPK pathways in lipopolysaccharide-induced RAW264.7 macrophages and BV2 microglia. *Molecules* **2014**, *19*, 18073–18089. [CrossRef] [PubMed]

102. Manzo, E.; Ciavatta, M.L.; Melck, D.; Schupp, P.; de Voogd, N.; Gavagnin, M. Aromatic cyclic peroxides and related keto-compounds from the *Plakortis* sp. component of a sponge consortium. *J. Nat. Prod.* **2009**, *72*, 1547–1551. [CrossRef] [PubMed]

103. Carbone, M.; Núñez-Pons, L.; Castelluccio, F.; Avila, C.; Gavagnin, M. Illudalane sesquiterpenoids of the alcyopterosin series from the Antarctic marine soft coral *Alcyonium grandis*. *J. Nat. Prod.* **2009**, *72*, 1357–1360. [CrossRef]

104. Palermo, J.A.; Brasco, M.F.; Spagnuolo, C.; Seldes, A.M. Illudalane sesquiterpenoids from the soft coral *Alcyonium paessleri*: The first natural nitrate esters. *J. Org. Chem.* **2000**, *65*, 4482–4486. [CrossRef]

105. Brasco, M.F.R.; Seldes, A.M.; Palermo, J.A. Paesslerins A and B: Novel tricyclic sesquiterpenoids from the soft coral Alcyonium paessleri. *Org. Lett.* **2001**, *3*, 1415–1417. [CrossRef]

106. Mellado, G.G.; Zubia, E.; Ortega, M.J.; Lopez-Gonzalez, P.J. Steroids from the Antarctic octocoral *Anthomastus bathyproctus*. *J. Nat. Prod.* **2005**, *68*, 1111–1115. [CrossRef]

107. Thomas, S.A.L.; Sanchez, A.; Kee, Y.; Wilson, N.G.; Baker, B.J. Bathyptilones: Terpenoids from an Antarctic sea pen, *Anthoptilum grandiflorum* (Verrill, 1879). *Mar. Drugs* **2019**, *17*, 513. [CrossRef]

108. Mellado, G.G.; Zubia, E.; Ortega, M.J.; Lopez-Gonzalez, P.J. New polyoxygenated steroids from the Antarctic octocoral *Dasystenella Acanthina*. *Steroids* **2004**, *69*, 291–299. [CrossRef]

109. Koljak, R.; Lopp, A.; Pehk, T.; Varvas, K.; Muurisepp, A.-M.; Jarving, I.; Samel, N. New cytotoxic sterols from the soft coral Gersemia fruticosa. *Tetrahedron* **1998**, *54*, 179–186. [CrossRef]

110. Lopp, A.; Pihlak, A.; Paves, H.; Samuel, K.; Koljak, R.; Samel, N. The effect of 9,11-secosterol, a newly discovered compound from the soft coral Gersemia fruticosa, on the growth and cell cycle progression of various tumor cells in culture. *Steroids* **1994**, *59*, 274–281. [CrossRef]

111. Angulo-Preckler, C.; Genta-Jouve, G.; Mahajan, N.; de la Cruz, M.; de Pedro, N.; Reyes, F.; Iken, K.; Avila, C.; Thomas, O.P. Gersemiols A–C and eunicellol A, diterpenoids from the Arctic soft coral *Gersemia fruticosa*. *J. Nat. Prod.* **2016**, *79*, 1132–1136. [CrossRef]

112. Thomas, S.A.L.; von Salm, J.L.; Clark, S.; Nemani, P.; Ferlita, S.; Wilson, N.G.; Baker, B.J. Keikipukalides, furanocembranoid aldehydes from the deep sea Antarctic coral *Plumerella delicatissima*. *J. Nat. Prod.* **2018**, *81*, 117–123. [CrossRef] [PubMed]

113. Von Salm, J.L.; Wilson, N.G.; Vesely, B.A.; Kyle, D.E.; Cuce, J.; Baker, B.J. Shagenes A and B, new tricyclic sesquiterpenes produced by an undescribed Antarctic octocoral. *Org. Lett.* **2014**, *16*, 2630–2633. [CrossRef] [PubMed]

114. Tadesse, M.; Tabudravu, J.N.; Jaspars, M.; Strom, M.B.; Hansen, E.; Andersen, J.H.; Kristiansen, P.E.; Haug, T. The antibacterial ent-eusynstyelamide B and eusynstyelamides D, E, and F from the Arctic bryozoan *Tegella* cf. *spitzbergensis*. *J. Nat. Prod.* **2011**, *74*, 837–841. [CrossRef]

115. Michael, P.; Hansen, E.; Isaksson, J.; Andersen, J.H.; Hansen, K.Ø. Dendrobeaniamine A, a new alkaloid from the Arctic marine bryozoan *Dendrobeania murrayana*. *Nat. Prod. Res.* **2019**, *34*, 2059–2064. [CrossRef]

116. Maschek, J.A.; Mevers, E.M.; Diyabalanage, T.; Chen, L.; Ren, Y.; Amsler, C.D.; McClintock, J.B.; Wu, J.; Baker, B.J. Palmadorin chemodiversity from the Antarctic nudibranch *Austrodoris kerguelenensis* and inhibition of Jak2/STAT5-dependent HEL leukemia cells. *Tetrahedron* **2012**, *68*, 9095–9104. [CrossRef]

117. Rodriguez, J.; Castro, R.; Riguera, R. Holothurinosides: New antitumour non-sulphated triterpenoid glycosides from the sea cucumber *Holothuria forskalii*. *Tetrahedron* **1991**, *47*, 4753–4762. [CrossRef]

118. Maier, M.S.; Roccatagliata, A.J.; Kuriss, A.; Chludil, H.; Seldes, A.M.; Pujol, C.A.; Damonte, E.B. Two new cytotoxic and virucidal trisulfated triterpene glycosides from the Antarctic sea cucumber *Staurocucumis liouvillei*. *J. Nat. Prod.* **2001**, *64*, 732–736. [CrossRef]

119. Kicha, A.A.; Ivanchina, N.V.; Malyarenko, T.V.; Kalinovsky, A.I.; Popov, R.S.; Stonik, V.A. Six new polyhydroxylated steroids conjugated with taurine, microdiscusols A-F, from the Arctic starfish *Asterias microdiscus*. *Steroids* **2019**, *150*, 108458. [CrossRef]

120. D'Auria, M.V.; Paloma, L.G.; Minale, L.; Riccio, R.; Zampella, A.; Morbidoni, M. Isolation and structure characterization of two novel bioactive sulphated polyhydroxysteroids from the Antarctic ophiuroid *Ophioderma longicaudum*. *Nat. Prod. Lett.* **1993**, *3*, 197–201. [CrossRef]

121. Comin, M.J.; Maier, M.S.; Roccatagliata, A.J.; Pujol, C.A.; Damonte, E.B. Evaluation of the antiviral activity of natural sulfated polyhydroxysteroids and their synthetic derivatives and analogs. *Steroids* **1999**, *64*, 335–340. [CrossRef]

122. Ma, W.S.; Mutka, T.; Vesley, B.; Amsler, M.O.; McClintock, J.B.; Amsler, C.D.; Perman, J.A.; Singh, M.P.; Maiese, W.M.; Zaworotko, M.J.; et al. Norselic acids A-E, highly oxidized anti-infective steroids that deter mesograzer predation, from the Antarctic sponge *Crella* sp. *J. Nat. Prod.* **2009**, *72*, 1842–1846. [CrossRef] [PubMed]

123. Ankisetty, S.; Amsler, C.D.; McClintock, J.B.; Baker, B.J. Further membranolide diterpenes from the antarctic sponge *Dendrilla membranosa*. *J. Nat. Prod.* **2004**, *67*, 1172–1174. [CrossRef] [PubMed]

124. Shilling, A.J.; Witowski, C.G.; Maschek, J.A.; Azhari, A.; Vesely, B.; Kyle, D.E.; Amsler, C.D.; McClintock, J.B.; Baker, B.J. Spongian diterpenoids derived from the Antarctic sponge *Dendrilla antarctica* are potent inhibitors of the *Leishmania* parasite. *J. Nat. Prod.* **2020**, *83*, 1553–1562. [CrossRef]

125. Von Salm, J.L.; Witowski, C.G.; Fleeman, R.M.; McClintock, J.B.; Amsler, C.D.; Shaw, L.N.; Baker, B.J. Darwinolide, a new diterpene scaffold thati inhibits methicillin-resistant *Staphylococcus aureus* biofilm from the Antarctic sponge *Dendrilla membranosa*. *Org. Lett.* **2016**, *18*, 2596–2599. [CrossRef]

126. Bory, A.; Shilling, A.J.; Allen, J.; Azhari, A.; Roth, A.; Shaw, L.N.; Kyle, D.E.; Adams, J.H.; Amsler, C.D.; McClintock, J.B.; et al. Bioactivity of spongian diterpenoid scaffolds from the Antarctic sponge *Dendrilla Antarctica*. *Mar. Drugs.* **2020**, *18*, 327. [CrossRef]

127. Ciaglia, E.; Malfitano, A.M.; Laezza, C.; Fontana, A.; Nuzzo, G.; Cutignano, A.; Abate, M.; Pelin, M.; Sosa, S.; Bifulco, M.; et al. Immuno-modulatory and anti-inflammatory effects of dihydrogracilin A, a terpene derived from the marine sponge *Dendrilla membranosa*. *Int. J. Mol. Sci.* **2017**, *18*, 1643. [CrossRef]

128. Lidgren, G.; Bohlin, L.; Bergman, J. Studies of Swedish marine organisms. VII. A novel biologically active indole alkaloid from the sponge Geodia baretti. *Tetrahedron Lett.* **1986**, *27*, 3283–3284. [CrossRef]

129. Lind, K.F.; Hansen, E.; Østerud, B.; Eilertsen, K.-E.; Bayer, A.; Engqvist, M.; Leszczak, K.; Jørgensen, T.Ø.; Andersen, J.H. Antioxidant and anti-inflammatory activities of barettin. *Mar. Drugs* **2013**, *11*, 2655–2666. [CrossRef] [PubMed]

130. Volk, C.A.; Koeck, M. Viscosaline: New 3-alkyl pyridinium alkaloid from the Arctic sponge Haliclona viscosa. *Org. Biomol. Chem.* **2004**, *2*, 1827–1830. [CrossRef] [PubMed]

131. Lippert, H.; Brinkmeyer, R.; Mulhaupt, T.; Iken, K. Antimicrobial activity in sub-Arctic marine invertebrates. *Polar Biol.* **2003**, *26*, 591–600. [CrossRef]

132. Perry, N.B.; Ettouati, L.; Litaudon, M.; Blunt, J.W.; Munro, M.H.G.; Parkin, S.; Hope, H. Alkaloids from the antarctic sponge Kirkpatrickia varialosa. Part 1: Variolin B, a new antitumor and antiviral compound. *Tetrahedron* **1994**, *50*, 3987–3992. [CrossRef]

133. Trimurtulu, G.; Faulkner, D.J.; Perry, N.B.; Ettouati, L.; Litaudon, M.; Blunt, J.W.; Munro, M.H.G.; Jameson, G.B. Alkaloids from the antarctic sponge Kirkpatrickia varialosa. Part 2: Variolin A and *N*(3')-methyl tetrahydrovariolin B. *Tetrahedron* **1994**, *50*, 3993–4000. [CrossRef]

134. Na, M.; Ding, Y.; Wang, B.; Tekwani, B.L.; Schinazi, R.F.; Franzblau, S.; Kelly, M.; Stone, R.; Li, X.-C.; Ferreira, D.; et al. Anti-infective discorhabdins from a deep-water Alaskan sponge of the senus *Latrunculia*. *J. Nat. Prod.* **2010**, *73*, 383–387. [CrossRef]

135. Cutignano, A.; De Palma, R.; Fontana, A. A chemical investigation of the Antarctic sponge *Lyssodendoryx flabellata*. *Nat. Prod. Res.* **2012**, *26*, 1240–1248. [CrossRef]

136. Cutignano, A.; Nuzzo, G.; D'Angelo, D.; Borbone, E.; Fusco, A.; Fontana, A. Mycalol: A natural lipid with promising cytotoxic properties against human anaplastic thyroid carcinoma cells. *Angew. Chem. Int. Ed.* **2013**, *52*, 9256–9260. [CrossRef]

137. Holzwarth, M.; Trendel, J.-M.; Albrecht, P.; Maier, A.; Michaelis, W. Cyclic peroxides derived from the marine sponge Plakortis simplex. *J. Nat. Prod.* **2005**, *68*, 759–761. [CrossRef]

138. Kong, F.; Andersen, R.J. Polymastiamide A, a novel steroid/amino acid conjugate isolated from the Norwegian marine sponge Polymastia boletiformis (Lamarck, 1815). *J. Org. Chem.* **1993**, *58*, 6924–6927. [CrossRef]

139. Kong, F.; Andersen, R.J. Polymastiamides B-F, novel steroid/amino acid conjugates isolated from the Norwegian marine sponge Polymastia boletiformis. *J. Nat. Prod.* **1996**, *59*, 379–385. [CrossRef]

140. Smyrniotopoulos, V.; Rae, M.; Soldatou, S.; Ding, Y.; Wolff, C.W.; McCormack, G.; Coleman, C.M.; Ferreira, D.; Tasdemir, D. Sulfated steroid-amino acid conjugates from the Irish marine sponge *Polymastia boletiformis*. *Mar. Drugs* **2015**, *13*, 1632–1646. [CrossRef] [PubMed]

141. Hanssen, K.O.; Andersen, J.H.; Stiberg, T.; Engh, R.A.; Svenson, J.; Geneviere, A.M.; Hansen, E. Antitumoral and mechanistic studies of ianthelline isolated from the Arctic sponge *Stryphnus fortis*. *Anticancer Res.* **2012**, *32*, 4287–4297. [PubMed]

142. Shin, J.; Seo, Y.; Rho, J.R.; Baek, E.; Kwon, B.-M.; Jeong, T.S.; Bok, S.H. Suberitenones A and B: Sesterterpenoids of an unprecedented skeletal class from the Antarctic sponge Suberites sp. *J. Org. Chem.* **1995**, *60*, 7582–7588. [CrossRef]

143. Gompel, M.; Leost, M.; De Joffe, E.B.K.; Puricelli, L.; Franco, L.H.; Palermo, J.; Meijer, L. Meridianins, a new family of protein kinase inhibitors isolated from the ascidian *Aplidium meridianum*. *Bioorg. Med. Chem. Lett.* **2004**, *14*, 1703–1707. [CrossRef] [PubMed]

144. Appleton, D.R.; Chuen, C.S.; Berridge, M.V.; Webb, V.L.; Copp, B.R. Rossinones A and B, biologically active meroterpenoids from the Antarctic ascidian, *Aplidium* species. *J. Org. Chem.* **2009**, *74*, 9195–9198. [CrossRef]

145. Steffan, B. Lepadin A, a decahydroquinoline alkaloid from the tunicate Clavelina lepadiformis. *Tetrahedron* **1991**, *47*, 8729–8732. [CrossRef]

146. Kubanek, J.; Williams, D.E.; de Silva, E.D.; Allen, T.; Andersen, R.J. Cytotoxic alkaloids from the flatworm Prostheceraeus villatus and its tunicate prey Clavelina lepadiformis. *Tetrahedron Lett.* **1995**, *36*, 6189–6192. [CrossRef]

147. Diyabalanage, T.; Amsler, C.D.; McClintock, J.B.; Baker, B.J. Palmerolide A, a cytotoxic macrolide from the Antarctic tunicate *Synoicum adareanum*. *J. Am. Chem. Soc.* **2006**, *128*, 5630–5631. [CrossRef]

148. Noguez, J.H.; Diyabalanage, T.; Miyata, Y.; Xie, X.-S.; Valeriote, F.A.; Amsler, C.D.; McClintock, J.B.; Baker, B.J. Palmerolide macrolides from the Antarctic tunicate *Synoicum adareanum*. *Bioorg. Med. Chem.* **2011**, *19*, 6608–6614. [CrossRef]

149. Miyata, Y.; Diyabalanage, T.; Amsler, C.D.; McClintock, J.B.; Valeriote, F.A.; Baker, B.J. Ecdysteroids from the Antarctic tunicate *Synoicum adareanum*. *J. Nat. Prod.* **2007**, *70*, 1859–1864. [CrossRef] [PubMed]

150. Tadesse, M.; Strøm, M.B.; Svenson, J.; Jaspars, M.; Milne, B.F.; Tørfoss, V.; Andersen, J.H.; Hansen, E.; Stensvåg, K.; Haug, T. Synoxazolidinones A and B: Novel bioactive alkaloids from the ascidian *Synoicum pulmonaria*. *Org. Lett.* **2010**, *12*, 4752–4755. [CrossRef]

151. Tadesse, M.; Svenson, J.; Jaspars, M.; Strøm, M.B.; Abdelrahman, M.H.; Andersen, J.H.; Hansen, E.; Kristiansen, P.E.; Stensvåg, K.; Haug, T. Synoxazolidinone C; a bicyclic member of the synoxazolidinone family with antibacterial and anticancer activities. *Tetrahedron Lett.* **2011**, *52*, 1804–1806. [CrossRef]

152. Tadesse, M.; Svenson, J.; Sepcic, K.; Trembleau, L.; Engqvist, M.; Andersen, J.H.; Jaspars, M.; Stensvag, K.; Haug, T. Isolation and synthesis of pulmonarins A and B, acetylcholinesterase inhibitors from the colonial ascidian *Synoicum pulmonaria*. *J. Nat. Prod.* **2014**, *77*, 364–369. [CrossRef] [PubMed]

153. Pettus, J.A., Jr.; Wing, R.M.; Sims, J.J. Marine natural products. XII. Isolation of a family of multihalogenated gamma-methylene lactones from the red seaweed Delisea fimbriata. *Tetrahedron Lett.* **1977**, *18*, 41–44. [CrossRef]

154. Shilling, A.J.; Von Salm, J.L.; Sanchez, A.R.; Kee, Y.; Amsler, C.D.; McClintock, J.B.; Baker, B.J. Anverenes B-E, new polyhalogenated monoterpenes from the Antarctic red alga *Plocamium cartilagineum*. *Mar. Drugs* **2019**, *17*, 230. [CrossRef]

155. Ankisetty, S.; Nandiraju, S.; Win, H.; Park, Y.C.; Amsler, C.D.; McClintock, J.B.; Baker, J.A.; Diyabalanage, T.K.; Pasaribu, A.; Singh, M.P.; et al. Chemical investigation of predator-deterred macroalgae from the Antarctic peninsula. *J. Nat. Prod.* **2004**, *67*, 1295–1302. [CrossRef]

156. Nedialkova, D.; Naidenova, M. Screening the antimicrobial activity of actinomycetes strains isolated from Antarctica. *J. Cult. Collect.* **2005**, *4*, 29–35.

157. Silber, J.; Ohlendorf, B.; Labes, A.; Erhard, A.; Imhoff, J.F. Calcarides A-E, antibacterial macrocyclic and linear polyesters from a *Calcarisporium* strain. *Mar. Drugs* **2013**, *11*, 3309–3323. [CrossRef]

158. Encheva-Malinova, M.; Stoyanova, M.; Avramova, H.; Pavlova, Y.; Gocheva, B.; Ivanova, I.; Moncheva, P. Antibacterial potential of streptomycete strains from Antarctic soils. *Biotechnol. Biotechnol. Equip.* **2014**, *28*, 721–727. [CrossRef]

159. Giudice, A.L.; Fani, R. Antimicrobial potential of cold-adapted bacteria and fungi from polar regions. In *Biotechnology of Extremophiles: Advances and Challenges*; Rampelotto, P.H., Ed.; Springer: Cham, Switzerland, 2016.

160. Parrilli, E.; Papa, R.; Carillo, S.; Tilotta, M.; Casillo, A.; Sannino, F.; Cellini, A.; Artini, M.; Selan, L.; Corsaro, M.M.; et al. Anti-biofilm activity of *Pseudoalteromonas haloplanktis* tac125 against *Staphylococcus epidermidis* biofilm: Evidence of a signal molecule involvement? *Int. J. Immunopathol. Pharmacol.* **2015**, *28*, 104–113. [CrossRef]

161. Bosi, E.; Fondi, M.; Orlandini, V.; Perrin, E.; Maida, I.; de Pascale, D.; Tutino, M.L.; Parrilli, E.; Lo Giudice, A.; Filloux, A.; et al. The pangenome of (Antarctic) *Pseudoalteromonas* bacteria: Evolutionary and functional insights. *BMC Genom.* **2017**, *18*, 93. [CrossRef] [PubMed]

162. Lipotec Antarcticine. Available online: https://www.lipotec.com/en/products/antarcticine-reg-marine-ingredient/ (accessed on 15 March 2020).

163. Lipotec SeaCode. Available online: https://www.lipotec.com/en/products/seacode-trade-marine-ingredient/ (accessed on 15 March 2020).

164. Knestrick, M.A.; Wilson, N.G.; Roth, A.; Adams, J.H.; Baker, B.J. Friomaramide, a highly modified linear hexapeptide from an Antarctic sponge, inhibits *Plasmodium falciparum* liver-stage development. *J. Nat. Prod.* **2019**, *82*, 2354–2358. [CrossRef]

165. Ebada, S.S.; Lin, W.; Proksch, P. Bioactive sesterterpenes and triterpenes from marine sponges: Occurrence and pharmacological significance. *Mar. Drugs* **2010**, *8*, 313–346. [CrossRef] [PubMed]

166. Elissawy, A.M.; El-Shazly, M.; Ebada, S.S.; Singab, A.B.; Proksch, P. Bioactive terpenes from marine-derived fungi. *Mar. Drugs* **2015**, *13*, 1966–1992. [CrossRef] [PubMed]

167. Hegazy, M.E.; Mohamed, T.A.; Alhammady, M.A.; Shaheen, A.M.; Reda, E.H.; Elshamy, A.I.; Aziz, M.; Pare, P.W. Molecular architecture and biomedical leads of terpenes from red sea marine invertebrates. *Mar. Drugs* **2015**, *13*, 3154–3181. [CrossRef]

168. Wu, Q.; Sun, J.; Chen, J.; Zhang, H.; Guo, Y.W.; Wang, H. Terpenoids from marine soft coral of the genus *Lemnalia*: Chemistry and biological activities. *Mar. Drugs* **2018**, *16*, 320. [CrossRef]

169. McNaught, A.D.; Wilkinson, A. *IUPAC Compendium of Chemical Terminology*; International Union of Pure and Applied Chemistry: Hoboken, NJ, USA, 1997.

170. Ruzicka, L. The isoprene rule and the biogenesis of terpenic compounds. *Experientia* **1953**, *9*, 357–367. [CrossRef]

171. Davis, E.M.; Croteau, R. Cyclization enzymes in the biosynthesis of monoterpenes, sesquiterpenes, and diterpenes. In *Biosynthesis: Aromatic Polyketides, Isoprenoids, Alkaloids*; Leeper, F.J., Vederas, J.C., Eds.; Springer: Berlin/Heidelberg, Germany, 2000; pp. 53–95.

172. Leal, M.C.; Madeira, C.; Brandao, C.A.; Puga, J.; Calado, R. Bioprospecting of marine invertebrates for new natural products-a chemical and zoogeographical perspective. *Molecules* **2012**, *17*, 9842–9854. [CrossRef]

173. Amsler, C.D.; McClintock, J.B.; Baker, B.J. Chemical mediation of mutualistic interactions between macroalgae and mesograzers structure unique coastal communities along the western Antarctic Peninsula. *J. Phycol.* **2014**, *50*, 1–10. [CrossRef]

174. Kladi, M.; Vagias, C.; Roussis, V. Volatile halogenated metabolites from marine red algae. *Phytochem. Rev.* **2004**, *3*, 337–366. [CrossRef]

175. Amsler, C.D.; Iken, K.B.; McClintock, J.B.; Amsler, M.O.; Peters, K.J.; Hubbard, J.M.; Furrow, F.B.; Baker, B.J. A comprehensive evaluation of the palatability and chemical defenses of subtidal macroalgae from the Antarctic peninsula. *Mar. Ecol. Prog. Ser.* **2005**, *294*, 141–159. [CrossRef]

176. McClintock, J.B.; Baker, B.J. *Marine Chemical Ecology*; CRC Press: New York, NY, USA, 2001.

177. Pérez, M.J.; Falqué, E.; Domínguez, H. Antimicrobial action of compounds from marine seaweed. *Mar. Drugs* **2016**, *14*, 52. [CrossRef] [PubMed]

178. MarinLit, a Database of the Marine Natural Products Literature. Available online: http://pubs.rsc.org/marinlit/ (accessed on 3 March 2020).

179. Fraga, B.M. Natural sesquiterpenoids. *Nat. Prod. Rep.* **2013**, *30*, 1226–1264. [CrossRef]

180. Duran-Pena, M.J.; Ares, J.M.B.; Hanson, J.R.; Collado, I.G.; Hernandez-Galan, R. Biological activity of natural sesquiterpenoids containing a gem-dimethylcyclopropane unit. *Nat. Prod. Rep.* **2015**, *32*, 1236–1248. [CrossRef]

181. Choudhary, A.; Naughton, L.M.; Montanchez, I.; Dobson, A.D.W.; Rai, D.K. Current status and future prospects of marine natural products (MNPs) as antimicrobials. *Mar. Drugs* **2017**, *15*, 272. [CrossRef]

182. Bideau, F.L.; Kousara, M.; Chen, L.; Wei, L.; Dumas, F. Tricyclic sesquiterpenes from marine origin. *Chem. Rev.* **2017**, *117*, 6110–6159. [CrossRef]

183. Kato, S.; Shindo, K.; Kawai, H.; Odagawa, A.; Matsuoka, M.; Mochizuki, J. Pyrrolostatin, a novel lipid peroxidation inhibitor from *Streptomyces chrestomyceticus*. Taxonomy, fermentation, isolation, structure elucidation and biological properties. *J. Antibiot. (Tokyo)* **1993**, *46*, 892–899. [CrossRef]

184. Liu, D.Z.; Liang, B.W. Structural elucidation and NMR assignments of two new pyrrolosesquiterpenes from *Streptomyces* sp. Hd7–21. *Magn. Reson. Chem.* **2014**, *52*, 57–59. [CrossRef]

185. Bradfield, A.E.; Penfold, A.R.; Simonsen, J.L. 414. The constitution of eremophilone and of two related hydroxy-ketones from the wood oil of *Eremophila mitchelli*. *J. Chem. Soc.* **1932**, 2744–2759. [CrossRef]

186. Riche, C.; Pascard-Billy, C.; Devys, M.; Gaudemer, A.; Barbier, M.; Bousquet, J.-F. Structure cristalline et moleculaire de la phomenone, phytotoxine produite par le champignon *Phoma exigua var. non oxydabilis*. *Tetrahedron Lett.* **1974**, *15*, 2765–2766. [CrossRef]

187. Yuyama, K.T.; Fortkamp, D.; Abraham, W.R. Eremophilane-type sesquiterpenes from fungi and their medicinal potential. *Biol. Chem.* **2018**, *399*, 13–28. [CrossRef] [PubMed]

188. Wang, L.; Li, M.; Tang, J.; Li, X. Eremophilane sesquiterpenes from a deep marine-derived fungus, *Aspergillus* sp. SCSIOW2, cultivated in the presence of epigenetic modifying agents. *Molecules* **2016**, *21*, 473. [CrossRef]

189. McDonald, L.A.; Barbieri, L.R.; Bernan, V.S.; Janso, J.; Lassota, P.; Carter, G.T. 07H239-A, a new cytotoxic eremophilane sesquiterpene from the marine-derived *Xylariaceous fungus* LL-07H239. *J. Nat. Prod.* **2004**, *67*, 1565–1567. [CrossRef] [PubMed]

190. Song, Y.; Wang, J.; Huang, H.; Ma, L.; Wang, J.; Gu, Y.; Liu, L.; Lin, Y. Four eremophilane sesquiterpenes from the mangrove endophytic fungus *Xylaria* sp. BL321. *Mar. Drugs* **2012**, *10*, 340–348. [CrossRef]

191. Torres, P.; Ayala, J.; Grande, C.; Anaya, J.; Grande, M. Furanoeremophilane derivatives from *Senecio flavus*. *Phytochemistry* **1999**, *52*, 1507–1513. [CrossRef]

192. Mohamed, A.E.-H.; Ahmed, H. Eremophilane-type sesquiterpene derivatives from *Senecio aegyptius* var. *discoideus*. *J. Nat. Prod.* **2005**, *68*, 439–442. [CrossRef]

193. Tori, M.; Otose, K.; Fukuyama, H.; Murata, J.; Shiotani, Y.; Takaoka, S.; Nakashima, K.; Sono, M.; Tanaka, M. New eremophilanes from *Farfugium japonicum*. *Tetrahedron* **2010**, *66*, 5235–5243. [CrossRef]

194. Harinantenaina, L.; Matsunami, K.; Otsuka, H. Chemical and biologically active constituents of *Pteris multifida*. *J. Nat. Med.* **2008**, *62*, 452–455. [CrossRef]

195. Fabian, K.; Lorenzen, K.; Anke, T.; Johansson, M.; Sterner, O. Five new bioactive sesquiterpenes from the fungus *Radulomyces confluens* (Fr.) Christ. *Z. Naturforsch. C. J. Biosci.* **1998**, *53*, 939–945. [CrossRef]

196. Mogi, Y.; Inanaga, K.; Tokuyama, H.; Ihara, M.; Yamaoka, Y.; Yamada, K.-i.; Takasu, K. Rapid assembly of protoilludane skeleton through tandem catalysis: Total synthesis of paesslerin A and its structural revision. *Org. Lett.* **2019**, *21*, 3954–3958. [CrossRef] [PubMed]

197. Iken, K.B.; Baker, B.J. Ainigmaptilones, sesquiterpenes from the Antarctic gorgonian coral *Ainigmaptilon antarcticus*. *J. Nat. Prod.* **2003**, *66*, 888–890. [CrossRef]

198. Cano, L.P.P.; Manfredi, R.Q.; Pérez, M.; García, M.; Blustein, G.; Cordeiro, R.; Pérez, C.D.; Schejter, L.; Palermo, J.A. Isolation and antifouling activity of azulene derivatives from the Antarctic gorgonian *Acanthogorgia laxa*. *Chem. Biodivers.* **2018**, *15*, e1700425. [CrossRef] [PubMed]

199. Iken, K.; Avila, C.; Ciavatta, M.L.; Fontana, A.; Cimino, G. Hodgsonal, a new drimane sesquiterpene from the mantle of the Antarctic nudibranch Bathydoris hodgsoni. *Tetrahedron Lett.* **1998**, *39*, 5635–5638. [CrossRef]

200. Avila, C. Terpenoids in marine heterobranch molluscs. *Mar. Drugs* **2020**, *18*, 162. [CrossRef]

201. Hortelano, S. Molecular basis of the anti-inflammatory effects of terpenoids. *Inflamm. Allergy Drug Targets* **2009**, *8*, 28–39.

202. Gao, S.S.; Li, X.M.; Zhang, Y.; Li, C.S.; Wang, B.G. Conidiogenones H and I, two new diterpenes of cyclopiane class from a marine-derived endophytic fungus *Penicillium chrysogenum* QEN-24S. *Chem. Biodivers.* **2011**, *8*, 1748–1753. [CrossRef]

203. Du, L.; Li, D.; Zhu, T.; Cai, S.; Wang, F.; Xiao, X.; Gu, Q. New alkaloids and diterpenes from a deep ocean sediment derived fungus *Penicillium* sp. *Tetrahedron* **2009**, *65*, 1033–1039. [CrossRef]

204. Roncal, T.; Cordobes, S.; Sterner, O.; Ugalde, U. Conidiation in *Penicillium cyclopium* is induced by conidiogenone, an endogenous diterpene. *Eukaryot. Cell* **2002**, *1*, 823–829. [CrossRef]

205. Cheng, Z.; Li, Y.; Xu, W.; Liu, W.; Liu, L.; Zhu, D.; Kang, Y.; Luo, Z.; Li, Q. Three new cyclopiane-type diterpenes from a deep-sea derived fungus *Penicillium* sp. YPGA11 and their effects against human esophageal carcinoma cells. *Bioorg. Chem.* **2019**, *91*, 103129. [CrossRef]

206. Gonzalez, Y.; Torres-Mendoza, D.; Jones, G.E.; Fernandez, P.L. Marine diterpenoids as potential anti-inflammatory agents. *Mediat. Inflamm.* **2015**, *2015*, 263543. [CrossRef] [PubMed]

207. Dayton, P.K.; Bobilliard, G.A.; Paine, R.T.; Dayton, L.B. Biological accomodation in the benthic community at McMurdo Sound, Antarctica. *Ecol. Monograph.* **1974**, *44*, 105–128. [CrossRef]

208. Nunez-Pons, L.; Carbone, M.; Vazquez, J.; Gavagnin, M.; Avila, C. Lipophilic defenses from alcyonium soft corals of Antarctica. *J. Chem. Ecol.* **2013**, *39*, 675–685. [CrossRef]

209. Cattaneo-Vietti, R.; Chiantore, M.; Schiaparelli, S.; Albertelli, G. Shallow- and deep-water mollusc distribution at Terra Nova Bay (Ross Sea, Antarctica). *Polar Biol.* **2000**, *23*, 173–182. [CrossRef]

210. Gavagnin, M.; De Napoli, A.; Castelluccio, F.; Cimino, G. Austrodorin-A and -B: First tricyclic diterpenoid 2'-monoglyceryl esters from an Antarctic nudibranch. *Tetrahedron Lett.* **1999**, *40*, 8471–8475. [CrossRef]

211. Davies-Coleman, M.T.; Faulkner, D.J. New diterpenoic acid glycerides from the Antarctic nudibranch Austrodoris kerguelensis. *Tetrahedron* **1991**, *47*, 9743–9750. [CrossRef]

212. Arens, J.; Bergs, D.; Mewes, M.; Merz, J.; Schembecker, G.; Schulz, F. Heterologous fermentation of a diterpene from *Alternaria brassisicola*. *Mycology* **2014**, *5*, 207–219. [CrossRef]

213. Xu, R.; Fazio, G.C.; Matsuda, S.P. On the origins of triterpenoid skeletal diversity. *Phytochemistry* **2004**, *65*, 261–291. [CrossRef]

214. Lednicer, D. *Steroid Chemistry at a Glance*; Wiley: Hoboken, NJ, USA, 2010.

215. Mijts, B.N.; Lee, P.C.; Schmidt-Dannert, C. Engineering carotenoid biosynthetic pathways. *Methods Enzymol.* **2004**, *388*, 315–329.

216. Fiedor, J.; Burda, K. Potential role of carotenoids as antioxidants in human health and disease. *Nutrients* **2014**, *6*, 466–488. [CrossRef]

217. Lohr, M. Carotenoids. In *The Chlamydomonas Sourcebook*, 2nd ed.; Harris, E.H., Stern, D.B., Witman, G.B., Eds.; Academic Press: London, UK, 2009; Chapter 21; pp. 799–817.

218. Gonzalez, S.; Gilaberte, Y.; Philips, N.; Juarranz, A. Current trends in photoprotection-A new generation of oral photoprotectors. *Open Dermatol. J.* **2011**, *5*, 6–14. [CrossRef]

219. Galasso, C.; Corinaldesi, C.; Sansone, C. Carotenoids from marine organisms: Biological Ffunctions and industrial applications. *Antioxidants* **2017**, *6*, 96. [CrossRef]

220. Matsuno, T. Aquatic animal carotenoids. *Fish. Sci.* **2001**, *67*, 771–783. [CrossRef]

221. Liao, L.; Su, S.; Zhao, B.; Fan, C.; Zhang, J.; Li, H.; Chen, B. Biosynthetic potential of a novel Antarctic actinobacterium *Marisediminicola antarctica* ZS314(T) revealed by genomic data mining and pigment characterization. *Mar. Drugs* **2019**, *17*, 388. [CrossRef] [PubMed]

222. Rodrigues, D.B.; Flores, E.M.M.; Barin, J.S.; Mercadante, A.Z.; Jacob-Lopes, E.; Zepka, L.Q. Production of carotenoids from microalgae cultivated using agroindustrial wastes. *Food Res. Int.* **2014**, *65*, 144–148. [CrossRef]

223. Da Silva, N.A.; Rodrigues, E.; Mercadante, A.Z.; de Rosso, V.V. Phenolic compounds and carotenoids from four fruits native from the Brazilian Atlantic forest. *J. Agri. Food Chem.* **2014**, *62*, 5072–5084. [CrossRef] [PubMed]

224. Gharibzahedi, S.M.T.; Razavi, S.H.; Mousavi, S.M. Characterization of bacteria of the genus *Dietzia*: An updated review. *Ann. Microbiol.* **2014**, *64*, 1–11. [CrossRef]

225. Guyomarc'h, F.; Binet, A.; Dufosse, L. Production of carotenoids by *Brevibacterium linens*: Variation among strains, kinetic aspects and HPLC profiles. *J. Ind. Microbiol. Biotech.* **2000**, *24*, 64–70. [CrossRef]

226. Wagener, S.; Volker, T.; De Spirt, S.; Ernst, H.; Stahl, W. 3,3'-Dihydroxyisorenieratene and isorenieratene prevent UV-induced DNA damage in human skin fibroblasts. *Free Radic. Biol. Med.* **2012**, *53*, 457–463. [CrossRef]

227. Kalinin, V.I.; Silchenko, A.S.; Avilov, S.A.; Stonik, V.A.; Smirnov, A.V. Sea cucumbers triterpene glycosides, the recent progress in structural elucidation and chemotaxonomy. *Phytochem. Rev.* **2005**, *4*, 221–236. [CrossRef]

228. Antonov, A.S.; Avilov, S.A.; Kalinovsky, A.I.; Anastyuk, S.D.; Dmitrenok, P.S.; Evtushenko, E.V.; Kalinin, V.I.; Smirnov, A.V.; Taboada, S.; Ballesteros, M.; et al. Triterpene glycosides from Antarctic Sea cucumbers. 1. Structure of liouvillosides A1, A2, A3, B1, and B2 from the sea cucumber *Staurocucumis liouvillei*: New procedure for separation of highly polar glycoside fractions and taxonomic revision. *J. Nat. Prod.* **2008**, *71*, 1677–1685. [CrossRef] [PubMed]

229. Howes, F.N. Fish-poison plants. *Bull. Misc. Infor. (R. Bot. Gard. Kew)* **1930**, *1930*, 129–153. [CrossRef]

230. Augustin, J.M.; Kuzina, V.; Andersen, S.B.; Bak, S. Molecular activities, biosynthesis and evolution of triterpenoid saponins. *Phytochemistry* **2011**, *72*, 435–457. [CrossRef] [PubMed]

231. Lorent, J.H.; Quetin-Leclercq, J.; Mingeot-Leclercq, M.P. The amphiphilic nature of saponins and their effects on artificial and biological membranes and potential consequences for red blood and cancer cells. *Org. Biomol. Chem.* **2014**, *12*, 8803–8822. [CrossRef] [PubMed]

232. Sun, H.X.; Xie, Y.; Ye, Y.P. Advances in saponin-based adjuvants. *Vaccine* **2009**, *27*, 1787–1796. [CrossRef]

233. Bahrami, Y.; Zhang, W.; Chataway, T.; Franco, C. Structure elucidation of five novel isomeric saponins from the viscera of the sea cucumber *Holothuria lessoni*. *Mar. Drugs* **2014**, *12*, 4439–4473. [CrossRef]

234. Antonov, A.S.; Avilov, S.A.; Kalinovsky, A.I.; Anastyuk, S.D.; Dmitrenok, P.S.; Kalinin, V.I.; Taboada, S.; Bosh, A.; Avila, C.; Stonik, V.A. Triterpene glycosides from Antarctic sea cucumbers. 2. Structure of achlioniceosides A(1), A(2), and A(3) from the sea cucumber *Achlionice violaecuspidata* (=*Rhipidothuria racowitzai*). *J. Nat. Prod.* **2009**, *72*, 33–38. [CrossRef] [PubMed]

235. Silchenko, A.S.; Kalinovsky, A.I.; Avilov, S.A.; Andryjashchenko, P.V.; Fedorov, S.N.; Dmitrenok, P.S.; Yurchenko, E.A.; Kalinin, V.I.; Rogacheva, A.V.; Gebruk, A.V. Kolgaosides A and B, two new triterpene glycosides from the Arctic deep water sea cucumber *Kolga hyalina* (Elasipodida: Elpidiidae). *Nat. Prod. Commun.* **2014**, *9*, 1259–1264. [CrossRef]

236. Menna, M.; Imperatore, C.; D'Aniello, F.; Aiello, A. Meroterpenes from marine invertebrates: Structures, occurrence, and ecological implications. *Mar. Drugs* **2013**, *11*, 1602–1643. [CrossRef]

237. Hemmings, A. *A question of politics: Bioprospecting and the Antarctic Treaty System. Antarctic Bioprospecting*; University of Canterbury: Christchurch, New Zealand, 2005; pp. 98–129.

238. Hemmings, A. Re-Justifying the Antarctic Treaty System for the 21st Century: Rights, Expectations and Global Equity. In *Polar Geopolitics: Knowledges, Resources and Legal Regimes*; Edward Elgar Publishing: Cheltenham, UK, 2014; pp. 55–73.

239. Greiber, T. Access and Benefit Sharing in Relation to Marine Genetic Resources from Areas Beyond National Jurisdiction: A Possible Way Forward; Study in Preparation of the Informal Workshop on Conservation of Biodiversity Beyond National Jurisdiction. In *Research Project of the Federal Agency for Nature Conservation*; Bundesamt für Naturschutz: Bonn, Germany, 2011.

240. Greiber, T. *An Explanatory Guide to the Nagoya Protocol on Access and Benefit-Sharing*; IUCN: Gland, Switzerland, 2012.

241. Wilkinson, M.D.; Dumontier, M.; Aalbersberg, I.J.; Appleton, G.; Axton, M.; Baak, A.; Blomberg, N.; Boiten, J.-W.; da Santos, L.B.S.; Bourne, P.E.; et al. The FAIR Guiding Principles for scientific data management and stewardship. *Sci. Data* **2016**, *3*, 160018. [CrossRef]

242. Thompson, C.C.; Kruger, R.H.; Thompson, F.L. Unlocking marine biotechnology in the developing world. *Trends Biotech.* **2017**, *35*, 1119–1121. [CrossRef] [PubMed]

243. Blasiak, R.; Jouffray, J.-B.; Wabnitz, C.C.C.; Sundström, E.; Österblom, H. Corporate control and global governance of marine genetic resources. *Sci. Adv.* **2018**, *4*, eaar5237. [CrossRef] [PubMed]

244. Hofland, P. Plitidepsin–A novel anti-cancer agent possibly active against COVID19. *Onco'Zine.* 2020. Available online: https://www.oncozine.com/plitidepsin-a-novel-anti-cancer-agent-possibly-active-against-covid18/ (accessed on 27 July 2020).

245. Andrade-Ochoa, S.; Nevárez-Moorillón, G.V.; Sánchez-Torres, L.E.; Villanueva-García, M.; Sánchez-Ramírez, B.E.; Rodríguez-Valdez, L.M.; Rivera-Chavira, B.E. Quantitative structure-activity relationship of molecules constituent of different essential oils with antimycobacterial activity against *Mycobacterium tuberculosis* and *Mycobacterium bovis. BMC Complement. Altern. Med.* **2015**, *15*, 332. [CrossRef] [PubMed]

 *marine drugs*

Review

Bioactive Compounds from Marine Heterobranchs

Conxita Avila [1,*] and Carlos Angulo-Preckler [1,2]

[1] Department of Evolutionary Biology, Ecology, Environmental Sciences, and Biodiversity Research Institute (IrBIO), Faculty of Biology, University of Barcelona, Av. Diagonal 643, 08028 Barcelona, Catalonia, Spain; carlos.a.preckler@uit.no

[2] Norwegian College of Fishery Science, UiT The Arctic University of Norway, Hansine Hansens veg 18, 9019 Tromsø, Norway

* Correspondence: conxita.avila@ub.edu

Received: 21 November 2020; Accepted: 7 December 2020; Published: 21 December 2020

Abstract: The natural products of heterobranch molluscs display a huge variability both in structure and in their bioactivity. Despite the considerable lack of information, it can be observed from the recent literature that this group of animals possesses an astonishing arsenal of molecules from different origins that provide the molluscs with potent chemicals that are ecologically and pharmacologically relevant. In this review, we analyze the bioactivity of more than 450 compounds from ca. 400 species of heterobranch molluscs that are useful for the snails to protect themselves in different ways and/or that may be useful to us because of their pharmacological activities. Their ecological activities include predator avoidance, toxicity, antimicrobials, antifouling, trail-following and alarm pheromones, sunscreens and UV protection, tissue regeneration, and others. The most studied ecological activity is predation avoidance, followed by toxicity. Their pharmacological activities consist of cytotoxicity and antitumoral activity; antibiotic, antiparasitic, antiviral, and anti-inflammatory activity; and activity against neurodegenerative diseases and others. The most studied pharmacological activities are cytotoxicity and anticancer activities, followed by antibiotic activity. Overall, it can be observed that heterobranch molluscs are extremely interesting in regard to the study of marine natural products in terms of both chemical ecology and biotechnology studies, providing many leads for further detailed research in these fields in the near future.

Keywords: marine natural products; Mollusca; Gastropoda; chemical ecology

1. Background

Marine heterobranch molluscs are a well-known source of marine natural products (MNPs) that have been studied in depth over the years [1–3]. MNPs from heterobranchs show an amazing structural diversity and display a wide variety of biological activities, as reported in previous reviews [1–4]. In general, MNPs have been demonstrated to be crucial in many ecological interactions among marine organisms, regulating several aspects of reproduction, development, settlement, growth, defense, and others [2,5–7]. Some general reviews have reported a significant amount of detailed information on the structure of MNPs, marine chemical ecology, and marine chemistry, or have analyzed some particular mollusc compounds [4,8–18]. The yearly reports by Blunt and collaborators [5,6] have provided very accurate information on new marine natural products. Previous reviews have also dealt with the different chemical structures found in heterobranchs, the origin and anatomical allocation of their compounds, their biosynthesis, biogeography, and their evolutionary patterns [1,2,19–29]. Therefore, all of these topics will not be considered again here.

Furthermore, MNPs have been described to be potentially useful as drugs, and some of them are already available on the market [7,8,10,12,30–33]. Remarkably, many MNPs possess unique chemical structures that are totally absent in terrestrial or freshwater environments [32,34–37].

Five drugs, at least, have been isolated from marine invertebrates and are approved for different (mostly anticancer) purposes, including cytarabine (Ara-C), eribulin mesylate, ziconotide, brentuximab vedotin, and trabectedin, obtained from two sponges, two molluscs, and a tunicate, respectively [31,33,38]. These molecules include very different chemical structures, from nucleosides to peptides, alkaloids, macrolides, and antibody–drug conjugates (ADCs). Many other compounds are currently in phase III, phase II, and phase I clinical trials, including several heterobranch compounds, and could soon be on the market [31]. Moreover, many studies deal with MNPs bioactivity, mechanisms of action, virtual screening, synthesis, derivatives, ADMET (absorption, distribution, metabolism, excretion, and toxicity), and others in an attempt to increase the chances of finding new useful drugs [31–43]. Some databases are also very good tools to search the details of MNPs described to date, such as MarinLit (http://pubs.rsc.org/marinlit/). In cancer research, for example, NPs are considered very relevant as potential drug leads, and approximately 80% of the approved chemotherapeutic drugs and more than 50% of all drugs are based on bioactive natural products, while almost 90% of human diseases are treated with natural products or their derivatives [39–43]. Thus, many MNPs are being tested as antitumor agents because of their potent growth inhibition against human tumor cells, both in vitro and in vivo in murine models (and others), as well as in cancer clinical trials [39,42,43].

In fact, marine organisms are still considered an underexplored source of NPs, displaying specific biological activities, with biomedically interesting applications to be potentially used as drugs [2,5,6,8,10,29–31,44]. Many compounds found in heterobranchs are also promising drugs and are being tested under clinical trials [36,43,45,46]. However, as far as we know, there has not yet been a comprehensive published review on the bioactivity of MNPs from heterobranch molluscs, despite the fact that this is one of the most chemodiverse invertebrate groups [2,4]. For this reason, we summarize here all the ecological and pharmacological activities reported in heterobranch molluscs, trying to emphasize in the assays carried out, whether they are or not ecologically and biomedically significant, and their potential interest, since it seemed timely and necessary now. As previously mentioned, this review does not cover other ecological or evolutionary aspects that are already covered in previous reviews [1,2], nor the chemical synthesis of the MNPs. The aim of this review is, therefore, to showcase the main ecological and pharmacological bioactivities of the chemical compounds found in heterobranch molluscs, describing in which groups they are found and their particular bioactivities with all of the information we have been able to compile up to June 2020.

Heterobranch molluscs are soft-bodied and mostly shell-less animals that live all around the planet at all latitudes and depths [2]. These animals are often protected by chemical strategies, although they may also present behavioral and/or morphological strategies to combine them with [1]. As a result of the most recent evolutionary, phylogenetic, and taxonomical studies on the group, heterobranch gastropods now comprise the classical "Opisthobranchia" and the marine "Pulmonata" together with several other groups, reaching a total of more than 33,000 species, although the most well-known groups account for only ca. 9000 species [47–51]. Among these, only about 400 species have been chemically analyzed, and, therefore, a lot of compounds remain to be potentially discovered [1,2,5,6,52]. Among the chemically studied heterobranch species, a wide variety of compounds has been described, many of them being bioactive at the ecological and/or pharmacological level [2,8]. At the ecological level, some NPs are used for protection against potential predators and competitors, enhancing their ecological performance, while others may have a role in their reproduction, development, growth, and feeding behavior [1,2,8]. In heterobranch molluscs, NPs may be de novo biosynthesized by the animals, obtained from their diet (biotransformed or not), or perhaps even produced by symbionts [1,2]. In any case, all of them are considered in this review because they are found in and used by the molluscs.

This review analyzes the bioactive compounds by activity (ecological and pharmacological, and different subtopics within them) and by taxonomical groups. Heterobranchs classically include eight major taxa: Nudibranchia, Pleurobranchoidea (or Pleurobranchida), Tylodinoidea (or Umbraculida), Cephalaspidea, Anaspidea (or Aplysiida), Pteropoda, Sacoglossa, and Pulmonata (Table 1) [47–50]. All of these taxa have different morphological and anatomical characteristics; different diet, behavioral, and ecological traits; and different chemical strategies [1,2]. Nudibranchs (sea slugs) are carnivorous and comprise Doridacea, Dendronotida, Euarminida, and Aeolidida, and are considered the most diverse group, with Doridacea feeding on porifera (sponges), bryozoans, tunicates, or other "opisthobranchs", Dendronotids prey on cnidarians (usually octocorals or hydrozoans) or some small animals (crustaceans or turbellarians), Euarminida feed on octocoral cnidarians or bryozoans, and Aeolidida are mainly cnidarian feeders [1]. All of them lack a shell in adult stage, and they possess interesting chemistry that may be de novo biosynthesized or obtained from their diet of the above-mentioned prey [1,2]. Pleurobranchoidea (side-gill slugs) are usually ascidian feeders or generalist scavengers, while Tylodinoidea (false limpets) feed on sponges, and Cephalaspidea (head-shielded slugs and snails) may be algal feeders or voracious predators of other animals (other "opisthobranchs", including other cephalaspideans), sponges, annelids, and others [1]. Anaspideans (sea hares) are herbivorous, feeding on different kinds of algae, but also on sea grasses, or even cyanobacteria. On the other hand, pelagic Pteropods (sea angels) are planktonic and feed on phytoplankton or other pteropods, while Sacoglossans and Pulmonates are herbivorous that feed on different types of algae [1,2].

Table 1. Species number and natural products numbers (NPs) for the different heterobranch groups [2,49,50]. * Accepted species number obtained from WoRMs (www.marinespecies.org), accessed on 11 November 2020). ** Natural products' number, main types of molecules, and diet according to Avila et al. [2]. *** Only marine pulmonata are considered here. # Number.

Phylum Mollusca Class Gastropoda Subclass Heterobranchia	Species #* 80548 33193	NPs #**	Main Types of Molecules **	Main Diet **
Nudibranchia	2462	~250	Terpenoids, alkaloids, macrolides, peptides, acidic secretions, etc.	Porifera, bryozoa, tunicata, cnidaria, other heterobranchs, crustacea, turbellaria
Pleurobranchoidea	96	25	Terpenoids, alkaloids, peptides, acidic secretions, etc.	Tunicates, other animals
Tylodinoidea	12	6	Alkaloids, diacylglycerols, etc.	Porifera
Cephalaspidea	875	40	Polyketides, polypropionates, polyacetates, ethers, acidic secretions, etc.	Algae, other heterobranchs, porifera, other animals
Anaspidea	94	~200	Polyketides, terpenoids, peptides, etc.	Algae, sea grasses, cyanobacteria
Pteropoda	409	5	Polypropionates, etc.	Phytoplankton, other pteropods
Sacoglossa	362	~120	Terpenoids, polypropionates, etc.	Algae
Pulmonata	500 ***	~75	Polypropionates, terpenoids, peptides, etc.	Algae

2. Ecological Activity

2.1. Predation

Heterobranch mollusc are protected against predation by a vast array of defensive strategies, many of which are combined with or include the use of natural products (Figures 1–5) [2]. These chemical strategies may, in fact, be useful against many different kinds of predators, which can usually be grouped into three main types: fish, crabs, and sea stars, although other potential predators, such as anemones, sea spiders, etc., have also been reported (Table 2) [1,2]. Whether defensive strategies used against one predator are also effective against another potential predator is seldom reported in the literature. Furthermore, when laboratory assays are carried out using non-sympatric potential predators, the presumed ecological roles become highly speculative, because laboratory results cannot and should not be directly extrapolated to the field. The possibility that chemical compounds are used in the field against a wider range of predators than those usually tested in the laboratory remains to be proven in most cases [1,2]. In general, as reported below, very few studies have been conducted in the field against sympatric predators, and, thus, the ecological role of NPs in the field should be carefully considered.

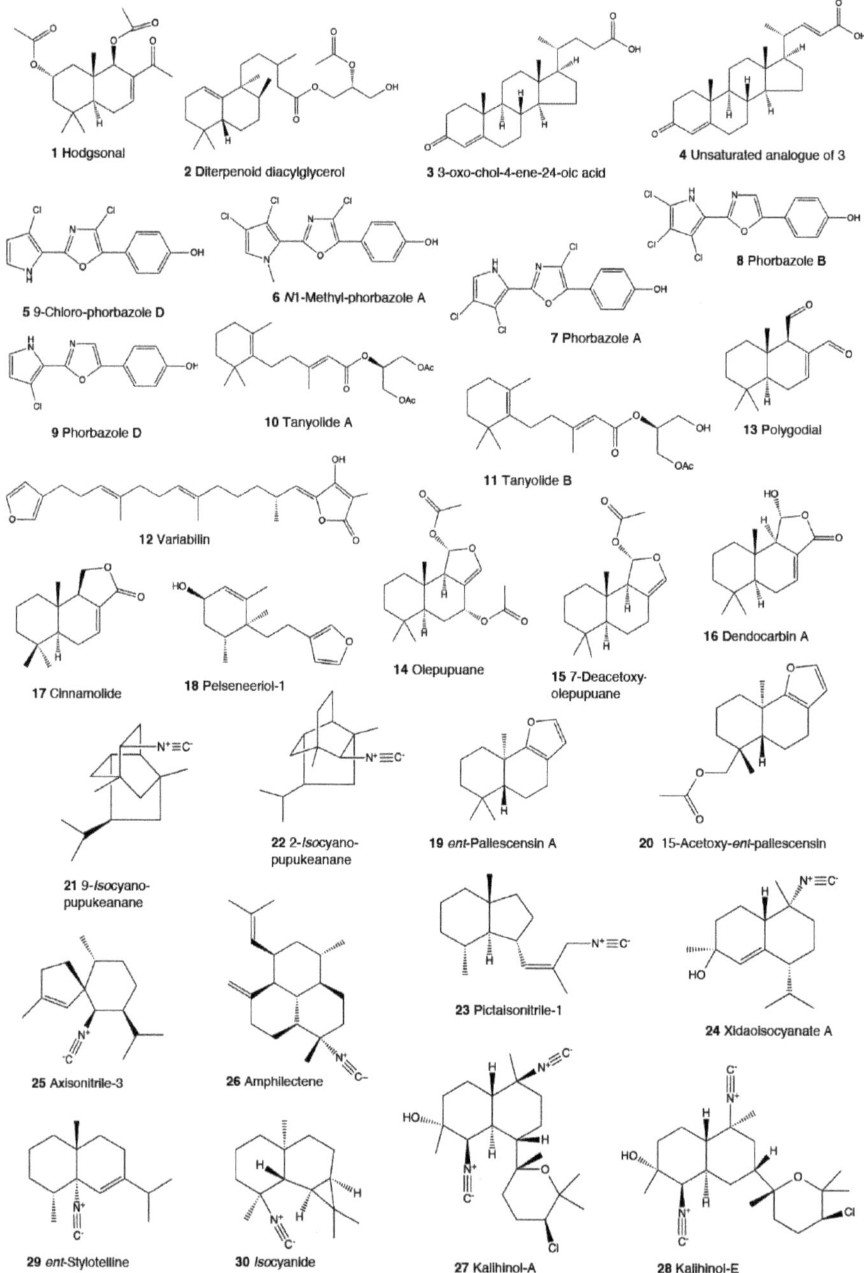

Figure 1. Structures of selected compounds used against predation in some Doridacea. These molecules may also display other activities, as reported in the text.

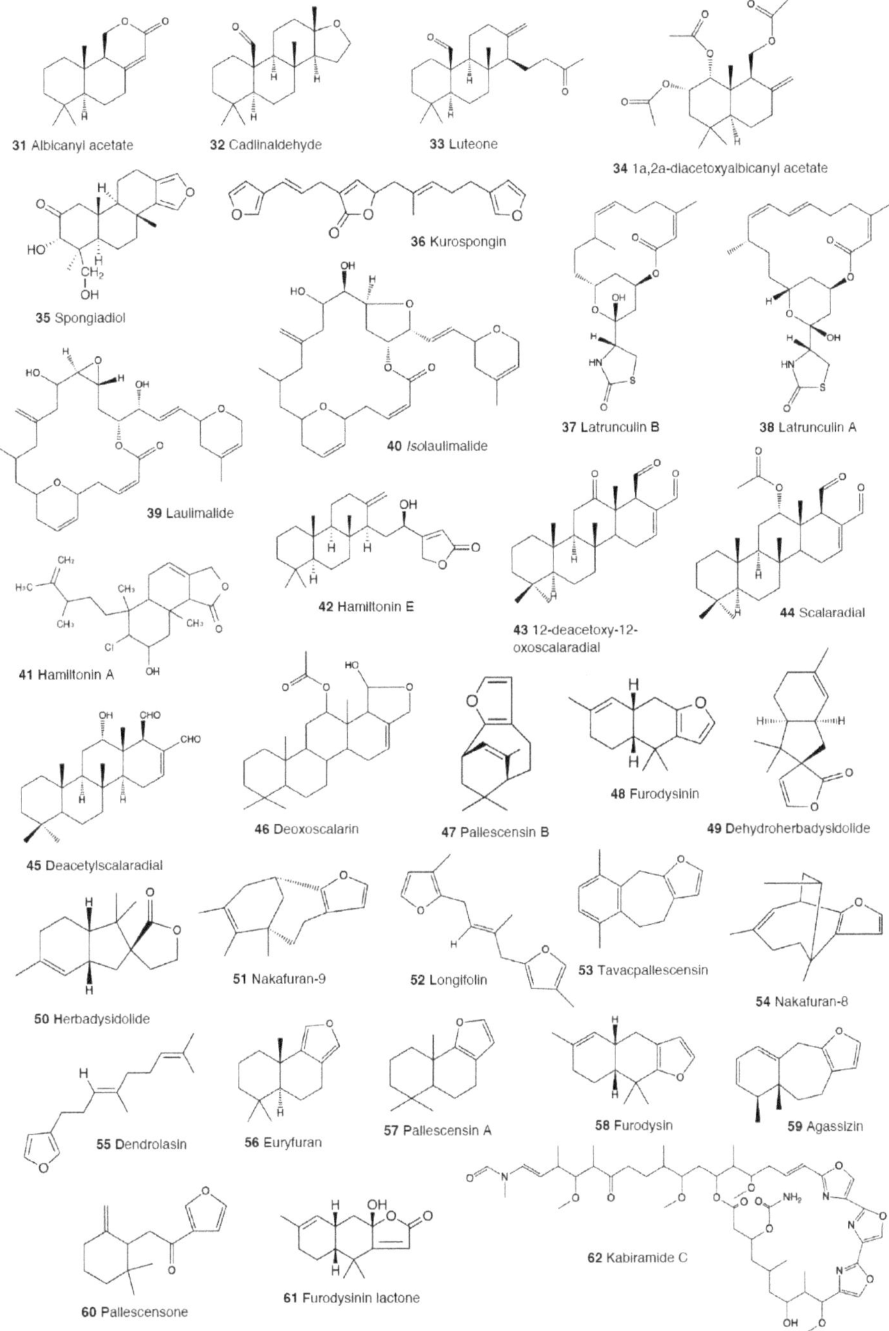

Figure 2. Structures of selected compounds used against predation in some Doridacea. These molecules may also display other activities, as reported in the text.

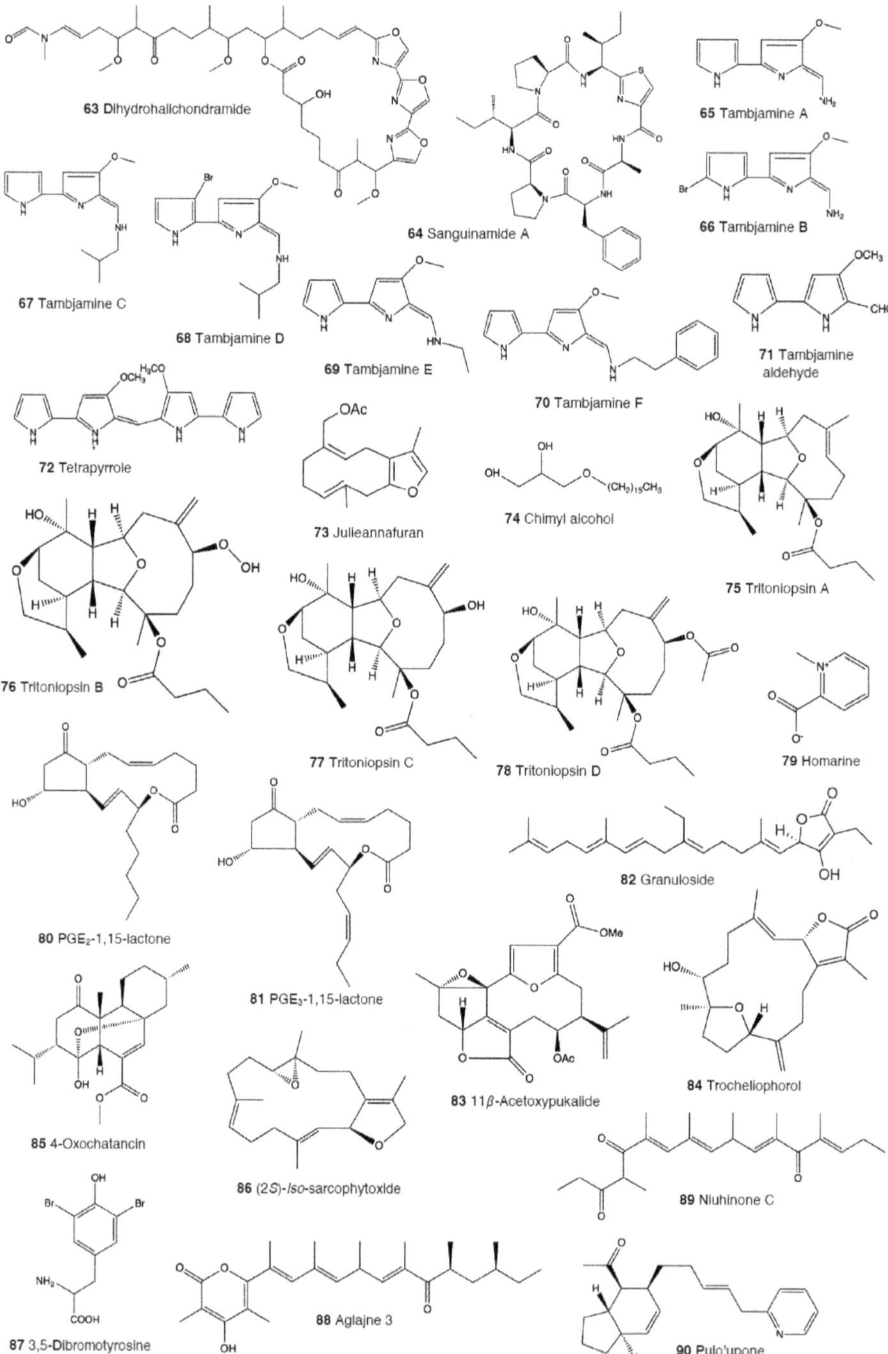

Figure 3. Structures of selected compounds used against predation in some Doridacea, Dendronotida, Euarminida, Aeolidida, Pleurobranchoidea, Tylodinoidea, and some Cephalaspidea. These molecules may also display other activities, as reported in the text.

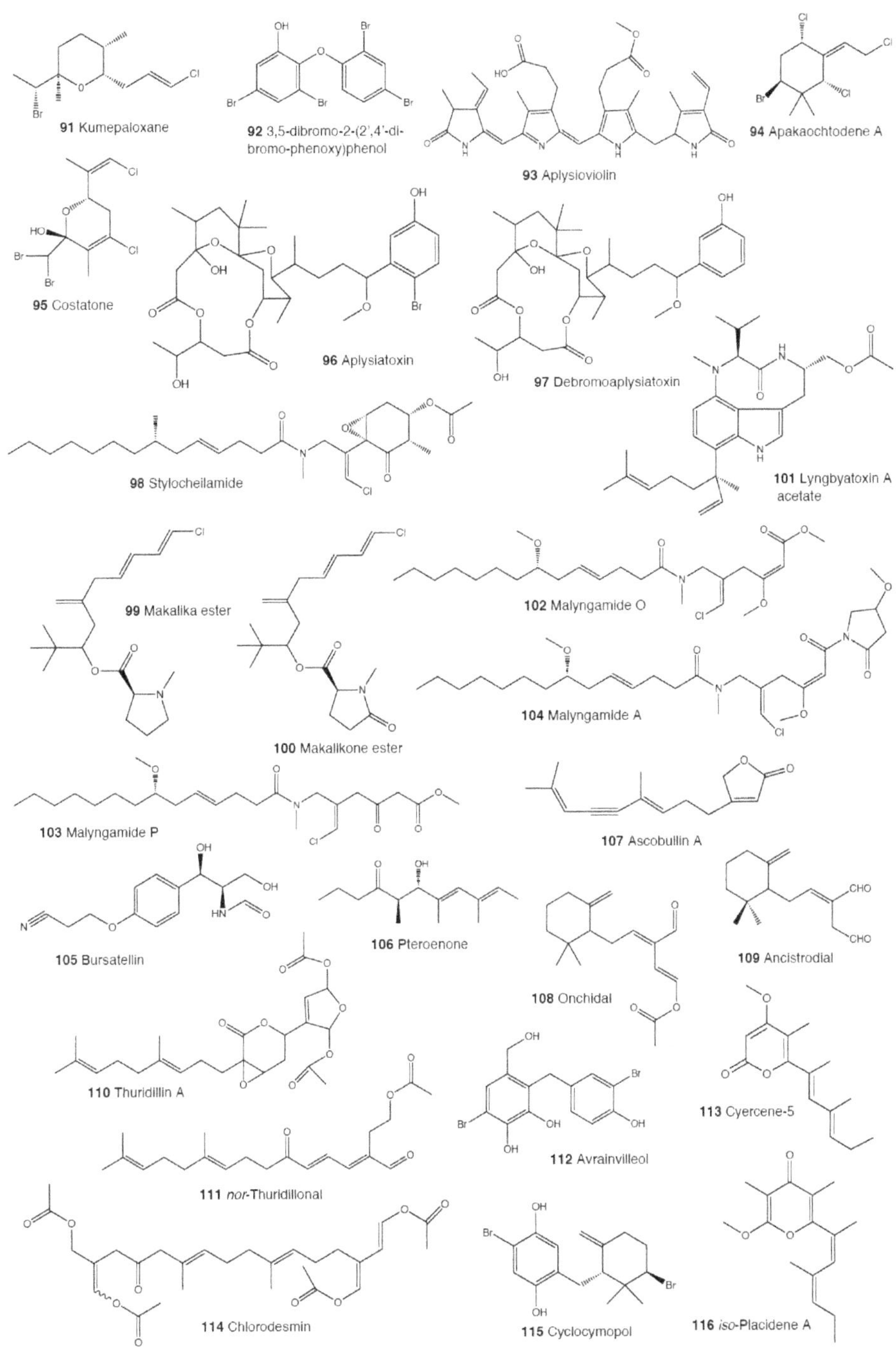

Figure 4. Structures of selected compounds used against predation in some Cephalaspidea, Anaspidea, Pteropoda, and Sacoglossa. These molecules may also display other activities, as reported in the text.

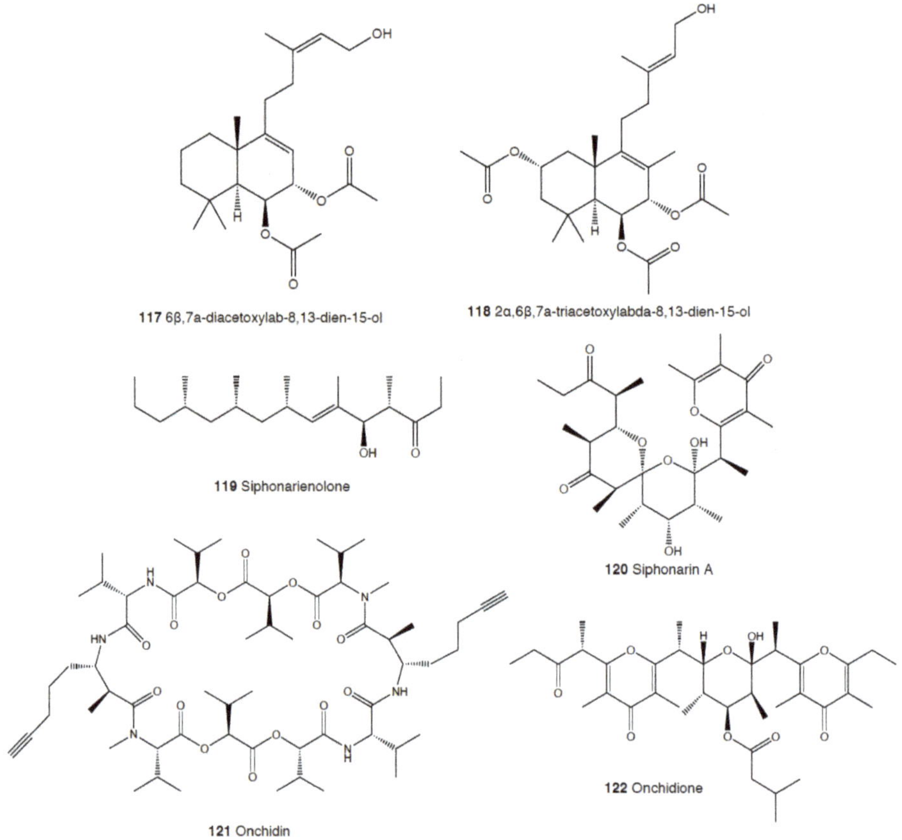

117 6β,7a-diacetoxylab-8,13-dien-15-ol

118 2a,6β,7a-triacetoxylabda-8,13-dien-15-ol

119 Siphonarienolone

120 Siphonarin A

121 Onchidin

122 Onchidione

Figure 5. Structures of selected compounds used against predation in Pulmonata. These molecules may also display other activities, as reported in the text.

2.1.1. Nudibranchia

Doridacea

This is the most studied group of heterobranchs regarding compounds against predation (Figures 1–3). Even the most basal species are protected against potential predators, such as the Antarctic *Bathydoris hodgsoni* [53,54]. This large slug presents the drimane sesquiterpene hodgsonal (**1**), which is located in its mantle and dorsal papillae, and which is suggested to be de novo biosynthesized. Hodgsonal (**1**) was the first described 2-substituted drimane sesquiterpene from a marine organism [55,56]. While *B. hodgsoni* is chemically protected against sympatric predators, such as the sea star *Odontaster validus* and the anemone *Epiactis* sp., its egg masses seem to rely only on physical defenses [54,57]. The related Antarctic species, *Prodoris (Bathydoris) clavigera* also possesses chemical defenses against *O. validus*, but the compounds behind this activity have not been yet described (C Avila and K Iken, unpublished results; [2]).

The most studied group within Doridacea are the Doridoidei, comprising the well-known dorids, phyllids, and chromodorids, among others. The Antarctic *Doris (Austrodoris) kerguelenensis* possesses a series of diterpene diacylglycerides (**2**) along with monoacylglycerides, and monoacylglycerides of regular fatty acids, which are located in the mantle and deter sympatric predators, such as sea stars (*O. validus*) and anemones (*Epiactis* sp.) [1,58–61]. This slug possesses many other molecules that may not be involved in defense against predators, including additional diterpene glycerides with different skeletons, such as *ent*-labdane, labdane, halimane, clerodane, and isocopalane diterpenes, as well as norsesquiterpenes [18,58,59,62–67]. Cryptic speciation has been reported in *D. kerguelenensis,* and this could be behind their chemical variability, even at the intrapopulation level, as well as perhaps the presence of different terpene synthase variants involved in their de novo biosynthesis [61,67–70]. Since these compounds occur in complex mixtures in the slug, it seems difficult to trace the bioactivity to the individual compounds. *Doris (Archidoris)* species also present similar glycerid compounds [1,71].

Several species have been reported to use steroids against potential predators. This is the case of *Aldisa sanguinea*, and perhaps also the Brazilian *Doris* aff. *verrucosa* [1,72]. The steroidal acids, 3-oxo-chol-4-ene-24-oic acid (**3**) and its unsaturated analogue (**4**) were reported from *Aldisa sanguinea* (*A. cooperi*), probably originated from some related inactive compounds from its diet of the sponge *Anthoarcuata graceae* [73]. The 3-oxo-chol-4-ene-24-oic acid (**3**) deterred feeding in the common freshwater goldfish (*Carassius auratus*) in laboratory assays [73]. Similarly, a progesterone homologue was found in the mantle of *Aldisa smaragdina* from Spain [74]. Another species, *A. andersoni* from India, is protected against predators by two phorboxazoles, 9-chloro-phorbazole D (**5**) and *N*1-methyl-phorbazole A (**6**), and the phorbazoles A (**7**), B (**8**), and D (**9**) located in their mantle and viscera [55,75,76]. The phorbazoles are chlorinated phenyl-pyrrolyloxazoles that were previously found in the sponge *Phorbas* aff. *clathrata*, and, therefore, a dietary origin from a sponge has been suggested [55,56,75,76]. The two phorboxazoles (**5,6**) and phorbazole A (**7**) were tested in the laboratory at 1 mg/mL against the shrimp *Palaemon elegans* and showed to be deterrent, although they were not in their natural concentration [75,77].

The Pacific slug *Sclerodoris tanya* presents the sesquiterpene glyceride esters tanyolides A (**10**) and B (**11**) in its mantle, reported to be effective deterrents against sympatric fish predators, such as *Gibbonsia elegans* and *Paraclinus integrippinis* at 1 mg/pellet [78]. The Mediterranean *Paradoris (Discodoris) indecora* incorporates furanosesterterpenes, including variabilin (**12**), from its sponge preys *Ircinia variabilis* and *I. fasciculata* [79] as deterrents against fish predation [79]. Variabilin (**12**) was tested in the laboratory at 300 µg/cm^2 against freshwater and marine fishes [79].

Dendrodoris species are well studied, with polygodial (**13**) from *D. limbata* being the first example of de novo biosynthesis in nudibranchs [80,81]. Polygodial (**13**), a drimane sesquiterpene, was first described in plants, where it is a deterrent against herbivores [82], and it is a deterrent in the slug against predation by marine and freshwater fish [80]. Polygodial (**13**) was found to be transformed from olepupuane (**14**) once secreted from the mantle cells, since it is not present in vivo in the slug tissues [80,83,84]. Furthermore, some fatty acid-esterified sesquiterpenoids were also found in *D. limbata*, and later in other species, generally found in the reproductive organs and egg masses and possibly with other functions, or perhaps just being stored as putative precursors of polygodial (**13**) [85]. Further studies with many other *Dendrodoris* species around the planet have yielded similar drimane sesquiterpenes located in the mantle, such as in *D. arborescens, D. carbunculosa, D. denisoni, D. grandiflora, D. carbunculosa, D. krebsii, D. nigra,* and *D. tuberculosa*, which are suggested to be used as feeding deterrents against predators [1,2,81,86–94]. In particular, *D. arborescens* presents 7-deacetoxyolepupuane (**15**) [87], *D. carbunculosa* possesses dendrocarbins A–N (**16**) [86], *D. krebsi* also has drimane sesquiterpenes and esters [89,90], and *D. denisoni* has cinnamolide (**17**), olepupuane (**14**), and polygodial (**13**) in its mantle [88].

Doriopsilla species also present similar metabolites to the related genus *Dendrodoris*. The Atlantic *Doriopsilla pelseneeri* presents the furanosesquiterpene alcohols pelseneeriols-1 and -2 (**18**) in the mantle [81,85,95–97]. *D. albopunctata* and *D. areolata* also have drimane sesquiterpenes and *ent*-pallescensin A (**19**) [89]. Other *Doriopsilla* species studied possess also drimane sesquiterpenoids and sesquiterpenoids with the *ent*-pallescensin A (**19**) skeleton in the mantle, including *D. janaina* and *D. pharpa* [81,89,95–98]. These natural products are de novo biosynthesized by the slugs, such as 15-acetoxy-*ent*-pallescensin (**20**) via the mevalonic pathway in *D. areolata* and *Doriopsilla* sp. [81,96,97,99]. It has been suggested that these compounds are used for defense against predators, but very few assays have been reported [81,96]. These include only the extracts of *D. pharpa* presenting polygodial (**13**), which deter feeding of the blenny fish *Chasmodes bosquianus* and the mummichog fish *Fundulus heteroclitus*, which even learned to avoid food items with extracts of slugs, and also deter the crabs *Callinectes similus* and *Panopeus herbstii* in the field [98].

The group of phyllidids has also been well studied over the last years [1,4]. These are usually brightly colored tropical animals, very specious, and quite similar in their external morphology, which has often resulted in some misidentifications [2,4,100]. These slugs are characterized by presenting isocyanate compounds that display a wide array of activities, apart from avoiding predation (see below) [1,101–105]. The first species studied was *Phyllidia varicosa* from Hawai'i, where a toxic compound, 9-*iso*cyanopupukeanane (**21**), and a tricyclic sesquiterpene isocyanide were described almost 50 years ago [106]. The compound was also found in its prey, the sponge *Ciocalypta (Hymeniacidon)* sp. [106], and a related compound was subsequently reported in the slug, 2-*iso*cyanopupukeanane (**22**) [107]. The extracts of Palauan *P. varicosa* deterred feeding by sympatric reef fish at natural concentration [108]. Similarly, the extracts from other species from Guam of the related genus *Phyllidia*, *Phyllidiella*, *Phyllidiopsis*, and *Fryeria* are deterrent to the sympatric crabs *Leptodius* sp., the mantle extracts being more deterrent than the viscera extracts [2]. A fast transformation of the secreted compounds was reported and was related to the loss of the deterrent activity [2]. The analysis of the sesquiterpene isocyanides that these slugs present suggests a broad diet of different demosponges, indicating a wide feeding variability [22]. Some experiments with agar-based food combined with different color patterns were also conducted, and the results showed that phyllidiids were defended against fish predators [109]. *P. varicosa* also possesses two 9-thiocyanatopupukeanane sesquiterpenes found in epimeric mixture; these were traced to its prey, the demosponge *Axinyssa aculeata* [110]. One of them is located in the mantle and is probably related to defense, but both compounds are found in the viscera, indicating their dietary origin. *Phyllidia coelestis* from Thailand also contains two pupukeanane sesquiterpenoids suggested to be used as for defense against predators [2,109,111]. *Phyllidia elegans* from Guam was a deterrent against reef fish, although the natural products have not been yet identified [109]. Other *Phyllidia* species contain related compounds, such as *Phyllidia picta* from Bali yielding two axane sesquiterpenoids, pictaisonitrile-1 (**23**) and pictaisonitrile-2, and *Phyllidia* sp. From Sri Lanka presenting the sponge-related 3-*iso*cyano-theonellin (similar to a cyanide from *Axinyssa*), together with some nitrogenous bisabolene sesquiterpenes [112–115].

Phyllidia varicosa, P. ocellata, Phyllidiella pustulosa, and *Phillidiopsis krempfi* from Australia also present three more sesquiterpene isonitriles, 10-*epi*-axisonitrile-3, 10-*iso*cyano-4-cadinene, and 2-*iso*cyanotrachyopsane, and the peroxide 1,7-*epi*dioxy-5-cadinene, together with some more sesquiterpene isonitriles [102,116]. Moreover, *Phyllidia ocellata* and *Phyllidiella pustulosa* contain stereoisomers of 10-*iso*cyano-4-amorphene and of 4-*iso*cyano-9-amorphene, respectively [102,116]. *Phyllidia coelestis* and *Phyllidiella pustulosa* from South China and their potential prey *Acanthella cavernosa* contain a nitrogenous cadinane-type sesquiterpenoid, xidaoisocyanate A (**24**), together with other sesquiterpenoids and diterpenoids [117]. *P. pustulosa* from Fiji possesses axisonitrile-3 (**25**), an isothiocyanate, and some minor related sesquiterpenes [118]. In China and Vietnam, *P. pustulosa* also presents sesquiterpene isocyanides, isothiocyanate, as well as some sterols, some of them also reported in *Acanthella* sponges, while in Japan, a sesquiterpene isonitrile is reported [103,119–122]. Samples from Hainan island present diterpenes together with sesquiterpenes, with the diterpenes amphilectene (**26**), kalihinol-A (**27**), and kalihinol-E (**28**) being previously found in sponges, and the sesquiterpene *ent*-stylotelline (**29**) being the enantiomer of the sponge compound stylotellin [120,123]. Amphilectene (**26**), kalihinol-A (**27**), and kalihinol-E (**28**) display deterrence in the laboratory against the allopatric goldfish *C. auratus* at 50 μg/cm^2 [120]. *P. pustulosa* is therefore a chemically rich species, containing a wide variety of compounds, perhaps related to its unrestricted sponge diet, or to the presence of unknown cryptic species, but only a few of their metabolites have been tested against predation. Moreover, in field experiments, living *Phyllidiella granulatus* were offered to fish but were never consumed, while crude lipophilic extracts of three species of phyllidiids were shown to be effective against fish predation [109]. These were *Phyllidia varicosa* from Palau, *P. elegans* from Guam, and *Phyllidiella pustulosa* from Palau, where crude extracts at natural concentrations deterred feeding by sympatric reef fish, such as *Abudefduf sexfasciatus, A. vaigiensis, Cheilinus fasciatus, Thalassoma lutescens, T. hardwickii, Naso vlamingii*, and *Bodianus axillaris*, although *P. pustulosa* extracts from Guam did not [109]. In this study, the authors reported that visual and chemical cues are more effective against fish when used together than either of them alone [109].

Another exhaustively studied group is that of "chromodoridids", which possess a huge diversity of compounds from their diet of demosponges, often accumulating them in mantle dermal formations (MDFs) [1,4,124]. This group was recently the subject of important taxonomical revisions that resulted in changes in several genus names [125]. One of the first species studied was *Cadlina luteomarginata*, where natural mixtures of three isocyanides and three isothiocyanates from its sponge prey were found, with the isocyanides (**30**) being deterrent in laboratory assays against goldfish at 10 μg/mL and both mixtures being deterrent against the woolly sculpin *Clinocottus analis* [126,127]. Some terpenoids from *C. luteomarginata* are de novo biosynthesized, while others are obtained from its sponge diet [128]. Specimens from British Columbia present de novo produced albicanyl acetate (**31**), cadlinaldehyde (**32**) and luteone (**33**) [128]. Albicanyl acetate (**31**), which is concentrated in mantle and mucus, was shown to be deterrent [129]. The related 1a,2a-diacetoxyalbicanyl acetate (**34**) was found in their egg masses and was suggested to be involved in defense against predators based on structural similarity [128,130].

Chromodoris is among the most studied heterobranch genus, although many studies were published using different names [76,131–165]. These slugs accumulate mostly terpenoids from their diet sponges, and many different structures have been reported, including sesquiterpenes, diterpenes and nor-diterpenes, sesterterpenes, macrolides, and bromophenols [131–133,135–165]. Previous studies analyzed the chemistry in the Mediterranean species *C. luterorosea, C. purpurea, C. krohni*, and *C. britoi* [1,2,4], containing diterpenoids from *Spongilla* sponges, while tropical species such as *C. mandapamensis* from India contain spongiadiol (**35**), previously found in sponges from Australia, within a mixture of related spongiane compounds [166]. In the Red Sea, *C. africana* presents the furanoterpene kurospongin (**36**), as well as a 14-membered macrolide with an attached 2-thiazolidinone unit, latrunculin B (**37**) [167–170]. Kurospongin (**36**) was obtained also from a *Spongia* sp. in Okinawa and reported to be deterrent [167–169]. Latrunculin B (**37**) was also found in *C. (Glossodoris) quadricolor* [171] and in the sponge *Latrunculia magnifica* [169,170]. In fact, also latrunculin A (**38**) is a sponge compound initially found in *L. magnifica* and reported in the MDFs of several *Chromodoris* species [136,141,153,164,169]. Other macrolides, such as laulimalide (**39**) and *iso*laulimalide (**40**), were reported in *C. lochi* and its sponge prey, *Hyattella* sp. [142,172–174]. *C. hamiltoni* from South Africa presents hamiltonins A–D (**41,42**), atypical chlorinated homoditerpenes, as well as the sesterterpene hamiltonin E (**42**) and latrunculins A and B (**37,38**), while specimens from Mozambique possess two spongian diterpene lactones in addition to latrunculin B (**37**) [153,155]. Many other compounds have been described in this genus, often located in the MDFs and suggesting a defensive role, but unfortunately very few tests for deterrence have been carried out [1,4].

In the genus *Glossodoris*, *G. vespa* and *G. averni* from Australia, as well as *G. pallida* from China, contain 12-deacetoxy-12-oxoscalaradial (**43**), while *G. pallida* from Guam contains some sesquiterpenes, such as scalaradial (**44**), deacetylscalaradial (**45**), and deoxoscalarin (**46**) [175–177]. The sesquiterpenes from *G. pallida* from Guam, located in their MDFs, have been proven to act as deterrents against sympatric reef fish (*Abudefduf sexfasciatus*, among others) and crabs (*Leptodius* sp.) at natural concentrations [176,177]. Further studies with *G. vespa* showed high concentrations of sesquiterpenes in mantle rim tissues that were more unpalatable to the allopatric palaemonid shrimp *Palaemon serenus* than metabolites from the viscera, suggesting selective accumulation of dietary compounds or perhaps even biotransformation to more potent defenses [178].

As taxonomical studies progress, many *Chromodoris* and *Glossodoris* species have been renamed, such as *Goniobranchus, Ardeadoris, Doriprismatica, Felimare*, and *Felimida*, respectively [171,175,179–185]. *Goniobranchus collingwoodi* presents six spongian-16-one diterpenes in the mantle, and the extract of the whole body displayed deterrence against the allopatric palaemonid shrimp *P. serenus* [185]. *G. reticulatus* from Australia contains a dialdehyde sesquiterpene and its ring-closed acetal, also reported in *G. sinensis* from China, where they are described to be deterrents against *Palaemon elegans* [186]. Specimens of *G. splendidus* from different localities in east Australia were described to present different abundances, types, and richness of natural products in addition to high individual variation between specimens from the same population [187]. These variations resulted in different potencies when deterring feeding in the allopatric, generalist rock-pool shrimp *P. serenus*, but in all cases, the specimens showed deterrent activity [187,188]. Other *Goniobranchus* species, such as *G. albonarus*, present diterpenes and nor-diterpenes obtained from their sponge prey, but they have not been tested for feeding deterrence [189–192].

Another interesting genus within this chromodorid group is *Ceratosoma*, because these species present a dorsal protuberance containing MDFs loaded with furanosesquiterpenoids. Although a defensive role has been suggested and it seems highly probable, it still remains to be demonstrated using sympatric predators [22,193]. These species include *C. trilobatum* and *C. gracillimum* from China, which possess pallescensin-B (**47**), (–)-furodysinin (**48**), (–)-dehydroherbadysidolide (**49**), and (–)-herbadysidolide (**50**) previously reported for *Dysidea* sponges [22,193–197]. From them, (–)-furodysinin (**48**) shows deterrent activity against the goldfish *Carassius auratus* in the laboratory [194]. Another compound, nakafuran-9 (**51**), present in *C. gracillimum* specimens from Hainan, was also reported as a deterrent [131]. In Australia, *C. trilobatum* possesses furodysinin (**48**), furodysin (**58**), and dendrolasin (**55**) in the viscera and, additionally, agassizin (**59**) and dehydroherbadysidolide (**49**) in the mantle, while *C. brevicaudatum* presents mixtures of the same compounds along with some unidentified metabolites [178].

Hypselodoris is another well-studied genus, although some species are now named *Felimare* or even *Risbecia* [125,131,165,198–208]. All of these species possess diet-derived furanosesquiterpenes, among other terpenoids, located in their MDFs [131,165,198–209]. Longifolin (**52**) is one of the main furanosesquiterpenes found in these groups, is located in MDFs, and is a deterrent in the lab against the goldfish *Carassius auratus*, like several other compounds of theirs [131,198,201,208]. Many of these molecules are obtained from *Dysidea* sponge species [165,200,206,207]. Some of the studied species include the Mediterranean *F. picta webbi*, *F. villafranca*, *F. cantabrica*, *F. tricolor*, *F. fontandraui*, and others, presenting longifolin (**52**) and some related compounds [2,124]. In the laboratory, the crude extracts of *F. cantabrica* displayed stronger deterrence against the allopatric shrimp *Palaemon elegans* than extracts from their prey sponge, *Dysidea fragilis*, suggesting a selective accumulation of compounds [206]. The main chemical behind the deterrence was nakafuran-9 (**51**). The Mediterranean and North Atlantic species mentioned above have aposematic colorations and conform Müllerian mimicry groups [1,210]. *F. fontandraui*, however, does not present MDFs and presents tavacpallescensin (**53**) in its mantle rim [6,205,210,211]. Tavacpallescensin (**53**) is a deterrent against the allopatric shrimp *Palaemon elegans* at 1 mg/mL in the laboratory, a very low concentration compared to that reported in its mantle (25.98 ± 1.41 mg/mL) [205]. In the Atlantic, *F. picta webbi* presents longifolin (**52**) and tavacfuran, while *F. picta azorica* also presents microcionin-1 [212]. *Hypselodoris capensis* presents the feeding deterrents nakafuran-8 (**54**) and -9 (**51**), which are active against the reef fishes *Chaetodon* spp., together with the sesterterpene 22-deoxy-23-hydroxymethyl-variabilin and other sesquiterpenes and sesterterpenes from its presumed prey, the sponges *Fasciospongia* sp. and *Dysidea* sp. [213]. The Australian *H. obscura* contains dendrolasin (**55**), (–)-euryfuran (**56**), and (+)-pallescensin A (**57**), while *H. whitei* presents (–)-euryfuran (**56**), (–)-furodysin (**58**), (–)-furosydinin (**48**), and dendrolasin (**55**), some of which are deterrents against the shrimp *P. elegans*, as previously mentioned [186]. *H. infucata* from Hawai'i also possesses nakafuran-8 (**54**) and -9 (**51**), probably obtained from *Dysidea fragilis* [157]. In Bali, *H. infucata* presents (–)-furodysinin (**48**), and its crude extract is repellent against the sympatric shrimp *Penaeus vannamei* at natural concentration [214]. In Hawai'i, *H. infucata* (*Chromodoris maridadilus*) contains a 3:1 mixture of nakafuran-8 (**54**) and nakafuran-9 (**51**), like its sponge prey *Dysidea fragilis*, both reported to be deterrent [165]. *H. bennetti* and *H. obscura* from Australia contain euryfuran (**56**), but *H. obscura* also has furodysinin (**48**), furodysin (**58**), and dendrolasin (**55**), while *H. bennetti* presents agassizin (**59**), dehydroherbadysidolide (**49**), and pallescensone (**60**) [178]. In addition, in Australia, *H. tryoni* presents dehydroherbadysidolide (**49**), furodysinin (**48**), nakafuran-9 (**51**), and dendrolasin (**55**) [178]. In India, *H. kanga* and its prey sponge *Dysidea* sp. also present furodysinin (**48**) [166]. In Brazil, *H. lajensis* presents furodysinin lactone (**61**), also originated from *Dysidea* species [207]. Other *Hypselodoris* species such as *H. jacksoni* contain similar or related compounds, but no activity against potential predators has been shown [209]. Similarly, the related *Mexichromis festiva* has euryfuran (**56**) and dendrolasin (**55**), while *M. mariei* presents only euryfuran (**56**) [178]. Other chromodoridid genera like *Tyrinna* contain interesting compounds, but none of them have been demonstrated to be used against predation to date [131,179,215,216].

The genus *Hexabranchus* mainly contains macrolides. In several locations around the Pacific and the Indo-Pacific, *H. sanguineus* presents several macrocyclic lactones, but only kabiramides and halichondramide derivatives have been proved to be deterrents against the sympatric fish *Thalassoma lunare* and the crab *Dardanus megistos* [217–220]. Active compounds consist mainly of kabiramide C (**62**) and halichondramide derivatives, such as dihydrohalichondramide (**63**) [217,218,220,221]. These macrolides are found in the slugs and even at higher concentrations in their spawn, suggesting a defensive role [218,222]. Since these compounds are found in mantle and viscera, they are suggested to be obtained and biotransformed from their diet of *Halichondria* sponges [218,221]. *H. sanguineus* from Fiji contains also macrolides along with two thiazole cyclic peptides, sanguinamides A (**64**) and B [219].

Finally, within the group of nembrothids, the tambjamines (**65–71**) are alkaloids obtained from their diet of several species [223]. *Tambja abdere* and *T. eliora* in the east Pacific accumulate tambjamines (**65–71**) from the bryozoan *Sessibugula translucens,* and they are in turn preyed on by another nembrothid slug, *Roboastra tigris* [61,224,225]. In Micronesia, *Nembrotha* species present tambjamines (**65–71**) from their ascidian prey, *Atapozoa* sp. [157,226,227]. These compounds include mixtures of tambjamine A (**65**), B (**66**), C (**67**), D (**68**), E (**69**), and F (**70**); a tambjamine aldehyde (**71**); and a blue tetrapyrrol (**72**) [226]. Crude extracts and mixtures containing tambjamine C (**67**) and F (**70**) and the tetrapyrrol (**72**) are reported to be deterrents against fish at (or below) natural concentrations, while tambjamines A (**65**) and E (**69**) are not deterrents [61,226]. *R. tigris* feeds on *T. abdere* and *T. eliora,* accumulating tambjamines A–D (**65–68**) [223]. Both *Tambja* species and *R. tigris* are able to detect the tambjamines released into the mucus by chemoreception and thus chemically locate their prey [61,223]. When the concentration of tambjamines is very high, *R. tigris* may reject its prey [61,223]. Similarly, tambjamines have also been reported in *T. ceutae* and *T. stegosauriformis* and their bryozoan prey, *Bugula dentata* [207,228].

Dendronotida

In Florida, *Tritonia hamnerorum* presents julieannafuran (**73**), a furano-germacrene obtained from its diet, the sea fan *Gorgonia ventalina* [229]. Julieannafuran (**73**) has been shown in reliable field assays to be a deterrent at natural concentrations against sympatric reef fish, such as *Thalassoma bifasciatum,* as well as in the laboratory [229]. The Antarctic *Tritonia challengeriana,* instead, has been proved to be chemically protected against feeding by the sympatric sea stars *Odontaster validus,* but no compounds have been identified from it to date ([2], Avila and K Iken, unpublished results). Furthermore, in Antarctica, *Tritoniella belli* sequesters 1-O-hexadecyl glycerol (chimyl alcohol) (**74**) from its diet, the stoloniferan coral *Clavularia frankliniana* [230–232]. This compound provides protection against the potential sympatric predator, the sea star *O. validus,* which is also deterred by the mantle tissue of the slug ([2,230–232], Avila and Iken, unpublished results). The spawn of *T. belli* is also chemically defended against predators [232,233].

Tritoniopsis elegans presents the sesquiterpenes tritoniopsins A–D (**75–78**) in the mantle, which are obtained from its diet of the soft coral *Cladiella krempfi* [234]. Tritoniopsins A (**75**) and B (**76**) are the major compounds, with tritoniopsin A (**75**) more abundant in the slug and tritoniopsin B (**76**) in the soft coral, thus suggesting a selective accumulation by the slug, which incorporates it in its mantle possibly for protection against potential predators [234].

The Mediterranean *Marionia blainvillea* presents homarine (**79**), a widespread zwitterionic natural product described to be a feeding deterrent, but it has not been tested against sympatric predators of the slug [235]. Homarine (**79**) has been suggested to derive from its cnidarian diet and could be the only defense of this slug that has no nematocysts [235]. Furthermore, homarine (**79**) has been found in other molluscs, for example, in Antarctica (*Marseniopsis mollis*), where it was described to deter feeding in the seastar *Odontaster validus* [235,236].

The colorful *Tethys fimbria* was described to de novo biosynthesize a series of prostaglandins (PG) and PG–lactones [1,237–240]. These compounds are well known in many organisms as promotors of hormonal responses [28]. Different PGEs, such as PGE_2-1,15-lactone (**80**) and PGE_3-1,15-lactone (**81**) are found in *T. fimbria* cerata [237], while PGFs are present in the reproductive system of the slugs [239]. Since cerata are detached when the animal is disturbed, together with a copious amount of mucus and a strong antero-posterior waving movement, PGEs are suggested to be involved somehow in defense, autotomy, and/or tissue protection, as well as further regeneration of cerata, while PGE–lactones (**80,81**) are converted to the free acid forms PGE_2 and PGE_3, respectively [237]. Similarly, *Melibe viridis* contains one of these prostaglandin lactones (**80**) in its mucus and cerata, suggested to be used for defense against predators [77].

Euarminida

Only one species has been suggested to use defensive compounds in this group [1,2], the Antarctic *Charcotia granulosa* [241,242], although no experiments have proved this yet. This species possesses a unique linear homosesterterpene lactone, granuloside (**82**), probably stored in its MDF-like structures [242]. Granuloside (**82**) was isolated from the lipophilic extract of the mantle of the slug, while it was absent in the gut and digestive gland as well as in the prey of the nudibranch, the bryozoan *Beania erecta*, strongly supporting its de novo biosynthetic origin. Sesterterpenes are known in nudibranchs [4], but, to date, granuloside (**82**) is the only known linear homosesterterpene in nature.

Aeolidida

Homarine (**79**), previously mentioned above, has been also found in the Atlantic aeolidids *Cratena pilata* and *Cuthona gymnota*, the Pacific *Hermissenda crassicornis*, the Australian *Phestilla lugubris*, and the Mediterranean *Cuthona coerulea* [2,92,235]. It has been suggested that the slugs obtain homarine (**79**) from their diet of hydrozoans or other cnidarians [235]. *Flabellina exoptata*, *F. ischitana*, *F. pedate*, and *F. affinis* also contain homarine (**79**) [235,243]. Despite the fact that homarine (**79**) has not been tested specifically for these species, its potential deterrent role cannot be overruled (see above) and may complement their cnidocyst defenses.

Phyllodesmium species do not to present functional cnidocysts, and, thus, their chemical defenses become their only protective shield, together with their cryptic behavior [244–248]. *P. magnum* from China presents an uncommon asteriscane sesquiterpene related to 11β-acetoxypukalide (**83**), as well as some other sesquiterpenes [249]. 11β-acetoxypukalide (**83**) was previously reported to be the chemical defense of *P. guamensis* from Guam, which accumulate it in their cerata, and it was suggested to be obtained from feeding on *Sinularia* soft corals [246]. 11β-Acetoxypukalide (**83**) was shown to deter feeding by the sympatric omnivorous pufferfish *Canthigaster solandri* at concentrations at least an order of magnitude lower than those found in their cerata (0.5% of dry mass in artificial food) [246]. Previously, trocheliophorol (**84**) was also found to be accumulated in the cerata of the Australian *P. longicirrum* and in its prey, the soft coral *Sarcophyton trocheliophorum* [245]. Four more polycyclic diterpenes and other compounds were described from *P. longicirrum*, some of them (for example, 4-oxochatancin (**85**), (2S)-*iso*sarcophytoxide (**86**), and cembranoid bisepoxide 12) being deterrent also to the pufferfish *C. solandri* [250,251]. The 4-oxochatancin (**85**) is probably obtained from a diet of *Sarcophyton* corals [28,250,251]. *P. longicirrum* also possesses many other compounds, including steroids, cembranoid diterpenes, biscembranoids, and the above-mentioned chatancin diterpenes [251].

Other *Phyllodesmium* species have been reported to contain other interesting natural products, but its role in deterring potential predators has not been proved to date [244,248].

2.1.2. Pleurobranchoidea

This group is well known for presenting acidic secretions that may deter putative predators [1,2]. Examples, with pHs as low as 1–2 include *Pleurobranchaea californica*, *Berthellina citrina*, and *Pleurobranchus strongi* from the Pacific, as well as *Berthella plumula* and *Pleurobranchus membranaceus* from the North Atlantic. In addition, *Berthella* sp. 1 from the Mediterranean and *Berthella* sp. 2 from Antarctica display pH ~1 [2]. *P. californica* and *P. membranaceus* have also been described to possess buccal acid glands [124,252]. Both *Berthella* and *Berthellina* are usually consumers of demosponges and occasionally of calcareous sponges and corals [253], and no chemical defenses have been described for them besides the acid secretions mentioned above. Similarly, the Antarctic *Bathyberthella antarctica* presents defensive acid secretions in its mantle [254,255].

2.1.3. Tylodinoidea

Tylodina species seem to be protected from predation by using sponge compounds and crypsis. *Tylodina fungina* from the Pacific contains an ester derivative of the brominated isoxazoline alkaloid 3,5-dibromotyrosine (**87**), which is a known feeding deterrent in sponges of the genus *Aplysina* [256]. *T. perversa* from the Mediterranean possesses similar metabolites from the sponge *Aplysina aerophoba* [257]. Finally, *T. corticalis* from Australia selectively accumulates several bromotyrosine-derived alkaloids from its sponge diet, *Pseudoceratina purpurea*, which contains a larger variety of these compounds [258]. In all cases, the natural products are sequestered by the molluscs and can then be found in the mantle, mucus, reproductive organs, and egg masses [259,260]. In the case of *T. perversa*, they feed preferentially on the symbiotic tissues of sponge prey loaded with cyanobacteria [261]. Furthermore, the slugs combine chemical defense with crypsis, while their mimetic yellow color (as well as that of their egg masses) on *Aplysina* species is due to uranidine, a phenolic pigment that becomes dark by oxidation when exposed to air, and it is also derived from the sponge [262,263].

2.1.4. Cephalaspidea

Species of the genus *Philine* often secrete sulfuric acid from subepithelial notal glands, and this is supposed to be a defense against predators, similarly to acid-secreting nudibranchs and pleurobranchoids [124,264]. *P. quadripartita* from the Mediterranean, Atlantic, South Africa, and Indo-Pacific is an example, possessing sulfuric and hydrochloric acid in acidic glands [265,266]. Some other cephalaspideans are able to de novo biosynthesize their own chemical defenses, such as *Bulla striata*, a generalist algal feeder found in the Atlantic and the Mediterranean [267,268]. Remarkably, cephalaspideans, such as the voracious predator *Philinopsis depicta*, are able to prey on *B. striata*, thus obtaining chemical defenses from them—in this case, the polypropionates aglajnes 1–3 (**88**), using them for their own defense, with aglajne-1 being the most deterrent [269–271]. Similarly, the Pacific species *P. speciosa* contains the polypropionates niuhinones A and B (**89**), as well as a pyridine derivate pulo'upone (**90**) reported to be deterrent, and although their origin is not yet known, *P. speciosa* probably also relies on other cephalaspideans [272,273]. In fact, niuhinones A and B (**89**) have also been found in the Atlantic species *B. occidentalis*, along with the acyclic polypropionate, niuhinone C (**89**) [274]. *P. speciosa* also presents other compounds, such as the depsipeptide kulolide-1, a linear tetrapeptide (see below), pupukeamide, additional peptides, and the macrolide tolytoxin-23-acetate [275–277]. Similarly, *Bulla gouldiana* possesses an isomer of pulo'upone (**90**) which is further found in its cephalaspidean predator, *Navanax inermis*, and suggested to be used for its own protection [278]. Moreover, *Nakamigawaia spiralis* from Guam has been reported to chemically deter sympatric reef fish, but the active compounds have not been identified to date [279].

Homarine (**79**), again, could be used against predators in this group of heterobranchs, since it has been found in the Mediterranean *Aglaja tricolorata*, probably from its diet of sea slugs, such as dendronotaceans and/or aeolidids [235].

Another interesting group is that of *Haminoea* species. In Guam, *H. cymbalum* uses a halogenated polyacetate, kumepaloxane (**91**), which it secretes when it is disturbed and which deters porcupine fish [280]. Similarly, a chemically related brominated tetrahydropyran has been found in the same species from India, as well as in *H. cyanomarginata* from the Mediterranean, strongly deterring predation by the generalist crustacean *Palaemon elegans* [77,166]. Moreover, the spawn of *H. virescens* from the Pacific has been shown to deter feeding in decapod crustaceans, although the compound(s) has not yet been identified [281].

In Guam, *Sagaminopteron* species concentrate polybrominated diphenyl ethers, probably for defense against potential predators, although this has not yet been demonstrated. *S. nigropunctatum* and *S. psychedelicum* both feed on the sponge *Dysidea granulosa* and sequester the sponge-polybrominated diphenyl ethers, concentrating them in their mantle and parapodia [282]. One of the compounds, 3,5 dibromo-2-(2′,4′-dibromo-phenoxy)phenol (**92**), is found at higher concentrations in the slug's parapodia (8–10%) than in the sponge or the rest of tissues of the slug (2–4%), thus supporting a potential defensive role [282].

2.1.5. Anaspidea

Although sea hares are among the most studied heterobranch groups and many compounds have been described, not so many studies have focused on metabolites used to avoid predation [1,4,283]. Usually, sea hares obtain natural products from their red algal food and are often able to biotransform them [284–286]. Surprising reports on sea hares include specimens of *Aplysia fasciata* (*A. brasiliana*) being rejected by sharks, even when hidden in fish fillets [287]. The sharks avoided all of the pieces, except for the buccal mass, presumably containing no defensive metabolites [287]. In fact, it is well known that sea hares present glandular structures containing deterring compounds, which may be secreted or stored in their external tissues. *A. juliana* is known to use opaline and ink secretions to deter crabs, while *A. californica*, *A. dactylomela*, and *A. parvula* present aplysioviolin (**93**) and phycoerythrobilin, biotransformed from their algal food in the ink gland and used to avoid blue crabs' predation [288–290]. Enzymatic interactions between opaline and ink secretions in *A. californica* involving escapin result in hydrogen peroxide production, and this induces deterrence against crabs, spiny lobsters, fishes, and anemones, as widely described in the literature [2,291–297]. Significant deterrence was also described when *A. californica* was fed on *Ulva* (green algae) and on *Plocamium* (red algae) and given to kelp bass (*Paralabrax clathratus*), and the effect proved to be stronger when the sea hare had fed on *Plocamium* (richer in natural products) [298]. *A. parvula* from Guam accumulates apakaochtodenes A (**94**) and B, two halogenated monoterpenes, from their red algal food *Portieria hornemanii*, using them as repellents against potential sympatric reef fish predators at natural concentrations [299]. In New Zealand, the same species contains several brominated and chlorinated terpenoids from the red algae *Plocamium costatum*, among which costatone (**95**) is found 14 times more concentrated in the slug than in the algae, supporting a potential defensive role [88,300].

Stylocheilus feeds on cyanobacteria using compounds from their diet to deter predators [301]. In Hawai'i, *S. longicauda* presents aplysiatoxin (**96**), debromoaplysiatoxin (**97**), stylocheilamide (**98**) and some complex proline esters (makalika ester (**99**) and makalikone ester (**100**)) together with lyngbyatoxin A acetate (**101**) [302–305]. Stylocheilamide (**98**) was later considered to be identical to acetyl malyngamide I, previously described from the Hawaiian cyanobacteria *Lyngbya majuscula* [306]. Moreover, the alkaloids malyngamides O (**102**) and P (**103**) were also found in the sea hare, being also structurally related to *L. majuscula* compounds [307]. Malyngamides A (**104**) and B were first found in *Microcoleus lyngbyaceus* (probably *L. majuscula*) [308]. In Guam, *S. longicauda* contains malyngamydes from the cyanobacteria and biotransforms malyngamyde B into an acetate. It has been proved that *S. longicauda* compounds are deterrents against sympatric fish (such as the pufferfish

Canthigaster solandri), amphipods, crabs (*Leptodius* spp.), and even the herbivorous cephalaspidean *Diniatys dentifer* [309,310].

Bursatella leachii plei from Puerto Rico presents bursatellin (**105**), a diol nitrile alkaloid, structurally related to chloramphenicol, while *B. leachii* from the Mediterranean possesses the (+) and (−) isomers of bursatellin (**105**), in their external extracts, but no deterrent activity has been reported to date [311,312].

2.1.6. Pteropoda

The amazing case of the Antarctic pelagic slug *Clione limacina* is worth mentioning here. *C. limacina* possesses a polypropionate-derived compound, pteroenone (**106**), which is a strong feeding deterrent against fish predators, such as *Pagothenia borchgrevincki* and *Pseudotrematomas bernachii* [313]. Pteroenone seems to be de novo biosynthesized, since it is not found in the prey of *C. limacina*, the thecosomate *Limacina helicina* [314]. The pelagic hyperiid crustacean *Hyperiella dilatata* captures and carries the chemically protected pteropods on its dorsum, thus increasing its chances of survival [315].

2.1.7. Sacoglossa

Despite the fact that the variety of compounds described in sacoglossa is huge [2], very few studies have tested deterrence at natural concentrations and against sympatric predators. The shelled sacoglossa *Ascobulla ulla* presents ascobullin A (**107**) and B, structurally related to oxytoxins (see below), but with less reactive molecules [316]. *Elysia crispata* from Venezuela contains, among other compounds, crispatenine and onchidal (**108**), the latter also found in the pulmonate *Onchidella* (see below) where it is presumably used to deter potential predators in its active form, ancistrodial (**109**) [316–319]. *Elysia translucens* contains udoteal as a main component from the green algae *Udotea petiolata*, which induces significant avoidance in the fish *Pomacentrus coeruleus* at 800 ppm [320].

Among the shell-less sacoglossans, the Mediterranean *Thuridilla hopei* contains the diterpenoids thuridillins (**110**), possessing a central α,β-epoxy-δ-lactone ring which is substituted by an uncyclized or cyclized isoprenoid chain and a 2,5-diacetoxy-2,5-dihydrofuran unit [321,322]. *T. hopei* also possesses *nor*-thuridillonal (**111**), the epoxylactone from the algae *Pseudochlorodesmis furcellata* [323], considered the putative precursor of thuridillins (**110**), and which is active in laboratory feeding deterrence tests against the shrimp *Palaemon elegans* at a concentration of 5.0 mg/mL [322]. *Thuridilla splendens* from Australia also presents thuridillins (**110**), but contrastingly, these thuridillins did not deter feeding by the sympatric shrimp *Palaemon serenus* in the laboratory [186,324].

The Caribbean *Costasiella ocellifera* (*C. lilianae*) contains avrainvilleol (**112**), a brominated diphenylmethane dietary algal derivative, from feeding on the algae *Avrainvillea longicaulis* [316,325]. Avrainvilleol (**112**) possesses deterrent properties against the tropical damselfish *Pomatocentrus coeruleus* at 100 ppm [316,325].

The Mediterranean *Cyerce cristallina* presents cyercene polypropionates (**113**) [326]. This slug has unknown feeding habits and may autotomize its cerata [326,327]. Cyercenes (**113**) are also found in the Australian *C. nigricans*, which feeds on *Chlorodesmis* algae and presents the algal diterpenoid chlorodesmin (**114**) [328]. The Atlantic *Mourgona germaineae* secrets a toxic mucus when disturbed and may also autotomize the cerata [329]. *M. germaineae* retains active chloroplasts form its algal diet, the calcareous green alga *Cymopolia barbata*, from which it also accumulates prenylated bromohydroquinones, such as cyclocymopol (**115**) [330]. Cyclocymopol (**115**) is similar to the deterrent avrainvilleol (**112**) mentioned above [325]. *Caliphylla mediterranea*, instead, seems to rely only on a defensive cryptic behavior to avoid predators, lacking propionates or other defensive chemistry [331]. This species captures chloroplasts from the algae *Bryopsis plumula* for camouflage and does not autotomize [331]. Contrastingly, *Placida dendritica* possesses polypropionate γ-pyrones such as *iso*-placidene A (**116**) that are probably used for deterrence; this species also uses crypsis as a defensive mechanism but does not autotomize [332].

2.1.8. Pulmonata

While many different compounds have been described in pulmonates, very few have been appropriately tested using natural concentrations and against sympatric predators [2]. *Trimusculus costatus* from South Africa presents the diterpenoid labdanes 6β,7a-diacetoxylab-8,13-dien-15-ol (**117**) and 2α,6β,7a-triacetoxylabda-8,13-dien-15-ol (**118**), which produce feeding deterrence against the predatory fish *Pomadasys commersonnii* [333]. *T. reticulatus* from New Zealand, instead, possesses some deterrent diterpenes, such as 6β-*iso*valeroxylabda-8,13-dien-7α,15-diol and 2α,7α-diacetoxy-6/3-*iso*valeroxylabda-8,13-dien-15-ol, which are located in the mantle and foot are effective against sea star predators [334]. Other species of this genus also display antifeeding activities, such as *T. costatus* from Chile and *T. peruvianus* from South Africa [333,335–337].

Contrastingly, species of the genus *Siphonaria* present two different classes of polypropionates, some of which are found in the mucus and mantle border, thus indicating some sort of deterrent role, and are considered to be de novo biosynthesized [338,339]. The first type of polypropionates is represented by acyclic compounds with a 2-pyrone and furanone rings, such as siphonarienolone (**119**), structurally related to the polypropionates of the cephalaspideans (see above). This type of polypropionate is found in some species from Australia, Atlantic Ocean, and South Africa [340–346]. The second type possesses variable lengths in the alkyl chain, producing a polyoxygenated network that often cyclizes, for example siphonarin A (**120**), similar to polypropionates from actinomycetes, and found in *Siphonaria* from Australia, New Zealand, Pacific Ocean, and South Africa [347–352]. The species that have been analyzed to date include *S. capensis*, *S. concinna*, *S. cristatus*, and *S. serrata*, and some of their polypropionates are deterrents against fish [353].

The Onchidiidae possess repugnatorial glands which may contain sesquiterpenoids, depsipeptide acetates, or propionates. *Onchidella binneyi* presents onchidal (**108**), which is secreted as ancistrodial (**109**), its active form, to deter potential predators [319]. Many species of *Onchidella* present variable amounts of natural products at different geographical locations, all of them being deterrent for sea stars, such as the sympatric *Leptasterias hexactis* for *Onchidella borealis* [354,355]. *Peronia peronii* and several *Onchidium* species present polypropionates similar to those of *Siphonaria* mentioned above [356,357], as well as some depsipeptides, such as onchidin (**121**) [358,359]. Finally, *Onchidium* sp. From China presents onchidione (**122**) in the mucus and mantle [360], with a potential defensive role, as well as onchidiol and 4-*epi*-onchidiol (see below) [361,362].

Table 2. Natural products used against predation in the different heterobranch groups. In brackets: number of species with antipredatory compounds, number of the compounds in figures, and reference numbers. # Number.

Species (#)	Compounds (#)	Predator(s) Tested	References (#)
		Nudibranchia (68)	
Bathydoris hodgsoni	Hodgsonal (**1**)	Sea star *Odontaster validus*, anemone *Epiactis* sp.	[54–56]
Doris (Austrodoris) kerguelenensis	Diterpene diacylglycerides (**2**)	Sea star *Odontaster validus*, anemone *Epiactis* sp.	[58–60]
Aldisa sanguinea	3-Oxo-chol-4-ene-24-oic acid (**3**), unsaturated analogue (**4**)	Goldfish (*Carassius auratus*)	[73]
Aldisa andersoni	9-Chloro-phorbazole D (**5**), N1-methyl-phorbazole A (**6**), phorbazoles A (**7**), B (**8**), and D (**9**)	Shrimp	[54,56,75–77]
Sclerodoris tanya	Tanyolides A (**10**) and B (**11**)	Fishes (*Gibbonsia elegans* and *Paraclinus integrippinis*)	[78]
Paradoris (Discodoris) indecora	Variabilin (**12**)	Marine and freshwater fishes	[79]
Dendrodoris limbata	Polygodial (**13**), olepupuane (**14**)	Marine and freshwater fishes	[80,81,83,84]
Dendrodoris arborescens	7-Deacetoxyolepupuane (**15**)	Feeding deterrence	[87]

Table 2. *Cont.*

Species (#)	Compounds (#)	Predator(s) Tested	References (#)
Dendrodoris carbunculosa	Dendrocarbins A–N (**16**)	Feeding deterrence	[86]
Dendrodoris denisoni	Cinnamolide (**17**), olepupuane (**14**), polygodial (**13**)	Fish	[88]
Doriopsilla pelseneeri	Pelseneeriols 1 (**18**) and 2, polygodial (**13**)	Feeding deterrence	[95,99]
Doriopsilla albopunctata, D. areolata, D. janaina, D. pharpa	*ent*-pallescensin A (**19**), 15-acetoxy-*ent*-pallescensin (**20**)	Feeding deterrence	[96,97,99]
Doriopsilla pharpa	Polygodial (**13**)	Fishes (*Chasmodes bosquianus, Fundulus heteroclitus*), crabs (*Callinectes similus, Panopeus herbstii*)	[98]
Phyllidia varicosa	9-*Iso*cyanopupukeanane (**21**), 2-isocyanopupukeanane (**22**)	Fish	[106–108]
Phyllidia coelestis, Phyllidiella pustulosa	Xidaoisocyanate A (**24**)	Fish	[117]
P. pustulosa	Axisonitrile-3 (**25**), amphilectene (**26**), kalihinol A (**27**), kalihinol E (**28**), *ent*-stylotelline (**29**)	Goldfish (*C. auratus*)	[118,120,123]
Cadlina luteomarginata	Isocyanides (**30**), albicanyl acetate (**31**), cadlinaldehyde (**32**), luteone (**33**), 1a,2a-diacetoxyalbicanyl acetate (**34**)	Fishes (*Carassius auratura, Clinocottus analis*)	[126–130]
Chromodoris africana, C. (Glossodoris) quadricolor	Kurospongin (**36**), latrunculin B (**37**)	Fish (*Tilapia mosambica*)	[167,168,170,171]
Chromodoris hamiltoni	Latrunculins A (**38**) and B (**37**), hamiltonins A–E (**41,42**)	Feeding deterrence	[153,155]
Glossodoris vespa, G. averni, G. pallida	12-Deacetoxy-12-oxoscalaradial (**43**)	Shrimp (*Palaemon serenus*)	[175–177]
Glossodoris pallida	Scalaradial (**44**), deacetylscalaradial (**45**), deoxoscalarin (**46**)	Crabs (*Leptodius* sp.), fish (*Abudefduf sexfasciatus*)	[176–178]
Ceratosoma trilobatum, C. gracillimum	Pallescensin B (**47**), (–)-furodysinin (**48**), (–)-dehydroherbadysidolide (**49**), (–)-herbadysidolide (**50**), nakafuran-9 (**51**), dendrolasin (**55**), furodysin (**58**), agassizin (**59**)	Goldfish (*C. auratus*)	[131,193–196]
Felimare (Hypselodoris) picta webbi, F. (Hypselodoris) villafranca, F. (Hypselodoris) cantabrica, F. (Hypselodoris) tricolor, F. (Hypselodoris) fontandraui	Longifolin (**52**)	Shrimp (*P. elegans*)	[124,165]
Felimare (Hypselodoris) fontandraui	Tavacpallescensin (**53**)	Shrimp (*P. elegans*)	[205,211]
Hypselodoris capensis	Nakafuran-8 and -9 (**54,51**)	Feeding deterrence	[213]
Hypselodoris obscura	Dendrolasin (**55**), (–)-euryfuran (**56**), (+)-pallescensin A (**57**), (–)-furodysinin (**48**), (–)-furodysin (**58**)	Feeding deterrence	[186]
Hypselodoris whitei	(–)-Euryfuran (**56**), (–)-furodysin (**58**), (–)-furosydinin (**48**), dendrolasin (**55**)	Feeding deterrence	[186]
Hypselodoris infucata	Nakafuran-8 and -9 (**54,51**), (–)-furodysinin (**48**)	Shrimp (*Penaeus vannamei*)	[157,165,186]
Hypselodoris benneti	Euryfuran (**56**), agassizin (**59**), dehydroherbadysidolide (**49**), pallescensone (**60**)	Feeding deterrence	[178]
Hypselodoris (Risbecia) tryoni	Dehydroherbadysidolide (**49**), furodysinin (**48**), nakafuran-9 (**51**), dendrolasin (**55**)	Feeding deterrence	[178]
Hypselodoris kanga	Furodysinin (**48**)	Feeding deterrence	[166]
Hypselodoris lajensis	Furodysinin lactone (**61**)	Feeding deterrence	[207]
Mexichromis festiva	Euryfuran (**56**), dendrolasin (**55**)	Feeding deterrence	[178]
Mexichromis marieri	Euryfuran (**56**)	Feeding deterrence	[178]
Hexabranchus sanguineus	Kabiramide C (**62**), dihydrohalichondramide (**63**), sanguinamides A (**64**) and B	Fish (*Thalassoma lunare*), crab (*Dardanus megistos*)	[217–222]

Table 2. *Cont.*

Species (#)	Compounds (#)	Predator(s) Tested	References (#)
Tambja abdere, T. eliora	Tambjamines A–F (**65–70**), tambjamine aldehyde (**71**)	Fish	[160,224,225]
Roboastra tigris, Nembrotha spp.	Tambjamines A–F (**65–70**), tambjamine aldehyde (**71**), tetrapyrrol (**72**)	Fish	[137,157,223,226,227]
Tritonia hamnerorum	Julieannafuran (**73**)	Fish	[229]
Tritoniella belli	1-O-hexadecyl glycerol (**74**)	Seastar (*O. validus*)	[230]
Tritoniopsis elegans	Tritoniopsins A–D (**75–78**)	Feeding deterrent	[234]
Marionia blainvillea	Homarine (**79**)	Feeding deterrent	[235]
Tethys fimbria, Melibe viridis	PGE$_2$-1,15-lactone (**80**), PGE$_3$-1,15-lactone (**81**)	Feeding deterrent	[237,238]
Charcotia granulosa	Granuloside (**82**)	Seastar (*O. validus*)	[241,242]
Cratena pilata, Cuthona gymnota, Hermissenda crassicornis, Phestilla lugubris, Cuthona coerulea, Flabellina exoptata, F. ischitana, F. pedata, F. affinis	Homarine (**79**)	Feeding deterrent	[235]
Phyllodesmium magnum, Phyllodesmium guamensis	11β-Acetoxypukalide (**83**)	Pufferfish (*Canthigaster solandri*)	[236,249]
Phyllodesmium longicirrum	Trocheliophorol (**84**), 4-oxochatancin (**85**), (2S)-*iso*sarcophytoxide (**86**), cembranoid bisepoxide 12	Pufferfish (*Canthigaster solandri*)	[245,250,251]
Tylodinoidea (3)			
Tylodina fungina, T. perversa	3,5-Dibromotyrosine (**87**)	Feeding deterrent	[256,257]
Tylodina corticalis	Bromotyrosine-derived alkaloids	Feeding deterrent	[258]
Cephalaspidea (9)			
Bulla striata, Philinopsis depicta	Aglajnes 1–3 (**88**)	Fish (*C. auratus*)	[269]
Bulla gouldiana, Navanax inermis	Pulo'upone (**90**)	Feeding deterrent	[278]
Aglaja tricolorata	Homarine (**79**)	Reef fish	[235]
Haminoea cymbalum	Kumepaloxane (**91**), tetrahydropyran	Porcupine fish	[280]
Haminoea cyanomarginata	Tetrahydropyran	Shrimp (*P. elegans*)	[77]
Sagaminopteron nigropunctatum, S. psychedelicum	3,5 Dibromo-2-(2′,4′-dibromo-phenoxy)phenol (**92**)	Feeding deterrent	[282]
Anaspidea (5)			
Aplysia californica, A. dactylomela, A. parvula	Aplysioviolin (**93**), phycoerythrobilin	Blue crabs, lobsters	[289,290,292]
Aplysia parvula	Apakaochtodene A (**94**) and B, costatone (**95**)	Fish	[88,299,300]
Stylocheilus longicauda	Aplysiatoxin (**96**), debromoaplysiatoxin (**97**), stylocheilamide (**98**), makalika ester (**99**), makalikone ester (**100**), lyngbyatoxin A acetate (**101**), malyngamide A (**104**), malyngamide B, malyngamide O (**102**), and malyngamide P (**103**)	Fish, amphipods, crabs, cephalaspidean	[302,303]
Bursatella leachii	Bursatellin (**105**)	Fish (*Oreochromis mossambicus* and *Caffragobius gilchristi*)	[311,312]
Pteropoda (1)			
Clione limacina	Pteroenone (**106**)	Fish	[314]
Sacoglossa (9)			
Ascobulla ulla	Ascobullin A (**107**) and B	Feeding deterrent	[316]
Elysia crispata	Crispatenine, onchidal (**108**)	Feeding deterrent	[131]
Elysia translucens	Udoteal	Fish (*Pomacentrus coeruleus*)	[320]
Thuridilla hopei	Thuridillins (**110**), *nor*-thuridillonal (**111**), epoxylactone	Shrimp (*P. elegans*)	[123,321,322]
Costasiella ocellifera	Avrainvilleol (**112**)	Fish	[316,325]
Cyerce cristallina, C. nigricans	Cyercenes (**113**), chlorodesmin (**114**)	Mosquito fish (*Gambusia affinis*)	[326–328]
Mourgona germaineae	Cyclocymopol (**115**)	Fish	[330]

Table 2. *Cont.*

Species (#)	Compounds (#)	Predator(s) Tested	References (#)
Placida dendritica	Polypropionate γ-pyrones (**116**)	Feeding deterrent	[332]
Pulmonata (11)			
Onchidella binneyi	Onchidal (**108**), ancistrodial (**109**)	Fish and crabs	[319]
Peronia peronii, Onchidium ssp.	Onchidin (**121**), onchidione (**122**), onchidiol, 4-*epi*-onchidiol	Sea stars	[343,358,359,361,362]
Trimusculus costatus	Labdanes 6β,7a-diacetoxylab-8,13-dien-15-ol (**117**), 2α,6β,7a-triacetoxylabda-8,13-dien-15-ol (**118**)	Fish (*Pomadasys commersonnii*)	[333]
Trimusculus reticulatus, T. costatus, T. peruvianus	6β-*iso*valeroxylabda-8,13-dien-7α, 15-diol, 2α,7α-diacetoxy-6/3-*iso*valeroxylabda-8,13-dien-15-ol	Sea stars	[334–337]
Siphonaria capensis, S. concinna, S. cristatus, S. serrata	Siphonarienolone (**119**), siphonarin A (**120**), diemenensins A and B	Fish	[341,342,344,347,350]

2.2. Toxicity

Toxicity was the first described activity in heterobranch molluscs, when the mucus secretion of *Phyllidia varicosa* was reported to be toxic to fish and crustaceans [106,109]. All nudibranchs except aeolidids, and all the other groups except pleurobranchoideans and pteropods, have been described to use toxic compounds for protection and survival (Figures 6 and 7). Toxicity may affect putative macropredators, such as fish, crabs, or others; small micropredators, such as amphipods or other crustaceans; and even gametes and early embryos of potential competitors or predators (Table 3). As mentioned above, the problem of assays that use species that are not sympatric puts in question the ecological validity of some of the results.

2.2.1. Nudibranchia

Doridacea

Species of the genus *Archidoris* present de novo biosynthesized ichthyotoxic diterpene glycerides (**123**) [71,363–367]. In the Atlantic, *A. pseudoargus* locates them in the mantle and egg masses [363]. Their compounds include a wide variety of terpenoids and related compounds (sesquiterpenoic and diterpenoic acid glycerides and glyceryl ether), although not all of them have been tested for ichthyotoxicity [367–373]. *Doris verrucosa* also presents ichthyotoxic diterpenoid acid glycerides, the verrucosins (**124**), active in the laboratory against *Gambusia affinis*, and most probably biosynthesized [370–372].

Phyllidia varicosa accumulates sponge compounds and secretes them in the mucus, producing toxicity in fish and crustaceans [106,109]. Among several other bioactive compounds, 9-*iso*cyanopupukeanane (**21**) and 2-*iso*cyanopupukeanane (**22**) are obtained from the demosponge *Ciocalypta* (*Hymeniacidon*) [106,107]. When 9-*iso*cyanopupukanane (**21**) was tested using the killifish *Oryzias latipes*, it was more toxic than its 9-epi-isomer, while 2-*iso*cyanoallopupukeanane (**125**) was toxic at 10 µg/mL [101,185]. In Indonesia, *P. varicosa* feeds on *Axinyssa aculeata* sequestering two epimeric 9-thiocyanatopupekeanane sesquiterpenes (**126**), which, together with 9-*iso*cyanopupukeanane (**21**), are mildly toxic to brine shrimp (LC_{50} 5 ppm) in the laboratory [110]. *P. pulitzeri* and its sponge food, *Axinella cannabina*, possess axisonitrile-1 (**127**), which was toxic against the marine fish *Chromis chromis* and the freshwater fish *Carassius carassius* [184]. Many other phyllidid species (*P. rosans* (*P. bourguini*), *P. coelestis*, *P. ocellata*, *Phyllidia* sp, *Phyllidiella pustulosa*, *Phyllidiopsis krempfi*, etc.) contain a wide variety of these and other nitrogenated compounds, but these have not been tested for toxicity [101–103,111,114,118–120,122,373–377].

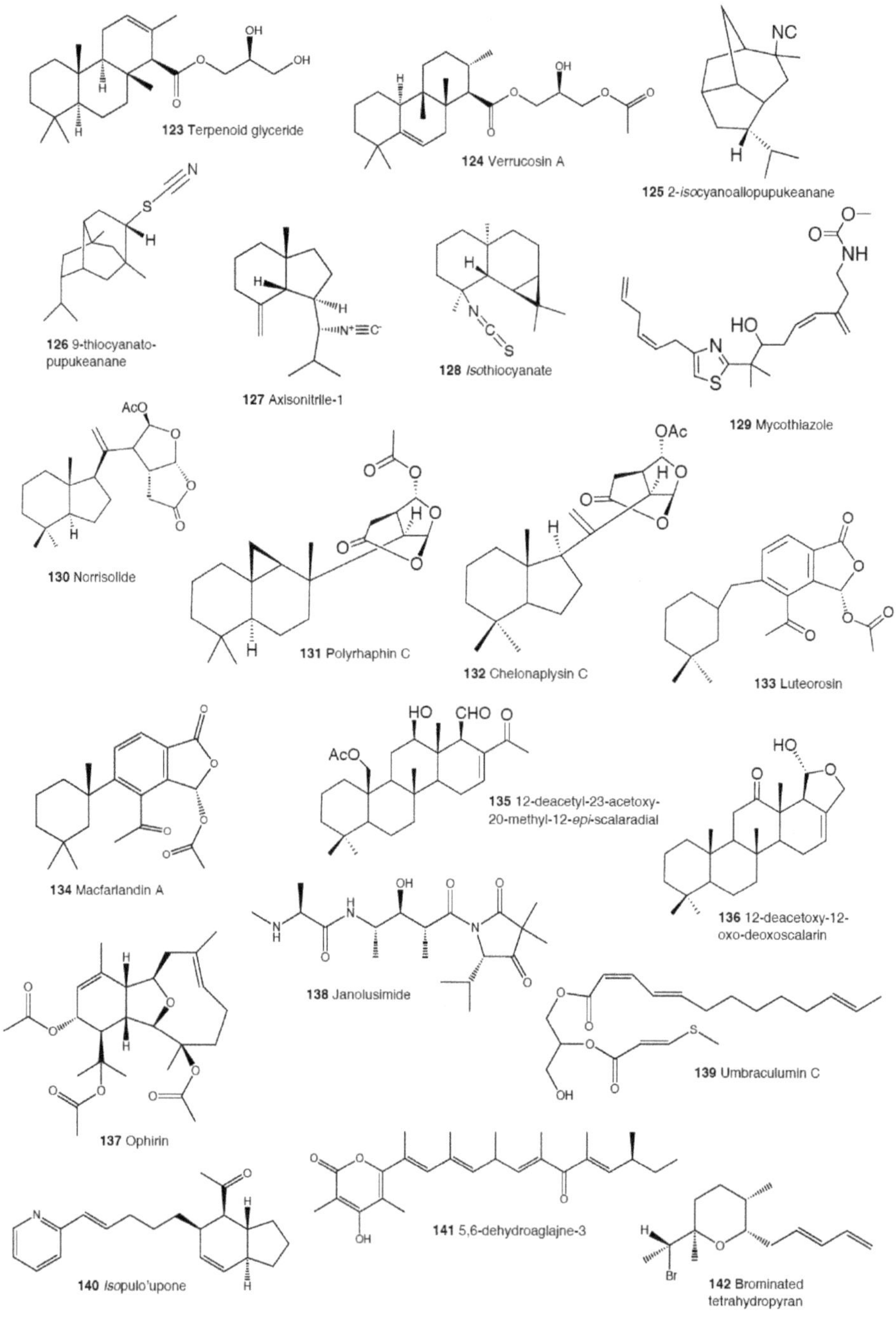

Figure 6. Structures of selected compounds displaying toxicity in Doridacea, Dendronotida, Euarminida, Tylodinoidea, and Cephalaspidea. These molecules may also display other activities, as reported in the text.

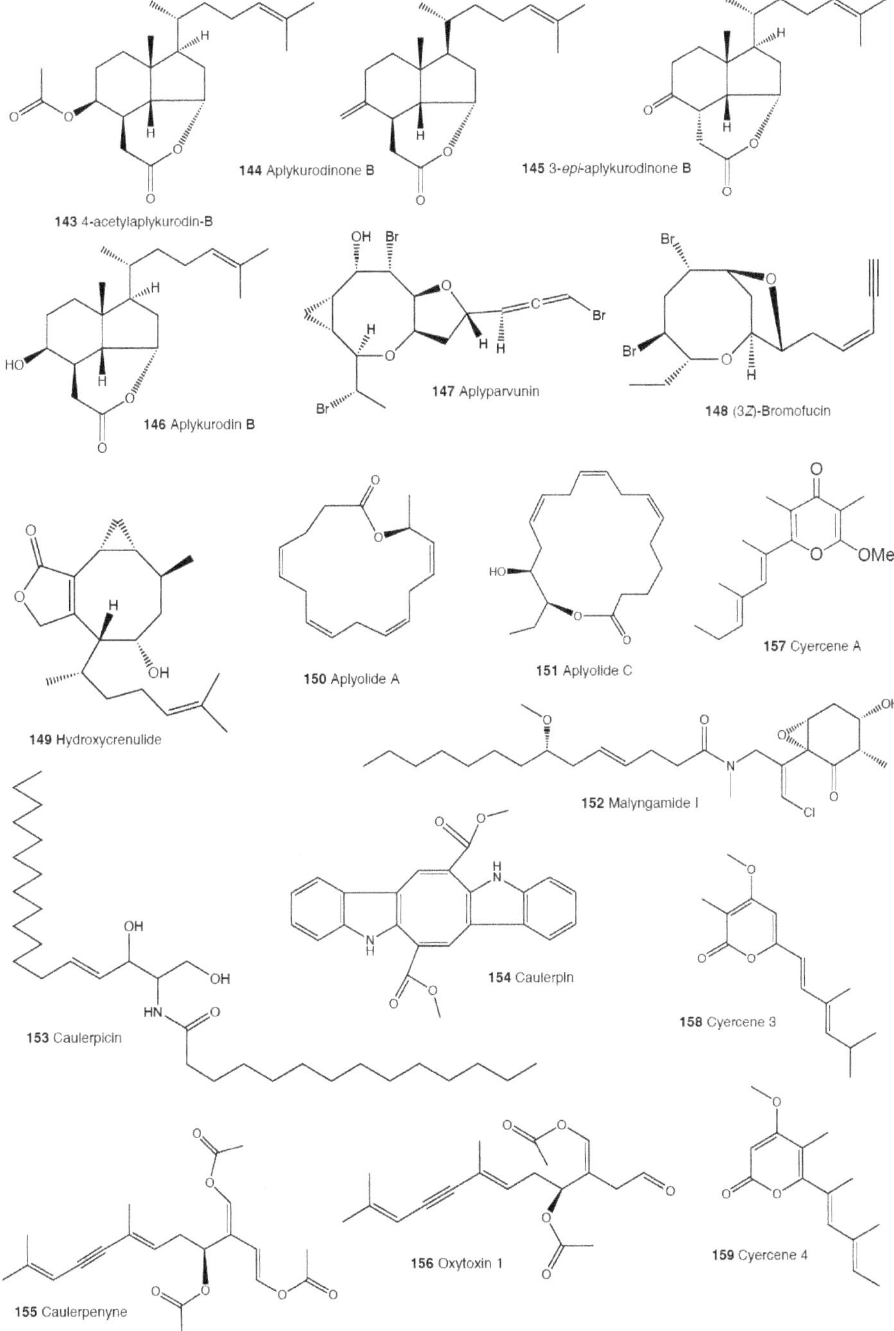

Figure 7. Structures of selected compounds displaying toxicity in Anaspidea, Sacoglossa, and Pulmonata. These molecules may also display other activities, as reported in the text.

Within chromodoridids, *Cadlina luteomarginata* presents three isocyanides (**30**) and three isothiocyanates (**128**) obtained from its sponge diet [126,127]. These metabolites resulted toxic in laboratory at 100 µg/mL, but no studies at natural concentrations and sympatric species are reported [126,127]. Further, as previously mentioned, the well-studied genus *Chromodoris* possess toxic compounds [1,6]. Kurospongin (**36**), a furanoterpene found in *C. africana* from the Red Sea, was obtained from an Okinawan *Spongia* sp. and reported to be strongly ichthyotoxic to the freshwater goldfish (*C. auratus*) at 5 µg/mL [167]. *C. hamiltoni* from South Africa and Mozambique presents one or both latrunculins A and B (**38**,**37**), among other compounds, as does *C. africana* from the Red Sea, and *C. quadricolor* (*Glossodoris quadricolor*) [153,168–171]. Latrunculin B (**37**) has been reported to be ichthyotoxic and was described from the sponge *Latrunculia* magnifica [168,169]. The Mediterranean *Felimida (Chromodoris) luteorosea* contains many ichthyotoxic sponge-derived diterpenes tested in the laboratory, including norrisolide (**130**), polyrhaphin C (**131**), chelonaplysin C (**132**), luteorosin (**133**), macfarlandin A (**134**), and closely related compounds [149]. Although many other *Chromodoris* species possess interesting chemicals, they have not been tested for toxicity.

Among the scalarane sesterterpenes described in *Doriprismatica (Glossodoris) sedna* from Costa Rica, 12-deacetyl-23-acetoxy-20-methyl-12-*epi*-scalaradial (**135**) was ichthyotoxic to the allopatric fish *Gambusia affinis* at 0.1 ppm [183]. *Goniobranchus splendidus* from Australia contains many sponge compounds, mainly spongian diterpenes, rearranged diterpenes, and nor-diterpenes [187]. Its chemical extracts have been proven to be toxic to brine shrimp (*Artemia* sp.) at natural concentrations, with potency depending on the mixture of chemicals present in each population analyzed, from no activity to toxicity [187]. *Doriprismatica (Glossodoris) atromarginata* presents furanoditerpenoids and scalarane sesterterpenes from its dietary sponges *Spongia (Hyatella)* sp. and *Hyrtios* spp., and these compounds display ichthyotoxicity against the mosquito fish, *G. affinis*—particularly, the activity of 12-deacetoxy-12-oxodeoxoscalarin (**136**) is noticeable [92,175,180,378–386]. Other NPs from chromodoridids were analyzed for ichthyotoxicity against *G. affinis*, including homoscalarane and scalarane compounds from *Felimida (Glossodoris) dalli*, *Glossodoris rufomarginata*, *Glossodoris pallida*, *Glossodoris vespa*, and *Ardeadoris (Glossodoris) averni*, and 12-deacetyl-23-acetoxy-20-methyl-12-*epi*-scalaradial (**135**) was the most potent of them [175,183,383].

Ceratosoma trilobatum and *C. gracillimum* from China contain the furanosesquiterpenes pallescensin B (**47**), (–)-furodysinin (**48**), (–)-dehydroherbadysidolide (**49**), and (–)-herbadysidolide (**50**), previously found in *Dysidea* sponges. These were tested for toxicity in the laboratory against mosquito fish and were all observed to be non-toxic except (–)-furodysinin (**48**) [22,131,193].

Dendronotida

The Mediterranean species *Tethys fimbria* contains a variety of de novo synthesized prostaglandins with diverse functions [1,240], among which is a prostaglandin lactone, PGE$_2$-1,15-lactone (**80**), later also found in *Melibe viridis* [77]. This prostaglandin lactone (**80**) is located in the mucus and cerata of *T. fimbria* and is ichthyotoxic in the laboratory against the mosquito fish [77].

Euarminida

Two euarminid species are reported to present toxic compounds. In China, *Dermatobranchus ornatus* has been reported to possess compounds inhibiting cell division in fertilized starfish eggs [9]. *D. ornatus* possesses four diterpenoids of the eunicellin class in the mantle, ophirin (**137**), calicophirin B, 13-deacetoxyl calicophirin B, and 13-deacetoxyl-3-deacetyl calicophirin B, two of them probably from its diet on the gorgonian *Muricella sinensis*, and another one previously found in an unidentified soft coral from the Pacific Ocean [22,387]. Among them, ophirin (**137**) is reported to induce brine shrimp (*Artemia* sp.) lethality. The second case is that of *Janolus cristatus*, which possesses janolusimide (**138**), a toxic tripeptide which is toxic to mice at LD 5 mg/kg [388,389]. The N-methyl analogue, janolusimide B, has been further isolated from *Bugula flabellata*, a bryozoan from New Zealand, thus suggesting a putative dietary origin for janolusimide (**138**) [390].

2.2.2. Tylodinoidea

The Mediterranean *Umbraculum mediterraneum* contains diacylglycerid fatty acid esters that are ichthyotoxic to the mosquito fish in the laboratory [391–393]. These natural products, umbraculumins A, B, and C (**139**), are suggested to be obtained from their sponge prey [263].

2.2.3. Cephalaspidea

Several compounds from *Bulla* species, such as niuhinone-B, isopulo'upone (**140**), and 5,6-dehydroaglajne-3 (**141**), are polypronionates described to be toxic to fish and shrimp [274,278]. Niuhinone-B is found in the Pacific *B. gouldiana* and the Mexican *B. occidentalis* [274,278]. In the Pacific Ocean, *Navanax inermis* also uses these compounds after ingesting *B. gouldiana* specimens, while in Hawai'i, *Philinopsis depicta* probably obtains niuhinone-B from other cephalaspideans [272,273,278]. *N. inermis* also contains *iso*pulo'upone (**140**), which is reported to be a strong ichthyotoxin that significantly affects the mosquito fish *Gambusia affinis* at 10 ppm and *Artemia salina* at 2 ppm in the laboratory [271,394]. The Mediterranean *P. depicta* contains aglajne-3 (**88**), a polypropionate toxic to *Artemia salina* (LD_{50} < 35 ppm) and *Gambusia affinis* [270].

Haminoea species also possess some toxic compounds. In the Mediterranean, *H. cyanomarginata* presents a brominated tetrahydropyran (**142**) reported to be highly toxic to the mosquito fish *G. affinis* at 1 ppm in the laboratory [77]. This tetrahydropyran (**142**) was also found in the Indian *H. cymbalum*, where it could play the same role, and it is structurally similar to kumepaloxane (**91**) from conspecifics of Guam [280].

2.2.4. Anaspidea

Several sea hares are reported to use toxic compounds. In the Mediterranean, *Aplysia fasciata* presents different compounds in different locations, with polyhalogenated monoterpenes similar to those of *Plocamium* red algae in some places [395], but some degraded sterols in other localities, such as 4-acetylaplykurodin-B (**143**), aplykurodinone B (**144**), and 3-*epi*-aplykurodinone B (**145**), which are located in the mantle and are described to be ichthyotoxic to the mosquito fish *G. affinis* in the laboratory [396]. These compounds are also related to the steroids found in the Atlantic *A. fasciata* [397] and to aplykurodin B (**146**) from the Pacific *A. kurodai* [398]. In Japan, instead, *A. parvula* possesses the ichthyotoxic brominated acetogenin dicyclic ether, aplyparvunin (**147**), which possesses strong activity (LC_{100} 3 ppm in 24h) against *G. affinis* in the laboratory [399], while specimens from South East Africa present (3*Z*)-bromofucin (**148**), a halogenated cyclic acetogenin obtained from its red algal food, *Laurencia implicata* [400]. *A. vaccaria* from the Pacific Ocean presents also ichthyotoxic compounds, in this case, the crenulides (**149**), non-halogenated diterpenoids obtained from its brown algal food, *Dictyota crenulata*, and located in their digestive gland [401,402]. Crenulides (**149**) are toxic to the reef-dwelling fish *Eupomacentrus leucosticus* at 10 µg/mL [401,402]. *A. depilans* also possesses ichthyotoxic fatty acid lactones, the aplyolides A−E (**150,151**), which are toxic in the laboratory to the mosquito fish *G. affinis* at 10 ppm [403]. In the Caribbean, *A. argus* presents ichthyotoxic biotransformed compounds from its diet, the brown algae *Stypopodium zonale*, but it possesses the bioactive diphenyl ether 2-(2′,4′dibromophenoxy)-dibromoanisole from the green alga *Cladophora vagabunda* in the digestive gland when it feeds on it [404,405].

Several bioactive compounds have also been isolated from *Stylocheilus*, mostly related to cyanobacterial metabolites [301–310]. However, only the related acetyl malyngamide I (**152**) from the Hawaiian *Lyngbya majuscula* was found to be ichthyotoxic [306], being structurally similar to stylocheilamide (**98**), a non-toxic amide from the Hawaiian *S. longicauda* [304].

2.2.5. Sacoglossa

The first toxic species reported in this group was *Oxynoe panamensis* from California, containing caulerpicin (**153**) and caulerpin (**154**) from its green algal food, *Caulerpa sertularioides* [406].

Later, in the Mediterranean, the shelled sacoglossans *Oxynoe olivacea* and *Ascobulla* (*Cylindrobulla*) *fragilis* were described to biotransform the sesquiterpenoid caulerpenyne (**155**) from its green algal food (*Caulerpa prolifera*) into the more potent ichthyotoxic aldehydes, oxytoxin-1 (**156**) and oxytoxin-2 [316,407]. In particular, oxytoxin-1 (**156**) is toxic to the mosquito fish *G. affinis* at >10 μg/mL in the laboratory, while oxytoxin-2 is toxic at 1 μg/mL. These animals are able to transport the compounds from the digestive gland to the mantle and secrete them into toxic whitish mucus [407]. Similarly, *Lobiger serradifalci*, also feeding on *C. prolifera*, presents only oxytoxin-1 (**156**) in its parapodial lobes and defensive secretion [407,408]. In the Caribbean species *Ascobulla ulla* (eating *Caulerpa fastigiata*), *Oxynoe antillarum* (eating *Caulerpa* sp.), and *Lobiger souberveii* (eating *Caulerpa racemosa*), also caulerpenyne (**155**) is also found [316]. In fact, only caulerpenyne (**155**) is detected in *L. souberveii*, while the rest of species modify it to oxytoxins (**156**) [316]. Caulerpenyne (**155**) is also found in *Volvatella* sp. in India [409].

Some shell-less species use the same system, transforming caulerpenyne (**155**) from *Caulerpa* species into oxytoxins (**156**) [410]. The Caribbean *Elysia subornata* feeds on *Caulerpa prolifera*, while *E. patina* and *E. nisbeti* feed on *Caulerpa* sp., and they all present caulerpenyne (**155**) and oxytoxin-1 (**156**) [316]. In India, *E.* cf. *expansa* also contains caulerpenyne (**155**), along with dihydrocaulerpenyne and expansinol, some minor reduced derivatives, similar to *Ascobulla ulla* compounds (see above) [411]. In *A. ulla*, ascobullin A (**107**) and ascobullin B have replaced oxytoxins, being structurally related but less reactive compounds detoxification process [316,411].

Avrainvilleol (**112**) from *Costasiella ocellifera* (*C. lilianae*) from the Caribbean is toxic to sympatric reef fishes at 10 μg/mL [325].

Cyercenes (**113**) are pyrone compounds found in several shell-less sacoglossans, displaying a very strong ichthyotoxicity against the mosquito fish, *G. affinis* in the laboratory [326,327]. The Mediterranean *Cyerce cristallina* de novo biosynthesizes the α- and γ-pyrones cyercene A (**157**) and B, as well as cyercenes 1–5 (**158,159**) [326,327]. In the toxicity assays, the most active compounds were cyercene A (**157**), cyercene-3 (**158**), and cyercene-4 (**159**), all at 10 μg/mL [326,327]. Although many other compounds of interest have been described in this group [19,412–419], they have not been proven to be toxic against sympatric species.

Table 3. Number of toxic compounds in the different heterobranch groups. In brackets: number of species with toxic compounds, number of the compounds in figures, and reference numbers. # NumberSpecies (#).

	Compounds (#)	Activity	References (#)
	Nudibranchia (22)		
Archidoris pseudoargus	Diterpenoic acid glycerides (**123**)	Ichthyotoxicity	[363,366]
Doris verrucosa	Verrucosins A (**124**) and B	Ichthyotoxicity, potent activators of PKC, and promotion of tentacle regeneration in the freshwater hydrozoan *Hydra vulgaris*	[371,420]
Phyllidia varicosa	2-*Isocyanopupukeanane* (**22**), 9-*isocyanopupukeanane* (**21**), 2-*isocyanoallopupukeanane* (**125**), 9-Thiocyanatopupekeanane (**126**)	Toxic to brine shrimp, killifish (*Oryzias latipes*), and crustaceans	[106,110,123]
Phyllidia pulitzeri	Axisonitrile-1 (**127**)	Toxic to fish (*Chromis chromis* and *Carassius carassius*)	[184]
Phyllidiella rosans (*P. bourguini*)	9-*Isocyanopupukeanane* (**21**), *epi*-9-*isocyanopupukeanane*	Ichthyotoxic to killifish *Oryzias latipes*	[373]
Cadlina luteomarginata	Isocyanides (**30**), isothiocyanates (**128**)	Toxic to goldfish (*Carassius auratus*)	[126,127]
Chromodoris africana	Kurospongin (**36**)	Ichthyotoxicity	[167–169]
Chromodoris africana, *C. quadricolor*	Latrunculins A (**38**) and B (**37**), kurospongin (**36**), 2-thiazolidinone	Ichthyotoxicity	[167,170,171]
Felimida (*Chromodoris*) *luteorosea*	Norrisolide (**130**), polyrhaphin C (**131**), chelonaplysin C (**132**), luteorosin (**133**), macfarlandin A (**134**)	Ichthyotoxicity	[149]

Table 3. *Cont.*

	Compounds (#)	Activity	References (#)
Doriprismatica (Glossodoris) sedna	12-Deacetyl-23-acetoxy-20-methyl-12-*epi*-scalaradial (**135**), 12-deacetyl-23-acetoxy-20-methyl-12-*epi*-deoxoscalarin, 12-deacetyl-20-methyl-12-*epi*-deoxoscalarin	Ichthyotoxic to *Gambusia affinis*	[183,421]
Doriprismatica (Glossodoris) atromarginata	12-Deacetoxy-12-oxodeoxoscalarin (**136**)	Ichthyotoxic to mosquito fish (*G. affinis*)	[380,381]
Felimida (Glossodoris) dalli, Glossodoris rufomarginata, Glossodoris pallida, Glossodoris vespa, Ardeadoris (Glossodoris) averni	Homoscalarane, scalarane, 12-deacetyl-23-acetoxy-20-methyl-12-*epi*scalaradial (**135**)	Ichthyotoxic to mosquito fish (*G. affinis*)	[175,183,383]
Ceratosoma trilobatum, C. gracillimum	Pallescensin B (**47**), (-)-furodysinin (**48**), (-)-dehydroherbadysidolide (**49**), (-)-herbadysidolide (**50**), nakafuran-9 (**51**)	Ichthyotoxicity	[22,131,193]
Tethys fimbria, Melibe viridis	Prostaglandin-1,15-lactones (**80**)	Ichthyotoxic to mosquito fish (*G. affinis*)	[77,240]
Dermatobranchus ornatus	Ophirin (**137**), calicophirin B, 13-deacetoxyl calicophirin B, 13-deacetoxyl-3-deacetyl calicophirin B	Inhibitory activity against the growth of silkworm *Bombyx mori*, and inhibition of cell division in fertilized starfish eggs	[387,422,423]
Janolus cristatus	Janolusimide (**138**)	Toxic to mice	[388,390]
Tylodinoidea (1)			
Umbraculum mediterraneum	Umbraculumins A–C (**139**)	Ichthyotoxic to mosquito fish (*G. affinis*)	[263,391–393]
Cephalaspidea (7)			
Bulla gouldiana	Niuhinone B, *iso*pulo'upone (**140**)	Ichthyotoxicity and shrimp toxicity	[278]
Bulla occidentalis	Niuhinone B	Ichthyotoxicity and shrimp toxicity	[274]
Navanax inermis	Niuhinone-B, *iso*pulo'upone (**140**), 5,6-dehydroagajne-3 (**141**)	Ichthyotoxicity and shrimp toxicity	[278]
Philinopsis depicta	Niuhinone B, aglajne 3 (**88**)	Toxic to *G. affinis* and *Artemia salina*	[270]
Philinopsis speciosa	Niuhinone A, B, pulo'upone (**90**), kulolide-1 (**271**), pupukeamide, tolytoxin-23-acetate	Ichthyotoxicity and shrimp toxicity	[272,273,275–277]
Haminoea cyanomarginata	Brominated tetrahydropyran (**142**)	Ichthyotoxic to mosquito fish (*G. affinis*)	[77]
Haminoea cymbalum	Brominated tetrahydropyran (**142**), kumepaloxane (**91**)	Ichthyotoxic to mosquito fish (*G. affinis*)	[280]
Anaspidea (7)			
Aplysia fasciata	4-Acetylaplykurodin-B (**143**), aplykurodinone B (**144**), 3-*epi*-aplykurodinone B (**145**)	Ichthyotoxicity	[396]
Aplysia juliana	Pyropheophorbides a and b, halogenated diterpenoid lactone, julianin-S	Lethal to crabs	[288]
Aplysia kurodai	Aplykurodin B (**146**)	Ichthyotoxicity	[398]
Aplysia parvula	Aplyparvunin (**147**), (3Z)-bromofucin (**148**)	Ichthyotoxicity	[399,400]
Aplysia vaccaria	Crenulides (**149**)	Ichthyotoxicity	[401,402]
Aplysia depilans	Aplyolides A–E (**150–151**)	Ichthyotoxicity	[403]
Stylocheilus longicauda	Makalika ester (**99**), makalikone ester (**100**), malyngamide I (**152**), malyngamide O (**102**), malyngamide P (**103**), lyngbyatoxin A acetate (**101**)	Ichthyotoxicity	[302–304]
Sacoglossa (15)			
Oxynoe panamensis	Caulerpicin (**153**), caulerpin (**154**)	Toxic to rats and mice	[406]
Oxynoe olivacea, Ascobulla fragilis	Caulerpenyne (**155**), oxytoxin 1 (**156**) and 2	Ichthyotoxicity	[316,407]
Lobiger serradifalci	Oxytoxin 1 (**156**)	Ichthyotoxicity	[316,407]
Ascobulla ulla	Ascobullin A (**107**) and B		[316]
Ascobulla ulla, Oxynoe antillarum, Lobiger souverbeii, Volvatella sp., Elysia subornata, E. patina, E. nisbeti	Caulerpenyne (**155**), oxytoxin-1 (**156**)	Ichthyotoxicity	[316,409]

Table 3. *Cont.*

	Compounds (#)	Activity	References (#)
Elysia expansa	Caulerpenyne (**155**), dihydrocaulerpenyne, expansinol	Ichthyotoxicity	[411]
Costasiella ocellifera (*C. lilianae*)	Avrainvilleol (**112**)	Toxic to sympatric reef fishes	[325]
Placida dendritica	Iso-placidene-A (**116**)	Strong ichthyotoxicity against *Gambusia affinis*	[332]
Cyerce cristallina	Cyercene A (**157**) and B, cyercenes 1–5 (**158,159**)	Strong ichthyotoxicity against *G. affinis*	[326,327]
Pulmonata (2)			
Siphonaria maura	Vallartanones B	In laboratory assays against krill and fish (*Thallasoma lunare*)	[350]
Trimusculus costatus	6β,7a-Diacetoxylab-8,13-dien-15-ol (**117**), 2α,6β,7a-triacetoxylabda-8,13-dien-15-ol (**118**)	Toxic to brine shrimp (*Artemia salina*)	[333]

2.2.6. Pulmonata

Trimusculus costatus from South Africa presents the labdanes 6β,7a-diacetoxylab-8,13-dien-15-ol (**117**) and 2α,6β,7a-triacetoxylabda-8,13-dien-15-ol (**118**), both toxic to the brine shrimp *Artemia salina* in the laboratory [333]. *Siphonaria* species present two different types of polypropionates, some of them located in the mucus and mantle border and reported to be ichthyotoxic [27,350]. *Siphonaria maura* from Mexico presents Vallartanone B, which in laboratory assays was rejected when applied to krill at 100 µg/mg and offered to the fish *Thallasoma lunare* [350].

2.3. Antimicrobials

Many marine organisms possess compounds to avoid microbial infections, and heterobranchs are no exception. Antimicrobial compounds against marine microorganisms described in heterobranchs are reported here (Figure 8, Table 4). To the best of our knowledge, however, euarminids, pleurobranchoids, tylodinoids, pteropods, and sacoglossans have not been studied for this activity to date.

2.3.1. Nudibranchia

Doridacea

Notodoris citrina from the Red Sea presents several imidazole alkaloids, among which *iso*naamidine-A (**160**) has been reported to strongly inhibit the AI-2 channel of the marine pathogen *Vibrio harveyi*, acting as a quorum sensing inhibitor [424,425]. Some of the compounds of *N. citrina* have been also found in the calcareous sponge *Leucetta chagosensis*, which is the slug diet at different geographical localities [424,425]. *Iso*naamidine-A (**160**) has also been found in *Notodoris gardineri* from the Philippines [426].

Several species of the colorful Phyllidids have been reported to contain isocyanate compounds with diverse bioactive properties [1,101–105]. As previously mentioned, this is a particularly difficult group to study since many species and genera are similar in shape and color, resulting in many misidentifications over the years [99], although some species have been studied in depth [101–105,427]. *Phyllidiella pustulosa* presents compounds obtained from the sponge *Acanthella cavernosa* [119]. *Acanthella* sponges are the dietary origin for different sesquiterpene isocyanides and related compounds in specimens from China and Vietnam [119–122]. Recent chemical analysis of the South China Sea nudibranchs, *P. pustulosa* and *Phyllidia coelestis*, as well as *A. cavernosa*, reported a nitrogenous cadinane-type sesquiterpenoid, xidaoisocyanate A (**24**), among other sesqui- and di-terpenoids [117]. Moreover, axisonitrile-3 (**25**) and several minor related sesquiterpenes were isolated from the same species, *P. pustulosa*, from Fiji [118]. Moreover, *P. pustulosa* and *Phyllidia ocellata* from Australia also present some stereoisomers of 4-isocyano-9-amorphene and of 10-isocyano-4-amorphene, respectively, while *Phyllidia picta* from Bali contains the axane sesquiterpenoids pictaisonitrile-1 (**23**) and pictaisonitrile-2 [112]. *Phyllidia* sp. from Sri Lanka contains 3-*iso*cyano-theonellin (**161**), closely related to a cyanide obtained from the demosponge *Axinyssa* [113]. *P. varicosa* presents two 9-thiocyanatopupukeanane sesquiterpenes (**126**), found also in its demosponge prey *Axinyssa aculeata* [110]. Several of these compounds are reported to have an antimicrobial role.

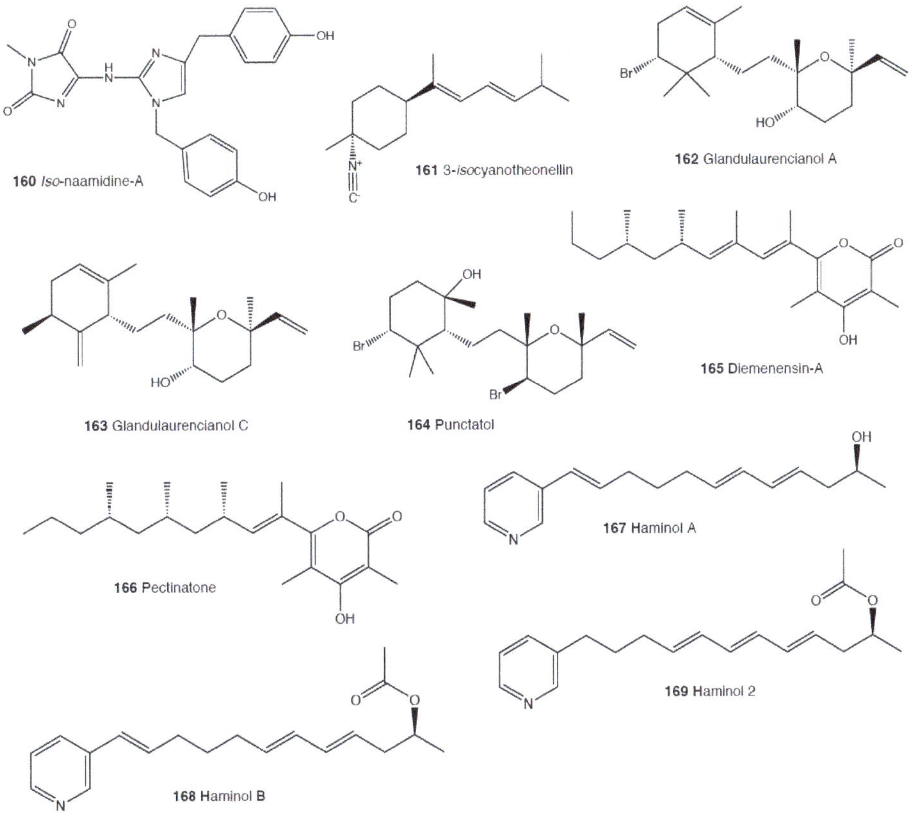

Figure 8. Structures of selected antimicrobial compounds in heterobranch molluscs. These molecules may also display other activities, as reported in the text.

Dendronotida

Only homarine (**79**) has been described in this group as a potential antimicrobial [428], and it is found in the species *Marionia blainvillea* [235].

Aeolidida

Again, homarine (**79**) has been found in nine aeolid species as mentioned above (Table 2), and it has been reported as a potential antimicrobial, probably derived from their cnidarian food sources [235,428].

2.3.2. Cephalaspidea

In cephalaspideans, homarine (**79**) has also been described in *Aglaja tricolorata*, originating probably from their diet of other heterobranchs [235].

2.3.3. Anaspidea

Regarding sea hares, *Aplysia punctata* possesses three brominated diterpenes, glandulaurencianols A–C (**162,163**), obtained from the red algae *Laurencia glandulifera*, along with punctatol (**164**) [429,430].

All these compounds showed a laurencianol skeleton, known for its antibacterial activity and a common algal dietary source [431]. Moreover, the cosmopolitan *Aplysia juliana* presents two toxic chlorophyll derivatives, pyropheophorbides *a* and *b*, and a halogenated diterpenoid lactone, while its purple secretion also includes an antibacterial and cytotoxic peptide, julianin-S, and their egg masses are protected from microbial infections by unsaturated fatty acids [288,432–434].

Table 4. Antimicrobial compounds in the different heterobranch groups. In brackets: number of species with antimicrobial compounds, number of the compounds in figures, and reference numbers. # Numbers.

Species (#)	Compounds (#)	Activity	References (#)
Nudibranchia (11)			
Notodoris citrina, Notodoris gardineri	Iso-naamidine-A (**160**)	Inhibits the AI-2 channel of the marine pathogen *Vibrio harveyi*	[123,426]
Phyllidiella pustulosa, Phyllidia coelestis, Phyllidia varicosa, Phyllidia sp.	Xidaoisocyanate A (**24**), axisonitrile-3 (**25**), 4-*iso*cyano-9-amorphene, 9-thiocyanato-pupukeanane (**126**), 3-*iso*cyanotheonellin (**161**)	Antimicrobial	[103,110,114,115,117,118,427]
Marionia blainvillea, Phestilla lugubris, Cuthona caerulea	Homarine (**79**)	Antimicrobial	[235]
Cephalaspidea (1)			
Aglaja tricolorata	Homarine (**79**)	Antibacterial	[235]
Anaspidea (3)			
Aplysia punctata	Glandulaurencianols A–C (**162,163**), punctatol (**164**)	Antibacterial	[429,430]
Aplysia juliana	Pyropheophorbides a and b, halogenated diterpenoid lactone, julianin-S	Antibacterial	[288,432,433]
Dolabella auricularia	Dolabellanin A	Antibacterial	[435]
Pulmonata (4)			
Siphonaria spp.	Siphonarienolone (**119**), diemenensins A (**165**) and B, siphonarin A (**120**), Vallartanones A and B	Antimicrobial	[27,340,344,348]
Siphonaria diemenensis	Diemenensin A (**165**)	Antibacterial	[341]
Siphonaria pectinata	Pectinatone (**166**)	Antimicrobial	[341,343]

Dolabella auricularia is another anaspidean known for protecting their eggs from bacterial pathogens, with a de novo biosynthesized glycoprotein, dolabellanin A, located in the albumen gland, showing antibacterial activity [435].

2.3.4. Pulmonata

Some *Siphonaria* species possess polypropionates in their mucus and mantle border [27]. Compounds with a 2-pyrone and furanone rings, such as siphonarienolone (**119**), structurally related to the polypropionates of cephalaspideans, are present in several species from Australia, the West and East Atlantic, and South Africa [27,340]. Both *S. diemenensis* and *S. pectinata* display antimicrobial activity due to the presence of diemenensin-A (**165**) and pectinatone (**166**), respectively [340,341,343].

2.4. Antifouling

Potentially, all surfaces under water are possible substrates for fouling colonization. Marine organisms have developed an amazing array of mechanisms to avoid fouling, and these include the use of chemicals [436]. In heterobranch molluscs, all nudibranchs except euarminids, as well as cephalaspideans, have been reported to possess antifouling compounds (Figure 9, Table 5).

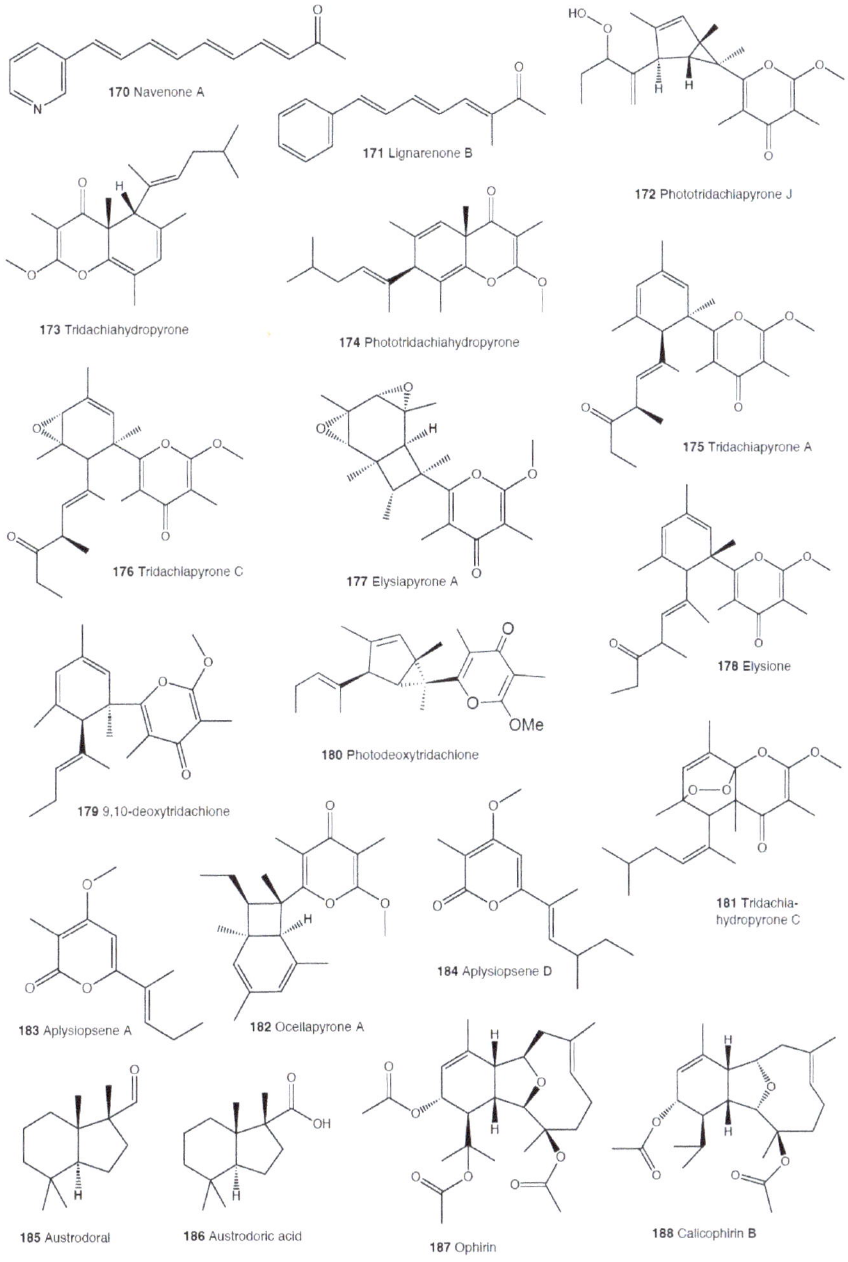

Figure 9. Structures of selected antifouling compounds, pheromones, tissue regeneration compounds, sunscreens, and other ecological activities in heterobranch molluscs. These molecules may also display other activities, as reported in the text.

2.4.1. Nudibranchia

Doridacea

Since the isolation of 9-*iso*cyanopupukeanane (**21**) from *Phillidia varicosa* [106], phyllidids have been shown to be chemically rich, presenting many nitrogenous mono-, bi- and tri-cyclic sesquiterpenes, usually traced back to their sponge prey [1,4,22,101–107,110,111,114,116,118–120,122,374–377,437]. Some of these compounds are potent antifouling agents, effective against barnacle larvae, such as the bisabolene 3-*iso*cyanotheonellin (**161**) of *P. varicosa* from Sri Lanka, and a sesquiterpene isonitrile from the Japanese *Phyllidiella pustulosa* [102,103,114–116,437]. Moreover, from *Phyllidia* sp. collected at Sri Lanka, some nitrogenous bisabolene sesquiterpenes exhibited a potent in vitro antifouling activity against barnacle larvae [114,115]. Different studies on *Phyllidia ocelata*, *P. varicosa*, *Phyllidiella pustulosa*, and *Phillidiopsis krempfi* with the aim of identifying antifouling activity reported three sesquiterpene isonitriles, namely, 10-*epi*-axisonitrile-3, 10-*iso*cyano-4-cadinene, and 2-*iso*cyanotrachyopsane, as well as a peroxide, 1,7-*epi*dioxy-5-cadinene, among others [102,116]. These molecules display potent antifouling activity against larvae of the barnacle *Balanus amphitrite* (EC_{50} = 0.14 µg/mL), while axisonitrile-3 (**25**) has an EC_{50} value of 3,2 µg/mL [437]. In fact, these natural products are present in many phyllidid species, such as *P. varicosa*, *P. coelestis*, *P. ocellata*, *P. picta*, *Phyllidia* sp., *Phillidiopsis krempfi*, *Phyllidiella pustulosa*, and *P. rosans (P. bourguini)* [102,103,116,119–122,373,437]. Moreover, *Reticulidia fungia* presents sesquiterpenes such as reticulidin A (**215**) with antifouling activity [438].

Dendronotida

As reported above, some species present homarine (**79**), such as the Mediterranean *Marionia blainvillea*, a compound that has also been proven to display potent antifouling activity [235,428,439]. This activity was previously reported for the gorgonian *Eunicella singularis* and the soft coral *Gersemia antarctica*, and homarine (**79**) has been suggested to be incorporated in dendronotids from their octocoral cnidarian food prey [235,428,439]. The presence of homarine (**79**) in the mucus secretion of the slugs would inhibit the growth of microorganisms in the mucus [235].

Aeolidida

Homarine (**79**) is also found in this group, particularly in *Cratena pilata* and *Cuthona gymnota* from the Atlantic, *Hermissenda crassicornis* from the Pacific (locations mistaken in [235]), *Cuthona coerulea* from the Mediterranean, and *Phestilla lugubris* from Australia [235,440]. In this case, homarine (**79**) is originated from their hydrozan cnidarian prey and may also be used to avoid microbial colonization of the slug mucus [235,441,442].

2.4.2. Cephalaspidea

Again, homarine (**79**) is reported here, although in a different context. The Mediterranean *Aglaja tricolorata* feeds on other heterobranchs, and may, therefore, accumulate their natural products, in this case homarine (**79**) [235,442].

Another compound has been reported to present antifouling activity in the mantle of *Sagaminopteron nigropunctatum* and *S. psychedelicum* [282]. This is the polybrominated diphenyl ether 3,5-dibromo-2-(2′,4′-dibromo-phenoxy)-phenol (**92**), which is also found in their prey, the demosponge *Dysidea granulosa*, together with other polybrominated diphenyl ethers [282]. This compound (**92**) is selectively accumulated in the parapodia of the slugs at a concentration of 8–10%, while it is present in other tissues of the slug and in the sponge at only 2–4% [282]. Its antifouling activity has been reported to be very high against marine bacteria, diatoms, barnacle larvae, and mussel juveniles [443].

Haminoea cyanomarginata from the Mediterranean is also protected against fouling by a brominated tetrahydropyran (**142**) [77]. This compound was previously found in an Australian sponge and later reported in *H. cymbalum* from India [77]. Haminols are 3-alkylpyridines also found in *Haminoea* species, which, among other activities, have been tested as antifouling molecules. These compounds include haminols A–C (**167–168**) and haminols 1–6 (**169**) [444], which are structurally related to the navenones (**170**) mentioned below. Haminol 2 (**169**) was the most potent compound, with low toxicity and strong activity against larvae of the barnacle *Amphibalanus amphitrite* [445]. De novo biosynthesis of haminols was proven in the Mediterranean *H. orbignyana* [446].

Table 5. Antifouling compounds in the different heterobranch groups. In brackets: number of species with antifouling compounds, number of the compounds in figures, and reference numbers. # Numbers.

Species (#)	Compounds (#)	Activity	References (#)
Nudibranchia (15)			
Phyllidia varicosa, Phyllidia rosans (P. bourguini)	9-Isocyanopupukeanane (**21**), 3-isocyanotheonellin (**161**)	Antifouling against barnacle larvae	[93,106]
Phyllidia sp.	3-Isocyanotheonellin (**161**)	Antifouling activity against barnacle larvae	[114,115]
Phyllidiella pustulosa	Sesquiterpene isonitrile	Antifouling against barnacle larvae	[101]
Phyllidia ocelata, P. varicosa, P. coelestis, P. picta, Phyllidiella pustulosa, Phillidiopsis krempfi	10-epi-Axisonitrile-3, 10-isocyano-4-cadinene, 2-isocyanotrachyopsane, 1,7-epidioxy-5-cadinene, 4-isocyano-9-amorphene and 10α-isocyano-4-amorphene, 9-thiocyanatopupukeanane sesquiterpenes	Antifouling activity against barnacle larvae	[110,112]
Reticulidia fungía	Reticulidin A (**215**)	Antifouling activity	[438]
Marionia blainvillea, Phestilla lugubris, Cratena pilata, Cuthona caerulea, Cuthona gymnota, Hermissenda crassicornis	Homarine (**79**)	Antifouling activity, prevents microbial colonization of the slug mucus	[235,428,439,441,442]
Cephalaspidea (9)			
Aglaja tricolorata	Homarine (**79**)	Antifouling activity	[235,441]
Sagaminopteron nigropunctatum, S. psychedelicum	3,5 Dibromo-2-(2′,4′-dibromo-phenoxy)phenol (**92**)	Antifouling activity against marine bacteria, diatoms, barnacle larvae, and mussel juveniles	[282,443]
Haminoea cyanomarginata, H. cymbalum	Brominated tetrahydropyran (**142**)	Antifouling activity	[77]
Haminoea orteai	Haminol A,B,C (**167–168**)	Antifouling activity	[444]
Haminoea orbignyana	Haminol 1-2 (**169**), haminol A and B (**167–168**)	Antifouling activity against larvae of the barnacle Amphibalanus amphitrite	[444–446]
Haminoea fusari	Haminol 1–6 (**169**)	Antifouling activity	[445]

2.5. Trail Following and Alarm Pheromones

In nature, behavioral and chemical defenses of heterobranchs are often combined to increase survival chances [1,2]. Heterobranchs display a high number of strategies, among which alarm pheromones are used to communicate and induce behavioral changes (Figure 9, Table 6) [1,2]. Doridacean nudibranchs and cephalaspideans, and perhaps some anaspideans, display trail-following behavior, which also implies some chemoreception mechanisms.

2.5.1. Nudibranchia

Doridacea

A trail-following behavior has been reported in some chromodoridids, such as *Hypselodoris* (*Risbecia*) *tryoni*, but the potential chemicals involved have not yet been described [447]. Whether this is related to defense or to reproduction remains to be further investigated.

Tambja and *Roboastra* are protected by the potent alkaloids tambjamines (**65–70**), which are obtained from their prey, for example, the bryozoan *Sessibugula translucens* for *Tambja abdere* and *T. eliora*, and these two slug species in turn for the voracious *Roboastra tigris* [61,226–228]. All these slugs are able to detect the tambjamines (**65–70**) secreted in mucous trails through chemoreception mechanisms, thus being able to find conspecifics and food (bryozoan) for *Tambja* species or detect food (slugs) for *Roboastra* [61,226]. When *Tambja abdere* specimens are disturbed, they release higher amounts of tambjamines (**65–70**), and, thus, they are also considered to be alarm pheromones [61,226]. Similar examples include *Tambja ceutae* and the bryozoan *Bugula dentata*, *Tambja stegosauriformis* and *B. dentata*, as well as some *Nembrotha* species and their ascidian prey, *Atapozoa* sp. [207,229–231].

2.5.2. Cephalaspidea

Several genera in this group have been reported to secrete alarm pheromones and display trail-following behavior. Navenones A–C (**170**) were isolated long ago from *Navanax inermis* and reported to be used as alarm signals [394]. Navenones (**170**) are accumulated in a specialized ventral gland and are released within its yellow slime trail when the animals are disturbed, inducing an avoidance alarm response in the conspecifics that follow them [269,394,448]. Navenones (**170**) are suggested to be de novo biosynthesized [448]. These slugs may also be cannibalistic, but, interestingly, the secretion is not released when a small slug is attacked by a larger conspecific [318].

Haminoea species similarly release alarm pheromones that induce escape behavior in the trail-following conspecifics [271,449]. The Mediterranean species *H. exigua*, *H. fusari*, *H. orbignyana*, *H. orteai*, and *H. navicula* present haminols (**167–169**), oxygenated 3-alkylpyridines that are secreted in their mucous trail [19,446]. Haminols (**167–169**) are biosynthesized through sequential addition of acetic acid units to nicotinic acid [446,450]. Haminols consist of several structures similar to navenones (**170**), the haminols A–C (**167–168**) and the haminols 1–6 (**169**) [444]. Similarly, *H callidegenita* presents haminols 7–11, which are also suggested to act as alarm pheromones [271,451]. The Pacific *H. japonica* also presents a series of alkylphenols, which are structurally similar to navenone-C and are suggested to act as alarm pheromones [271,451].

The widespread *Scaphander lignarius* lives in sandy bottoms, feeding usually on foraminiferans [452,453]. This species presents the ω-arylmethylketones, lignarenones (**171**), de novo biosynthesized through a polyketide pathway from benzoic acid and accumulated in the Blochmann's gland [454–456]. Lignarenones (**171**) are released within the bright yellow slime that *S. lignarius* secretes when disturbed and are also suggested to be used as alarm pheromones [454–456].

Table 6. Alarm signal compounds in the different heterobranch groups. In brackets: number of species with alarm signal compounds, number of the compounds in figures, and reference numbers. # Numbers.

Species (#)	Compounds (#)	Activity	References (#)
Nudibranchia (7)			
Tambja abdere, T. eliora, Roboastra tigris	Tambjamines (**65–70**)	Alarm pheromones and cues	[223]
Tambja ceutae, Tambja stegosauriformis, Nembrotha spp.	Tambjamines (**65–70**)	Alarm pheromones and cues	[226–228]
Cephalaspidea (10)			
Navanax inermis	Navenones A–C (**170**)	Alarm pheromones	[394,448]
Haminoea exigua, H. fusari, H. orbignyana, H. orteai, H. navicula	Haminols (**167–169**)	Alarm pheromones employed during cross copulation, escape reaction in conspecifics	[19,449]
Haminoea navicula	Haminols A and B (**167,168**)	Alarm pheromones	[449]
Haminoea orteai	Haminols A–C (**167,168**)	Alarm pheromones	[444]
Haminoea fusari, H. hydatis	Haminols 1-6 (**169**)	Alarm pheromones	[271,445]
Haminoea japonica	Navenone-C (**170**)	Alarm pheromones	[271,445]
Haminoea callidegenita	Haminols 7-11	Alarm pheromones	[271,451]
Scaphander lignarius	Lignarenones (**171**)	Alarm pheromones	[453–455]
Anaspidea (1)			
Aplysia californica	Aplysiapalythines A–C (mycosporine-like amino acids, MAAs), asterina, palythine	Alarm cues, causing avoidance behaviors in neighboring conspecifics	[457,458]

2.5.3. Anaspidea

Mycosporine-like amino acids (MAAs) have been suggested to act as alarm cues in the sea hare *Aplysia californica* [457].

2.6. Sunscreens and UV Protection

Avoiding UV radiation (UVR) is very relevant for marine organisms that live in shallow waters, since the damaging effects of UVR on cells and tissues are potentially huge [459]. In heterobranch molluscs, those living close to the surface (plankton and pteropods) or living on algae (for food and camouflage; sea hares and sacoglossans) are the most susceptible to UVR, and UV-protecting compounds have indeed been described in anaspideans, pteropods, and sacoglossa (Figure 9, Table 7).

2.6.1. Anaspidea

As mentioned above, sea hares are generalist herbivores. Among them, *Aplysia* species possess many dietary compounds from their algal food, such as many halogenated terpenoids and some carotenoids, which may provide UVR protection [459,460]. Furthermore, *A. californica* contains mycosporine-like amino acids (MAAs), the most common photoprotective molecules reported in marine organisms [457,459].

2.6.2. Pteropoda

The sea butterflies *Limacina helicina* and *Clione limacina* also possess MAAs originated from phytoplanktonic microalgal species, which provide them with UV protection [461].

2.6.3. Sacoglossa

Polypropionates from sacoglossans have been reported to be used as sunscreens in several species [412,461]. These compounds are de novo biosynthesized by the slugs, and their biosynthesis is affected by UVR [462,463]. *Elysia* species possess γ-pyrone propionates, such as phototridachiapyrone J (**172**) in the Atlantic *E. patagonica* [412], and tridachiahydropyrone (**173**) and phototridachiahydropyrone (**174**) in the Caribbean *Elysia crispata* [78,317]. Phototridachiahydropyrone (**174**) seems to originate from a rearrangement mechanism during the photochemical electrocyclic formation of tridachiahydropyrone (**173**) under prolonged UVR exposition [464]. Similar γ-pyrone compounds, tridachiapyrones A–F (**175,176**), as well as elysiapyrones (**177**), along with other compounds, were found in the Pacific *E. (Tridachiella) diomedea* [413,415–419]. *E. viridis* feeds on the green algae *Codium vermiliara* and biosynthesizes the polypropionate (+) elysione (**178**) [463,465]. Elysione (**178**) is also found in *E. chlorotica*, which feeds on *Cladophora* algae [414]. Tridachiapyrone A (**175**) is the enantiomer of (+) elysione (**178**), and like tridachiapyrone C (**176**) and crispatene, it has been suggested that these metabolites represent the protected forms of (+) elysione (**178**) under strong light irradiation [414,466]. All of these polyene γ-pyrone compounds are localized in cell membranes, where they may act as sunscreens, thus protecting the photosynthetic apparatus of the chloroplasts [466,467].

Placobranchus ocellatus and *Placobranchus* sp. from the Pacific Ocean also present propionate-derived γ-pyrones, such as 9,10-deoxytridachione (**179**), photodeoxytridachione (**180**), tridachiahydropyrone B and C (**181**), and *iso*-9,10-deoxy-tridachione, probably also used as sunscreens [19,412,466,468,469]. In specimens from India, the elysiapyrone-related compounds ocellapyrones (**182**) were also found [468].

2.7. Tissue Regeneration

As in terrestrial organisms, vagile marine organisms, as well as heterobranchs, may use autotomy and tissue regeneration as a mechanism to distract predators and escape [1,2]. In fact, many nudibranchs and sacoglossa may lose part of their mantle or their cerata [1,2]. The known compounds involved in these mechanisms are reported here (Figure 9, Table 8).

Table 7. Photoprotective compounds in the different heterobranch groups. In brackets: number of species with photoprotective compounds, number of the compounds in figures, and reference numbers.

Species (#)	Compounds (#)	Activity	References (#)
Anaspidea (1)			
Aplysia californica	Aplysiapalythines A, B, C (mycosporine-like amino acids, MAAs), asterine, palythine	Sunscreens	[457]
Pteropoda (2)			
Limacina helicina, Clione limacina	Mycosporine-like amino acids (MAAs)	UV photoprotection	[461]
Sacoglossa (8)			
Cyerce cristallina	Cyercene A (**157**) and B, cyercenes 1–5 (**158,159**)	Protection against sunlight-induced damage	[326,327]
Elysia patagonica	Phototridachiapyrone J (**172**)	Sunscreens	[412]
Elysia crispata	Tridachiahydropyrone (**173**), phototridachiahydropyrone (**174**)	Sunscreens	[317,464]
Elysia (Tridachiella) diomedea	Tridachiapyrones A–F (**175,176**), elysiapyrones (**177**)	Sunscreens, photoprotection	[417–419]
Elysia viridis, E. chlorotica	Elysione (**178**)	Sunscreens	[463,465]
Placobranchus ocellatus, Placobranchus sp.	9,10-Deoxy-tridachione (**179**), photodeoxytridachione (**180**), tridachiahydropyrone B and C (**181**), *iso*-9,10-deoxy-tridachione, ocellapyrones (**182**)	Sunscreens	[466,468,469]

2.7.1. Nudibranchia

Doridacea

Peltodoris atromaculata feeds on the sponge *Petrosia ficiformis*, accumulating polyacetylenes, petroformynes, and other compounds, although the autotomy mechanism of the slug does not seem to be related with this chemistry, and, in fact, may be related to stress or senescence instead [1,210,470].

Dendronotida

Tethys fimbria presents de novo biosynthesized prostaglandin (PG) lactones (**80–81**) involved in predator distraction by the release of their cerata [237–239]. When *T. fimbria* is molested, cerata are detached together with a copious amount of mucus and continue to move for some time. Although PGEs and PGFs have been suggested to be involved in defense and reproduction, respectively, a role in tissue regeneration has also been suggested, since they are common hormonal compounds [237–240,471,472]. In fact, PGE_2-1,15-lactone (**80**) and PGE_3-1,15-lactone (**81**) are found in the cerata, detached when disturbed to distract potential predators and reverting to the free acid forms, PGE_2 and PGE_3, which are suggested to be the defensive compounds [237].

2.7.2. Sacoglossa

Many sacoglossans also release their cerata when disturbed. Among them, the Mediterranean *Ercolania viridis* presents γ- and α-pyrone polypropionates, including cyercenes (**113, 157–159**) [473]. Cyercene polypropionates (**113, 157–159**) are also found in the Mediterranean *Cyerce cristallina*, which also displays cerata autotomy [326,327], as well as in *C. nigricans* from Australia [328]. Moreover, in *Aplysiopsis formosa* from the Atlantic Ocean, the α-pyrones aplysiopsenes A–D (**183,184**) have been found [325]. The implication of polypropionates in autotomy and further cerata regeneration has been further demonstrated by experiments with *Hydra* [474]. However, some sacoglossa may also present cerata autotomy without possessing polypropionates, such as the Atlantic *Mourgona germaineae*, which instead possesses prenylated bromohydroquinones obtained from the calcareous green alga *Cymopolia barbata* [330], similarly to *Costasiella ocellifera* from *Avrainvillea longicaulis* [325]. In contrast, some species may present polypropionates without displaying autotomy behavior, such as *Placida dendritica*, which possesses polypropionate γ-pyrones (**116**) with no regenerative activity on the *Hydra* assay [332]. Evolutionary trends are probably behind this behavioral and chemical variability [2].

Table 8. Tissue-regenerator compounds in the different heterobranch groups. In brackets: number of species with tissue-regenerator compounds, number of the compounds in figures, and reference numbers. # Numbers.

Species (#)	Compounds (#)	Activity	References (#)
Nudibranchia (1)			
Tethys fimbria	PGE2-1,15-lactone (**80**), PGE3-1,15-lactone (**81**)	Predator distraction by the release of their cerata, cerata regeneration	[237,238]
Sacoglossa (6)			
Ercolania viridis	Cyercenes (**113, 157–159**)	Cerata autotomy, cerata regeneration	[332]
Cyerce cristallina, C. nigricans	Cyercenes (**113, 157–159**)	Cerata autotomy, cerata regeneration	[326–328]
Aplysiopsis formosa	Aplysiopsenes A–D (**183,184**)	Cerata autotomy and cerata regeneration	[475]
Mourgona germaineae, Costasiella ocellifera	Prenylated bromohydroquino-nes	Cerata autotomy	[325,330]

2.8. Other Ecological Activities

Some nudibranchs have been reported to display activities other than those mentioned above [1,2]. These may include metabolites that may be considered stress metabolites, as well as compounds involved in reproduction or in hormonal responses, as reported below (Figure 9, Table 9).

2.8.1. Nudibranchia

Doridacea

The Antarctic slug *Doris kerguelenensis* possesses diterpenoid glycerols (**2**) and terpenoid glyceryl esters of different types (*ent*-labdane, halimane, isocopalane diterpenoids) in its mantle many [59,62,64,370,381,476]. *D. kerguelenensis* may also present some nor-sesquiterpenes in its mantle, austrodoral (**185**) and its oxidized derivative austrodoric acid (**186**), which have been detected in higher amounts after the animals were kept in aquarium for 15 days before freezing [64]. This is the reason why austrodoral (**185**) has been suggested to be a stress metabolite, although perhaps some physiological alterations may also be behind these higher amounts of austrodoral (**185**) [64].

Table 9. Other ecological activities in the different heterobranch groups. In brackets: number of species with active compounds, number of the compounds in figures, and reference numbers. # Number.

Species (#)	Compounds (#)	Activity	References (#)
Nudibranchia (3)			
Doris kerguelenensis	Austrodoral (**185**), austrodoric acid (**186**)	Stress metabolites	[64,65]
Tethys fimbria	PGs–lactones (**80,81**)	Role in reproduction	[237,238]
Dermatobranchus ornatus	Eunicellin, ophirin (**187**), calicophirin B (**188**), 13-deacetoxycalicophirin B, 13-deacetoxy-3-deacetylcalicophirin	Inhibition of cell division in fertilized starfish eggs	[22,243,387,477]

Dendronotida

Prostaglandins (PGs) (**80–81**) from *Tethys fimbria* have been reported to be de novo biosynthesized, having different roles in the species, and being located in different body parts of the slug [237–239]. As previously mentioned, PGE$_2$-1,15-lactone (**80**) and PGE$_3$-1,15-lactone (**81**) are found in the cerata, reverting to the free acid forms PGE$_2$ and PGE$_3$ probably for defense [237]. In their reproductive system, instead, PGF–lactones are found, and these have been suggested to have a role in reproduction [239].

Euarminida

Some species have been reported to inhibit cell division in embryos or eggs of sympatric species to avoid competition and/or further predation by their adults, for example, against sea urchins or sea stars [1,2]. An example is *Dermatobranchus ornatus* from China, which presents several types of diterpenes that inhibit cell division in fertilized starfish eggs [22]. These compounds include four eunicellin diterpenes, such as ophirin (**187**). In fact, two of these compounds are probably taken from its gorgonian prey, *Muricella sinensis* [22], while another one was previously found in an unidentified Pacific soft coral [387]. Furthermore, *D. ornatus* possesses a calicophirin diterpenoid (**188**) probably from a gorgonian prey, *Muricella* sp. [477].

3. Pharmacological Activity

Marine natural products may have many diverse applications that are beneficial for humans and, thus, may become potentially very useful drugs [5,6,8,10,29–31,33]. Some recent reviews offer wide summaries of the different compounds and activities, as well as a historical perspective [31–33,35–40, 42,43,123,478–482]. Overall, it remains clear that natural products are still the best option to find novel bioactive entities and potentially be modified to find leads to fight several human diseases.

3.1. Cytotoxicity and Antitumoral Activity

Many reviews have dealt with this topic in depth, and, therefore, we include here only a brief summary of all the activities described [31,42,479]. Among the most active compounds, it is worth mentioning jorumycin (**189**) from the doridacean *Jorunna funebris*, ulapualides (**190**) from *Hexabranchus sanguineus*, kabiramides (**62**) from *Hexabranchus* sp., aplyronines (**191**) and dolastatins (**192**) from the anaspidean *Dolabella auricularia*, and bursatellanins (**193**) from *Bursatella leachii*, as well as kahalalides (**194**) from the sacoglossan *Elysia rufescens*, all of which are under clinical trials for antitumoral activity [36,43,45,46,479]. It is important to remember that cytotoxic, anticancer, and antitumoral compounds are strongly needed, since cancer remains one of the deadliest diseases worldwide [42]. In 2018, about 18 million new cancer cases were reported globally, producing around 10 million deaths [483]. Thus, many studies are being developed to find new compounds with novel modes of action, and marine organisms, such as heterobranch molluscs, are considered to be reservoirs of new bioactive compounds that are potentially useful [42]. In fact, about 30% of the top twenty drugs currently on the market originate from a natural source (mostly plants), while approximately 50% of the marketed drugs are naturally derived or based on natural compounds [31,480,484]. Comparing marine and terrestrial natural products, it has been reported by the NCI (USA) that more than 1% of marine compounds present antitumoral activity compared to the 0.01% for terrestrial compounds [481]. Many natural compounds present cytotoxic effects against macromolecules expressed by cancer cells, such as in oncogenic signal transduction pathways, and display growth inhibition of human tumor cells [39,42,43,482]. Moreover, many studies on the mechanism of action of marine natural compounds inhibiting tumor growth indicate that it involves processes of apoptosis, necrosis, and lysis of the tumor cells, both in vitro and in vivo [42,485]. All heterobranch groups reviewed here possess some anticancer, antitumoral, and/or cytotoxic compounds, except pteropods (Figures 10–23, Table 10). The natural products described include a wide variety of compounds, from terpenoids and steroids, to peptides, polyketides, as well as nitrogen-containing metabolites.

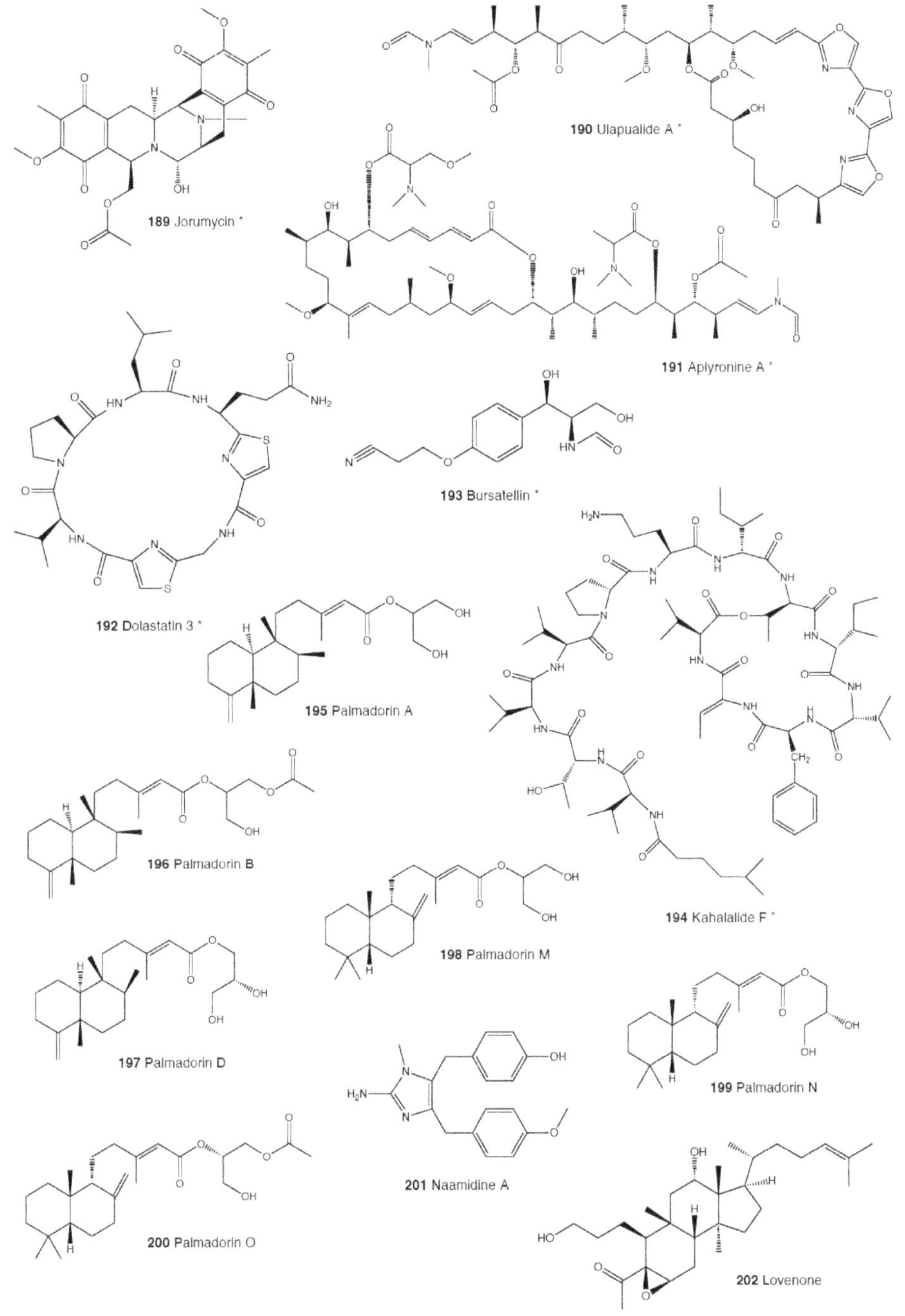

Figure 10. Structures of selected cytotoxic and antitumoral compounds in some heterobranch molluscs. These molecules may also display other activities, as reported in the text. * = compounds in clinical trials.

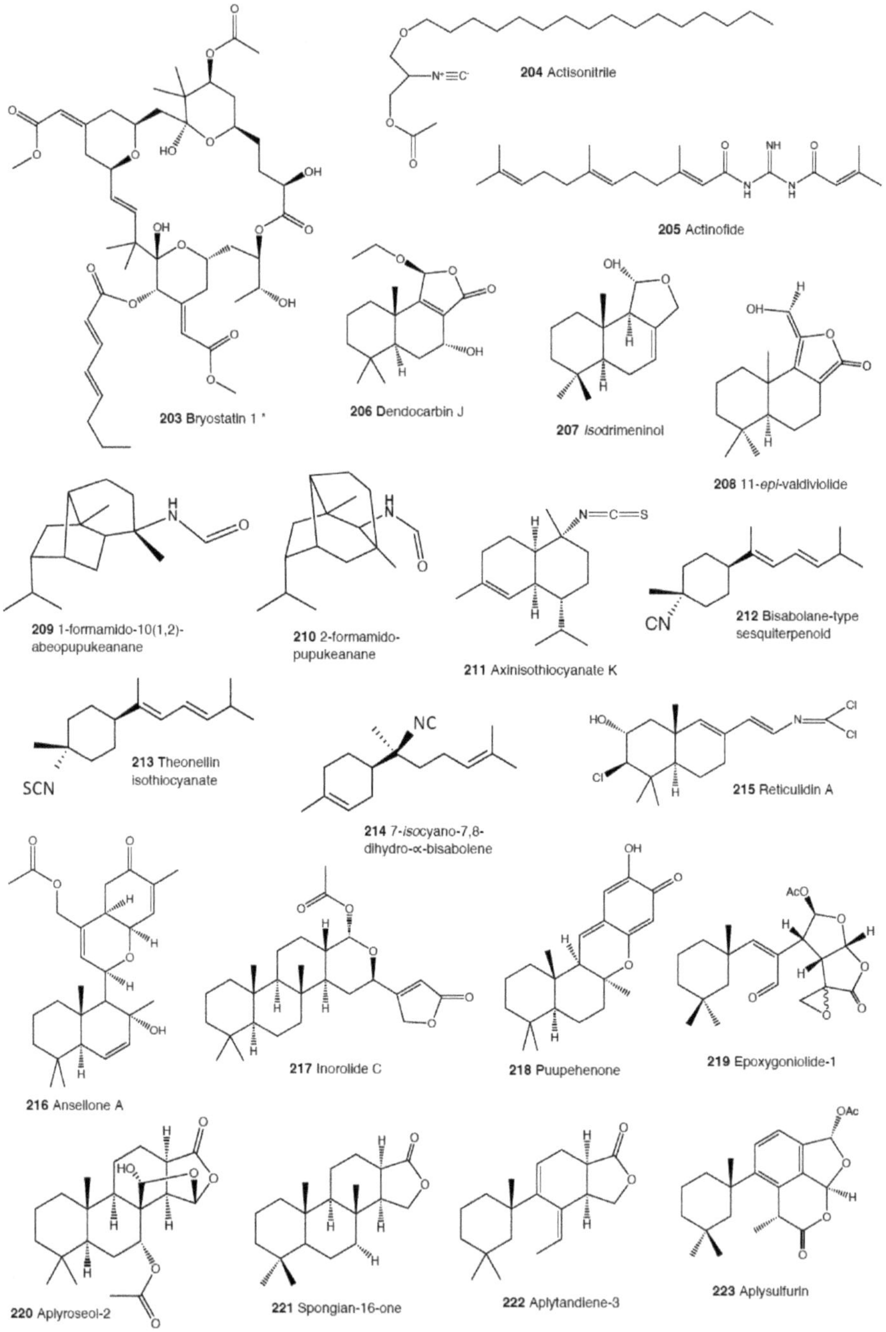

Figure 11. Structures of selected cytotoxic and antitumoral compounds in some Doridacea nudibranchs. These molecules may also display other activities, as reported in the text. * = compounds in clinical trials.

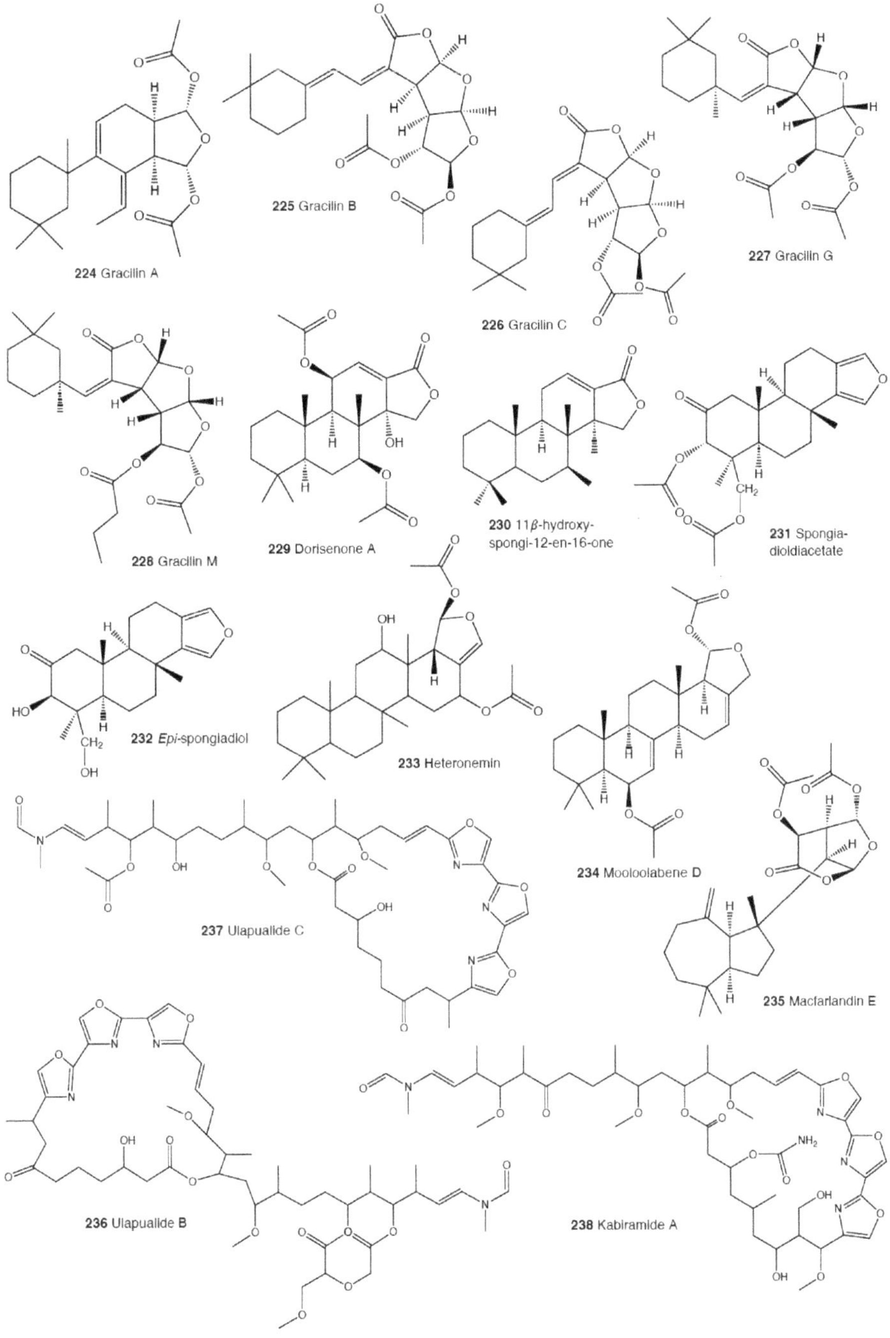

Figure 12. Structures of selected cytotoxic and antitumoral compounds in some Doridacea nudibranchs. These molecules may also display other activities, as reported in the text.

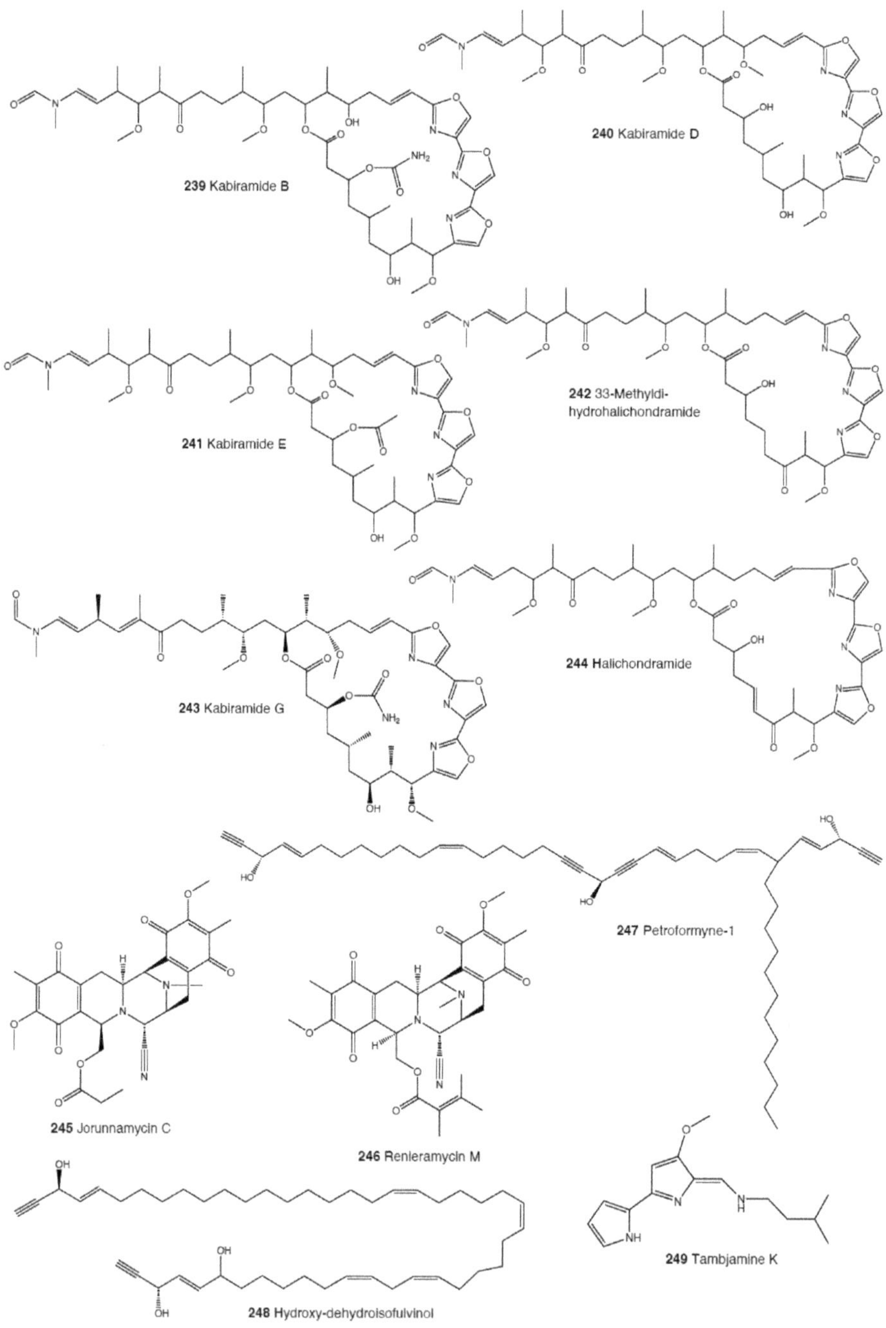

Figure 13. Structures of selected cytotoxic and antitumoral compounds in some Doridacea nudibranchs. These molecules may also display other activities, as reported in the text.

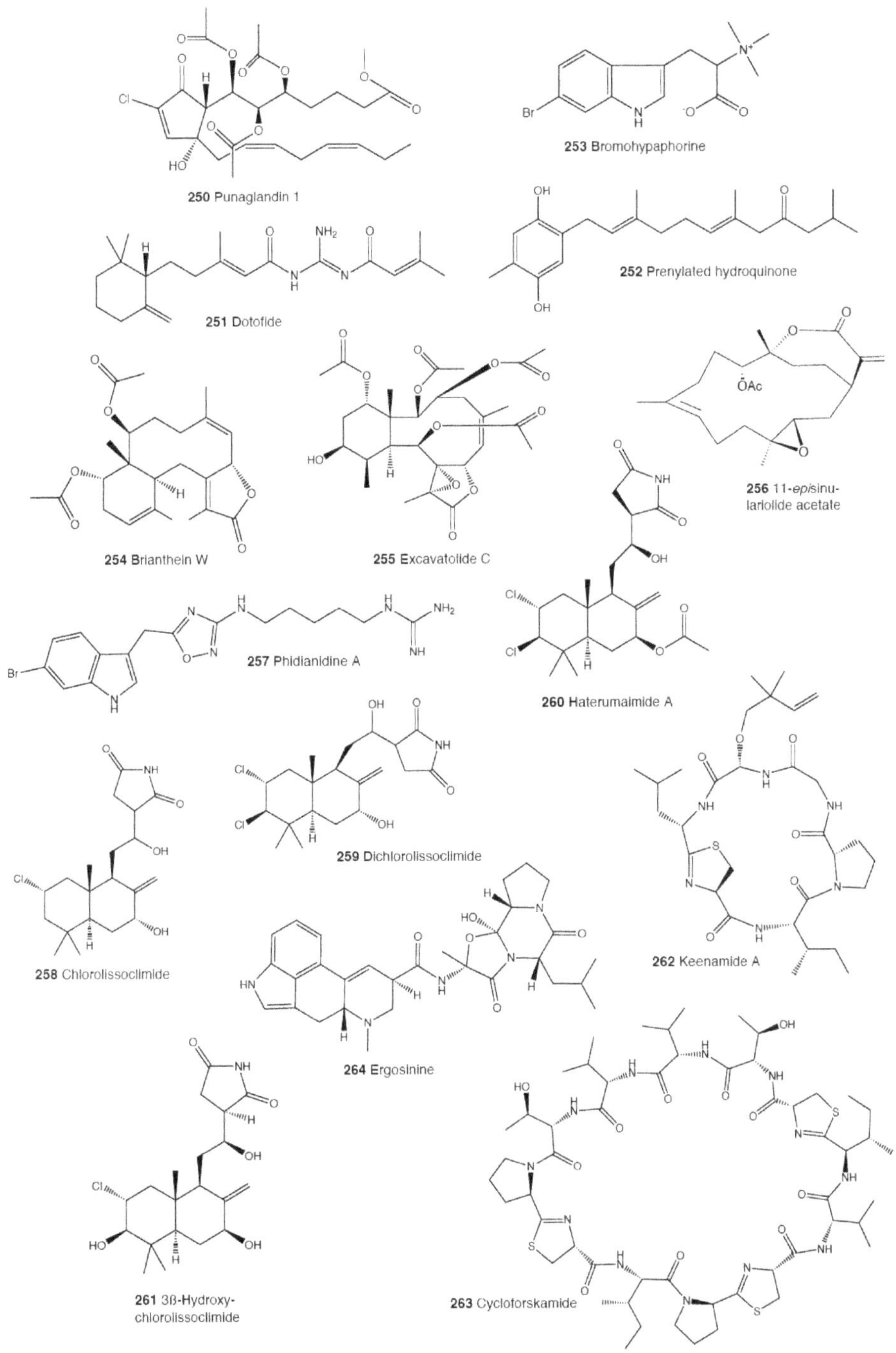

Figure 14. Structures of selected cytotoxic and antitumoral compounds in some Dendronotida, Euarminida, and Aeolidida nudibranchs, and Pleurobranchoidea. These molecules may also display other activities, as reported in the text.

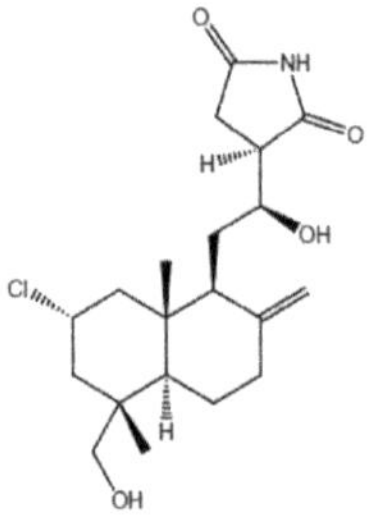

265 Haterumaimide J

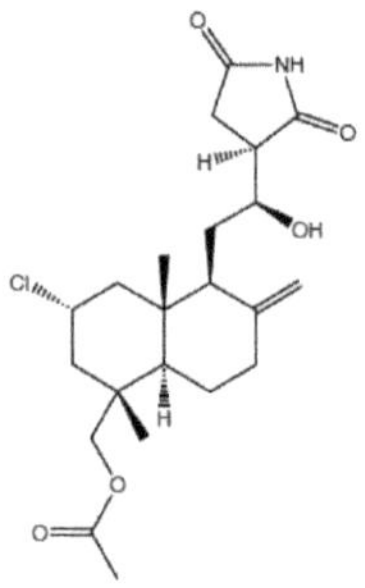

266 Haterumaimide K

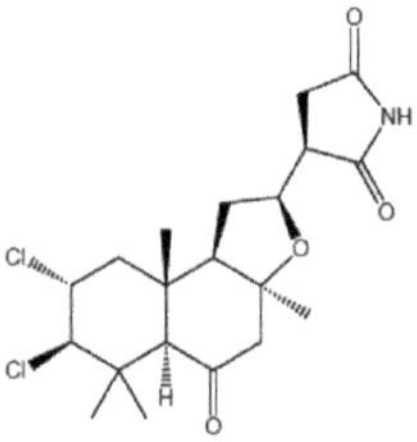

267 Haterumaimide C

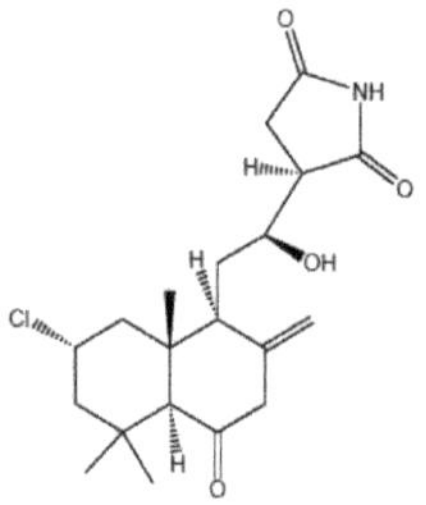

268 Haterumaimide G

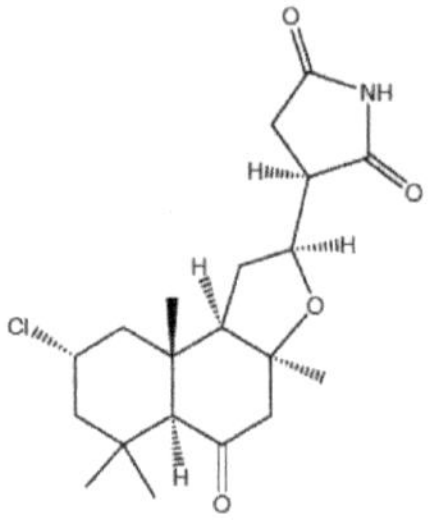

269 Haterumaimide I

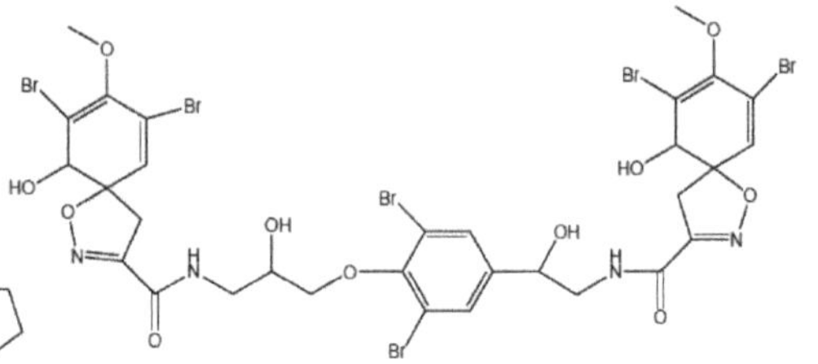

270 *Iso*-fistularin-3

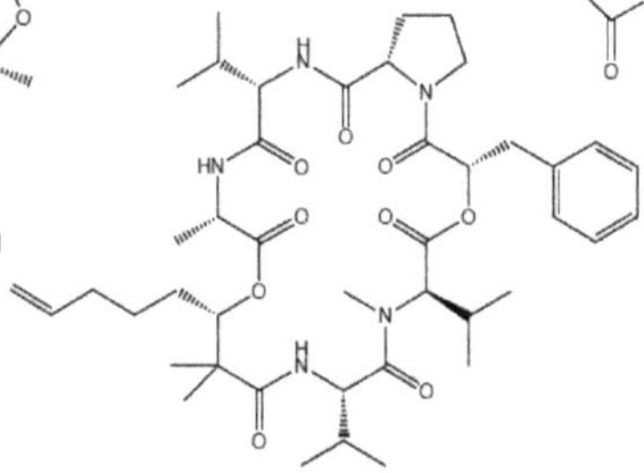

272 Kulolide-2

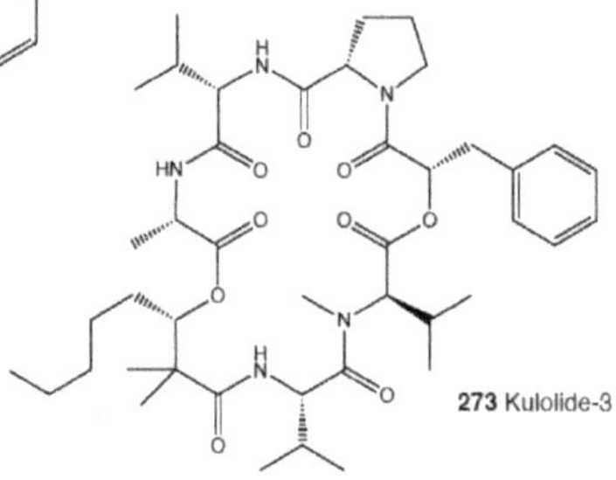

273 Kulolide-3

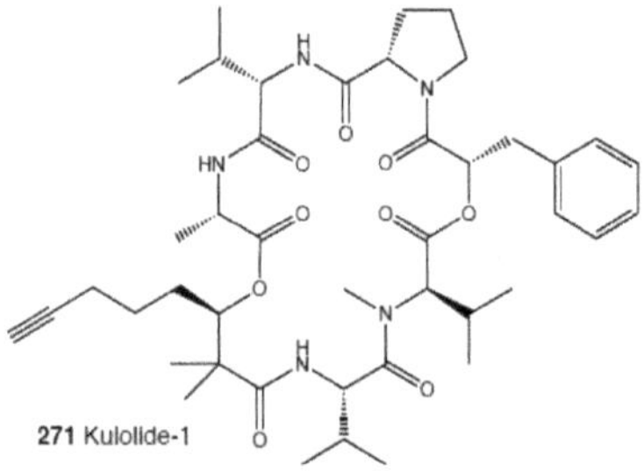

271 Kulolide-1

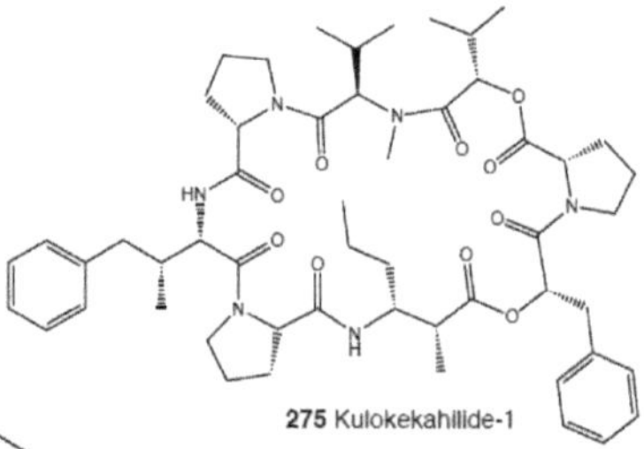

275 Kulokekahilide-1

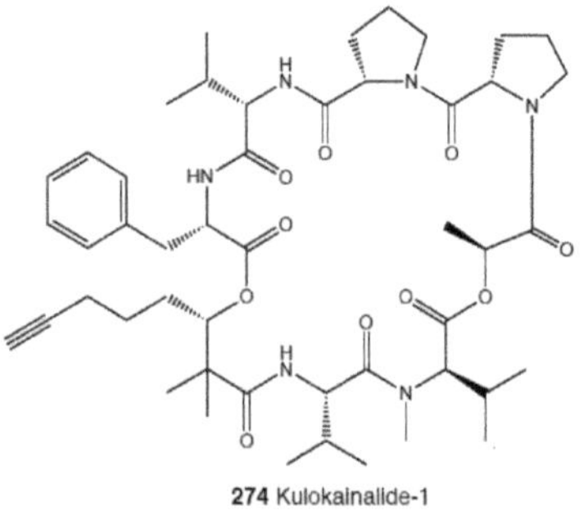

274 Kulokainalide-1

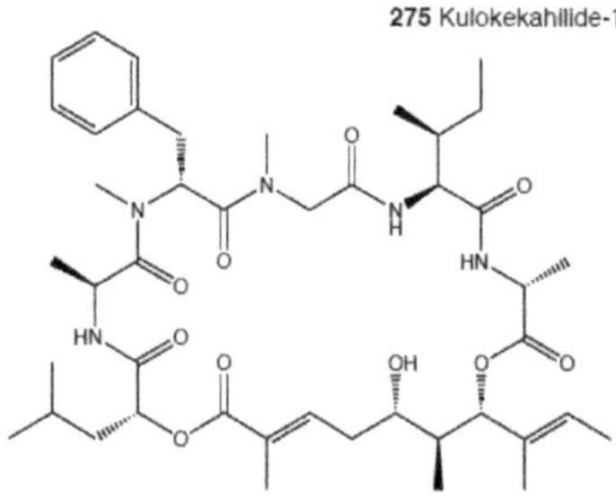

276 Kulokekahilide-2

Figure 15. Structures of selected cytotoxic and antitumoral compounds in some Pleurobranchoidea, Tylodinoidea, and Cephalaspidea. These molecules may also display other activities, as reported in the text.

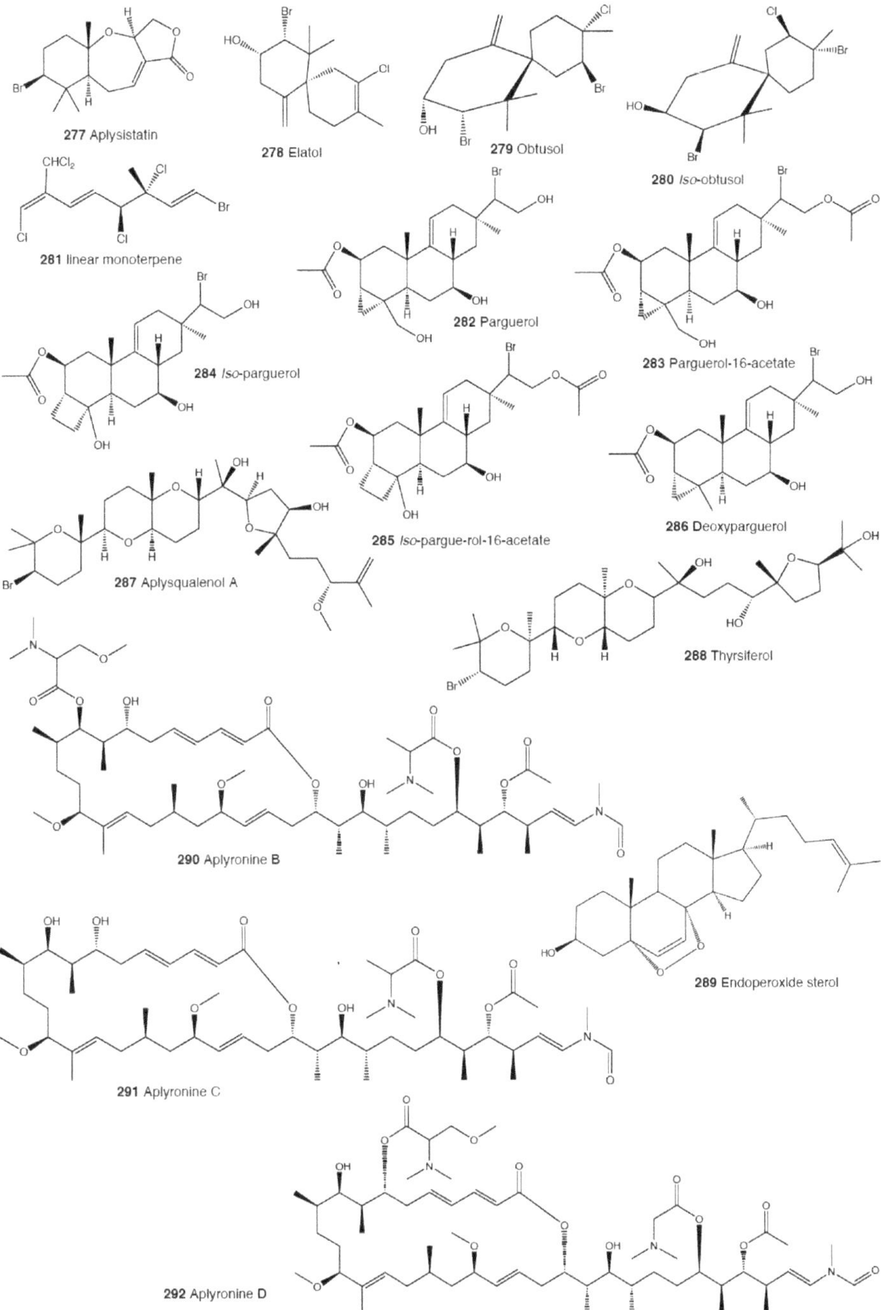

Figure 16. Structures of selected cytotoxic and antitumoral compounds in some Anaspidea. These molecules may also display other activities, as reported in the text.

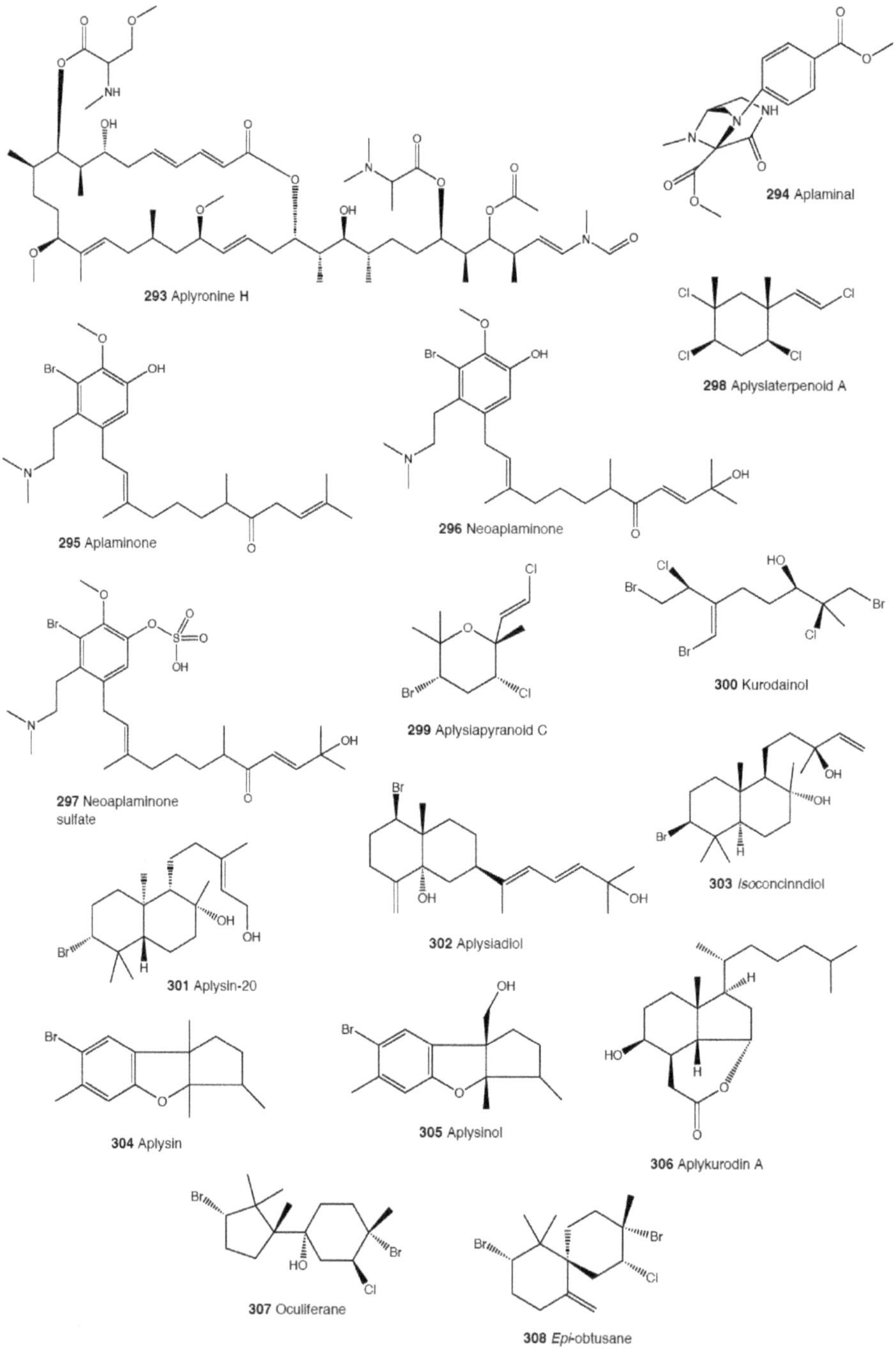

Figure 17. Structures of selected cytotoxic and antitumoral compounds in some Anaspidea. These molecules may also display other activities, as reported in the text.

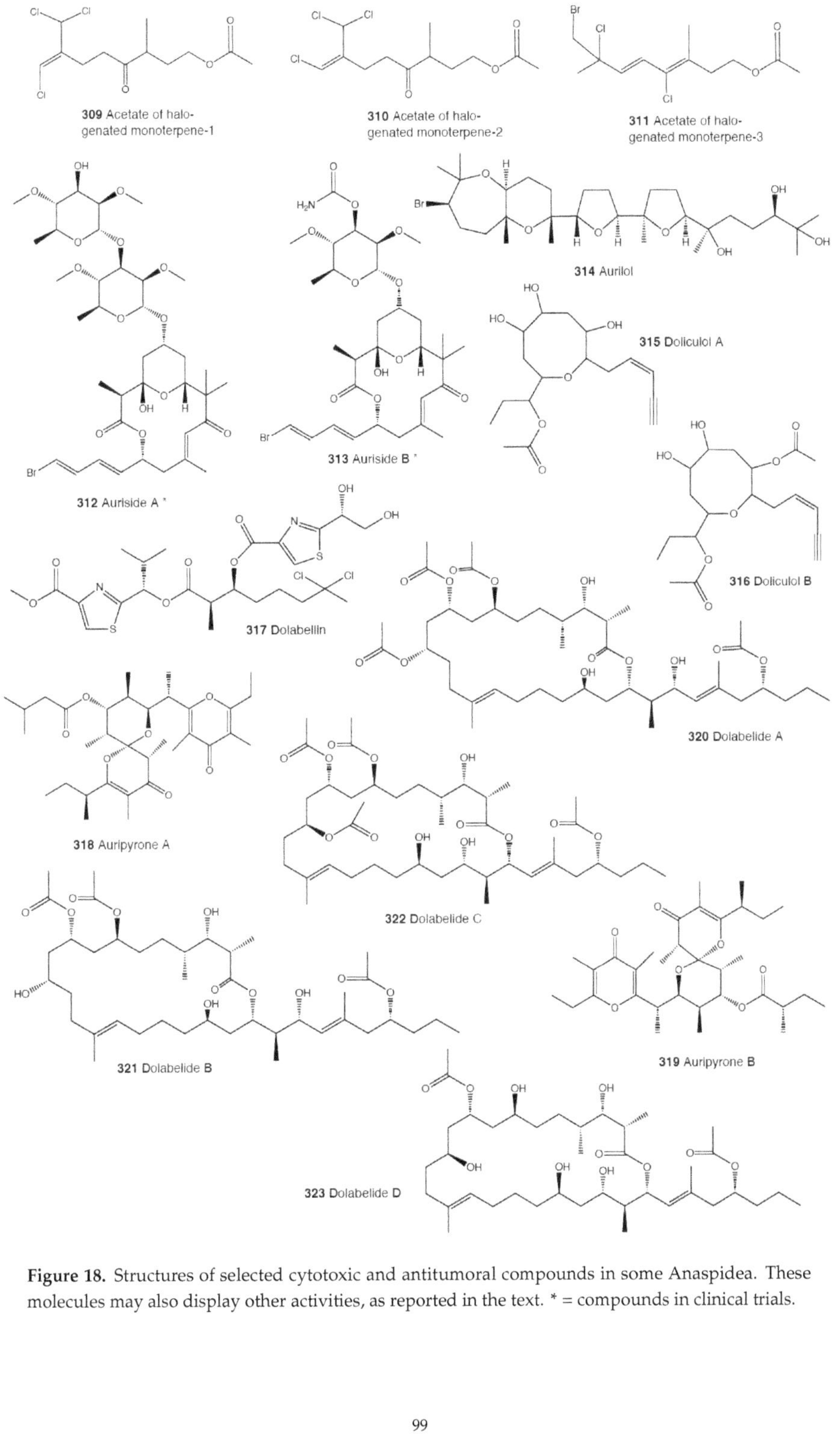

Figure 18. Structures of selected cytotoxic and antitumoral compounds in some Anaspidea. These molecules may also display other activities, as reported in the text. * = compounds in clinical trials.

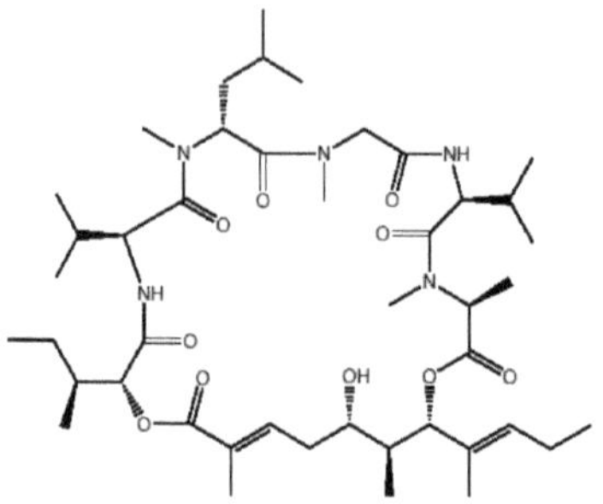

324 Aurilide

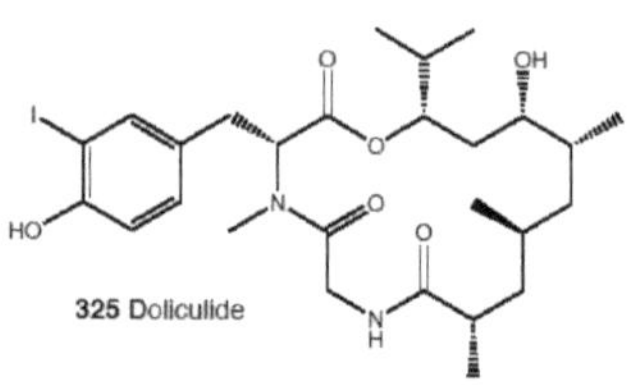

325 Doliculide

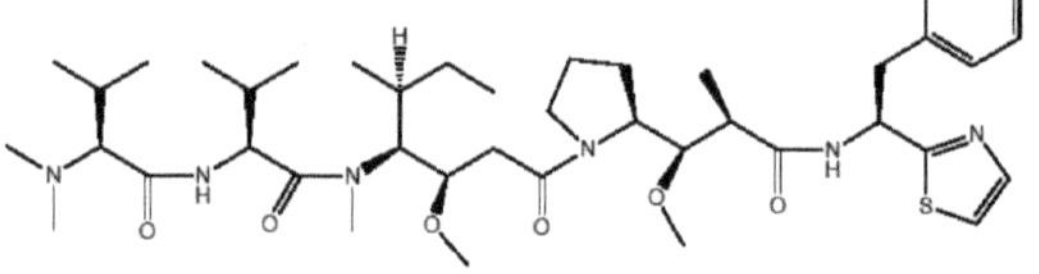

326 Dolastatin 10 *

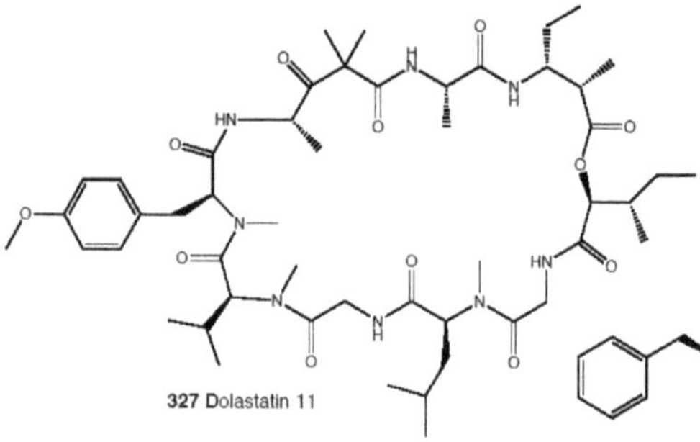

327 Dolastatin 11

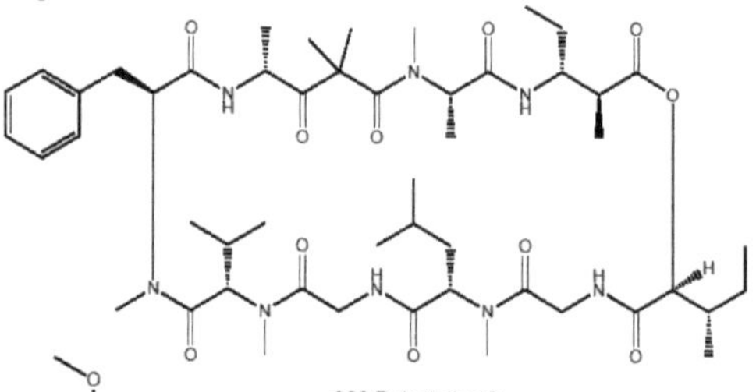

328 Dolastatin 12

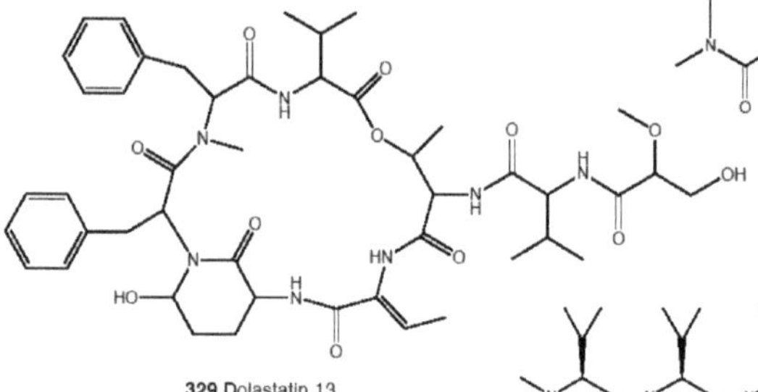

329 Dolastatin 13

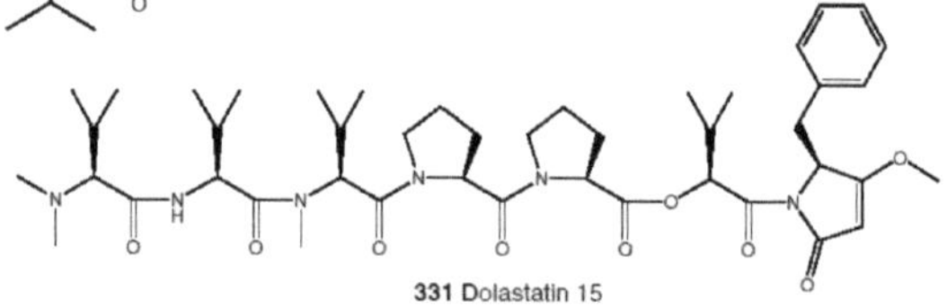

331 Dolastatin 15

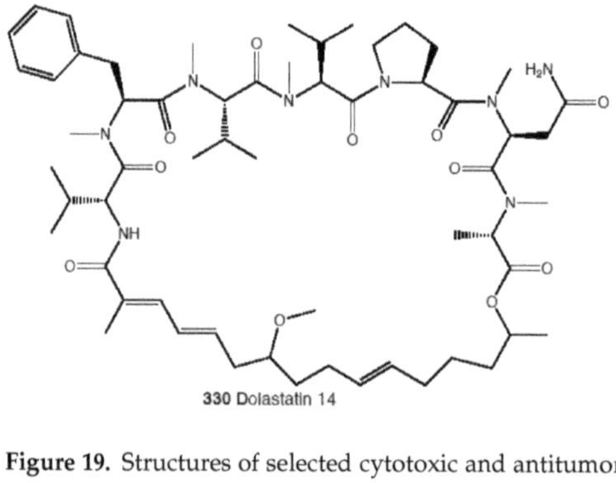

330 Dolastatin 14

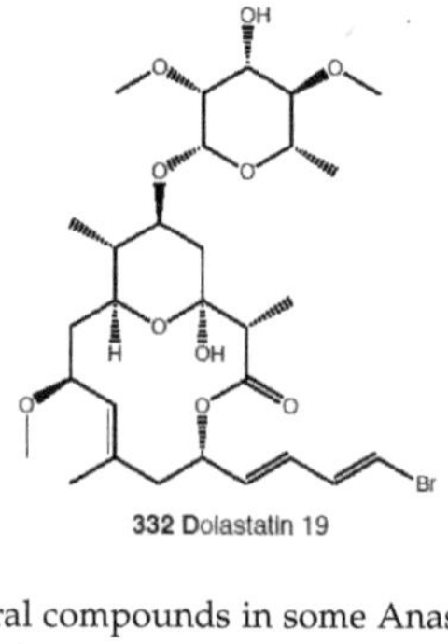

332 Dolastatin 19

Figure 19. Structures of selected cytotoxic and antitumoral compounds in some Anaspidea. These molecules may also display other activities, as reported in the text. * = compounds in clinical trials.

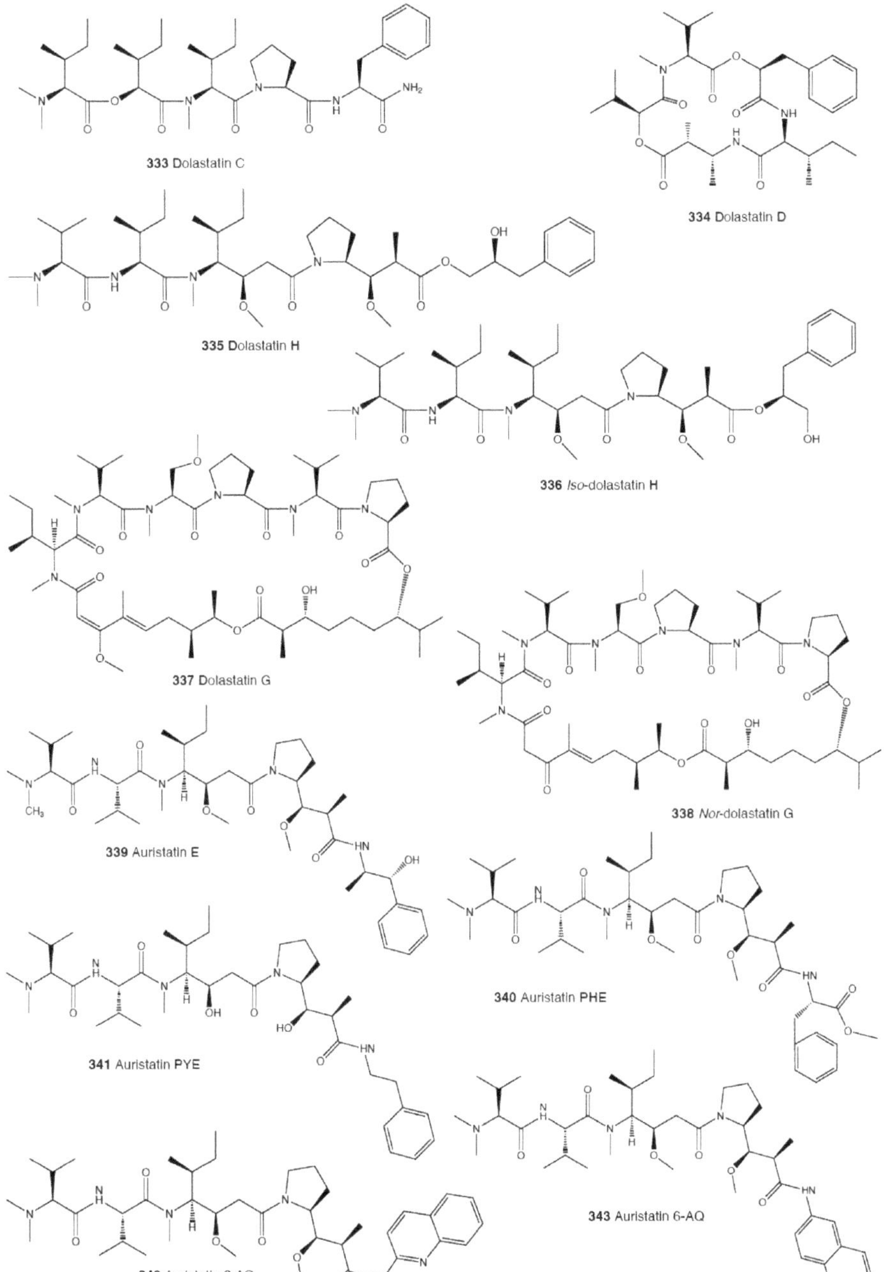

333 Dolastatin C

334 Dolastatin D

335 Dolastatin H

336 *Iso*-dolastatin H

337 Dolastatin G

338 *Nor*-dolastatin G

339 Auristatin E

340 Auristatin PHE

341 Auristatin PYE

342 Auristatin 2-AQ

343 Auristatin 6-AQ

Figure 20. Structures of selected cytotoxic and antitumoral compounds in some Anaspidea. These molecules may also display other activities, as reported in the text.

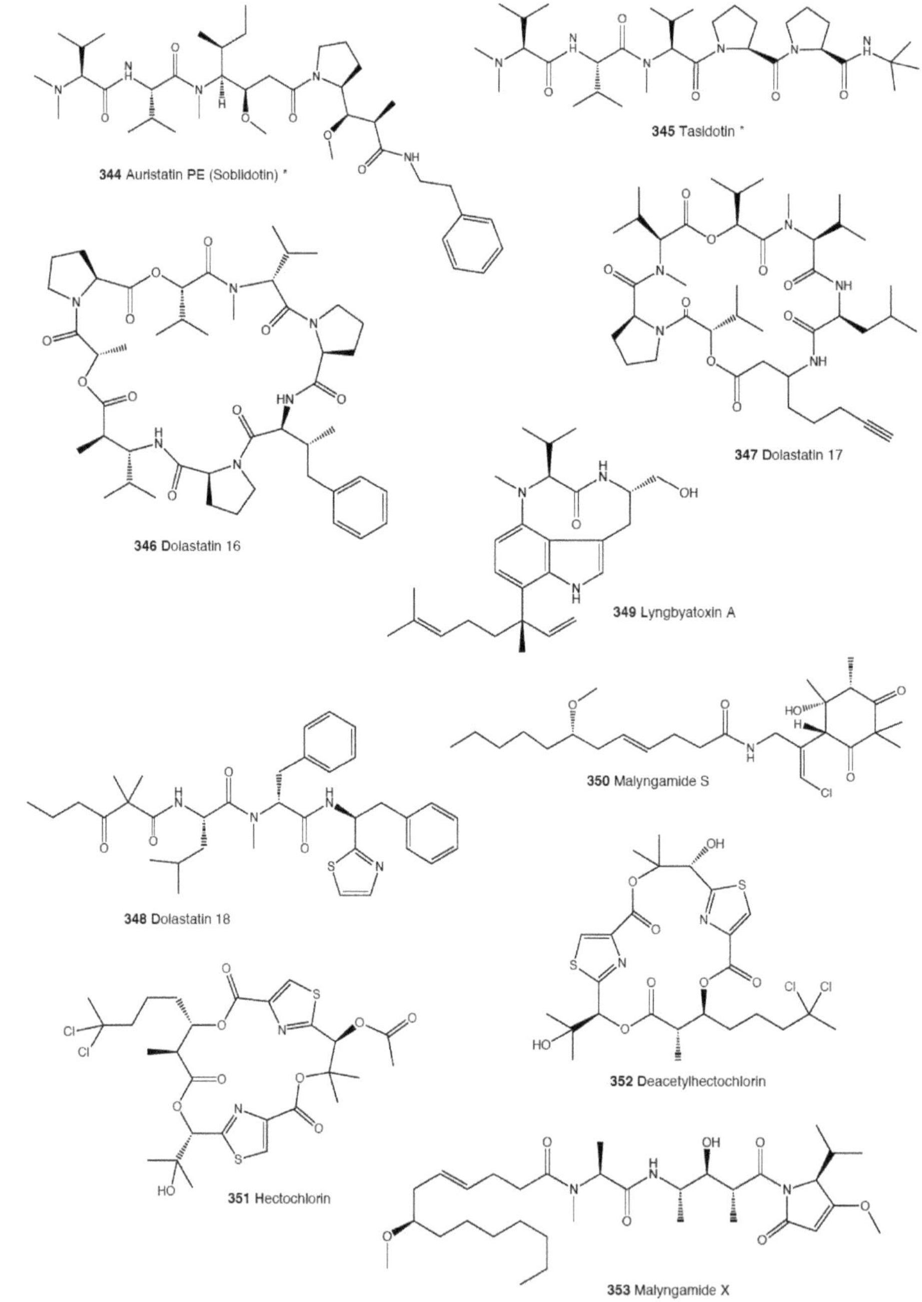

Figure 21. Structures of selected cytotoxic and antitumoral compounds in some Anaspidea. These molecules may also display other activities, as reported in the text. * = compounds in clinical trials.

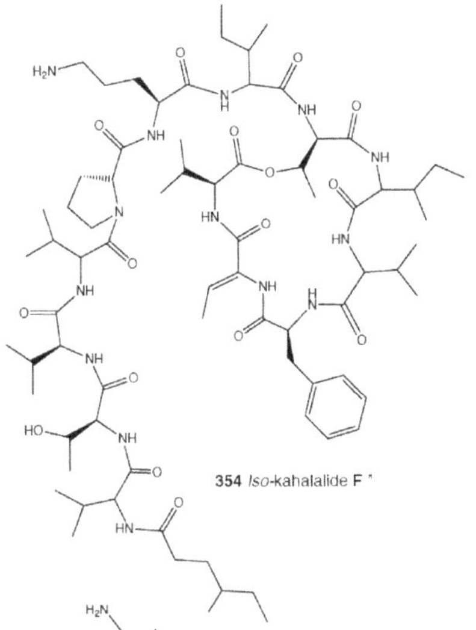

354 *Iso*-kahalalide F *

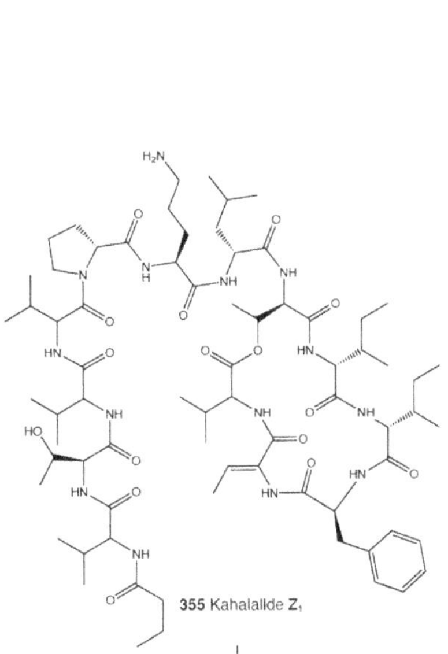

355 Kahalalide Z_1

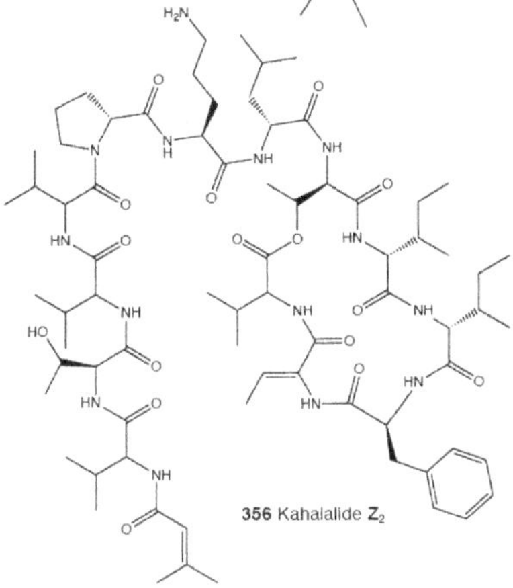

356 Kahalalide Z_2

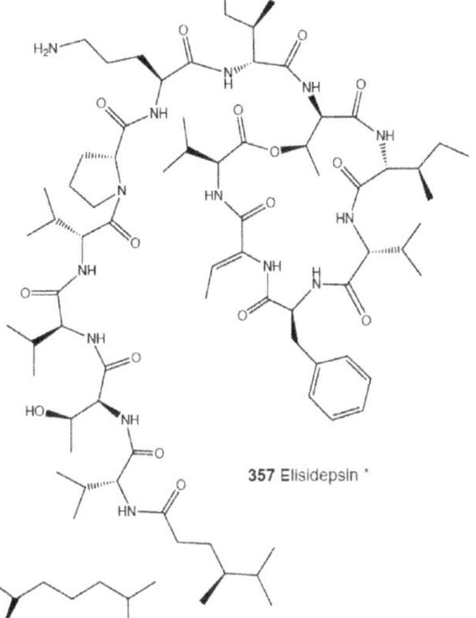

357 Elisidepsin *

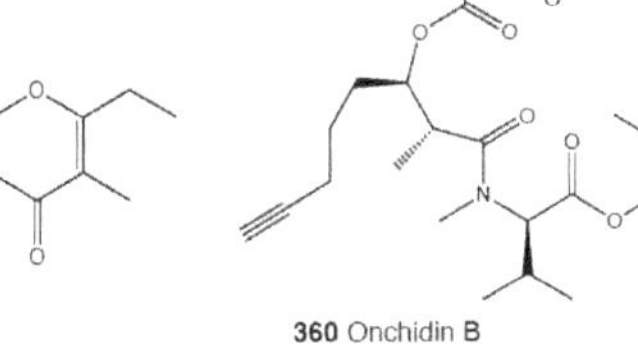

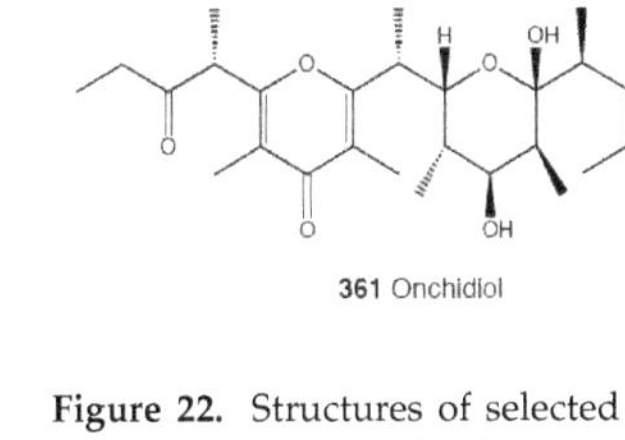

358 C-21 Hydroxylated sterol

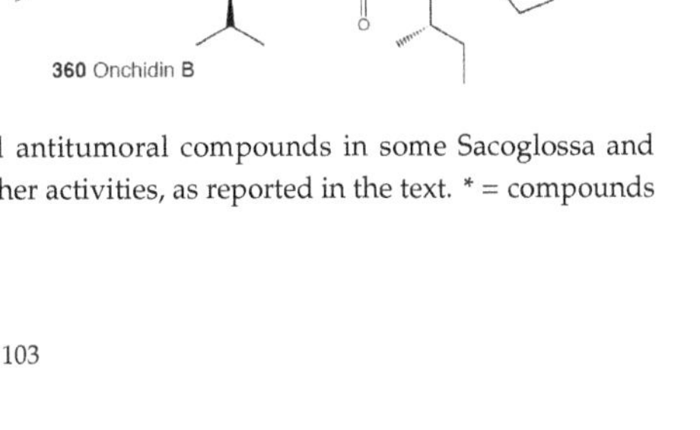

359 Secosterol

361 Onchidiol

360 Onchidin B

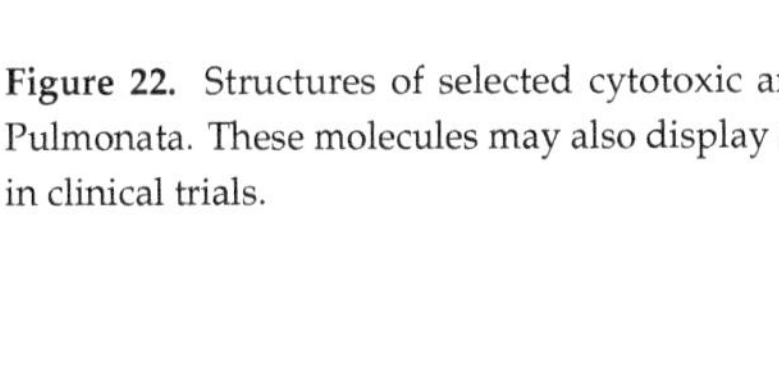

Figure 22. Structures of selected cytotoxic and antitumoral compounds in some Sacoglossa and Pulmonata. These molecules may also display other activities, as reported in the text. * = compounds in clinical trials.

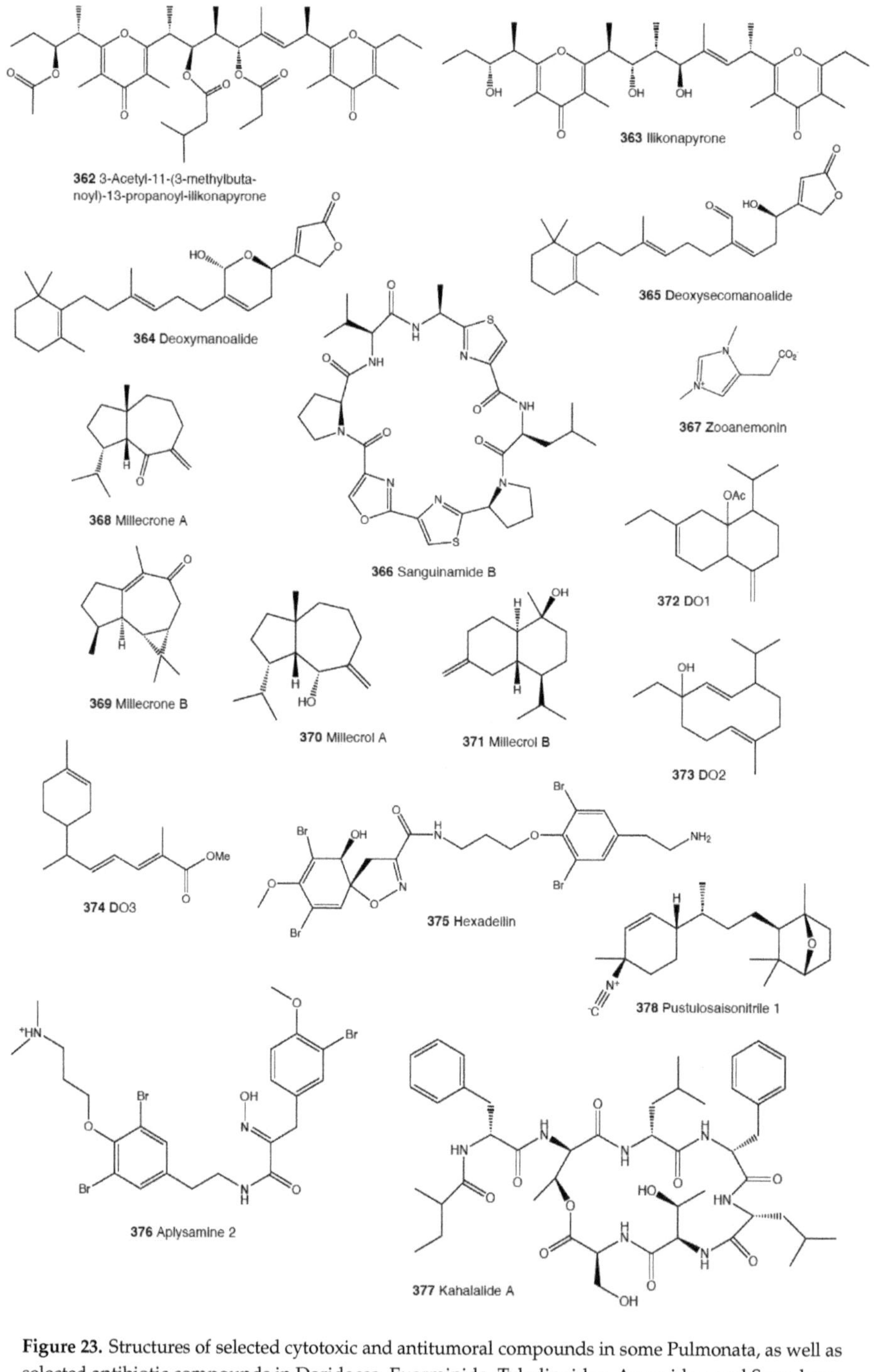

Figure 23. Structures of selected cytotoxic and antitumoral compounds in some Pulmonata, as well as selected antibiotic compounds in Doridacea, Euarminida, Tylodinoidea, Anaspidea, and Sacoglossa, and an antiparasitic compound from a Doridacea. These molecules may also display other activities, as reported in the text.

3.1.1. Nudibranchia

Doridacea

The Antarctic slug *Doris kerguelenensis* presents clerodane and labdane diterpenes, such as palmadorins (**195–200**), among other compounds [66,67]. Palmadorins A (**195**), B (**196**), D (**197**), M (**198**), N (**199**), and O (**200**) have been described to inhibit human erythroleukemia cells (HEL) at low IC_{50} (micromolar), and palmadorin M (**198**) has been reported to inhibit Jak2, STAT5, and Erk1/2 activation in HEL cells, producing apoptosis at 5 μM [67]. The Mediterranean and Atlantic *Doris verrucosa*, contains the de novo biosynthesized verrucosins, diterpenoid acid glycerides, among which verrucosins A (**124**) and B are potent activators of protein kinase C, and they promote tentacle regeneration in the freshwater hydrozoan *Hydra vulgaris* [370,371,486]. *Notodoris* nudibranchs feed on *Leucetta* calcareous sponges, presenting sponge-derived imidazole alkaloids [424,426,487,488]. In the Red Sea, *N. citrina* and *N. gardineri* obtain their chemicals from *Leucetta chagosensis*, presenting among others, naamidine A (**201**) and *iso*-naamidine-A (**160**) [424,426]. Naamidine A (**201**) was later tested from the sponge (from different localities) as a selective inhibitor of the epidermal growth factor (EGF) and was found to inhibit human tumor xenografts in mice, as well as displaying antitumour activity that promotes caspase-dependent apoptosis in tumor cells [489,490].

Adalaria loveni from the North Sea presents lovenone (**202**), a degraded triterpenoid suggested to come from an unidentified bryozoan prey [491]. Lovenone (**202**) was reported to present modest cytotoxicity to two HTCLs (human tumor cell lines) [491]. Another bryozoan-feeder, *Polycera atra* accumulates bryostatins (**203**) from *Bugula neritina*, also including them in their spawn [340,492–494]. Bryostatins (**203**) are polyketide macrolides known to be biosynthesized by a microbial symbiont, *Endobugula sertula* [495]. Bryostatins (**203**) are highly bioactive compounds, with bryostatin 1 (**203**), for example, being investigated in more than 20 clinical trials (two phase I trials) against multiple carcinomas and Alzheimer's disease [31,42,123,496,497]. Bryostatin 1 (**203**) modulates the Bcl-2 and p53 oncoproteins in human diffuse large-cell lymphoma WSU-DLCL2, inducing a decrease in the expression of Bcl-2 [498]. Furthermore, bryostatin 1 (**203**) showed a notable activity of modulating the paclitaxel inhibitor of protein kinase C (PKC) [499–501], as well as inducing ubiquitination and proteasome degradation of Bcl-2 in lymphoblastic leukemia, allowing for the growth of progenitor cells from bone marrow [502]. Moreover, bryostatins (**203**) are strong activators of PKC, regulating the activation, growth, and differentiation of cells [503].

Another interesting species is the Chinese *Actinocyclus papillatus*, which presents the mildly cytotoxic (–)-actisonitrile (**204**) in its mantle, along with actinofide (**205**), a terpenoid diacylguanidine [361,504,505]. Both enantiomers, (–)- and (+)-actisonitrile (**204**), were tested for cytotoxicity against tumor and non-tumor mammalian cells, resulting in a parallel concentration-dependent toxic profile at a micromolar concentration [504]. On the other hand, actinofide (**205**), a guanidine moiety acylated by two terpenoid acid units, allowed for the synthesis of a series of structural analogues which were tested for growth inhibitory activity against some cancer cell lines in vitro [506–508]. Actinofide (**205**) and some of its analogues were tested against six cancer cell lines: two human carcinoma cancer cell lines (MCF-7 (breast) and A549 (non–small-cell lung cancer, NSCLC)) of epithelial origin, two human cancer cell lines from glial origin (Hs683 oligodendroglioma and U373 glioblastoma of astrocytic origin), and two melanoma models (mouse B16F10 and human SKMEL-28 cell lines), resulting in many relevant activities [479]. Actinofide (**205**) presents GI_{50} values of 8.3 ± 1.8 for Hs683, 15.7 ± 10.1 for U373, 23.4 ± 5.5 for A549, 23.4 ± 5.9 for MCF-7, 24.2 ± 8.2 for SKMEL-28, and 7.5 ± 3.1 for B16F10 [479,505]. The most active compounds are reported to be those with one or two N-C_{15} residues, and a preliminary correlation between structure and activity was proposed [479].

Aldisa andersoni from the Indo-Pacific contains some phorbazoles, peculiar chlorinated phenyl-pyrrolyloxazoles, such as 9-chloro-phorbazole D (**5**) and *N*1-methyl-phorbazole A (**6**), along with phorbazoles A (**7**), B (**8**), and C described in the sponge *Phorbas* aff. *clathrata* [55,56,75,76].

Both 9-chloro-phorbazole D (**5**) and *N*1-methyl-phorbazole A (**6**) produce cytostatic effects in vitro against several HTCLs [75]. More specifically, *N*1-methyl-phorbazole A (**6**) inhibits human SKMEL-28 melanoma and U373 glioblastoma cells [75].

Dendrodoris carbunculosa from Japan possesses drimane sesquiterpenoids, the dendocarbins A–N (**16,206**), along with *iso*drimeninol (**207**) and 11-*epi*valdiviolide (**208**), some of them displaying cytotoxicity against murine leukemia P388 cell lines [87]. In particular, the IC_{50} values against adriamycin (ADR)- and vincristine (VCR)-resistant P388 cells (P388/ADR and P388/VCR, respectively), as well as those against sensitive P388 strain (P388/S) were reported for all compounds, with dendocarbin J (**206**) and 11-*epi*valdiviolide (**208**) showing moderate cytotoxicity against both sensitive and resistant cell strains [86]. The IC_{50} values for dendocarbin J (**206**) were 17 μg/mL for P388/S, 4 μg/mL for P388/VCR(-), 4 μg/mL for P388/VCR(+), 11 μg/mL for P388/ADR(-), and 8 μg/mL for P388/ADR(+), while for 11-*epi*valdiviolide (**208**) the values were 3.2, 2.5, 2.5, 2.5, and 2.5 μg/mL, respectively [86]. The origin of 11-*epi*valdiviolide (**208**) could be dietary, since it was first described in the sponge *Dysidea fusca* [509].

Regarding phyllidids, several species are described to present bioactive compounds. In Thailand, *Phyllidia coelestis* presents two cytotoxic pupukeanane sesquiterpenoids, 1-formamido-10(1,2)-abeopupukeanane (**209**) and 2-formamidopupukeanane (**210**), which show in vitro growth inhibitory activity against four human cancer cell lines [111]. In particular, this activity was tested against HeLa (cervical), MCF-7 (breast), KB (oral cavity), and HT-29 (colon) cancer cell lines with IC_{50} values between 0.05 and 10 μM [111,510]. Furthermore, both molecules (**209,210**) present selectivity, weakly inhibiting the growth of human gingival fibroblasts by 65% and 25% at 20 μM, respectively [111,510]. *Phyllidiella pustulosa* also contains many sesquiterpenoids, as abovementioned, and specimens from Okinawa, for example, possess the moderately cytotoxic substituted axinisothiocyanate K derivative (**211**) and an isocyano compound [377]. In Fiji, *P. pustulosa* contains the isothiocyanate axisonitrile-3 (**25**), which is weakly cytotoxic (IC_{50} > 20 μg/mL), in addition to some related sesquiterpenes [118,427]. In southern China, both *P. coelestis* and *P. pustulosa* present several nitrogenous terpenoids from their demosponge prey, *Acanthella cavernosa* [117]. Among these compounds, a bisabolane-type sesquiterpenoid (**212**), a theonellin *iso*thiocyanate (**213**), and 7-*iso*cyano-7,8-dihydro-∝-bisabolene (**214**) display cytotoxicity against several HCCLs [117]. In particular, they all show strong cytotoxicity against HCCL SNU-398 with IC_{50} values of 0.50, 2.15, and 0.50 μM, respectively [117]. Furthermore, the bisabolane sesquiterpenoid (**212**) presents broad cytotoxicity, being active against HCCLs A549, HT-29, and Capan-1, with IC_{50} values of 8.60, 3.35, and 1.98 μM, respectively [117]. Contrastingly, *Reticulidia fungia* from Okinawa presents the cytotoxic carbonimidic dichlorides, reticulidins A (**215**) and B, which are two uncommon sponge sesquiterpenes, probably obtained from their diet of *Pseudaxinyssa* sponges [438,511,512]. Both compounds are moderately cytotoxic in vitro against KB cells, with IC_{50} values of ~1 μM for both, and against mouse L1210 leukemia cells, with IC_{50} values of ~2 and ~0.3 μM, respectively [438].

Chromodoridids are also reported to possess cytotoxic and anticancer compounds. In the Paficic, *Cadlina luteomarginata* and its sponge prey *Phorbas* sp. Contain, among other compounds, ansellone A (**216**), a sesterterpenoid that moderately activates the cyclic adenosine monophosphate (cAMP) signaling pathway, with an EC_{50} value of 14 mM in the HEK293 cell-based test [513]. The modulation of the cAMP signaling pathway is used in stem cell techniques, and it is relevant to treat diseases such as cancer, heart failure, and neurodegenerative diseases [514]. On the other hand, several *Chromodoris* species possess latrunculin A (**38**), a PKS-NRPS-derived macrolide, reported to be cytotoxic [141]. Latrunculin A (**38**) was found in *C. lochi* and its sponge prey *Spongia* (*Cacospongia*) *mycofijiensis* in Fiji [141]. In fact, latrunculins A (**38**) and B (**37**) were first described from the Red Sea *Negombata* (*Latrunculia*) *magnifica* and later in other sponges (*Hyattela* sp.) [168]. Latrunculins (**37,38**) were then reported in several *Chromodoris* species at different localities, including *C. africana*, *C. annae*, *C. elisabethina*, *C. hamiltoni*, *C. kuiteri*, *C. magnifica*, and *C. quadricolor* [153,155,164,170,171]. *C.* (*Glossodoris*) also obtains latrunculin B (**37**) from the demosponge *Latrunculia magnifica* [171]. Latrunculins (**37,38**) interfere with the cytoskeleton, disrupting the organization of cell microfilaments, and inhibit the proliferation of cancer cells due to their strong actin binding properties [153,164,515–517]. A PKS-NRPS-derived mycothiazole (**129**) found in *C. lochi* from Vanuatu and its prey sponge (see above) possesses selective cytotoxicity, inhibits the hypoxia-inducible factor-1 (HIF-1), and also suppresses the mitochondrial respiration at complex I in sensitive cell lines (IC$_{50}$ values of 0.36–13.8 nM for HeLa, P815, RAW 264.7, MDCK, HeLa S3, 4T1, B16, and CD4/CD8 T cells) [103–106]. Latrunculins A (**38**) and B (**37**) display antimigratory activity against highly metastatic human prostate cancer PC-3M-CT$^+$ cells and murine brain-metastatic melanoma B16B15b cells [515,518]. Latrunculin A (**38**) presented IC$_{50}$ values of ~0.5 µM against murine P388 leukemia, human HT-29 colon cancer, and human A549 NSCLC, with more than a fivefold in vitro growth inhibitory effects against A549 NSCLC than to P388 leukemia [519]. Latrunculin A (**38**), coded NSC613011 on the NCI database, has an IC$_{50}$ mean value of ~0.7 µM in the 60 cancer cell line panel, with a more than twofold log magnitude difference between the most sensitive and the most resistant cancer cell lines, being as active against MDR NCI/ADR-RES as against cells without the MDR phenotype [47]. Finally, latrunculin A (**38**) induces apoptosis in cancer cells via activation of the caspase-3/caspase-7 pathway and displays strong anticancer effects in peritoneal dissemination models of MKN45 and NUGC-4 human gastric cancer in mice [455]. In vivo anticancer assays using A549 NSCLC xenografts also show that latrunculin A (**38**) increased the life span of treated tumor-bearing mice by 46% compared to controls [520].

Chromodoris lochi from Indonesia, instead, presents sponge-derived polyketides, such as laulimalide (**39**) and *iso*laulimalide (**40**) from the sponge *Hyattella* sp., both compounds being cytotoxic due to their microtubule-stabilizing action at a different binding site to taxanes, located on two adjacent β-tubulin units between tubulin protofilaments of a microtubule [521,522]. Therefore, these compounds (**39–40**), with IC$_{50}$ values of 15 ng/mL in the KB cell line, are being tested as potential antitumor agents [142,523,524]. Laulimalide (**39**) inhibits growth in more than ten cancer cell lines at low nanomolar concentrations, while *iso*laulimalide (**40**) is effective at low micromolar values [142,525,526]. As previously mentioned, laulimalide (**39**) is a microtubule stabilizer, like the plant compounds taccalonolide and paclitaxel, but laulimalide (**39**) has been reported to cause the formation of aberrant, structurally distinct mitotic spindles, differently from the other two molecules [527]. Moreover, laulimalide (**39**) has an effect in P-gp-overexpressing cancer cells and against cell lines resistant to paclitaxel and epothilones [525,528]. Further studies have shown that in ovarian cancer cells, the increased expression of βII- and βIII tubulin isotypes induces resistance to laulimalide (**39**), as does the downregulation of vimentin expression in human ovarian carcinoma cells [529,530]. Moreover, assays in vivo tested the anticancer activity of laulimalide (**39**) in two xenograft models, the human MDA-MB-435 breast cancer and the human HT-1080 fibrosarcoma models, describing little tumor growth inhibition accompanied by a strong toxicity and mortality, contrasting to paclitaxel [525].

Furthermore, inorolides A–C (**217**), sesquiterpenoids found in the Japanese *Chromodoris inornata* (*C. aspersa*) and other scalarane terpenoids are cytotoxic against murine L1210 leukemia and human epidermoid carcinoma KB cell lines [156]. Particularly, inorolides A–C (**217**) display IC$_{50}$ values of ~7, ~5, and ~4 µM, respectively, while sesterterpenoids like deoxoscalarin (**46**) and its analogues, deoxoscalarin-3-one, 21-hydroxydeoxoscalarin, 21-acetoxydeoxoscalarin, and 12-O-acetyl-16-O-deacetyl-12,16-episcalarolbutenolide, display values of ~3, ~2, ~9, ~1, and ~5 µM, respectively [156]. Some of these metabolites have been reported also in *Hyrtios* sponges and reviewed for their bioactivity [156,531].

Similarly, cytoxicity has been reported in several compounds from other *Chromodoris* species, such as a sponge diterpene found in *C. petechialis* from Hawai'i, puupehenone (**218**), an oxygenated diterpene of *C. elisabethina* from Australia, a spongian diterpene of *C. kunei* from Okinawa, furanoditerpenoids found in *C. reticulata* from China, and the mildly toxic diterpenes from a *Chromodoris* sp. from Australia [22,154,158,160,186]. Puupehenone (**218**) is active at IC$_{50}$ values of 1 µg/mL to P388, 0.1–1 µg/mL to A-549 and MCF-7, as well as 1–10 µg/mL to HCT-8 [22,154,158,160,186].

Goniobranchus species also possess some cytotoxic compounds. In Australia, *G. splendidus* has a cytotoxic spiroepoxide lactone, epoxygoniolide-1 (**219**), suspected to originate from its dietary sponge prey [532]. Epoxygoniolide-1 (**219**) shows moderate cytotoxicity to NCIH-460, SW60, and HepG2 cancer cells [532]. *G. (Chromodoris) sinensis* from China presents aplyroseol-2 (**220**) [131]. *G. reticulatus* from Australia presents a dialdehyde sesquiterpene, together with the ring-closed acetal, both bioactive against P388 mouse leukemia cells [161,186]. *G. reticulatus* contains spongian-16-one (**221**), aplytandiene-3 (**222**), aplysulfurin (**223**), aplyroseol-2 (**220**), and gracilins A (**224**), B (**225**), C (**226**), G (**227**), and M (**228**), all of which display cytotoxicity against HeLa S3 cells [190]. Moreover, gracilins B (**225**) and C (**226**), as well as some isomers, obtained from the demosponge *Spongionella* sp. are cytotoxic against a wider panel of HTCLs, and have also been reported as cyclosporine A mimics and BACE1 and ERK inhibitors (see below) [533,534]. Gracilins M–Q (**228**) showed a significant potency against the HeLa S3 cell line [113]. In Japan, *G. (Chromodoris) obsoletus* contains dorisenones A–D (**229**), cytotoxic sponge diterpenoids, along with related compounds, such as 11β-hydroxyspongi-12-en-16-one (**230**), spongian-16-one (**221**) [154]. All of these compounds are active against murine lymphoma L1210 and human epidermoid carcinoma KB cells at IC$_{50}$ submicromolar (as low as IC$_{50}$ 0.2 μg/mL) and low micromolar values, respectively, and although dorisenone D and a related compound have also been tested in vivo against P388 leukemia, they show no activity [154]. Similarly, in *G. collingwoodii*, some spongian-16-one diterpenes are reported to be inactive to a range of HTCLs [186].

The spongian diterpenes of *Glossodoris cincta* from Egypt and Sri Lanka have also been reported to be cytotoxic [22,166,380,381,535–537]. The compound 12-*epi*-scalaradial found in *G. cincta* and *G. hikuerensis* is the most active, inhibiting the epidermal growth factor receptor (EGFR) implicated in many cancers, and also inhibiting the human recombinant PLA2 at 0.02 μM. *Doriprismatica (Glossodoris) atromarginata* presents furanoditerpenoids and scalarane sesterterpenes originating from its dietary sponges *Spongia (Hyatella)* sp. and *Hyrtios* spp., depending on the geographical location (Australia, Sri Lanka, India) [93,176,181,386–388]. The most active metabolites were spongiadiol (**35**), spongiadiol diacetate (**231**), *epi*-spongiadiol (**232**), 12-deacetoxy-12-oxodeoxoscalarin (**136**), heteronemin (**233**), and mooloolabene D (**234**) [180,381,383–386,519,535,538–540]. In particular, spongiadiol (**35**) is active against P388 murine leukemia cells at an IC$_{50}$ of 0.5 μg/mL. Other species, such as *Felimida (Glossodoris) dalli*, *Doriprismatica (Glossodoris) sedna*, *Glossodoris rufomarginata*, *Glossodoris pallida*, *Glossodoris vespa*, and *Ardeadoris (Glossodoris) averni*, also present homoscalarane and scalarane metabolites, among which 12-deacetyl-23-acetoxy-20-methyl-12-*epi*-scalaradial (**135**) moderately inhibits mammalian phospholipase A2 (IC$_{50}$ = 18 μM) [175,183,383]. Heteronemin (**233**), also found in several chromodoridid species and derived from dietary sponges, such as *Heteronema erecta*, blocks tumor cell intravasation through the lymph-endothelial barrier in a three-dimensional (3D) cell culture model, using spheroids of the MCF-7 breast cancer cell [541,542].

Hypselodoris infucata from Bali presents (–)-furodysinin (**48**), which is active against the HeLa cell line with an IC$_{50}$ of 102.7 μg/mL [214]. *Felimida (Chromodoris) macfarlandi* from California presents macfarlandines A–E (**134,235**), from which macfarlandin E (**235**) displays unique Golgi-modifying properties [139,140,149,543].

Hexabranchus sanguineus is a chemically rich species, transferring several compounds to their egg masses, including the macrolides ulapualides A (**190**) and B (**236**) in Hawai'i, presenting three contiguous oxazole rings with the attached lipid-like side chain ending in the N-methyl-N-alkenylformamide group [222]. Ulapualide A (**190**) has also been found in sponges and, thus, is suggested to be of dietary origin in the slug [544]. Both compounds are reported to display activity against murine L1210 leukemia cells, with IC$_{50}$ values of ~10 and ~30 nM, respectively [222]. Ulapualide A (**190**) has been described to be a potent actin-depolymerizer [545]. Later, ulapualides C–E (**237**) have been found in egg masses from Hawai'i, and ulapualide C (**237**) is reported to display cytotoxicity against several HTCLs, although it is 2–4 times less potent than ulapualides A (**190**) and B (**236**) [546]. A different study found several kabiramides, including kabiramide A (**238**), B (**239**), D (**240**), and E (**241**), along with the halichondramide derivatives, dihydrohalichondramide (**63**) and 33-methylhalichondramide (**242**), in egg masses of the same species from a different locality [217–221]. Kabiramides (**238–241**) contain a macrolide ring with contiguous trisoxazole rings [547]. Kabiramide C (**62**) was traced to a *Halichondria* sponge and further reported in the sponge *Pachastrissa nux* together with other kabiramides, as well as to some adult slug specimens at a lower concentration [221,222,548]. Kabiramide C (**62**) presents growth inhibitory effects in vitro against human MCF-7 breast cancer cells (IC$_{50}$ ~0.5 µM), ten times higher than for human fibroblasts (IC$_{50}$ ~8 µM), displaying bioselectivity [548]. The mechanism of action has been described as kabiramide C (**62**) binding to actin and its actin complex, which is achieved through a two-step binding reaction and forms a very stable and long-lived complex [549]. Kabiramide G (**243**) presents an even stronger bioselectivity, with IC$_{50}$ values of 0.02 µM against MCF-7 cancer cells and >2 µM for human fibroblasts [548]. Kabiramides A (**238**) and B (**239**) were active with IC$_{50}$ values of ~10 nM, and kabiramides D (**240**) and E (**241**) with IC$_{50}$ values of ~30 nM, against the murine L1210 leukemia cell line [220]. On the other hand, dihydrohalichondramide (**63**) and 33-methylhalichondramide (**242**) inhibit growth in the murine L1210 leukemia cell line with IC$_{50}$ values of ~40 and ~60 nM, respectively, in this case by disrupting actin microfilaments [220,550]. Halichondramide (**244**) presents also a wide array of bioactivities, including cytotoxic and cytostatic activities and antiproliferative and antimigratory effects in vitro, as reported elsewhere [479,551,552]. Most of these compounds probably originate from a sponge diet [218]. A mechanism of detoxification has been suggested for the slug to deal with halichondramide and transform it into less toxic compounds [218,221].

Another interesting species is *Jorunna funebris*, which in India presents the cytotoxic isoquinoline alkaloid jorumycin (**189**) [166,553,554]. Jorumycin (**189**) is cytotoxic at very low concentrations, with an IC$_{50}$ of 12.5 ng/mL against some cancer cell lines, such as P388, A549, HT29, and MEL28, including cells resistant to apopototic stimuli, and having a saframycin-like structure similar to ecteinascidin 743, one of the most active marine-derived antitumor agents isolated from the tunicate *Ecteinascidia turbinata* and an approved drug already on the market [43,553,555–557]. A synthetic compound derived from jorumycin (**189**) is in phase II clinical studies for endometrial and cervical cancer, as well as solid human tumors and hematological diseases (Ewing sarcoma, urothelial carcinoma, and multiple myeloma) [556,558,559]. Specimens from Thailand, instead, present jorunnamycins A–C (**245**), along with renieramycins (**246**), which are also cytotoxic, while specimens from Sri Lanka presented several isoquinoline–quinone metabolites from the sponge *Xestospongia* sp. [22,560]. Jorunnamycins A–C (**245**) are reported to be obtained after treatment of the samples with potassium cyanide, yielding more stable compounds while still conserving high cytotoxicity against HTCLs [560]. Jorunnamycin C (**245**) and renieramycin M (**246**) present IC$_{50}$ values at low nanomolar ranges against human colon (HCT-116) and breast (MDA-MB-435) cancer cells [561]. Their mechanism of action has been reported as the downregulation of protein tyrosine phosphatase receptor type K (PTPRK) in vitro, since PTPRK is a tumor suppressor gene product that may be involved in colon cancer [516,561–563]. Some structurally related compounds were also found in specimens from South China, such as the fennebricins A and B, and other molecules, probably also from a *Xestospongia* sp. [564,565].

The Mediterranean *Peltodoris atromaculata*, and the sponges on which it feeds on, *Petrosia ficiformis* and *Haliclona fulva*, contain cytotoxic long-chain fulvinol-like polyacetylenes, namely, petroformynes (**247**) [566–569]. These compounds are structurally very similar to the sponge compounds neopetroformynes and are active against murine P388 leukemia cells with IC$_{50}$ values of 0.09–0.45 µM [570]. Further metabolites from this slug include some other polyacetylenes, such as the hydroxy-dehydroisofulvinol (**248**), very similar to fulvinol, which is active against four cell lines, i.e., murine P388 leukemia, A549 NSCLC, HT-29 colon cancer, and SKMEL-28, at IC50 ~2 µM [571]. Hydroxy-dehydroisofulvinol (**248**) presents an IC$_{50}$ value of ~3 µM against the SKMEL28 melanoma cell line [571].

Finally, nembrothids also present some cytoxocity. The alkaloids tambjamines (**65–70**), as mentioned above, are found in several *Tambja* species (*T. capensis, T. ceutae, T. eliora, T. morosa, T. stegosauriformis, T. verconis*) along with their bryozoan prey (*Bugula dentata* or *Sessibugula translucens*) in different geographical localities [21,88,207,223,228,572,573]. *Roboastra* species feeding on *Tambja* species also present tambjamines (**65–70**), and *R. tigris* obtains them from *T. abdere* and *T. eliora* [223]. Tambjamines (**65–70**) have also been found in *Nembrotha* spp. and the tunicate *Atapozoa* sp. [226,227]. Some tambjamines (**65–70**), which are similar to the bacterial compounds prodigiosins, have been described to cause DNA damage and induce apoptosis [223,574–577]. Tambjamine D (**68**) is active against several tumor cell lines by intercalating into DNA, as well as by promoting single-strand DNA oxidative cleavage, although a lack of selectivity was described [228,578–586]. Tambjamine K (**249**) and the tetrapyrrole (**72**) display concentration-dependent cytotoxicity against tumor and non-tumor mammalian cells, with IC_{50} values between ~0.004 and 15 µM and IC_{50} ~19 µM against mouse 3T3-L1 fibroblasts for tambjamine K (**249**) [232,584,587]. In fact, tambjamine K (**249**) is very selective, showing a ~4000-fold differential sensitivity between human Caco-2 colon cancer cells and HeLa cervix cancer cells [229]. Tambjamine C (**67**), instead, is a good transmembrane anion transporter, similar to prodigiosins, which are relevant in cancer cell biology and cancer cell migration, and they are expressed differently in diverse cancer cells [581–586].

Dendronotida

Within this group, several species have been described to present cytotoxic compounds. Punaglandins (**250**) are cytotoxic PGs obtained by *Tritonia* sp. From its octocoral prey, *Telesto riisei*, active at an IC_{50} of 0.03 µM to mouse leukemia cells [587]. Dotofide (**251**) is found in *Doto pinnatifida* and is active against several cell lines using the MTT colorimetric assay [505]. Against human glioma, dotofide (**251**) shows a GI_{50} value of 18.1 µM for Hs683 oligodendroglioma (ATCC HTB-138), and of 28.8 µM for U373 glioblastoma (ECACC 08061901) [505]. For human carcinoma, dotofide (**251**) displays GI_{50} values of 29.4 µM for A549 NSCLC (DSMZ ACC107), and 28.1 µM for MCF-7 breast carcinoma (DSMZ ACC115) [505]. Finally, for melanoma, it displays GI_{50} values of 60.5 µM for human SKMEL-28 (ATCC HTB-72) and 9.6 µM for mouse B16F10 (ATCC CRL-6475) [505].

The sesquiterpenes tritoniopsins A–D (**75–78**) are found in *Tritoniopsis elegans* and its soft coral prey *Cladiella krempfi*, where the slug accumulates tritoniopsin A (**75**) in its mantle at higher concentrations than the coral or other slug tissues [238]. Rat cell lines were used to test the cytotoxicity of tritoniopsin A (**75**), resulting in a weak to moderated activity [234].

Euarminida

The previously mentioned eunicellin diterpenes (**187**) from *Dermatobranchus ornatus* display moderate cytotoxicity to A-549, SKOV-3, SK-MEL-2, and HCT-15, along with inhibition of cell division in fertilized starfish eggs [22]. The South African slug *Leminda millecra* possesses sesquiterpenes and prenylquinones from its diet of octocorals, mainly of the genus *Alcyonium* and gorgonians such as *Leptogorgia palma* [588–590]. Among them, a prenylated hydroquinone (**252**) possesses moderate inhibitory activity, with values of GI_{50} around 6–9 µM against WHCO1 and WHCO6 esophageal cancer cell lines, inducing apoptosis via generation of reactive oxygen moieties [589,590].

Aeolidida

Hermissenda crassicornis presents L-6-bromohypaphorine (**253**), reported to be an agonist of human a7 nicotinic acetylcholine receptor [591]. In *Phyllodesmium briareum*, brianthein W (**254**) is reported to have an ED_{50} of 0.76 µg/mL against P-388, while excavatolide C (**255**) displays an ED_{50} of 0.3µg/mL for P-388, and an ED_{50} of 1.9µg/mL for KB, A-549, HT-29 [248]. Both compounds were traced to the diet the octocoral *Briareum* sp. [248]. Similarly, *P. magnum* possesses 11-*epi*sinulariolide acetate (**256**) with an ED_{50} of 1.2µg/mL for P-388, ED_{50} of 1.9µg/mL for HT-29, and ED_{50} of 0.8µg/mL for HL-60 [248]. Moreover, the previously mentioned diterpene trocheliophorol (**84**) from *Phyllodesmium longicirrum* has been reported as cytotoxic [245].

Phidianidines A (**257**) and B are bromoindole alkaloids from the Chinese slug *Phidiana militaris* [361,592]. Phidianidines (**257**) are the only known marine source of the 1,2,4-oxadiazole system, and their interesting structure promoted their synthesis, as well as that of several analogues [593–605]. Phidianidines (**257**) show cytotoxicity against several cell lines, such as C6 and HeLa tumor cells at nanomolar concentrations [592]. Their IC_{50} values stand from ~0.4 to >100 µM in three cancer cell lines, with no selectivity against mouse 3T3-L1 fibroblasts and rat H9c2 cardiomyocytes (IC_{50}: ~0.1 and ~5 µM) [592,606]. Human HeLa cervix cancer cells are very sensitive to the growth inhibitory effects of phidianidines (**257**), contrarily to Caco-2 colon cancer cells [606]. Phidianidines (**257**) are also selective and potent ligands with partial agonist activity against the µ opiod receptor (when compared to δ- and κ-opiod receptors), which is involved in cancer progression [593,607]. Virtual screening allowed for the identification of phidianidine A (**257**) as a potential ligand for CXCR4, a chemokine receptor involved in several diseases, including cancer progression, metastasis, and immunodeficiency disorders, and competing with natural ligand CXCL12 as observed by molecular docking, proving that it is a CXCR4 antagonist [608]. Moreover, phidianidine A (**257**) significantly reduced the CXCL12-induced migration at 50 µM in a rat cell line of pituitary adenoma [609].

3.1.2. Pleurobranchoidea

Pleurobranchus albiguttatus and *P. forskalii* from the Philippines contain the cytotoxic chlorinated diterpenes chlorolissoclimide (**258**), dichlorolissoclimide (**259**), haterumaimides (**260**), and 3β-hydroxychlorolissoclimide (**261**), which are acquired from its tunicate prey *Lissoclinum* [610]. These were tested against the NCI panel of 60 tumor cell lines [610]. In Indonesia, *P. forskalii* presents a cyclic hexapeptide instead, keenamide A (**262**), containing a thiazoline and an isoprene residue, which has been reported active against four cancer cell lines (P388, A549, MEL20, and HT29) with IC_{50} values of 2.5–8 µM [611,612]. In Japan, *P. forskalii* contains a macrocyclic dodecapeptide, cycloforskamide (**263**), contaning three thiazoline heterocycles [612]. Cycloforskamide (**263**) displays an IC_{50} value of 5.8 µM against murine P388 leukemia cells [387]. Unfortunately, the low sensitivity of this model resulted in the NCI not pursuing this compound further [613,614]. In Japan, *P. forskalii* also presented ergosinine (**264**), an alkaloid only described previously in terrestrial higher plants and fungi, which is a known activator of caspase-3 [615,616]. The tunicate-derived compounds chlorolissoclimide (**258**) and dichlorolissoclimide (**259**) have been further tested, resulting in IC_{50} values of 0.7 and 1.25 µM against the NCI 60 cell lines panel, respectively, and also displaying selectivity toward melanoma cell lines [610,613,617,618]. Furthermore, both compounds (**258,259**) inhibit growth in the Corbett–Valeriote soft agar disk diffusion test, but with no selectivity for solid tumors, while the related 3β-hydroxychlorolissoclimide (**261**) presents some selectivity for solid tumors in the same test [610]. In particular, chlorolissoclimide (**258**) and dichlorolissoclimide (**259**) were active against murine P388 leukemia cells resistant to adriamycin by blocking the translation elongation via inhibition of translocation, thus producing an accumulation of ribosomes on mRNA [610,618,619]. On the other hand, haterumaimides A–Q (**260,265–269**) also present interesting activities, as do their synthetic derivatives, against the NCI 60 cell lines panel, with IC_{50} values of 0.08–1 µM for haterumaimide D, and from 0.5 nM to >20 µM for haterumaimides J (**265**) and K (**266**) and haterumaimides C (**267**), G (**268**), and I (**269**), respectively [610,617,620–624].

3.1.3. Tylodinoidea

As previously mentioned, *Tylodina perversa* obtains brominated isoxazoline alkaloids from sponges of the genus *Aplysina* [257,625]. Among them, *iso*-fistularin-3 (**270**) presents a IC_{50} value of ~9 µM against human HeLa cervix carcinoma cells [625].

3.1.4. Cephalaspidea

The most active species in this group is *Philinopsis speciosa*. Because its molecules are structurally similar to cyanobacterial compounds, it has been hypothesized that *P. speciosa* obtains them through its diet of the sea hares *Stylocheilus longicauda* and *Dolabella auricularia*, which in turn feed on cyanobacteria [276]. In Hawai'i, *P. speciosa* presents the cyclodepsipeptide kulolide-1 (**271**), which displays potent antitumoral activity against L-1210 leukemia and P388 murine leukemia cell lines, with IC_{50} values of 0.7 and 2.1 µg/mL, respectively, although these assays are highly sensitive to proapoptotic stimuli [277,614]. Kulolide-1 (**271**) also produced morphological changes in rat 3Y1 fibroblast cells at 50 µM [277]. Further chemical analysis of the species yielded some more related peptides, such as kulolide-2 (**272**), kulolide-3 (**273**), kulokainalide-1 (**274**) and the bidepsipeptides kulokekahilide-1 (**275**) and kulokekahilide-2 (**276**) [276,626,627]. Kulokainalide-1 (**274**) and kulokekahilide-1 (**275**) are reported to be moderately cytotoxic, while kulokekahilide-2 (**276**) is highly effective. Kulokekahilide-1 (**275**) shows growth inhibitory effects with an IC_{50} of ~2 µM in murine P388 leukemia cells, while kulokekahilide-2 (**276**) presents a higher potency with an IC_{50} of 4.2 nM [626,627]. Kulokekahilide-2 (**276**) also displays activity against human SK-OV-3 ovarian and tMDA-MB-435 breast cancer cell lines, with IC_{50} values of 7.5 and 14.6 nM, respectively [627]. For A-10 (non-transformed rat (*Rattus norvegicus*) aortic cells), kulokekahilide-2 (**276**) shows an IC_{50} of 59.1 nM, acting selectively [627]. In other cancer cell lines, namely, human A549 NSCLC, K562 chronic myelogenous leukemia, and MCF-7 breast cancer, kulokekahilide-2 (**276**) shows IC_{50} values of ~0.2 nM [628]. In a methodologically different study, the IC_{50} values for kulokekahilide-2 (**276**) were not the same, with ~19 and ~4 nM for murine P388 leukemia and human HeLa cervix carcinoma cells, respectively [629]. Some very potent analogues have also been synthesized, with IC_{50} values of ~0.001 and ~0.008 nM, respectively, allowing the authors to ascertain the best substituents for cytotoxic activity [628].

Moreover, several polyunsaturated fatty acids with cytotoxic activity against a range of human cancer cells lines, including HT-29, MCF7, and A2058, were described from Arctic specimens of *Scaphander lignarius* [630].

3.1.5. Anaspidea

Sea hares have been proven to contain many interesting cytotoxic compounds, many of them derived from the algae or the cyanobacteria that they feed on [1,2]. Within the genus *Aplysia*, many species have been studied, including *A. angasi, A. dactylomela, A. depilans, A. fasciata, A. juliana, A. kurodai, A. oculifera,* and *A. punctata* [1,2]. *A. angasi* (also reported as *A. dactylomela*), which showed growth inhibitory activity against murine lymphocytic leukemia P338 cells due to the presence of the brominated tricyclic 6-7-5-fused sesquiterpene, aplysistatin (**277**) [631]. Aplysistatin (**277**) showed an IC_{50} value of ~8 µM for P388 leukemia cells and human KB cancer cells [613,631].

A. dactylomela presents a wide variety of metabolites [631–634]. Some dietary halogenated chamigrene sesquiterpenes were described in specimens from the Canary Islands, such as elatol (**278**) and obtusol (**279**), previously found in the red algae *Laurencia elata* and *L. microcladia*, while *iso*-obtusol (**280**) was found in *L. obtusa* [631–634]. Elatol (**278**) is a potent, non-selective, cytotoxic natural product, with IC_{50} values between 1–10 µM against ten cancer cell lines in several studies [631,635,636]. Remarkably, elatol (**278**) is proapoptotic in murine B16F10 melanoma cells by decreasing Bcl-x and increasing Bak, caspase-9 and p53 expression, although it is known that B16F10 melanoma cells are very sensitive to these stimuli [636–638]. Obtusol (**279**) and *iso*-obtusol (**280**) show a much weaker effect inhibiting growth in vitro [631,633,635]. A linear halogenated monoterpene (**281**) was also found, showing growth inhibitory effects towards HM02 (gastric carcinoma), HEP-G2 (liver carcinoma), and MCF-7 (breast carcinoma) cancer cells with IC_{50} values of ~1 µM [631]. Some tricyclic monobromoditerpenes were also found in *A. dactylomela* specimens, such as parguerol (**282**), parguerol-16-acetate (**283**), *iso*-parguerol (**284**), *iso*-parguerol-16-acetate (**285**), and deoxyparguerol (**286**), which were also present in red algae, such as *Jania rubens* [639,640]. In Bahamas and Puerto Rico, however, specimens of this sea hare contain different chemistry [639]. Parguerol (**282**) and the previously mentioned related compounds present in vitro growth inhibitory effects at low micromolar values on murine P388 leukemia, are highly sensitive to proapoptotic stimuli as previously discussed, and are also significantly sensitive to Ehrlich ascite carcinoma [614,639,640]. These activities were more potent with *iso*-parguerol (**284**) derivatives [639,640]. In the Caribbean, some other compounds are present, such as the bromotriterpene polyether aplysqualenol A (**287**), which displays potent IC_{50} values in the NCI assay (60 cancer cell lines panel), particularly an IC_{50} of ~0.4 µM in human SNB-19 central nervous system cancer and T-47D breast cancer cells [641]. Aplysqualenol A (**287**) has been described to be a ligand for DYNLL1 (light chain of dynein type 1), also indicating some anticancer potential [642–645]. In China, specimens of *A. dactylomela* also present a brominated triterpene polyether, thyrsiferol (**288**), previously reported in the red algae *Laurencia thyrsifera* [646,647]. Thyrsiferol (**288**) has been reported to display a strong growth inhibitory in vitro activity against mouse P388 leukemia cells, and only moderate activity in solid tumors cell lines, with IC_{50} values of ~0.02 and ~17 μM for P388 and A549 NSCLC cancer cell lines, respectively [614,648,649]. Apparently, thyrsiferol (**288**) inhibits hypoxia-induced HIF-1 activation in T47D human breast tumor cells and suppresses hypoxic induction of HIF-1 target genes (VEGF and GLUT-1) at the mRNA level, also suppressing mitochondrial respiration in complex I [648].

A. depilans presents an endoperoxide sterol (**289**) with an IC_{50} value of 2.5 μM for human HCT-116 colorectal cancer cells [606,650,651]. *A. fasciata* from Spain presents a degraded sterol, 3-*epi*-aplykurodinone B (**145**) with an IC_{50} value of ~8 μM against mouse P388 leukemia, human A549 NSCLC, HT-29 colon cancer, and SKMEl-28 melanoma [397]. *A. juliana* contains two chlorophyll derivatives, pyropheophorbides a and b, and a halogenated diterpenoid lactone, derived from its diet of green algae, as well as a cytotoxic peptide, julianin-S, which is secreted within its purple ink [288,432–434].

A. kurodai has been widely studied over the years, containing a series of compounds that include polyketide macrolides, halogenated and brominated mono- and di-terpenes, brominated sesquiterpenoids, sterols, alkaloids, peptides, and others [1,2]. Aplyronines A–H (**191,290–293**) are polyketide macrolides found in Pacific specimens along with aplaminal (**294**), some of them being tested in antitumor clinical trials [46,652–656]. For human HeLa-S3 cancer cells, IC_{50} values were 0.5 nM for aplyronine A (**191**), 3 nM for aplyronine B (**290**), 22 nM for aplyronine C (**291**), 0.08 nM for aplyronine D (**292**), and 10 nM for aplyronine H (**293**) [653,657,658]. However, in a different study with the same cell line, IC_{50} values were ~0.4 nM for aplyronine A (**191**), ~4 nM for aplyronine B (**290**), and ~20 nM for aplyronine C (**291**) [652]. It has been suggested that the methylated amino acids (*N,N,O*-trimethyl-serine or *N,N*-dimethyl-alanine) in position 22 are important for the inhibition activity of these compounds [653]. Aplyronine A (**191**) is registered at the NCI database as NSC687160, with a mean IC_{50} value of ~0.2 nM in the 60 cancer cell lines, but being selective against some cell lines and most active against MDR cancer cell line NCI/ADR-RES, with an IC_{50} value of ~0.2 nM, as well as for P388 murine leukemia, Lewis murine lung carcinoma, Erhlich murine carcinoma, colon 26 murine carcinoma, and B16 murine melanoma, but very moderated against HOP-92 (NSCLC), OVCAR-4 (ovarian cancer), TK-10 and UO-31 (renal cancers), and BT-549 and T47-D (breast cancers) [652]. Aplyronine A (**191**) also presents proapoptotic effects in cancer cells [46]. Aplyronine A (**191**) was in fact suggested to inhibit the actin microfilaments, since it can depolymerize F-actin and inhibit actin polymerization, forming a complex with monomeric actin (1:1), in a similar way to the well-known tubulin inhibitors vincristine and vinblastine currently employed in some cancer treatments [479,659,660]. Later, other studies suggested that aplyronine A (**191**) forms a 1:1:1 heterotrimeric complex with actin and tubulin, and this is what actually inhibits tubulin polymerization; thus, synthesis is being carried out [661–664]. Furthermore, an analogue of aplyronine D (**292**) is being analyzed as an antibody–drug conjugate (ADC) and for bearing a linker suitable for bioconjugation [665]. Overall, aplyronines (**191,290–293**) are extremely toxic molecules with huge potential as leads, but they are not yet known to be attached to a suitable mAb [666]. Aplaminal (**294**) is a triazabicyclo-[3.2.1]-octane, displaying an IC_{50} value of ~2 μM against human HeLa S3 cervix carcinoma cells [654].

In Japan, *A. kurodai* presents aplaminone (**295**), neoaplaminone (**296**), and neoaplaminone sulfate (**297**), formed from a bromine-containing dopamine and a sesquiterpenoid [667]. Both aplaminone (**295**) and neoaplaminone sulfate (**297**) show IC_{50} values of ~1 μM for human HeLa cervix cancer cells, while neoaplaminone (**296**) is active at IC_{50} ~1 nM [667]. In Japan, *A. kurodai* specimens also contain aplysiaterpenoid A (**298**) and aplysiapyranoids A–D (**299**), displaying moderate cytotoxicities against Vero, MDCK, and B16 cell lines (IC_{50} 19–96 μg/mL) [668,669]. Moreover, mono- and di-terpenes, such as kurodainol (**300**), aplysiaterpenoids A–D (**298**), aplysin-20 (**301**), *iso*-aplysin-20, aplysiadiol (**302**), *epi*-aplysin-20, and *ent*-isoconcinndiol (**303**), are found in *A. kurodai* specimens and are suggested to originate from isoconncindiol of the red algae *Laurencia snyderae* [668–674]. *A. kurodai* also contains other compounds, including aplydilactone, a dieicosanoid lactone, aplysepsine, and a 1,4-benzoidiasepine alkaloid [675,676]. The egg masses and albumen gland moreover contain cytotoxic peptides, such as aplysianin A or aplysianin E, the latter highly tumor-lytic at 10 ng/mL against MM 46 and MM 48 mice tumor cells [294,677–679].

Furthermore, *A. kurodai* presents some brominated sesquiterpenoids, such as (-)-aplysin (**304**), aplysinol (**305**), and aplykurodin A (**306**) and B (**146**) [398,680]. (-)-Aplysin (**304**), originating from its diet on algae, was in fact one of the first halogenated compounds found in marine organisms, and it shows IC$_{50}$ values of 4–8 μM for several cancer cells, with a mean IC$_{50}$ value of ~30 μM at the NCI 60 cell line panel and no selectivity [680,681]. (-)-Aplysin (**304**) was active against the human A549 NSCLC xenograft with 18% of tumor growth reduction in vivo as compared to the control, supposedly by acting as a sensitizer for tumor necrosis factor-related apoptosis, producing TRAIL-induced apoptosis in cancer cells via the P38 MAPK/survivin pathway [682]. (-)-Aplysin (**304**) is also effective against human glioma cells by increasing miR-181 expression, sensitizing the cytotoxic effects of the alkylating drug temozolomide, and inducing cell cycle arrest and apoptosis through the inhibition of the PI3K/Akt signaling, which is relevant in the survival of glioma cells [683].

In Egypt, *A. oculifera* contains two halogenated sesquiterpenes, oculiferane (**307**) and *epi*-obtusane (**308**), which displayed IC$_{50}$ values between 2 and 8 μM against human PC-3 prostate cancer, A549 NSCLC, MCF-7 breast cancer, HepG2 liver cancer, and HCT116 colon cancer [684]. *A. punctata* from Spanish coasts present atypical acetates of halogenated monoterpenes, among which three compounds (**309–311**) show IC$_{50}$ values between 1.5 and 2.5 μM against P388, HT-29, A-549, and MEL-28 cancer cell lines [685].

Another sea hare, *Dolabella auricularia*, is one of the most studied marine invertebrates, with many bioactive natural products, including polyketides, halogenated terpenes, and peptides, along a wide geographical range [1,2]. Several excellent reviews deal with *D. auricularia* most bioactive compounds [46,478,479], and, therefore, only a brief summary will be included here. This sea hare is able to modify dietary molecules from brown and red algae and also to de novo biosynthesize peptides and polypropionates, while it may also contain cyanobacterial metabolites [686]. We will summarize the cytotoxic activities of aurisides A and B (**312,313**), aurilol (**314**), doliculols A and B (**315,316**), dolabellin (**317**), auripyrones A and B (**318,319**), dolabelides A–D (**320–323**), aurilide (**324**), doliculide (**325**), auristatins (**339–344**), and finally, dolastatins (**192,326–338,346–348**). In Japan, among the dietary compounds from red algae, the macrolide glycosides aurisides A and B (**312,313**) and aurilol (**314**), a polyether bromotriterpene, are cytotoxic to HeLa tumor cell line [687,688]. Aurisides A and B (**312,313**) present a carbon backbone 5,7,13-trihydroxy-3,9-dioxoheptadecanoic acid with a bromine-substituted conjugated diene moiety, a cyclic hemiacetal, and a 14-membered lactone [688]. Auriside A (**312**) displays IC$_{50}$ values of ~0.2 μM against human HeLa S3 cervix cancer cells, while auriside B (**313**) is less potent (IC$_{50}$ 2 μM), and both are being studied for their synthesis [688–690]. Aurilol (**314**) presents IC$_{50}$ values of ~7 μM against human HeLa S3 cancer cells, and it is structurally similar to enshuol, a bromo triterpenic polyether with a dioxabicyclo-(5.4.0)-undecane ring system described in *Laurencia*, thus supporting a dietary origin in the slug [687]. Doliculols A and B (**315,316**), non-halogenated acetylenic cyclic ethers similar to *Laurencia* ethers, are moderately cytotoxic macrolides [691–694]. Dolabellin (**317**), a bisthiazole metabolite found in Indian specimens, showed IC$_{50}$ values of ~10 μM against human HeLa S3 cervix carcinoma cells [695].

In *D. auricularia* from Japan, auripyrones A and B (**318,319**) were described to have IC$_{50}$ values of ~0.5 μM against human HeLa S3 carcinoma cells and are being synthesized [696–700]. Furthermore, Japanese specimens contain the macrolides dolabelides A–D (**320–323**), displaying IC$_{50}$ values of ~8 and ~2 μM, for dolabelides A and B (**320,321**) respectively, while dolabelides C and D (**322,323**) showed both values of ~2 μM against human HeLa S3 cervix cancer cells, and synthesis studies are being carried out [692,693,701,702].

The cyclic depsipeptide aurilide (**324**) was found in Japanese *D. auricularia* specimens, while aurilides B and C were found in the cyanobacterium *Lyngbya majuscula* from Papua New Guinea [703–705]. Aurilide (**324**) displayed an IC_{50} mean value of ~0.01 μM in the NCI 60 cell line panel and was selective for renal, ovarian, and prostate cancer cell lines. It was active in the in vivo NCI hollow fiber assay, but inactive due to high toxicity in a xenograft model [705,706]. The mechanism of action of aurilide (**324**) seems to involve microtubule stabilization, since it does not interact with tubulin, thus being different from taxol [705]. Furthermore, aurilide (**324**) has been reported to selectively bind to prohibitin 1 (PHB1) in the mitochondria, activating the proteolytic processing of optic atrophy 1 (OPA1) and resulting in mitochondria-induced apoptosis [707].

Doliculide (**325**) is a mixed peptide–polyketide-originated compound found in Japanese specimens of *D. auricularia*, possessing an iodo-N-Me-tyrosine and a glycine, which inhibits growth of human HeLa S3 cervix carcinoma cells (IC_{50} ~2 nM) [708]. Its action mechanism consists of binding to actin and stopping cancer cells at the G2/M phase of the cell cycle, thus interfering in normal actin assemblage and producing the hyperassemblage of purified actin in the form of F-actin [709]. Synthesis and computational studies are being developed [710–712]. Doliculide (**325**) inhibits proliferation and impairs the migratory potential of human MCF-7 and MDA-MB-231 breast cancer cells, while modifying senescence-related genes at non-toxic concentrations in p53 wild-type cancer cells by up to 13% [713].

The most famous compounds of all heterobranchs are the diverse cytotoxic linear and cyclic peptides of *D. auricularia*, the dolastatins (**192,326–338,346–348**), which were probably used by the Romans [46,197]. Dolastatins (**192,326–338,346–348**) include many different active molecules, including linear and cyclic peptides, depsipeptides, peptides containing thiazole and oxazole heterocycles, and macrolides [479]. Dolastatins (**192,326–338,346–348**) are found in small amounts in sea hares and are suggested to originate from their diet, particularly from cyanobacteria of the genera *Symploca*, *Caldora* and *Lyngbya* [714–716]. Among the most known dolastatins, dolastatin 3 (**192**) and dolastatins 10–15 (**326–331**) display mild to strong biological activities and were further studied, such as the macrocyclic lactone dolastatin 19 (**332**); dolastatins C (**333**), D (**334**), H (**335**) and *iso*-dolastatin H (**336**); dolastatin G (**337**); and *nor*-dolastatin G (**338**) [694,717–724]. Dolastatin G (**337**) and *nor*-dolastatin G (**338**) show moderate cytotoxicity and are analogs of lyngbyastatin 2 and *nor*-lyngbyastatin 2 described in the cyanobacterium *Lyngbya majuscula* from Guam [725]. Dolastatin 19 (**332**) is structurally similar to the previously mentioned aurisides A (**312**) and B (**313**). In general, these compounds did not pass phase II trials alone, and several studies are being developed to use them in combination with other structures [478,479,726].

Dolastatin 3 (**192**), a cyclic peptide with two thiazole rings, was also found in *D. auricularia* from Japan, and further synthesized, showing an $IC_{50} < 1$ μM in P388 murine leukemia cells [718,727]. Dolastatin 3 (**192**) was also found in *Lyngbya majuscula* from Palau [728]. Dolastatin 3 (**192**) induces a 78% life extension in vivo in the murine P388 lymphocytic leukemia model, and a 52% life extension in murine colon carcinoma 38 [728].

Dolastatin 10 (**326**) is a linear pentapeptide with four unique residues described in 1987 from *D. auricularia* collected at the Indian Ocean, and later found in the cyanobacterium *Symploca hydnoides* together with its methyl derivative, symplostatin 1, and has been often reviewed in the literature [478,714,726,729–731]. Dolastatin 10 (**326**) was tested in phase I trials by the NCI but failed later in phase II for advanced and metastatic soft tissue sarcoma, advanced hepatobiliary cancers, pancreatic cancers, and others because of its side effects [478,479,726,731,732]. However, in 2011, brentuximab vedotin (Adcetris®), an antibody-dolastatin 10 conjugate (ADC), was approved by the FDA (Food and Drugs Administration) for the treatment of Hodgkin's lymphoma [43]. This ADC is composed of a highly toxic "warhead" derived from dolastatin 10 (**326**) which is attached to a specific targeting moiety, a monoclonal antibody (mAb) directed to a particular epitope on the cancer cell [726]. Dolastatin 10 (**326**) inhibits microtubule assembly, causing cells to accumulate in metaphase, but it produced bone marrow toxicity in initial trials, local irritation at the injection site, and some mild peripheral neuropathy [733,734]. Dolastatin 10 (**326**) exhibits cytotoxic effects against human lung and breast cancer cell lines via both Bcl-2 phosphorylation and caspase-3 protein activation, and it modulates p53 oncoproteins in human diffuse large-cell lymphoma [735,736]. Other studies describe dolastatin 10 (**326**) activity against ovarian carcinoma xenografts as well as mouse P388 and L1210 leukemia, B16 melanoma, M5076 sarcoma, human LOX melanoma, and MX-1 breast cancer xenografts [736–738]. Dolastatin 10 (**326**) is not only an inhibitor of tubulin polymerization, it also inhibits tubulin-dependent GTP hydrolysis as well as the binding of vinblastine, maytansine, and vincristine to tubulin, although the binding site on tubulin is not the same as that of the vinca alkaloids [733,739].

Auristatins (**339–344**) are peptides related to dolastatin 10 (**326**) (see below), approved by the FDA as microtubule-destabilizing agents (MDA), and used as antibody–drug conjugates (ADCs) [740–743]. Auristatins (**339–344**) were synthesized by Pettit's group in the 1990s while working on dolastatins, and they included auristatin E (**339**), auristatin PHE (**340**), auristatin PYE (**341**), and two aminoquinoline derivatives, auristatin -2-AQ (**342**) and auristatin-6-AQ (**343**), all active against tumor lines (between 10–100 pM) [32,744–746]. Several derivatives of auristatins are (or are expected to be soon) in preclinical trials as "ADC warheads", some reaching phase I, such as DZ-2384 [478,726]. Auristatin PE (**344**), also named soblidotin, TZT-1027, or YHI-501, is a microtubule active drug that exerts a considerable antivascular effect along with a potent cytotoxic effect in several models, including murine P388 leukemia, colon 26 cancer, Lewis lung carcinoma, B16 melanoma, and M5076 sarcoma, as well as human MX-1 breast cancer and LX-1 and SBC-3 SCLC xenografts [747,748]. Auristatin PE (**344**) entered clinical trials for advanced and metastatic soft tissue sarcomas, NSCLCs, and others, but did not proceeded beyond phase II due to toxicity and/or a lack of efficacy in the trials [726]. More than 30 ADCs in clinical trials are based on auristatins [43,555,717,749].

Several excellent reviews, such as those of Newman, have dealt with dolastatin-10 (**326**) and all its derivatives, their evolution as ADC warheads, the auristatin-based ADCs, and the approved and/or tested drugs, such as brentuximab vedotin, polatuzumab vedotin, enfortumab vedotin, ladiratuzumab vedotin, lifastuzumab vedotin, PSMA-ADC, RC-48, telisotuzumab vedotin, tisotumab vedotin, BA-3021, CX-2029, HuMax-AXL-107 (enapotamab vedotin), pinatuzumab vedotin, ABBV-085, AGS67E, ALT-P7, CDX-014, losatuxizumab vedotin, SGN-CD48A, rituximab-MC-vc-MMAE (TRS-005), GM-103, HT-1511, OBI-999, depatuxizumab mafadotin, AGS-16C3F, GSK-2857916 (belantamab mafodotin), W-0101, cofetuzumab pelidotin, NG-HER2 ADC, XMT-1536, ASN-004, ARX-788, lupartumab amadotin, AGS-62P1, and ZW-49, among others [478,666,726]. We strongly recommend these reviews for further details on the different status of the many derivatives that are currently being tested and those that are in clinical trials.

Dolastatin 11 (**327**), 12 (**328**), 13 (**329**), and 14 (**330**) are further depsipeptides isolated from *D. auricularia*, while dolastatin 12 (**328**) was also reported in *Lyngbya majuscula–Schizothrix calcicola* cyanobacterial assemblages [750–753]. In the NCI cell line panel, dolastatin 11 (**327**) showed an IC_{50} mean value of ~0.07 µM, while dolastatin 12 (**328**) displayed different IC_{50} values in several assays, such as 1 nM for human NCI-H460 NSCLC, 30 nM for human SF-295 CNS cancer, ~0.1 µM for mouse neuro-2a neuroblastoma and 1 µM for P388 leukemia cells [750,754,755]. Dolastatin 11 (**327**) produces a massive rearrangement of the actin filament network in cells, inducing a cytoplasmic retraction and further cell division arrest at the level of cytokinesis [239,241]. Dolastatin 12 (**328**) also targets actin microfilaments [753]. The values of IC_{50} for dolastatin 13 (**329**) and 14 (**330**) against the murine P388 leukemia cell line are reported to be 14 and 20 nM, respectively [751,752].

Dolastatin 15 (**331**) is another linear peptide from *D. auricularia* widely used as a potential warhead, with an IC_{50} of 3–5 nM, and with many derivatives being tested after chemical modifications, for example, replacing the C-terminal (S)-dolapyrrolidinone unit with some diverse amides while maintaining its anti-tubulin activity [756–761]. Dolastatin 15 (**331**) also produces microtubule depolymerization in vitro, possibly binding to the vinca domain of tubulin, and it is a classical inducer of apoptosis in cancer cells, acting as a conventional proapoptotic cytotoxic agent [762–764]. Dolastatin 15 (**331**) presents an IC_{50} against the NCI panel about ten times higher than that of dolastatin

10 (**326**), that is, 2 vs. 0.2 nM, and it is three to four times more potent than vincristine, a clinically used common antiproliferative agent [765]. Tasidotin (**345**), an analog of dolastatin 15 (**331**) where the carboxyl-terminal ester group is replaced by a tert-butyl amide, is also a proapoptotic cytotoxic compound, tested against many different cancer lines, but it did not go beyond phase II clinical trials because of its lack of efficacy [766–768].

Dolastatin 16 (**346**), instead, is a cyclic depsipeptide with two amino acids, dolamethylleuine and dolaphenvaline, described in *D. auricularia* from Papua New Guinea, and found also in *Lyngbya majuscula* from Madagascar and *Symploca* cf. *hydnoides* from Guam [769–771]. In the NCI cell line panel, dolastatin 16 (**346**) displayed an IC_{50} mean value of ~0.3 µM [770,772]. Specimens of *D. auricularia* from Papua New Guinea also contained dolastatin 17 (**347**), another cyclodepsipeptide with a novel acetylenic β-amino acid named dolayne (Doy), similar to that of onchidin (**121**), which has submicromolar values of IC_{50} against four cancer cell lines [773,774]. Dolastatin 18 (**348**) was found in *D. auricularia* from the Indian Ocean, contains a thiazole ring, and shows submicromolar values of IC_{50} against mouse P388 lymphocytic leukemia and human NCI-H460 NSCLC cell lines [775]. *D. auricularia* from California contains dolastatin 19 (**332**), a macrocyclic lactone related to the previously mentioned aurisides (**312,313**), showing in vitro growth inhibition for breast MCF-7 and colon KM20L2 cancer cells with IC_{50} values of ~1 µM [776]. Similarly, other dolastatins from specimens from Japan and the Indian Ocean, such as dolastatin D (**334**), dolastatin G (**337**), *nor*-dolastatin G (**338**), dolastatin H (**335**), and *iso*-dolastatin H (**336**), also display activity against some cancer cell lines, such as human HeLa S3 cancer cells [721,723,724].

Stylocheilus and *Bursatella* are also cyanobacterial feeders containing interesting cytotoxic compounds [1,2,4]. Aplysiatoxin (**96**) and debromoaplysiatoxin (**97**) were found in *S. longicauda* from Hawai'i, originating from *Lyngbya majuscula* [302,303,777]. The mixture of these two compounds was toxic to mice (LD_{100} 0.3 mg/kg), and both compounds are potent PKC activators that are being tested as anticancer lead structures (along with some derivatives) based on their anti-proliferative activity while removing their tumor-promoting activities [778–780]. Further studies on different populations of *S. longicauda* also from Hawai'i reported complex proline esters, makalika (**99**) and makalikone (**100**), together with lyngbyatoxin A (**349**) with antitumor properties, again from a diet of *L. majuscula* [304,305]. Makalikone (**100**) shows moderate activity against P388, A549, and HTB38 cancer cell lines, with IC_{50} values between 2.5 and 5 µg/mL [305]. Lyngbyatoxin A (**349**) is toxic to mice (LD_{100} 0.3 mg/kg), and it has been reported to act as a tumor promoter [781]. Lyngbyatoxin A acetate (**101**), also found in the sea hare and its cyanobacterial prey in Hawai'i, displays very potent toxicity against several cancer cell lines, with IC_{50} values ~0.05 µg/mL [305]. Furthermore, some alkaloids such as malyngamides O (**102**) and P (**103**) were found in *S. longicauda* and *L. majuscula* [307,781,782]. Malyngamide O (**102**) shows IC_{50} values of 2 µg/mL against the cancer cell lines P388, A549, and HT29 [123]. In Guam, *S. longicauda* accumulates malyngamydes and transforms malyngamyde B into an acetate [307]. In fact, more than 30 malyngamides have been isolated from cyanobacteria and sea hares, and they have been observed to be N-substituted amides of long-chain methoxylated fatty acids, which are characterized by presenting a trans double bond and a 7S configuration of the oxygen-bearing carbon [479].

Bursatella leachii also feeds on "*Lyngbya*", accumulating lyngbyatoxin A (**349**) and debromoaplysiatoxin (**97**) in the digestive gland [783]. In New Zealand, *B. leachii* presents an alkaloid derived also from cyanobacteria, malyngamide S (**350**), with cytotoxic properties, while in Thailand, it presents hectochlorin (**351**) and deacetylhectochlorin (**352**), which are also cytotoxic compounds previously isolated from *Lyngbya majuscula* and structurally similar to dolabellin from *Dolabella auricularia* [784–786]. Malyngamide S (**350**) and malyngamide X (**353**), both found in *B. leachii*, possess in vitro growth inhibitory properties against several cancer cell lines, with IC_{50} values between ~4 and ~8 µM against murine P388 leukemia and human A549 NSCLC, NCI-H187 (SCLC), HT-29 colon cancer, HL60 leukemia, KB and BC breast cancer lines [307,786,787]. Hectochlorin (**351**) and deacetylhectochlorin (**352**) show growth inhibitory effects in vitro against human KB, NCI-H187

SCLCL, and BC breast cancer cell lines, with deacetylhectochlorin (**352**) displaying a mean IC$_{50}$ value of ~1 µM, and hectochlorin (**351**) a mean IC$_{50}$ value of ~5 µMin the NCI cell panel (being more potent against colon, melanoma, ovarian, and renal cell lines) [784,785]. Additionally, hectochlorin (**351**) seems to be cytostatic rather than cytotoxic in regard to the obtained dose–response curves [785].

3.1.6. Sacoglossa

Within this group, caulerpenyne (**155**)—a sesquiterpene found in *Elysia* spp. and other species (see above), as well as in its diet of the green algae *Caulerpa*—is active against several cancer cell lines at IC$_{50}$ ~10 µM, while it has an IC$_{50}$ ~40 µM in the NCI panel [316,411,788–790]. Caulerpenyne (**155**) is not selective for normal (hamster fibroblasts, human keratinocytes, and melanocytes) and cancer cells [790,791]. Caulerpenyne (**155**) induces tubulin aggregation, inhibiting the polymerization of tubulin and bundling of the residual microtubules, but it does not bind to colchicine, taxol, or vinca-alkaloid binding domains [790,792]. It has been shown that caulerpenyne (**155**) may block the stimulation of mitogen-activated protein kinase (MAPK), thus affecting the control of cell proliferation, differentiation, or death [790,793].

Kahalalides (**194,354–356**) are cyclodepsipeptides found in *Elysia* species (*E. rufescens, E. ornata, and E. grandifolia*) and their algal food, *Bryopsis pennata* [331,794–797]. Kahalalides (**194,354–356**) include more than 20 structurally diverse molecules, ranging from a C-31 tripeptide to a C-77 tridecapeptide, where each peptide contains a different fatty acid chain [794]. Among them, kahalalide F (**194**), a cyclic peptide connected by an amidic bond to a short fatty acid chain, is the most potent, being reported to show antitumour activity and tested in phase I trials in patients with hormone-refractory prostate cancer [797,798]. Treating cancer cells with kahalalide F (**194**) resulted in critical changes in lysosomal membranes and large vacuoles, producing cell swelling, while it is also reported to display specific interactions with cell membrane proteins [482,799]. Kahalalide F (**194**) inhibits the PI3K–AKT signaling pathway in the breast cancer cell lines SKBR3 and BT474 [800]. The IC$_{50}$ values of kahalalide F (**194**) in the NCI 60 cell line panel were from 0.2 to 10 µM, with hormone-independent prostate cancer cells being the most sensitive [267]. Kahalalide F (**194**) also displays in vivo activity against human prostate hormone-independent xenograft models and in the hollow fiber test [794,801–803]. Kahalalide F (**194**) was tested in several oncological clinical trials and was taken to phase II, although it failed to be effective [804,805]. Kahalalide F (**194**) is found in nature as a mixture with *iso*-kahalalide F (**354**), also possessing interesting bioactivities [331,806]. *Iso*-kahalalide F (**354**) also entered phase II clinical trials for liver cancer, melanoma, and NSCLC patients, but was also ineffective [807]. The origin of these compounds has been suggested to be *Mycoplasma* spp. Or *Vibrio* spp. Bacteria, since they are affiliated with *E. rufescens* and its mucus [808]. In *E. ornata* from India, two more compounds, kahalalide Z$_1$ (**355**) and kahalalide Z$_2$ (**356**), were found, differing from kahalalide F (**194**) in the *N*-terminal acid moiety and some aminoacid units of the peptide chain, and displaying a bioactivity profile comparable with kahalalide F (**194**) [795]. Furthermore, elisidepsin trifluoroacetate (PM02734, IrvalecR), a kahalalide-derived synthetic cyclic depsipeptide (**357**), displays cytotoxic activity, causing cell death by inhibiting the AKT/mTOR pathway [331,809]. Elisidepsin (**357**) also underwent clinical development after showing IC$_{50}$ values between 0.4 and ~9 µM in a 23 cancer cell lines, including breast, colon, head and neck, lung, ovary, pancreas, prostate, and melanoma types [810–812]. Elisidepsin (**357**) acts at the cell membrane level, interacting directly with glycosylceramides in the membrane of cancer cells, while inducing necrosis-like cell death in the yeast *Saccharomyces cerevisiae* [813–815]. Elisidepsin (**357**) is active in vivo against human melanoma, liver, pancreas, breast, and prostate cancer xenografts [816]. However, in clinical trials, elisidepsin (**357**) has been ineffective to date [811,817].

3.1.7. Pulmonata

Trimusculus species present a single type of labdane diterpenoids, such as *T. costatus* and *T. reticulatus* from different geographic localities [333,334,818]. In Chile and South Africa, *T. costatus* and *T. peruvianus* metabolites present cytotoxicity [333,335–337]. An atypical C-21 hydroxylated sterol (**358**) from

T. peruvianus presented IC$_{50}$ values of ~6 µM when tested against human HCT-116 and HT29 colon cancer cells [819]. A secosterol (**359**) from *T. costatus* was also active, with an IC$_{50}$ value of ~3 µM against the WHCO1 esophageal cancer cell line [337].

Siphonaria species, *S. capensis*, *S. concinna*, *S. cristatus*, and *S. serrata*, contain de novo biosynthesized polypropionates, and some of which also present cytotoxic activity [27,339–352]. In particular, *iso*pectinatone, siphonarienolone (**119**) and others are active at 2.5 µg/mL against P388, A549, HT29, and MEL28, while pectinatone (**166**) and siphonarienfuranone are active at 5 µg/mL, and siphonarienedione, siphonarienone, and *iso*siphonarienolone are active at 10 µg/mL [27,339–352]. Onchidiidids present sesquiterpenoids, depsipeptide acetates, and propionates with 32 carbon atoms, two γ-pyrone rings, and a number of hydroxyl groups [12]. *Onchidium* species possess cytotoxic cyclic depsipeptides, such as onchidin (**121**), and the tropical *Onchidium* sp. also possesses cytotoxic acetates and propionates [358,359,820]. Onchidin (**121**) and onchidin B (**360**) are active against murine P388 leukemia and human KB oral cancer cells at IC$_{50}$ values of ~ 7 µM [358,359]. In China, *Onchidium* sp. presents bis-γ-pyrone polypropionates, such as onchidione (**122**) and related compounds, such as onchidiol (**361**) and ilikonapyrones (**362,363**) lacking the hemiketal ring, in different populations [12,357,360,362,821–823]. Both kinds of compounds were tested against HCCLs, resulting in 3-acetyl-11-(3-methylbutanoyl)-13-propanoyl-ilikonapyrone (**362**) being active, inhibiting growth in all tested cell lines with IC$_{50}$ between 3 and 9 µM (A549 NSCLC, MCF-7 breast cancer, PC-3 prostate cancer, Hs683 oligodendroglioma, U373 glioblastoma, and SKMEL-28 melanoma), and being comparable to etoposide and camptothecin [357]. This compound (**362**) seems to be active against cancer cells that present resistance to proapoptotic stimuli [357]. Moreover, onchidione (**122**) and the related 3-acetyl-onchidionol and 4-*epi*-onchidione were reported to have significant effects on the splicing of XBP1 mRNA, which is an important regulator of some genes related to the growth of tumors [821].

Table 10. Cytotoxic and antitumoral compounds in the different heterobranch groups. In brackets: number of species with these compounds, number of the compounds in figures, and reference numbers. # Numbers.

Species (#)	Compounds (#)	Activity	References (#)
Nudibranchia (59)			
Doris kerguelenensis	Palmadorins A (**195**), B (**196**), D (**197**), M (**198**), N (**199**), and O (**200**)	Inhibition of human erythroleukemia cells (HEL), inhibition of Jak2, STAT5, and Erk1/2 activation in HEL cells	[66,67,824]
Doris verrucosa	Verrucosins A (**124**) and B	Activation of protein kinase C	[371,420]
Notodoris citrina, N. gardineri	Naamidine A (**201**), iso-naamidine-A (**160**)	Inhibition of the epidermal growth factor (EGF), inhibition of human tumor xenografts in mice, and promotion of caspase-dependent apoptosis in tumor cells	[424,489,490]
Adalaria loveni	Lovenone (**202**)	Cytotoxic to two HTCLs	[491]
Polycera atra	Bryostatins (**203**)	Cytotoxic to P388 lymphocytic leukemia and Alzheimer's disease cells	[424,486,487]
Actinocyclus papillatus	(–)-Actisonitrile (**204**), actinofide (**205**)	Cytotoxic to tumor and non-tumor cells	[361,479,503,504]
Aldisa andersoni	9-Chloro-phorbazole D (**5**), N1-methyl-phorbazole A (**6**), phorbazoles A (**7**), B (**8**), and C	Cytostatic effects in vitro against several HTCLs (human SKMEL-28 melanoma and U373 glioblastoma cells)	[55,75,76]
Dendrodoris carbunculosa	Dendocarbins A–N (**16**, **206**), isodrimeninol (**207**), 11-epivaldiviolide (**208**)	Cytotoxic to murine leukemia P388 cell lines	[86,509]
Phyllidiella coelestis	1-Formamido-10(1→2)-abeopupukeanane (**209**), 2-formamidopupukeanane (**210**)	Cytotoxic to HeLa, MCF-7, KB, HT-29 cancer cell lines	[111]
Phyllidiella pustulosa	Axinisothiocyanate K (**211**), isothiocyanate axisonitrile-3 (**25**)	Cytotoxic to NBT-2 cells	[123,377]
Phyllidiella coelestis, P. pustulosa	Bisabolane-type sesquiterpenoid (**212**), theonellin isothiocyanate (**213**), 7-isocyano-7,8-dihydro-α-bisabolene (**214**)	Cytotoxic to several HCCLs	[117]

Table 10. *Cont.*

Species (#)	Compounds (#)	Activity	References (#)
Reticulidia fungia	Reticulidins A (**215**) and B	Cytotoxic in vitro to KB cells and mouse L1210 leukemia cells	[438,511,512]
Cadlina luteomarginata	Ansellone A (**216**)	Activation of the cyclic adenosine monophosphate (cAMP) signaling pathway	[513]
Chromodoris elisabetina, C. hamiltoni, C. lochi, C. africana, C. annae, C. kuiteri, C. magnifica, C. quadricolor	Latrunculins A (**38**) and B (**37**)	Disruption of normal cell organization and function	[153,155,168–171]
Chromodoris lochi	Mycothiazole (**129**)	Inhibition of the hypoxia-inducible factor-1 (HIF-1), and suppression of the mitochondrial respiration at complex I	[378,379,825,826]
Chromodoris lochi	Laulimalide (**39**), isolaulimalide (**40**)	Cytotoxic to the KB cell line	[142,523,524]
Chromodoris inornata	Inorolides A–C (**217**)	Cytotoxic to murine L1210 leukemia and human epidermoid carcinoma KB cell lines	[156,531]
Chromodoris petechialis	Puupehenone (**218**)	Human peripheral blood mononuclear (PBM) cells	[827]
Goniobranchus splendidus	Epoxygoniolide-1 (**219**)	Cytotoxic	[532]
Goniobranchus (Chromodoris) sinensis	Aplyroseol-2 (**220**)	Cytotoxic to HeLa S3 cells	[131]
Goniobranchus reticulatus	Spongian-16-one (**221**), aplytandiene-3 (**222**), aplysulfurin (**223**), aplyroseol-2 (**220**), gracilins A (**224**), B (**225**), C (**226**), G (**227**), and M (**228**)	Cytotoxic to P388 mouse leukemia, HTCLs cell lines, and BACE1 and ERK inhibition	[161,190]
Goniobranchus (Chromodoris) obsoleta	Dorisenones A–D (**229**), 11β-hydroxyspongi-12-en-16-one (**230**), spongian-16-one (**221**)	Cytotoxic to murine lymphoma L1210 and KB cells	[154]
Doriprismatica (Glossodoris) atromarginata	Spongiadiol (**35**), spongiadiol diacetate (**231**), epispongiadiol (**232**), 12-deacetoxy-12-oxodeoxoscalarin (**136**), heteronemin (**233**), mooloolabene D (**234**)	Cytotoxic to MCF-7 breast cancer cells	[180,182,381,828]
Felimida (Glossodoris) dalli, Doriprismatica (Glossodoris) sedna, Glossodoris rufomarginata, G. pallida, G. vespa, Ardeadoris (Glossodoris) averni	12-deacetyl-23-acetoxy-20-methyl-12-epi-scalaradial (**135**)	Inhibition of mammalian phospholipase A2	[175,381]
Hypselodoris infucata	(−)-Furodysinin (**48**)	Cytotoxic to HeLa cell	[214]
Felimida (Chromodoris) macfarlandi	Macfarlandin E (**235**)	Golgi-modifying properties	[139,140,149,543]
Hexabranchus sanguineus	Ulapualides A (**190**), B (**236**) and C (**237**), kabiramides A (**238**), B (**239**), C (**62**), D (**240**), E (**241**) and G (**243**), dihydrohalichondramide (**63**), 33-methylhalichondramide (**242**), halichondramide (**244**), Hurghadin	Cytotoxic to murine L1210 leukemia cells, cytotoxic to several HTCLs, and cytotoxic to human MCF-7 breast cancer cells	[545]
Jorunna funebris	Jorumycin (**189**), jorunnamycins A–C (**245**), renieramycin M (**246**)	Cytotoxic to cancer cell lines P388, A549, HT29, and MEL28, and cytotoxic to human colon (HCT-116) and breast (MDA-MB-435) cancer cells	[560]
Jorunna funebris	Fennebricins A and B, N-formyl-1,2-dihydrorenierol	Strong NF-κB inhibition, and cytotoxic to A549 and HL-60 tumor cell lines	[564,565]
Peltodoris atromaculata	Petroformynes (**247**), hydroxyl-dehydroisofulvinol (**248**), fulvinol	Cytotoxic to murine P388 leukemia cells, A549 NSCLC, HT-29 colon cancer and SKMEL-28 melanoma cells	[567,568]
Halgerda aurantiomaculata	Zooanemonin (**367**)	Antineoplastic	[829]
Tambja capensis, T. ceutae, T. eliora, T. morosa, T. stegosauriformis, T. verconis, Roboastra tigris, Nembrotha spp.	Tambjamines (**65–70**), tambjamine K (**249**), tetrapyrrole (**72**)	Cytotoxic to several tumor cell lines (Caco-2 colon cancer cells, HeLa cervix cancer cells)	
Tritonia sp.	Punaglandins (**250**)	Cytotoxic	[587]
Doto pinnatifida	Dotofide (**251**)	Cytotoxic to Hs683 oligodendroglioma, U373 glioblastoma, A549 NSCLC human carcinoma, MCF-7 breast carcinoma, SKMEL-28, and mouse B16F10 cells	[243,505]
Tritoniopsis elegans	Tritoniopsins A–D (**75–78**)	Cytotoxic to rat cell lines	[234]
Dermatobranchus ornatus	Ophirin (**187**)	Cytotoxic	[22]
Leminda millecra	Prenylated hydroquinone (**252**)	Cytotoxic to WHCO1, WHCO6 esophageal cancer cell lines	[588–590]
Hermissenda crassicornis	L-6-bromohypaphorine (**253**)	Cytotoxic to human a7 nicotinic acetylcholine receptor (nAChR subtype)	[591]
Phyllodesmium briareum	Brianthein W (**254**), excavatolide C (**255**)	Cytotoxic to cancer cell line P-388	[248]
Phyllodesmium magnum	11-episinulariolide acetate (**256**)	Cytotoxic to cancer cell line P-388	[248]
Phyllodesmium longicirrum	Trocheliophorol (**84**)	Cytotoxic	[245]
Phidiana militaris	Phidianidines A (**257**) and B	Cytotoxic to C6 and HeLa tumor cells	[592,593,595]

Table 10. *Cont.*

Species (#)	Compounds (#)	Activity	References (#)
Pleurobranchoidea (2)			
Pleurobranchus albiguttatus, P. forskalii	Chlorolissoclimide (**258**), dichlorolissoclimide (**259**), haterumaimides A (**260**), C (**267**), J (**265**), K (**266**), G (**268**), I (**269**), 3ß-hydroxylissoclimide (**261**)	Cytotoxic to melanoma cells	[610,619]
P. forskalii	Keenamide A (**262**), cycloforskamide (**263**), ergosinine (**264**)	Cytotoxic to P-388, A-549, MEL-20, and HT-29 tumor cell lines	[611,612]
Tylodinoidea (1)			
Tylodina perversa	Iso-fistularin 3 (**270**)	Cytotoxic to human HeLa cervix carcinoma cells	[257,625]
Cephalaspidea (2)			
Philinopsis speciosa	Kulolides 1 (**271**), 2 (**272**) and 3 (**273**), kulokainalide 1 (**274**), lulokekahilides 1 (**275**) and 2 (**276**)	Cytotoxic to L-1210, P388 leukemia, human SK-OV-3 ovarian, tMDA-MB-435 breast cancer, human A549 NSCLC, K562 chronic myelogenous leukemia, HeLa cervix carcinoma, and MCF-7 breast cancer cell lines	[277,626–628]
Scaphander lignarius	ARA, EPA, HTA (fatty acids)	Cytotoxic to a set of cancer and normal cell lines	[630]
Anaspidea (11)			
Aplysia angasi, A. dactylomela, A. depilans, A. fasciata, A. juliana, A. kurodai, A. oculifera, A. punctata	Aplysistatin (**277**)	Cytotoxic to mouse P388 leukemia, human KB cancer, and HeLa cervix carcinoma cells	[613,631,820]
Aplysia dactylomela	Elatol (**278**), obtusol (**279**), iso-obtusol (**280**), linear halogenated monoterpene (**281**)	Cytotoxic to ten cancer cell lines, B16F10 melanoma, HM02 gastric carcinoma, HEP-G2 liver carcinoma, and MCF-7 breast carcinoma cancer cells	[631–633,633]
A. dactylomela	Parguerol (**282**), parguerol-16-acetate (**283**), iso-parguerol (**284**), iso-parguerol-16-acetate (**285**), deoxyparguerol (**286**)	Cytotoxic to P388 leukemia and Ehrlich ascite carcinoma cells	[611,636,640]
A. dactylomela	Aplysqualenol A (**287**)	Cytotoxic to 60 cancer cell lines	[641,642]
A. dactylomela	Thyrsiferol (**288**)	Cytotoxic to P388 leukemia and T47D human breast tumor cells, and suppression of hypoxic induction of HIF-1 target genes	[647–649]
Aplysia depilans	Endoperoxide sterol (**289**)	Cytotoxic to human HCT-116 colorectal cancer cells	[650,651]
Aplysia fasciata	3-epi-aplykurodinone B (**145**)	Cytotoxic to mouse P388 leukemia, human A549 NSCLC, HT-29 colon cancer, and SKMEl-28 melanoma	[397]
Aplysia juliana	Pyropheophorbides a and b, julianin S	Cytotoxic	[288,432]
Aplysia kurodai	Aplyronines A (**191**), B (**290**), C (**291**), D (**292**) and H (**293**), aplaminal (**294**)	Cytotoxic to human HeLa S3 cervix carcinoma cells	[653,654,657,658]
A. kurodai	Aplaminone (**295**), neoaplaminone (**296**), neoaplaminone sulfate (**297**)	Cytotoxic to human HeLa S3 cervix carcinoma cells	[667]
A. kurodai	Aplysiaterpenoid A (**298**), aplysiapyranoids A–D (**299**)	Cytotoxic to Vero, MDCK, and B16 cells	[668,669]
A. kurodai	Kurodainol (**300**), aplysiaterpenoids A–D (**298**), aplysin-20 (**301**), iso-aplysin-20, aplysiadiol (**302**), epi-aplysin-20, ent-isoconcinndiol (**303**), aplysianin A	Induction of growth inhibitory effects in various cancer cell lines	[668–674]
A. kurodai	(-)-Aplysin (**304**), aplysinol (**305**), aplykurodins A (**306**) and B (**146**)	Cytotoxic to various cancer cell lines, human A549 NSCLC, and human glioma cells	[680,682,683]
Aplysia oculifera	Oculiferane (**308**), epi-obtusane (**309**)	Cytotoxic to PC-3 prostate, A549 NSCLC, MCF-7 breast, HepG2 liver, and HCT116 colon cancer cell lines	[684]
Aplysia punctata	Halogenated monoterpenes (**310–312**)	Cytotoxic to four tumor cell lines	[685]
Dolabella auricularia	Dolabellanin A	Antineoplastic	[435]
D. auricularia	Auripyrones A (**319**) and B, dolabelides A (**320**), B (**321**), C (**322**) and D (**323**)	Cytotoxic to human HeLa S3 cancer cells	[687,696,701,702]
D. auricularia	Doliculide (**325**)	Cytotoxic to human HeLa S3, MCF-7, and MDA-MB-231 breast cancer cells	[708]
D. auricularia	Dolastatins 3 (**192**), 10 (**326**), 11 (**327**), 12 (**328**), 13 (**329**), 14 (**330**), 15 (**331**), 16 (**346**), 17 (**347**), 18 (**348**), 19 (**332**), C (**333**), D (**334**), G (**337**) and H (**335**), iso-dolastatin H (**336**), debromoaplysiatoxin (**97**), anhydrodebromo-aplysiatoxin, aurilide (**324**), nor-dolastatin G (**338**), auristatins E (**339**), PHE (**340**), PYE (**341**), 2-AQ (**342**), 6-AQ (**343**) and PE (**344**), tasidotin (**345**)	Cytotoxic to renal, ovarian, prostate, hepatobiliary, pancreatic cancer cell lines, P388 murine leukemia, colon 26 cancer, Lewis lung carcinoma, B16 melanoma, M5076 sarcoma, human MX-1 breast cancer, LX-1, MCF-7, colon KM20L2 cancer, and SBC-3 SCLC cell lines	[46,688,694,703,705, 717–719,721,722, 724,744,750,751, 766]

Table 10. *Cont.*

Species (#)	Compounds (#)	Activity	References (#)
Stylocheilus longicauda	Aplysiatoxin (**96**), debromoaplysiatoxin (**97**), makalika (**99**), makalikone (**100**), lyngbyatoxin A (**349**), lyngbyatoxin A acetate (**101**), malyngamide B, O (**102**) and P (**103**)	Cytotoxic to P388, A549, HT29, and HTB38 cancer cell lines, and toxic to mice	[302,304,305,307, 777–779]
Bursatella leachii	Lyngbyatoxin A (**349**), debromoaplysiatoxin (**97**), hectochlorin (**351**), deacetylhectochlorin (**352**), malyndamides S (**350**) and X (**353**)	Cytotoxic to murine P388 leukemia, human A549, NSCLC, NCI-H187 (SCLC), HT-29 colon cancer, HL60 leukemia, KB, and BC breast cancer	[783–787]
Sacoglossa (3)			
Elysia subornata	Caulerpenyne (**155**)	Cytotoxic to neuroblastoma SK-N-SH cell line	[316,411,788–790]
Elysia rufescens	Kahalide F (**194**), iso-kahalalide F (**354**)	Cytotoxic to A549 and Hs683 cell lines, and breast cancer cell lines SKBR3 and BT474	[331,797,800]
Elysia ornata	Kahalides F (**194**), Z_1 (**356**), Z_2 (**355**), elisidepsin (**356**)	Cytotoxic to A549 and Hs683, breast, colon, head, neck, lung, ovary, pancreas, prostate, and melanoma cell lines	[795,797,810–812]
Pulmonata (9)			
Trimusculus peruvianus	Hydroxylated sterol (**358**)	Cytotoxic to human HCT-116 and HT29 colon cancer cell lines	[819]
Trimusculus costatus	Secosterol (**359**)	Cytotoxic to WHCO1 esophageal cancer cell line	[337]
Siphonaria capensis, S. concinna, S. cristatus, S. serrata	Siphonarienfuranon, capensinone, denticulatins	Cytotoxic	[339,346,352]
Siphonaria spp.	Siphonarienolone (**119**), diemensins A (**165**) and B, siphonarin A (**120**), vallartanones A and B	Cytotoxic	[27,134,340–344, 348,351]
Onchidium sp.	onchidin (**121**), onchidin B (**360**), onchidione (**122**), onchidiol (**361**), ilikonapyrones (**362,363**), onchidionol	Cytotoxic to murine P388 and KB oral cancer cells, and regulation of some genes related to tumor growth	[131,358,359,821]

3.2. Antibiotic Activity

Most groups of heterobranchs present compounds with antibiotic activity, the exceptions being dendronotacean and aeolidacean nudibranchs, pleurobranchoideans, cephalaspideans, and pteropods (Figure 23, Table 11) [1,2]. This is an open field for research, since bacterial strains are becoming resistant or multiresistant to known antibiotics, and new molecules are strongly needed to target them [31,830].

3.2.1. Nudibranchia

Doridacea

Both olepupuane (**14**) and polygodial (**13**), previously mentioned above, are found in different dendrodorid nudibranchs and show antifungal activity against *Saccharomyces cerevisiae* IFO 0203 and *Hansenula anomala* IFO 0136 [81,831]. Moreover, extracts of the egg masses of *Dendrodoris fumata* show antibacterial activity against *Escherichia coli*, *Staphylococcus aureus* and *Pseudomonas aeruginosa*, thus protecting embryos against bacterial infection [832].

Phyllidids, well-studied colorful nudibranchs, often contain isocyanate compounds [1,2]. These isocyanates display a wide array of activities, including antibiotic activity, and have been well studied in recent years [101–105]. One of the best-known species is *Phyllidiella pustulosa*, with compounds obtained from its dietary sponges, mainly *Acanthella cavernosa*, from different geographical localities [119–122]. *P. pustulosa* from Fiji presents axisonitrile-3 (**25**), displaying strong growth inhibition against *Mycobacterium tuberculosis* (MIC 2 μg/mL) [118,427]. Similarly, *Phyllidia picta* from Bali contains the axane sesquiterpenoids pictaisonitrile-1 (**23**) and pictaisonitrile-2, while *Phyllidia coelestis* and *P. pustulosa* from China, as well as their probable sponge prey *A. cavernosa*, present a nitrogenous cadinane-type sesquiterpenoid, xidaoisocyanate A (**24**), among other terpenoids [112,113,117]. Furthermore, *Phyllidia varicosa* presents two 9-thiocyanatopupukeanane sesquiterpenes (**126**), sequestered from the sponge *Axinyssa aculeata* [110]. Altogether, 9-isocyanopupukeanane (**21**) and *epi*-9-isocyanopupukeanane, as well as their thiocyano derivatives, are moderately antibacterial against *Bacillus subtilis* and antifungal against *Candida albicans*, and they were also further isolated from *Phyllidiella rosans* (*P. bourguini*) [110,373].

Doriprismatica (Glossodoris) atromarginata from Australia, Sri Lanka, and India, possesses different compounds from its demosponge preys (*Spongia (Hyatella)* sp., *Hyrtios* spp., and *Hyattella cribriformis*), two scalarane sesterterpenes in their MDFs, some pentacyclic scalaranes, and heteronemin (**233**) [22,92,166,176,180,380–382]. Heteronemin (**233**), a scalarin-skeleton sesterterpene first reported in the sponge *Heteronema erecta*, displays antibacterial activity towards *M. tuberculosis* H$_{37}$Rv, with a MIC of 6.25 µg/mL [85,180,833]. Other *Glossodoris* species from Australia, *G. hikuerensis* and *G. vespa*, also contain heteronemin (**233**) in their viscera, along with scalaradial (**44**), 12-deacetoxy-12-oxoscalaradial (**43**), and 12-deacetoxy-12-oxo-deoxoscalarin (**136**) in their mantle [178]. In particular, 12-*epi*-scalaradial of *G. hikuerensis* and *G. cincta* shows antimicrobial activity at 10 µg/disk against *S. aureus*, *B. subtilis*, and *C. albicans* [166].

Some compounds described in *Chromodoris willani* collected in Okinawa contain two sesterterpenes, deoximanoalide (**364**) and deoxysecomanoalide (**365**), biotransformed from manoalide and secomanoalide, respectively, possessing antimicrobial activity against *E. coli* and *B. subtilis*, and showing inhibition of snake venom phospholipase A2 at 0.2–0.5 µg [159]. On the other hand, nakafuran-8 (**54**) and nakafuran-9 (**51**), found in several chromodorid species as reported above, were tested against *E. coli*, *S. aureus*, *P. aeriginosa* and *B. subtilis* in a disk diffusion assay, displaying no antibacterial activity [157,165]. In contrast, in *Chromodoris petechialis*, puupehenone (**218**) shows a MIC of 3 µg/mL against *C. albicans* [158].

Hawaiian specimens of the bright red "Spanish dancer" nudibranch, *Hexabranchus sanguineus*, yielded several bioactive macrolides (see above) [222]. Specimens of *H. sanguineus* from Fiji additionally yielded two thiazole cyclic peptides, sanguinamide A (**64**) and B (**366**) [219]. Both sanguinamide A (**64**) and B (**366**) were found at very low concentrations, 0.0023 and 0.011% dry weight, respectively [219]. Sanguinamide B (**366**) showed a moderate antibacterial activity against *P. aeruginosa*, reducing twitching motility [834,835]. *H. sanguineus* specimens from different locations showed different compounds, most of them probably from dietary sponges (*Halichondria*, *Axinella*, and *Dysidea*) [22,219,545,836–838]. In Indonesia, Hawai'i and Japan, *H. sanguineus* presents two trisoxazole macrolides, ulapualide A (**190**) and B (**236**), that inhibit the growth of *C. albicans*, while halichondramides (**244**) and kabiramides A–E (**62,238–241,243**) inhibited several fungi [218,222,546,547,836].

The Indian *Jorunna funebris* possesses, among other related metabolites, the isoquinoline alkaloid jorumycin (**189**), which is very similar to renieramycin E from the sponge *Reniera* sp. [166,553,839]. Jorumycin (**189**) presents antimicrobial activity against *Bacillus subtilis* and *Staphylococcus aureus* [557]. Jorumycin (**189**) was found in the mucus secretion, and, thus, a defensive role was proposed for it [166]. *J. funebris* and its prey, *Xestospongia* spp., present several isoquinolinequinones and bistetrahydroisoquinolines, among which some isoquinolinequinones also display antibacterial activity [553,560,564,565,840].

Although the metabolites have not yet been described, the organic extract of *Halgerda stricklandi* displays modest activity against *Staphylococcus aureus*, but no activity against a range of several other bacteria and fungi [840]. *Halgerda aurantiomaculata* contains a tryptophane derivative called zooanemonin (**367**), previously isolated from several sponges and the sea anemone *Anemonia sulcata*, which is also reported to show antibacterial activity [841].

Finally, within the Nembrothidae, the previously mentioned alkaloids tambjamines (**65–70,249**) also possess antimicrobial activity from their diet [223]. Particularly, their blue tetrapyrrol (**72**), presumably derived from a diet of ascidians, is active against *B. subtilis* at 5 µg/disc [157].

Euarminida

Chemical studies of the South African *Leminda millecra* described some sesquiterpenes from dietary origin, including millecrones A and B (**368,369**) and millecrols A and B (**370,371**) [588,589]. Millecrone A (**368**), originating from the soft coral *Alcyonium fauri*, inhibited the growth of *Candida albicans* at 50 µg/disk, while millecrone B (**369**) from the gorgonian *Leptogorgia palma* was active against both *Staphylococcus aureus* and *Bacillus subtilis* at 50 µg/disk [21,589]. In contrast, millecrol B (**371**) was only active against *B. subtilis* at 50 µg/disk [21].

Furthermore, *Dermatobranchus otome* from Japan presents the germacrane sesquiterpenoids DO1 (**372**), DO2 (**373**), and DO3 (**374**), displaying antibacterial activity against *B. subtilis* [842].

Extracts of *Armina babai* also display antimicrobial activity against *Pseudomonas* sp. and *Proteus mirabilis*, although the compounds have not yet been described [843]. *A. babai* from India possesses a ceramide also found in the gorgonian *Acabaria undulata* [842,844].

3.2.2. Tylodinoidea

As mentioned above, the Australian *Tylodina corticalis* selectively accumulates some bromotyrosine-derived alkaloids from the sponge *Pseudoceratina purpurea*, which contains a larger variety of these compounds [258]. Among them, hexadellin (**375**) and aplysamine 2 (**376**) display mild antibiotic activity against *E. coli* and *S. aureus* at concentrations of 125–250 µg/mL [262].

3.2.3. Anaspidea

Aplysia, as mentioned above, is one of the most studied genera, with many NPs displaying a wide range of activities around the world [1,2]. In particular, the brominated diterpenes, glandulaurencianols A–C (**162,163**), as well as punctatol (**164**) from *Aplysia punctata*, probably from the red algae *Laurencia glandulifera*, possess the laurencianol skeleton, a known antibacterial diterpene from *Laurencia glandulifera* that is active against *Escherichia coli* and *Bacillus subtilis* [429–431]. Moreover, a purple secretion of the sea hare *A. juliana* contains julianin-S, an antibacterial peptide suggested to protect its egg masses from microbial infections, together with some unsaturated fatty acids [288,434]. Similarly, aplysianin E from *A. kurodai* eggs shows antifungal activity against *C. albicans* at IC_{50} >16 µg/mL [672–674].

As mentioned above, *Dolabella auricularia*, possesses the glycoprotein dolabellanin A, probably de novo biosynthesized [435]. Besides its antineoplastic activity, dolabellanin A also shows antibacterial activity against *E. coli*, which may protect the egg masses from bacterial pathogens [435].

Finally, bursatellin (**105**), a diol nitrile alkaloid found in *Bursatella leachii plei* from Puerto Rico, is structurally related to the well-known antibiotic chloramphenicol [311]. In the Mediterranean, both the + and – isomers of bursatellin (**105**) are found in the external extracts of *B. leachii leachii* and *B. leachii savignyana* [312].

3.2.4. Sacoglossa

In this group, the previously mentioned cyclodepsipeptides kahalalides A (**377**) and F (**194**) from *Elysia rufescens* and its algal food, *Bryopsis* sp., are active against mycobacteria, with kahalalide A (**377**) inhibiting *M. tuberculosis* H37Rv by 83% at 12.5 µg/mL, and kahalalide F (**194**) by 67% at 12.5 µg/mL [331,411,794,797]. Kahalalide F (**194**) also inhibited *Mycobacterium intracellulare* at a MIC of 25 µg/mL [331]. Kahalalides (**194,354–356**) are also found in *E. rufescens*, *E. ornata* and *E. grandifolia*, and their algal diet *Bryopsis pennata* [794,795], with kahalalide F (**194**) always being the most active compound [794,795]. Kahalalide F (**194**) is found in a mixture together with its isomer *iso*-kahalalide F (**354**), which also shows relevant bioactivities [331,806]. Both compounds have been suggested to originate from bacterial symbionts, although more research is needed to prove this [808].

Chlorodesmin (**114**), from the Australian *Cyerce nigricans*, is a diterpenoid previously known from the green algae *Chlorodesmis fastigiata* with antibacterial and antifungal activity [845,846].

3.2.5. Pulmonata

Several species of *Siphonaria* (*S. capensis*, *S. concinna*, *S. cristatus*, and *S. serrata*) possess different types of polypropionates in their mantle and mucous secretion, affecting gramm + bacteria [338]. Species from Australia, West and East Atlantic, and South Africa displayed antimicrobial activity due to acyclic compounds with a 2-pyrone and furanone rings (type I), such as siphonarienolone (**119**), which were similar to the polypropionates of cephalaspideans [340–346]. On the other hand, polypropionates with a profuse polyoxygenated network that frequently cyclizes (Type II), such as siphonarin A (**120**), similar to those of actinomycetes, are found in *Siphonaria* species from Australia, New Zealand, North-East Pacific, Pacific Islands, and South Africa [347–352]. *S. diemenensis* and

S. pectinata present diemenensin-A (**165**) and pectinatone (**166**) [340,341,343]. Diemenensin-A (**165**) inhibited *S. aureus* and *B. subtilis* at 1 µg/disc and 5 µg/disc, respectively, while pectinatone (**166**) inhibited *S. aureus*, *B. subtilus*, *C. albicans*, and *S. cerevisiae* [341,343].

Table 11. Number of antibiotic compounds in the different heterobranch groups. In brackets: number of species with antibiotic compounds, number of the compounds in figures, and reference numbers. # Number.

Species (#)	Compounds (#)	Activity	References (#)
Nudibranchia (21)			
Phyllidiella pustulosa	Axisonitrile-3 (**25**)	Antimycobacterial activity against *Mycobacterium tuberculosis*	[427]
P. pustulosa, P. coelestis	Xidaoisocyanate A (**24**)	Antibiotic activity	[117]
Phyllidia picta	Pictaisonitrile 1 (**23**) and 2	Antibiotic activity	[112]
Phyllidia varicosa	9-Thiocyanatopupukeanane (**126**)	Antibiotic activity	[247]
Phyllidiella rosans	9-Isocyanopupukeanane (**21**), epi-9-isocyanopupukeanane	Antibacterial activity against *Bacillus subtilis* and *Candida albicans*	[110]
Doriorismatica (Glossodoris) atromarginata	Scalaranes, heteronemin (**233**)	Antimycobacterial activity against *M. tuberculosis* H$_{37}$Rv	[833]
Glossodoris hikuerensis, G. vespa, G. cincta	Heteronemin (**233**), scalaradial (**44**), 12-deacetoxy-12-oxoscalaradial (**43**), 12-deacetoxy-12-oxo-deoxoscalarin (**136**), 12-epi-scalaradial	Antibiotic activity	[178]
Felimida (Chromodoris) macfarlandi	Macfarlandines D and E (**235**)	Antibacterial activity against *B. subtilis* in the disk assay system at 10 gg per disk, and activity against *Vibrio anguillarum* and *Beneckea harveyi* at 100 gg per disk	[139,140,149,543]
Chromodoris willani	Deoximanoalide (**364**), deoxysecomanoalide (**365**)	Antimicrobial activity against *Escherichia coli* and *B. subtilis*, and inhibitor of snake venom phospholipase A2	[159]
Chromodoris spp.	Nakafuran-8 (**54**), nakafuran-9 (**51**), puupehenone (**218**)	Antibacterial activity against *E. coli*, *Staphylococcus aureus*, *Pseudomonas aeruginosa*, *B. subtilis*, and antifungal activity against *C. albicans*	[157,158,165]
Hexabranchus sanguineus	Kabiramides A–E (**238, 239, 62, 240, 241**), sanguinamides A (**64**), B (**366**), halichondriamides (**244**), ulapualides A (**190**) and B (**236**)	Antibacterial activity against *P. aeruginosa*, and antifungal activity against *C. albicans*	[208,219,221,222,545,834–837]
Jorunna funebris	Jorumycin (**189**), jorunnamycins A–C (**245**)	Antimicrobial activity against *B. subtilis* and *S. aureus*	[166,553,560]
Halgerda aurantiomaculata	Zooanemonin (**367**)	Antibacterial	[841]
Roboastra tigris, Tambja abdere, T. eliora	Tambjamines (**65–70, 249**), tetrapyrrole (**72**)	Antibacterial activity against *B. subtilis*	[223]
Leminda millecra	Millecrones A (**368**) and B (**369**), millecrols A (**370**) and B (**371**)	Antibiotic activity against *C. albicans*, *S. aureus* and *B. subtilis*	[588]
Dermatobranchus otome	DO1 (**372**), DO2 (**373**), DO3 (**374**)	Antibacterial activity against *B. subtilis*	[842]
Armina babai	Extracts	Antibacterial activity against *Pseudomonas* sp. and *Proteus mirabilis*	[843]
Tylodinoidea (1)			
Tylodina corticalis	Hexadellin (**375**), aplysamine 2 (**376**)	Antibacterial activity against *E. coli* and *S. aureus*	[262]
Anaspidea (5)			
Aplysia punctata	Glandulaurencianols A–C (**162,163**), punctatol (**164**)	Antibacterial activity against *B. subtilis* and *E. coli*	[429–431]
Aplysia juliana	Julianin-S	Antibacterial activity	[288]
Aplysia kurodai	Aplysianin E	Antifungal activity against *C. albicans*	[672–674]
Dolabella auricularia	Dolabellanin A	Antibacterial activity against *E. coli*	[435]
Bursatella leachii plei, B. leachii savignyana	Bursatellin (**105**)	Antibiotic activity	[311,312]
Sacoglossa (4)			
Elysia rufescens	Kahalalides A (**377**) and F (**194**), iso-kahalalide F (**354**)	Antimycobacterial activity against *Mycobacterium tuberculosis* and *M. intracellulare*	[794]
Elysia ornata, E. grandifolia	Kahalalide F (**194**)	Antimycobacterial activity against *M. tuberculosis* and *M. intracellulare*	[331,411,797]
Cyerce nigricans	Chlorodesmin (**114**)	Antibacterial and antifungal activity	[845,846]
Pulmonata (7)			
Siphonaria australis, S. diemensis, S. capensis, S. concinna, S. cristatus, S. serrata, S. pectinata	Siphonarienolone (**119**), siphonarin A (**120**), pectinatone (**166**)	Antimicrobial activity	[340–343,346]

3.3. Antiparasitic Activity

Currently, another important need is to identify antiparasitic compounds, although the antiparasitic activities included here mainly comprise compounds related to antiplasmodial effects. Within heterobranchs, antimalarial compounds have been described in several doridacean nudibranchs, while only one species of sacoglossa has been cited to possess antileishmanial activity [1,2] (Figures 23 and 24, Table 12).

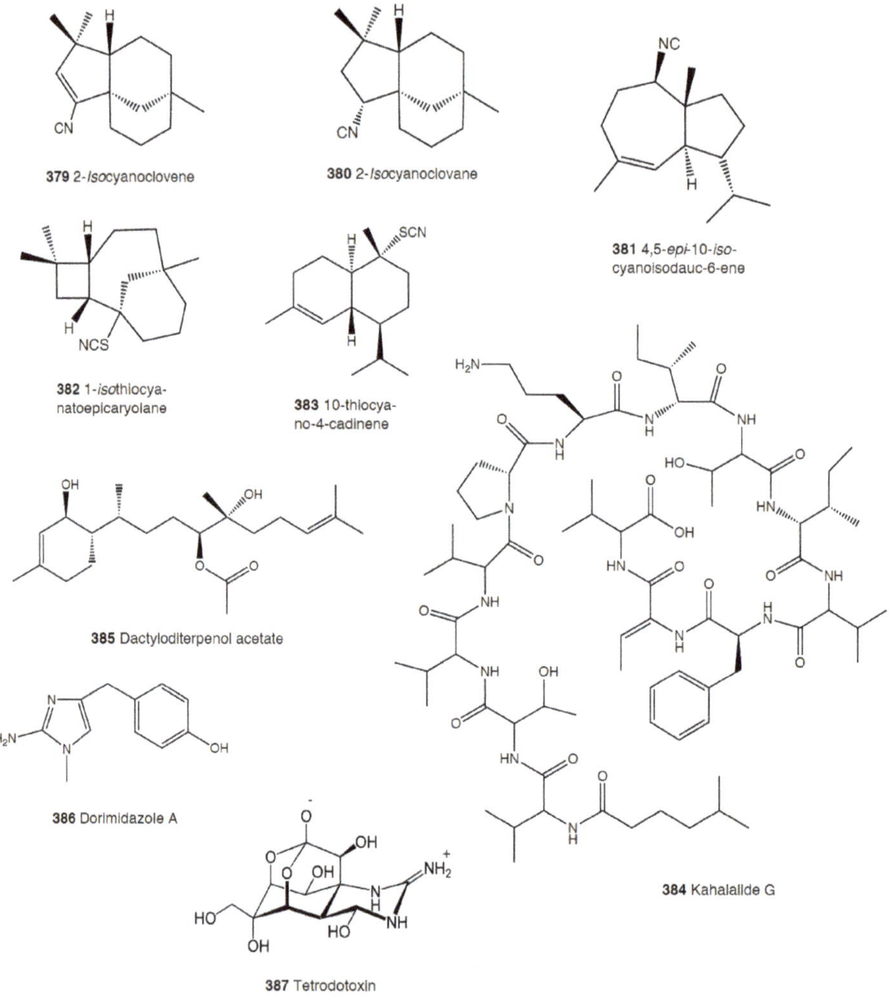

Figure 24. Structures of selected antiparasitic compounds and antivirals from Doridacea and Sacoglossa, anti-inflammatory compounds from Anaspidea, and compounds with other pharmacological activities in heterobranch molluscs. These molecules may also display other activities, as reported in the text.

3.3.1. Nudibranchia

Doridacea

The doridacean nudibranch *Notodoris gardineri* from the Philippines presents the imidazole alkaloids *iso*-naamidine-A (**160**) and dorimidazole-A (**386**), the latter exhibiting anthelminthic activity against the nematode parasite *Nippostrongylus brasiliensis* at 50 µg/mL [424,426]. Moreover, *Chromodoris lochi* from Vanuatu contains the PKS-NRPS-derived mycothiazole (**129**) which is described to possess anthelminthic activity against the nematode parasite *N. brasiliensis* at 50 µg/mL [378]. Mycothiazole (**129**) has also been found in the prey sponge of the slug, the sponge *Spongia (Cacospongia) mycofijiensis* [379].

Among the numerous nitrogenated compounds reported from phyllidid nudibranchs, mostly obtained from their sponge prey, several have been found to have antiplasmodial activity [44,847]. Axisonitrile-3 (**25**) shows an IC_{50} towards *Plasmodium falciparum* of 16.5 ng/mL for chloroquine-resistant strain W2 and no associated cytotoxicity [848]. Axisonitrile-3 (**25**) has been reported to interfere with the detoxification of heme, a degradation product of hemoglobin digestion within infected erythrocytes, and to form a binary complex with iron in protoporphyrin IX, producing heme accumulation, which results in toxicity to the malaria parasite [849].

Table 12. Antiparasitic compounds in the different heterobranch groups. In brackets: number of species with antiparasitic compounds, number of the compounds in figures, and reference numbers. # Number.

Species (#)	Compounds (#)	Activity	References (#)
Nudibranchia (4)			
Phyllidiella pustulosa	Axisonitrile-3 (**25**), pustulosaisonitrile-1 (**378**), 10-thiocyano-4-cadinene (**383**)	Activity against *Plasmodium falciparum*	[118,123,848–850]
Phyllidia ocellata	2-*Iso*cyanoclovene (**379**), 2-*iso*cyanoclovane (**380**), 4,5-*epi*-10-isocyanoisodauc-6-ene (**381**), 1-*iso*thiocyanatoepicaryolane (**382**)	Activity against *Plasmodium falciparum*	[376]
Notodoris gardineri	Iso-naamidine-A (**160**), dorimidazole A (**386**)	Anthelminthic activity	[424,426]
Chromodoris lochi	Mycothiazole (**129**)	Anthelminthic and toxic activity	[378]
Sacoglossa (3)			
Elysia rufescens, E. ornata, E. grandifolia	Kahalalides (**194,354–356,377**)	Antileishmanial activity	[794–797]

Pustulosaisonitrile-1 (**378**) from *Phyllidiella pustulosa* from Australia presents moderate levels of in vitro antimalarial activity [850]. Among the diverse nitrogenous mono-, bi- and tri-cyclic sesquiterpenes found in *Phyllidia ocelata* and *Phyllidiella pustulosa* from different geographical locations, several are reported to possess antimalarial activity against *Plasmodium falciparum* [102,118–120,122,375–377]. In particular, in *P. ocellata* from Australia, 2-*iso*cyanoclovene (**379**), its dihydro analogue 2-*iso*cyanoclovane (**380**) and 4,5-*epi*-10-isocyanoisodauc-6-ene (**381**) present IC_{50} values of 0.26–0.30 µM, while 1-*iso*thiocyanatoepicaryolane (**382**) has an IC_{50} > 10 µM [376]. In *P. pustulosa* from Fiji, 10-thiocyano-4-cadinene (**383**) shows moderate antiplasmodial activity [118,123].

3.3.2. Sacoglossa

The depsipeptides kahalalides (**194,354–356,377**) found in different *Elysia* species, such as *E. rufescens, E. ornata, E. grandifolia*, and their algal food *Bryopsis pennata*, have been reported to possess antileishmanial properties [794–797]. Kahalalides (**194,354–356,377**) are active against *Leishmania* spp. At micromolar ranges, and their lethality is linked to the alteration of the plasmatic membrane of the protozoan [794–797].

3.4. Antiviral Activity

Marine organisms are considered an underexplored source of antiviral compounds [851–853]. Many of the drugs currently employed produce strong side effects and develop resistances [853]. Viral diseases cause a huge number of deaths annually; for example, human immunodeficiency virus (HIV) is one of the top five most deadly diseases worldwide [853]. Furthermore, new viruses are appearing with extreme virulence, such as COVID-19, with no known treatment to date [854]. Therefore, the need for new antiviral drugs is clear, and heterobranchs, with their amazing biodiversity and chemodiversity, may perhaps contribute to this. To date, only doridacean nudibranchs, sea hares, and sacoglossans have been reported to possess antiviral compounds (Figure 24, Table 13).

Table 13. Antiviral compounds in the different heterobranch groups. In brackets: number of species with antiviral compounds, number of the compounds in figures, and reference numbers. # Number.

Species (#)	Compounds (#)	Activity	References (#)
Nudibranchia (7)			
Cadlina luteomarginata	Ansellone A (**216**)	Activation of the latent proviral HIV-1 gene expression	[855]
Chromodoris mandapamensis, Glossodoris cincta	Spongiadiol (**35**), *epi*-spongiadiol (**232**)	Activity against herpes simplex virus, type 1 (HSV-1) and P388 murine leukemia cells	[166,535]
Chromodoris hamiltoni	Latrunculins A (**38**) and B (**37**)	Activity against HIV-1	[153,155]
Chromodoris africana, C. quadricolor	Latrunculin B (**37**)	Activity against HIV-1	[155,853]
Chromodoris petechialis	Puupehenone (**218**)	Anti-HIV-1	[797]
Anaspidea (1)			
Dolabella auricularia	Dolastatin 3 (**192**)	Activity against HIV life cycle	[718,728,853]
Sacoglossa (3)			
Elysia rufescens, E. grandifolia, E. ornata	Kahalalide F (**194**), *iso*-kahalalide F (**354**)	Activity against herpes simplex virus II	[331,794,795,797,853]

3.4.1. Nudibranchia

Doridacea

The chromodoridid *Cadlina luteomarginata* presents compounds with the tricyclic ansellane carbon skeleton, among other compounds, obtained from its sponge prey *Phorbas* sp. [513]. Among them, ansellone A (**216**) from the sponge was tested for the "shock and kill" approach to a sterilizing HIV-1 cure [855]. Ansellone A (**216**), together with other sponge compounds, was found to activate the latent proviral HIV-1 gene expression, as well as to possess LRA profiles comparable to prostratin, which is in phase I as a potential HIV treatment [855].

Several chromodorid species contain spongiadiol (**35**) from the sponges they feed on, among other spongian diterpenes [22,166,380,381,535,536]. This is the case of *Chromodoris mandapamensis* and *Glossodoris cincta* (*G. atromarginata*) specimens from different localities, which present these compounds in their mantle and digestive gland [22,166,380,381,535,536]. Spongiadiol (**35**) is active against the herpes simplex virus, showing an IC_{50} of 0.25 µg/mL against herpes simplex virus type I [535,537]. *Chromodoris hamiltoni* presents a wide arsenal of chemicals, among which the 2-thiazolidinone macrolides latrunculins A (**38**) and B (**37**) from its diet of sponges are found at different localities [154,156]. Latrunculin B (**37**) is also present in *C. africana* and *C. quadricolor* [170,171]. Latrunculin B (**37**), a very active compound, as previously mentioned, was reported in the sponge *Latrunculia magnifica* [168–170]. The EC_{50} of latrunculin B (**37**) against HIV-1 is 16.4 µM, thus showing a moderate activity while being non-cytotoxic [853]. *Doriprismatica (Glossodoris) atromarginata* also presents furanoditerpenoids and scalarane sesterterpenes originating from its dietary sponges *Spongia (Hyatella)* sp. and *Hyrtios* spp., depending on the geographical location (Australia, Sri Lanka, and India) [92,175,180,381,382]. Some of these compounds are reported as antivirals, particularly spongiadiol (**35**) and *epi*-spongiadiol (**232**) [180,383–386,519,535,539]. Puupehenone (**218**) from *Chromodoris petechialis* is also active against HIV-1 [797].

3.4.2. Anaspidea

Dolabella auricularia presents some diet-derived cyclic depsipeptides reported to exhibit a wide range of activity against different stages of the HIV life cycle [718]. Dolastatin 3 (**192**) from *D. auricularia* was further isolated from the circumtropical cyanobacterium *Lyngbya majuscula* from Palau [718,728]. Dolastatin-3 (**192**) inhibits HIV-1 integrase at relatively high concentrations, with EC_{50} of 5 mM for the terminal cleavage and 4.1 mM for the strand-transfer reactions [728]. The activity of dolastatin-3 (**192**) was lost after some time in the laboratory, since it was a difficult-to-handle molecule and also presented some cytotoxicity; therefore, it was not taken forward for further investigation [728].

3.4.3. Sacoglossa

In this group, the previously mentioned depsipeptides kahalalides (**194,354–356,377**) are found in *Elysia rufescens*, *E. grandifolia*, and *E. ornata*, as well as in the green algae *Bryopsis pennata in* their diet [47,49]. Among them, kahalalide F (**194**) is the most bioactive compound, although all kahalalides possess many activities as reported above. Kahalalide F (**194**) presents antiviral properties against herpes simplex virus II, while kahalalides A (**377**) and G (**384**) are inactive [331,794–797]. Kahalalide F (**194**) is found along with its isomer, *iso*-kahalalide F (**354**), both displaying a wide array of bioactivities and suggested to be of bacterial origin [331,806,808]. Kahalalide F (**194**) also exhibits moderate activity against HIV-1, with an EC_{50} of 14.2 µM, while it is not cytotoxic against human peripheral blood mononuclear (PBM) cells [853]. Kahalalides A (**377**) and G (**384**), contrastingly, are not active against HIV-1 [853].

3.5. Anti-Inflammatory Activity

Only a few nudibranchs and some sea hares are known to possess anti-inflammatory compounds, while there have been no studies to date regarding the remaining groups (Figure 24, Table 14).

Table 14. Anti-inflammatory compounds in the different heterobranch groups. In brackets: number of species with anti-inflammatory compounds, number of the compounds in figures, and reference numbers. # Number.

Species (#)	Compounds (#)	Activity	References (#)
Nudibranchia (11)			
Glossodoris rufomarginata, G. pallida, G. vespa, G. averni, G. hikuerensis, G. atromarginata, G. cincta	Scalaradial (**44**)	Potent inhibition of PLA_2, and potent anti-inflammatory activity	[175,177,381,383,856–858]
Goniobranchus splendidus	Gracilins (**224–228**)	Cyclosporine A mimics, BACE1 and ERK inhibition	[190,533,534]
Tethys fimbria, Melibe viridis	Prostaglandin E-1,15-lactones (**80, 81**)	Reduction of inflammation after autotomy and tissue regeneration	[77,240]
Tritonia sp.	Punaglandins (**250**)	Anti-inflammatory activity	[587]
Anaspidea (3)			
Aplysia depilans	Carotenoids, polyunsaturated fatty acids	Anti-inflammatory activity	[859]
Aplysia dactylomela	Dactyloditerpenol acetate (**385**)	Anti-neuroinflammatory activity	[860,861]
Bursatella leachii	Malyngamide S (**350**)	Anti-inflammatory activity	[786]

3.5.1. Nudibranchia

Doridacea

As reported above, several species of *Glossodoris* present scalaradial (**44**) and other scalarane compounds derived from the sponges they feed on [22,166,175,180,381,383,856]. These include *Glossodoris rufomarginata*, *G. pallida*, *G. vespa*, *G. averni*, *G. hikuerensis*, *G. atromarginata*, and *G. cincta* from different geographical locations [22,166,175,178,380,381,383,536,856]. Scalaradial (**44**) is a potent anti-inflammatory compound [856], but it is also toxic to slugs, and, thus, after feeding, they quickly transform scalaradial (**44**) into its 12-deacetyl derivative or other related scalaranes in a detoxification process, locating them in MDFs in their mantle rims [2,4,22,166,176,177,383]. Scalaradial (**44**) was first found in the Mediterranean sponge *Cacospongia mollior* [857] and has been reported to display a potent inhibition of PLA$_2$ [858]. Similarly, *Goniobranchus* species usually present spongian cyclic diterpenes, which are often cytotoxic, as reported above, and obtained from their diet of *Spongionella* sponges [154,190,533]. Among these compounds, *G. splendidus* contains gracilins (**224–228**), some of which have been tested from the sponge and possess a high anti-inflammatory potential, such as cyclosporine A mimics and as BACE1 and ERK inhibitors [534].

Dendronotida

The invasive species *Melibe viridis* presents a prostaglandin lactone in its mucus and cerata which had been previously reported in *Tethys fimbria* [77,240]. In fact, *T. fimbria*, presents a wide array of de novo biosynthesized prostaglandins (**80,81**) with different roles, which may include reducing inflammation in their tissues after autotomy and tissue regeneration [240].

Moreover, punaglandins (**250**) from *Tritonia* sp. show anti-inflammatory activity, and a synthetic 10-thiomethyl derivative enhances in vivo mineralization in human osteoblasts [587].

3.5.2. Anaspidea

Several species of sea hares have been studied to date for anti-inflammatory activity. *Aplysia depilans* presents 8 carotenoids and 22 polyunsaturated fatty acids obtained from their algal diet and found in the digestive gland which possess anti-inflammatory activity [859]. *Aplysia dactylomela* possesses dactyloditerpenol acetate (**385**); this is probably derived from laurenditerpenol from *Laurencia intricata*, which is reported to have a significant in vitro anti-neuroinflammatory activity [860,861].

Bursatella leachii feeds on cyanobacteria and accumulates its natural products usually in its digestive gland, using them for its own defense as previously discussed [311,312,783,784]. Among these compounds, the alkaloid malyngamide S (**350**) from New Zealand specimens presents anti-inflammatory properties [786].

3.6. Against Neurodegenerative Diseases

Activity against neurodegenerative diseases has been described for several marine natural compounds [862]. In heterobranch molluscs, compounds from several species have been tested, providing some interesting results (Figure 24, Table 15). The doridacean nudibranch *Polycera atra* feeds on the bryozoan *Bugula neritina*, accumulating the polyketide macrolides bryostatins (**203**) and transferring them to its spawn, as mentioned above [492–494]. Bryostatins (**203**) have been further traced to the symbiont Candidatus *Endobugula sertula*, where the biosynthetic genes have been described [495]. Among them, bryostatin 1 (**203**) is the most studied molecule as a potential treatment for many diseases, including cancer and Alzheimer disease (AD), and it is in phase I trials for AD [495,497].

The doridacean nudibranchs *Goniobranchus obsoletus* and *G. splendidus* from Australia possess many cyclic diterpenes of the spongian type, including gracilins (**224–228**) which are accumulated from feeding on *Spongiella* sponges [154,190,533]. Gracilins (**224–228**), as previously mentioned, possess several interesting properties as drug candidates and also show a potential role against neurodegenerative diseases, such as AD, which is also being tested [534,863].

The chromodoridid *Cadlina luteomarginata* obtains ansellone A (**216**) from its diet of the sponge *Phorbas* sp. [513]. Ansellone A (**216**) shows cAMP activation (EC_{50} = 14 mM) comparable to that of forskolin in the HEK293 cell-based assay [513]. This activity is very useful in stem cell techniques, because modulating the cAMP signaling pathway is crucial for treating many diseases, such as cancer and heart failure, as well as neurodegenerative diseases [513].

In cephalaspideans, the cylichnidae *Scaphander lignarius* lives in soft bottom, muddy areas, usually feeding on foraminiferans [229,231]. As previously mentioned, *S. lignarius* specimens from the Mediterranean and East Atlantic present the so-called lignarenones (**171**), which are de novo biosynthesized and secreted in the Blochmann's gland [245]. These compounds are suggested to be used as alarm pheromones, similarly to other cephalaspidean species mentioned above. Interestingly, recent studies suggest that lignarenone B (**171**) could also be used as a possible therapeutic candidate for the treatment of GSK3β-involved pathologies, such as AD [864]. In silico binding studies revealed that lignarenone B (**171**) can act over the ATP and/or substrate binding regions of GSK3β [864]. The predicted inhibitory potential of lignarenone B (**171**) was experimentally validated by an in vitro assay showing a ~50% increase in Ser9 phosphorylation levels of GSK3β, while it also potentiates structural neuronal plasticity in vitro using neuronal primary cultures [865]. Future studies are aimed to test lignarenones in preclinical mouse models of AD.

Table 15. Compounds used against neurodegenerative diseases in the different heterobranch groups. In brackets: number of species with these compounds, number of the compounds in figures, and reference numbers. # Number.

Species (#)	Compounds (#)	Activity	References (#)
Nudibranchia (4)			
Polycera atra	Bryostatin 1 (**203**)	Alzheimer disease (AD)	[492–494,866]
Goniobranchus obsoletus, G. splendidus	Gracilins (**224–228**)	Potential against neurodegenerative diseases	[533,534,863]
Cadlina luteomarginata	Ansellone A (**216**)	cAMP activation (neurodegenerative diseases)	[513]
Cephalaspidea (1)			
Scaphander lignarius	Lignarenone B (**171**)	Alzheimer disease (AD)	[864,867]

3.7. Other Pharmacological Activities

Other activities that were not included in the previous sections comprise those of a couple of nudibranch species and a pleurobranchoidea (Figure 24, Table 16). No other activities have been described in the remaining groups.

Janolusimide (**138**) is a tripeptide described in the Mediterranean euarminid nudibranch *Janolus cristatus* [388]. Janolusimide (**138**) is toxic to mice (LD 5 mg/kg) and affects acetylcholine receptors, thus having a neurotoxic action at lower concentrations [388]. A N-methyl analogue, janolusimide B, was further described in the New Zealand bryozoan *Bugula flabellata* [390], suggesting a dietary origin for janolusimide (**138**), since *J. cristatus* has been reported to feed on bryozoans, including *B. flabellata* [390].

The pleurobranchoid genus *Pleurobranchaea* is often used as a model for neurobiology investigations because of its peculiar escape swimming behavior, which is achieved by alternating dorsal and ventral body flexions [868]. Furthermore, it is also interesting because *P. maculata* from New Zealand possesses tetrodotoxin (TTX) (**387**). TTX (**387**) is found in its adult tissues, gonads, and egg masses, thus suggesting a defensive role [869,870]. TTX (**387**) is a very potent neurotoxin that inhibits action potential in nerve cells, and it has been found in many poisonous animals, such as flatworms, arrow worms, ribbon worms, snails, blue-ringed octopus, xanthid crabs, sea stars, fish, and toads [871,872]. In some of these cases, it has been demonstrated that TTX (**387**) is produced by symbiotic bacteria from the *Pseudoalteromonas*, *Pseudomonas*, *Vibrio*, and other strains, and that it is bioaccumulated along the food chain. However, the bacterial origin of TTX (**387**) has not been proved in all cases [871,872].

Table 16. Other pharmacological activities in compounds from different heterobranch groups. In brackets: number of species with active compounds, number of the compounds in figures, and reference numbers. # Number.

Species (#)	Compounds (#)	Activity	References (#)
	Nudibranchia (1)		
Janolus cristatus	Janolusimide (**138**)	Toxic to mice	[388,390]
	Pleurobranchoidea (1)		
Pleurobranchaea maculata	Tetrodotoxin (TTX) (**387**)	Neurotoxin	[869,870]

4. Concluding Remarks

Despite the fact that only a small proportion of heterobranch molluscs has been reported to date, they represent a particularly rich group of natural products. Their NPs display an astonishing variation in bioactivity, both ecological and pharmacological, reflecting the huge chemodiversity they possess (Figure 25). Biodiversity and chemodiversity correlate here to offer a huge amount of bioactive molecules in these peculiar group of molluscs. Heterobranchs indeed comprise a very diverse group of organisms that present almost all classes of natural products described to date, but not all of these NPs have been tested for potential bioactivities [1,2,28]. Thus, many more studies are expected to find not only new NPs but also their potential bioactivities.

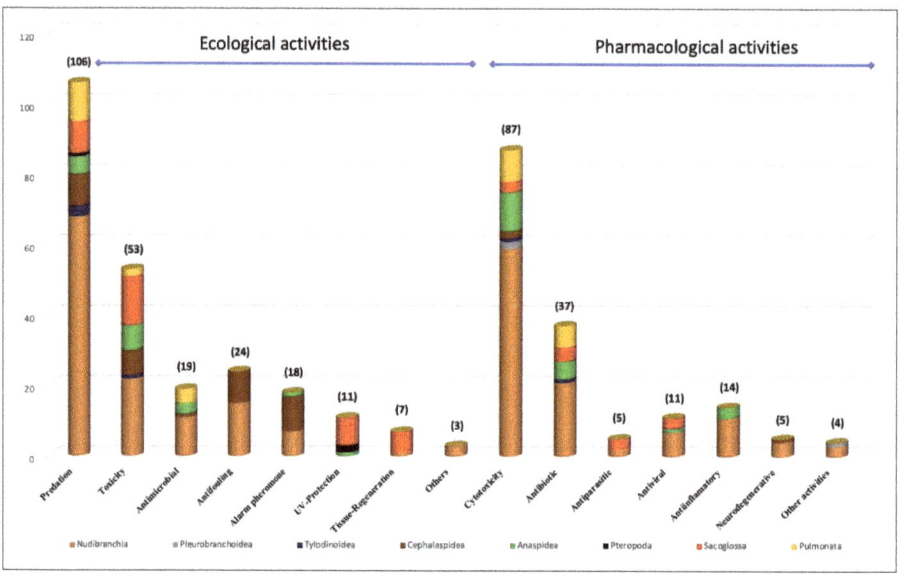

Figure 25. Number of species with bioactive compounds in the different heterobranch groups.

Regarding ecological activity, in fact, very few compounds from the total NPs described have been tested, as described above, and this keeps the door open for many other ecological interactions to be identified in the future. NPs from heterobranchs have been shown to be ecologically relevant, as in the case of those from other marine organisms, although many experiments should still be performed to complete the available information. Some molecules are shown to display multiple roles, as is the case in other marine invertebrates, while similar structures are shown to display similar roles in geographically distant localities by phylogenetically related species [1,2]. The most studied activity is feeding deterrence, followed by toxicity (Figure 25), although the NPs are not usually tested against sympatric species, raising doubts about their real effects in the habitat where the molluscs live. Reliable field data are scarce, and, therefore, the ecological significance of many compounds remains to be demonstrated.

The pharmacological activity of heterobranch NPs is still underexplored because we are only aware of a small part of their chemical arsenal. However, their potential is obvious from the molecules reported above, which are proven to be of interest in many fields, with some compounds being promising drugs, such as dolastatins, ulapualides, kabiramides, latrunculins, doliculides, and others. Further research is needed to develop these compounds. Overall, the most studied activities are cytotoxicity and anticancer activity, followed by antibiotic activity (Figure 25).

In this review, we discussed more than 450 metabolites isolated from ca. 400 species of heterobranch molluscs. Heterobranch molluscs are thus an important source of bioactive NPs, even if not all of them are produced by the molluscs themselves. John Faulkner once said that in order to find the most bioactive compounds in an ecosystem, heterobranchs would be the best shot to find them, because they have already selected them along evolution. This continues to appear to be true. While symbionts may be behind some of the NP syntheses, heterobranchs have evolved to use them for their own benefit [1,2]. In some cases, dietary cyanobacteria or other bacteria have been proven to be the source of compounds, but, in general, origin from symbionts remains difficult to prove [481]. In any case, if NPs can be traced to a microorganism, this may help to solve the supply problems for many of the bioactive NPs, either by culturing, by isolating the BGCs (biosynthetic gene clusters), or by using other metagenomic techniques [873,874]. Culturing the molluscs is a difficult-but-not-impossible task, which could also be useful for some heterobranch species. Moreover, possible strategies to improve MNP selection include many dereplication strategies described in the literature, in addition to the many chemical techniques used to obtain derivatives, as well as synthesis methods [44,761,875]. Furthermore, virtual screening, computational chemistry, as well as more studies on molecular targets are needed to overcome the limitations of studying MNPs. The use of nanotechnology to deliver drugs is also a promising field that requires further investigation; ADCs, for instance, show considerable potential [481]. As an example, kahalalide F (**194**) conjugated to 40 nm gold nanoparticles resulted in higher cell growth inhibition in HeLa cervical carcinoma cells than the compound alone [481,876]. Further research should also be devoted to this field. Overall, we have seen that heterobranch molluscs are extremely interesting in regard to the study of marine natural products in terms of both chemical ecology and biotechnology studies, providing many leads for further detailed research in these fields in the near future.

Author Contributions: Conceptualization and design C.A.; literature review, investigation, and data curation C.A. and C.A.-P.; writing—original draft preparation, review and editing, C.A. and C.A.-P.; funding acquisition, C.A. All authors have read and agreed to the published version of the manuscript.

Funding: Support for this work was provided by BLUEBIO grant to C.A. (CTM2016-78901/ANT) and the Ramon Areces Foundation to C.A.-P.

Acknowledgments: This is an AntECO (SCAR) contribution.

Conflicts of Interest: The authors declare no conflict of interest.

Abbreviations

AD	Alzheimer Disease
ADCs	Antibody-drug conjugates
ADMET	Absortion, Distribution, Metabolism, Excretion, and Toxicity
ADR	Adriamycin resistant
BGCs	Biosynthetic Gene Clusters
cAMP	Cyclic Adenosine Monophosphate
EC_{50}	Half maximal Effective Concentration
ED_{50}	Half effective dose
EGF	Epidermal Groth Factor
EGFR	Epidermal Groth Factor Receptor
ERK	Extracellular signal-regulated kinases
FDA	Food and Drugs Administration
GI_{50}	Maximal inhibition of cell proliferation
HCCLs	Human Colon Cancer Cell Lines
HEL	Human Erythroleukemia cells
HeLa	Henrietta_Lacks cell line from cervical cancer cells
HIF-1	Hypoxia Inducible Factor 1
HTCLs	Human Tumor Cell Lines
KB	Subline of the KERATIN-forming tumor cell line HeLa
IC_{50}	Half minimal Inhibitory Concentration
LD	Lethal Dose
LRA	Latency Reversal Agent
MAAs	Mycosporine-like Amino Acids
MAPK	Mitogen-Activated Protein Kinase
MDA	Microtubule-Desestabilizing Agent
MDFs	Mantle Dermal Formations
MDR	Multidrug resistant variant
MIC	Minimum Inhibitory Concentration
MNPs	Marine Natural Products
MTT	Dimethyl Thiazolyl Diphenyl Tetrazolium Bromide
NCI	National Cancer Institute
NPs	Natural Products
NSCLC	Nonsmall Cell Lung Cancer
PBM	Peripheral Blood Mononuclear
PG	Prostaglandins
PKC	Protein Kinase C
PSMA-ADC	Prostate-specific membrane antigen antibody–drug conjugate
PTPRK	Protein Tyrosine Phosphatase Receptor type K
TRAIL	Tumor necrosis factor-related apoptosis-inducing ligand
TTX	Tetrodotoxin
UVR	Ultra-Violet Radiation
VCR	Vincristine resistant

References

1. Avila, C.; Núñez-Pons, L.; Moles, J. From the tropics to the poles: Chemical defensive strategies in sea slugs (Mollusca: Heterobranchia). In *Chemical Ecology: The Ecological Impacts of Marine Natural Products*; Puglisi, M.P., Becerro, M.A., Eds.; CRC Press: Boca Raton, FL, USA; Taylor and Francis: Abingdon, UK, 2018.
2. Avila, C. Natural products of opisthobranch molluscs: A biological review. *Oceanogr. Mar. Biol. Annu. Rev.* **1995**, *33*, 487–559.
3. Avila, C. A preliminary catalogue of natural substances of opistobranch molluscs from Western Mediterranean and near Atlantic. *Sci. Mar.* **1992**, *56*, 373–382.
4. Avila, C. Terpenoids in Marine Heterobranch Molluscs. *Mar. Drugs* **2020**, *8*, 162. [CrossRef] [PubMed]

5. Blunt, J.W.; Carroll, A.R.; Copp, B.R.; Davis, R.A.; Keyzers, R.A.; Prinsep, M.R. Marine natural products. *Nat. Prod. Rep.* **2018**, *35*, 8–53. [CrossRef]

6. Carroll, A.R.; Copp, B.R.; Davis, R.A.; Keyzers, R.A.; Prinsep, M.R. Marine natural products. *Nat. Prod. Rep.* **2020**, *37*, 175–223. [CrossRef]

7. Puglisi, M.P.; Becerro, M.A. *Chemical Ecology: The Ecological Impacts of Marine Natural Products*; CRC Press: Boca Raton, FL, USA; Taylor and Francis: Abingdon, UK, 2018.

8. McClintock, J.B.; Baker, P.J. *Marine Chemical Ecology*; CRC Marine Science Series Press: Boca Raton, FL, USA, 2001.

9. Ianora, A.; Boersma, M.; Casotti, R.; Fontana, A.; Harder, J.; Hoffmann, F.; Pavia, H.; Potin, P.; Poulet, S.A.; Toth, G. New trends in marine chemical ecology. *Estuar. Coasts* **2006**, *29*, 531–551. [CrossRef]

10. Puglisi, M.P.; Sneed, J.M.; Sharp, K.H.; Ritson-Williams, R.; Paul, V.J. Marine chemical ecology in benthic environments. *Nat. Prod. Rep.* **2014**, *31*, 1510–1553. [CrossRef]

11. Tian, Y.; Li, Y.-L.; Zhao, F.-C. Secondary metabolites from polar organisms. *Mar. Drugs* **2017**, *15*, 28. [CrossRef]

12. Cimino, G.; Gavagnin, M. (Eds.) *Progress in Molecular and Subcellular Biology*; Subseries Marine Molecular Biotechnology; Springer: Berlin/Heidelberg, Germany, 2006; Volume 43.

13. Benkendorff, K. Molluscan biological and chemical diversity: Secondary metabolites and medicinal resources produced by marine molluscs. *Biol. Rev.* **2010**, *85*, 757–775. [CrossRef]

14. Kandyuk, R.P. Sterols and their functional role in Mollusks (a review). *Hydrobiol. J.* **2006**, *42*, 56–66. [CrossRef]

15. Garson, M. Marine natural products as antifeedants. In *Comprehensive Natural Products II. Chemistry and Biology*; Mander, L., Liu, H.W., Eds.; Elsevier Science: Amsterdam, The Netherlands, 2010; pp. 503–537.

16. Cimino, G.; Ghiselin, M.T. Marine natural products chemistry as an evolutionary narrative. In *Marine Chemical Ecology*; McClintock, J.B., Baker, P.J., Eds.; CRC Marine Science Series Press: Boca Raton, FL, USA, 2001; pp. 115–154.

17. McClintock, J.B.; Amsler, C.D.; Baker, B.J. Overview of the chemical ecology of benthic marine invertebrates along the Western Antarctic peninsula. *Integr. Comp. Biol.* **2010**, *50*, 967–980. [CrossRef] [PubMed]

18. Núñez-Pons, L.; Avila, C. Natural products mediating ecological interactions in Antarctic benthic communities: A mini-review of the known molecules. *Nat. Prod. Rep.* **2015**, *32*, 1114–1130. [CrossRef] [PubMed]

19. Cimino, G.; Ghiselin, M.T. Chemical defense and the evolution of opisthobranch gastropods. *Proc. Calif. Acad. Sci.* **2009**, *60*, 175–422.

20. Garson, M.J. Marine mollusks from Australia and New Zealand: Chemical and ecological studies. In *Progress in Molecular and Subcellular Biology*; Cimino, G., Gavagnin, M., Eds.; Springer: Berlin/Heidelberg, Germany, 2006; Volume 43, pp. 159–174.

21. Davies-Coleman, M.T. Secondary metabolites from the marine gastropod molluscs of Antarctica, Southern Africa and South America. In *Progress in Molecular and Subcellular Biology*; Cimino, G., Gavagnin, M., Eds.; Springer: Berlin/Heidelberg, Germany, 2006; Volume 43, pp. 133–157.

22. Wahidulla, S.; Guo, Y.W.; Fakhr, I.M.I.; Mollo, E. Chemical diversity in opisthobranch molluscs from scarcely investigated Indo-Pacific areas. In *Progress in Molecular and Subcellular Biology*; Cimino, G., Gavagnin, M., Eds.; Springer: Berlin/Heidelberg, Germany, 2006; Volume 43, pp. 175–198.

23. Miyamoto, T. Selected bioactive compounds from Japanese anaspideans and nudibranchs. In *Progress in Molecular and Subcellular Biology*; Cimino, G., Gavagnin, M., Eds.; Springer: Berlin/Heidelberg, Germany, 2006; Volume 43, pp. 199–214.

24. Wang, J.R.; He, W.F.; Guo, Y.W. Chemistry, chemoecology, and bioactivity of the South China Sea opisthobranch molluscs and their dietary organisms. *J. Asian Nat. Prod. Res.* **2013**, *15*, 185–197. [CrossRef] [PubMed]

25. Andersen, R.J.; Desjardine, K.; Woods, K. Skin chemistry of nudibranchs from the West Coast of North America. In *Progress in Molecular and Subcellular Biology*; Cimino, G., Gavagnin, M., Eds.; Springer: Berlin/Heidelberg, Germany, 2006; Volume 43, pp. 277–301.

26. Kamiya, H.; Sakai, R.; Jimbo, M. Bioactive molecules from sea hares. In *Progress in Molecular and Subcellular Biology*; Cimino, G., Gavagnin, M., Eds.; Springer: Berlin/Heidelberg, Germany, 2006; Volume 43, pp. 215–239.

27. Darias, J.; Cueto, M.; Díaz-Marrero, A.R. The chemistry of marine pulmonate gastropods. In *Progress in Molecular and Subcellular Biology*; Cimino, G., Gavagnin, M., Eds.; Springer: Berlin/Heidelberg, Germany, 2006; Volume 43, pp. 105–131.

28. Dean, L.J.; Prinsep, M.R. The chemistry and chemical ecology of nudibranchs. *Nat. Prod. Rep.* **2017**, *34*, 1359–1390. [CrossRef] [PubMed]

29. Avila, C.; Taboada, S.; Núñez-Pons, L. Antarctic marine chemical ecology: What is next? *Mar. Ecol.* **2008**, *29*, 1–71. [CrossRef]

30. Paul, V.J. *Ecological Roles of Marine Natural Products*; Comstock Publishications Association: Ithaka, NY, USA, 1992.

31. Ruiz-Torres, V.; Encinar, J.A.; Herranz-López, M.; Pérez-Sánchez, A.; Galiano, V.; Barrajón-Catalán, E.; Micol, V. An updated review on marine anticancer compounds: The use of virtual screening for the discovery of small-molecule cancer drugs. *Molecules* **2017**, *22*, 1037. [CrossRef]

32. Newman, D.J.; Cragg, G.M. Marine natural products and related compounds in clinical and advanced preclinical trials. *J. Nat. Prod.* **2004**, *67*, 1216–1238. [CrossRef]

33. Baker, B.J. *Marine Biomedicine: From Beach to Bedside*; CRC Press: Boca Raton, FL, USA; Taylor and Francis: Abingdon, UK, 2015.

34. Haefner, B. Drugs from the deep: Marine natural products as drug candidates. *Drug Discov. Today* **2003**, *8*, 536–544. [CrossRef]

35. Clardy, J.; Walsh, C. Lessons from natural molecules. *Nature* **2004**, *432*, 829–837. [CrossRef]

36. Molinski, T.F.; Dalisay, D.S.; Lievens, S.L.; Saludes, J.P. Drug development from marine natural products. *Nature Rev. Drug Discov.* **2009**, *8*, 69–85. [CrossRef] [PubMed]

37. Cragg, G.M.; Grothaus, P.G.; Newman, D.J. New horizons for old drugs and drug leads. *J. Nat. Prod.* **2014**, *77*, 703–723. [CrossRef] [PubMed]

38. Mudit, M.; El Sayed, K.A. Cancer control potential of marine natural product scaffolds through inhibition of tumor cell migration and invasion. *Drug Discov. Today* **2016**, *21*, 1745–1760. [CrossRef] [PubMed]

39. Newman, D.J.; Cragg, G.M. Natural products as sources of new drugs over the 30 years from 1981 to 2010. *J. Nat. Prod.* **2012**, *75*, 311–335. [CrossRef]

40. Newman, D.J.; Cragg, G.M. Natural products as sources of new drugs from 1981 to 2014. *J. Nat. Prod.* **2016**, *79*, 629–661. [CrossRef] [PubMed]

41. White, J. Drug Addiction: From Basic Research to Therapy. *Drug Alcohol. Rev.* **2009**, *28*, 455. [CrossRef]

42. Khalifa, S.A.; Elias, N.; Farag, M.A.; Chen, L.; Saeed, A.; Hegazy, M.E.; Moustafa, M.S.; El-Wahed, A.; Al-Mousawi, S.M.; Musharraf, S.G.; et al. Marine natural products: A source of novel anticancer drugs. *Mar. Drugs* **2019**, *17*, 491. [CrossRef]

43. Newman, D.J.; Cragg, G.M. Marine-sourced anti-cancer and cancer pain control agents in clinical and late preclinical development. *Mar. Drugs* **2014**, *12*, 255–278. [CrossRef]

44. Gross, H.; König, G.M. Terpenoids from marine organisms: Unique structures and their pharmacological potential. *Phytochem. Rev.* **2006**, *5*, 115–141. [CrossRef]

45. Faircloth, G.; Cuevas, M.C. Kahalalide F and ES285: Potent anticancer agents from marine molluscs. In *Progress in Molecular and Subcellular Biology*; Cimino, G., Gavagnin, M., Eds.; Springer: Berlin/Heidelberg, Germany, 2006; Volume 43, pp. 363–379.

46. Kigoshi, H.; Kita, M. Antitumor effects of sea hare-derived compounds in cancer. In *Handbook of Anticancer Drugs from Marine Origin*; Kim, S.K., Ed.; Springer International Publishing: Cham, Switzerland, 2015; pp. 701–739.

47. Schrödl, M.; Jörger, K.M.; Klussmann-Kolb, A.; Wilson, N.G. Bye bye 'Opisthobranchia'! A review on the contribution of mesopsammic sea slugs to euthyneuran systematics. *Thalassas* **2011**, *27*, 101–112.

48. Medina, M.; Lal, S.; Vallès, Y.; Takaoka, T.L.; Dayrat, B.A.; Boore, J.L.; Gosliner, T. Crawling through time: Transition of snails to slugs dating back to the Paleozoic, based on mitochondrial phylogenomics. *Mar. Genom.* **2011**, *4*, 51–59. [CrossRef] [PubMed]

49. Wägele, H.; Klussmann-Kolb, A.; Verbeek, E.; Schrödl, M. Flashback and foreshadowing; a review of the taxon Opisthobranchia. *Org. Div. Evol.* **2014**, *14*, 133–149. [CrossRef]

50. Zapata, F.; Wilson, N.G.; Howison, M.; Andrade, S.C.; Jörger, K.M.; Schrödl, M.; Goetz, F.E.; Giribet, G.; Dunn, C.W. Phylogenomic analyses of deep gastropod relationships reject Orthogastropoda. *Proc. R. Soc. B Biol. Sci.* **2014**, *281*, 20141739. [CrossRef] [PubMed]

51. WoRMS. World Register of Marine Species, Database. Available online: http://www.marinespecies.org (accessed on 11 November 2020).

52. Avila, C. Molluscan natural products as biological models: Chemical ecology, histology, and laboratory culture. In *Progress in Molecular and Subcellular Biology*; Cimino, G., Gavagnin, M., Eds.; Springer: Berlin/Heidelberg, Germany, 2006; Volume 43, pp. 1–23.

53. Iken, K.; Avila, C.; Ciavatta, M.L.; Fontana, A.; Cimino, G. Hodgsonal, a new drimane sesquiterpene from the mantle of the Antarctic nudibranch *Bathydoris hodgsoni*. *Tetrahedron Lett.* **1998**, *39*, 5635–5638. [CrossRef]

54. Avila, C.; Iken, K.; Fontana, A.; Cimino, G. Chemical ecology of the Antarctic nudibranch *Bathydoris hodgsoni* Eliot, 1907: Defensive role and origin of its natural products. *J. Exp. Mar. Biol. Ecol.* **2000**, *252*, 27–44. [CrossRef]

55. Loughlin, W.A.; Muderawan, I.W.; McCleary, M.A.; Volter, K.E.; King, M.D. Studies towards the synthesis of phorbazoles A–D: Formation of the pyrrole oxazole skeleton. *Aust. J. Chem.* **1999**, *52*, 231–234. [CrossRef]

56. Radspieler, A.; Liebscher, J. Total synthesis of phorbazole C. *Tetrahedron* **2001**, *57*, 4867–4871. [CrossRef]

57. Moles, J.; Wägele, H.; Cutignano, A.; Fontana, A.; Ballesteros, M.; Avila, C. Giant embryos and hatchlings of Antarctic nudibranchs (Mollusca: Gastropoda: Heterobranchia). *Mar. Biol.* **2017**, *164*, 114. [CrossRef]

58. Gavagnin, M.; De Napoli, A.; Castelluccio, F.; Cimino, G. Austrodorin-A and-B: First tricyclic diterpenoid 2′-monoglyceryl esters from an Antarctic nudibranch. *Tetrahedron Lett.* **1999**, *40*, 8471–8475. [CrossRef]

59. Gavagnin, M.; De Napoli, A.; Cimino, G.; Iken, K.; Avila, C.; Garcia, F.J. Absolute configuration of diterpenoid diacylglycerols from the Antarctic nudibranch *Austrodoris kerguelenensis*. *Tetrahedron Asym.* **1999**, *10*, 2647–2650. [CrossRef]

60. Iken, K.; Avila, C.; Fontana, A.; Gavagnin, M. Chemical ecology and origin of defensive compounds in the Antarctic nudibranch *Austrodoris kerguelenensis* (Opisthobranchia: Gastropoda). *Mar. Biol.* **2002**, *141*, 101–109.

61. Carte, B.; Faulkner, D. Role of Secondary Metabolites in Feeding Associations between a Predatory Nudibranch, 2 Grazing Nudibranchs, and a Bryozoan. *J. Chem. Ecol.* **1986**, *12*, 795–804. [CrossRef] [PubMed]

62. Davies-Coleman, M.T.; Faulkner, D.J. New diterpenoic acid glycerides from the Antarctic nudibranch *Austrodoris kerguelensis*. *Tetrahedron* **1991**, *47*, 9743–9750. [CrossRef]

63. Gavagnin, M.; Trivellone, E.; Castelluccio, F.; Cimino, G.; Cattaneo-Vietti, R. Glyceryl ester of a new halimane diterpenoic acid from the skin of the antarctic nudibranch *Austrodoris kerguelenensis*. *Tetrahedron Lett.* **1995**, *36*, 7319–7322. [CrossRef]

64. Gavagnin, M.; Carbone, M.; Mollo, E.; Cimino, G. Austrodoral and austrodoric acid: Nor-sesquiterpenes with a new carbon skeleton from the Antarctic nudibranch *Austrodoris kerguelenensis*. *Tetrahedron Lett.* **2003**, *44*, 1495–1498. [CrossRef]

65. Gavagnin, M.; Carbone, M.; Mollo, E.; Cimino, G. Further chemical studies on the Antarctic nudibranch *Austrodoris kerguelenensis*: New terpenoid acylglycerols and revision of the previous stereochemistry. *Tetrahedron* **2003**, *59*, 5579–5583. [CrossRef]

66. Diyabalanage, T.; Iken, K.B.; McClintock, J.B.; Amsler, C.D.; Baker, B.J. Palmadorins A–C, diterpene glycerides from the Antarctic nudibranch *Austrodoris kerguelenensis*. *J. Nat. Prod.* **2010**, *73*, 416–421. [CrossRef] [PubMed]

67. Maschek, J.A.; Mevers, E.; Diyabalanage, T.; Chen, L.; Ren, Y.; McClintock, J.B.; Amsler, C.D.; Wu, J.; Baker, B.J. Palmadorin chemodiversity from the Antarctic nudibranch *Austrodoris kerguelenensis* and inhibition of Jak2/STAT5-dependent HEL leukemia cells. *Tetrahedron* **2012**, *68*, 9095–9104. [CrossRef]

68. Cutignano, A.; Zhang, W.; Avila, C.; Cimino, G.; Fontana, A. Intrapopulation variability in the terpene metabolism of the Antarctic opisthobranch mollusc *Austrodoris kerguelenensis*. *Eur. J. Org. Chem.* **2011**, 5383–5389. [CrossRef]

69. Wilson, N.G.; Maschek, J.A.; Baker, B.J. A species flock driven by predation? Secondary metabolites support diversification of slugs in Antarctica. *PLoS ONE* **2013**, *8*, e80277. [CrossRef]

70. Gavagnin, M.; Fontana, A.; Ciavatta, M.L.; Cimino, G. Chemical studies on Antarctic nudibranch molluscs. *Italian J. Zool.* **2000**, *1*, 101–109. [CrossRef]

71. Graziani, E.I.; Andersen, R.J.; Krug, P.J.; Faulkner, D.J. Stable isotope incorporation evidence for the de novo biosynthesis of terpenoic acid glycerides by dorid nudibranchs. *Tetrahedron* **1996**, *52*, 6869–6878. [CrossRef]

72. Granato, A.C.; Berlinck, R.G.S.; Magalhaes, A.; Schefer, A.B.; Ferreira, A.G.; De Sanctis, B.; De Freitas, J.C.; Hajdu, E.; Migotto, A.E. Natural products from the marine sponges *Aaptos* sp. and *Hymeniacidon* aff. *heliophila*, and from the nudibranch *Doris* aff. *verrucosa*. *Quim. Nov.* **2000**, *23*, 594–599.

73. Ayer, S.W.; Andersen, R.J. Steroidal antifeedants from the dorid nudibranch *Aldisa sanguinea cooperi*. *Tetrahedron Lett.* **1982**, *23*, 1039–1042. [CrossRef]

74. Gavagnin, M.; Ungur, N.; Mollo, E.; Templado, J.; Cimino, G. Structure and synthesis of a progesterone homologue from the skin of the dorid nudibranch *Aldisa smaragdina*. *Eur. J. Org. Chem.* **2002**, *9*, 1500–1504. [CrossRef]

75. Nuzzo, G.; Ciavatta, M.L.; Kiss, R.; Mathieu, V.; Leclercqz, H.; Manzo, E.; Villani, G.; Mollo, E.; Lefranc, F.; D'Souza, L.; et al. Chemistry of the nudibranch *Aldisa andersoni*: Structure and biological activity of phorbazole metabolites. *Mar. Drugs* **2012**, *10*, 1799–1811. [CrossRef]

76. Rudi, A.; Stein, Z.; Green, S.; Goldberg, I.; Kashman, Y.; Benayahu, Y.; Schleyer, M. Phorbazoles A–D, novel chlorinated phenylpyrrolyloxazoles from the marine sponge *Phorbas* aff. *clathrata*. *Tetrahedron Lett.* **1994**, *35*, 2589–2592. [CrossRef]

77. Mollo, E.; Gavagnin, M.; Carbone, M.; Castelluccio, F.; Pozone, F.; Roussis, V.; Templado, J.; Ghiselin, M.T.; Cimino, G. Factors promoting marine invasions: A chemoecological approach. *Proc. Natl. Acad. Sci. USA* **2008**, *105*, 4582–4586. [CrossRef]

78. Krug, P.J.; Boyd, K.G.; Faulkner, D.J. Isolation and synthesis of tanyolides A and B, metabolites of the nudibranch *Sclerodoris tanya*. *Tetrahedron* **1995**, *51*, 11063–11074. [CrossRef]

79. Marín, A.; López-Belluga, M.D.; Scognamiglio, G.; Cimino, G. Morphological and chemical camouflage of the Mediterranean nudibranch *Discodoris indecora* on the sponges *Ircinia variabilis* and *Ircinia fasciculata*. *J. Mollus. Stud.* **1997**, *63*, 431–439. [CrossRef]

80. Cimino, G.; De Rosa, S.; De Stefano, S.; Sodano, G.; Villani, G. Dorid nudibranch elaborates its own chemical defense. *Science* **1983**, *219*, 1237–1238. [CrossRef] [PubMed]

81. Gavagnin, M.; Mollo, E.; Castelluccio, F.; Ghiselin, M.T.; Calado, G.; Cimino, G. Can molluscs biosynthesize typical sponge metabolites? The case of the nudibranch *Doriopsilla areolata*. *Tetrahedron* **2001**, *57*, 8913–8916. [CrossRef]

82. Kubo, I.; Nakanishi, K. Insect antifeedants and repellents from African plants. In *Host Plant Resistance to Pests*; ACS Symposium Series; American Chemical Society: Washington, DC, USA, 1977; Volume 62, pp. 165–178.

83. Cimino, G.; De Rosa, S.; De Stefano, S.; Sodano, G. Observations on the toxicity and metabolic relationships of polygodial, the chemical defense of the nudibranch *Dendrodoris limbata*. *Experientia* **1985**, *41*, 1335–1336. [CrossRef]

84. Cimino, G.; Sodano, G.; Spinella, A. Occurrence of olepupuane in two Mediterranean nudibranchs: A protected form of polygodial. *J. Nat. Prod.* **1988**, *51*, 1010–1011. [CrossRef] [PubMed]

85. Cimino, G.; De Rosa, S.; De Stefano, S.; Sodano, G. Novel sesquiterpenoid esters from the nudibranch *Dendrodoris limbata*. *Tetrahedron Letters*. **1981**, *22*, 1271–1272. [CrossRef]

86. Sakio, Y.; Hirano, Y.J.; Hayashi, M.; Komiyama, K.; Ishibashi, M. Dendocarbins A–N, new drimane sesquiterpenes from the nudibranch *Dendrodoris carbunculosa*. *J. Nat. Prod.* **2001**, *64*, 726–731. [CrossRef]

87. Fontana, A.; Ciavatta, M.L.; Miyamoto, T.; Spinella, A.; Cimino, G. Biosynthesis of drimane terpenoids in dorid molluscs: Pivotal role of 7-deacetoxyolepupuane in two species of *Dendrodoris* nudibranchs. *Tetrahedron* **1999**, *55*, 5937–5946. [CrossRef]

88. Grkovic, T.; Appleton, D.R.; Copp, B.R. Chemistry and chemical ecology of some of the common opisthobranch molluscs found on the shores of NE New Zealand. *Chem. N. Z.* **2005**, *69*, 12–15.

89. Gavagnin, M.; Mollo, E.; Calado, G.; Fahey, S.; Ghiselin, M.T.; Ortea, J.; Cimino, G. Chemical studies of porostome nudibranchs: Comparative and ecological aspects. *Chemoecology* **2001**, *11*, 131–136. [CrossRef]

90. Avila, C.; Cimino, G.; Crispino, A.; Spinella, A. Drimane sesquiterpenoids in Mediterranean Dendrodoris nudibranchs: Anatomical distribution and biological role. *Experientia* **1991**, *47*, 306–310. [CrossRef]

91. Okuda, R.K.; Scheuer, P.J.; Hochlowski, J.E.; Walker, R.P.; Faulkner, D.J. Sesquiterpenoid constituents of eight porostome nudibranchs. *J. Org. Chem.* **1983**, *48*, 1866–1869. [CrossRef]

92. Karuso, P. Chemical ecology of the nudibranchs. In *BioorganicMarine Chemistry*; Scheuer, P.P.J., Ed.; Bioorganic Marine Chemistry; Springer: Berlin/Heidelberg, Germany, 1987; pp. 31–60.

93. Cimino, G.; Ghiselin, M.T. Chemical defense and evolutionary trends in biosynthetic capacity among Dorid nudibranchs (Mollusca: Gastropoda: Opisthobranchia). *Chemoecology* **1999**, *9*, 187–207. [CrossRef]

94. Faulkner, D.J. Marine Natural Products. *Nat. Prod. Rep.* **2001**, *18*, 1–49. [CrossRef] [PubMed]

95. Gaspar, H.; Gavagnin, M.; Calado, G.; Castelluccio, F.; Mollo, E.; Cimino, G. Pelseneeriol-1 and-2: New furanosesquiterpene alcohols from porostome nudibranch *Doriopsilla pelseneeri*. *Tetrahedron* **2005**, *61*, 11032–11037. [CrossRef]

96. Fontana, A.; Tramice, A.; Cutignano, A.; d'Ippolito, G.; Gavagnin, M.; Cimino, G. Terpene biosynthesis in the nudibranch *Doriopsilla areolata*. *J. Org. Chem.* **2003**, *68*, 2405–2409. [CrossRef]

97. Spinella, A.; Alvarez, L.A.; Avila, C.; Cimino, G. New acetoxy-ent-pallescensin-A sesquiterpenoids from the skin of the porostome nudibranch *Doriopsilla areolata*. *Tetrahedron Lett.* **1994**, *35*, 8665–8668. [CrossRef]

98. Long, J.D.; Hay, M.E. Fishes learn aversions to a nudibranch's chemical defense. *Mar. Ecol. Progr. Ser.* **2006**, *307*, 199–208. [CrossRef]

99. Gaspar, H.; Cutignano, A.; Ferreira, T.; Calado, G.; Cimino, G.; Fontana, A. Biosynthetic evidence supporting the generation of terpene chemodiversity in marine mollusks of the genus *Doriopsilla*. *J. Nat. Prod.* **2008**, *71*, 2053–2056. [CrossRef]

100. Brunckhorst, D.J. The systematics and phylogeny of phyllidiid nudibranchs (Doridoidea). *Rec. Aust. Mus. Suppl.* **1993**, *16*, 1–107. [CrossRef]

101. Fusetani, N.; Wolstenholme, H.J.; Matsunaga, S.; Hirota, H. Two new sesquiterpene isonitriles from the nudibranch *Phyllidia pustulosa*. *Tetrahedron Lett.* **1991**, *32*, 7291–7294. [CrossRef]

102. Okino, T.; Yoshimura, E.; Hirota, H.; Fusetani, N. New antifouling sesquiterpenes from four nudibranchs of the family Phyllidiidae. *Tetrahedron* **1996**, *52*, 9447–9454. [CrossRef]

103. Hirota, H.; Okino, T.; Yoshimura, E.; Fusetani, N. Five new antifouling sesquiterpenes from two marine sponges of the genus *Axinyssa* and the nudibranch *Phyllidia pustulosa*. *Tetrahedron* **1998**, *54*, 13971–13980. [CrossRef]

104. Cimino, G.; Fontana, A.; Gavagnin, M. Marine opisthobranch molluscs: Chemistry and ecology in sacoglossan and dorids. *Curr. Org. Chem.* **1999**, *3*, 327–372.

105. Garson, M.J.; Simpson, J.S. Marine isocyanides and related natural products—Structure, biosynthesis and ecology. *Nat. Prod. Rep.* **2004**, *21*, 164–179. [CrossRef]

106. Burreson, B.J.; Scheuer, P.J.; Finer, J.; Clardy, J. 9-Isocyanopupukeanane, a marine invertebrate allomone with a new sesquiterpene skeleton. *J. Am. Chem. Soc.* **1975**, *97*, 4763–4764. [CrossRef]

107. Hagadone, M.R.; Burreson, B.J.; Scheuer, P.J.; Finer, J.S.; Clardy, J. Defense allomones of the nudibranch *Phyllidia varicosa* Lamarck 1801. *Helv. Chim. Acta* **1979**, *62*, 2484–2494. [CrossRef]

108. Ungur, N.; Gavagnin, M.; Fontana, A.; Cimino, G. Absolute stereochemistry of natural sesquiterpenoid diacylglycerols. *Tetrahedron Asymmetry* **1999**, *10*, 1263–1273. [CrossRef]

109. Ritson-Williams, R.; Paul, V.J. Marine benthic invertebrates use multimodal cues for defense against reef fish. *Mar. Ecol. Progr. Ser.* **2007**, *340*, 29–39. [CrossRef]

110. Yasman, Y.; Edrada, R.A.; Wray, V.; Proksch, P. New 9-thiocyanatopupukeanane sesquiterpenes from the nudibranch *Phyllidia varicosa* and its sponge-prey *Axinyssa aculeata*. *J. Nat. Prod.* **2003**, *66*, 1512–1514. [CrossRef]

111. Jaisamut, S.; Prabpai, S.; Tancharoen, C.; Yuenyongsawad, S.; Hannongbua, S.; Kongsaeree, P.; Plubrukarn, A. Bridged tricyclic sesquiterpenes from the tubercle nudibranch *Phyllidia coelestis* Bergh. *J. Nat. Prod.* **2013**, *76*, 2158–2161. [CrossRef]

112. Sim, D.C.-M.; Mudianta, I.W.; White, A.M.; Martiningsih, N.W.; Loh, J.J.M.; Cheney, K.L.; Garson, M.J. New sesquiterpenoid isonitriles from three species of phyllid nudibranchs. *Fitoterapia* **2018**, *126*, 69–73. [CrossRef] [PubMed]

113. Iwashima, M.; Terada, I.; Iguchi, K.; Yamori, T. New biologically active marine sesquiterpenoid and steroid from the Okinawan sponge of the genus *Axinyssa*. *Chem. Pharm. Bull.* **2002**, *50*, 1286–1289. [CrossRef] [PubMed]

114. Gulavita, N.K.; De Silva, E.D.; Hagadone, M.R.; Karuso, P.; Scheuer, P.J.; van Duyne, G.D.; Clardy, J. Nitrogenous bisabolene sesquiterpenes from marine invertebrates. *J. Org. Chem.* **1986**, *51*, 5136–5139. [CrossRef]

115. Kitano, Y.; Ito, T.; Suzuki, T.; Nogata, Y.; Shinshima, K.; Yoshimura, E.; Chiba, K.; Tada, M.; Sakaguchi, I. (2002) Synthesis and antifouling activity of 3-isocyanotheonellin and its analogues. *J. Chem. Soc. Perkin Trans.* **2002**, *1*, 2251–2255. [CrossRef]

116. Fusetani, N.; Hirota, H.; Okino, T.; Tomono, Y.; Yoshimura, E. Antifouling activity of isocyanoterpenoids and related compounds isolated from a marine sponge and nudibranchs. *J. Nat. Toxins* **1996**, *5*, 249–259.

117. Wu, Q.; Chen, W.-T.; Li, S.-W.; Ye, J.-Y.; Huan, X.-J.; Gavagnin, M.; Yao, L.-G.; Wang, H.; Miao, Z.-H.; Li, X.-W.; et al. Cytotoxic nitrogenous terpenoids from two South China Sea nudibranchs *Phyllidiella pustulosa*, *Phyllidia coelestis*, and their sponge-prey *Acanthella cavernosa*. *Mar. Drugs* **2019**, *17*, 56. [CrossRef]

118. Wright, A.D. GC-MS and NMR analysis of *Phyllidiella pustulosa* and one of its dietary sources, the sponge *Phakellia carduus*. *Comp. Biochem. Physiol.* **2003**, *134A*, 307–313. [CrossRef]

119. Dumdei, E.J.; Flowers, A.E.; Garson, M.J.; Moore, C.J. The biosynthesis of sesquiterpene isocyanides and isothiocyanates in the marine sponge *Acanthella cavernosa* (Dendy); evidence for dietary transfer to the dorid nudibranch *Phyllidiella pustulosa*. *Comp. Biochem. Physiol.* **1997**, *118*, 1385–1392. [CrossRef]

120. Manzo, E.; Ciavatta, M.L.; Gavagnin, M.; Mollo, E.; Guo, Y.-W.; Cimino, G. Isocyanide terpene metabolites of *Phyllidiella pustulosa*, a nudibranch from the South China Sea. *J. Nat. Prod.* **2004**, *67*, 1701–1704. [CrossRef]

121. Shimomura, M.; Miyaoka, H.; Yamada, Y. Absolute configuration of marine diterpenoid kalihinol A. *Tetrahedron Lett.* **1999**, *40*, 8015–8017. [CrossRef]

122. Lyakhova, E.G.; Kolesnikova, S.A.; Kalinovskii, A.I.; Stonik, V.A. Secondary metabolites of the Vietnamese nudibranch mollusc *Phyllidiella pustulosa*. *Chem. Nat. Comp.* **2010**, *46*, 534–538. [CrossRef]

123. Fisch, K.M.; Hertzer, C.; Böhringer, N.; Wuisan, Z.G.; Schillo, D.; Bara, R.; Kaligis, F.; Wägele, H.; König, G.M.; Schäberle, T.F. The potential of Indonesian heterobranchs found around Bunaken Island for the production of bioactive compounds. *Mar. Drugs* **2017**, *15*, 384. [CrossRef] [PubMed]

124. Wägele, H.; Ballesteros, M.; Avila, C. Defensive glandular structures in opisthobranch molluscs-from histology to ecology. *Oceanogr. Mar. Biol.* **2006**, *44*, 197.

125. Johnson, R.F.; Gosliner, T.M. Traditional taxonomic groupings mask evolutionary history: A molecular phylogeny and new classification of the chromodorid nudibranchs. *PLoS ONE* **2012**, *7*, e33479. [CrossRef]

126. Thompson, J.E.; Walker, R.P.; Wratten, S.J.; Faulkner, D.J. A chemical defense mechanism for the nudibranch *Cadlina luteomarginata*. *Tetrahedron* **1982**, *38*, 1865–1873. [CrossRef]

127. Faulkner, D.J.; Molinski, T.F.; Andersen, R.J.; Dumdei, E.J.; De Silva, E.D. Geographical variation in defensive chemicals from Pacific coast dorid nudibranchs and some related marine molluscs. *Comp. Biochem. Physiol. C Toxicol. Pharmacol.* **1990**, *97*, 233–240. [CrossRef]

128. Kubanek, J.; Graziani, E.I.; Andersen, R.J. Investigations of terpenoid biosynthesis by the dorid nudibranch *Cadlina luteomarginata*. *J. Org. Chem.* **1997**, *62*, 7239–7246. [CrossRef]

129. Hellou, J.; Andersen, R.J.; Thompson, J.E. Terpenoids from the dorid nudibranch *Cadlina luteomarginata*. *Tetrahedron* **1982**, *38*, 1875–1879. [CrossRef]

130. Dumdei, E.J.; Kubanek, J.; Coleman, J.E.; Pika, J.; Andersen, R.J.; Steiner, J.R.; Clardy, J. New terpenoid metabolites from the skin extracts, an egg mass, and dietary sponges of the Northeastern Pacific dorid nudibranch *Cadlina luteomarginata*. *Can. J. Chem.* **1997**, *75*, 773–789. [CrossRef]

131. Carbone, M.; Gavagnin, M.; Haber, M.; Guo, Y.-W.; Fontana, A.; Manzo, E.; Genta-Jouve, G.; Tsoukatou, M.; Rudman, W.B.; Cimino, G.; et al. Packaging and delivery of chemical weapons: A defensive trojan horse stratagem in Chromodorid nudibranchs. *PLoS ONE* **2013**, *8*, e62075. [CrossRef]

132. Hochlowski, J.E.; Faulkner, D.J. Chemical constituents of the nudibranch *Chromodoris marislae*. *Tetrahedron Lett.* **1981**, *22*, 271–274. [CrossRef]

133. Schulte, G.R.; Scheuer, P.J. Defense allomones of some marine mollusks. *Tetrahedron* **1982**, *38*, 1857–1863. [CrossRef]

134. Hochlowski, J.E.; Faulkner, D.J.; Matsumoto, G.K.; Clardy, J. Norrisolide, a novel diterpene from the dorid nudibranch *Chromodoris norrisi*. *J. Org. Chem.* **1983**, *48*, 1141–1142. [CrossRef]

135. Bergquist, P.R.; Bowden, B.F.; Cambie, R.C.; Craw, P.A.; Karuso, P.; Poiner, A.; Taylor, W.C. The constituents of marine sponges. VI. Diterpenoid metabolites of the New Zealand sponge *Chelonaplysilla violacea*. *Aust. J. Chem.* **1993**, *46*, 623–632. [CrossRef]

136. Okuda, R.K.; Scheuer, P.J. Latrunculin-A, ichthyotoxic constituent of the nudibranch *Chromodoris elisabethina*. *Experientia* **1985**, *41*, 1355–1356. [CrossRef]

137. Carte, B.; Kernan, M.R.; Barrabee, E.B.; Faulkner, D.J.; Matsumoto, G.K.; Clardy, J. Metabolites of the nudibranch *Chromodoris funerea* and the singlet oxygen oxidation products of furodysin and furodysinin. *J. Org. Chem.* **1986**, *51*, 3528–3532. [CrossRef]

138. Kimura, J.; Hyosu, M. Two new sesterterpenes from the marine sponge, *Coscinoderma mathewsi*. *Chem. Lett.* **1999**, *28*, 61–62. [CrossRef]

139. Molinski, T.F.; Faulkner, D.J. Aromatic norditerpenes from the nudibranch *Chromodoris macfarlandi*. *J. Org. Chem.* **1986**, *51*, 2601–2603. [CrossRef]

140. Molinski, T.F.; Faulkner, D.J.; He, C.H.; Van Duyne, G.D.; Clardy, J. Three new rearranged spongian diterpenes from *Chromodoris macfarlandi*: Reappraisal of the structures of dendrillolides A and B. *J. Org. Chem.* **1986**, *51*, 4564–4567. [CrossRef]

141. Kakou, Y.; Crews, P.; Bakus, G.J. Dendrolasin and latrunculin A from the Fijian sponge *Spongia mycofijiensis* and an associated nudibranch *Chromodoris lochi*. *J. Nat. Prod.* **1987**, *50*, 482–484. [CrossRef]

142. Corley, D.G.; Herb, R.; Moore, R.E.; Scheuer, P.J.; Paul, V.J. Laulimalides. New potent cytotoxic macrolides from a marine sponge and a nudibranch predator. *J. Org. Chem.* **1988**, *53*, 3644–3646. [CrossRef]

143. Kernan, M.R.; Barrabee, E.B.; Faulkner, D.J. Variation of the metabolites of *Chromodoris funerea*: Comparison of specimens from a Palauan marine lake with those from adjacent waters. *Comp. Biochem. Physiol. B Comp. Biochem.* **1988**, *89*, 275–278. [CrossRef]

144. Bobzin, S.C.; Faulkner, D.J. Diterpenes from the marine sponge *Aplysilla polyrhaphis* and the dorid nudibranch *Chromodoris norrisi*. *J. Org. Chem.* **1989**, *54*, 3902–3907. [CrossRef]

145. Dumdei, E.J.; De Silva, E.D.; Andersen, R.J.; Choudhary, M.I.; Clardy, J. Chromodorolide A, a rearranged diterpene with a new carbon skeleton from the Indian ocean nudibranch *Chromodoris cavae*. *J. Am. Chem. Soc.* **1989**, *111*, 2712–2713. [CrossRef]

146. Cimino, G.; Crispino, A.; Gavagnin, M.; Sodano, G. Diterpenes from the nudibranch *Chromodoris luteorosea*. *J. Nat. Prod.* **1990**, *53*, 102–106. [CrossRef]

147. de Silva, E.D.; Morris, S.A.; Miao, S.; Dumdei, E.; Andersen, R.J. Terpenoid metabolites from skin extracts of four Sri Lankan Nudibranchs in the Genus *Chromodoris*. *J. Nat. Prod.* **1991**, *54*, 993–997. [CrossRef]

148. Morris, S.A.; Silva, E.D.D.; Andersen, R.J. Chromodorane diterpenes from the tropical dorid nudibranch *Chromodoris cavae*. *Can. J. Chem.* **1991**, *69*, 768–771. [CrossRef]

149. Gavagnin, M.; Vardaro, R.R.; Avila, C.; Cimino, G.; Ortea, J. Ichthyotoxic diterpenoids from the Cantabrian nudibranch *Chromodoris luteorosea*. *J. Nat. Prod.* **1992**, *55*, 368–371. [CrossRef]

150. Chi, Y.; Hashimoto, F.; Nohara, T.; Nakamura, M.; Yoshizawa, T.; Yamashita, M.; Marubayashi, N. Tennen Yuki Kagobutsu Toronkai Koen Yoshishu. Available online: https://www.scienceopen.com/document?vid=294f7cd1-f803-4b53-907a-bc3edf8ab46e (accessed on 11 November 2020).

151. Miyamoto, T.; Sakamoto, K.; Amano, H.; Higuchi, R.; Komori, T.; Sasaki, T. Three new cytotoxic sesterterpenoids, inorolide A, B, and C from the nudibranch *Chromodoris inornata*. *Tetrahedron Lett.* **1992**, *33*, 5811–5814. [CrossRef]

152. Puliti, R.A.; Gavagnin, M.A.; Cimino, G.U.; Mattia, C.A.; Mazzarella, L.E. Structure of chelonaplysin C: A spongian diterpenoid from nudibranch *Chromodoris luteorosea*. *Acta Crystallogr. C* **1992**, *48*, 2145–2147. [CrossRef]

153. Pika, J.; Faulkner, D.J. Unusual chlorinated homo-diterpenes from the South African nudibranch *Chromodoris hamiltoni*. *Tetrahedron* **1995**, *51*, 8189–8198. [CrossRef]

154. Miyamoto, T.; Sakamoto, K.; Arao, K.; Komori, T.; Higuchi, R.; Sasaki, T. Dorisenones, cytotoxic spongian diterpenoids, from the nudibranch *Chromodoris obsoleta*. *Tetrahedron* **1996**, *52*, 8187–8198. [CrossRef]

155. McPhail, K.L.; Davies-Coleman, M.T. New spongiane diterpenes from the East African nudibranch *Chromodoris hamiltoni*. *Tetrahedron* **1997**, *53*, 4655–4660. [CrossRef]

156. Miyamoto, T.; Sakamoto, K.; Amano, H.; Arakawa, Y.; Nagarekawa, Y.; Komori, T.; Higuchi, R.; Sasaki, T. New cytotoxic sesterterpenoids from the nudibranch *Chromodoris inornata*. *Tetrahedron* **1999**, *55*, 9133–9142. [CrossRef]

157. Karuso, P.; Scheuer, P.J. Natural products from three nudibranchs: *Nembrotha kubaryana*, *Hypselodoris infucata* and *Chromodoris petechialis*. *Molecules* **2002**, *7*, 1–6. [CrossRef]

158. Yong, K.W.; Salim, A.A.; Garson, M.J. New oxygenated diterpenes from an Australian nudibranch of the genus *Chromodoris*. *Tetrahedron* **2008**, *64*, 6733–6738. [CrossRef]

159. Uddin, M.H.; Otsuka, M.; Muroi, T.; Ono, A.; Hanif, N.; Matsuda, S.; Higa, T.; Tanaka, J. Deoxymanoalides from the nudibranch *Chromodoris willani*. *Chem. Pharm. Bull.* **2009**, *57*, 885–887. [CrossRef]

160. Agena, M.; Tanaka, C.; Hanif, N.; Yasumoto-Hirose, M.; Tanaka, J. New cytotoxic spongian diterpenes from the sponge *Dysidea* cf. *arenaria*. *Tetrahedron* **2009**, *65*, 1495–1499. [CrossRef]

161. Suciati, S.; Lambert, L.K.; Garson, M.J. Structures and anatomical distribution of oxygenated diterpenes in the Australian nudibranch *Chromodoris reticulata*. *Aust. J. Chem.* **2011**, *64*, 757–765. [CrossRef]

162. Katavic, P.L.; Jumaryatno, P.; Hooper, J.N.; Blanchfield, J.T.; Garson, M.J. Oxygenated terpenoids from the Australian sponges *Coscinoderma matthewsi* and *Dysidea* sp., and the nudibranch *Chromodoris albopunctata*. *Aust. J. Chem.* **2012**, *65*, 531–538. [CrossRef]

163. Katavic, P.L.; Jumaryatno, P.; Hooper, J.N.; Blanchfield, J.T.; Garson, M.J. Note of clarification about: Oxygenated Terpenoids from the Australian Sponges *Coscinoderma matthewsi* and *Dysidea* sp., and the Nudibranch *Chromodoris albopunctata*. *Aust. J. Chem.* **2013**, *66*, 1461.

164. Cheney, K.L.; White, A.; Mudianta, I.W.; Winters, A.E.; Quezada, M.; Capon, R.J.; Mollo, E.; Garson, M.J. Choose your weaponry: Selective storage of a single toxic compound, latrunculin A, by closely related nudibranch molluscs. *PLoS ONE* **2016**, *11*, e0145134. [CrossRef]

165. Schulte, G.; Scheuer, P.J.; McConnell, O.J. Two furanosesquiterpene marine metabolites with antifeedant properties. *Helv. Chim. Acta* **1980**, *63*, 2159–2167. [CrossRef]

166. Fontana, A.; Ciavatta, M.L.; D'Souza, L.; Mollo, E.; Naik, C.G.; Parameswaran, P.S.; Wahidulla, S.; Cimino, G. Selected chemo-ecological studies of marine opisthobranchs from Indian coasts. *J. Indian Inst. Sci.* **2001**, *81*, 403–415.

167. Tanaka, J.; Higa, T. The absolute configuration of kurospongin a new furanoterpene from a marine sponge, *Spongia* sp. *Tetrahedron* **1988**, *44*, 2805–2810. [CrossRef]

168. Kashman, Y.; Croweiss, A.; Shmueli, U. Latrunculin, a new 2-thiazolidinone macrolide from the marine sponge *Latrunculia magnifica*. *Tetrahedron Lett.* **1980**, *21*, 3629–3632. [CrossRef]

169. Kashman, Y.; Croweiss, A.; Kidor, R.; Blasberger, D.; Carmelya, S. Latrunculins: NMR study, two new toxins and a synthetic approach. *Tetrahedron* **1985**, *41*, 1905–1914. [CrossRef]

170. Guo, Y.W. Chemical Studies of the Novel Bioactive Secondary Metabolites from the Benthic Invertebrates: Isolation and Structure Characterization. Ph.D. Thesis, University of Naples, Naples, Italy, 1997.

171. Mebs, D. Chemical defense of a dorid nudibranch, *Glossodoris quadricolor*, from the Red Sea. *J. Chem. Ecol.* **1985**, *11*, 713–716. [CrossRef] [PubMed]

172. Jefford, C.W.; Bernardinelli, G.; Tanaka, J.I.; Higa, T. Structures and absolute configurations of the marine toxins, latrunculin A and laulimalide. *Tetrahedron Lett.* **1996**, *37*, 159–162. [CrossRef]

173. Ghosh, A.K.; Wang, Y. Total synthesis of (−)-laulimalide. *J. Am. Chem. Soc.* **2000**, *122*, 11027–11028. [CrossRef]

174. Gollner, A.; Mulzer, J. Total synthesis of neolaulimalide and isolaulimalide. *Organic Lett.* **2008**, *10*, 4701–4704. [CrossRef] [PubMed]

175. Manzo, E.; Gavagnin, M.; Somerville, M.J.; Mao, S.-C.; Ciavatta, M.L.; Mollo, E.; Schupp, P.J.; Garson, M.J.; Guo, V.; Cimino, G. Chemistry of *Glossodoris* nudibranchs: Specific occurrence of 12-keto scalaranes. *J. Chem. Ecol.* **2007**, *33*, 2325–2336. [CrossRef] [PubMed]

176. Rogers, S.D.; Paul, V.J. Chemical defenses of three *Glossodoris* nudibranchs and their dietary *Hyrtios* sponges. *Mar. Ecol. Progr. Ser.* **1991**, *77*, 221–232. [CrossRef]

177. Avila, C.; Paul, V.J. Chemical ecology of the nudibranch *Glossodoris pallida*: Is the location of diet-derived metabolites important for defense? *Mar. Ecol. Progr. Ser.* **1997**, *150*, 171–180. [CrossRef]

178. Winters, A.E.; White, A.M.; Dewi, A.S.; Mudianta, I.W.; Wilson, N.G.; Forster, L.C.; Garson, M.J.; Cheney, K.L. Distribution of defensive metabolites in nudibranch molluscs. *J. Chem. Ecol.* **2018**, *44*, 384–396. [CrossRef]

179. Zhukova, N.V. Lipids and fatty acids of nudibranch molluscs: Potential sources of bioactive compounds. *Mar. Drugs* **2014**, *12*, 4578–4592. [CrossRef]

180. Li, X.L.; Li, S.W.; Yao, L.G.; Mollo, E.; Gavagnin, M.; Guo, Y.W. The chemical and chemo-ecological studies on Weizhou nudibranch Glossodoris atromarginata. *Magn Reson Chem.* **2019**, 1–7. [CrossRef]

181. Somerville, M.J.; Mollo, E.; Cimino, G.; Rungprom, W.; Garson, M.J. Spongian diterpenes from Australian nudibranchs: An anatomically guided chemical study of *Glossodoris atromarginata*. *J. Nat. Prod.* **2007**, *70*, 1836. [CrossRef]

182. Yong, K.W.; Mudianta, I.W.; Cheney, K.L.; Mollo, E. : Blanchfield, J.T.; Garson, M.J. Isolation of norsesterterpenes and spongian diterpenes from *Dorisprismatica* (=*Glossodoris*) *atromarginata*. *J. Nat. Prod.* **2015**, *78*, 421–430. [CrossRef]

183. Fontana, A.; Mollo, E.; Ortea, J.; Gavagnin, M.; Cimino, G. Scalarane and homoscalarane compounds from the nudibranchs *Glossodoris sedna* and *Glossodoris dalli*: Chemical and biological properties. *J. Nat. Prod.* **2000**, *63*, 527–530. [CrossRef]

184. Cimino, G.; De Rosa, S.; De Stefano, S.; Sodano, G. The chemical defense of four Mediterranean nudibranchs. *Comp. Biochem. Physiol. B Comp. Biochem.* **1982**, *73*, 471–474. [CrossRef]

185. Forster, L.C.; Winters, A.E.; Cheney, K.L.; Dewapriya, P.; Capon, R.J.; Garson, M.J. Spongian-16-one diterpenes and their anatomical distribution in the Australian nudibranch *Goniobranchus collingwoodi*. *J. Nat. Prod.* **2017**, *80*, 670–675. [CrossRef] [PubMed]

186. Mudianta, W.; White, A.M.; Suciati, P.L.K.; Krishnaraj, R.R.; Winters, A.E.; Mollo, E.; Cheney, K.L.; Garson, M.J. Chemoecological studies on marine natural products: Terpene chemistry from marine mollusks. *Pure Appl. Chem.* **2014**, *86*, 995–1002. [CrossRef]

187. Winters, A.E.; White, A.M.; Cheney, K.L.; Garson, M.J. Geographic variation in diterpene-based secondary metabolites and level of defence in an aposematic nudibranch, *Goniobranchus splendidus*. *J. Moll. Stud.* **2019**, *85*, 133–142. [CrossRef]

188. Winters, A.E.; Green, N.F.; Wilson, N.G.; How, M.J.; Garson, M.J.; Marshall, N.J.; Cheney, K.L. Stabilizing selection on individual pattern elements of aposematic signals. *Proc. R. Soc. B* **2017**, *284*, 20170926. [CrossRef] [PubMed]

189. White, A.M.; Dewi, A.S.; Cheney, K.L.; Winters, A.E.; Blanchfield, J.T.; Garson, M.J. Oxygenated diterpenes from the Indo-Pacific nudibranchs *Goniobranchus splendidus* and *Ardeadoris egretta*. *Nat. Prod. Commun.* **2016**, *11*, 921–924. [CrossRef]

190. Hirayama, Y.; Katavic, P.L.; White, A.M.; Pierens, G.K.; Lambert, L.K.; Winters, A.E.; Kigoshi, H.; Kita, M.; Garson, M.J. New cytotoxic norditerpenes from the Australian nudibranchs *Goniobranchus splendidus* and *Goniobranchus daphne*. *Aust. J. Chem.* **2016**, *69*, 136–144. [CrossRef]

191. Mudianta, I.W.; White, A.M.; Garson, M.J. Oxygenated Terpenes from Indo-Pacific nudibranchs: Scalarane sesterterpenes from *Glossodoris hikuerensis* and 12-Acetoxy dendrillolide A from *Goniobranchus albonares*. *Nat. Prod. Commun.* **2015**, *10*, 865–868. [CrossRef]

192. White, A.M.; Pierens, G.K.; Forster, L.C.; Winters, A.E.; Cheney, K.L.; Garson, M.J. Rearranged diterpenes and norditerpenes from three Australian *Goniobranchus* mollusks. *J. Nat. Prod.* **2016**, *79*, 477–483. [CrossRef]

193. Mollo, E.; Gavagnin, M.; Carbone, M.; Guo, Y.-W.; Cimino, G. Chemical studies on Indopacific *Ceratosoma* nudibranchs illuminate the protective role of their dorsal horn. *Chemoecology* **2005**, *15*, 31–36. [CrossRef]

194. Cimino, G.; De Stefano, S.; Guerriero, A.; Minale, L. Furanosesquiterpenoids in sponges-III. Pallescensins AD from *Disidea pallescens*: New skeletal types. *Tetrahedron Lett.* **1975**, *16*, 1425–1428. [CrossRef]

195. Kazlauskas, R.; Murphy, P.T.; Wells, R.J.; Daly, J.J.; Schönholzer, P. Two sesquiterpene furans with new carbocyclic ring systems and related thiol acetates from a species of the sponge genus *Dysidea*. *Tetrahedron Lett.* **1978**, *19*, 4951–4954. [CrossRef]

196. Cameron, G.M.; Stapleton, B.L.; Simonsen, S.M.; Brecknell, D.J.; Garson, M.J. New sesquiterpene and brominated metabolites from the tropical marine sponge *Dysidea* sp. *Tetrahedron* **2000**, *56*, 5247–5252. [CrossRef]

197. Charles, C.; Braekman, J.C.; Daloze, D.; Tursch, B.; Declercq, J.P.; Germain, G.; Van Meerssche, M. Chemical studies of marine invertebrates. XXXIV. Herbadysidolide and herbasolide, two unusual sesquiterpenoids from the sponge *Dysidea herbacea*. *Bull. Soc. Chim. Belg.* **1978**, *87*, 481–486. [CrossRef]

198. Fontana, A.; Avila, C.; Martinez, E.; Ortea, J.; Trivellone, E.; Cimino, G. Defensive allomones in three species of *Hypselodoris* (gastropoda: Nudibranchia) from the Cantabrian sea. *J. Chem. Ecol.* **1993**, *19*, 339–356. [CrossRef]

199. Hochlowski, J.E.; Walker, R.P.; Ireland, C.; Faulkner, D.J. Metabolites of four nudibranchs of the genus *Hypselodoris*. *J. Org. Chem.* **1982**, *47*, 88–91. [CrossRef]

200. Grode, S.H.; Cardellina, J.H. Sesquiterpenes from the sponge *Dysidea etheria* and the nudibranch *Hypselodoris zebra*. *J. Nat. Prod.* **1984**, *47*, 76–83. [CrossRef]

201. García-Gómez, J.C.; Cimino, G.; Medina, A. Studies on the defensive behaviour of *Hypselodoris* species (Gastropoda: Nudibranchia): Ultrastructure and chemical analysis of mantle dermal formations (MDFs). *Mar. Biol.* **1990**, *106*, 245–250. [CrossRef]

202. Avila, C.; Cimino, G.; Fontana, A.; Gavagnin, M.; Ortea, J.; Trivellone, E. Defensive strategy of two *Hypselodoris* nudibranchs from Italian and Spanish coasts. *J. Chem. Ecol.* **1991**, *17*, 625–636. [CrossRef]

203. Cimino, G.; Fontana, A.; Giménez, F.; Marin, A.; Mollo, E.; Trivellone, E.; Zubia, E. Biotransformation of a dietary sesterterpenoid in the Mediterranean nudibranch *Hypselodoris orsini*. *Experientia* **1993**, *49*, 582–586. [CrossRef]

204. Fontana, A.; Trivellone, E.; Mollo, E.; Cimino, G.; Avila, C.; Martinez, E.; Ortea, J. Further chemical studies of Mediterranean and Atlantic *Hypselodoris* nudibranchs: A new furanosesquiterpenoid from *Hypsdodoris webbi*. *J. Nat. Prod.* **1994**, *57*, 510–513. [CrossRef]

205. Haber, M.; Cerfeda, S.; Carbone, M.; Calado, G.; Gaspar, H.; Neves, R.; Maharajan, V.; Cimino, G.; Gavagnin, M.; Ghiselin, M.T.; et al. Coloration and defense in the nudibranch gastropod *Hypselodoris fontandraui*. *Biol. Bull.* **2010**, *218*, 181–188. [CrossRef] [PubMed]

206. Da Cruz, J.F.; Gaspar, H.; Calado, G. Turning the game around: Toxicity in a nudibranch-sponge predator-prey association. *Chemoecology* **2012**, *22*, 47–53. [CrossRef]

207. Pereira, F.R.; Berlinck, R.G.S.; Rodrigues Filho, E.; Veloso, K.; Ferreira, A.G.; Padula, V. Metabólitos secundários dos nudibrânquios *Tambja stegosauriformis*, *Hypselodoris lajensis* e *Okenia zoobotryon* e dos briozoários *Zoobotryon verticillatum* e *Bugula dentata* da costa do Brasil. *Quim. Nova* **2012**, *35*, 2194–2201. [CrossRef]

208. Cimino, G.; De Stefano, S.; Minale, L.; Trivellone, E. Furanosesquiterpenoids in sponges-V: Spiniferins from *Pleraplysilla spinifera*. *Tetrahedron Lett.* **1975**, *16*, 3727–3730. [CrossRef]

209. Mudianta, I.W.; Challinor, V.L.; Winters, A.E.; Cheney, K.L.; De Voss, J.J.; Garson, M.J. Synthesis and determination of the absolute configuration of (−)-(5R, 6Z)-dendrolasin-5-acetate from the nudibranch *Hypselodoris jacksoni*. *Beilstein J. Org. Chem.* **2013**, *9*, 2925–2933. [CrossRef]

210. Avila, C. Substancias Naturales de Moluscos Opistobranquios: Estudio de su Estructura, Origen y Función en Ecosistemas Bentónicos. Ph.D. Thesis, University of Barcelona, Barcelona, Catalonia, Spain, 1993.

211. Avila, C.; Durfort, M. Histology of epithelia and mantle glands of selected species of doridacean mollusks with chemical defensive strategies. *Veliger* **1996**, *39*, 148–163.

212. Gaspar, H.; Rodrigues, A.I.; Calado, G. Comparative study of chemical defences from two allopatric north Atlantic subspecies of *Hypselodoris picta* (Mollusca: Opisthobranchia). *Açoreana* **2009**, *6*, 137–143.

213. McPhail, K.L.; Davies-Coleman, M.T.; Coetzee, P. A new furanosesterterpene from the South African nudibranch *Hypselodoris capensis* and a Dictyoceratida sponge. *J. Nat. Prod.* **1998**, *61*, 961–964. [CrossRef]

214. Mudianta, W.I.; Martiningsih, N.W.; Dodik Prasetia, I.N.; Nursid, M. Bioactive terpenoid from the balinese nudibranch *Hypselodoris infucata*. *Indones. J. Pharm.* **2016**, *27*, 104–110. [CrossRef]

215. Guella, G.; Mancini, I.; Guerriero, A.; Pietra, F. New furano-sesquiterpenoids from Mediterranean sponges. *Helv. Chim. Acta* **1985**, *68*, 1276–1282. [CrossRef]

216. Fontana, A.; Muniaín, C.; Cimino, G. First chemical study of patagonian nudibranchs: A new seco-11, 12-spongiane, tyrinnal, from the defensive organs of *Tyrinna nobilis*. *J. Nat. Prod.* **1998**, *61*, 1027–1029. [CrossRef] [PubMed]

217. Matsunaga, S.; Fusetani, N.; Hashimoto, K.; Koseki, K.; Noma, M. Bioactive marine metabolites. *J. Ame. Chem. Soc.* **1986**, *13*, 847–849. [CrossRef]

218. Pawlik, J.R.; Kernan, M.R.; Molinski, T.F.; Harper, M.K.; Faulkner, D.J. Defensive chemicals of the Spanish dancer nudibranch *Hexabranchus sanguineus* and its egg ribbons: Macrolides derived from a sponge diet. *J. Exp. Mar. Biol. Ecol.* **1988**, *119*, 99–109. [CrossRef]

219. Dalisay, D.S.; Rogers, E.W.; Edison, A.S.; Molinski, T.F. Trisoxazole macrolides and thiazole-containing cyclic peptides from the nudibranch *Hexabranchus sanguineus*. *J. Nat. Prod.* **2009**, *72*, 732–738. [CrossRef] [PubMed]

220. Matsunaga, S.; Fusetani, N.; Hashimoto, K.; Koseki, K.; Noguchi, H.; Noma, M.; Sankawa, U. Bioactive marine metabolites, part 25. Further kabiramides and halichondramides cytotoxic peptides from *Hexabranchus* egg masses. *J. Org. Chem.* **1989**, *54*, 1360–1363. [CrossRef]

221. Kernan, M.R.; Molinski, T.F.; Faulkner, D.J. Macrocyclic antifungal metabolites from the Spanish dancer nudibranch *Hexabranchus sanguineus* and sponges of the genus *Halichondria*. *J. Org. Chem.* **1988**, *53*, 5014–5020. [CrossRef]

222. Roesener, J.A.; Scheuer, P.J. Ulapualide A and B, extraordinary antitumor macrolides from nudibranch eggmasses. *J. Am. Chem. Soc.* **1986**, *108*, 846–847. [CrossRef]

223. Carté, B.; Faulkner, D.J. Defensive metabolites from three nembrothid nudibranchs. *J. Org. Chem.* **1983**, *48*, 2314–2318. [CrossRef]

224. Granato, A.C.; de Oliveira, J.H.; Seleghim, M.H.; Berlinck, R.G.; Macedo, M.L.; Ferreira, A.G.; Rocha, R.M.D.; Hajdu, E.; Peixinho, S.; Pessoa, C.O.; et al. Produtos naturais da ascidia *Botrylloides giganteum*, das esponjas *Verongula gigantea*, *Ircinia felix*, *Cliona delitrix* e do nudibrânquio *Tambja eliora*, da costa do Brasil. *Quim. Nova* **2005**, *28*, 192–198. [CrossRef]

225. Blackman, A.J.; Li, C.P. New tambjamine alkaloids from the marine bryozoan *Bugula dentata*. *Aust. J. Chem.* **1994**, *47*, 1625–1629. [CrossRef]

226. Paul, V.; Lindquist, N.; Fenical, W. Chemical defenses of the tropical ascidian *Atapozoa* sp. and its nudibranch predators *Nembrotha* spp. *Mar. Ecol. Prog. Ser.* **1990**, *59*, 109–111. [CrossRef]

227. Lindquist, N.; Fenical, W. New tamjamine class alkaloids from the marine ascidian *Atapozoa* sp. and its nudibranch predators. Origin of the tambjamines in *Atapozoa*. *Experientia* **1991**, *47*, 504–506. [CrossRef]

228. Carbone, M.; Irace, C.; Costagliola, F.; Castelluccio, F.; Villani, G.; Calado, G.; Padula, V.; Cimino, G.; Cervera, J.L.; Santamaria, R.; et al. A new cytotoxic tambjamine alkaloid from the Azorean nudibranch *Tambja ceutae*. *Bioorg. Med. Chem. Lett.* **2010**, *20*, 2668–2670. [CrossRef] [PubMed]

229. Cronin, G.; Hay, M.; Fenical, W.; Lindquist, N. Distribution, density, and sequestration of host chemical defenses by the specialist nudibranch *Tritonia hamnerorum* found at high densities on the sea fan *Gorgonia ventalina*. *Mar. Ecol. Prog. Ser.* **1995**, *119*, 177–189. [CrossRef]

230. McClintock, J.B.; Bryan, P.J.; Slattery, M.; Baker, B.J.; Yoshida, W.Y.; Hamann, M.; Heine, J.N. Chemical ecology of three Antarctic gastropods. *Antarct. J.* **1994**, *29*, 151–154.

231. McClintock, J.B.; Baker, B.J.; Slattery, M.; Heine, J.N.; Bryan, P.J.; Yoshida, W.; Davies-Coleman, M.T.; Faulkner, D.J. Chemical defense of common Antarctic shallow-water nudibranch *Tritoniella belli* Eliot (Mollusca: Tritonidae) and its prey, *Clavularia frankliniana* Rouel (Cnidaria: Octocorallia). *J. Chem. Ecol.* **1994**, *20*, 3361–3372. [CrossRef]

232. Bryan, P.J.; McClintock, J.B.; Baker, B.J. Population biology and antipredator defenses of the shallow-water Antarctic nudibranch *Tritoniella belli*. *Mar. Biol.* **1998**, *132*, 259–265. [CrossRef]

233. McClintock, J.B.; Baker, B.J. Palatability and chemical defense in the eggs, embryos and larvae of shallow-water Antarctic marine invertebrates. *Mar. Ecol. Progr. Ser.* **1997**, *154*, 121–131. [CrossRef]

234. Ciavatta, M.L.; Manzo, E.; Mollo, E.; Mattia, C.A.; Tedesco, C.; Irace, C.; Guo, Y.-W.; Li, X.-B.; Cimino, G.; Gavagnin, M. Tritoniopsins A–D, cladiellane-based diterpenes from the south china sea nudibranch *Tritoniopsis elegans* and its prey *Cladiella krempfi*. *J. Nat. Prod.* **2011**, *74*, 1902–1907. [CrossRef]

235. Affeld, S.; Wägele, H.; Avila, C.; Kehraus, S.; König, G.M. Distribution of homarine in some Opisthobranchia (Gastropoda: Mollusca). *Bonn. Zool. Beitr.* **2007**, *55*, 181–190.

236. McClintock, J.B.; Baker, B.J.; Hamann, M.T.; Yoshida, W.; Slattery, M.; Heine, J.N.; Bryan, P.J.; Jayatilake, G.S.; Moon, B.H. Homarine as a feeding deterrent in common shallow-water antarctic lamellarian gastropod *Marseniopsis mollis*: A rare example of chemical defense in a marine prosobranch. *J. Chem. Ecol.* **1994**, *20*, 2539–2549. [CrossRef] [PubMed]

237. Cimino, G.; Crispino, A.; Di Marzo, V.; Sodano, G.; Spinella, A.; Villani, G. A marine mollusc provides the first example of in vivo storage of prostaglandins: Prostaglandin-1, 15-lactones. *Experientia* **1991**, *47*, 56–60. [CrossRef] [PubMed]

238. Cimino, G.; Spinella, A.; Sodano, G. Naturally occurring prostaglandin-1, 15-lactones. *Tetrahedron Lett.* **1989**, *30*, 3589–3592. [CrossRef]

239. Cimino, G.; Crispino, A.; Di Marzo, V.; Spinella, A.; Sodano, G. Prostaglandin 1, 15-lactones of the F series from the nudibranch mollusk *Tethys fimbria*. *J. Org. Chem.* **1991**, *56*, 2907–2911. [CrossRef]

240. Di Marzo, V.; Cimino, G.; Crispino, A.; Minardi, C.; Sodano, G.; Spinella, A. A novel multifunctional metabolic pathway in a marine mollusc leads to unprecedented prostaglandin derivatives (prostaglandin 1, 15-lactones). *Biochem. J.* **1991**, *273*, 593–600. [CrossRef]

241. Cutignano, A.; Moles, J.; Avila, C.; Fontana, A. Granuloside, a unique linear homosesterterpene from the Antarctic nudibranch *Charcotia granulosa*. *J. Nat. Prod.* **2015**, *78*, 1761–1764. [CrossRef]

242. Moles, J.; Wägele, H.; Cutignano, A.; Fontana, A.; Avila, C. Distribution of granuloside in the Antarctic nudibranch *Charcotia granulosa* (Gastropoda: Heterobranchia: Charcotiidae). *Mar. Biol.* **2016**, *163*, 54–65. [CrossRef]

243. Putz, A.; König, G.M.; Wägele, H. Defensive strategies of Cladobranchia (Gastropoda, Opisthobranchia). *Nat. Prod. Rep.* **2010**, *27*, 1386–1402. [CrossRef]

244. Affeld, S.; Kehraus, S.; Wägele, H.; König, G.M. Dietary derived sesquiterpenes from *Phyllodesmium lizardensis*. *J. Nat. Prod.* **2009**, *72*, 298–300. [CrossRef]

245. Coll, J.; Bowden, B.; Tapiolas, D.; Willis, R.; Djura, P.; Streamer, M.; Trott, L. Studies of Australian soft corals XXXV: The terpenoid chemistry of soft corals and its implications. *Tetrahedron* **1985**, *41*, 1085–1092. [CrossRef]

246. Slattery, M.; Avila, C.; Starmer, J.; Paul, V.J. A sequestered soft coral diterpene in the aeolid nudibranch *Phyllodesmium guamensis*. *J. Exp. Mar. Biol. Ecol.* **1998**, *226*, 33–49. [CrossRef]

247. Edrada, R.A.; Wray, V.; Witte, L.; van Ofwegen, L.; Proksch, P. Bioactive terpenes from the soft coral *Heteroxenia* sp. from Mindoro, Philippines. *Z. Naturforsch.* **2000**, *55*, 82–86. [CrossRef] [PubMed]

248. Bogdanov, A.; Kehraus, S.; Bleidissel, S.; Preisfeld, G.; Schillo, D.; Piel, J.; Brachmann, A.O.; Wägele, H.; König, G.M. Defense in the Aeolidoidean Genus *Phyllodesmium* (Gastropoda). *J. Chem. Ecol.* **2014**, *40*, 1013–1024. [CrossRef]

249. Mao, S.C.; Gavagnin, M.; Mollo, E.; Guo, Y.-W. A new rare asteriscane sesquiterpene and other related derivatives from the Hainan aeolid nudibranch *Phyllodesmium magnum*. *Biochem. System. Ecol.* **2011**, *39*, 408–411. [CrossRef]

250. Bogdanov, A.; Hertzer, C.; Kehraus, S.; Nietzer, S.; Rohde, S.; Schupp, P.J.; Wägele, H.; König, G.M. Defensive diterpene from the Aeolidoidean *Phyllodesmium longicirrum*. *J. Nat. Prod.* **2016**, *79*, 611–615. [CrossRef]

251. Bogdanov, A.; Hertzer, C.; Kehraus, S.; Nietzer, S.; Rohde, S.; Schupp, P.J.; Wägele, H.; König, G.M. Secondary metabolome and its defensive role in the aeolidoidean *Phyllodesmium longicirrum* (Gastropoda, Heterobranchia, Nudibranchia). *Beilstein J. Org. Chem.* **2017**, *13*, 502–519. [CrossRef]

252. Gillette, R.; Saeki, M.; Huang, R.C. Defensive mechanisms in notaspid snails: Acid humor and evasiveness. *J. Exp. Biol.* **1991**, *156*, 335–347.

253. Willan, R.C. A review of the diets in the Notaspidea (Mollusca: Opisthobranchia). *J. Malacol. Soc. Aust.* **1984**, *6*, 125–142. [CrossRef]

254. Taboada, S.; Núñez-Pons, L.; Avila, C. Feeding repellence of Antarctic and sub-Antarctic benthic invertebrates against the omnivorous sea star *Odontaster validus* Koehler, 1906. *Pol. Biol.* **2013**, *36*, 13–25. [CrossRef]

255. Moles, J.; Núñez-Pons, L.; Taboada, S.; Figuerola, B.; Cristobo, J.; Avila, C. Anti-predatory chemical defences in Antarctic benthic fauna. *Mar. Biol.* **2015**, *162*, 1813–1821. [CrossRef]

256. Andersen, R.J.; Faulkner, D.J. Antibiotics from marine organisms of the Gulf of California. In Proceedings of the Abstracts from 3rd Conference on Food and Drugs from the Sea, Kingston, RI, USA, 20–23 August 1972; pp. 111–115.

257. Teeyapant, R.; Kreis, P.; Wray, V.; Witte, L.; Proksch, P. Brominated secondary compounds from the marine sponge *Verongia aerophoba* and the sponge feeding gastropod *Tylodina perversa*. *Z. Naturforsch.* **1993**, *48*, 630–644. [CrossRef]

258. Gotsbacher, M.P.; Karuso, P. New antimicrobial bromotyrosine analogues from the sponge *Pseudoceratina purpurea* and its predator *Tylodina corticalis*. *Mar. Drugs* **2015**, *13*, 1389–1409. [CrossRef] [PubMed]

259. Ebel, R.; Marin, A.; Proksch, P. Organ-specific distribution of dietary alkaloids in the marine opisthobranch *Tylodina perversa*. *Biochem. System. Ecol.* **1999**, *27*, 769–777. [CrossRef]

260. Thoms, C.; Wolff, M.; Padmakumar, K.; Ebel, R.; Proksch, P. Chemical defense of Mediterranean sponges *Aplysina cavernicola* and *Aplysina aerophoba*. *Z. Naturforsch.* **2004**, *59*, 113–122. [CrossRef] [PubMed]

261. Becerro, M.A.; Turon, X.; Uriz, M.J.; Templado, J. Can a sponge feeder be an herbivore? Tylodina perversa (Gastropoda) feeding on *Aplysina aerophoba*. *Biol. J. Linn. Soc.* **2003**, *78*, 429–438. [CrossRef]

262. Cimino G, De Rosa S, De Stefano S, Spinella A, Sodano G The zoochrome of the sponge *Verongia aerophoba* ("Uranidine"). *Tetrahedron Lett.* **1984**, *25*, 2925–2928. [CrossRef]

263. Cimino, G.; Sodano, G. Transfer of sponge secondary metabolites to predators. In *Sponges in Time and Space: Biology, Chemistry, Paleontology*; van Soest, R.W.M., van Kempen, T.M.G., Braekman, J.-C., Eds.; AA: Balkema, Rotterdam, 1994; pp. 459–472.

264. Thompson, T.E. Defensive acid-secretion in marine gastropods. *J. Mar. Biolog. Assoc.* **1960**, *39*, 115–122. [CrossRef]

265. Thompson, T.E. Investigation of the acidic allomone of the gastropod mollusc *Philine aperta* by means of ion chromatography and histochemical localisation of sulphate and chloride ions. *J. Mollus. Stud.* **1986**, *52*, 38–44. [CrossRef]

266. Moles, J.; Avila, C.; Malaquias, M.A.E. Unmasking Antarctic mollusc lineages: Novel evidence from philinoid snails (Gastropoda: Cephalaspidea). *Cladistics* **2019**, *35*, 487–513. [CrossRef]

267. Neves, R.; Gaspar, H.; Calado, G. Does a shell matter for defence? Chemical deterrence in two cephalaspidean gastropods with calcified shells. *J. Mollus. Stud.* **2009**, *75*, 127–131. [CrossRef]

268. Fontana, A.; Cutignano, A.; Giordano, A.; Coll, A.D.; Cimino, G. Biosynthesis of aglajnes, polypropionate allomones of the opisthobranch mollusc *Bulla striata*. *Tetrahedron Lett.* **2004**, *45*, 6847–6850. [CrossRef]

269. Cimino, G.; Sodano, G.; Spinella, A.; Trivellone, E. Aglajne-1, a polypropionate metabolite from the opisthobranch mollusk *Aglaja depicta*. *Tetrahedron Lett.* **1985**, *26*, 3389–3392. [CrossRef]

270. Cimino, G.; Sodano, G.; Spinella, A. New propionate-derived metabolites from *Aglaja depicta* and from its prey *Bulla striata* (opisthobranch mollusks). *J. Org. Chem.* **1987**, *52*, 5326–5331. [CrossRef]

271. Marín, A.; Álvarez, L.A.; Cimino, G.; Spinella, A. Chemical defence in cephalaspidean gastropods: Origin, anatomical location and ecological roles. *J. Mollus. Stud.* **1999**, *65*, 121–131. [CrossRef]

272. Coval, S.J.; Scheuer, P.J. An intriguing C16-alkadienone-substituted 2-pyridine from a marine mollusk. *J. Org. Chem.* **1985**, *50*, 3024–3025. [CrossRef]

273. Coval, S.J.; Schulte, G.R.; Matsumoto, G.K.; Roll, D.M.; Scheuer, P.J. Two polypropionate metabolites from the cephalaspidean mollusk *Philinopsis speciosa*. *Tetrahedron Lett.* **1985**, *26*, 5359–5362. [CrossRef]

274. Cutignano, A.; Calado, G.; Gaspar, H.; Cimino, G.; Fontana, A. Polypropionates from *Bulla occidentalis*: Chemical markers and trophic relationships in cephalaspidean molluscs. *Tetrahedron Lett.* **2011**, *52*, 4595–4597. [CrossRef]

275. Nakao, Y.; Yoshida, W.Y.; Scheuer, P.J. Pupukeamide, a linear tetrapeptide from a cephalaspidean mollusk *Philinopsis speciosa*. *Tetrahedron Lett.* **1996**, *37*, 8993–8996. [CrossRef]

276. Nakao, Y.; Yoshida, W.Y.; Szabo, C.M.; Baker, B.J.; Scheuer, P.J. More peptides and other diverse constituents of the marine mollusk *Philinopsis speciosa*. *J. Org. Chem.* **1998**, *63*, 3272–3280. [CrossRef]

277. Reese, M.T.; Gulavita, N.K.; Nakao, Y.; Hamann, M.T.; Yoshida, W.Y.; Coval, S.J.; Scheuer, P.J. Kulolide: A cytotoxic depsipeptide from a cephalaspidean mollusk, *Philinopsis speciosa*. *J. Am. Chem. Soc.* **1996**, *118*, 11081–11084. [CrossRef]

278. Spinella, A.; Álvarez, L.A.; Cimino, G. Predator-prey relationship between *Navanax inermis* and *Bulla gouldiana*: A chemical approach. *Tetrahedron* **1993**, *49*, 3203–3210. [CrossRef]

279. Cruz-Rivera, E. Evidence for chemical defence in the Cephalaspidean *Nakamigawaia spiralis* Kuroda and Habe, 1961. *J. Mollus. Stud.* **2011**, *77*, 95–97. [CrossRef]

280. Poiner, A.; Paul, V.J.; Scheuer, P.J. Kumepaloxane, a rearranged trisnor sesquiterpene from the bubble shell *Haminoea cymbalum*. *Tetrahedron* **1989**, *45*, 617–622.

281. Chang, E.S. *Possible Anti-Predation Properties of the Egg Masses of the Marine Gastropods Dialula sandiegensis, Doris montereyensis and Haminoea virescens (Mollusca, Gastropoda)*; Friday Harbor Laboratories Student Research Papers; Friday Harbor Laboratories: Washington, DC, USA, 2014; p. 528.

282. Becerro, M.A.; Starmer, J.A.; Paul, V.J. Chemical defenses of cryptic and aposematic gastropterid molluscs feeding on their host sponge *Dysidea granulosa*. *J. Chem. Ecol.* **2006**, *32*, 1491–1500. [CrossRef]

283. Pereira, R.B.; Andrade, P.B.; Valentão, P. Chemical diversity and biological properties of secondary metabolites from sea hares of *Aplysia* genus. *Mar. Drugs* **2016**, *14*, 39. [CrossRef]

284. Ellingson, R.A.; Krug, P.J. Evolution of poecilogony from planktotrophy: Cryptic speciation, phylogeography, and larval development in the gastropod genus *Alderia*. *Evolution* **2006**, *60*, 2293–2310. [CrossRef]

285. Hunt, B.P.V.; Pakhomov, E.A.; Hosie, G.W.; Siegel, V.; Ward, P.; Bernard, K. Pteropods in Southern Ocean ecosystems. *Prog. Oceanogr.* **2008**, *78*, 193–221. [CrossRef]

286. Jörger, K.M.; Norenburg, J.L.; Wilson, N.G.; Schrödl, M. Barcoding against a paradox? Combined molecular species delineations reveal multiple cryptic lineages in elusive meiofaunal sea slugs. *BMC Evol. Biol.* **2012**, *12*, 245. [CrossRef]

287. Kinnel, R.B.; Dieter, R.K.; Meinwald, J.; Van Engen, D.; Clardy, J.; Eisner, T.; Stallard, M.O.; Fenical, W. Brasilenyne and cis-dihydrorhodophytin: Antifeedant medium-ring haloethers from a sea hare (*Aplysia brasiliana*). *PNAS* **1979**, *76*, 3576–3579. [CrossRef]

288. Kamiya, H.; Muramoto, K.; Goto, R.; Sakai, M.; Endo, Y.; Yamazaki, M. Purification and characterization of an antibacterial and antineoplastic protein secretion of a sea hare, *Aplysia juliana*. *Toxicon* **1989**, *27*, 1269–1277. [CrossRef]

289. Kamio, M.; Grimes, T.V.; Hutchins, M.H.; van Dam, R.; Derby, C.D. The purple pigment aplysioviolin in sea hare ink deters predatory blue crabs through their chemical senses. *Anim. Behav.* **2010**, *80*, 89–100. [CrossRef]

290. Kamio, M.; Nguyen, L.; Yaldiz, S.; Derby, C.D. How to produce a chemical defense: Structural elucidation and anatomical distribution of aplysioviolin and phycoerythrobilin in the sea hare *Aplysia californica*. *Chem. Biodivers.* **2010**, *7*, 1183–1197. [CrossRef] [PubMed]

291. Derby, C.D. Escape by inking and secreting: Marine molluscs avoid predators through a rich array of chemicals and mechanisms. *Biol. Bull.* **2007**, *213*, 274–289. [CrossRef]

292. Kicklighter, C.E.; Shabani, S.; Johnson, P.M.; Derby, C.D. Sea hares use novel antipredatory chemical defenses. *Curr. Biol.* **2005**, *15*, 549–554. [CrossRef]

293. Sheybani, A.; Nusnbaum, M.; Caprio, J.; Derby, C.D. Responses of the sea catfish Ariopsis felis to chemical defenses from the sea hare *Aplysia californica*. *J. Exp. Mar. Biol. Ecol.* **2009**, *368*, 153–160. [CrossRef]

294. Johnson, P.M.; Kicklighter, C.E.; Schmidt, M.; Kamio, M.; Yang, H.; Elkin, D.; Michel, W.C.; Tai, P.C.; Derby, C.D. Packaging of chemicals in the defensive secretory glands of the sea hare *Aplysia californica*. *J. Exp. Biol.* **2006**, *209*, 78–88. [CrossRef]

295. Nusnbaum, M.; Derby, C.D. Effects of sea hare ink secretion and its escapin-generated components on a variety of predatory fishes. *Biol. Bull.* **2010**, *218*, 282–292. [CrossRef]

296. Kicklighter, C.E.; Derby, C.D. Multiple components in ink of the sea hare *Aplysia californica* are aversive to the sea anemone *Anthopleura sola*. *J. Exp. Mar. Biol. Ecol.* **2006**, *334*, 256–268. [CrossRef]

297. Kamio, M.; Ko, K.C.; Zheng, S.; Wang, B.; Collins, S.L.; Gadda, G.; Tai, P.C.; Derby, C.D. The chemistry of escapin: Identification and quantification of the components in the complex mixture generated by an L-amino acid oxidase in the defensive secretion of the sea snail *Aplysia californica*. *Chem. Eur. J.* **2009**, *15*, 1597–1603. [CrossRef]

298. Pennings, S.C. Multiple factors promoting narrow host range in the sea hare, *Aplysia californica*. *Oecologia* **1990**, *82*, 192–200. [CrossRef] [PubMed]

299. Ginsburg, D.W.; Paul, V.J. Chemical defenses in the sea hare *Aplysia parvula*: Importance of diet and sequestration of algal secondary metabolites. *Mar. Ecol. Progr. Ser.* **2001**, *215*, 261–274. [CrossRef]

300. Stierle, D.B.; Wing, R.M.; Sims, J.J. Marine natural products, XI Costatone and costatolide, new halogenated monoterpenes from the red seaweed, *Plocamium costatum*. *Tetrahedron Lett.* **1976**, *49*, 4455–4458. [CrossRef]

301. Paul, V.J.; Arthur, K.E.; Ritson-Williams, R.; Ross, C.; Sharp, K. Chemical defenses: From compounds to communities. *Biol. Bull.* **2007**, *213*, 226–251. [CrossRef] [PubMed]

302. Kato, Y.; Scheuer, P.J. Aplysiatoxin and debromoaplysiatoxin, constituents of the marine mollusk *Stylocheilus longicauda* (Quoy and Gaimard, 1824). *J. Am. Chem. Soc.* **1974**, *96*, 2245–2246. [CrossRef] [PubMed]

303. Kato, Y.; Scheuer, P.J. The aplysiatoxins. *Pure Appl. Chem.* **1975**, *41*, 1–14. [CrossRef]

304. Rose, A.F.; Scheuer, P.J.; Springer, J.P.; Clardy, J. Stylocheilamide, an unusual constituent of the sea hare *Stylocheilus longicauda*. *J. Am. Chem. Soc.* **1978**, *100*, 7665–7670. [CrossRef]

305. Gallimore, W.A.; Galario, D.L.; Lacy, C.; Zhu, Y.; Scheuer, P.J. Two complex proline esters from the sea hare *Stylocheilus longicauda*. *J. Nat. Prod.* **2000**, *63*, 1022–1026. [CrossRef]

306. Todd, J.S.; Gerwick, W.H. Malyngamide I from the tropical marine cyanobacterium *Lyngbya majuscula* and the probable structure revision of stylocheilamide. *Tetrahedron Lett.* **1995**, *36*, 7837–7840. [CrossRef]

307. Gallimore, W.A.; Scheuer, P.J. Malyngamides O and P from the sea hare *Stylocheilus longicauda*. *J. Nat. Prod.* **2000**, *63*, 1422–1424. [CrossRef]

308. Pennings, S.C.; Paul, V.J. Sequestration of dietary secondary metabolites by three species of sea hares: Location, specificity and dynamics. *Mar. Biol.* **1993**, *117*, 535–546. [CrossRef]

309. Paul, V.J.; Pennings, S.C. Diet-derived chemical defenses in the sea hare *Stylocheilus longicauda* (Quoy et Gaimard 1824). *J. Exp. Mar. Biol. Ecol.* **1991**, *151*, 227–243. [CrossRef]

310. Capper, A.; Cruz-Rivera, E.; Paul, V.J.; Tibbetts, I.R. Chemical deterrence of a marine cyanobacterium against sympatric and non-sympatric consumers. *Hydrobiologia* **2006**, *553*, 319–326. [CrossRef]

311. Gopichand, Y.; Schmitz, F.J. Bursatellin: A new diol dinitrile from the sea hare *Bursatella leachii pleii*. *J. Org. Chem.* **1980**, *45*, 5383–5385. [CrossRef]

312. Cimino, G.; Gavagnin, M.; Sodano, G.; Spinella, A.; Strazzullo, G. Revised structure of Bursatellin. *J. Org. Chem.* **1987**, *52*, 2303–2306. [CrossRef]

313. Yoshida, W.Y.; Bryan, P.J.; Baker, B.J.; Mcclintock, J.B. Pteroenone: A defensive metabolite of the abducted Antarctic pteropod *Clione antarctica*. *J. Org. Chem.* **1995**, *60*, 780–782. [CrossRef]

314. Bryan, P.J.; Yoshida, W.Y.; McClintock, J.B.; Baker, B.J. Ecological role for pteroenone, a novel antifeedant from the conspicuous Antarctic pteropod *Clione antarctica* (Gymnosomata: Gastropoda). *Mar. Biol.* **1995**, *122*, 271–277.

315. McClintock, J.B.; Janssen, J. Pteropod abduction as a chemical defence in a pelagic Antarctic amphipod. *Nature* **1990**, *346*, 462–464. [CrossRef]

316. Gavagnin, M.; Mollo, E.; Montanaro, D.; Ortea, J.; Cimino, G. Chemical studies of Caribbean sacoglossans: Dietary relationships with green algae and ecological implications. *J. Chem. Ecol.* **2000**, *26*, 1563–1578. [CrossRef]

317. Gavagnin, M.; Mollo, E.; Cimino, G.; Ortea, J. A new γ-dihydropyrone-propionate from the Caribbean sascoglossan *Tridachia crispata*. *Tetrahedron Lett.* **1996**, *37*, 4259–4262. [CrossRef]

318. Gavagnin, M.; Mollo, E.; Castelluccio, F.; Montanaro, D.; Ortea, J.; Cimino, G. A novel dietary sesquiterpene from the marine sacoglossan *Tridachia crispata*. *Nat. Prod. Lett.* **1997**, *10*, 151–156. [CrossRef]

319. Ireland, C.; Faulkner, D.J. The defensive secretion of the opisthobranch mollusc *Onchidella binneyi*. *Bioorg. Chem.* **1978**, *7*, 125–131. [CrossRef]

320. Paul, V.J.; Sun, H.H.; Fenical, W. Udoteal, a linear diterpenoid feeding deterrent from the tropical green alga *Udotea flabellum*. *Phytochemistry* **1982**, *21*, 468–469. [CrossRef]

321. Gavagnin, M.; Spinella, A.; Crispino, A.; Epifanio, R.D.A.; Marn, A.; Cimino, G. Chemical-components of the Mediterranean ascoglossan *Thuridilla hopei*. *Gazz. Chim. Ital.* **1993**, *123*, 205–208.

322. Carbone, M.; Ciavatta, M.L.; De Rinaldis, G.; Castelluccio, F.; Mollo, E.; Gavagnin, M. Identification of thuridillin-related aldehydes from Mediterranean ascoglossan mollusk *Thuridilla hopei*. *Tetrahedron* **2014**, *70*, 3770–3773. [CrossRef]

323. Paul, V.; Ciminiello, P.; Fenical, W. Diterpenoid feeding deterrents from the Pacific green-alga *Pseudochlorodesmis furcellata*. *Phytochemistry* **1988**, *27*, 1011–1014. [CrossRef]

324. Somerville, M.J.; Katavic, P.L.; Lambert, L.K.; Pierens, G.K.; Blanchfield, J.T.; Cimino, G.; Mollo, E.; Gavagnin, M.; Banwell, M.G.; Garson, M.J. Isolation of thuridillins D-F, diterpene metabolites from the Australian sacoglossan mollusk *Thuridilla splendens*; relative configuration of the epoxylactone ring. *J. Nat. Prod.* **2012**, *75*, 1618–1624. [CrossRef]

325. Hay, M.E.; Duffy, J.E.; Paul, V.J.; Renaud, P.E.; Fenical, W. Specialist herbivores reduce their susceptibility to predation by feeding on the chemically defended seaweed *Avrainvillea longicaulis*. *Limnol. Oceanogr.* **1990**, *35*, 1734–1747. [CrossRef]

326. Vardaro, R.R.; Di Marzo, V.; Crispino, A.; Cimino, G. Cyercenes, novel polypropionate pyrones from the autotomizing Mediterranean mollusc *Cyerce cristallina*. *Tetrahedron* **1991**, *47*, 5569–5576. [CrossRef]

327. Di Marzo, V.; Vardaro, R.R.; De Petrocellis, L.; Villani, G.; Minei, R.; Cimino, G. Cyercenes, novel pyrones from the ascoglossan mollusc *Cyerce cristallina*. Tissue distribution, biosynthesis and possible involvement in defense and regenerative processes. *Experientia* **1991**, *47*, 1221–1227. [CrossRef]

328. Roussis, V.; Pawlik, J.R.; Hay, M.E.; Fenical, W. Secondary metabolites of the chemically rich ascoglossan *Cyerce nigricans*. *Experientia* **1990**, *49*, 327–329. [CrossRef]

329. Jensen, K.R. Defensive behavior and toxicity of the ascoglossan opisthobranch *Mourgona germaineae* Marcus. *J. Chem. Ecol.* **1984**, *10*, 475–486. [CrossRef] [PubMed]

330. Högberg, H.E.; Thompson, R.H.; King, T.J. The cymopols, a group of prenylated bromohydroquinones from the green calcareous alga *Cymopolia barbata*. *J. Chem. Soc. Perkin Trans. I* **1976**, *1*, 1696–1701. [CrossRef]

331. Hamann, M.T.; Otto, C.S.; Scheuer, P.J.; Dunbar, D.C. Kahalalides: Bioactive peptides from a marine mollusk *Elysia rufescens* and its algal diet *Bryopsis* sp. *J. Org. Chem.* **1996**, *61*, 6594–6600. [CrossRef] [PubMed]

332. Vardaro, R.R.; Di Marzo, V.; Cimino, G. Placidenes: Cyercene-like polypropionate γ-pyrones from the Mediterranean ascoglossan mollusc *Placida dendritica*. *Tetrahedron Lett.* **1992**, *33*, 2875–2878. [CrossRef]

333. Gray, C.A.; Davies-Coleman, M.T.; McQuaid, C. Labdane diterpenes from the South African marine pulmonate *Trimusculus costatus*. *Nat. Prod. Lett.* **1998**, *12*, 47–53. [CrossRef]

334. Manker, D.C.; Faulkner, D.J. Investigation of the role of diterpenes produced by marine pulmonates *Trimusculus reticulatus* and *T. conica*. *J. Chem. Ecol.* **1996**, *22*, 23–35. [CrossRef]

335. Díaz-Marrero, A.R.; Dorta, E.; Cueto, M.; Rovirosa, J.; San-Martín, A.; Loyola, A.; Darias, J. Labdane diterpenes with a new oxidation pattern from the marine pulmonate *Trimusculus peruvianus*. *Tetrahedron* **2003**, *59*, 4805–4809. [CrossRef]

336. Rovirosa, J.; Quezada, E.; San-Martin, A. New diterpene from the mollusc *Trimusculus peruvianus*. *Bol. Soc. Chil. Quim.* **1992**, *37*, 143–145.

337. Van Wyk, A.W.W.; Gray, C.A.; Whibley, C.E.; Osoniyi, O.; Hendricks, D.T.; Caira-Mino, R.; Davies-Coleman, M.T. Bioactive metabolites from the South African marine mollusk *Trimusculus costatus*. *J. Nat. Prod.* **2008**, *71*, 420–425. [CrossRef]

338. Wiggering, B.; Neiber, M.T.; Gebauer, K.; Glaubrecht, M. One species, two developmental modes: a case of geographic poecilogony in marine gastropods. *BMC Evol. Biol.* **2020**, *20*, 76. [CrossRef] [PubMed]

339. Manker, D.C.; Garson, M.J.; Faulkner, D.J. *De novo* biosynthesis of polypropionate metabolites in the marine pulmonate *Siphonaria denticulata*. *J. Chem. Soc. Chem. Commun.* **1988**, *16*, 1061–1062. [CrossRef]

340. Paul, M.C.; Zubía, E.; Ortega, M.J.; Salvá, J. New polypropionates from *Siphonaria pectinata*. *Tetrahedron* **1997**, *53*, 2303–2308. [CrossRef]

341. Hochlowski, J.E.; Faulkner, D.J. Antibiotics from the marine pulmonate *Siphonaria diemenensis*. *Tetrahedron Lett.* **1983**, *24*, 1917–1920. [CrossRef]

342. Hochlowski, J.E.; Faulkner, D.J. Metabolites of the marine pulmonate *Siphonaria australis*. *J. Org. Chem.* **1984**, *49*, 3838–3840. [CrossRef]

343. Biskupiak, J.E.; Ireland, C.M. Pectinatone, a new antibiotic from the mollusc *Siphonaria pectinata*. *Tetrahedron Lett.* **1983**, *24*, 3055–3058. [CrossRef]

344. Norte, M.; Cataldo, F.; González, A.G.; Rodríguez, M.L.; Ruiz-Pérez, C. New metabolites from the marine mollusc *Siphonaria grisea*. *Tetrahedron* **1990**, *46*, 1669–1678. [CrossRef]

345. Norte, M.; Fernández, J.J.; Padilla, A. Isolation and synthesis of siphonarienal a new polypropionate from *Siphonaria grisea*. *Tetrahedron Lett.* **1994**, *35*, 3413–3416. [CrossRef]

346. Beukes, D.R.; Davies-Coleman, M.T. Novel polypropionates from the South African marine mollusc *Siphonaria capensis*. *Tetrahedron* **1999**, *55*, 4051–4056. [CrossRef]

347. Hochlowski, J.E.; Faulkner, D.J.; Matsumoto, G.K.; Clardy, J. The denticulatins, two propionate metabolites from the pulmonate *Siphonaria denticulata*. *J. Am. Chem. Soc.* **1983**, *105*, 7413–7415. [CrossRef]

348. Hochlowski, J.; Coll, J.; Faulkner, D.J.; Clardy, J. Novel metabolites of four *Siphonaria* species. *J. Am. Chem. Soc.* **1984**, *106*, 6748–6750. [CrossRef]

349. Roll, D.M.; Biskupiak, J.E.; Mayne, C.L.; Ireland, C.M. Muamvatin, a novel tricyclic spiro ketal from the Fijian mollusk *Siphonaria normalis*. *J. Am. Chem. Soc.* **1986**, *108*, 6680–6682. [CrossRef]

350. Manker, D.C.; Faulkner, D.J. Vallartanones A and B, polypropionate metabolites of *Siphonaria maura* from Mexico. *J. Org. Chem.* **1989**, *54*, 5374–5377. [CrossRef]

351. Manker, D.C.; Faulkner, D.J.; Stout, T.J.; Clardy, J. The baconipyrones. Novel polypropionates from the pulmonate *Siphonaria baconi*. *J. Org. Chem.* **1989**, *54*, 5371–5374. [CrossRef]

352. Brecknell, D.J.; Collett, L.A.; Davies-Coleman, M.T.; Garson, M.J.; Jones, D.D. New non-contiguous polypropionates from marine molluscs: A comment on their natural product status. *Tetrahedron* **2000**, *56*, 2497–2502. [CrossRef]

353. Mary J., G. The biosynthesis of marine natural products. *Chem. Rev.* **1993**, *93*, 1699–1733.

354. Abramson, S.N.; Radic, Z.; Manker, D.; Faulkner, D.J.; Taylor, P. Onchidal: A naturally occurring irreversible inhibitor of acetylcholinesterase with a novel mechanism of action. *Mol. Pharm.* **1989**, *36*, 349–354.

355. Young, C.M.; Greenwood, P.G.; Powell, C.J. The ecological role of defensive secretions in the intertidal pulmonate *Onchidella borealis*. *Biol. Bull.* **1986**, *171*, 391–404. [CrossRef]

356. Biskupiak, J.E.; Ireland, C.M. Cytotoxic metabolites from the mollusc *Peronia peronii*. *Tetrahedron Lett.* **1985**, *26*, 4307–4310. [CrossRef]

357. Carbone, M.; Ciavatta, M.L.; Wang, J.R.; Cirillo, I.; Mathieu, V.; Kiss, R.; Mollo, E.; Guo, Y.W.; Gavagnin, M. Extending the record of bis-γ-pyrone polypropionates from marine pulmonate mollusks. *J. Nat. Prod.* **2013**, *76*, 2065–2073. [CrossRef]

358. Rodríguez, J.; Fernández, R.; Quiñoá, E.; Riguera, R.; Debitus, C.; Bouchet, P. Onchidin: A cytotoxic depsipeptide with C 2 symmetry from a marine mollusc. *Tetrahedron Lett.* **1994**, *35*, 9239–9242. [CrossRef]

359. Fernández, R.; Rodríguez, J.; Quiñoá, E.; Riguera, R.; Muñoz, L.; Fernández-Suárez, M.; Debitus, C. Onchidin B: A new cyclodepsipeptide from the mollusc *Onchidium* sp. *J. Am. Chem. Soc.* **1996**, *118*, 11635–11643. [CrossRef]

360. Carbone, M.; Gavagnin, M.; Mattia, C.A.; Lotti, C.; Castelluccio, F.; Pagano, B.; Mollo, E.; Guo, Y.W.; Cimino, G. Structure of onchidione, a bis-γ-pyrone polypropionate from a marine pulmonate mollusk. *Tetrahedron* **2009**, *65*, 4404–4409. [CrossRef]

361. Guo, Y.W.; Gavagnin, M.; Carbone, M.; Mollo, E.; Cimino, G. Recent Sino-Italian collaborative studies on marine organisms from the South China Sea. *Pure Appl. Chem.* **2012**, *84*, 1391–1405. [CrossRef]

362. Wang, J.R.; Carbone, M.; Gavagnin, M.; Mándi, A.; Antus, S.; Yao, L.G.; Cimino, G.; Kurtán, T.; Guo, Y.W. Assignment of absolute configuration of bis-γ-pyrone polypropionates from marine pulmonate molluscs. *Eur. J. Org. Chem.* **2012**, *6*, 1107–1111. [CrossRef]

363. Young, R.M.; Baker, B.J. Defensive chemistry of the Irish nudibranch *Archidoris psuedoargus* (Gastropoda opisthobranchia). *Planta Med.* **2016**, *82*, 597. [CrossRef]

364. Zubia, E.; Gavagnin, M.; Crispino, A.; Martinez, E.; Ortea, J.; Cimino, G. Diasteroisomeric ichthyotoxic acylglycerols from the dorsum of two geographically distinct populations of *Archidoris* nudibranchs. *Experientia* **1993**, *49*, 268–271. [CrossRef]

365. Gustafson, K.; Andersen, R.J. Chemical studies of British Columbia nudibranchs. *Tetrahedron* **1985**, *41*, 1101–1108. [CrossRef]

366. Soriente, A.; Sodano, G.; Reed, K.C.; Todd, C. A new ichthyotoxic diacylglycerol from the nudibranch *Archidoris pseudoargus*. *Nat. Prod. Lett.* **1993**, *3*, 31–35. [CrossRef]

367. Cimino, G.; Crispino, A.; Gavagnin, M.; Trivellone, E.; Zubía, E.; Martínez, E.; Ortea, J. Archidorin: A new ichthyotoxic diacylglycerol from the Atlantic dorid nudibranch *Archidoris tuberculata*. *J. Nat. Prod.* **1993**, *56*, 1642–1646. [CrossRef]

368. Andersen, R.J.; Sum, F.W. Farnesic acid glycerides from the nudibranch *Archidoris odhneri*. *Tetrahedron Lett.* **1980**, *21*, 797–800. [CrossRef]

369. Gustafson, K.; Andersen, R.J.; Chen, M.H.; Clardy, J.; Hochlowski, J.E. Terpenoic acid glycerides from the dorid nudibranch *Archidoris montereyensis*. *Tetrahedron Lett.* **1984**, *25*, 11–14. [CrossRef]

370. Gavagnin, M.; Ungur, N.; Castelluccio, F.; Cimino, G. Novel verrucosins from the skin of the Mediterranean nudibranch *Doris verrucosa*. *Tetrahedron* **1997**, *53*, 1491–1504. [CrossRef]

371. Cimino, G.; Gavagnin, M.; Sodano, G.; Puliti, R.; Mattia, C.A.; Mazzarella, L. Verrucosin-A and-B, ichthyotoxic diterpenoic acid glycerides with a new carbon skeleton from the dorid nudibranch *Doris verrucosa*. *Tetrahedron* **1988**, *44*, 2301–2310. [CrossRef]

372. Avila, C.; Ballesteros, M.; Cimino, G.; Crispino, A.; Gavagnin, M.; Sodano, G. Biosynthetic origin and anatomical distribution of the main secondary metabolites in the nudibranch mollusc *Doris verrucosa*. *Comp. Biochem. Physiol. B Comp. Biochem.* **1990**, *97*, 363–368. [CrossRef]

373. Fusetani, N.; Wolstenholme, H.J.; Matsunaga, S. Co-occurrence of 9-isocyanopupukeanane and its C-9 epimer in the nudibranch *Phyllidia bourguini*. *Tetrahedron Lett.* **1990**, *31*, 5623–5624. [CrossRef]

374. Kassuhlke, K.E.; Potts, B.C.; Faulkner, D.J. New nitrogenous sesquiterpenes from two Philippine nudibranchs, *Phyllidia pustulosa* and *P. varicosa*, and from a Palauan sponge, *Halichondria* cf. *lendenfeldi*. *J. Org. Chem.* **1991**, *56*, 3747–3750. [CrossRef]

375. Fusetani, N.; Wolstenholme, H.J.; Shinoda, K.; Asai, N.; Matsunaga, S.; Onuki, H.; Hirota, H. Two sesquiterpene isocyanides and a sesquiterpene thiocyanate from the marine sponge *Acanthella* cf. *cavernosa* and the nudibranch *Phyllidia ocellata*. *Tetrahedron Lett.* **1992**, *33*, 6823–6826. [CrossRef]

376. White, A.M.; Pierens, G.K.; Skinner-Adams, T.; Andrews, K.T.; Bernhardt, P.V.; Krenske, E.H.; Mollo, E.; Garson, M.J. Antimalarial isocyano and isothiocyanato sesquiterpenes with tri-and bicyclic skeletons from the nudibranch *Phyllidia ocellata*. *J. Nat. Prod.* **2015**, *78*, 1422–1427. [CrossRef]

377. Jomori, T.; Shibutani, T.; Ahmadi, P.; Suzuka, T.; Tanaka, J. A New Isocyanosesquiterpene from the nudibranch *Phyllidiella pustulosa*. *Nat. Prod. Commun.* **2015**, *10*, 1913–1914. [CrossRef]

378. Crews, P.; Kakou, Y.; Quinoa, E. Mycothiazole, a polyketide heterocycle from a marine sponge. *J. Am. Chem. Soc.* **1988**, *110*, 4365–4368. [CrossRef]

379. Sonnenschein, R.N.; Johnson, T.A.; Tenney, K.; Valeriote, F.A.; Crews, P. A reassignment of (−)-mycothiazole and the isolation of a related diol. *J. Nat. Prod.* **2006**, *69*, 145–147. [CrossRef] [PubMed]

380. De Silva, E.D.; Scheuer, P.J. Furanoditerpenoids from the dorid nudibranch *Casella atromarginata*. *Heterocycles* **1982**, *17*, 167–170.

381. Fontana, A.; Cavaliere, P.; Ungur, N.; D'Souza, L.; Parameswaram, P.S.; Cimino, G. New scalaranes from the nudibranch *Glossodoris atromarginata* and its sponge prey. *J. Nat. Prod.* **1999**, *62*, 1367–1370. [CrossRef] [PubMed]

382. Gross, H.; Wright, A.D.; Reinscheid, U.; König, G.M. Three new spongian diterpenes from the Fijian marine sponge *Spongia* sp. *Nat. Prod. Commun.* **2009**, *4*, 315–322. [CrossRef] [PubMed]

383. Gavagnin, M.; Mollo, E.; Docimo, T.; Guo, Y.W.; Cimino, G. Scalarane metabolites of the nudibranch *Glossodoris rufomarginata* and its dietary sponge from the South China Sea. *J. Nat. Prod.* **2004**, *67*, 2104–2107. [CrossRef]

384. Betancur-Galvis, L.; Zuluaga, C.; Arnó, M.; González, M.A.; Zaragozá, R.J. Cytotoxic effect (on tumor cells) and in vitro antiviral activity against herpes simplex virus of synthetic spongiane diterpenes. *J. Nat. Prod.* **2002**, *65*, 189–192. [CrossRef]

385. Kamel, H.N.; Kim, Y.B.; Rimoldi, J.M.; Fronczek, F.R.; Ferreira, D.; Slattery, M. Scalarane sesterterpenoids: Semisynthesis and biological activity. *J. Nat. Prod.* **2009**, *72*, 1492–1496. [CrossRef]

386. Wu, S.Y.; Sung, P.J.; Chang, Y.L.; Pan, S.L.; Teng, C.M. Heteronemin, a spongean sesterterpene, induces cell apoptosis and autophagy in human renal carcinoma cells. *BioMed Res. Int.* **2015**, *2015*, 1–13. [CrossRef]

387. Hochlowski, J.E.; Faulkner, D.J. A diterpene related to cladiellin from a Pacific soft coral. *Tetrahedron Lett.* **1980**, *21*, 4055–4056. [CrossRef]

388. Sodano, S.; Spinella, A. Janolusimide, a lipophilic tripeptide toxin from the nudibranch mollusc *Janolus cristatus*. *Tetrahedron Lett.* **1986**, *27*, 2505–2508. [CrossRef]

389. Cimino, G.; De Rosa, S.; De Stefano, S.; Sodano, G. Marine natural products: New results from Mediterranean invertebrates. *Pure Appl. Chem.* **1986**, *58*, 375–386. [CrossRef]

390. Wang, J.; Prinsep, M.R.; Gordon, D.P.; Page, M.J.; Copp, B.R. Isolation and stereospecific synthesis of janolusimide B from a New Zealand collection of the bryozoan *Bugula flabellata*. *J. Nat. Prod.* **2015**, *78*, 530–533. [CrossRef] [PubMed]

391. Gavagnin, M.; Spinella, A.; Cimino, G.; Sodano, G. Stereochemistry of ichthyotoxic diacylglycerols from opisthobranch molluscs. *Tetrahedron Lett.* **1990**, *31*, 6093–6094. [CrossRef]

392. Cimino, G.; Crispino, A.; Spinella, A.; Sodano, G. Two ichthyotoxic diacylglycerols from the opisthobranch mollusc *Umbraculum mediterraneum*. *Tetrahedron Lett.* **1988**, *29*, 3613–3616. [CrossRef]

393. Cimino, G.; Spinella, A.; Scopa, A.; Sodano, G. Umbraculumin-B, an unusual 3- hydroxybutyric acid ester from the opisthobranch mollusc *Umbraculum mediterraneum*. *Tetrahedron Lett.* **1989**, *30*, 1147–1148. [CrossRef]

394. Sleeper, H.L.; Fenical, W. Navenones A–C: Trail-breaking alarm pheromones from the marine opisthobranch *Navanax inermis*. *J. Am. Chem. Soc.* **1977**, *99*, 2367–2368. [CrossRef]

395. Imperato, F.; Minale, L.; Riccio, R. Constituents of the digestive gland of molluscs of the genus *Aplysia*. II. Halogenated monoterpenes from *Aplysia limacina*. *Experientia* **1977**, *33*, 1273–1274. [CrossRef]

396. Spinella, A.; Gavagnin, M.; Crispino, A.; Cimino, G. 4-acetylaplykurodin B and aplykurodinone B, two ichthyotoxic degraded sterols from the Mediterranean mollusk *Aplysia fasciata*. *J. Nat. Prod.* **1992**, *5*, 989–993. [CrossRef]

397. Ortega, M.J.; Zubía, E.; Salvá, J. 3-epi-aplykurodinone B, a new degraded sterol from *Aplysia fasciata*. *J. Nat. Prod.* **1997**, *60*, 488–489. [CrossRef]

398. Miyamoto, T.; Higuchi, R.; Komori, T. Isolation and structures of aplykurodins A and B, two new isoprenoids from the marine mollusk *Aplysia kurodai*. *Tetrahedron Lett.* **1986**, *27*, 1153–1156. [CrossRef]

399. Miyamoto, T.; Ebisawa, Y.; Higuchi, R. Aplyparvunin, a bioactive acetogenin from the sea hare *Aplysia parvula*. *Tetrahedron Lett.* **1995**, *36*, 6073–6074. [CrossRef]

400. McPhail, K.L.; Davies-Coleman, M.T. (3Z)-Bromofucin from a South African sea hare. *Nat. Prod. Res.* **2005**, *19*, 449–452. [CrossRef] [PubMed]

401. Midland, S.L.; Wing, R.M.; Sims, J.J. New crenulides from the sea hare *Aplysia vaccaria*. *J. Org. Chem.* **1983**, *48*, 1906–1909. [CrossRef]

402. Sun, H.H.; McEnroe, F.J.; Fenical, W. Acetoxycrenulide, a new bicyclic cyclopropane-containing diterpenoid from the brown seaweed *Dictyota crenulata*. *J. Org. Chem.* **1983**, *48*, 1903–1906. [CrossRef]

403. Spinella, A.; Zubía, E.; Martînez, E.; Ortea, J.; Cimino, G. Structure and stereochemistry of aplyolides A– E, lactonized dihydroxy fatty acids from the skin of the marine mollusk *Aplysia depilans*. *J. Org. Chem.* **1997**, *62*, 5471–5475. [CrossRef]

404. Gerwick, W.H.; Fenical, W.; Fritsch, N.; Clardy, J. Stypotriol and stypoldione; ichthyotoxins of mixed biogenesis from the marine alga *Stypopodium zonale*. *Tetrahedron Lett.* **1979**, *2*, 145–148. [CrossRef]

405. Kuniyoshi, M.; Yamada, K.; Higa, T. A biologically active diphenyl ether from the green alga *Cladophora fascicularis*. *Experientia* **1985**, *41*, 523–524. [CrossRef]

406. Doty, M.S.; Aguilar-Santos, G. Transfer of toxic algal substances in marine food chains. *Pac. Sci.* **1970**, *24*, 351–355.

407. Gavagnin, M.; Marín, A.; Castelluccio, F.; Villani, G.; Cimino, G. Defensive relationships between *Caulerpa prolifera* and its shelled sacoglossan predators. *J. Exp. Mar. Biol. Ecol.* **1994**, *175*, 197–210. [CrossRef]

408. Cimino, G.; Crispino, A.; Di Marzo, V.; Gavagnin, M.; Ros, J.D. Oxytoxins, bioactive molecules produced by the marine opisthobranch mollusc *Oxynoe olivacea* from a diet-derived precursor. *Experientia* **1990**, *46*, 767–770. [CrossRef]

409. Fontana, A.; Ciavatta, M.L.; Mollo, E.; Naik, C.D.; Wahidulla, S.; D'Sousa, L.; Cimino, G. Volvatellin, cauerpenyne-related product from the sacoglossan *Volvatella* sp. *J. Nat. Prod.* **1999**, *62*, 931–933. [CrossRef] [PubMed]

410. Jensen, K.R. Evolution of the sacoglossa (Mollusca, Opisthobranchia) and the ecological associations with their food plants. *Evol. Ecol.* **1997**, *11*, 301–335. [CrossRef]

411. Ciavatta, M.L.; López Gresa, M.P.; Gavagnin, M.; Manzo, E.; Mollo, E.; D'Souza, L.; Cimino, G. New caulerpenyne-derived metabolites of an *Elysia* sacoglossan from the south Indian coast. *Molecules* **2006**, *11*, 808–816. [CrossRef] [PubMed]

412. Carbone, M.; Muniain, C.; Castelluccio, F.; Iannicelli, O.; Gavagnin, M. First chemical study of the sacoglossan *Elysia patagonica*: Isolation of a γ-pyrone propionate hydroperoxide. *Biochem. System. Ecol.* **2013**, *49*, 172–175. [CrossRef]

413. Gosliner, T.M. The genus *Thuridilla* (Opisthobranchia: Elysiidae) from thetropical indo-pacific, with a revision of the phylogeny of the elysiidae. *Calif. Acad. Sci.* **1995**, *59*, 1–54.

414. Dawe, R.D.; Wright, J.L.C. The major polypropionate metabolites from the sacoglossan mollusc *Elysia chlorotica*. *Tetrahedron Lett.* **1986**, *27*, 2559–2562. [CrossRef]

415. Ireland, C.D.; Faulkner, J.; Solheim, B.A.; Clardy, J. Tridachione, a propionate-derived metabolite of the opisthobranch mollusc *Tridachiella diomedea*. *J. Am. Chem. Soc.* **1978**, *100*, 1002–1003. [CrossRef]

416. Ireland, C.; Faulkner, J. The metabolites of the marine molluscs *Tridachiella diomedea* and *Tridachia crispata*. *Tetrahedron* **1981**, *37*, 233–240. [CrossRef]

417. Ksebati, M.B.; Schmitz, F.J. Tridachiapyrones: Propionate-derived metabolites from the sacoglossan mollusc *Tridachia crispata*. *J. Org. Chem.* **1985**, *50*, 5637–5642. [CrossRef]

418. Cueto, M.; D'Croz, L.; Maté, J.L.; San-Martín, A.; Darias, J. Elysiapyrones from *Elysia diomedea*. Do such metabolites evidence an enzymatically assisted electrocyclization cascade for the biosynthesis of their bicyclo[4.2.0]octane core? *Org. Lett.* **2005**, *7*, 415–418. [CrossRef] [PubMed]

419. Díaz-Marrero, A.R.; Cueto, M.; D'Croz, L.; Darias, J. Validating and endoperoxide as a key intermediate in the biosynthesis of elysiapyrones. *Org. Lett.* **2008**, *10*, 3057–3060. [CrossRef] [PubMed]

420. De Petrocellis, L.; Orlando, P.; Gavagnin, M.; Ventriglia, M.; Cimino, G.; Di Marzo, V. Novel diterpenoid diacylglycerols from marine molluscs: Potent morphogens and protein kinase C activators. *Experientia* **1996**, *52*, 874–877. [CrossRef] [PubMed]

421. Hochlowski, J.E.; Faulkner, D.J.; Bass, L.S.; Clardy, J. Metabolites of the dorid nudibranch *Chromodoris sedna*. *J. Org. Chem.* **1983**, *48*, 1738–1740. [CrossRef]

422. Fusetani, N.; Nagata, H.; Hirota, H.; Tsuyuki, T. Astrogorgiadiol and astrogorgin, inhibitors of cell division in fertilized starfish eggs, from a gorgonian *Astrogorgia* sp. *Tetrahedron Lett.* **1989**, *30*, 7079–7082. [CrossRef]

423. Ochi, M.; Yamada, K.; Shirase, K.; Kotsuki, H. Calicophirins A and B, two new insect growth inhibitory diterpenoids from a gorgonian coral *Calicogorgia* sp. *Heterocycles (Sendai)* **1991**, *32*, 19–21. [CrossRef]

424. Carmely, S.; Ilan, M.; Kashman, Y. 2-Amino Imidazole Alkaloids from the Marine Sponge *Leucetta chagosensis*. *Tetrahedron* **1989**, *45*, 2193–2200. [CrossRef]

425. Mai, T.; Tintillier, F.; Lucasson, A.; Moriou, C.; Bonno, E.; Petek, S.; Magré, K.; Al Mourabit, A.; Saulnier, D.; Debitus, C. Quorum sensing inhibitors from *Leucetta chagosensis* Dendy, 1863. *Lett. Appl. Microbiol.* **2015**, *61*, 311–317. [CrossRef]

426. Alvi, K.A.; Crews, P.; Loughhead, D.G. Structures and total synthesis of 2-aminoimidazoles from a *Notodoris* nudibranch. *J. Nat. Prod.* **1991**, *54*, 1509–1515. [CrossRef]

427. König, G.M.; Wright, A.D.; Franzblau, S.G. Assessment of antimycobacterial activity of a series of mainly marine derived natural products. *Planta Med.* **2000**, *66*, 337–342. [CrossRef]

428. Slattery, M.; Hamann, M.T.; McClintock, J.B.; Perry, T.L.; Puglisi, M.P.; Yoshida, W.Y. Ecological roles for water-borne metabolites from Antarctic soft corals. *Mar. Ecol. Progr. Ser.* **1997**, *161*, 133–144. [CrossRef]

429. Kladi, M.; Ntountaniotis, D.; Zervou, M.; Vagias, C.; Ioannou, E.; Roussis, V. Glandulaurencianols A–C, brominated diterpenes from the red alga, *Laurencia glandulifera* and the sea hare, *Aplysia punctata*. *Tetrahedron Lett.* **2014**, *55*, 2835–2837. [CrossRef]

430. Findlay, J.A.; Li, G. Novel terpenoids from the sea hare *Aplysia punctata*. *Can. J. Chem.* **2002**, *80*, 1697–1707. [CrossRef]

431. Caccamese, S.; Toscano, R.M.; Cerrini, S.; Gavuzzo, E. Laurencianol, a new halogenated diterpenoid from the marine alga *Laurencia obtusa*. *Tetrahedron Lett.* **1982**, *23*, 114–116. [CrossRef]

432. Kobayashi, M.; Kanda, F.; Kamiya, H. Occurrence of pyropheophorbides a and b in the viscera of the sea hare *Aplysia juliana*. *Nippon Suisan Gakk.* **1991**, *57*, 1983. [CrossRef]

433. Atta-ur-Rahman, K.A.; Abbas, S.A.; Sultana, T.; Shameel, M. A diterpenoid lactone from *Aplysia juliana*. *J. Nat. Prod.* **1991**, *54*, 886–888. [CrossRef]

434. Benkendorff, K.; Davis, A.R.; Rogers, C.N.; Bremner, J.B. Free fatty acids and sterols in the benthic spawn of aquatic molluscs, and their associated antimicrobial properties. *J. Exp. Mar. Biol. Ecol.* **2005**, *316*, 29–44. [CrossRef]

435. Kisugi, J.; Ohye, H.; Kamiya, H.; Yamazaki, M. Biopolymers from marine invertebrates. XIII. Characterization of an antibacterial protein, dolabellanin A, from the albumen gland of the sea hare, *Dolabella auricularia*. *Chem. Pharm. Bull.* **1992**, *40*, 1537–1539. [CrossRef]

436. Fusetani, N. Biofouling and antifouling. *Nat. Prod. Rep.* **2004**, *21*, 94–104. [CrossRef]

437. Nishikawa, K.; Nakahara, H.; Shirokura, Y.; Nogata, Y.; Yoshimura, E.; Umezawa, T.; Okino, T.; Matsuda, F. Total synthesis of 10-isocyano-4-cadinene and its stereoisomers and evaluations of antifouling activities. *J. Org. Chem.* **2011**, *76*, 6558–6573. [CrossRef]

438. Tanaka, J.; Higa, T. Two new cytotoxic carbonimidic dichlorides from the nudibranch *Reticulidia fungia*. *J. Nat. Prod.* **1999**, *62*, 1339–1340. [CrossRef] [PubMed]

439. Targett, N.M.; Bishop, S.S.; McConnell, O.J.; Yoder, J.A. Antifouling agents against the benthic marine diatom, *Navicula salinicola* Homarine from the gorgonians *Leptogorgia virgulata* and *L. setacea* and analogs. *J. Chem. Ecol.* **1983**, *9*, 817–829. [CrossRef] [PubMed]

440. Klein, D. A Proposed Definition of Mental Illness. Critical Issues in Psychiatric Diagnosis. Ph.D. Thesis, New York University, New York, NY, USA, 1978.

441. Berking, S. Is homarine a morphogen in the marine hydroid *Hydractinia*? *Roux's Arch. Dev. Biol.* **1986**, *195*, 33–38. [CrossRef] [PubMed]

442. Berking, S. Homarine (N-methylpicolinic acid) and trigonelline (N-methylnicotinic acid) appear to be involved in pattern control in a marine hydroid. *Development* **1987**, *99*, 211–220.

443. Ortlepp, S.; Pedpradap, S.; Dobretsov, S.; Proksch, P. Antifouling activity of sponge-derived polybrominated diphenyl ethers and synthetic analogues. *Biofouling* **2008**, *24*, 201–208. [CrossRef]

444. Spinella, A.; Alvarez, L.; Passeggio, A.; Cimino, G. New 3-alkylpyridines from 3 Mediterranean cephalaspidean mollusks; structure, ecological role and taxonomic relevance. *Tetrahedron* **1993**, *49*, 1307–1314. [CrossRef]

445. Blihoghe, D.; Manzo, E.; Villela, A.; Cutignano, A.; Picariello, G.; Faimali, M.; Fontana, A. Evaluation of the antifouling properties of 3-alyklpyridine compounds. *Biofouling* **2011**, *27*, 99–109. [CrossRef]

446. Cutignano, A.; Tramice, A.; De Caro, S.; Villani, G.; Cimino, G.; Fontana, A. Biogenesis of 3-alkylpyridine alkaloids in the marine mollusc *Haminoea orbignyana*. *Angew. Chem. Int. Ed.* **2003**, *42*, 2633–2636. [CrossRef]

447. Koehler, E. *'Trailing' Behaviour in Risbecia tryoni*. *Sea Slug Forum*; Australian Museum: Sydney, Australia, 1999; Available online: http://www.seaslugforum.net/find/760 (accessed on 20 November 2020).

448. Fenical, W.; Sleeper, H.L.; Paul, V.J.; Stallard, M.O.; Sun, H.H. Defensive chemistry of *Navanax* and related opisthobranch molluscs. *Pure Appl. Chem.* **1979**, *51*, 1865–1874. [CrossRef]

449. Cimino, G.; Passeggio, A.; Sodano, G.; Spinella, A.; Villani, G. Alarm pheromones from the Mediterranean opisthobranch *Haminoea navicula*. *Experientia* **1991**, *47*, 61–63. [CrossRef]

450. Cutignano, A.; Cimino, G.; Giordano, A.; d'Ippolito, G.; Fontana, A. Polyketide origin of 3-alkylpyridines in the marine mollusc *Haminoea orbignyana*. *Tetrahedron Lett.* **2004**, *45*, 2627–2629. [CrossRef]

451. Spinella, A.; Álvarez, L.A.; Cimino, G. Alkylphenols from the cephalaspidean mollusc *Haminoea callidegenita*. *Tetrahedron Lett.* **1998**, *39*, 2005–2008. [CrossRef]

452. Domènech, A.; Avila, C.; Ballesteros, M. Opisthobranch molluscs from the subtidal trawling grounds off Blanes (Girona, northeast Spain). *J. Mar. Biolog. Assoc.* **2006**, *86*, 383–389. [CrossRef]

453. Eilertsen, M.H.; Malaquias, M.A.E. Speciation in the dark: Diversification and biogeography of the deep-sea gastropod genus *Scaphander* in the Atlantic Ocean. *J. Biogeogr.* **2015**, *42*, 843–855. [CrossRef] [PubMed]

454. Cutignano, A.; Avila, C.; Rosica, A.; Romano, G.; Laratta, B.; Domenech-Coll, A.; Cimino, G.; Mollo, E.; Fontana, A. Biosynthesis and cellular localization of functional polyketides in the gastropod mollusc *Scaphander lignarius*. *Chembiochem.* **2012**, *13*, 1759–1766. [CrossRef] [PubMed]

455. Cutignano, A.; Avila, C.; Domenech-Coll, A.; d'Ippolito, G.; Cimino, G.; Fontana, A. First biosynthetic evidence on the phenyl-containing polyketides of the marine mollusc *Scaphander lignarius*. *Organic Lett.* **2008**, *10*, 2963–2966. [CrossRef]

456. Della Sala, G.; Cutignano, A.; Fontana, A.; Spinella, A.; Calabrese, G.; Domènech, A.; D'Ippolito, G.; Monica, C.D.; Cimino, G. Towards the biosynthesis of the aromatic products of the Mediterranean mollusc *Scaphander lignarius*: Isolation and synthesis of analogues of lignarenones. *Tetrahedron* **2007**, *63*, 7256–7263. [CrossRef]

457. Kamio, M.; Kicklighter, C.E.; Nguyen, L.; Germann, M.W.; Derby, C.D. Isolation and structural elucidation of novel mycosporine-like amino acids as alarm cues in the defensive ink secretion of the sea hare *Aplysia californica*. *Helv. Chim. Acta* **2011**, *94*, 1012–1018. [CrossRef]

458. Kicklighter, C.E.; Kamio, M.; Nguyen, L.; Germann, M.W.; Derby, C.D. Mycosporine-like amino acids are multifunctional molecules in sea hares and their marine community. *Proc. Nat. Acad. Sci. USA* **2011**, *108*, 11494–11499. [CrossRef]

459. Núñez-Pons, L.; Avila, C.; Romano, G.; Verde, C.; Giordano, D. UV-protective compounds in marine organisms from the Southern Ocean. *Mar. Drugs* **2018**, *16*, 336. [CrossRef]

460. Czeczuga, B. Investigations of carotenoids in some animals of the Adriatic Sea—VI Representatives of sponges, annelids, molluscs and echinodermates. *Comp. Biochem. Physiol. B Comp. Biochem.* **1984**, *78*, 259–264. [CrossRef]

461. Whitehead, K.; Karentz, D.; Hedges, J. Mycosporine-like amino acids (MAAs) in phytoplankton, a herbivorous pteropod (*Limacina helicina*), and its pteropod predator (*Clione antarctica*) in McMurdo Bay, Antarctica. *Mar. Biol.* **2001**, *139*, 1013–1019.

462. Gavagnin, M.; Carbone, M.; Ciavatta, M.L.; Mollo, E. Natural products from marine Heterobranchs: An overview of recent results. *Chem. J. Mold.* **2019**, *14*, 9–31. [CrossRef]

463. Cutignano, A.; Cimino, G.; Villani, G.; Fontana, A. Shaping the polypropionate biosynthesis in the solar-powered mollusc *Elysia viridis*. *ChemBioChem.* **2009**, *10*, 315–322. [CrossRef]

464. Gavagnin, M.; Mollo, E.; Cimino, G. Is phototridachiahydropyrone a true natural product? *Rev. Bras. Pharmacogn.* **2015**, *25*, 588–591. [CrossRef]

465. Gavagnin, M.; Marín, A.; Mollo, E.; Crispino, A.; Villani, G.; Cimino, G. Secondary metabolites from Mediterranean Elysioidea: Origin and biological role. *Comp. Biochem. Physiol. B Comp. Biochem.* **1994**, *108*, 107–115. [CrossRef]

466. Ireland, C.; Scheuer, P.J. Photosynthetic marine molluscs: In vivo 14C incorporation into metabolites of the sacoglossan *Placobranchus ocellatus*. *Science* **1979**, *205*, 922–923. [CrossRef]

467. Powell, K.J.; Richens, J.L.; Bramble, J.P.; Han, L.C.; Sharma, P.; O'Shea, P.; Moses, J.E. Photochemical activity of membrane-localised polyketide derived marine natural products. *Tetrahedron* **2017**, *74*, 1191–1198. [CrossRef]

468. Manzo, E.; Ciavatta, M.L.; Gavagnin, M.; Mollo, E.; Wahidulla, S.; Cimino, G. New γ-pyrone propionates from the Indian Ocean sacoglossan *Placobranchus ocellatus*. *Tetrahedron Lett.* **2005**, *46*, 465–468. [CrossRef]

469. Fu, X.; Hong, E.P.; Schmitz, F.J. New polypropionate pyrones from the Philippine sacoglossan mollusc *Placobranchus ocellatus*. *Tetrahedron* **2000**, *56*, 8989–8993. [CrossRef]

470. Avila, C. The growth of *Peltodoris atromaculata* Bergh, 1880 (Gastropoda: Nudibranchia) in the laboratory. *J. Molluscan Stud.* **1996**, *62*, 151–157. [CrossRef]

471. Di Marzo, V.; Minardi, C.; Vardaro, R.R.; Mollo, E.; Cimino, G. Prostaglandin F-1, 15-lactone fatty acyl esters: A prostaglandin lactone pathway branch developed during the reproduction and early larval stages of a marine mollusc. *Comp. Biochem. Physiol. B Biochem.* **1992**, *101*, 99–104. [CrossRef]

472. Marin, A.; Di Marzo, V.; Cimino, G. A histological and chemical study of the cerata of the opisthobranch mollusc *Tethys fimbria*. *Mar. Biol.* **1991**, *111*, 353–358. [CrossRef]

473. Vardaro, R.R.; Di Marzo, V.; Marín, A.; Cimino, G. A- and g-Pyrone-polypropionates from the Mediterranean ascoglossan mollusc *Ercolania funerea*. *Tetrahedron* **1992**, *48*, 9561–9566. [CrossRef]

474. Di Marzo, V.; Marín, A.; Vardaro, R.R.; De Petrocellis, L.; Villani, G.; Cimino, G. Histological and biochemical bases of defense mechanisms in four species of Polybranchoidea ascoglossan molluscs. *Mar. Biol.* **1993**, *117*, 367–380. [CrossRef]

475. Ciavatta, M.L.; Manzo, E.; Nuzzo, G.; Villani, G.; Cimino, G.; Cervera, J.L.; Malaquias, M.A.E.; Gavagnin, M. Aplysiopsenes: An additional example of marine polyketides with a mixed acetate/propionate pathway. *Tetrahedron Lett.* **2009**, *50*, 527–529. [CrossRef]

476. Ungur, N.; Gavagnin, M.; Fontana, A.; Cimino, G. Synthetic studies on natural diterpenoid glyceryl esters. *Tetrahedron* **2000**, *14*, 2503–2512. [CrossRef]

477. Zhang, W.; Gavagnin, M.; Guo, Y.-W.; Mollo, E.; Cimino, G. Chemical studies on the South China Sea nudibranch *Dermatobranchus ornatus* and its suggested prey gorgonian *Muricella* sp. *Chin. J. Org. Chem.* **2006**, *26*, 1667–1672.

478. Newman, D.J.; Cragg, G.M. Natural products as sources of new drugs over the nearly four decades from 01/1981 to 09/2019. *J. Nat. Prod.* **2020**, *83*, 770–803. [CrossRef]

479. Ciavatta, M.L.; Lefranc, F.; Carbone, M.; Mollo, E.; Gavagnin, M.; Betancourt, T.; Dasari, R.; Kornienko, A.; Kiss, R. Marine mollusk-derived agents with antiproliferative activity as promising anticancer agents to overcome chemotherapy resistance. *Med. Res. Rev.* **2017**, *37*, 702–801. [CrossRef]

480. Newman, D.J.; Cragg, G.M. Natural products as sources of new drugs over the last 25 years. *J. Nat. Prod.* **2007**, *70*, 461–477. [CrossRef] [PubMed]

481. Vinothkumar, S.; Parameswaran, P.S. Recent advances in marine drug research. *Biotechnol. Adv.* **2013**, *31*, 1826–1845. [CrossRef] [PubMed]

482. Nobili, S.; Lippi, D.; Witort, E.; Donnini, M.; Bausi, L.; Mini, E.; Capaccioli, S. Natural compounds for cancer treatment and prevention. *Pharmacol. Res.* **2009**, *59*, 365–378. [CrossRef] [PubMed]

483. Bray, F.; Ferlay, J.; Soerjomataram, I.; Siegel, R.L.; Torre, L.A.; Jemal, A. Global cancer statistics 2018: GLOBOCAN estimates of incidence and mortality worldwide for 36 cancers in 185 countries. *CA A Cancer J. Clin.* **2018**, *68*, 394–424. [CrossRef] [PubMed]

484. Howitz, K.T.; Sinclair, D.A. Xenohormesis: Sensing the chemical cues of other species. *Cell* **2008**, *133*, 387–391. [CrossRef] [PubMed]

485. Ang, K.K.H.; Holmes, M.J.; Higa, T.; Hamann, M.T.; Kara, U.A.K. In vivo antimalarial activity of the beta-carboline alkaloid manzamine A. Antimicrob. *Agents Chemother.* **2000**, *44*, 1645–1649. [CrossRef] [PubMed]

486. De Petrocellis, L.; Di Marzo, V.; Arca, B.; Gavagnin, M.; Minei, R.; Cimino, G. The effect of diterpenoidic diacylglycerols on tentacle regeneration in *Hydra vulgaris*. *Comp. Biochem. Physiol. C Comp. Pharmacol.* **1991**, *100C*, 603–607. [CrossRef]

487. Alvi, K.A.; Peters, B.M.; Lisa, H.M.; Phillip, C. 2-Aminoimidazoles and their zinc complexes from Indo-Pacific *Leucetta* sponges and *Notodoris* nudibranchs. *Tetrahedron* **1993**, *49*, 329–336. [CrossRef]

488. Carroll, A.R.; Bowden, B.F.; Coll, J.C. New imidazole alkaloids from the sponge *Leucetta* sp. and the associated predatory nudibranch *Notodoris gardineri*. *Aust. J. Chem.* **1993**, *46*, 1229–1234. [CrossRef]

489. Copp, B.R.; Fairchild, C.R.; Cornell, L.; Casazza, A.M.; Robinson, S.; Ireland, C.M. Naamidine A is an antagonist of the epidermal growth factor receptor and an in vivo active antitumor agent. *J. Med. Chem.* **1998**, *41*, 3909–3911. [CrossRef]

490. LaBarbera, D.V.; Modzelewska, K.; Glazar, A.I.; Gray, P.D.; Kaur, M.; Liu, T.; Grossman, D.; Harper, M.K.; Kuwada, S.K.; Moghal, N.; et al. The marine alkaloid Naamidine A promotes caspase-dependent apoptosis in tumor cells. *Anticancer Drugs* **2009**, *20*, 425–436. [CrossRef] [PubMed]

491. Graziani, E.I.; Allen, T.M.; Andersen, R.J. Lovenone, a cytotoxic degraded triterpenoid isolated from skin extracts of the North Sea dorid nudibranch *Adalaria loveni*. *Tetrahedron Lett.* **1995**, *36*, 1763–1766. [CrossRef]

492. Trindade-Silva, A.E.; Lim-Fong, G.E.; Sharp, K.H.; Haygood, M.G. Bryostatins: Biological context and biotechnological prospects. *Curr. Opin. Biotechnol.* **2010**, *21*, 834–842. [CrossRef] [PubMed]

493. Davidson, S.K. The Biology of the Bryostatins in the Marine Bryozoan *Bugula neritina*. Ph.D. Thesis, University of California, San Diego, CA, USA, 1999.

494. Lim, G.E. Bugula (Bryozoa) and Their Bacterial Symbionts: A Study in Symbiosis, Molecular Phylogenetics and Secondary Metabolism. Ph.D. Thesis, University of California, San Diego, CA, USA, 2004.

495. Sudek, S.; Lopanik, N.B.; Waggoner, L.E.; Hildebrand, M.; Anderson, C.; Liu, H.; Patel, A.; Sherman, D.H.; Haygood, M.G. Identification of the putative bryostatin polyketide synthase gene cluster from *Candidatus Endobugula sertula*, the uncultivated microbial symbiont of the marine bryozoan *Bugula neritina*. *J. Nat. Prod.* **2007**, *70*, 67–74. [CrossRef] [PubMed]

496. Zonder, J.A.; Shields, A.F.; Zalupski, M.; Chaplen, R.; Heilbrun, L.K.; Arlauskas, P.; Philip, P.A. A phase II trial of bryostatin 1 in the treatment of metastatic colorectal cancer. *Clin. Cancer Res.* **2001**, *7*, 38–42. [PubMed]

497. Mayer, A.M.S.; Glaser, K.B.; Cuevas, C.; Jacobs, R.S.; Kem, W.; Little, R.D.; McIntosh, J.M.; Newman, D.J.; Potts, B.C.; Shuster, D.E. The odyssey of marine pharmaceuticals: A current pipeline perspective. *Trends Pharmacol. Sci.* **2010**, *31*, 255–265. [CrossRef] [PubMed]

498. Maki, A.; Diwakaran, H.; Redman, B.; Al-Asfar, S.; Pettit, G.R.; Mohammad, R.M.; Al-Katib, A. The bcl-2 and p53 oncoproteins can be modulated by bryostatin 1 and dolastatins in human diffuse large cell lymphoma. *Anticancer Drugs* **1995**, *6*, 392–397. [CrossRef]

499. Skropeta, D.; Pastro, N.; Zivanovic, A. Kinase inhibitors from marine sponges. *Mar. Drugs* **2011**, *9*, 2131–2154. [CrossRef]

500. Kortmansky, J.; Schwartz, G.K. Bryostatin-1: A novel PKC inhibitor in clinical development. *Cancer Investig.* **2003**, *21*, 924–936. [CrossRef]

501. Kinnel, R.B.; Scheuer, P.J. 11-hydroxystaurosporine: A highly cytotoxic, powerful protein kinase C inhibitor from a tunicate. *J. Org. Chem.* **1992**, *57*, 6327–6329. [CrossRef]

502. Sekar, M.; Poomalai, S.; Gunasekaran, M.; Mani, P.; Krishnamurthy, A. Bioactive compounds from marine yeast inhibits lung cancer. *J. Appl. Pharm. Sci.* **2015**, *5*, 7–15.

503. Wall, N.R.; Mohammad, R.M.; Reddy, K.B.; Al-Katib, A.M. Bryostatin 1 induces ubiquitination and proteasome degradation of Bcl-2 in the human acute lymphoblastic leukemia cell line, Reh. *Int. J. Mol. Med.* **2000**, *5*, 165–236. [CrossRef] [PubMed]

504. Manzo, E.; Carbone, M.; Mollo, E.; Irace, C.; Di Pascale, A.; Li, Y.; Ciavatta, M.L.; Cimino, G.; Guo, Y.-W.; Gavagnin, M. Structure and synthesis of a unique isonitrile lipid isolated from the marine mollusk *Actinocyclus papillatus*. *Org. Lett.* **2011**, *13*, 1897–1899. [CrossRef] [PubMed]

505. Carbone, M.; Ciavatta, M.L.; Mathieu, V.; Ingels, A.; Kiss, R.; Pascale, P.; Mollo, E.; Ungur, N.; Guo, Y.-W.; Gavagnin, M. Marine terpenoid diacylguanidines: Structure, synthesis, and biological evaluation of naturally occurring actinofide and synthetic analogues. *J. Nat. Prod.* **2017**, *80*, 1339–1346. [CrossRef]

506. Parikh, M.; Riess, J.; Lara, P.N., Jr. New and emerging developments in extensive-stage small cell lung cancer therapeutics. *Curr. Opin. Oncol.* **2016**, *28*, 97–103. [CrossRef]

507. Stupp, R.; Hegi, M.E.; Mason, W.P.; Van Den Bent, M.J.; Taphoorn, M.J.; Janzer, R.C.; Ludwin, S.K.; Allgeier, A.; Fisher, B.; Belanger, K.; et al. Effects of radiotherapy with concomitant and adjuvant temozolomide versus radiotherapy alone on survival in glioblastoma in a randomised phase III study: 5-year analysis of the EORTC-NCIC trial. *Lancet Oncol.* **2009**, *10*, 459–466. [CrossRef]

508. Perret, G.Y.; Uzzan, B. An anticancer strategic dilemma: To kill or to contain. The choice of the pharmaceutical industry in 2009. *Fund. Clin. Pharmaccol.* **2011**, *25*, 283–295. [CrossRef]

509. Montagnac, A.; Martin, M.T.; Debitus, C.; Pais, M. Drimane sesquiterpenes from the sponge *Dysidea fusca*. *J. Nat. Prod.* **1996**, *59*, 866–868. [CrossRef]

510. Jaisamut, S. Terpenoids from the Nudibranch *Phyllidia coelestis* Bergh. Ph.D. Thesis, Prince of Songkla University, Hat Yai, Thailand, 2014.

511. Wratten, S.J.; Faulkner, D.J. Minor carbonimidic dichlorides from the marine sponge *Pseudaxinyssa pitys*. *Tetrahedron Lett.* **1978**, *19*, 1395–1398. [CrossRef]

512. Wratten, S.J.; John, D.; Van Engen, D.; Clardy, J. A vinyl carbonimidic dichloride from the marine sponge *Pseudaxinyssa pitys*. *Tetrahedron Lett.* **1978**, *19*, 1391–1394. [CrossRef]

513. Daoust, J.; Fontana, A.; Merchant, C.E.; De Voogd, N.J.; Patrick, B.O.; Kieffer, T.J.; Andersen, R.J. Ansellone A, a sesterterpenoid isolated from the nudibranch *Cadlina luteromarginata* and the sponge *Phorbas* sp., activates the cAMP signaling pathway. *Org. Lett.* **2010**, *12*, 3208–3211. [CrossRef] [PubMed]

514. Shirley, H.J.; Jamieson, M.L.; Brimble, M.A.; Bray, C.D. A new family of sesterterpenoids isolated around the Pacific Rim. *Nat. Prod. Rep.* **2018**, *35*, 210–219. [CrossRef] [PubMed]

515. El Sayed, K.A.; Youssef, D.T.A.; Marchetti, D. Bioactive natural and semisynthetic latrunculins. *J. Nat. Prod.* **2006**, *69*, 219–223. [CrossRef] [PubMed]

516. Spector, I.; Shochet, N.R.; Blasberger, D.; Kashman, Y. Latrunculins; novel marine macrolides that disrupt microfilament organization and affect cell growth: I. Comparison with cytochalasin D. *Cell Motil. Cytoskelet.* **1989**, *13*, 127–144. [CrossRef] [PubMed]

517. Spector, I.; Shochet, N.R.; Kashman, Y.; Groweiss, A. Latrunculins: Novel marine toxins that disrupt microfilament organization in cultured cells. *Science* **1983**, *219*, 493–495. [CrossRef] [PubMed]

518. Sayed, K.A.; Khanfar, M.A.; Shallal, H.M.; Muralidharan, A.; Awate, B.; Youssef, D.T.; Liu, Y.; Zhou, Y.D.; Nagle, D.G.; Shah, G. Latrunculin A and its C-17-O-carbamates inhibit prostate tumor cell invasion and HIF-1 activation in breast tumor cells. *J. Nat. Prod.* **2008**, *71*, 396–402. [CrossRef] [PubMed]

519. Longley, R.E.; McConnell, O.J.; Essich, E.; Harmody, D. Evaluation of marine sponge metabolites for cytotoxicity and signal transduction activity. *J. Nat. Prod.* **1993**, *56*, 915–920. [CrossRef]

520. Konishi, H.; Kikuchi, S.; Ochiai, T.; Ikoma, H.; Kubota, T.; Ichikawa, D.; Fujiwara, H.; Okamoto, K.; Sakakura, C.; Sonoyama, T.; et al. Latrunculin a has a strong anticancer effect in a peritoneal dissemination model of human gastric cancer in mice. *Anticancer Res.* **2009**, *29*, 2091–2097.

521. Prota, A.E.; Bargsten, K.; Northcote, P.T.; Marsh, M.; Altmann, K.H.; Miller, J.H.; Díaz, J.F.; Steinmetz, M.O. Structural basis of microtubule stabilization by laulimalide and peloruside A. Angew. *Chem. Int. Ed.* **2014**, *5*, 1621–1625. [CrossRef]

522. Churchill, C.D.; Klobukowski, M.; Tuszynski, J.A. The unique binding mode of laulimalide to two tubulin protofilaments. *Chem. Biol. Drug Des.* **2015**, *86*, 190–199. [CrossRef]

523. Mooberry, S.L.; Randall-Hlubek, D.A.; Leal, R.M.; Hegde, S.G.; Hubbard, R.D.; Zhang, L.; Wender, P.A. Microtubule-stabilizing agents based on designed laulimalide analogues. *Proc. Natl. Acad. Sci. USA* **2004**, *101*, 8803–8808. [CrossRef] [PubMed]

524. Churchill, C.D.M.; Klobukowski, M.; Tuszynski, J.A. Analysis of the binding mode of laulimalide to microtubules: Establishing a laulimalide-tubulin pharmacophore. *J. Biomol. Struct. Dyn.* **2016**, *34*, 1455–1469. [CrossRef] [PubMed]

525. Liu, J.; Towle, M.J.; Cheng, H.; Saxton, P.; Reardon, C.; Wu, J.; Murphy, E.A.; Kuznetsov, G.; Johannes, C.W.; Tremblay, M.R.; et al. In vitro and in vivo anticancer activities of synthetic (-)-laulimalide, a marine natural product microtubule stabilizing agent. *Anticancer Res.* **2007**, *27*, 1509–1518. [PubMed]

526. Mooberry, S.L.; Tien, G.; Hernandez, A.H.; Plubrukarn, A.; Davidson, B.S. Laulimalide and isolaulimalide, new paclitaxel-like microtubule-stabilizing agents. *Cancer Res.* **1999**, *59*, 653–660. [PubMed]

527. Rohena, C.C.; Peng, J.; Johnson, T.A.; Crews, P.; Mooberry, S.L. Chemically diverse microtubule stabilizing agents initiate distinct mitotic defects and dysregulated expression of key mitotic kinases. *Biochem. Pharmacol.* **2013**, *85*, 1104–1114. [CrossRef] [PubMed]

528. Pryor, D.E.; O'Brate, A.; Bilcer, G.; Díaz, J.F.; Wang, Y.; Wang, Y.; Kabaki, M.; Jung, M.K.; Andreu, J.M.; Ghosh, A.K.; et al. The microtubule stabilizing agent laulimalide does not bind in the taxoid site, kills cells resistant to paclitaxel and epothilones, and may not require its epoxide moiety for activity. *Biochemistry* **2002**, *41*, 9109–9115. [CrossRef]

529. Kanakkanthara, A.; Northcote, P.T.; Miller, J.H. βII-tubulin and βIII-tubulin mediate sensitivity to peloruside A and laulimalide, but not paclitaxel or vinblastine, in human ovarian carcinoma cells. *Mol. Cancer Ther.* **2012**, *11*, 393–404. [CrossRef]

530. Kanakkanthara, A.; Rawson, P.; Northcote, P.T.; Miller, J.H. Acquired resistance to peloruside A and laulimalide is associated with downregulation of vimentin in human ovarian carcinoma cells. *Pharm. Res.* **2012**, *29*, 3022–3032. [CrossRef]

531. Evidente, A.; Kornienko, A.; Lefranc, F.; Cimmino, A.; Dasari, R.; Evidente, M.; Mathieu, V.; Kiss, R. Sesterterpenoids with anticancer activity. *Curr. Med. Chem.* **2015**, *22*, 3502–3522. [CrossRef]

532. Forster, L.C.; Pierens, G.K.; White, A.M.; Cheney, K.L.; Dewapriya, P.; Capon, R.J.; Garson, M.J. Cytotoxic spiroepoxide lactone and its putative biosynthetic precursor from *Goniobranchus splendidus*. *ACS Omega* **2017**, *2*, 2672–2677. [CrossRef]

533. Rueda, A.; Losada, A.; Fernandez, R.; Cabanas, C.; Garcia-Fernandez, L.F.; Reyes, F.; Cuevas, C. Gracilins G-I, cytotoxic bisnorditerpenes from *Spongionella pulchella*, and the anti-adhesive properties of gracilin B. *Lett. Drug. Des. Discov.* **2006**, *3*, 753–760. [CrossRef]

534. Sanchez, J.A.; Alfonso, A.; Leiros, M.; Alonso, E.; Rateb, M.E.; Jaspars, M.; Houssen, W.E.; Ebel, R.; Tabudravu, J.; Botana, L.M. Identification of *Spongionella* compounds as cyclosporine A mimics. *Pharmacol. Res.* **2016**, *107*, 407–414. [CrossRef] [PubMed]

535. Kohmoto, S.; McConnell, O.J.; Wrigth, A.; Cross, S. Isospongiadiol, a cytotoxic and antiviral diterpene from a Caribbean deep water marine sponge, *Spongia* sp. *Chem. Lett.* **1987**, *16*, 1687–1690. [CrossRef]

536. Fontana, A.; Mollo, E.; Ricciardi, D.; Fakhr, I.; Cimino, G. Chemical studies of Egyptian opisthobranchs: Spongian diterpenoids from *Glossodoris atromarginata*. *J. Nat. Prod.* **1997**, *60*, 444–448. [CrossRef]

537. Cambie, R.C.; Craw, P.A.; Stone, M.J.; Bergquist, P.R. Chemistry of sponges, IV. Spongian diterpenes from *Hyatella intestinalis*. *J. Nat. Prod.* **1988**, *51*, 293–297. [CrossRef]

538. Agrawal, M. Isolation and Structural Elucidation of Cytotoxic Agents from Marine Invertebrates and Plants Sourced from the Great Barrier Reef, Australia. Ph.D. Thesis, James Cook University, Townsville, Australia, 2007.

539. Yong, K.W.; Garson, M.J.; Bernhardt, P.V. Absolute structures and conformations of the spongian diterpenes spongia-13(16), 14-dien-3-one, epispongiadiol and spongiadiol. *Acta Crystallogr. Sect. C Cryst. Struct. Commun.* **2009**, *65*, 167–170. [CrossRef] [PubMed]

540. Wonganuchitmeta, S.; Yuenyongsawad, S.; Keawpradub, N.; Plubrukarn, A. Antitubercular sesterterpenes from the Thai sponge *Brachiaster* sp. *J. Nat. Prod.* **2004**, *67*, 1767–1770. [CrossRef]

541. Kazlauskas, R.; Murphy, P.T.; Quinn, R.J.; Wells, R.J. Heteronemin, a new scalarin type sesterterpene from the sponge *Heteronema erecta*. *Tetrahedron Lett.* **1976**, *17*, 2631–2634. [CrossRef]

542. Kopf, S.; Viola, K.; Atanasov, A.G.; Jarukamjorn, K.; Rarova, L.; Kretschy, N.; Teichmann, M.; Vonach, C.; Saiko, P.; Giessrigl, B.; et al. In vitro characterisation of the anti-intravasative properties of the marine product heteronemin. *Arch. Toxicol.* **2013**, *87*, 1851–1861. [CrossRef]

543. Schnermann, M.J.; Beaudry, C.M.; Egorova, A.V.; Polishchuk, R.S.; Suetterlin, C.; Overman, L.E. Golgi-modifying properties of macfarlandin E and the synthesis and evaluation of its 2,7-dioxabicyclo[3.2.1]octan-3-one core. *Proc. Natl. Acad. Sci. USA* **2010**, *107*, 6158–6163. [CrossRef]

544. Chattopadhyay, S.K.; Pattenden, G. Total synthesis of ulapualide A, a novel tris-oxazole containing macrolide from the marine nudibranch *Hexabranchus sanguineus*. *Tetrahedron Lett.* **1998**, *39*, 6095–6098. [CrossRef]

545. Vincent, E.; Saxton, J.; Baker-Glenn, C.; Moal, I.; Hirst, J.D.; Pattenden, G.; Shaw, P.E. Effects of ulapualide A and synthetic macrolide analogues on actin dynamics and gene regulation. *Cell Mol. Life Sci.* **2007**, *64*, 487–497. [CrossRef] [PubMed]

546. Parrish, S.M.; Yoshida, W.; Yang, B.; Williams, P.G. Ulapualides C-E Isolated from a Hawaiian *Hexabranchus sanguineus* Egg Mass. *J. Nat. Prod.* **2017**, *80*, 726–730. [CrossRef] [PubMed]

547. Matsunaga, S.; Fusetani, N.; Hashimoto, K. Kabiramide C, a novel antifungal macrolide from nudibranch eggmasses. *J. Am. Chem. Soc.* **1986**, *108*, 847–849. [CrossRef]

548. Sirirak, T.; Kittiwisut, S.; Janma, C.; Yuenyongsawad, S.; Suwanborirux, K.; Plubrukarn, A. Kabiramides J and K, trisoxazole macrolides from the sponge *Pachastrissa nux*. *J. Nat. Prod.* **2011**, *74*, 1288–1292. [CrossRef]

549. Tanaka, J.; Yan, Y.; Choi, J.; Bai, J.; Klenchin, V.A.; Rayment, I.; Marriott, G. Biomolecular mimicry in the actin cytoskeleton: Mechanisms underlying the cytotoxicity of kabiramide C and related macrolides. *Proc. Natl. Acad. Sci. USA* **2003**, *100*, 13851–13856. [CrossRef]

550. Braet, F.; Spector, I.; Shochet, N.; Crews, P.; Higa, T.; Menu, E.; De Zanger, R.; Wisse, E. The new anti-actin agent dihydrohalichondramide reveals fenestrae-forming centers in hepatic endothelial cells. *BMC Cell Biol.* **2002**, *3*, 1–14. [CrossRef]

551. Shin, Y.; Kim, G.D.; Jeon, J.E.; Shin, J.; Lee, S.K. Antimetastatic effect of halichondramide, a trisoxazole macrolide from the marine sponge *Chondrosia corticata*, on human prostate cancer cells via modulation of epithelial-to-mesenchymal transition. *Mar. Drugs* **2013**, *11*, 2472–2485. [CrossRef]

552. Bae, S.Y.; Kim, G.D.; Jeon, J.E.; Shin, J.; Lee, S.K. Anti-proliferative effect of (19Z)-halichondramide, a novel marine macrolide isolated from the sponge *Chondrosia corticata*, is associated with G2/M cell cycle arrest and suppression of mTOR signaling in human lung cancer cells. *Toxicol. In Vitro* **2013**, *27*, 694–699. [CrossRef]

553. Fontana, A.; Cavaliere, P.; Wahidulla, S.; Naik, C.G.; Cimino, G. A new antitumor isoquinoline alkaloid from the marine nudibranch *Jorunna funebris*. *Tetrahedron* **2000**, *56*, 7305–7308. [CrossRef]

554. Oku, N.; Matsunaga, S.; van Soest, R.W.; Fusetani, N. Renieramycin J, a highly cytotoxic tetrahydroisoquinoline alkaloid, from a marine sponge *Neopetrosia* sp. *J. Nat. Prod.* **2003**, *66*, 1136–1139. [CrossRef] [PubMed]

555. Newman, D.J.; Cragg, G.M. Drugs and drug candidates from marine sources: An assessment of the current "State of Play". *Planta Med.* **2016**, *82*, 775–789. [CrossRef] [PubMed]

556. Petek, B.J.; Jones, R.L. PM00104 (Zalypsis®): A marine derived alkylating agent. *Molecules* **2014**, *19*, 12328–12335. [CrossRef] [PubMed]

557. Cimino, G.; Ciavatta, M.L.; Fontana, A.; Gavagnin, M. Metabolites of marine opisthobranchs: Chemistry and biological activity. In *Bioactive Compounds from Natural Sources; Isolation, Characterization and Biological Properties*; Tringali, C., Ed.; Taylor and Francis: London, UK, 2001; pp. 579–637.

558. Malve, H. Exploring the ocean for new drug developments: Marine pharmacology. *J. Pharm. Bioallied Sci.* **2016**, *8*, 83–91. [CrossRef]

559. Cuevas, C.; Perez, M.; Francesch, A.; Fernandez, C.; Chicharro, J.L.; Gallego, P.; Zarzuelo, M.; de la Calle, F.; Manzanares, I. Hemisynthetic method and intermediates thereof. PCT Application WO 2000069862 A2, 23 November 2000.

560. Charupant, K.; Suwanborirux, K.; Amnuoypol, S.; Saito, E.; Kubo, A.; Saito, N. Jorunnamycins A–C, new stabilized renieramycin-type bistetrahydroisoquinolines isolated from the Thai nudibranch *Jorunna funebris*. *Chem. Pharm. Bull.* **2007**, *55*, 81–86. [CrossRef]

561. Charupant, K.; Daikuhara, N.; Saito, E.; Amnuoypol, S.; Suwanborirux, K.; Owa, T.; Saito, N. Chemistry of renieramycins. Part 8: Synthesis and cytotoxicity evaluation of renieramycin M-jorunnamycin a analogues. *Bioorg. Med. Chem.* **2009**, *17*, 4548–4558. [CrossRef]

562. McPherson, J.R.; Ong, C.K.; Ng, C.C.Y.; Rajasegaran, V.; Heng, H.L.; Yu, W.S.S.; Tan, B.K.T.; Madhukumar, P.; Teo, M.C.C.; Ngeow, J.; et al. Whole-exome sequencing of breast cancer, malignant peripheral nerve sheath tumor and neurofibroma from a patient with neurofibromatosis type 1. *Cancer Med.* **2015**, *4*, 1871–1878. [CrossRef]

563. Storm, E.E.; Durinck, S.; e Melo, F.D.S.; Tremayne, J.; Kljavin, N.; Tan, C.; Ye, X.; Chiu, C.; Pham, T.; Hongo, J.A.; et al. Targeting PTPRK-RSPO3 colon tumours promotes differentiation and loss of stem-cell function. *Nature* **2016**, *529*, 97–100. [CrossRef]

564. He, W.F.; Li, Y.; Feng, M.T.; Gavagnin, M.; Mollo, E.; Mao, S.C.; Guo, Y.W. New isoquinolinequinone alkaloids from the South China Sea nudibranch *Jorunna funebris* and its possible sponge-prey *Xestospongia* sp. *Fitoterapia* **2014**, *96*, 109–114. [CrossRef]

565. Huang, R.Y.; Chen, W.T.; Kurtán, T.; Mándi, A.; Ding, J.; Li, J.; Guo, Y.W. Bioactive isoquinolinequinone alkaloids from the South China Sea nudibranch *Jorunna funebris* and its sponge-prey *Xestospongia* sp. *Fut. Med. Chem.* **2016**, *8*, 17–27. [CrossRef]

566. Castiello, D.; Cimino, G.; De Rosa, S.; De Stefano, S.; Sodano, G. High molecular weight polyacetylenes from the nudibranch *Peltodoris atromaculata* and the sponge *Petrosia ficiformis*. *Tetrahedron Lett.* **1980**, *21*, 5047–5050. [CrossRef]

567. Gemballa, S.; Schermutzki, F. Cytotoxic haplosclerid sponges preferred: A field study on the diet of the dotted sea slug *Peltodoris atromaculata* (Doridoidea: Nudibranchia). *Mar. Biol.* **2004**, *144*, 1213–1222. [CrossRef]

568. Ciavatta, M.L.; Nuzzo, G.; Takada, K.; Mathieu, V.; Kiss, R.; Villani, G.; Gavagnin, M. Sequestered fulvinol-related polyacetylenes in *Peltodoris atromaculata*. *J. Nat. Prod.* **2014**, *77*, 1678–1684. [CrossRef] [PubMed]

569. Castiello, D.; Cimino, G. de Rosa, S.; de Stefano, S.; Izzo, G.; Sodano, G. In Colloq. *Int. CNRS* **1979**, *291*, 413.

570. Ueoka, R.; Ise, Y.; Matsunaga, S. Cytotoxic polyacetylenes related to petroformyne-1 from the marine sponge *Petrosia* sp. *Tetrahedron* **2009**, *65*, 5204–5208. [CrossRef]

571. Ortega, M.G.; Zubia, E.; Carballo, J.L.; Salvá, J. Fulvinol, a new long-chain diacetylenic metabolite from the sponge *Reniera fulva*. *J. Nat. Prod.* **1996**, *59*, 1069–1071. [CrossRef]

572. Rapson, T.D. Bioactive 4-Methoxypyrrolic Natural Products from Two South African Marine Invertebrates. Master's Sc. Thesis, Rhodes University, Rhodes, South Africa, 2004.

573. Berlinck, R.G.; Hajdu, E.; da Rocha, R.M.; de Oliveira, J.H.; Hernández, I.L.; Seleghim, M.H.; Granato, A.C.; de Almeida, E.V.; Nuñez, C.V.; Muricy, G.; et al. Challenges and rewards of research in marine natural products chemistry in Brazil. *J. Nat. Prod.* **2004**, *67*, 510–522. [CrossRef]

574. Cavalcanti, B.C.; Júnior, H.V.; Seleghim, M.H.; Berlinck, R.G.; Cunha, G.M.; Moraes, M.O.; Pessoa, C. Cytotoxic and genotoxic effects of tambjamine D, an alkaloid isolated from the nudibranch *Tambja eliora*, on Chinese hamster lung fibroblasts. *Chem. Biol. Interact.* **2008**, *174*, 155–162. [CrossRef]

575. Pinkerton, D.M.; Banwell, M.G.; Garson, M.J.; Kumar, N.; de Moraes, M.O.; Cavalcanti, B.C.; Barros, F.W.A.; Pessoa, C. Antimicrobial and cytotoxic activities of synthetically derived tambjamines C and E-J, BE-18591, and a related alkaloid from the marine bacterium *Pseudoalteromonas tunicata*. *Chem. Biodivers.* **2010**, *7*, 1311–1324. [CrossRef]

576. Melvin, M.S.; Ferguson, D.C.; Lindquist, N.; Manderville, R.A. DNA binding by 4-methoxypyrrolic natural products. Preference for intercalation at AT sites by tambjamine E and prodigiosin. *J. Org. Chem.* **1999**, *64*, 6861–6869. [CrossRef]

577. Aldrich, L.N.; Stoops, S.L.; Crews, B.C.; Marnett, L.J.; Lindsley, C.W. Total synthesis and biological evaluation of tambjamine K and a library of unnatural analogs. *Bioorg. Med. Chem. Lett.* **2010**, *20*, 5207–5211. [CrossRef] [PubMed]

578. Kazlauskas, R.; Marwood, J.F.; Murphy, P.T.; Wells, R.J. A blue pigment from a compound ascidian. *Aust. J. Chem.* **1982**, *35*, 215–217. [CrossRef]

579. Melvin, M.S.; Wooton, K.E.; Rich, C.C.; Saluta, G.R.; Kucera, G.L.; Lindquist, N.; Manderville, R.A. Copper-nuclease efficiency correlates with cytotoxicity for the 4-methoxypyrrolic natural products. *J. Inorg. Biochem.* **2001**, *87*, 129–135. [CrossRef]

580. Matsunaga, S.; Fusetani, N.; Hashimoto, K. Bioactive marine metabolites. VIII. Isolation of an antimicrobial blue pigment from the bryozoan *Bugula dentata*. *Experientia* **1986**, *42*, 84. [CrossRef]

581. Hernández, P.I.; Moreno, D.; Javier, A.A.; Torroba, T.; Pérez-Tomás, R.; Quesada, R. Tambjamine alkaloids and related synthetic analogs: Efficient transmembrane anion transporters. *ChemComm.* **2012**, *48*, 1556–1558. [CrossRef]

582. Xu, W.Q.; Song, L.J.; Liu, Q.; Zhao, L.; Zheng, L.; Yan, Z.W.; Fu, G.H. Expression of anion exchanger 1 is associated with tumor progress in human gastric cancer. *J. Cancer Res. Clinical Oncol.* **2009**, *135*, 1323–1330. [CrossRef]

583. Liedauer, R.; Svoboda, M.; Wlcek, K.; Arrich, F.; Jäger, W.; Toma, C.; Thalhammer, T. Different expression patterns of organic anion transporting polypeptides in osteosarcomas, bone metastases and aneurysmal bone cysts. *Oncol. Rep.* **2009**, *22*, 1485–1492.

584. Buxhofer-Ausch, V.; Secky, L.; Wlcek, K.; Svoboda, M.; Kounnis, V.; Briasoulis, E.; Tzakos, A.G.; Jaeger, W.; Thalhammer, T. Tumor-specific expression of organic anion-transporting polypeptides: Transporters as novel targets for cancer therapy. *J. Drug Deliv.* **2013**, *2013*, 1–12. [CrossRef]

585. AbuAli, G.; Grimm, S. Isolation and characterization of the anticancer gene organic cation transporter like-3 (ORCTL3). In *Anticancer Genes*; Springer: London, UK, 2014; pp. 213–227.

586. Liu, T.; Li, Q. Organic anion-transporting polypeptides: A novel approach for cancer therapy. *J. Drug. Target.* **2014**, *22*, 14–22. [CrossRef]

587. Baker, B.; Scheuer, P. The punaglandins: 10-chloroprostanoids from the octocoral *Telesto riisei*. *J. Nat. Prod.* **1994**, *57*, 1346–1353. [CrossRef]

588. Pika, J.; Faulkner, D.J. Four sesquiterpenes from the South African nudibranch *Leminda millecra*. *Tetrahedron* **1994**, *50*, 3065–3070. [CrossRef]

589. McPhail, K.L.; Davies-Coleman, M.T.; Starmer, J. Sequestered chemistry of the arminacean nudibranch *Leminda millecra* in Algoa Bay, South Africa. *J. Nat. Prod.* **2001**, *64*, 1183–1190. [CrossRef] [PubMed]

590. Whibley, C.E.; McPhail, K.L.; Keyzers, R.A.; Maritz, M.F.; Leaner, V.D. : Birrer, M.J.; Davies-Coleman, M.T.; Hendricks, D.T. Reactive oxygen species mediated apoptosis of esophageal cancer cells induced by marine triprenyl toluquinones and toluhydroquinones. *Mol. Cancer Ther.* **2007**, *6*, 2535–2543. [CrossRef] [PubMed]

591. Kasheverov, I.E.; Shelukhina, I.V.; Kudryavtsev, D.S.; Makarieva, T.N.; Spirova, E.N.; Guzii, A.G.; Stonik, V.A.; Tsetlin, V.I. 6-Bromohypaphorine from marine nudibranch mollusk *Hermissenda crassicornis* is an agonist of human α7 nicotinic acetylcholine receptor. *Mar. Drugs* **2015**, *13*, 1255–1266. [CrossRef]

592. Carbone, M.; Li, Y.; Irace, C.; Mollo, E.; Castelluccio, F.; Di Pascale, A.; Cimino, G.; Santamaria, R.; Guo, Y.W.; Gavagnin, M. Structure and cytotoxicity of phidianidines A and B: First finding of 1, 2, 4-oxadiazole system in a marine natural product. *Organic Lett.* **2011**, *13*, 2516–2519. [CrossRef]

593. Brogan, J.T.; Stoops, S.L.; Lindsley, C.W. Total synthesis and biological evaluation of phidianidines A and B uncovers unique pharmacological profiles at CNS targets. *ACS Chem. Neurosci.* **2012**, *3*, 658–664. [CrossRef]

594. Lin, H.Y.; Snider, B.B. Synthesis of phidianidines A and B. *J. Org. Chem.* **2012**, *77*, 4832–4836. [CrossRef]

595. Manzo, E.; Pagano, D.; Carbone, M.; Ciavatta, M.L.; Gavagnin, M. Synthesis of phidianidine B, a highly cytotoxic 1, 2, 4-oxadiazole marine metabolite. *Arkivoc* **2012**, *9*, 220–228. [CrossRef]

596. Buchanan, J.C.; Petersen, B.P.; Chamberland, S. Concise total synthesis of phidianidine A and B. *Tetrahedron Lett.* **2013**, *54*, 6002–6004. [CrossRef]

597. Maftei, C.V.; Fodor, E.; Jones, P.G.; Franz, M.H.; Kelter, G.; Fiebig, H.; Neda, I. Synthesis and characterization of novel bioactive 1, 2, 4-oxadiazole natural product analogs bearing the N-phenylmaleimide and N-phenylsuccinimide moieties. *Beilstein J. Org. Chem.* **2013**, *9*, 2202–2215. [CrossRef]

598. Jiang, C.S.; Fu, Y.; Zhang, L.; Gong, J.X.; Wang, Z.Z.; Xiao, W.; Zhang, H.Y.; Guo, Y.W. Synthesis and biological evaluation of novel marine-derived indole-based 1, 2, 4-oxadiazoles derivatives as multifunctional neuroprotective agents. *Bioorg. Med. Chem. Lett.* **2015**, *25*, 216–220. [CrossRef] [PubMed]

599. Zhang, L.; Jiang, C.S.; Gao, L.X.; Gong, J.X.; Wang, Z.H.; Li, J.Y.; Li, J.; Li, X.W.; Guo, Y.W. Design, synthesis and in vitro activity of phidianidine B derivatives as novel PTP1B inhibitors with specific selectivity. *Bioorg. Med. Chem. Lett.* **2016**, *26*, 778–781. [CrossRef] [PubMed]

600. Khan, I.; Ibrar, A.; Abbas, N. Oxadiazoles as privileged motifs for promising anticancer leads: Recent advances and future prospects. *Archiv. Pharm. Chem. Life Sci.* **2014**, *347*, 1–20. [CrossRef] [PubMed]

601. Ding, D.; Boudreau, M.A.; Leemans, E.; Spink, E.; Yamaguchi, T.; Testero, S.A.; O'Daniel, P.I.; Lastochkin, E.; Chang, M.; Mobashery, S. Exploration of the structure-activity relationship of 1,2,4-oxadiazole antibiotics. *Bioorg. Med. Chem. Lett.* **2015**, *25*, 4854–4857. [CrossRef] [PubMed]

602. Lukin, A.; Karapetian, R.; Ivanenkov, Y.; Krasavin, M. Privileged 1,2,4-oxadiazoles in anticancer drug design: Novel 5-aryloxymethyl-1,2,4-oxadiazole leads for prostate cancer therapy. *Lett. Drug Des. Discov.* **2016**, *13*, 198–204. [CrossRef]

603. Pitasse-Santos, P.; Sueth-Santiago, V.; Lima, M.E.F. 1,2,4- and 1,3,4-oxadiazoles as scaffolds in the development of antiparasitic agents. *J. Brazil Chem. Soc.* **2018**, *29*, 435–456. [CrossRef]

604. Chawla, G. 1,2,4-Oxadiazole as a privileged scaffold for anti-inflammatory and analgesic activities: A review. *Mini-Rev. Med. Chem.* **2018**, *18*, 1536–1547. [CrossRef]

605. Manzo, E.; Pagano, D.; Ciavatta, M.L.; Carbone, M.; Gavagnin, M. 1, 2, 4-Oxadiazol Derivatives, Process for Their Preparation and Use Thereof as Intermediates in the Preparation of Indolic Alkaloids. U.S. Patent 20150051405A1, 19 February 2015.

606. Carroll, A.R.; Scheuer, P.J. Kuanoniamines A, B, C, and D: Pentacyclic alkaloids from a tunicate and its prosobranch mollusk predator *Chelynotus semperi*. *J. Org. Chem.* **1990**, *55*, 4426–4431. [CrossRef]

607. Singleton, P.A.; Moss, J.; Karp, D.D.; Atkins, J.T.; Janku, F. The mu opioid receptor: A new target for cancer therapy? *Cancer* **2015**, *121*, 2681–2688. [CrossRef]

608. Vitale, R.M.; Gatti, M.; Carbone, M.; Barbieri, F.; Felicità, V.; Gavagnin, M.; Florio, T.; Amodeo, P. Minimalist hybrid ligand/receptor-based pharmacophore model for CXCR4 applied to a small-library of marine natural products led to the identification of phidianidine A as a new CXCR4 ligand exhibiting antagonist activity. *ACS Chem. Biol.* **2013**, *8*, 2762–2770. [CrossRef]

609. Barbieri, F.; Thellung, S.; Wurth, R.; Gatto, F.; Corsaro, A.; Villa, V.; Nizzari, M.; Albertelli, M.; Ferone, D.; Florio, T. Emerging targets in pituitary adenomas: Role of the CXCL12/CXCR4-R7 system. *Int. J. Endocrinol.* **2014**, *2014*, 753524. [CrossRef] [PubMed]

610. Fu, X.; Palomar, A.J.; Hong, E.P.; Schmitz, F.J.; Valeriote, F.A. Cytotoxic lissoclimide-type diterpenes from the molluscs *Pleurobranchus albiguttatus* and *Pleurobranchus forskalii*. *J. Nat. Prod.* **2004**, *67*, 1415–1418. [CrossRef] [PubMed]

611. Wesson, K.J.; Hamann, M.T. Keenamide A, a bioactive cyclic peptide from the marine mollusk *Pleurobranchus forskalii*. *J. Nat. Prod.* **1996**, *59*, 629–631. [CrossRef] [PubMed]

612. Tan, K.C.; Wakimoto, T.; Takada, K.; Ohtsuki, T.; Uchiyama, N.; Goda, Y.; Abe, I. Cycloforskamide, a cytotoxic macrocyclic peptide from the sea slug *Pleurobranchus forskalii*. *J. Nat. Prod.* **2013**, *76*, 1388–1391. [CrossRef]

613. Shoemaker, R.H. The NCI60 human tumour cell line anticancer drug screen. *Nat. Rev. Cancer* **2006**, *6*, 813–823. [CrossRef]

614. Darro, F.; Decaestecker, C.; Gaussin, J.F.; Mortier, S.; Van Ginckel, R.; Kiss, R. Are syngeneic mouse tumor models still valuable experimental models in the field of anti-cancer drug discovery? *Int. J. Oncol.* **2005**, *27*, 607–616.

615. Wakimoto, T.; Tan, K.C.; Abe, I. Ergot Alkaloid from the Sea Slug *Pleurobranchus forskalii*. *Toxicon* **2013**, *72*, 1–4. [CrossRef]

616. Mulac, D.; Humpf, H.-U. Cytotoxicity and accumulation of ergot alkaloids in human primary cells. *Toxicology* **2011**, *282*, 112–121. [CrossRef]

617. Boyd, M.R.; Paull, K.D. Some practical considerations and applications of the National Cancer Institute in vitro anticancer drug discovery screen. *Drug. Dev. Res.* **1995**, *34*, 91–109. [CrossRef]

618. Robert, F.; Gao, H.Q.; Donia, M.; Merrick, W.C.; Hamann, M.T.; Pelletier, J. Chlorolissoclimides: New inhibitors of eukaryotic protein synthesis. *RNA* **2006**, *12*, 717–725. [CrossRef]

619. Malochet-Grivois, C.; Roussakis, C.; Robillard, N.; Biard, J.F.; Riou, D.; Debitus, C.; Verbist, J.F. Effects in vitro of two marine substances, chlorolissoclimide and dichlorolissoclimide, on a non-small-cell bronchopulmonary carcinoma line (NSCLC-N6). *Anticancer Drug Des.* **1992**, *7*, 493–505. [PubMed]

620. Uddin, M.J.; Kokubo, S.; Suenaga, K.; Ueda, K.; Uemura, D. Haterumaimides AE, five new dichlorolissoclimide-type diterpenoids from an ascidian, *Lissoclinum* sp. *Heterocycles* **2001**, *54*, 1039–1047.

621. Uddin, M.J.; Kokubo, S.; Ueda, K.; Suenaga, K.; Uemura, D. Haterumaimides F– I, four new cytotoxic diterpene alkaloids from an ascidian *Lissoclinum* species. *J. Nat. Prod.* **2001**, *64*, 1169–1173. [CrossRef] [PubMed]

622. Uddin, M.J.; Kokubo, S.; Ueda, K.; Suenaga, K.; Uemura, D. Haterumaimides J and K, potent cytotoxic diterpene alkaloids from the ascidian *Lissoclinum* species. *Chem. Lett.* **2002**, *31*, 1028–1029. [CrossRef]

623. Uddin, J.; Ueda, K.; Siwu, E.R.; Kita, M.; Uemura, D. Cytotoxic labdane alkaloids from an ascidian *Lissoclinum* sp.: Isolation, structure elucidation, and structure-activity relationship. *Bioorg. Med. Chem.* **2006**, *14*, 6954–6961. [CrossRef] [PubMed]

624. González, M.A.; Romero, D.; Zapata, B.; Betancur-Galvis, L. First synthesis of lissoclimide-type alkaloids. *Lett. Org. Chem.* **2009**, *6*, 289–292. [CrossRef]

625. Thoms, C.; Ebel, R.; Proksch, P. Sequestration and possible role of dietary alkaloids in the sponge-feeding mollusk *Tylodina perversa*. In *Molluscs*; Springer: Berlin/Heidelberg, Germany, 2006; pp. 261–275.

626. Kimura, J.; Takada, Y.; Inayoshi, T.; Nakao, Y.; Goetz, G.; Yoshida, W.Y.; Scheuer, P.J. Kulokekahilide-1, a cytotoxic depsipeptide from the cephalaspidean mollusk *Philinopsis speciosa*. *J. Org. Chem.* **2002**, *67*, 1760–1767. [CrossRef]

627. Nakao, Y.; Yoshida, W.Y.; Takada, Y.; Kimura, J.; Yang, L.; Mooberry, S.L.; Scheuer, P.J. Kulokekahilide-2, a cytotoxic depsipeptide from a cephalaspidean mollusk *Philinopsis speciosa*. *J. Nat. Prod.* **2004**, *67*, 1332–1340. [CrossRef]

628. Umehara, M.; Negishi, T.; Tashiro, T.; Nakao, Y.; Kimura, J. Structure-related cytotoxic activity from kulokekahilide-2, a cyclodepsipeptide of Hawaiian marine mollusk. *Bioorg. Med. Chem. Lett.* **2012**, *22*, 7422–7425. [CrossRef]

629. Takada, Y.; Umehara, M.; Katsumata, R.; Nakao, Y.; Kimura, J. The total synthesis and structure-activity relationships of a highly cytotoxic depsipeptide kulokekahilide-2 and its analogs. *Tetrahedron* **2012**, *68*, 659–669. [CrossRef]

630. Vasskog, T.; Andersen, J.H.; Hansen, E.; Svenson, J. Characterization and cytotoxicity studies of the rare 21: 4 n-7 acid and other polyunsaturated fatty acids from the marine opisthobranch *Scaphander lignarius*, isolated using bioassay guided fractionation. *Mar. Drugs* **2012**, *10*, 2676–2690. [CrossRef] [PubMed]

631. Wessels, M.; König, G.M.; Wright, A.D. New Natural Product Isolation and Comparison of the Secondary Metabolite Content of Three Distinct Samples of the Sea Hare *Aplysia dactylomela* from Tenerife. *J. Nat. Prod.* **2000**, *63*, 920–928. [CrossRef] [PubMed]

632. Sims, J.J.; Lin, G.H.; Wing, R.M. Marine natural products X elatol, a halogenated sesquiterpene alcohol from the red alga *Laurencia elata*. *Tetrahedron Lett.* **1974**, *15*, 3487–3490. [CrossRef]

633. Lang, K.L.; Silva, I.T.; Zimmermann, L.A.; Lhullier, C.; Mañalich Arana, M.V.; Palermo, J.A.; Falkenberg, M.; Simões, C.M.; Schenkel, E.P.; Durán, F.J. Cytotoxic activity of semi-synthetic derivatives of elatol and isoobtusol. *Mar. Drugs* **2012**, *10*, 2254–2264. [CrossRef] [PubMed]

634. Gonzalez, A.G.; Darias, J.; Diaz, A.; Fourneron, J.D.; Martin, J.D.; Perez, C. Evidence for the biogenesis of halogenated chamigrenes from the red alga *Laurencia obtusa*. *Tetrahedron Lett* **1976**, *17*, 3051–3054. [CrossRef]

635. Dias, T.; Brito, I.; Moujir, L.; Paiz, N.; Darias, J.; Cueto, M. Cytotoxic Sesquiterpenes from *Aplysia dactylomela*. *J. Nat. Prod.* **2005**, *68*, 1677–1679. [CrossRef]

636. Campos, A.; Souza, C.B.; Lhullier, C.; Falkenberg, M.; Schenkel, E.P.; Ribeiro-do-Valle, R.M.; Siqueira, J.M. Anti-tumour effects of elatol, a marine derivative compound obtained from red algae *Laurencia microcladia*. *J. Pharm. Pharmacol.* **2012**, *64*, 1146–1154. [CrossRef]

637. Van Goietsenoven, G.; Hutton, J.; Becker, J.P.; Lallemand, B.; Robert, F.; Lefranc, F.; Pirker, C.; Vandenbussche, G.; Van Antwerpen, P.; Evidente, A.; et al. Targeting of eEF1A with Amaryllidaceae isocarbostyrils as a strategy to combat melanomas. *FASEB J.* **2010**, *24*, 4575–4584. [CrossRef]

638. Mathieu, V.; Le Mercier, M.; De Neve, N.; Sauvage, S.; Gras, T.; Roland, I.; Lefranc, F.; Kiss, R. Galectin-1 knockdown increases sensitivity to temozolomide in a B16F10 mouse metastatic melanoma model. *J. Invest. Dermatol.* **2007**, *127*, 2399–2410. [CrossRef]

639. Schmitz, F.J.; Michaud, D.P.; Schmidt, P.G. Marine natural products: Parguerol, deoxyparguerol, and isoparguerol. New brominated diterpenes with modified pimarane skeletons from the sea hare *Aplysia dactylomela*. *J. Am. Chem. Soc.* **1982**, *104*, 6415–6423. [CrossRef]

640. Awad, N.E. Bioactive brominated diterpenes from the marine red alga *Jania rubens* (L.) Lamx. *Phytother. Res.* **2004**, *18*, 275–279. [CrossRef] [PubMed]

641. Vera, B.; Rodríguez, A.D.; Avilés, E.; Ishikawa, Y. Aplysqualenols A and B: Squalene-derived polyethers with antitumoral and antiviral activity from the caribbean sea slug *Aplysia dactylomela*. *Eur. J. Org. Chem.* **2009**, *31*, 5327–5336. [CrossRef] [PubMed]

642. Vera, B.; Rodríguez, A.D.; La Clair, J.J. Aplysqualenol A binds to the light chain of dynein type 1 (DYNLL1). *Angew. Chem.* **2011**, *123*, 8284–8288. [CrossRef]

643. Zhu, G.; Yang, F.; Balachandran, R.; Höök, P.; Vallee, R.B.; Curran, D.P.; Day, B.W. Synthesis and biological evaluation of purealin and analogues as cytoplasmic dynein heavy chain inhibitors. *J. Med. Chem.* **2006**, *49*, 2063–2076. [CrossRef]

644. Alberti, C. Cytoskeleton structure and dynamic behaviour: Quick excursus from basic molecular mechanisms to some implications in cancer chemotherapy. *Eur. Rev. Med. Pharmacol. Sci.* **2009**, *13*, 13–21.

645. Wong, D.M.; Li, L.; Jurado, S.; King, A.; Bamford, R.; Wall, M.; Walia, M.K.; Kelly, G.L.; Walkley, C.R.; Tarlinton, D.M.; et al. The transcription factor ASCIZ and its target DYNLL1 are essential for the development and expansion of MYC-driven B cell lymphoma. *Cell Rep.* **2016**, *14*, 1488–1499. [CrossRef]

646. Manzo, E.; Gavagnin, M.; Bifulco, G.; Cimino, P.; Di Micco, S.; Ciavatta, M.L.; Guo, Y.-W.; Cimino, G. Aplysiols A and B, squalene-derived polyethers from the mantle of the sea hare *Aplysia dactylomela*. *Tetrahedron* **2007**, *63*, 9970–9978. [CrossRef]

647. Blunt, J.W.; Hartshorn, M.P.; McLennan, T.J.; Munro, M.H.G.; Robinson, W.T.; Yorke, S.C. Thyrsiferol: A squalene-derived metabolite of *Laurencia thyrsifera*. *Tetrahedron Lett.* **1978**, *19*, 69–72. [CrossRef]

648. Mahdi, F.; Falkenberg, M.; Ioannou, E.; Roussis, V.; Zhou, Y.D.; Nagle, D.G. Thyrsiferol inhibits mitochondrial respiration and HIF-1 activation. *Phytochem. Lett.* **2011**, *4*, 75–78. [CrossRef]

649. Fernández, J.; Souto, M.L.; Norte, M. Evaluation of the cytotoxic activity of polyethers isolated from *Laurencia*. *Bioorg. Med. Chem.* **1998**, *6*, 2237–2243. [CrossRef]

650. Jiménez, C.; Quiñoá, E.; Castedo, L.; Riguera, R. Epidioxy sterols from the tunicates *Dendrodoa grossularia* and *Ascidiella aspersa* and the gastropoda *Aplysia depilans* and *Aplysia punctata*. *J. Nat. Prod.* **1986**, *49*, 905–909. [CrossRef]

651. Mun, B.; Wang, W.; Kim, H.; Hahn, D.; Yang, I.; Won, D.H.; Kim, E.H.; Lee, J.; Han, C.; Kim, H.; et al. Cytotoxic 5α, 8α-epidioxy sterols from the marine sponge *Monanchora* sp. *Arch. Pharm. Res.* **2015**, *38*, 18–25. [CrossRef] [PubMed]

652. Yamada, K.; Ojika, M.; Ishigaki, T.; Yoshida, Y.; Ekimoto, H.; Arakawa, M. Aplyronine A, a potent antitumor substance, and the congeners aplyronines B and C isolated from the sea hare *Aplysia kurodai*. *J. Am. Chem. Soc.* **1993**, *115*, 11020–11021. [CrossRef]

653. Ojika, M.; Kigoshi, H.; Suenaga, K.; Imamura, Y.; Yoshikawa, K.; Ishigaki, T.; Sakakura, A.; Mutou, T.; Yamada, K. Aplyronines D-H from the sea hare *Aplysia kurodai*: Isolation, structures, and cytotoxicity. *Tetrahedron* **2012**, *68*, 982–987. [CrossRef]

654. Kuroda, T.; Kigoshi, H. Aplaminal: A novel cytotoxic aminal isolated from the sea hare *Aplysia kurodai*. *Organic Lett.* **2008**, *10*, 489–491. [CrossRef]

655. Kigoshi, H.; Suenaga, K.; Takagi, M.; Akao, A.; Kanematsu, K.; Kamei, N.; Okugawa, Y.; Yamada, K. Cytotoxicity and actin-depolymerizing activity of aplyronine A, a potent antitumor macrolide of marine origin, and its analogs. *Tetrahedron* **2002**, *58*, 1075–1102. [CrossRef]

656. Ojika, M.; Kigoshi, H.; Yoshida, Y.; Ishigaki, T.; Nisiwaki, M.; Tsukada, I.; Arakawa, M.; Ekimoto, H.; Yamada, K. Aplyronine A, a potent antitumor macrolide of marine origin, and the congeners aplyronines B and C: Isolation, structures, and bioactivities. *Tetrahedron* **2007**, *63*, 3138–3167. [CrossRef]

657. Yamada, K.; Ojika, M.; Kigoshi, H.; Suenaga, K. Aplyronine A, a potent antitumour macrolide of marine origin, and the congeners aplyronines B-H: Chemistry and biology. *Nat. Prod. Rep.* **2009**, *26*, 27–43. [CrossRef]

658. Kita, M.; Kigoshi, H. Marine natural products that interfere with multiple cytoskeletal protein interactions. *Nat. Prod. Rep.* **2015**, *32*, 534–542. [CrossRef]

659. Ohno, O.; Morita, M.; Kitamura, K.; Teruya, T.; Yoneda, K.; Kita, M.; Kigoshi, H.; Suenaga, K. Apoptosis-inducing activity of the actin-depolymerizing agent aplyronine A and its side-chain derivatives. *Bioorg. Med. Chem. Lett.* **2013**, *23*, 1467–1471. [CrossRef]

660. Hirata, K.; Muraoka, S.; Suenaga, K.; Kuroda, T.; Kato, K.; Tanaka, H.; Yamamoto, M.; Takata, M.; Yamada, K.; Kigoshi, H. Structure basis for antitumor effect of aplyronine A. *J. Mol. Biol.* **2006**, *356*, 945–954. [CrossRef] [PubMed]

661. Kita, M.; Hirayama, Y.; Sugiyama, M.; Kigoshi, H. Development of highly cytotoxic and actin-depolymerizing biotin derivatives of aplyronine A. *Angew. Chem. Int. Ed.* **2011**, *50*, 9871–9874. [CrossRef] [PubMed]

662. Kita, M.; Hirayama, Y.; Yoneda, K.; Yamagishi, K.; Chinen, T.; Usui, T.; Sumiya, E.; Uesugi, M.; Kigoshi, H. Inhibition of microtubule assembly by a complex of actin and antitumor macrolide aplyronine A. *J. Am. Chem. Soc.* **2013**, *135*, 18089–18095. [CrossRef] [PubMed]

663. Hong, W.P.; Noshi, M.N.; El-Awa, A.; Fuchs, P.L. Synthesis of the C1-C20 and C15-C27 segments of aplyronine A. *Org. Lett.* **2011**, *13*, 6342–6345. [CrossRef] [PubMed]

664. Paterson, I.; Fink, S.J.; Lee, L.Y.; Atkinson, S.J.; Blakey, S.B. Total synthesis of aplyronine C. *Org. Lett.* **2013**, *15*, 3118–3121. [CrossRef]

665. Anžiček, N.; Williams, S.; Housden, M.P.; Paterson, I. Toward aplyronine payloads for antibody-drug conjugates: Total synthesis of aplyronines A and D. *Org. Biomol. Chem.* **2018**, *16*, 1343–1350. [CrossRef]

666. Newman, D.J.; Cragg, G.M. Current status of marine-derived compounds as warheads in anti-tumor drug candidates. *Mar. Drugs* **2017**, *15*, 99. [CrossRef]

667. Kigoshi, H.; Imamura, Y.; Yoshikawa, K.; Yamada, K. Three new cytotoxic alkaloids, aplaminone, neoaplaminone and neoaplaminone sulfate from the marine mollusc *Aplysia kurodai*. *Tetrahedron Lett.* **1990**, *31*, 4911–4914. [CrossRef]

668. Kusumi, T.; Uchida, H.; Inouye, Y.; Ishitsuka, M.; Yamamoto, H.; Kakisawa, H. Novel cytotoxic monoterpenes having a halogenated tetrahydropyran from *Aplysia kurodai*. *J. Org. Chem.* **1987**, *52*, 4597–4600. [CrossRef]

669. Miyamoto, T.; Higuchi, R.; Marubayashi, N.; Komori, T. Two new polyhalogenated monoterpenes from the sea hare *Aplysia kurodai*. *Liebigs Ann. Chem.* **1988**, *12*, 1191–1193. [CrossRef]

670. Katayama, A.; Ina, K.; Nozaki, H.; Nakayama, M. Structural elucidation of kurodainol, a novel halogenated monoterpene from sea hare (*Aplysia kurodai*). *Agr. Biol. Chem.* **1982**, *46*, 859–860. [CrossRef]

671. Yamamura, S.; Hirata, Y. A naturally-occurring bromo-compound, aplysin-20 from *Aplysia kurodai*. *Bull. Chem. Soc. Jpn.* **1971**, *44*, 2560–2562. [CrossRef]

672. Yamamura, S.; Terada, Y. Isoaplysin-20, a natural bromine-containing diterpene, from *Aplysia kurodai*. *Tetrahedron Lett.* **1977**, *25*, 2171–2172. [CrossRef]

673. Ojika, M.; Yoshida, Y.; Okumura, M.; Ieda, S.; Yamada, K. Aplysiadiol, a new brominated diterpene from the marine mollusc *Aplysia kurodai*. *J. Nat. Prod.* **1990**, *53*, 1619–1622. [CrossRef]

674. Ojika, M.; Kigoshi, H.; Yoshikawa, K.; Nakayama, Y.; Yamada, K. A new bromo diterpene, epi-aplysin-20, and ent-isoconcinndiol from the marine mollusc *Aplysia kurodai*. *Bull. Chem. Soc. Jpn.* **1992**, *65*, 2300–2302. [CrossRef]

675. Ojika, M.; Yoshida, Y.; Nakayama, Y.; Yamada, K. Aplydilactone, a novel fatty acid metabolite from the marine mollusc *Aplysia kurodai*. *Tetrahedron Lett.* **1990**, *31*, 4907–4910. [CrossRef]

676. Ojika, M.; Yoshida, T.; Yamada, K. Aplysepine, a novel 1,4-benzodiazepine alkaloid from the sea hare *Aplysia kurodai*. *Tetrahedron Lett.* **1993**, *34*, 5307–5308. [CrossRef]

677. Iijima, R.; Kisugi, J.; Yamazaki, M. Antifungal activity of aplysianin E, a cytotoxic protein of sea hare (*Aplysia kurodai*) eggs. *Dev. Comp. Immunol.* **1995**, *19*, 13–19.

678. Iijima, R.; Kisugi, J.; Yamazaki, M. A novel antimicrobial peptide from the sea hare *Dolabella auricularia*. *Dev. Comp. Immunol.* **2003**, *27*, 305–311. [CrossRef]

679. Kaviarasan, T.; Siva, S.R.; Yogamoorthi, A. Antimicrobial secondary metabolites from marine gastropod egg capsules and egg masses. *Asian Pac. J. Trop Biomed.* **2012**, *2*, 916–922. [CrossRef]

680. Yamamura, S.; Hirata, Y. Structures of aplysin and aplysinol, naturally occurring bromo-compounds. *Tetrahedron* **1963**, *19*, 1485–1496. [CrossRef]

681. Ryu, G.S.; Park, S.H.; Choi, B.W.; Lee, N.H.; Hwang, H.J.; Ryu, S.Y.; Lee, B.H. Cytotoxic activities of brominated sesquiterpenes from the red alga *Laurencia okamurae*. *Nat. Prod. Sci.* **2002**, *8*, 103–107.

682. Liu, J.; Ma, L.; Wu, N.; Liu, G.; Zheng, L.; Lin, X. Aplysin sensitizes cancer cells to TRAIL by suppressing P38 MAPK/survivin pathway. *Mar. Drugs* **2014**, *12*, 5072–5088. [CrossRef] [PubMed]

683. Gong, A.J.; Gong, L.L.; Yao, W.C.; Ge, N.; Lu, L.X.; Liang, H. Aplysin induces apoptosis in glioma cells through HSP90/AKT pathway. *Exp. Biol. Med.* **2015**, *240*, 639–644. [CrossRef]

684. Hegazy, M.E.F.; Moustfa, A.Y.; Mohamed, A.E.H.H.; Alhammady, M.A.; Elbehairi, S.E.I.; Ohta, S.; Paré, P.W. New cytotoxic halogenated sesquiterpenes from the Egyptian sea hare, Aplysia oculifera. *Tetrahedron Lett.* **2014**, *55*, 1711–1714. [CrossRef]

685. Ortega, M.J.; Zubía, E.; Salvá, J. New polyhalogenated monoterpenes from the sea hare *Aplysia punctata*. *J. Nat. Prod.* **1997**, *60*, 482–484. [CrossRef]

686. Pennings, S.C.; Paul, V.J.; Dunbar, D.C.; Hamann, M.T.; Lumbang, W.; Novack, B.; Jacobs, R.S. Unpalatable compounds in the marine gastropod *Dolabella auricularia*: Distribution and effect of diet. *J. Chem. Ecol.* **1999**, *25*, 735–755. [CrossRef]

687. Suenaga, K.; Shibata, T.; Takada, N.; Kigoshi, H.; Yamada, K. Aurilol, a cytotoxic bromotriterpene isolated from the sea hare *Dolabella auricularia*. *J. Nat. Prod.* **1998**, *61*, 515–518. [CrossRef]

688. Sone, H.; Kigoshi, H.; Yamada, K. Aurisides A and B, cytotoxic macrolide glycosides from the Japanese sea hare *Dolabella auricularia*. *J. Org. Chem.* **1996**, *61*, 8956–8960. [CrossRef]

689. Paterson, I.; Florence, G.J.; Heimann, A.C.; Mackay, A.C. Stereocontrolled total synthesis of (−)-aurisides A and B. *Angew. Chem. Int. Ed.* **2005**, *44*, 1130–1133. [CrossRef]

690. Tello-Aburto, R.; Olivo, H.F. A formal synthesis of the auriside aglycon. *Organic Lett.* **2008**, *10*, 2191–2194. [CrossRef] [PubMed]

691. Ojika, M.; Nemoto, T.; Yamada, K. Doliculols A and B, the non-halogenated C15 acetogenins with cyclic ether from the sea hare *Dolabella auricularia*. *Tetrahedron Lett* **1993**, *34*, 3461–3462. [CrossRef]

692. Ojika, M.; Nagoya, T.; Yamada, K. Dolabelides A and B, cytotoxic 22-membered macrolides isolated from the sea hare *Dolabella auricularia*. *Tetrahedron Lett.* **1995**, *36*, 7491–7494. [CrossRef]

693. Suenaga, K.; Nagoya, T.; Shibata, T.; Kigoshi, H.; Yamada, K. Dolabelides C and D, cytotoxic macrolides isolated from the sea hare *Dolabella auricularia*. *J. Nat. Prod.* **1997**, *60*, 155–157. [CrossRef]

694. Yamada, K.; Ojika, M.; Kigoshi, H.; Suenaga, K. Cytotoxic substances from two species of Japanese sea hares: Chemistry and bioactivity. *Proc. Japan Acad. Ser. B Phys. Biol. Sci.* **2010**, *86*, 176–189. [CrossRef]

695. Sone, H.; Kondo, T.; Kiryu, M.; Ishiwata, H.; Ojika, M.; Yamada, K. Dolabellin, a cytotoxic bisthiazole metabolite from the sea hare *Dolabella auricularia*: Structural determination and synthesis. *J. Org. Chem.* **1995**, *60*, 4774–4781. [CrossRef]

696. Suenaga, K.; Kigoshi, H.; Yamada, K. Auripyrones A and B, cytotoxic polypropionates from the sea hare *Dolabella auricularia*: Isolation and structures. *Tetrahedron Lett.* **1996**, *37*, 5151–5154. [CrossRef]

697. Lister, T.; Perkins, M.V. Total synthesis of auripyrone A. Angew. *Chem. Int. Ed.* **2006**, *45*, 2560–2564. [CrossRef]

698. Jung, M.E.; Chaumontet, M.; Salehi-Rad, R. Total synthesis of auripyrone B using a non-aldol aldolcuprate opening process. *Org. Lett.* **2010**, *12*, 2872–2875. [CrossRef]

699. Jung, M.E.; Salehi-Rad, R. Total synthesis of auripyrone A using a tandem non-aldol aldol/Paterson aldol process as a key step. *Angew. Chem. Int. Ed.* **2009**, *48*, 8766–8769. [CrossRef]

700. Hayakawa, I.; Takemura, T.; Fukasawa, E.; Ebihara, Y.; Sato, N.; Nakamura, T.; Suenaga, K.; Kigoshi, H. Total synthesis of auripyrones A and B and determination of the absolute configuration of auripyrone B. *Angew. Chem. Int. Ed.* **2010**, *49*, 2401–2405. [CrossRef] [PubMed]

701. Park, P.K.; O'Malley, S.J.; Schmidt, D.R.; Leighton, J.L. Total synthesis of dolabelide D. *J. Am. Chem. Soc.* **2006**, *128*, 2796–2797. [CrossRef] [PubMed]

702. Hanson, P.R.; Chegondi, R.; Nguyen, J.; Thomas, C.D.; Waetzig, J.D.; Whitehead, A. Total synthesis of dolabelide C: A phosphate-mediated approach. *J. Org. Chem.* **2011**, *76*, 4358–4370. [CrossRef] [PubMed]

703. Suenaga, K.; Mutou, T.; Shibata, T.; Itoh, T.; Kigoshi, H.; Yamada, K. Isolation and stereostructure of aurilide, a novel cyclodepsipeptide from the Japanese sea hare *Dolabella auricularia*. *Tetrahedron Lett.* **1996**, *37*, 6771–6774. [CrossRef]

704. Han, B.; Gross, H.; Goeger, D.E.; Mooberry, S.L.; Gerwick, W.H. Aurilides B and C, cancer cell toxins from a Papua New Guinea collection of the marine cyanobacterium *Lyngbya majuscula*. *J. Nat. Prod.* **2006**, *69*, 572–575. [CrossRef]

705. Suenaga, K.; Mutou, T.; Shibata, T.; Itoh, T.; Fujita, T.; Takada, N.; Hayamizu, K.; Takagi, M.; Irifune, T.; Kigoshi, H.; et al. Aurilide, a cytotoxic depsipeptide from the sea hare *Dolabella auricularia*: Isolation, structure determination, synthesis, and biological activity. *Tetrahedron* **2004**, *60*, 8509–8527. [CrossRef]

706. Hollingshead, M.G.; Alley, M.C.; Camalier, R.F.; Abbott, B.J.; Mayo, J.G.; Malspeis, L.; Grever, M.R. In vivo cultivation of tumor cells in hollow fibers. *Life Sci.* **1995**, *57*, 131–141. [CrossRef]

707. Sato, S.I.; Murata, A.; Orihara, T.; Shirakawa, T.; Suenaga, K.; Kigoshi, H.; Uesugi, M. Marine natural product aurilide activates the OPA1-mediated apoptosis by binding to prohibitin. *Chem. Biol.* **2011**, *18*, 131–139. [CrossRef]

708. Ishiwata, H.; Nemoto, T.; Ojika, M.; Yamada, K. Isolation and stereostructure of doliculide, a cytotoxic cyclodepsipeptide from the Japanese sea hare *Dolabella auricularia*. *J. Org. Chem.* **1994**, *59*, 4710–4711. [CrossRef]

709. Bai, R.; Covell, D.G.; Liu, C.; Ghosh, A.K.; Hamel, E. (−)-Doliculide, a new macrocyclic depsipeptide enhancer of actin assembly. *J. Biol. Chem.* **2002**, *277*, 32165–32171. [CrossRef]

710. Ishiwata, H.; Sone, H.; Kigoshi, H.; Yamada, K. Enantioselective total synthesis of doliculide, a potent cytotoxic cyclodepsipeptide of marine origin and structure-cytotoxicity relationships of synthetic doliculide congeners. *Tetrahedron* **1994**, *50*, 12853–12882. [CrossRef]

711. Ghosh, A.K.; Liu, C. Total synthesis of antitumor depsipeptide (-)-doliculide. *Org. Lett.* **2001**, *3*, 635–638. [CrossRef] [PubMed]

712. Matcha, K.; Madduri, A.V.; Roy, S.; Ziegler, S.; Waldmann, H.; Hirsch, A.K.; Minnaard, A.J. Total synthesis of (−)-Doliculide, structure-activity relationship studies and its binding to F-actin. *Chem. Bio. Chem.* **2012**, *13*, 2537–2548. [CrossRef] [PubMed]

713. Foerster, F.; Braig, S.; Chen, T.; Altmann, K.H.; Vollmar, A.M. Pharmacological characterization of actin-binding (−)-Doliculide. *Bioorg. Med. Chem.* **2014**, *22*, 5117–5122. [CrossRef] [PubMed]

714. Harrigan, G.G.; Luesch, H.; Yoshida, W.Y.; Moore, R.E.; Nagle, D.G.; Paul, V.J.; Mooberry, S.L.; Corbett, T.H.; Valeriote, F.A. Symplostatin 1: A Dolastatin 10 analogue from the marine cyanobacterium *Symploca hydnoides*. *J. Nat. Prod.* **1998**, *61*, 1075–1077. [CrossRef]

715. Luesch, H.; Harrigan, G.G.; Goetz, G.; Horgen, F.D. The cyanobacterial origin of potent anticancer agents originally isolated from sea hares. *Curr. Med. Chem.* **2002**, *9*, 1791–1806. [CrossRef]

716. Engene, N.; Tronholm, A.; Salvador-Reyes, L.A.; Luesch, H.; Paul, V.J. *Caldora penicillata* gen. nov., comb. nov. (Cyanobacteria), a pantropical marine species with biomedical relevance. *J. Phycol.* **2015**, *51*, 670–681. [CrossRef]

717. Maderna, A.; Leverett, C.A. Recent advances in the development of new auristatins: Structural modifications and application in antibody drug conjugates. *Mol. Pharm.* **2015**, *12*, 1798–1812. [CrossRef]

718. Pettit, G.R.; Kamano, Y.; Brown, P.; Gust, D.; Inoue, M.; Herald, C.L. Structure of the cyclic peptide dolastatin 3 from *Dolabella auricularia*. *J. Am. Chem. Soc.* **1982**, *104*, 905–907. [CrossRef]

719. Pettit, G.R.; Kamano, Y.; Herald, C.L.; Fujii, Y.; Kizu, H.; Boyd, M.R.; Boettner, F.E.; Doubek, D.L.; Schmidt, J.M.; Chapuis, J.-C.; et al. Isolation of dolastatins 10-15 from the marine mollusc *Dolabella auricularia*. *Tetrahedron* **1993**, *49*, 9151–9170. [CrossRef]

720. Yamada, K.; Kigoshi, H. Bioactive compounds from the sea hares of two genera: *Aplysia* and *Dolabella*. *Bull. Chem. Soc. Jpn.* **1997**, *70*, 1479–1489. [CrossRef]

721. Sone, H.; Nemoto, T.; Ishiwata, H.; Ojika, M.; Yamada, K. Isolation, structure, and synthesis of dolastatin D, a cytotoxic cyclic depsipeptide from the sea hare *Dolabella auricularia*. *Tetrahedron Lett.* **1993**, *34*, 8449–8452. [CrossRef]

722. Sone, H.; Nemoto, T.; Ojika, M.; Yamada, K. Isolation, structure, and synthesis of dolastatin C, a cytotoxic cyclic depsipeptide from the sea hare *Dolabella auricularia*. *Tetrahedron Lett.* **1993**, *34*, 8445–8448. [CrossRef]

723. Sone, H.; Shibata, T.; Fujita, T.; Ojika, M.; Yamada, K. Dolastatin H and isodolastatin H, potent cytotoxic peptides from the sea hare *Dolabella auricularia*: Isolation, stereostructures, and synthesis. *J. Am. Chem. Soc.* **1996**, *118*, 1874–1880. [CrossRef]

724. Mutou, T.; Kondo, T.; Ojika, M.; Yamada, K. Isolation and stereostructures of dolastatin G and nordolastatin G, cytotoxic 35-membered cyclodepsipeptides from the Japanese sea hare *Dolabella auricularia*. *J. Org. Chem.* **1996**, *61*, 6340–6345. [CrossRef]

725. Luesch, H.; Yoshida, W.Y.; Moore, R.E.; Paul, V.J. Lyngbyastatin 2 and norlyngbyastatin 2, analogues of dolastatin G and nordolastatin G from the marine cyanobacterium *Lyngbya majuscula*. *J. Nat. Prod.* **1999**, *62*, 1702–1706. [CrossRef]

726. Newman, D.J. The "utility" of highly toxic marine-sourced compounds. *Mar. Drugs* **2019**, *17*, 324. [CrossRef]

727. Pettit, G.R.; Kamano, Y.; Holzapfel, C.W.; Van Zyl, W.J.; Tuinman, A.A.; Herald, C.L.; Baczynskyj, L.; Schmidt, J.M. Antineoplastic agents. 150. The structure and synthesis of dolastatin 3. *J. Am. Chem. Soc.* **1987**, *109*, 7581–7582. [CrossRef]

728. Mitchell, S.S.; Faulkner, D.J.; Rubins, K.; Bushman, F.D. Dolastatin 3 and two novel cyclic peptides from a Palauan collection of *Lyngbya majuscula*. *J. Nat. Prod.* **2000**, *63*, 279–282. [CrossRef]

729. Pettit, G.R.; Kamano, Y.; Herald, C.L.; Tuinman, A.A.; Boettner, F.E.; Kizu, H.; Schmidt, J.M.; Baczynskyj, L.; Tomer, K.B.; Bontems, R.J. The isolation and structure of a remarkable marine animal antineoplastic constituent: Dolastatin 10. *J. Am. Chem. Soc.* **1987**, *109*, 6883–6885. [CrossRef]

730. Pettit, G.R.; Singh, S.B.; Hogan, F.; Lloyd-Williams, P.; Herald, D.L.; Burkett, D.D.; Clewlow, P.J. The absolute configuration and synthesis of natural (-)-dolastatin 10. *J. Am. Chem. Soc.* **1989**, *111*, 5463–5466. [CrossRef]

731. Singh, R.; Sharma, M.; Joshi, P.; Rawat, D.S. Clinical status of anti-cancer agents derived from marine sources. *Anticancer Agents Med. Chem.* **2008**, *8*, 603–617. [CrossRef] [PubMed]

732. Poncet, J. The dolastatins, a family of promising antineoplastic agents. *Cur. Pharm. Des.* **1999**, *5*, 139–162.

733. Bai, R.; Pettit, G.R.; Hamel, E. Dolastatin 10, a powerful cytostatic peptide derived from a marine animal. Inhibition of tubulin polymerization mediated through the vinca alkaloid binding domain. *Biochem. Pharmacol.* **1990**, *39*, 1941–1949. [CrossRef]

734. Pathak, S.; Multani, A.S.; Ozen, M.; Richardson, M.A.; Newman, R.A. Dolastatin-10 induces polyploidy, telomeric associations and apoptosis in a murine melanoma cell line. *Oncol. Rep.* **1998**, *5*, 373–379. [CrossRef]

735. Mooberry, S.L.; Leal, R.M.; Tinley, T.L.; Luesch, H.; Moore, R.E.; Corbett, T.H. The molecular pharmacology of symplostatin 1: A new antimitotic dolastatin 10 analog. *Int. J. Cancer* **2003**, *104*, 512–521. [CrossRef]

736. Kalemkerian, G.P.; Ou, X.; Adil, M.R.; Rosati, R.; Khoulani, M.M.; Madan, S.K.; Pettit, G.R. Activity of dolastatin 10 against small-cell lung cancer in vitro and in vivo: Induction of apoptosis and bcl-2 modification. *Cancer Chemother. Pharmacol.* **1999**, *43*, 507–515. [CrossRef]

737. Mohammad, R.M.; Pettit, G.R.; Almatchy, V.P.; Wall, N.; Varterasian, M.; Al-Katib, A. Synergistic interaction of selected marine animal anticancer drugs against human diffuse large cell lymphoma. *Anticancer Drugs* **1998**, *9*, 149–156. [CrossRef]

738. Aherne, G.W.; Hardcastle, A.; Valenti, M.; Bryant, A.; Rogers, P.; Pettit, G.R.; Srirangam, J.K.; Kelland, L.R. Antitumour evaluation of dolastatins 10 and 15 and their measurement in plasma by radioimmunoassay. *Cancer Chemother. Pharmacol.* **1996**, *38*, 225–232. [CrossRef]

739. Watanabe, J.; Natsume, T.; Fujio, N.; Miyasaka, K.; Kobayashi, M. Induction of apoptosis in human cancer cells by TZT-1027, an antimicrotubule agent. *Apoptosis* **2000**, *5*, 345–353. [CrossRef]

740. Verdier-Pinard, P.; Kepler, J.A.; Pettit, G.R.; Hamel, E. Sustained intracellular retention of dolastatin 10 causes its potent antimitotic activity. *Mol. Pharmacol.* **2000**, *57*, 180–187. [PubMed]

741. Doronina, S.O.; Toki, B.E.; Torgov, M.Y.; Mendelsohn, B.A.; Cerveny, C.G.; Chace, D.F.; DeBlanc, R.L.; Gearing, R.P.; Bovee, T.D.; Siegall, C.B.; et al. Development of potent monoclonal antibody auristatin conjugates for cancer therapy. *Nature Biotech.* **2003**, *21*, 778–784. [CrossRef] [PubMed]

742. Steinmetz, M.O.; Prota, A.E. Microtubule-targeting agents: Strategies to hijack the cytoskeleton. *Trends Cell Biol.* **2018**, *28*, 776–792. [CrossRef] [PubMed]

743. Johansson, M.P.; Maaheimo, H.; Ekholm, F.S. New insight on the structural features of the cytotoxic auristatins MMAE and MMAF revealed by combined NMR spectroscopy and quantum chemical modelling. *Sci. Rep.* **2017**, *7*, 1–10. [CrossRef] [PubMed]

744. Flahive, E.; Srirangam, J. The Dolastatins. In *Anticancer Agents from Natural Products*, 2nd ed.; Cragg, G.M., Kingston, D.G.I., Newman, D.J., Eds.; Taylor and Francis: Boca Raton, FL, USA, 2011; pp. 263–290.

745. Salvador-Reyes, L.A.; Luesch, H. Biological targets and mechanisms of action of natural products from marine cyanobacteria. *Nat. Prod. Rep.* **2015**, *32*, 478–503. [CrossRef] [PubMed]

746. Pettit, G.R.; Barkoczy, J.; Kantoci, D. Human Cancer Inhibitory Pentapeptide Amides. US Patent 5410024, 25 April 1995.

747. Watanabe, J.; Minami, M.; Kobayashi, M. Antitumor activity of TZT-1027 (Soblidotin). *Anticancer Res.* **2006**, *26*, 1973–1981.

748. Natsume, T.; Watanabe, J.; Ogawa, K.; Yasumura, K.; Kobayashi, M. Tumor-specific antivascular effect of TZT-1027 (Soblidotin) elucidated by magnetic resonance imaging and confocal laser scanning microscopy. *Cancer Sci.* **2007**, *98*, 598–604. [CrossRef]

749. Martins, A.; Vieira, H.; Gaspar, H.; Santos, S. Marketed marine natural products in the pharmaceutical and cosmeceutical industries: Tips for success. *Mar. Drugs* **2014**, *12*, 1066–1101. [CrossRef]

750. Pettit, G.R.; Kamano, Y.; Kizu, H.; Dufresne, C.; Herald, C.L.; Bontems, R.J.; Schmidt, J.M.; Boettner, F.E.; Nieman, R.A. Isolation and structure of the cell growth inhibitory depsipeptides dolastatins 11 and 12. *Heterocycles* **1989**, *28*, 553–558. [CrossRef]

751. Pettit, G.R.; Kamano, Y.; Herald, C.L.; Dufresne, C.; Cerny, R.L.; Herald, D.L.; Schmidt, J.M.; Kizu, H. Isolation and structure of the cytostatic depsipeptide dolastatin 13 from the sea hare *Dolabella auricularia*. *J. Am. Chem. Soc.* **1989**, *111*, 5015–5017. [CrossRef]

752. Pettit, G.R.; Kamano, Y.; Herald, C.L.; Dufresne, C.; Bates, R.B.; Schmidt, J.M.; Cerny, R.L.; Kizu, H. Antineoplastic agents. 190. Isolation and structure of the cyclodepsipeptide dolastatin 14. *J. Org. Chem.* **1990**, *55*, 2989–2990. [CrossRef]

753. Harrigan, G.G.; Yoshida, W.Y.; Moore, R.E.; Nagle, D.G.; Park, P.U.; Biggs, J.; Paul, V.J.; Mooberry, S.L.; Corbett, T.H.; Valeriote, F.A. Isolation, structure determination, and biological activity of dolastatin 12 and lyngbyastatin 1 from *Lyngbya majuscula/Schizothrix calcicola* cyanobacterial assemblages. *J. Nat. Prod.* **1998**, *61*, 1221–1225. [CrossRef] [PubMed]

754. Ali, M.A.; Bates, R.B.; Crane, Z.D.; Dicus, C.W.; Gramme, M.R.; Hamel, E.; Marcischak, J.; Martinez, D.S.; McClure, K.J.; Nakkiew, P.; et al. Dolastatin 11 conformations, analogues and pharmacophore. *Bioorg. Med. Chem.* **2005**, *13*, 4138–4152. [CrossRef] [PubMed]

755. Thornburg, C.C.; Thimmaiah, M.; Shaala, L.A.; Hau, A.M.; Malmo, J.M.; Ishmael, J.E.; Youssef, D.T.; McPhail, K.L. Cyclic depsipeptides, grassypeptolides D and E and ibu-epidemethoxylyngbyastatin 3, from a Red Sea *Leptolyngbya* cyanobacterium. *J. Nat. Prod.* **2011**, *74*, 1677–1685. [CrossRef] [PubMed]

756. Pettit, G.R.; Kamano, Y.; Dufresne, C.; Cerny, R.L.; Herald, C.L.; Schmidt, J.M. Isolation and structure of the cytostatic linear depsipeptide dolastatin 15. *J. Org. Chem.* **1989**, *54*, 6005–6006. [CrossRef]

757. Pettit, G.R.; Herald, D.L.; Singh, S.B.; Thornton, T.J.; Mullaney, J.T. Antineoplastic agents. 220. Synthesis of natural (-)-dolastatin 15. *J. Am. Chem. Soc.* **1991**, *113*, 6692–6693. [CrossRef]

758. Akaji, K.; Hayashi, Y.; Kiso, Y.; Kuriyama, N. Convergent synthesis of dolastatin 15 by solid phase coupling of an N-methylamino acid. *J. Org. Chem.* **1999**, *64*, 405–411. [CrossRef]

759. Yokosaka, A.; Izawa, A.; Sakai, C.; Sakurada, E.; Morita, Y.; Nishio, Y. Synthesis and evaluation of novel dolastatin 10 derivatives for versatile conjugations. *Bioorg. Med. Chem.* **2018**, *26*, 1643–1652. [CrossRef]

760. Newman, D.J.; Cragg, G.M. Advanced preclinical and clinical trials of natural products and related compounds from marine sources. *Curr. Med. Chem.* **2004**, *11*, 1693–1713. [CrossRef]

761. Miller, J.H.; Field, J.J.; Kanakkanthara, A.; Owen, J.G.; Singh, A.J.; Northcote, P.T. Marine invertebrate natural products that target microtubules. *J. Nat. Prod.* **2018**, *81*, 691–702. [CrossRef]

762. Bai, R.; Friedman, S.J.; Pettit, G.R.; Hamel, E. Dolastatin 15, a potent antimitotic depsipeptide derived from *Dolabella auricularia*: Interaction with tubulin and effects on cellular microtubules. *Biochem. Pharmacol.* **1992**, *43*, 2637–2645. [CrossRef]

763. Ali, M.A.; Rosati, R.; Pettit, G.; Kalemkerian, G.P. Dolastatin 15 induces apoptosis and BCL-2 phosphorylation in small cell lung cancer cell lines. *Anticancer Res.* **1998**, *18*, 1021–1026. [PubMed]

764. Sato, M.; Sagawa, M.; Nakazato, T.; Ikeda, Y.; Kizaki, M. A natural peptide, dolastatin 15, induces G2/M cell cycle arrest and apoptosis of human multiple myeloma cells. *Int. J. Oncol.* **2007**, *30*, 1453–1459. [CrossRef] [PubMed]

765. Beckwith, M.; Urba, W.J.; Longo, D.L. Growth inhibition of human lymphoma cell lines by the marine products, dolastatins 10 and 15. *JNCI J. Nat. Cancer Ins.* **1993**, *85*, 483–488. [CrossRef]

766. Bai, R.; Edler, M.C.; Bonate, P.L.; Copeland, T.D.; Pettit, G.R.; Luduena, R.F.; Hamel, E. Intracellular activation and deactivation of tasidotin, an analog of dolastatin 15: Correlation with cytotoxicity. *Mol. Pharmacol.* **2009**, *75*, 218–226. [CrossRef]

767. Ray, A.; Okouneva, T.; Manna, T.; Miller, H.P.; Schmid, S.; Arthaud, L.; Luduena, R.; Jordan, M.A.; Wilson, L. Mechanism of action of the microtubule-targeted antimitotic depsipeptide tasidotin (formerly ILX651) and its major metabolite tasidotin C-carboxylate. *Cancer Res.* **2007**, *67*, 3767–3776. [CrossRef]

768. Garg, V.; Zhang, W.; Gidwani, P.; Kim, M.; Kolb, E.A. Preclinical analysis of tasidotin HCl in Ewing's sarcoma, rhabdomyosarcoma, synovial sarcoma, and osteosarcoma. *Clin. Cancer Res.* **2007**, *13*, 5446–5454. [CrossRef]

769. Salvador, L.A.; Biggs, J.S.; Paul, V.J.; Luesch, H. Veraguamides A– G, cyclic hexadepsipeptides from a dolastatin 16-producing cyanobacterium *Symploca* cf. *hydnoides* from Guam. *J. Nat. Prod.* **2011**, *74*, 917–927. [CrossRef]

770. Pettit, G.R.; Xu, J.P.; Hogan, F.; Williams, M.D.; Doubek, D.L.; Schmidt, J.M.; Cerny, R.L.; Boyd, M.R. Isolation and structure of the human cancer cell growth inhibitory cyclodepsipeptide dolastatin 16. *J. Nat. Prod.* **1997**, *60*, 752–754. [CrossRef]

771. Nogle, L.M.; Gerwick, W.H. Isolation of four new cyclic depsipeptides, antanapeptins A– D, and dolastatin 16 from a Madagascan collection of *Lyngbya majuscula*. *J. Nat. Prod.* **2002**, *65*, 21–24. [CrossRef]

772. Monks, A.; Scudiero, D.; Skehan, P.; Shoemaker, R.; Paull, K.; Vistica, D.; Hose, C.; Langley, J.; Cronise, P.; Vaigro-Wolff, A.; et al. Feasibility of a high-flux anticancer drug screen using a diverse panel of cultured human tumor cell lines. *JNCI J. Natl. Cancer I.* **1991**, *83*, 757–766. [CrossRef] [PubMed]

773. Pettit, G.R.; Xu, J.P.; Hogan, F.; Cerny, R.L. Isolation and structure of dolastatin 17. *Heterocycles* **1998**, *47*, 491–496. [CrossRef]

774. Kobayashi, S.; Kobayashi, J.; Yazaki, R.; Ueno, M. Toward the total synthesis of onchidin, a cytotoxic cyclic depsipeptide from a mollusc. *Chem. Asian J.* **2007**, *2*, 135–144. [CrossRef] [PubMed]

775. Pettit, G.R.; Xu, J.P.; Williams, M.D.; Hogan, F.; Schmidt, J.M.; Cerny, R.L. Antineoplastic agents 370. Isolation and structure of dolastatin 18. *Bioorg. Chem. Med. Lett.* **1997**, *7*, 827–832. [CrossRef]

776. Pettit, G.R.; Xu, J.P.; Doubek, D.L.; Chapuis, J.C.; Schmidt, J.M. Antineoplastic Agents. 510. Isolation and structure of dolastatin 19 from the Gulf of California sea hare *Dolabella auricularia*. *J. Nat. Prod.* **2004**, *67*, 1252–1255. [CrossRef]

777. Moore, R.E. Toxins, anticancer agents, and tumor promoters from marine prokaryotes. *Pure Appl. Chem.* **1982**, *54*, 1919–1934. [CrossRef]

778. Watson, M. Midgut gland toxins of Hawaiian sea hares. I. Isolation and preliminary toxicological observations. *Toxicon* **1973**, *11*, 259–267. [CrossRef]

779. Watson, M.; Rayner, M.D. Midgut gland toxins of Hawaiian sea hares. II. A preliminary pharmacological study. *Toxicon* **1973**, *11*, 269–276. [CrossRef]

780. Ashida, Y.; Yanagita, R.C.; Takahashi, C.; Kawanami, Y.; Irie, K. Binding mode prediction of aplysiatoxin, a potent agonist of protein kinase C, through molecular simulation and structure-activity study on simplified analogs of the receptor-recognition domain. *Bioorg. Med. Chem.* **2016**, *24*, 4218–4227. [CrossRef]

781. Cardellina, J.H.; Marner, F.J.; Moore, R.E. Seaweed dermatitis: Structure of Lyngbyatoxin A. *Science* **1979**, *204*, 193–195. [CrossRef]

782. Cardellina, J.H.; Marner, F.J.; Moore, R.E. Malyngamide A, a novel chlorinated metabolite of the marine cyanophyte *Lyngbya majuscula*. *J. Am. Chem. Soc.* **1979**, *101*, 240–242. [CrossRef]

783. Capper, A.; Tibbetts, I.R.; O'Neil, J.M.; Shaw, G.R. The fate of *Lyngbya majuscula* toxins in three potential consumers. *J. Chem. Ecol.* **2005**, *31*, 1595–1606. [CrossRef] [PubMed]

784. Suntornchashwej, S.; Chaichit, N.; Isobe, M.; Suwanborirux, K. Hectochlorin and morpholine derivatives from the Thai sea hare, *Bursatella leachii*. *J. Nat. Prod.* **2005**, *68*, 951–955. [CrossRef] [PubMed]

785. Marquez, B.L.; Watts, K.S.; Yokochi, A.; Roberts, M.A.; Verdier-Pinard, P.; Jimenez, J.I.; Hamel, E.; Scheuer, P.J.; Gerwick, W.H. Structure and absolute stereochemistry of hectochlorin, a potent stimulator of actin assembly. *J. Nat. Prod.* **2002**, *65*, 866–871. [CrossRef] [PubMed]

786. Appleton, D.R.; Sewell, M.A.; Berridge, M.V.; Copp, B.R. A new biologically active malyngamide from a New Zealand collection of the sea hare *Bursatella leachii*. *J. Nat. Prod.* **2002**, *65*, 630–631. [CrossRef]

787. Suntornchashwej, S.; Suwanborirux, K.; Koga, K.; Isobe, M. Malyngamide X: The first (7R)-lyngbic acid that connects to a new tripeptide backbone from the Thai sea hare *Bursatella leachii*. *Chem. Asian J.* **2007**, *2*, 114–122. [CrossRef]

788. Fischel, J.L.; Lemee, R.; Formento, P.; Caldani, C.; Moll, J.L.; Pesando, D.; Meinesz, A.; Grelier, P.; Pietra, P.; Guerriero, A. Cell growth inhibitory effects of caulerpenyne, a sesquiterpenoid from the marine algae *Caulerpa taxifolia*. *Anticancer Res.* **1995**, *15*, 2155–2160.

789. Cavas, L.; Baskin, Y.; Yurdakoc, K.; Olgun, N. Antiproliferative and newly contributed apoptotic activities from an invasive marine alga: *Caulerpa racemosa var. cylindracea*. *J. Exp. Mar. Biol. Ecol.* **2006**, *339*, 111–119. [CrossRef]

790. Barbier, P.; Guise, S.; Huitorel, P.; Amade, P.; Pesando, D.; Briand, C.; Peyrot, V. Caulerpenyne from *Caulerpa taxifolia* has an antiproliferative activity on tumor cell line SK-N-SH and modifies the microtubule network. *Life Sci.* **2001**, *70*, 415–429. [CrossRef]

791. Parent-Massin, D.; Fournier, V.; Amade, P.; Lemee, R.; Durand-Clement, M.; Delescluse, C.; Pesando, D. Evaluation of the toxicological risk to humans of caulerpenyne using human hematopoietic progenitors, melanocytes, and keratinocytes in culture. *J. Toxicol. Environ. Health* **1996**, *47*, 47–59. [CrossRef]

792. Bourdron, J.; Barbier, P.; Allegro, D.; Villard, C.; Lafitte, D.; Commeiras, L.; Parrain, J.L.; Peyrot, V. Caulerpenyne binding to tubulin: Structural modifications by a non conventional pharmacological agent. *Med. Chem.* **2009**, *5*, 182–190. [CrossRef] [PubMed]

793. Pesando, D.; Pesci-Bardon, C.; Huitorel, P.; Girard, J.P. Caulerpenyne blocks MBP kinase activation controlling mitosis in sea urchin eggs. *Eur. J. Cell. Biol.* **1999**, *78*, 903–910. [CrossRef]

794. Gao, J.; Hamann, M.T. Chemistry and biology of kahalalides. *Chem. Rev.* **2011**, *111*, 3208–3235. [CrossRef] [PubMed]

795. Ciavatta, M.L.; Devi, P.; Carbone, M.; Mathieu, V.; Kiss, R.; Casapullo, A.; Gavagnin, M. Kahalalide F analogues from the mucous secretion of Indian sacoglossan mollusc *Elysia ornata*. *Tetrahedron* **2016**, *72*, 625–631. [CrossRef]

796. Ashour, M.; Edrada, R.; Ebel, R.; Wray, V.; Wätjen, W.; Padmakumar, K.; Müller, W.E.G.; Lin, W.H.; Proksch, P. Kahalalide derivatives from the Indian dacoglossan mollusk *Elysia grandifolia*. *J. Nat. Prod.* **2006**, *69*, 1547–1553. [CrossRef]

797. Hamann, M.T.; Scheuer, P.J. Kahalalide F: A bioactive depsipeptide from the sacoglossan mollusk *Elysia rufescens* and the green alga *Bryopsis* sp. *J. Am. Chem. Soc.* **1993**, *115*, 5825–5826. [CrossRef]

798. Kan, Y.; Fujita, T.; Sakamoto, B.; Hokama, Y.; Nagai, H. A new cyclic depsipeptide from the Hawaiian green alga *Bryopsis* species. *J. Nat. Prod.* **1999**, *62*, 1169–1172. [CrossRef]

799. Suárez, Y.; González, L.; Cuadrado, A.; Berciano, M.; Lafarga, M.; Muñoz, A. Kahalalide F, a new marine-derived compound, induces oncosis in human prostate and breast cancer cells. *Mol. Cancer Ther.* **2003**, *2*, 863–872.

800. Janmaat, M.L.; Rodriguez, J.A.; Jimeno, J.; Kruyt, F.A.E.; Giaccone, G. Kahalalide F induces necrosis-like cell death that involves depletion of ErbB3 and inhibition of Akt signaling. *Mol. Pharmacol.* **2005**, *68*, 502–510. [CrossRef]

801. Pardo, B.; Paz-Ares, L.; Tabernero, J.; Ciruelos, E.; García, M.; Salazar, R.; López, A.; Blanco, M.; Nieto, A.; Jimeno, J.; et al. Phase I clinical and pharmacokinetic study of kahalalide F administered weekly as a 1-hour infusion to patients with advanced solid tumors. *Clin. Cancer Res.* **2008**, *14*, 1116–1123. [CrossRef]

802. Rademaker-Lakhai, J.M.; Horenblas, S.; Meinhardt, W.; Stokvis, E.; de Reijke, T.M.; Jimeno, J.M.; Lopez-Lazaro, L.; Martin, J.A.L.; Beijnen, J.H.; Schellens, J.H. Phase I clinical and pharmacokinetic study of kahalalide F in patients with advanced androgen refractory prostate cancer. *Clin. Cancer Res.* **2005**, *11*, 1854–1862. [CrossRef] [PubMed]

803. Sewell, J.M.; Mayer, I.; Langdon, S.P.; Smyth, J.F.; Jodrell, D.I.; Guichard, S.M. The mechanism of action of Kahalalide F: Variable cell permeability in human hepatoma cell lines. *Eur. J. Cancer* **2005**, *41*, 1637–1644. [CrossRef] [PubMed]

804. Miguel-Lillo, B.; Valenzuela, B.; Peris-Ribera, J.E.; Soto-Matos, A.; Pérez-Ruixo, J.J. Population pharmacokinetics of kahalalide F in advanced cancer patients. *Cancer Chemother. Pharmacol.* **2015**, *76*, 365–374. [CrossRef] [PubMed]

805. Martín-Algarra, S.; Espinosa, E.; Rubió, J.; López, J.J.L.; Manzano, J.L.; Carrión, L.A.; Plazaola, A.; Tanovic, A.; Paz-Ares, L. Phase II study of weekly Kahalalide F in patients with advanced malignant melanoma. *Eur. J. Cancer* **2009**, *45*, 732–735. [CrossRef]

806. Wang, B.; Waters, A.L.; Valeriote, F.A.; Hamann, M.T. An efficient and cost-effective approach to Kahalalide F N-terminal modifications using a nuisance algal bloom of *Bryopsis pennata*. *Biochim. Biophys. Acta (BBA) Gen. Subj.* **2015**, *1850*, 1849–1854. [CrossRef]

807. Shilabin, A.G.; Hamann, M.T. In vitro and in vivo evaluation of select kahalalide F analogs with antitumor and antifungal activities. *Bioorg. Med. Chem.* **2011**, *19*, 6628–6632. [CrossRef]

808. Davis, J.; Fricke, W.F.; Hamann, M.T.; Esquenazi, E.; Dorrestein, P.C.; Hill, R.T. Characterization of the bacterial community of the chemically defended hawaiian sacoglossan *Elysia rufescens*. *Appl. Environ. Microbiol.* **2013**, *79*, 7073–7081. [CrossRef]

809. Ling, Y.H.; Miguel, A.; Zou, Y.; Yuan, Z.; Lu, B.; José, J.; Ana, M.C.; Perez-Soler, R. PM02734 (elisidepsin) induces caspase-independent cell death associated with features of autophagy, inhibition of the Akt/mTOR signaling pathway, and activation of death-associated protein kinase. *Clin. Cancer Res.* **2011**, *17*, 5353–5366. [CrossRef]

810. Serova, M.; de Gramont, A.; Bieche, I.; Riveiro, M.E.; Galmarini, C.M.; Aracil, M.; Jimeno, J.; Faivre, S.; Raymond, E. Predictive factors of sensitivity to elisidepsin, a novel Kahalalide F-derived marine compound. *Mar. Drugs* **2013**, *11*, 944–959. [CrossRef]

811. Goldwasser, F.; Faivre, S.; Alexandre, J.; Coronado, C.; Fernandez-Garcia, E.M.; Kahatt, C.M.; Paramio, P.G.; Dios, J.L.; Miguel-Lillo, B.; Raymond, E. Phase I study of elisidepsin (IrvalecR) in combination with carboplatin or gemcitabine in patients with advancedmalignancies. *Invest. New Drugs* **2014**, *32*, 500–509. [CrossRef]

812. Ratain, M.J.; Geary, D.; Undevia, S.D.; Coronado, C.; Alfaro, V.; Iglesias, J.L.; Schilsky, R.L.; Miguel-Lillo, B. First-in-human, phase I study of elisidepsin (PM02734) administered as a 30-min or as a 3-hour intravenous infusion every three weeks in patients with advanced solid tumors. *Invest. New Drugs* **2015**, *33*, 901–910. [CrossRef] [PubMed]

813. Herrero, A.B.; Astudillo, A.M.; Balboa, M.A.; Cuevas, C.; Balsinde, J.; Moreno, S. Levels of SCS7/FA2H-mediated fatty acid 2-hydroxylation determine the sensitivity of cells to antitumor PM02734. *Cancer Res.* **2008**, *68*, 9779–9787. [CrossRef] [PubMed]

814. Váradi, T.; Roszik, J.; Lisboa, D.; Vereb, G.; Molina-Guijarro, J.M.; Galmarini, C.M.; Szöllősi, J.; Nagy, P. ErbB protein modifications are secondary to severe cell membrane alterations induced by elisidepsin treatment. *Eur. J. Pharmacol.* **2011**, *667*, 91–99. [CrossRef] [PubMed]

815. Molina-Guijarro, J.M.; García, C.; Macías, Á.; García-Fernández, L.F.; Moreno, C.; Reyes, F.; Martínez-Leal, J.F.; Fernández, R.; Martínez, V.; Valenzuela, C.; et al. Elisidepsin interacts directly with glycosylceramides in the plasma membrane of tumor cells to induce necrotic cell death. *PLoS ONE* **2015**, *10*, e0140782. [CrossRef]

816. Salazar, R.; Jones, R.J.; Oaknin, A.; Crawford, D.; Cuadra, C.; Hopkins, C.; Gil, M.; Coronado, C.; Soto-Matos, A.; Cullell-Young, M.; et al. A phase I and pharmacokinetic study of elisidepsin (PM02734) in patients with advanced solid tumors. *Cancer Chemother. Pharmacol.* **2012**, *70*, 673–681. [CrossRef] [PubMed]

817. Petty, R.; Anthoney, A.; Metges, J.P.; Alsina, M.; Gonçalves, A.; Brown, J.; Montagut, C.; Gunzer, K.; Laus, G.; Dios, J.L.I.; et al. Phase Ib/II study of elisidepsin in metastatic or advanced gastroesophageal cancer (IMAGE trial). *Cancer Chemother. Pharmacol.* **2016**, *77*, 819–827. [CrossRef]

818. Ciavatta, M.L.; Villani, G.; Trivellone, E.; Cimino, G. Two new labdane aldehydes from the skin of the notaspidean *Pleurobranchaea meckelii*. *Tetrahedron Lett.* **1995**, *36*, 8673–8676. [CrossRef]

819. Díaz-Marrero, A.R.; Dorta, E.; Cueto, M.; Rovirosa, J.; San Martín, A.; Loyola, L.A.; Darias, J. New polyhydroxylated steroids from the marine pulmonate *Trimusculus peruvianus*. *Arkivoc J. Org. Chem.* **2003**, *10*, 107–117. [CrossRef]

820. Rodriguez, J.; Riguera, R.; Debitus, C. The natural polypropionate-derived esters of the mollusk *Onchidium* sp. *J. Org. Chem.* **1992**, *57*, 4624–4632. [CrossRef]

821. Zhou, Z.F.; Li, X.L.; Yao, L.G.; Li, J.; Gavagnin, M.; Guo, Y.W. Marine bis-γ-pyrone polypropionates of onchidione family and their effects on the XBP1 gene expression. *Bioorg. Med. Chem. Lett.* **2018**, *28*, 1093–1096. [CrossRef]

822. Ireland, C.M.; Biskupiak, J.E.; Hite, G.J.; Rapposch, M.; Scheuer, P.J.; Ruble, J.R. Ilikonapyrone esters, likely defense allomones of the mollusk *Onchidium verruculatum*. *J. Org. Chem.* **1984**, *49*, 559–561. [CrossRef]

823. Arimoto, H.; Cheng, J.-F.; Nishiyama, S.; Yamamura, S. Synthetic studies on fully substituted γ-pyrone-containing natural products: The absolute configurations of ilikonapyrone and peroniatriols I and II. *Tetrahedron Lett.* **1993**, *34*, 5781–5784. [CrossRef]

824. Maschek, J.A. Chemical Investigation of the Antarctic Marine Invertebrates *Austrodoris kerguelenensis* & *Dendrilla membranosa* and the Antarctic Red Alga *Gigartina skottsbergii*. Ph.D. Thesis, University of South Florida, Tampa, FL, USA, 2011.

825. Morgan, J.B.; Mahdi, F.; Liu, Y.; Coothankandaswamy, V.; Jekabsons, M.B.; Johnson, T.A.; Sashidhara, K.V.; Crews, P.; Nagle, D.G.; Zhou, Y.D. The marine sponge metabolite mycothiazole: A novel prototype mitochondrial complex I inhibitor. *Bioorg. Med. Chem.* **2010**, *18*, 5988–5994. [CrossRef] [PubMed]

826. Meyer, K.J.; Singh, A.J.; Cameron, A.; Tan, A.S.; Leahy, D.C.; O'Sullivan, D.; Joshi, P.; La Flamme, A.C.; Northcote, P.T.; Berridge, M.V.; et al. Mitochondrial genome-knockout cells demonstrate a dual mechanism of action for the electron transport complex I inhibitor mycothiazole. *Mar. Drugs* **2012**, *10*, 900–917. [CrossRef] [PubMed]

827. Hamann, M.T.; Scheuer, P.J.; Kelly-Borges, M. Biogenetically diverse, bioactive constituents of a sponge, order Verongida: Bromotyramines and sesquiterpene-shikimate derived metabolites. *J. Org. Chem.* **1993**, *58*, 6565–6569. [CrossRef]

828. Kazlauskas, R.; Murphy, P.T.; Wells, R.J.; Noack, K.; Oberhansli, W.E.; Schonholzer, P. A new series of diterpenes from Australian Spongia species. *Australian J. Chem.* **1979**, *32*, 867–880. [CrossRef]

829. Pettit, G.R.; Herald, C.L.; Allen, M.S.; Von Dreele, R.B.; Vanell, L.D.; Kao, J.P.; Blake, W. Antineoplastic agents. 48. The isolation and structure of aplysistatin. *J. Am. Chem. Soc.* **1977**, *99*, 262–263. [CrossRef]

830. Fischbach, M.A.; Walsh, C.T. Antibiotics for emerging pathogens. *Science* **2009**, *325*, 1089–1093. [CrossRef]

831. Kubo, I.; Taniguchi, M. Polygodial, an antifungal potentiator. *J. Nat. Prod.* **1988**, *51*, 22–29. [CrossRef]

832. Benkendorff, K.; Davis, A.R.; Bremner, J.B. Chemical defense in the egg masses of benthic invertebrates: An assessment of antibacterial activity in 39 mollusks and 4 polychaetes. *J. Invertebr. Pathol.* **2001**, *78*, 109–118. [CrossRef]

833. El Sayed, K.A.; Bartyzel, P.; Shen, X.; Perry, T.L.; Zjawiony, J.K.; Hamann, M.T. Marine natural products as anti-tuberculosis agents. *Tetrahedron* **2000**, *56*, 949–953. [CrossRef]

834. Singh, E.K.; Ramsey, D.M.; McAlpine, S.R. Total synthesis of trans,trans-sanguinamide B and conformational isomers. *Organic Lett.* **2012**, *14*, 1198–1201. [CrossRef] [PubMed]

835. Wahyudi, H.; Tantisantisom, W.; Liu, X.; Ramsey, D.M.; Singh, E.K.; McAlpine, S.R. Synthesis, structure-activity analysis, and Biological Evaluation of Sanguinamide B Analogues. *J. Org. Chem.* **2012**, *77*, 10596–10616. [CrossRef] [PubMed]

836. Doralyn S., D.; Evan W., R.; Arthur S., E.; Tadeusz F., M. Structure Elucidation at the Nanomole Scale. 1. Trisoxazole Macrolides and Thiazole-Containing Cyclic Peptides from the Nudibranch Hexabranchus sanguineus. *J. Nat. Prod.* **2009**, *72*, 732–738.

837. Zhang, W.; Gavagnin, M.; Guo, Y.W.; Mollo, E.; Ghiselin, M.T.; Cimino, G. Terpenoid metabolites of the nudibranch *Hexabranchus sanguineus* from the South China Sea. *Tetrahedron* **2007**, *63*, 4725–4729. [CrossRef]

838. Guo, Y.; Gavagnin, M.; Mollo, E.; Trivellone, E.; Cimino, G.; Fakhr, I. Structure of the pigment of the Red Sea nudibranch *Hexabranchus sanguineus*. *Tetrahedron Lett.* **1998**, *39*, 2635–2638. [CrossRef]

839. He, H.; Faulkner, D.J. Renieramycins E and F from the sponge *Reniera* sp. Reassignment of the stereochemistry of the renieramycins. *J. Org. Chem.* **1989**, *54*, 5822–5824. [CrossRef]

840. Reddy, K.V.; Mohanraju, R.; Murthy, K.N.; Ramesh, C.; Karthick, P. Antimicrobial properties of nudibranchs tissues extracts from South Andaman, India. *J. Coast. Life Med.* **2015**, *3*, 582–584.

841. Fahey, S.J.; Carroll, A.R. Natural products isolated from species of *Halgerda* Bergh, 1880 (Mollusca: Nudibranchia) and their ecological and evolutionary implications. *J. Chem. Ecol.* **2007**, *33*, 1226–1234. [CrossRef]

842. Ishibashi, M.; Yamaguchi, Y.; Hirano, Y.J. Bioactive natural products from nudibranchs. In *Biomaterials from Aquatic and Terrestrial Organisms*; Fingerman, M., Nagabhushanam, R., Eds.; Science Publishers: Enfield, NH, USA, 2006; pp. 513–535.

843. Ramya, M.S.; Sivasubramanian, K.; Ravichandran, S.; Anbuchezhian, R. Screening of antimicrobial compound from the sea slug *Armina babai*. *Bangladesh J. Pharmacol.* **2014**, *9*, 268–274. [CrossRef]

844. Shin, J.; Seo, Y. Isolation of new ceramides from the gorgonian *Acabaria undulata*. *J. Nat. Prod.* **1995**, *58*, 948–953. [CrossRef]

845. Hay, M.E.; Pawlik, J.R.; Duffy, J.E.; Fenical, W. Seaweed-herbivore-predator interactions: Host-plant specialization reduces predation on small herbivores. *Oecologia* **1989**, *81*, 418–427. [CrossRef] [PubMed]

846. Paul, V.J.; Fenical, W. Chemical defense in tropical green algae, order Caulerpales. *Mar. Ecol. Prog. Ser.* **1986**, *34*, 157–169. [CrossRef]

847. Wright, A.D.; König, G.M.; Angerhofer, C.K.; Greenidge, P.; Linden, A.; Desqueyroux-Faundez, R. Antimalarial activity: The search for marine-derived natural products with selective antimalarial activity. *J. Nat. Prod.* **1996**, *59*, 710–716. [CrossRef] [PubMed]

848. Angerhofer, C.K.; Pezzuto, J.M.; König, G.M.; Wright, A.D.; Sticher, O. Antimalarial activity of sesquiterpenes from the marine sponge *Acanthella klethra*. *J. Nat. Prod.* **1992**, *55*, 1787–1789. [CrossRef] [PubMed]

849. Wright, A.D.; Wang, H.; Gurrath, M.; König, G.M.; Kocak, G.; Neumann, G.; Loria, P.; Foley, M.; Tilley, L. Inhibition of heme detoxification processes underlies the antimalarial activity of terpene isonitrile compounds from marine sponges. *J. Med. Chem.* **2001**, *44*, 873–885. [CrossRef]

850. White, A.M.; Dao, K.; Vrubliauskas, D.; Könst, Z.A.; Pierens, G.K.; Mándi, A.; Andrews, K.T.; Skinner-Adams, T.S.; Clarke, M.E.; Narbutas, P.T.; et al. Catalyst-controlled stereoselective synthesis secures the structure of the antimalarial isocyanoterpene pustulosaisonitrile-1. *J. Org. Chem.* **2017**, *82*, 13313–13323. [CrossRef]

851. Yang, S.S.; Cragg, G.M.; Newman, D.J.; Bader, J.P. Natural product-based anti-HIV drug discovery and development facilitated by the NCI developmental therapeutics program. *J. Nat. Prod.* **2001**, *64*, 265–277. [CrossRef]

852. Newman, D.J.; Cragg, G.M.; Snader, K.M. The influence of natural products upon drug discovery. *Nat. Prod. Rep.* **2000**, *17*, 175–285. [CrossRef]

853. Gochfeld, D.J.; El Sayed, K.A.; Yousaf, M.; Hu, J.; Bartyzel, P.; Dunbar, D.C.; Wilkins, S.P.; Zjawiony, J.K.; Schinazi, R.F.; Schlueter-Wirtz, S.; et al. Marine natural products as lead anti-HIV agents. *Mini-Rev. Med. Chem.* **2003**, *3*, 401–424. [CrossRef]

854. Mitjà, O.; Clotet, B. Use of antiviral drugs to reduce COVID-19 transmission. *Lancet Glob. Health* **2020**, *8*, e639–e640. [CrossRef]

855. Wang, M.; Tietjen, I.; Chen, M.; Williams, D.E.; Daoust, J.; Brockman, M.A.; Andersen, R.J. Sesterterpenoids isolated from the sponge *Phorbas* sp. activate latent HIV-1 provirus expression. *J. Org. Chem.* **2016**, *81*, 11324–11334. [CrossRef] [PubMed]

856. Jacobs, R.S.; Bober, M.A.; Pinto, I.; Williams, A.B.; Jacobson, P.B.; de Carvalho, M.S. Pharmacological studies of marine novel marine metabolites. In *Advances in Marine Biotechnology*; Attaway, D.H., Zaborsky, O.R., Eds.; Plenum Press: New York, NY, USA, 1993; Volume 1, pp. 77–99.

857. Cimino, G.; De Stefano, S.; Minale, L. Scalaradial, a third sesterterpene with the tetracarbocyclic skeleton of scalarin, from the sponge *Cacospongia mollior*. *Experientia* **1974**, *30*, 846–847.

858. De Carvalho, M.S.; Jacobs, R.S. Two-step inactivation of bee venom phospholipase A2 by scalaradial. *Biochem. Pharm.* **1991**, *42*, 1621–1626. [CrossRef]

859. Oliveira, A.P.; Lobo-da-Cunha, A.; Taveira, M.; Ferreira, M.; Valentão, P.; Andrade, P.B. Digestive gland from *Aplysia depilans* Gmelin: Leads for inflammation treatment. *Molecules* **2015**, *20*, 15766–15780. [CrossRef]

860. Jiménez-Romero, C.; Mayer, A.M.; Rodriguez, A.D. Dactyloditerpenol acetate, a new prenylbisabolane-type diterpene from *Aplysia dactylomela* with significant in vitro anti-neuroinflammatory activity. *Bioorg. Med. Chem. Lett.* **2014**, *24*, 344–348. [CrossRef]

861. Mohammed, K.A.; Hossain, C.F.; Zhang, L.; Bruick, R.K.; Zhou, Y.D.; Nagle, D.G. Laurenditerpenol, a new diterpene from the tropical marine alga *Laurencia intricata* that potently inhibits HIF-1 mediated hypoxic signaling in breast tumor cells. *J. Nat. Prod.* **2004**, *67*, 2002–2007. [CrossRef]

862. Choi, D.Y.; Choi, H. Natural products from marine organisms with neuroprotective activity in the experimental models of Alzheimer's disease, Parkinson's disease and ischemic brain stroke: Their molecular targets and action mechanisms. *Arch. Pharm. Res.* **2015**, *38*, 139–170. [CrossRef]

863. Leiros, M.; Alonso, E.; Rateb, M.E.; Houssen, W.E.; Ebel, R.; Jaspars, M.; Alfonso, A.; Botana, L.M. Gracilins: Spongionella-derived promising compounds for Alzheimer disease. *Neuropharmacol.* **2015**, *93*, 285–293. [CrossRef]

864. Llorach-Pares, L.; Rodriguez-Urgelles, E.; Nonell-Canals, A.; Alberch, J.; Avila, C.; Sanchez-Martinez, M.; Giralt, A. Meridianins and Lignarenone B as Potential GSK3β Inhibitors and Inductors of Structural Neuronal Plasticity. *Biomolecules* **2020**, *10*, 639. [CrossRef]

865. Meijer, L.; Flajolet, M.; Greengard, P. Pharmacological inhibitors of glycogen synthase kinase 3. *Trends Pharmacol. Sci.* **2004**, *25*, 471–480. [CrossRef] [PubMed]

866. Sun, M.K.; Alkon, D.L. Bryostatin-1: Pharmacology and Therapeutic Potential as a CNS Drug. *CNS Drug Rev.* **2006**, *12*, 1–8. [CrossRef] [PubMed]

867. Cimino, G.; Spinella, A.; Sodano, G. Potential alarm pheromones from the Mediterranean opisthobranch *Scaphander lignarius*. *Tetrahedron Lett.* **1989**, *30*, 5003–5004. [CrossRef]

868. Davis, W.J.; Mpitsos, G.J. Behavioral choice and habituation in the marine mollusk *Pleurobranchaea californica* MacFarland (Gastropoda, Opisthobranchia). *Z. Vgl. Physiol.* **1971**, *75*, 207–232.

869. McNabb, P.; Selwood, A.I.; Munday, R.; Wood, S.A.; Taylor, D.I.; MacKenzie, L.A.; van Ginkel, R.; Rhodes, L.L.; Cornelisen, C.; Heasman, K.; et al. Detection of tetrodotoxin from the grey side-gilled sea slug *Pleurobranchaea maculata*, and associated dog neurotoxicosis on beaches adjacent to the Hauraki Gulf, Auckland, New Zealand. *Toxicon* **2010**, *56*, 466–473. [CrossRef]

870. Salvitti, L.R.; Wood, S.A.; Winsor, L.; Cary, S.C. Intracellular immunohistochemical detection of tetrodotoxin in *Pleurobranchaea maculata* (Gastropoda) and *Stylochoplana* sp. (Turbellaria). *Mar. Drugs* **2015**, *13*, 756–769. [CrossRef]
871. Chau, R.; Kalaitzis, J.A.; Neilan, B.A. On the origins and biosynthesis of tetrodotoxin. *Aquat. Toxicol.* **2011**, *104*, 61–72. [CrossRef]
872. Salvitti, L.R.; Wood, S.A.; Fairweather, R.; Cary, S.C. In situ accumulation of tetrodotoxin in non-toxic *Pleurobranchaea maculata* (Opisthobranchia). *Aquat. Sci.* **2016**, *79*, 1–10. [CrossRef]
873. Böhringer, N.; Fisch, K.M.; Schillo, D.; Bara, R.; Hertzer, C.; Grein, F.; Eisenbarth, J.-H.; Kaligis, F.; Schneider, T.; Wägele, H.; et al. Antimicrobial potential of bacteria associated with marine sea slugs from North Sulawesi, Indonesia. *Front. Microbiol.* **2017**, *8*, 1092. [CrossRef]
874. Fisch, K.M.; Schäberle, T.F. Toolbox for antibiotics discovery from microorganisms. *Arch. Pharm.* **2016**, *349*, 683–691. [CrossRef]
875. Newman, D.J. Developing natural product drugs: Supply problems and how they have been overcome. *Pharmacol. Ther.* **2016**, *162*, 1–9. [CrossRef] [PubMed]
876. Hosta, L.; Pla-Roca, M.; Arbiol, J.; López-Iglesias, C.; Samitier, J.; Cruz, L.J.; Kogan, M.J.; Albericio, F. Conjugation of Kahalalide F with gold nanoparticles to enhance in vitro antitumoral activity. *Bioconjugate Chem.* **2009**, *20*, 138–146. [CrossRef] [PubMed]

marine drugs

Article

Bioactivity Screening of Antarctic Sponges Reveals Anticancer Activity and Potential Cell Death via Ferroptosis by Mycalols

Gennaro Riccio [1], Genoveffa Nuzzo [2], Gianluca Zazo [3], Daniela Coppola [1], Giuseppina Senese [2], Lucia Romano [2], Maria Costantini [1,4], Nadia Ruocco [1], Marco Bertolino [5], Angelo Fontana [2,6], Adrianna Ianora [1], Cinzia Verde [1,4], Daniela Giordano [1,4] and Chiara Lauritano [1,*]

[1] Department of Marine Biotechnology, Stazione Zoologica Anton Dohrn, Villa Comunale, 80121 Napoli, Italy; gennaro.riccio@szn.it (G.R.); daniela.coppola@szn.it (D.C.); maria.costantini@szn.it (M.C.); nadia.ruocco@szn.it (N.R.); adrianna.ianora@szn.it (A.I.); c.verde@ibp.cnr.it (C.V.); daniela.giordano@ibbr.cnr.it (D.G.)
[2] Istituto di Chimica Biomolecolare, Consiglio Nazionale delle Ricerche, Via Campi Flegrei 34, 80078 Pozzuoli, Italy; nuzzo.genoveffa@icb.cnr.it (G.N.); giusi.senese@icb.cnr.it (G.S.); l.romano@icb.cnr.it (L.R.); angelo.fontana@icb.cnr.it (A.F.)
[3] Research Infrastructure for Marine Biological Resources Department, Stazione Zoologica Anton Dohrn, Villa Comunale, 80121 Napoli, Italy; gzazo@inogs.it
[4] Institute of Biosciences and BioResources (IBBR), National Research Council (CNR), Via Pietro Castellino 111, 80131 Napoli, Italy
[5] Dipartimento di Scienze della Terra, dell'Ambiente e della Vita (DISTAV), Università degli Studi di Genova, Corso Europa 26, 16132 Genova, Italy; marco.bertolino@unige.it
[6] Laboratory of Bio-Organic Chemistry and Chemical Biology, Department of Biology, Università di Napoli "Federico II", Via Cupa Nuova Cinthia 21, 80126 Napoli, Italy
* Correspondence: chiara.lauritano@szn.it

Citation: Riccio, G.; Nuzzo, G.; Zazo, G.; Coppola, D.; Senese, G.; Romano, L.; Costantini, M.; Ruocco, N.; Bertolino, M.; Fontana, A.; et al. Bioactivity Screening of Antarctic Sponges Reveals Anticancer Activity and Potential Cell Death via Ferroptosis by Mycalols. *Mar. Drugs* **2021**, *19*, 459. https://doi.org/10.3390/md19080459

Academic Editor: Bill J. Baker

Received: 13 July 2021
Accepted: 10 August 2021
Published: 14 August 2021

Abstract: Sponges are known to produce a series of compounds with bioactivities useful for human health. This study was conducted on four sponges collected in the framework of the XXXIV Italian National Antarctic Research Program (PNRA) in November-December 2018, i.e., *Mycale (Oxymycale) acerata*, *Haliclona (Rhizoniera) dancoi*, *Hemimycale topsenti*, and *Hemigellius pilosus*. Sponge extracts were fractioned and tested against hepatocellular carcinoma (HepG2), lung carcinoma (A549), and melanoma cells (A2058), in order to screen for antiproliferative or cytotoxic activity. Two different chemical classes of compounds, belonging to mycalols and suberitenones, were identified in the active fractions. Mycalols were the most active compounds, and their mechanism of action was also investigated at the gene and protein levels in HepG2 cells. Of the differentially expressed genes, ULK1 and GALNT5 were the most down-regulated genes, while MAPK8 was one of the most up-regulated genes. These genes were previously associated with ferroptosis, a programmed cell death triggered by iron-dependent lipid peroxidation, confirmed at the protein level by the down-regulation of GPX4, a key regulator of ferroptosis, and the up-regulation of NCOA4, involved in iron homeostasis. These data suggest, for the first time, that mycalols act by triggering ferroptosis in HepG2 cells.

Keywords: Antarctica; sponges; drug discovery; mycalols; marine biotechnology

1. Introduction

Marine organisms represent an excellent source of natural products with bioactivities useful for the treatment and prevention of human pathologies, such as cancer, inflammation, and infections. In recent years, the chemistry of natural products derived from marine organisms has received growing interest in the scientific community. In particular, there are fourteen approved pharmaceutical products in clinical use and more than 20 marine natural products in various stages of clinical development (i.e., four compounds in Phase III, twelve in Phase II, and seven in Phase I clinical trials), especially in the field of oncology (https:// www.midwestern.edu/departments/marinepharmacology/clinical-pipeline.xml). These

Mar. Drugs **2021**, *19*, 459. https://doi.org/10.3390/md19080459 https://www.mdpi.com/journal/marinedrugs

compounds have been identified mainly from invertebrates, such as sponges, mollusks, bryozoans, and ascidians. The last two drugs approved in 2020 are Belantamabmafodotin-blmf (Blenrep[TM]) from a mollusk/cyanobacterium for the treatment of relapsed/refractory multiple myeloma and Lurbinectedin (Zepzelca[TM]) from a tunicate for metastatic small cell lung cancer treatment.

It has been known for a long time that sponges may produce interesting compounds with bioactivities useful for human health [1–4]. In 1907, Richter outlined that the active component of the roasted bath sponge was rich in iodine and was already used by Roger cosmetics against struma [1]. The first marine-derived anticancer compound was cytarabine or Ara-C (Cytosar-U®), which was developed for clinical use as a synthetic analogue of a C-nucleoside isolated from the Caribbean sponge *Tectitethya crypta* (de Laubenfels, 1949) (ex *Tethya crypta*) [5]. Ara-C was approved in 1969 and is still used to treat acute myelocytic leukemia and non-Hodgkin's lymphoma [6]. Successively, a synthetic analog of spongouridine, vidarabine or Ara-A (Vira-A®), from the sponge *Tectitethya crypta* was approved as an antiviral (1976), while a synthetic analog of halichondrin A, Eribulin (Halaven®), from the sponge *Halichondria (Halichondria) okadai* (Kadota, 1922) was approved for the treatment of drug-refractory breast cancer (2010). Other compounds from sponges in clinical trials are Plocabulin (PM184), actually in Phase II clinical trials and MORAb-202 in Phase I. Both compounds are undergoing trials for the treatment of solid tumors (https://www.midwestern.edu/departments/marinepharmacology/clinical-pipeline.xml).

The current investigation focuses on antiproliferative bioactivity screening of four sponges collected from two different sites in the framework of the XXXIV Italian National Antarctic Research Program (PNRA) in November–December 2018. Three of them have been previously characterized by morphological analysis of spicules and molecular marker (i.e., 18S, 28S, ITS, and CO1) amplification and were identified as *Mycale (Oxymycale) acerata* Kirkpatrick, 1907, *Haliclona (Rhizoniera) dancoi* (Topsent, 1901), and *Hemigellius pilosus* (Kirkpatrick, 1907) [7].

Extreme environments, such as the poles, represent an almost untapped source of marine natural products which is still largely unexplored compared to more accessible sites. Studies in extreme environments are rare due to logistic problems and expedition costs. The Southern Ocean represents 9.6% of the world's oceans and extends approximately 35 million km^2. The Antarctic region is strongly affected by snow and ice-cover changes, extreme photoperiods, and low temperatures [8]. Due to these harsh characteristics, Antarctic organisms have evolved various physiological and behavioral adaptations [9]. For example, a longer period of larval development has been observed for sponges in Antarctica [10], where they represent the major component of the Antarctic zoobenthos (counting about 400 species; [10,11]). Previous studies have shown that Antarctic sponges may produce bioactive compounds for possible human applications [2,12–19]. This study aims to further explore the Antarctic region and to give an overview of sponge bioactivity from different sites, their chemical composition, and the mechanism of action of the active principles. Herein we report a bioassay-guided fractionation of four Antarctic sponges that led to the identification of two different classes of compounds, suberitenone A (**1**) and B (**2**) and mycalols, as bioactive compounds present in the identified cytotoxic fractions. The mixture containing the marine metabolite mycalol (**3**) and its analogues (**4–9**), already reported to possess anticancer activity on anaplastic thyroid carcinoma cells (ATC) [20,21], were identified as the most promising fraction. Thus, the mechanism of action of the mycalol mixture was further investigated at the gene and protein levels (Figure 1).

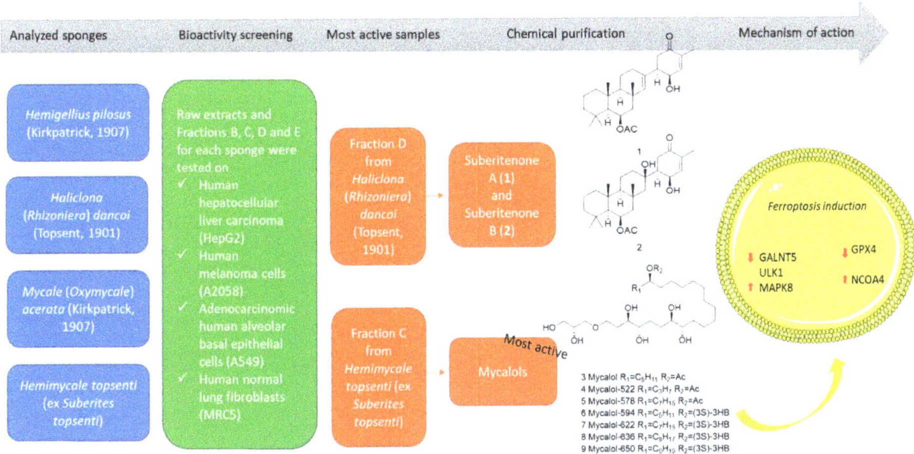

Figure 1. Work pipeline reporting the various experimental steps, the chemical structures of suberitenone A (**1**) and B (**2**) and mycalols, and the main results on the mechanism of action of mycalols, which were identified as the most active compounds in the current study.

2. Results

2.1. Species Identification of Specimen C7

The C7 specimen belongs to the species *Hemimycale topsenti* (Burton, 1929) due to the characteristic areolated surface—presence of a plumose skeleton formed by columns of subtilostyles characterized by oval elongated heads.

The 663 bp PCR fragment (using primer pair dgLCO1490/dgHCO2198) displayed a 100% of pairwise sequence similarity (85% of query cover) to the species *Hemimycale topsenti* (ex *Suberites topsenti*) mitochondrial partial COI gene voucher NIWA28884 (Accession Number: LN850246.1) (Figure S1). The other three sponges were identified in our previous paper [7] (Table 1).

Table 1. Collected species and relative sample IDs and MNA codes.

Species Name	Abbreviation	Sample IDs	MNA Code
Hemigellius pilosus (Kirkpatrick, 1907)	*H. p.*	D4	13266
Haliclona (*Rhizoniera*) *dancoi* (Topsent, 1901)	*H. d.*	C6	13265
Mycale (*Oxymycale*) *acerata* (Kirkpatrick, 1907)	*M. a.*	B4	13264
Hemimycale topsenti (ex *Suberites topsenti*)	*H. t.*	C7	13860

2.2. Chemical Fractionation and Cell Viability on Cancer Cell Lines

Different concentrations (1, 10, and 100 µg mL^{-1}) of total MeOH extract (TE) of sampled Antarctic sponges, and their enriched extracts B–E, obtained by solid-phase extraction (SPE) fractionation [22], were screened for their capability to affect the viability of the human cancer cell lines HepG2 (hepatocellular carcinoma), A549 (lung adenocarcinoma) and A2058 (melanoma), and the normal lung cell line MRC5 (Figure S2).

Bioactivity screening on A2058 cells showed that fraction C, D, and E of the sponge *M. a.* were active only at 100 µg/mL ($p < 0.1$ for fraction C and $p < 0.001$ for fraction D and E), as well as fraction C and D of *H. d.* ($p < 0.05$ for fraction C and $p < 0.001$ for fraction D). For *H. t.*, fraction C was active at both 10 and 100 µg/mL ($p < 0.001$ for both) while total extract and fraction D were active only at 100 µg/mL ($p < 0.001$ for both) (Figure S2). For *H. p.*, total extract, fraction C and fraction E were active only at 100 µg/mL ($p < 0.01$ for total extract and fraction C, while $p < 0.05$ for fraction E; Figure S2).

Regarding A549 cells, *M. a.* fraction C, D, and E were able to significantly reduce A549 cell proliferation only at 100 μg/mL ($p < 0.05$ for fraction C and $p < 0.001$ for fraction D and E,). For *H. d.*, total extract and fraction D were active at 100 μg/mL ($p < 0.01$ for total extract and $p < 0.01$ for fraction D), while fraction C was active both at 10 and 100 μg/mL ($p < 0.05$ and $p < 0.001$, respectively). For *H. t.*, total extract, fraction C and fraction D were active at 100 μg/mL ($p < 0.001$ for all). For *H. p.*, fraction E was the only one active at 100 μg/mL ($p < 0.001$) (Figure S2).

Bioactivity on HepG2 cells showed that fraction D and E of *M. a.* were active at 100 μg/mL ($p < 0.01$ and $p < 0.001$, respectively). Fraction C of *H. d.* was active at both 10 and 100 μg/mL ($p < 0.001$ for both), total extract and fraction D were active only at 100 μg/mL ($p < 0.001$ for both). For *H. t.*, total extract, fraction C, D and E were active only at 100μg/mL ($p < 0.001$ for all). *H. p.* total extract and fraction D were active at 100 μg/mL (*p*, <, 0001 and $p < 0.01$, respectively).

2.3. Identification of Compounds in Most Active Fractions

Preliminary ^{1}H NMR of the active fraction D of the sponge *H. t.* (Figure S3) showed the methyl pattern that clearly indicated the presence of a sesterterpene. Further purification on silica column led to the isolation of two known compounds, suberitenone A (**1**) and B (**2**) (Figure 2), whose identification was obtained by comparing the ^{1}H NMR data (Figures S4 and S5) with the literature [23], and confirmed by ESI$^+$ MS analysis (Figures S6 and S7).

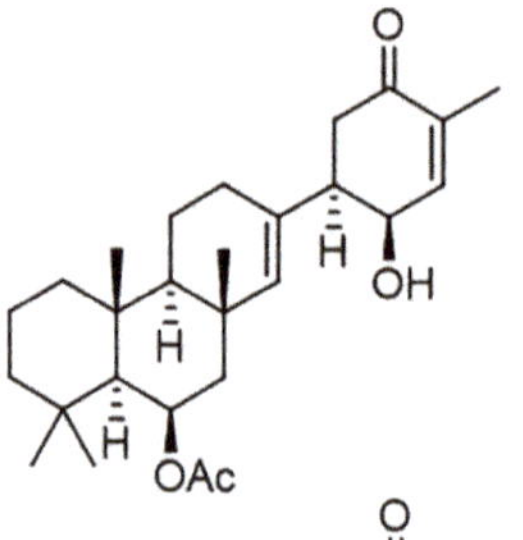

1

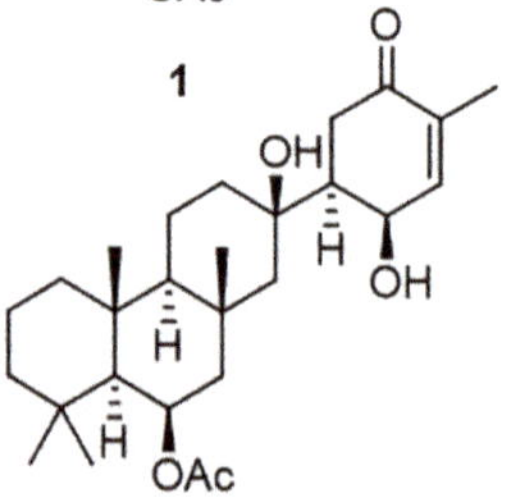

2

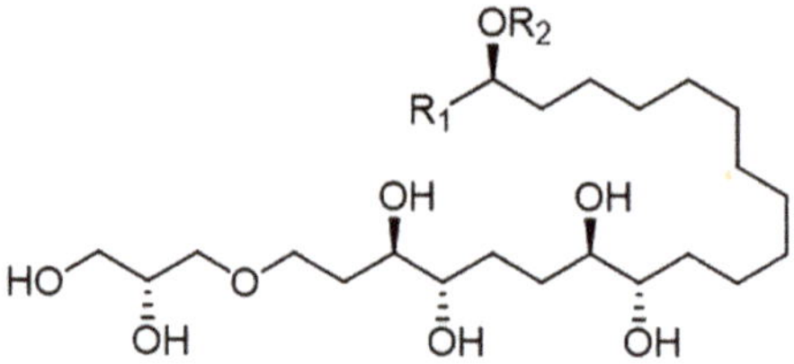

3 Mycalol-550 R_1=C_5H_{11} R_2=Ac
4 Mycalol-522 R1=C_3H_7 R_2=Ac
5 Mycalol-578 R1=C_7H_{15} R_2=Ac
6 Mycalol-594 R1=C_5H_{11} R_2=(3*S*)-3HB
7 Mycalol-622 R1=C_7H_{15} R_2=(3*S*)-3HB
8 Mycalol-636 R1=C_8H_{17} R_2=(3*S*)-3HB
9 Mycalol-650 R1=C_9H_{19} R_2=(3*S*)-3HB

Figure 2. Molecular structure of suberitenones A (**1**) and B (**2**), and mycalols (**3–9**).

Chemical investigation of fraction C resulting from the SPE-fractionation of *H. d.* extract suggested the presence of polioxygenated compounds (Figure S8). Through a normal phase purification, we isolated and identified the main products family. Proton NMR spectrum (Figure S9) contained all the diagnostic signals of mycalol and its derivatives (Figure 2), previously isolated from the Antarctic sponge *Mycale (Oxymycale) acerata* [20,21,24]. HR ESI$^+$-MS analysis confimed our hypothesis and allowed us to define the species composition, establishing the following percentage mycalol-522/mycalol-550/mycalo-578/mycalol-594/mycalol-622/mycalol-636/mycalol-650 as 2/54/15/3/20/2/4, respectively (Figure S10).

2.4. Bioactivity of Suberitenones A and B from Hemimycale topsenti and Mycalols from Haliclona (Rhizoniera) dancoi on Cancer Cell Lines

The mixture of mycalols, and suberitenones A and B purified from the most active fractions were tested on cancer cell lines (A549, A2058 and HepG2) and a normal cell line (MRC5) to confirm the cytotoxic activity. In order to evaluate the IC_{50} values, the compounds were challenged at different concentrations (0.05, 0.10, 0.19, 0.39, 0.78, 1.56, 3.12, 6.25, 12.5, 25, 50, 100 µM) (Figure 3 for mycalols and Figure S11 for suberitenone A and B).

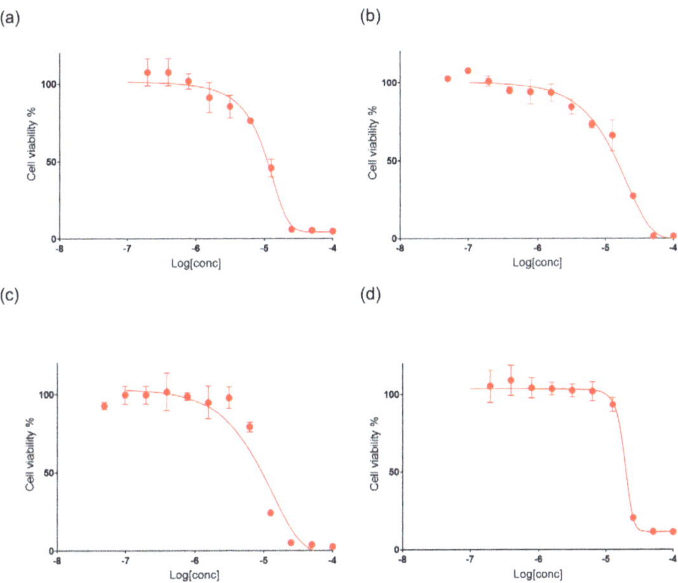

Figure 3. Cell viability assay. The figure shows the anti-proliferative effects of mycalols (red line) on A549 (**a**), A2058 (**b**), HepG2 (**c**), and MRC5 (**d**) cell lines, at increasing concentrations (0.05, 0.10, 0.19, 0.39, 0.78, 1.56, 3.12, 6.25, 12.5, 25, 50, 100 µM). Control cells were incubated with complete cell medium and DMSO. Results are expressed as percent survival after 72 h exposure.

IC_{50} values of mycalols on A549, A2058, HepG2, and MRC5 cell lines were 10.1, 15.3, 9.0 and 21.3 µM, respectively (Table 2), which were roughly in line with the activity reported for the pure compounds on anaplastic thyroid carcinoma-derived FRO cells [21]. IC_{50} values of suberitenones A (**1**) on A549, A2058, HepG2, and MRC5 cells were 28.5, 10.2, 17.6, and 7.4 µM, respectively. IC_{50} values of suberitenones B (**2**) on A549, A2058, HepG2, and MRC5 were 80.7, 14.6, 19.2, and 8.5 µM, respectively (Table 2, Figure S10). The mycalols showed higher activity against cancer cell lines with respect to normal cell lines, and, in particular, they had the lowest IC_{50} when tested against HepG2 cells. On the contrary, **1** and **2** showed lower activity against cancer cell lines with respect to the normal ones (Table 2).

Table 2. IC_{50} values of mycalols, suberitenone A and B bioactivity against A549, A2058, HepG2, and MRC5 cell lines. Values are expressed in µM.

Compounds	A549	A2058	HepG2	MRC5
Mycalols	10.1	15.3	9.0	21.3
Suberitenones A (**1**)	28.5	10.2	17.6	7.4
Suberitenones B (**2**)	80.7	14.6	19.2	8.5

2.4.1. Mechanism of Action

In order to elucidate the cell death metabolic pathway induced by mycalols, expression levels of selected genes by using a PCR array were analyzed in HepG2 cells treated in the presence of mycalols at 9 µM for 48 h. Gene transcription was considered affected by compounds if expression values were greater than two-fold difference with respect to the control (DMSO alone). Both differentially up-regulated and down-regulated genes are reported in Table 3. Gene expression analyses indicated that Unc-51-like kinase 1 (ULK1) and UDP-N-acetyl-alpha-D-galactosamine:polypeptide N-acetylgalactosaminyltransferase 5 (GALNT5) were the most down-regulated genes (−12.03 and −10.71-fold expression, respectively). ULK1 is involved in mammalian autophagic signaling [25] in hepatic cells and is known to be a regulator of different metabolic pathways/processes, such as mevalonate/cholesterol biosynthesis pathway [26] and lipotoxicity. Lipotoxicity is known to induce tissue damage and inflammation in metabolic disorders [27] and has been found to be related to ferroptosis [28,29], an iron-dependent programmed cell death triggered by the accumulation of lethal lipid species [30]. GALNT5 is involved in carcinogenesis and progression of choloangiocarcinoma, and activates the AKT/Erk (extracellular signal-regulated kinases) pathway [31]. In addition, a decrease in its substrate concentration (UDP-N-acetylglucosamine) has been related to ferroptosis in hepatoma cells [32].

Table 3. Transcriptional modulation of a subset of genes involved in human cell death signaling pathways in mycalols-treated HepG2 cells. Gene transcription is considered unaffected by compound treatment if fold regulation is in the range ± 2.0.

Unigene	RefSeq	Symbol	Description	Fold	SD
Genes down-regulated by mycalols treatment					
Hs.47061	NM_003565	ULK1	Unc-51-like kinase 1 (*C. elegans*)	−12.03	0.025
Hs.269027	NM_014568	GALNT5	UDP-N-acetyl-alpha-D-galactosamine:polypeptide N-acetylgalactosaminyltransferase 5 (GalNAc-T5)	−10.71	0.001
Hs.2490	NM_033292	CASP1	Caspase 1, apoptosis-related cysteine peptidase (interleukin 1, beta, convertase)	−6.85	0.005
Hs.484111	NM_002546	TNFRSF11	Tumor necrosis factor receptor superfamily, member 11b	−5.57	0.001
Hs.81791	NM_014592	KCNIP1	Kv channel interacting protein 1	−2.57	0.001
Hs.160562	NM_000618	IGF1	Insulin-like growth factor 1 (somatomedin C)	−2.41	0.002
Hs.552567	NM_001160	APAF1	Apoptotic peptidase activating factor 1	−2.03	0.026
Genes up-regulated by mycalols treatment					
Hs.513667	NM_003946	NOL3	Nucleolar protein 3 (apoptosis repressor with CARD domain)	7.40	0.029
Hs.227817	NM_004049	BCL2A1	BCL2-related protein A1	5.14	0.002
Hs.587290	NM_003900	SQSTM1	Sequestosome 1	4.75	1.256
Hs.442337	NM_176823	S100A7A	S100 calcium binding protein A7A	4.15	0.001
Hs.553833	NM_001004467	OR10J3	Olfactory receptor, family 10, subfamily J, member 3	4.05	0.001
Hs.202676	NM_014258	SYCP2	Synaptonemal complex protein 2	3.74	0.001
Hs.138211	NM_002750	MAPK8	Mitogen-activated protein kinase 8	3.71	0.132
Hs.519680	NM_001145805	IRGM	Immunity-related GTPase family, M	3.19	0.001
Hs.643440	NM_002361	MAG	Myelin associated glycoprotein	3.17	0.006
Hs.181301	NM_004079	CTSS	Cathepsin S	2.97	0.004
Hs.32949	NM_005218	DEFB1	Defensin, beta 1	2.74	0.005
Hs.29169	NM_024610	HSPBAP1	HSPB (heat shock 27kDa) associated protein 1	2.28	0.03

In the current study, the mitogen-activated protein kinase 8 (MAPK8; also known as JNK1), which has been directly related to ferroptosis as well [33], was up-regulated (3.71-fold regulation). On the other hand, gene expression analyses showed that the proapoptotic genes caspase 1 (CASP1), tumor necrosis factor receptor superfamily, member 11b (TNFRSF11; known to promote apoptosis in osteoclasts, https://www.uniprot.org/uniprot/O00300) and apoptotic peptidase activating factor 1 (APAF) were down-regulated (-6.85, -5.57, and -2.03 fold regulation, respectively), while the anti-apoptotic genes nucleolar protein 3 (apoptosis repressor with CARD domain; NOL3) and B-cell lymphoma 2 A1 (BCL2A1) were the most up-regulated genes (7.40 and 5.14-fold regulation, respectively), suggesting that apoptosis was not involved in the mechanism of action of mycalol toxicity.

To confirm whether mycalols triggered ferroptosis, the levels of glutathione peroxidase 4 (GPX4) and nuclear receptor coactivator 4 (NCOA4) were evaluated in hepatocellular carcinoma at the protein level. Both GPX4 and NCOA4 have an important role in ferroptosis processes. GPX4 inhibition has been associated with the accumulation of lipid peroxides, which leads to ferroptosis [29]. On the contrary, NCOA4 overexpression has been associated with an improved sensitivity to ferroptosis [34]. In the current study, the HepG2 cell line was treated in the presence or in the absence of mycalols. As shown in Figure 4a,b, the level of GPX4 protein was strongly reduced in HepG2 treated in the presence of 9 µM mycalols with respect to cells treated in the presence of the DMSO alone ($p < 0.05$). On the contrary, NCOA4 protein level was increased after mycalols treatment ($p < 0.01$; Figure 4c,d) with respect to the treatment in the presence of the vehicle alone. The reduction in GPX4 and the increase in NCOA4 protein levels suggested that mycalols were able to induce ferroptosis in hepatocarcinoma cells.

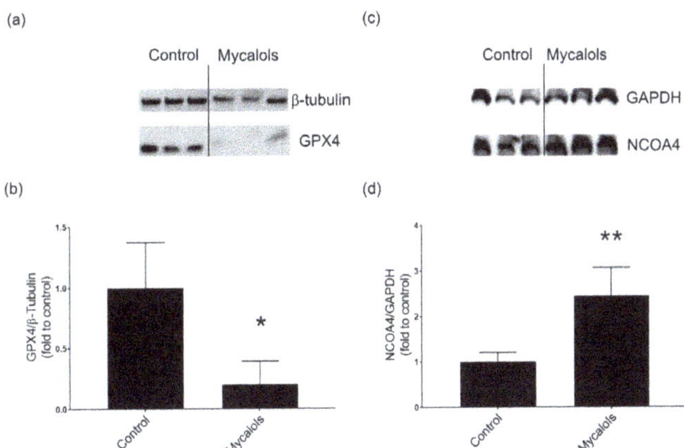

Figure 4. Effects of the mycalol mixture on ferroptosis-related proteins glutathione peroxidase 4 (GPX4) and nuclear receptor coactivator 4 (NCOA4). Western blotting analyses with anti-GPX4 (**a**) or anti-NCOA4 (**b**) antibody of the extracts from HepG2 treated with the DMSO alone (control) or in the presence of mycalols. The intensity of the bands of GPX4 (**c**) and NCOA4 (**d**) were normalized, respectively, with β-tubulin or glyceraldehyde 3-phosphate dehydrogenase (GAPDH), used as standard proteins, and are reported as relative to GPX4 and NCOA4 levels (* $p < 0.05$; ** $p < 0.01$).

3. Discussion

Sponges represent an important reservoir of bioactive marine natural compounds. Abdelaleem et al. [35] reported the number of isolated bioactive compounds in a taxon of the order Demospongiae between 2013 and 2019, showing that 195 new compounds and 189 known compounds were isolated. These compounds have been reported to have a broad range of bioactivities, such as antimicrobial, cytotoxic, antioxidant, antiviral, anti-allergic, anti-parasitic, and anti-inflammatory [35].

In the current study, we have screened four Antarctic sponges belonging to the class of Demospongiae, i.e., *Mycale (Oxymycale) acerata*, *Haliclona (Rhizoniera) dancoi*, *Hemimycale topsenti*, and *Hemigellius pilosus*, for their possible cytotoxic activity against different cancer cell lines. Two of these sponges, *H. dancoi* and *H. topsenti*, have shown high cytotoxic activity. Bioactivity-guided fractionation brought to the identification of mycalols and suberitenones A and B from *H. dancoi* and *H. topsenti*, respectively.

Mycalols, polyoxygenated glyceryl alkyl ethers, have been already isolated from the Antarctic sponge *M. (Oxymycale) acerata* [20,21,24], and they have been shown to possess cytotoxic activity against thyroid cancer cells [20]. Suberitenones A and B are sesterterpenoids isolated for the first time in the Antarctic sponge *Suberites* sp. [23] and, successively, in the Antarctic sponge *Phorbas areolatus* (Thiele, 1905) [36]. Suberitenones B has shown cytotoxic activity against lung adenocarcinoma cells (A549) and hepatocellular carcinoma cells (HepG2) [36].

Here, we report a new cytotoxic activity of mycalols against A549, HepG2, and melanoma cells (A2058), other than thyroid cancer cells. In addition to A549 and HepG2, suberitenones B also showed activity against melanoma cells. Suberitenones A was shown, for the first time, to be active against lung carcinoma, hepatocellular carcinoma, and melanoma (A549, HepG2, and A2058) cells. However, both suberitenones A and B also showed high toxicity against normal cells (MRC-5, lung fibroblasts). On the contrary, as previously reported on thyroid cells [20,21], cytoxicity of mycalols was weaker against normal cells than tumor cell lines. Particularly, mycalols showed the highest activity against hepatocarcinoma cells. It is worth noting that the members of this family of products show an in vitro activity that is dependent on the length and the substitution of the alkyl skeleton, with a good correlation with the lipophilicity of the products [21]. Reasonably, this suggests that the activity of one of these products may be significantly stronger than that we recorded with the mixture. In this study, we further investigated the cell death pathway triggered by the pool of mycalols in hepatocellular carcinoma cells. Expression levels of 84 genes, involved in different death pathways, were studied. Results showed that the differentially expressed genes could be associated with ferroptosis. In particular, ULK1, GALNT5 down-regulation and MAPK8 (also known as JNK1) up-regulation have been previously related to ferroptosis processes [28,32,33,37]. Ferroptosis is a programmed cell death triggered by iron accumulation and iron-dependent lipid peroxidation. Ferroptosis has been associated to many diseases, such as cancer, kidney injury, and neurological diseases [38]. Thus, activating or blocking ferroptosis could be a strategy to treat various diseases [29,39]. GPX4 has a key role in ferroptosis and is involved in the inhibition of lipid peroxides. GPX4 converts glutathione (GSH) into its oxidized form (GSSG) and reduces lipid peroxides [29]. A marine terpenoid, named heteronemin, was recently isolated from *Hippospongia* sp., which induces apoptosis and ferroptosis in hepatocarcinoma cells, reducing GPX4 expression as well [40]. Ferroptosis has also been related to NCOA4 levels [41]. In fact, NCOA4 over-expression in cellular models increases sensitivity to ferroptosis [34]. In order to confirm that mycalols from *Haliclona (Rhizoniera) dancoi* triggered the ferroptosis pathway, we analyzed the protein levels of both GPX4 and NCOA4, in the hepatocarcinoma cell line. In cells treated in the presence of mycalols, GPX4 protein levels were significantly reduced, while NCOA4 levels increased, confirming our hypothesis.

4. Conclusions

The aim of this study was to screen different sponges collected in Antarctic waters, in the framework of the XXXIV Italian National Antarctic Research Program (PNRA) in November–December 2018, in order to find bioactive natural products. Four different species of sponges have been collected and for each species, both total extracts and their four SPE-fractions have been prepared and tested on different tumor cell lines. A mixture of mycalols as well as suberitenones A and B were found in the most active fractions. Mycalols showed the highest activity. Thus, we studied the mechanism of action of the mycalol mixture at the gene and protein level, and the results suggested the activation of the ferroptosis death pathway in the hepatocarcinoma cell line. Several cancer cells have been found to modulate lipid metabolism in order to reduce ferroptosis sensitivity [42], thus targeting ferroptosis and its mediators could become a promising strategy in cancer therapy [43].

5. Materials and Methods

5.1. Sampling

Three analyzed sponge specimens were previously identified by morphological and molecular approaches in our previous paper [7]. All samples belonged to the class Demospongiae and to the following two orders: Haplosclerida with two species, *Hemigellius pilosus* (Kirkpatrick, 1907) and *Haliclona (Rhizoniera) dancoi* (Topsent, 1901), and Poecilosclerida with one species, *Mycale (Oxymycale) acerata* (Kirkpatrick, 1907). The species of sample C7 (MNA code 13860) was identified as *Hemimycale topsenti* in the present work (Table 1). Sponges were collected by scuba divers in November–December 2018 at two sites of Tethys Bay: (1) *M. (Oxymicale) acerata* at 26 m of depth (74°42.0670′ S, 164°02.5180′ E) and (2) *H. (Rhizoniera) dancoi*, *H. pilosus* and *H. topsenti* at 28 m of depth (74°40.5370′ S, 164°04.1690′ E). Samples were immediately washed with filtered and sterilized seawater to remove loosely attached bacteria and/or debris [7,44]. A small fragment of each sponge was stored in 70% ethanol for taxonomic identification, another fragment was preserved in RNA*later*™ into sterile tubes and stored at −20 °C until use for DNA extraction, and the remaining sample was frozen and stored at −80 °C for chemical extraction procedures. In addition, sponge slides of spicules were deposited at the Italian National Antarctic Museum (MNA, Section of Genoa, Italy). The MNA voucher codes of each sample are reported in Table 1.

5.2. Morphological Analysis and Polymerase Chain Reaction (PCR) of 18S/28S rRNA and CO1 Markers

For the taxonomic identification, small fragments of each sponge were heat-dissolved in nitric acid, rinsed in water, and dehydrated in ethanol. Then, spicules were mounted on slides for microscopic analyses, following standard methods [45]. The skeletal architecture was examined under a light microscope and hand-cut sections of sponge portions were made as described in Hooper [46]. The taxonomic classification follows the updated nomenclature reported in the World Porifera Database (WPD) [47].

About 10 mg of tissue, stored in RNA*later* at −20 °C, were excised and used for DNA extraction using the QIAamp® DNA Micro kit (Qiagen, Hilden, Germany), according to the manufacturer's instructions. DNA quantity (ng/μL) and quality (A260/A280; A260/A230) were evaluated by a NanoDrop spectrophotometer. PCR reactions were performed on a C1000 Touch Thermal Cycler (BioRad, Hercules, CA, USA) in a 30 μL reaction mixture by adding 1 μL of genomic DNA (starting concentration = ~100 ng/μL) from serial dilution (1:1, 1:10, 1:50 and 1:100), 6 μL of 5× Buffer GL (GeneSpin Srl, Milan, Italy), 0.6 μL of dNTPs (10 mM each), 1 μL of each forward and reverse primer (20 pmol/μL), and 0.2 μL of Xtra Taq Polymerase (5 U/μL, GeneSpin Srl, Milan, Italy). The PCR cycles were set as follows:

i. for 18S and 28S, a denaturation step at 95 °C for 2 min, [35 cycles denaturation step at 95 °C for 1 min, annealing step at 52 °C (18S1/18S2; [48]), 55 °C (18S-AF/18S-BR, NL4F/NL4R; [49,50]), 57 °C (C2/D2; [51]), for 1 min and 72 °C of primer extension for 2 min], a final extension step at 72 °C for 10 min;

ii. CO1 primers (dgLCO1490/dgHCO2198, COX1-R1/COX1-D2; [52,53]), a first denaturation at 94 °C for 3 min, [35 cycles of denaturation at 94 °C for 30 s, annealing at 45 °C for 30 s and primer extension at 72 °C for 1 min].

PCR products were separated on 1.5% agarose gel electrophoresis in TAE buffer (40 mM Tris-acetate, 1 mM EDTA, pH 8.0) using a 100 bp DNA ladder (GeneSpin Srl, Milan, Italy) and purified using the QIAquick Gel Extraction Kit (Qiagen, Hilden, Germany) according to the manufacturer's instructions. PCR amplicons were then sequenced in both strands through Applied Biosystems (Life Technologies, Waltham, MA, USA) 3730 Analyzer (48 capillaries). The total 18S, 28S, and CO1 regions were aligned to the nucleotide collection (GenBank, EMBL, DDBJ, PDB, RefSeq sequences) of Basic Local Alignment

Search Tool (BLAST [54]) and then aligned with highly similar sequences using MultiAlin (http://multalin.toulouse.inra.fr/multalin/ accessed on June 2021) [55]).

5.3. Chemical Extraction and SPE Fractionation

After lyophilization, the organic material was extracted with methanol ((Merk Life Science S.r.l., Milano, Italy) at room temperature and a small amount (about 50 mg) of raw extract (hereafter referred to as TE) was subjected to SPE on a GX-271 ASPEC Gilson apparatus by using CHROMABOND$^®$ HRX cartridges (6 mL/500 mg, MACHEREY-NAGEL, Düeren, Germany) as reported by Cutignano et al. [22]. Briefly, this extraction yielded five fractions (A, B, C, D, and E) eluted with H_2O, CH_3CN/H_2O 7:3, CH_3CN, and CH_2Cl_2/CH_3OH 9:1, respectively. The raw extract and the SPE fractions B-E were tested on A549, A2058 and HepG2 cell lines. Fraction A mainly full of sea salt was not further analyzed. SPE fractions were monitored by TLC revealed by spraying with $Ce(SO_4)_2$.

5.4. Purification and Characterization of Active Compounds

5.4.1. Haliclona (Rhizoniera) dancoi (H. d.)

The active SPE-HRX fraction C (5.1 mg) of *H. dancoi* was purified by chromatography on a silica Pasteur pipette (SiO_2) eluted with a gradient of chloroform in methanol, to obtain 1.2 mg of mycalols mixture. ^{1}H NMR in pyr-d$_5$ (Figure S9) and HR ESI$^+$-MS were acquired to define the chemical composition (Figure 2).

5.4.2. Hemimycale topsenti (H. t.)

The active SPE-HRX fraction D (2.3 mg) of *H. topsenti* was further purified by chromatography on a silica Pasteur pipette (SiO_2) eluted with a gradient of light petroleum ether (EP) in Et_2O (EE). Suberitenone A (**1**, 1 mg) and B (**2**, 0.4 mg) were eluted with EP/EE 8:2 and 7:3, respectively. ^{1}H NMR in $CDCl_3$ and HR ESI$^+$-MS were acquired to define the chemical characterization (Figures S4–S7).

5.5. Cell Lines

Human cells were bought at ATCC (https://www.lgcstandards-atcc.org/ accessed on June 2021). Human hepatocellular liver carcinoma (HepG2; ATCC$^®$ HB-8065™) and human normal lung fibroblasts (MRC5; ATCC$^®$ CCL-171™) were cultured in EMEM medium, human melanoma cells (A2058; ATCC$^®$ CRL-11147TM) were cultured in DMEM, adenocarcinomic human alveolar basal epithelial cells (A549; ATCC$^®$ CL-185TM) were cultured in F-12K medium. The media were supplemented with 10% fetal bovine serum, 50 U/mL penicillin, and 50 µg/mL streptomycin.

5.6. Antibody

The following antibodies were used: the rabbit monoclonal anti-glutathione peroxidase 4 (GPX4; 52455s, Cell Signaling Technology, Danvers, MA, USA), the rabbit monoclonal anti-β-tubulin (9F3, 2118s, Cell Signaling Technology, Danvers, MA, USA), the anti-rabbit IgG, HRP-linked antibody (7074s, Cell Signaling Technology, Danvers, MA, USA), the rabbit monoclonal anti-nuclear receptor coactivator 4 (NCOA4; E8H8Z, 66849s, Cell Signaling Technology, Danvers, MA, USA), the rabbit monoclonal anti-glyceraldehyde 3-phosphate dehydrogenase (GAPDH; 14C10, 2118s, Cell signaling Technology, Danvers, MA, USA), and the anti-mouse IgG, HRP-linked antibody (7074s, Cell Signaling Technology, Danvers, MA, USA).

5.7. In Vitro Cell Viability Studies

To evaluate the in vitro effects of the extracts or fractions on the cell viability of HepG2, A2058, A549, and MRC5 cell lines were seeded in 96-well microtiter plates at a density of 1×10^4 cells/well and incubated at 37 °C to allow for cell adhesion in the plates. After 16 h, the medium was replaced with fresh medium containing increasing concentrations of extracts, fractions (0.01, 0.1, 1, 10, and 100 µg/mL) or purified compounds (0.05, 0.10, 0.19,

0.39, 0.78, 1.56, 3.12, 6.25, 12.5, 25, 50, 100 μM) dissolved in dimethyl sulfoxide (DMSO), for 72 h. The maximum concentration of DMSO used was 1% (*v/v*). Each concentration was tested at least in triplicate. After 72 h, the 3-(4,5-dimethyl-2-thizolyl)-2,5-diphenyl-2H-tetrazolium bromide (MTT; A2231,0001, AppliChem Panreac Tischkalender, GmbH, Darmstadt, Germany) assay was carried out. Briefly, the medium was replaced with a medium containing MTT at 0.5 mg/mL and the plates were incubated for 3 h at 37 °C. After incubation, the cells were treated with isopropyl alcohol (used as MTT solvent) for 30 min at room temperature. Absorbance was measured at OD = 570 nm by a microplate reader (Multiskan™ FC Microplate Photometer, Thermo Fisher Scientific, Waltham, MA, USA). Cell survival was expressed as a percentage of viable cells in the presence of the tested samples, with respect to untreated control cultures with only DMSO.

5.8. RNA Extraction and Reverse Transcription-Quantitative Polymerase Chain Reaction (RT-qPCR)

HepG2 cells to be used for RNA extraction were seeded in 6 wells plates (at 500,000 per well) and kept 16 h for attachment. The seeded cells were then treated in the presence of mycalols at 9 μM (molarity was evaluated based on the average molecular weights of the mycalols into the mixture) for 48 h at 37 °C. Cells were washed by adding phosphate-buffered saline (PBS 1×).

Cells were lysed by adding 1 mL of Trisure Reagent (Meridian bioscience, Memphis, TN, USA). RNA was isolated as previously described [56]. RNA concentration, quality, and purity were assessed using an ND-1000 UV-Vis spectrophotometer (NanoDrop Technologies, Thermo Fisher Scientific, Waltham, MA, USA), monitoring the absorbance at 260 nm, and the 260/280 nm and 260/230 nm ratios (both ratios were about 2.0). RNA quality was evaluated by gel electrophoresis that showed intact RNA, with sharp ribosomal bands. About 500 ng of RNA was subjected to reverse transcription reaction using the RT2 first strand kit (cat.330401, Qiagen, Hilden, Germany) according to the manufacturer's instructions. The RT-qPCR analysis was performed in duplicate using the RT2 Profiler PCR Array kit (cat. PAHS-212ZE-4, Qiagen, Hilden, Germany) to analyze the expression of 84 genes involved in cell death signaling pathways. Plates were run on a ViiA7 (Applied Biosystems, Foster City, CA, USA, 384-well blocks). The PCR program consisted of a denaturation step at 95 °C for 20 s followed by 40 cycles at 95 °C for 15 s and 60 °C for 1 min. The cycle threshold (Ct)-values were analyzed with PCR array data analysis online software (GeneGlobe Data Analysis Center http://pcrdataanalysis.sabiosciences.com/pcr/arrayanalysis.php accessed on May 2021, Qiagen, Hilden, Germany). Real-time data were expressed as the fold of expression, describing the changes in gene expression between cells treated in the presence of mycalols and cells treated in the presence of DMSO alone. Only expression values greater than a two-fold difference with respect to the controls were considered significant.

5.9. Protein Extraction and Western Blotting Analyses

The HepG2 cells to be used for protein extraction were seeded in 6-well plates (at 500,000 per well) and kept 16 h for attachment. The seeded cells were then treated in the presence of mycalols at 9 μM for 48 h at 37 °C. Cells were washed by adding phosphate-buffered saline (PBS 1×). Cells were lysed in RIPA Buffer (Cell signaling technology, Danvers, MA, USA) 1% Triton X-100, supplemented with protease inhibitors (#310A7779, AppliChem GmbH, Darmstadt, Germany). The lysates were centrifuged at 14,000 rpm for 30 min at 4 °C. The total amount of proteins in each lysate was measured using the Bradford assay. Before sodium dodecyl sulfate-polyacrylamide gel electrophoresis (SDS-PAGE), each lysate was diluted in Laemmli sample buffer (#161-0747, Biorad, Hercules, CA, USA) containing β-mercapto-ethanol and then boiled for 5 min at 95 °C. For Western blotting analysis, proteins were transferred onto polyvinylidene difluoride (PVDF) membranes (#1704156, Trans-blot Turbo transfer pack-Biorad) using the Trans blot turbo transfer system (Biorad, Hercules, CA, USA). After the transfer, PVDF membranes were incubated in a blocking solution (1× Tris-buffered saline, TBS, 5% BSA) for 1 h at 25 °C and then incubated with primary and secondary antibodies supplemented with 5% BSA. ECL (#170-5060, Biorad, Hercules, CA, USA) reactions were performed as per the manufacturer's instructions and immunoreactive bands were detected by chemiluminescence using ChemiDoc MP imaging system (Biorad, Hercules, CA, USA). The obtained immunoreactive bands were quantitated using Image Lab v6.0 (Biorad, Hercules, CA, USA). Different housekeeping proteins were used to normalize immunoreactive bands depending on the molecular weight of the detected proteins. B-tubulin and GAPDH were used to normalize GPX4 and NCOA4, respectively.

Supplementary Materials: The following are available online at https://www.mdpi.com/article/10.3390/md19080459/s1, Figure S1: MultiAlin output of CO1 PCR product indicated as C7_dgLCO1490/dgHCO21 aligned to the BLAST highly similar sequence of the strain *Hemimycale topsenti* (ex *Suberites topsenti*; LN850246.1); Figure S2: Cell viability assay. The figure shows the effects on cell viability of fractions (B green bars; C blue bars; D yellow bars; E red bars) or total extracts (ex, black bars) of the samples (a) *M. a.*, (b) *H. d.*, (c) *H.t.* and (d) *H. p.* on A2058, A549, and HepG2 cell lines at increasing concentrations (10 and 100 ng/mL and 1, 10, and 100 µg/mL). Cell viability was normalized using cells with only DMSO as a control sample. Results are expressed as percent survival after 72 h exposure (n = 3; * for $p < 0.05$; ** for $p < 0.01$ and *** for $p < 0.001$, Student's *t*-test); Figure S3: ^{1}H NMR spectrum of fraction D of the sponge *H. t.* (600 MHz, CDCl3); Figure S4: ^{1}H NMR spectrum of suberitenone A **1** (600 MHz, CDCl3); Figure S5: ^{1}H NMR spectrum of suberitenone B **2** (600 MHz, CDCl3); Figure S6: ESI$^+$-MS spectrum of suberitenone A **1** (m/z [M+Na]$^+$ 451.29); Figure S7: ESI$^+$-MS spectrum of suberitenone B **2** (m/z [M+Na]$^+$ 469.30); Figure S8: ^{1}H NMR spectrum of fraction C of the sponge *H. d.* (600 MHz, CD$_3$OD); Figure S9: ^{1}H NMR spectrum of purified fraction containing mycalols (600 MHz, C$_6$D$_5$N); Figure S10: ESI+-MS of fraction containing mycalols (**3–9**); Figure S11: Cell viability antiproliferative effects induced by suberitenones A and suberitenone B. The figure shows the anti-proliferative effects of (a) suberitenones A (**1**, green line) and (b) suberitenones B (**2**, blue line) on the cell viability of A549, A2058, HepG2, and MRC5 cell lines, at increasing concentrations (0.05, 0.10, 0.19, 0.39, 0.78, 1.56, 3.12, 6.25, 12.5, 25, 50, 100 µM). The control sample contained only DMSO. Results are expressed as percent survival after 72 h exposure.

Author Contributions: Conceptualization, G.R. and C.L.; collection of sponge samples in the framework of XXXIV Italian National Antarctic Research Program (PNRA) expedition, G.Z.; morphological analysis and classification of sponges, M.B.; methodology, G.R., G.N., D.C., G.S., L.R., M.C., N.R. and C.L.; formal analysis, data curation, G.R., G.N., C.L.; writing—original draft preparation, G.R., G.N., C.L.; writing—review and editing, G.R., G.N., D.C., M.C., A.F., A.I., C.V., D.G. and C.L.; funding acquisition, A.F., A.I., C.V. and D.G. All authors have read and agreed to the published version of the manuscript.

Funding: This research was supported by PNRA16_00043, Cosmeceuticals And Nutraceuticals From Antarctic Biological REsources (CAN FARE). It was carried out in the framework of the SCAR Programme "Antarctic Thresholds–Ecosystem Resilience and Adaptation" (AnT-ERA). Authors acknowledge the financial support of the project "Antitumor Drugs and Vaccines from the Sea (ADViSE)" (CUP B43D18000240007–SURF 17061BP000000011; PG/2018/0494374) funded by POR Campania FESR 2014–2020 "Technology Platform for Therapeutic Strategies against Cancer"-Action 1.1.2 and 1.2.2.

Acknowledgments: Authors thank Servier Medical Art (SMART) website (https://smart.servier.com/) by Servier for the cell element of Figure 1. D.G. and C.V. wish to thank Valentina Brasiello and Chiara Nobile for technical support and assistance.

Conflicts of Interest: The authors declare no conflict of interest.

References

1. Thakur, N.L.; Thakur, A.N.; Müller, W.E.G. Article RP 18 Marine natural products in drug discovery. *NISCAIR Online Period. Repos.* **2005**, *4*, 471–477.
2. Berne, S.; Kalauz, M.; Lapat, M.; Savin, L.; Janussen, D.; Kersken, D.; Ambrožič Avguštin, J.; Zemljič Jokhadar, Š.; Jaklič, D.; Gunde-Cimerman, N.; et al. Screening of the Antarctic marine sponges (Porifera) as a source of bioactive compounds. *Polar Biol.* **2016**, *39*, 947–959. [CrossRef]

3.   Giordano, D.; Costantini, M.; Coppola, D.; Lauritano, C.; Núñez Pons, L.; Ruocco, N.; di Prisco, G.; Ianora, A.; Verde, C. Biotechnological Applications of Bioactive Peptides From Marine Sources. In *Advances in Microbial Physiology*; Academic Press: London, UK, 2018; Volume 73, pp. 171–220, ISBN 9780128151907.

4.   Romano, G.; Ianora, A.; Costantini, M.; Sansone, C.; Lauritano, C.; Ruocco, N.; Ianora, A. Marine microorganisms as a promising and sustainable source of bioactive molecules. *Mar. Environ. Res.* **2016**, *128*, 58–69. [CrossRef] [PubMed]

5.   Sagar, S.; Kaur, M.; Minneman, K.P. Antiviral lead compounds from marine sponges. *Mar. Drugs* **2010**, *8*, 2619–2638. [CrossRef]

6.   Jaspars, M.; De Pascale, D.; Andersen, J.H.; Reyes, F.; Crawford, A.D.; Ianora, A. The marine biodiscovery pipeline and ocean medicines of tomorrow. *J. Mar. Biol. Assoc. UK* **2016**, *96*, 151–158. [CrossRef]

7.   Ruocco, N.; Esposito, R.; Bertolino, M.; Zazo, G.; Sonnessa, M.; Andreani, F.; Coppola, D.; Giordano, D.; Nuzzo, G.; Lauritano, C.; et al. A metataxonomic approach reveals diversified bacterial communities in antarctic sponges. *Mar. Drugs* **2021**, *19*, 173. [CrossRef]

8.   Lauritano, C.; Rizzo, C.; Giudice, A.L.; Saggiomo, M. Physiological and molecular responses to main environmental stressors of microalgae and bacteria in polar marine environments. *Microorganisms* **2020**, *8*, 1957. [CrossRef] [PubMed]

9.   Peck, L.S. Antarctic marine biodiversity: Adaptations, Environments and Responses to Change. *An Annu. Rev.* **2018**, *56*, 105–236.

10.   McClintock, J.B.; Amsler, C.D.; Baker, B.J.; Van Soest, R.W.M. Ecology of antarctic marine sponges: An overview. *Integr. Comp. Biol.* **2005**, *45*, 359–368. [CrossRef]

11.   de Broyer, C.; Koubbi, P.; Griffiths, H.; Raymond, B.; d'Acoz, C.d.; van de Putte, A.; Danis, B.; David, B.; Grant, S.; Gutt, J.; et al. *Biogeographic Atlas of the Southern Ocean*; XII; Scientific Committee on Antarctic Research: Cambridge, UK, 2014; ISBN 978-0-948277-28-3.

12.   Papaleo, M.C.; Fondi, M.; Maida, I.; Perrin, E.; Lo Giudice, A.; Michaud, L.; Mangano, S.; Bartolucci, G.; Romoli, R.; Fani, R. Sponge-associated microbial Antarctic communities exhibiting antimicrobial activity against *Burkholderia cepacia* complex bacteria. *Biotechnol. Adv.* **2012**, *30*, 272–293. [CrossRef]

13.   Von Salm, J.L.; Witowski, C.G.; Fleeman, R.M.; McClintock, J.B.; Amsler, C.D.; Shaw, L.N.; Baker, B.J. Darwinolide, a New Diterpene Scaffold That Inhibits Methicillin-Resistant *Staphylococcus aureus* Biofilm from the Antarctic Sponge *Dendrilla membranosa*. *Org. Lett.* **2016**, *18*, 2596–2599. [CrossRef]

14.   Di, X.; Rouger, C.; Hardardottir, I.; Freysdottir, J.; Molinski, T.F.; Tasdemir, D.; Omarsdottir, S. 6-Bromoindole derivatives from the icelandic marine sponge *Geodia barretti*: Isolation and anti-inflammatory activity. *Mar. Drugs* **2018**, *16*, 437. [CrossRef]

15.   Li, F.; Peifer, C.; Janussen, D.; Tasdemir, D. New discorhabdin alkaloids from the antarctic deep-sea sponge *Latrunculia biformis*. *Mar. Drugs* **2019**, *17*, 439. [CrossRef] [PubMed]

16.   Shilling, A.J.; Witowski, C.G.; Maschek, J.A.; Azhari, A.; Vesely, B.A.; Kyle, D.E.; Amsler, C.D.; McClintock, J.B.; Baker, B.J. Spongian Diterpenoids Derived from the Antarctic Sponge *Dendrilla antarctica* Are Potent Inhibitors of the Leishmania Parasite. *J. Nat. Prod.* **2020**, *83*, 1553–1562. [CrossRef] [PubMed]

17.   Núñez-Pons, L.; Shilling, A.; Verde, C.; Baker, B.J.; Giordano, D. Marine terpenoids from polar latitudes and their potential applications in biotechnology. *Mar. Drugs* **2020**, *18*, 401. [CrossRef] [PubMed]

18.   Giordano, D. Bioactive Molecules from Extreme Environments. *Mar. Drugs* **2020**, *18*, 640. [CrossRef]

19.   Ciaglia, E.; Malfitano, A.M.; Laezza, C.; Fontana, A.; Nuzzo, G.; Cutignano, A.; Abate, M.; Pelin, M.; Sosa, S.; Bifulco, M.; et al. Immuno-Modulatory and Anti-Inflammatory Effects of Dihydrogracilin A, a Terpene Derived from the Marine Sponge *Dendrilla membranosa*. *Int. J. Mol. Sci.* **2017**, *18*, 1643. [CrossRef]

20.   Cutignano, A.; Nuzzo, G.; D'Angelo, D.; Borbone, E.; Fusco, A.; Fontana, A. Mycalol: A natural lipid with promising cytotoxic properties against human anaplastic thyroid carcinoma cells. *Angew. Chemie Int. Ed.* **2013**, *52*, 9256–9260. [CrossRef]

21.   Cutignano, A.; Seetharamsingh, B.; D'Angelo, D.; Nuzzo, G.; Khairnar, P.V.; Fusco, A.; Reddy, D.S.; Fontana, A. Identification and Synthesis of Mycalol Analogues with Improved Potency against Anaplastic Thyroid Carcinoma Cell Lines. *J. Nat. Prod.* **2017**, *80*, 1125–1133. [CrossRef]

22.   Cutignano, A.; Nuzzo, G.; Ianora, A.; Luongo, E.; Romano, G.; Gallo, C.; Sansone, C.; Aprea, S.; Mancini, F.; D'Oro, U.; et al. Development and application of a novel SPE-method for bioassay-guided fractionation of marine extracts. *Mar. Drugs* **2015**, *13*, 5736–5749. [CrossRef]

23.   Shin, J.; Seo, Y.; Rho, J.R.; Baek, E.; Kwon, B.M.; Jeong, T.S.; Bok, S.H. Suberitenones a and B: Sesterterpenoids of an Unprecedented Skeletal Class from the Antarctic Sponge *Suberites* sp. *J. Org. Chem.* **1995**, *60*, 7582–7588. [CrossRef]

24.   Seetharamsingh, B.; Rajamohanan, P.R.; Reddy, D.S. Total synthesis and structural revision of mycalol, an anticancer natural product from the marine source. *Org. Lett.* **2015**, *17*, 1652–1655. [CrossRef] [PubMed]

25.   Lin, M.G.; Hurley, J.H. Structure and function of the ULK1 complex in autophagy. *Curr. Opin. Cell Biol.* **2016**, *39*, 61–68. [CrossRef]

26.   Rajak, S.; Iannucci, L.F.; Zhou, J.; Anjum, B.; George, N.; Singh, B.K.; Ghosh, S.; Yen, P.M.; Sinha, R.A. Loss of ULK1 Attenuates Cholesterogenic Gene Expression in Mammalian Hepatic Cells. *Front. Cell Dev. Biol.* **2020**, *8*, 523550. [CrossRef]

27.   Sinha, R.A.; Singh, B.K.; Zhou, J.; Xie, S.; Farah, B.L.; Lesmana, R.; Ohba, K.; Tripathi, M.; Ghosh, S.; Hollenberg, A.N.; et al. Loss of ULK1 increases RPS6KB1-NCOR1 repression of NR1H/LXR-mediated Scd1 transcription and augments lipotoxicity in hepatic cells. *Autophagy* **2017**, *13*, 169–186. [CrossRef] [PubMed]

28.   Sun, Y.; Chen, P.; Zhai, B.; Zhang, M.; Xiang, Y.; Fang, J.; Xu, S.; Gao, Y.; Chen, X.; Sui, X.; et al. The emerging role of ferroptosis in inflammation. *Biomed. Pharmacother.* **2020**, *127*, 110108. [CrossRef]

29. Li, J.; Cao, F.; Yin, H.l.; Huang, Z.j.; Lin, Z.t.; Mao, N.; Sun, B.; Wang, G. Ferroptosis: Past, present and future. *Cell Death Dis.* **2020**, *11*, 88. [CrossRef] [PubMed]

30. Yan, H.f.; Zou, T.; Tuo, Q.Z.; Xu, S.; Li, H.; Belaidi, A.A.; Lei, P. Ferroptosis: Mechanisms and links with diseases. *Signal Transduct. Target. Ther.* **2021**, *6*, 49. [CrossRef]

31. Detarya, M.; Sawanyawisuth, K.; Aphivatanasiri, C.; Chuangchaiya, S.; Saranaruk, P.; Sukprasert, L.; Silsirivanit, A.; Araki, N.; Wongkham, S.; Wongkham, C. The O-GalNAcylating enzyme GALNT5 mediates carcinogenesis and progression of cholangiocarcinoma via activation of AKT/ERK signaling. *Glycobiology* **2020**, *30*, 312–324. [CrossRef]

32. Jin, M.; Shi, C.; Li, T.; Wu, Y.; Hu, C.; Huang, G. Solasonine promotes ferroptosis of hepatoma carcinoma cells via glutathione peroxidase 4-induced destruction of the glutathione redox system. *Biomed. Pharmacother.* **2020**, *129*, 110282. [CrossRef]

33. Wu, C.; Zhao, W.; Yu, J.; Li, S.; Lin, L.; Chen, X. Induction of ferroptosis and mitochondrial dysfunction by oxidative stress in PC12 cells. *Sci. Rep.* **2018**, *8*, 574. [CrossRef]

34. Del Rey, M.Q.; Mancias, J.D. NCOA4-mediated ferritinophagy: A potential link to neurodegeneration. *Front. Neurosci.* **2019**, *13*, 238. [CrossRef]

35. Abdelaleem, E.R.; Samy, M.N.; Desoukey, S.Y.; Liu, M.; Quinn, R.J.; Abdelmohsen, U.R. Marine natural products from sponges (Porifera) of the order Dictyoceratida (2013 to 2019); a promising source for drug discovery. *RSC Adv.* **2020**, *10*, 34959–34976. [CrossRef]

36. Solanki, H.; Angulo-Preckler, C.; Calabro, K.; Kaur, N.; Lasserre, P.; Cautain, B.; de la Cruz, M.; Reyes, F.; Avila, C.; Thomas, O.P. Suberitane sesterterpenoids from the Antarctic sponge *Phorbas areolatus* (Thiele, 1905). *Tetrahedron Lett.* **2018**, *59*, 3353–3356. [CrossRef]

37. Li, X.; Wang, T.X.; Huang, X.; Li, Y.; Sun, T.; Zang, S.; Guan, K.L.; Xiong, Y.; Liu, J.; Yuan, H.X. Targeting ferroptosis alleviates methionine-choline deficient (MCD)-diet induced NASH by suppressing liver lipotoxicity. *Liver Int.* **2020**, *40*, 1378–1394. [CrossRef]

38. Kist, M.; Vucic, D. Cell death pathways: Intricate connections and disease implications. *EMBO J.* **2021**, *40*, e106700. [CrossRef] [PubMed]

39. Guzmán, E.A. Regulated cell death signaling pathways and marine natural products that target them. *Mar. Drugs* **2019**, *17*, 76. [CrossRef] [PubMed]

40. Chang, W.T.; Bow, Y.D.; Fu, P.J.; Li, C.Y.; Wu, C.Y.; Chang, Y.H.; Teng, Y.N.; Li, R.N.; Lu, M.C.; Liu, Y.C.; et al. A Marine Terpenoid, Heteronemin, Induces Both the Apoptosis and Ferroptosis of Hepatocellular Carcinoma Cells and Involves the ROS and MAPK Pathways. *Oxid. Med. Cell. Longev.* **2021**, *2021*, 7689045. [CrossRef]

41. Tang, M.; Chen, Z.; Wu, D.; Chen, L. Ferritinophagy/ferroptosis: Iron-related newcomers in human diseases. *J. Cell. Physiol.* **2018**, *233*, 9179–9190. [CrossRef] [PubMed]

42. Broadfield, L.A.; Pane, A.A.; Talebi, A.; Swinnen, J.V.; Fendt, S.-M. Lipid metabolism in cancer: New perspectives and emerging mechanisms. *Dev. Cell* **2021**, *56*, 1363–1393. [CrossRef] [PubMed]

43. Mishchenko, T.A.; Balalaeva, I.V.; Vedunova, M.V.; Krysko, D.V. Ferroptosis and Photodynamic Therapy Synergism: Enhancing Anticancer Treatment. *Trends Cancer* **2021**, *7*, 484–487. [CrossRef]

44. Mangano, S.; Michaud, L.; Caruso, C.; Brilli, M.; Bruni, V.; Fani, R.; Lo Giudice, A. Antagonistic interactions between psychrotrophic cultivable bacteria isolated from Antarctic sponges: A preliminary analysis. *Res. Microbiol.* **2009**, *160*, 27–37. [CrossRef]

45. Rützler, K. Sponges in coral reefs. In *Coral Reefs: Research Methods, Monographs on Oceanographic Methodology*; Stoddart, D.R., Johannes, R.E., Eds.; Unesco: Paris, France, 1978; pp. 299–313.

46. Hooper, J.N.A. '*Sponguide*'. *Guide to Sponge Collection and Identification*; Queensland Museum: South Brisbane, Australia, 2000.

47. Soest, V.; Rob, W.M.; Boury-Esnault, N.; Hooper, J.N.A.; Rützler, K.; de Voogd, N.J.; Alvarez, B.; Hajdu, E.; Pisera, A.B.; Manconi, R.; et al. *World Porifera Database*; World Register of Marine Species: Ostend, Belgium, 2014.

48. Manuel, M.; Borchiellini, C.; Alivon, E.; Le Parco, Y.; Vacelet, J.; Boury-Esnault, N. Phylogeny and evolution of calcareous sponges: Monophyly of calcinea and calcaronea, high level of morphological homoplasy, and the primitive nature of axial symmetry. *Syst. Biol.* **2003**, *52*, 311–333. [CrossRef]

49. Collins, A.G. Phylogeny of medusozoa and the evolution of cnidarian life cycles. *J. Evol. Biol.* **2002**, *15*, 418–432. [CrossRef]

50. Dohrmann, M.; Janussen, D.; Reitner, J.; Collins, A.G.; Wörheide, G. Phylogeny and evolution of glass sponges (*Porifera, Hexactinellida*). *Syst. Biol.* **2008**, *57*, 388–405. [CrossRef] [PubMed]

51. Chombard, C.; Boury-Esnault, N.; Tillier, S. Reassessment of homology of morphological characters in Tetractinellid sponges based on molecular data. *Syst. Biol.* **1998**, *47*, 351–366. [CrossRef] [PubMed]

52. Meyer, C.P.; Geller, J.B.; Paulay, G. Fine scale endemism on coral reefs: Archipelagic differentiation in turbinid gastropods. *Evolution* **2005**, *59*, 113–125. [CrossRef]

53. Rot, C.; Goldfarb, I.; Ilan, M.; Huchon, D. Putative cross-kingdom horizontal gene transfer in sponge (Porifera) mitochondria. *BMC Evol. Biol.* **2006**, *6*, 1–11. [CrossRef] [PubMed]

54. Altschul, S.F.; Madden, T.L.; Schäffer, A.A.; Zhang, J.; Zhang, Z.; Miller, W.; Lipman, D.J. Gapped BLAST and PSI-BLAST: A new generation of protein database search programs. *Nucleic Acids Res.* **1997**, *25*, 3389–3402. [CrossRef]

55.   Corpet, F. Multiple sequence alignment with hierarchical clustering. *Nucleic Acids Res.* **1988**, *16*, 10881–10890. [CrossRef]
56.   Riccio, G.; Bottone, S.; La Regina, G.; Badolati, N.; Passacantilli, S.; Rossi, G.B.; Accardo, A.; Dentice, M.; Silvestri, R.; Novellino, E.; et al. A Negative Allosteric Modulator of WNT Receptor Frizzled 4 Switches into an Allosteric Agonist. *Biochemistry* **2018**, *57*, 839–851. [CrossRef] [PubMed]

 *marine drugs*

Article

Antibacterial Cyclic Tripeptides from Antarctica-Sponge-Derived Fungus *Aspergillus insulicola* HDN151418

Chunxiao Sun [1], Ziping Zhang [1], Zilin Ren [1], Liu Yu [1], Huan Zhou [1], Yaxin Han [1], Mudassir Shah [1], Qian Che [1], Guojian Zhang [1,2], Dehai Li [1,2,3,*] and Tianjiao Zhu [1,*]

[1] Key Laboratory of Marine Drugs, Chinese Ministry of Education, School of Medicine and Pharmacy, Ocean University of China, Qingdao 266003, China; sunchunxiao93@163.com (C.S.); zhangziping1998@163.com (Z.Z.); rzl17854263950@163.com (Z.R.); yuliu906@163.com (L.Y.); zhouhuanchn@icloud.com (H.Z.); 17669475579@163.com (Y.H.); s84mudassir@gmail.com (M.S.); cheqian064@ouc.edu.cn (Q.C.); zhangguojian@ouc.edu.cn (G.Z.)

[2] Laboratory for Marine Drugs and Bioproducts, Pilot National Laboratory for Marine Science and Technology, Qingdao 266237, China

[3] Open Studio for Druggability Research of Marine Natural Products, Pilot National Laboratory for Marine Science and Technology, Qingdao 266237, China

[*] Correspondence: dehaili@ouc.edu.cn (D.L.); zhutj@ouc.edu.cn (T.Z.); Tel.: +86-532-82031619 (D.L.); +86-532-82031632 (T.Z.)

Received: 7 October 2020; Accepted: 24 October 2020; Published: 26 October 2020

Abstract: Three new aspochracin-type cyclic tripeptides, sclerotiotides M–O (**1–3**), together with three known analogues, sclerotiotide L (**4**), sclerotiotide F (**5**), and sclerotiotide B (**6**), were obtained from the ethyl acetate extract of the fungus *Aspergillus insulicola* HDN151418, which was isolated from an unidentified Antarctica sponge. Spectroscopic and chemical approaches were used to elucidate their structures. The absolute configuration of the side chain in compound **4** was elucidated for the first time. Compounds **1** and **2** showed broad antimicrobial activity against a panel of pathogenic strains, including *Bacillus cereus*, *Proteus species*, *Mycobacterium phlei*, *Bacillus subtilis*, *Vibrio parahemolyticus*, *Edwardsiella tarda*, MRCNS, and MRSA, with MIC values ranging from 1.56 to 25.0 µM.

Keywords: cyclic tripeptides; antibacterial; Antarctica sponge-derived fungus; *Aspergillus insulicola*

---

1. Introduction

Marine life is radically different from its terrestrial counterpart, resulting in an interesting difference between its metabolites [1,2]. Sponge-derived fungi isolated from the marine environment have shown great potential of producing diverse bioactive secondary metabolites [1–4]. Cyclic peptides are a class of essential metabolites that are widely present in marine tunicates [5], sponge [6], algae [7], bacteria [8], fungi [9,10], etc. The structure of aspochracin-type cyclic tripeptides usually contains a unique macro-cyclic ring and a polyketide side chain. For the macro-cyclic ring, the most common features are a 12-member (composed of Ala-Val-Orn) and 13-member (composed of Ala-Val-Lys) ring. Only 15 of the aspochracin-type cyclic tripeptides have been obtained from natural sources (Figure S1). Their structures were mainly different regarding the polyketide side chains, the constitution of amino acids, and the *N*-methylation level in the amino acid moieties. Those chemistry diversities were able to be generated by chemical transformations. For example, some analogs can be synthesized from JBIR-15 or aspochracin via photoisomerization-initiated radical reaction or air oxidation during the fermentation or subsequent isolation steps [10]. Some of the analogs exhibited antifungal or anti-inflammatory activities [10–13].

During our ongoing research on bioactive natural products from Antarctic marine-derived fungi [14–17], *Aspergillus insulicola* HDN151418, a fungal strain isolated from an unidentified sponge, was chosen based on its unique HPLC-UV profile (series UV absorption around 260 nm) (Figure S2) and its antibacterial activity against MRCNS and MRSA (detected by the paper diffusion method). Consequently, chemical investigation resulted in the identification of three new aspochracin-type cyclic tripeptides, sclerotiotides M–O (**1–3**), together with three known compounds, sclerotiotide L (**4**), sclerotiotide F (**5**), and sclerotiotide B (**6**) (Figure 1). Compounds **1** and **2** represent the first example of an aspochracin-type cyclic tripeptide that possess a hexa-2,4-dienedioic acid/methyl ester moiety. Here, we address the isolation, elucidation of the structure, and biological activities of the new aspochracin-type cyclic tripeptides, sclerotiotides M–O (**1–3**).

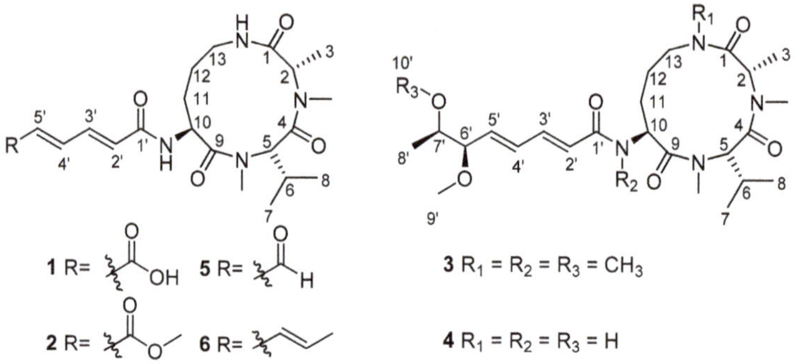

Figure 1. Structures of compounds **1–6**.

2. Results and Discussion

The fungus *A. insulicola* HDN151418 was cultured for 30 days under static condition (30 L). The crude extract (32.3 g) was fractionated and purified by LH-20, ODS, and HPLC, sequentially, yielding compounds **1** (10.2 mg), **2** (5.7 mg), **3** (6.5 mg), **4** (6.5 mg), **5** (20.2 mg), and **6** (12.0 mg).

Sclerotiotide M (**1**) was isolated as a pale yellow amorphous powder. The molecular formula was assigned as $C_{21}H_{32}N_4O_6$ based on the HRESIMS ion peak at *m/z* 435.2246 [M − H]$^-$ (calcd for $C_{21}H_{31}N_4O_6$, 435.2249). The IR spectrum showed absorption bands for amide groups at 3394 cm^{-1} and 1681 cm^{-1}. The ^{1}H and ^{13}C NMR spectra of **1** showed two amide NH protons (δ_H 7.47 and 8.39), two *N*-methyl protons (δ_H 2.83), and three characteristic α-methine signals (δ_H 4.50, 4.71, and 4.97). These features are characteristic of a tripeptide structure. Comprehensive analysis of the 1D NMR data of **1** (Tables 1 and 2) revealed that it is very similar to those of sclerotiotide F (**5**) [10]. The only difference between **1** and **5** was the presence of a carboxyl acid group ($\delta_{H/C}$ 12.5 brs/167.5) in **1** instead of an aldehyde group in **5**. The hexa-2,4-dienedioic acid side chain of **1** was further confirmed by the COSY correlations of H-2′/H-3′/H-4′/H-5′ and HMBC correlations from H-2′ to C-1′ and H-5′ to C-6′ (Figure 2). The geometric configurations of the two double bonds in the side chain are assigned as *E* on the basis of the large coupling constants ($J_{2'-3'}$ = 14.3 Hz, $J_{4'-5'}$ =14.4 Hz) and ROEs of H-2′/H-4′ and H-3′/H-5′. The absolute configurations of the α-carbons in the three amino acid units were determined by Marfey's method [18,19]. In detail, sclerotiotide M (**1**) was hydrolyzed into free amino acids, which were further derivatized with FDAA (1-fluoro-2-4-dinitrophenyl-5-L-alanine amide). HPLC analyses of FDAA derivatives of the hydrolysates and authentic samples revealed that the amino acid residues in **1** were L-*N*Me-Val, L-*N*Me-Ala, and L-Orn (Figure 3). Thus, sclerotiotide M (**1**) was established as (2′*E*,4′*E*)-*cyclo*-[(*N*Me-L-Ala) -(*N*Me-L-Val)-(N_α-5-carboxyhexa-2,4-dienoyl-L-Orn)].

Table 1. ^{1}H NMR (δ_H, J in Hz) spectroscopic data for compounds **1–3** in DMSO-d_6.

No.	1 [a]	2 [b]	3 [b]
2	4.50, q (7.1)	4.52, q (7.1)	4.65, q (6.4)
3	1.37, d (7.1)	1.39, d (7.1)	1.20, d (6.4)
N_{Ala}-CH$_3$	2.83, s	2.85, s	2.63, s
5	4.97, d (10.4)	4.99, d (10.3)	4.94, d (10.6)
6	2.21, m	2.22, m	2.28, m
7	0.63, d (6.7)	0.64, d (6.7)	0.74, d (6.7)
8	0.79, d (6.4)	0.80, d (6.3)	0.81, d (6.2)
N_{Val}-CH$_3$	2.83, s	2.85, s	2.91, s
10	4.71, m	4.73, m	5.33, t (5.5)
$N_{Orn\,(\alpha)}$H/ CH$_3$	8.39, d (7.6)	8.46, d (7.5)	3.08, s
11	1.61. m	1.71, m	1.70, m
	1.96, m	1.96, m	1.95, m
12	1.46, m	1.47, m	1.48, m
	1.67, m	1.63, m	1.66, m
13	2.86, m	2.89, m	2.97, m
	3.01, m	3.01, m	3.12, m
$N_{Orn\,(\omega)}$H/CH$_3$	7.47, t (6.0)	7.51, t (5.7)	2.83, s
2′	6.55, d (14.3)	6.60, d (14.6)	6.64, d (15.0)
3′	7.12, ov.	7.17, ov.	7.11, dd (15.0, 11.2)
4′	7.16, ov.	7.26, ov.	6.46, dd (15.5, 11.2)
5′	6.22, d (14.4)	6.36, d (14.8)	6.01, dd (15.5, 7.1)
6′			3.71, dd (7.1, 3.9)
7′		3.69, s	3.34, m
8′			1.02, d (6.4)
9′			3.23, s
10′			3.25, s
COOH	12.5, brs		

[a] Recorded at 500 MHz. [b] Recorded at 400 MHz. *ov.* Overlapped signal.

Table 2. ^{13}C NMR spectroscopic data for compounds **1–3** in DMSO-d_6.

No.	1 [a]	2 [b]	3 [c]
1	171.1	171.2	168.5
2	54.9	55.0	52.9
3	16.7	16.7	17.2
N_{Ala}-CH$_3$	30.1	30.1	28.7
4	169.5	169.6	168.2
5	58.0	58.1	57.6
6	26.8	26.9	26.3
7	18.1	18.2	18.1
8	20.2	20.3	19.9
N_{Val}-CH$_3$	30.2	30.3	29.7
9	171.9	172.0	171.8
10	50.1	50.2	52.7
$N_{Orn\,(\alpha)}$-CH$_3$			31.7
11	28.4	28.4	23.0
12	23.1	23.2	24.5
13	39.5	39.5	47.3
$N_{Orn\,(\omega)}$-CH$_3$			33.6
1′	163.8	163.8	166.6
2′	132.3	133.0	122.3
3′	141.8	142.5	141.7
4′	136.7	136.7	131.5
5′	128.2	126.8	138.8
6′	167.5	166.7	83.9
7′		52.1	78.9
8′			15.6
9′			57.1
10′			56.8

[a] Recorded at 125 MHz. [b] Recorded at 150 MHz. [c] Recorded at 100 MHz.

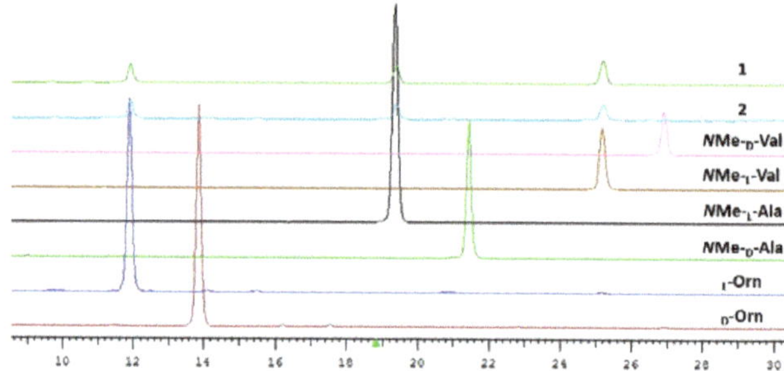

Figure 2. Key 2D NMR correlations of 1–3.

Figure 3. HPLC analysis of the FDAA derivatives of the compounds **1** and **2** and the standard amino acids.

Sclerotiotide N (**2**) was obtained as a pale yellow amorphous powder. The HRESIMS peak at *m/z* 451.2556 [M + H]⁺ indicated that its molecular formula was $C_{22}H_{34}N_4O_6$, which is 14 Da more than that of compound 1. The NMR data of **2** (Tables 1 and 2) were almost the same as those of **1** except for the additional signal of a methoxy group (δ_H 3.69, s). The methoxy group was determined to be linked to C-6′ by the HMBC correlation between H-7′ and C-6′ (Figure 2). The geometric configurations of the two double bonds in the side chain are both assigned as *E* on the basis of the coupling constants ($J_{2'-3'}$ = 14.6 Hz, $J_{4'-5'}$ =14.8 Hz) and ROEs of H-2′/H-4′ and H-3′/H-5′. Marfey's analysis was used to determine the absolute configuration of the amino acids present in the cyclic tripeptide [18]. The absolute configurations of the amino acid units of 2 were determined to be identical to 1. Acid hydrolysis and FDAA derivatization revealed ʟ-*N*Me-Ala, ʟ-*N*Me-Val, and ʟ-Orn by HPLC analysis (Figure 3). Thus, sclerotiotide N (**2**) was established as (2′*E*,4′*E*)-*cyclo*-[(*N*Me-ʟ-Ala)-(*N*Me-ʟ-Val)-(N_α-6-methoxy-6-oxohexa-2,4-dienoyl-ʟ-Orn)].

Sclerotiotide L (**4**) and sclerotiotide O (**3**) were isolated as pale yellow powders with the molecular formulas of $C_{24}H_{40}N_4O_6$ and $C_{27}H_{46}N_4O_6$, respectively, according to the analysis of HRESIMS data. The NMR data of **4** were identical to that of sclerotiotide L [12], indicating that they share the same planar structure (Table S1). Sclerotiotide L was first reported in 2018, while the stereochemistry of C-6′ and C-7′ remain unknown. Here, the absolute configuration of them was first determined using coupling constants analysis and Mosher's method. The small coupling constant ($^3J_{H-6', H-7'}$ = 4.9 Hz) between H-6′ and H-7′ indicates they are in a *gauche* conformation, which allowed focusing on two (**4a** and **4e**) of the six possible relative conformations (Figure 4). The relative configuration was further determined to be 6′*R** and 7′*R** by the ROESY correlations of H-6′/H-8′/H-5′ (Figure 4, **4a**). The absolute configuration of C-7′ was determined by Mosher's method [20]. Accordingly, compound **4** was derivatized into the esters **4g** and **4h** with (*R*)- and (*S*)-MPA (*α*-methoxyphenylacetic acid), respectively. The chemical shifts differences $\Delta\delta^{RS}$ suggested *R* configuration at C-7′ (Figure 5). Thus, compound **4** was established as (2′*E*,4′*E*)-*cyclo*-[(*N*Me-ʟ-Ala)-(*N*Me-ʟ-Val)-(N_α-(6*R*,7*R*)-7-hydroxy-6-methoxyocta-2,4-dienoyl-ʟ-Orn)]. Distinguished from **4**, compound **3** possessed three extra methyls (δ_H 2.83, s; 3.08, s; 3.25, s), which were assigned at $N_{Orn\,(\alpha)}$, $N_{Orn\,(\omega)}$, and 7′-OH on the basis of HMBC correlations from $N_{Orn\,(\alpha)}$-CH$_3$ to C-1 and C-13, from $N_{Orn\,(\omega)}$-CH$_3$ to C-9 and C-11, and from H-10′ to C-7′, respectively (Figure 2). The geometric configurations of the two double bonds in the side chain are both assigned as *E* on the basis of the coupling constants ($J_{2'-3'}$ = 15.0 Hz, $J_{4'-5'}$ =15.5 Hz) and ROEs of H-2′/H-4′ and H-3′/H-5′. Finally, the absolute configurations of C-6′ and C-7′ in **3** were assigned to be the same as **4** by the semisynthesis of **3** from **4**. Compound **4** was treated with sodium hydride in tetrahydrofuran to obtain **4A**. The identical NMR chemical shifts, ECD curves, and specific rotation values between **4A** and **3** indicated that **3** displayed the same stereochemistry with **4** (Figure S40). Thus, compound **3** was established as (2′*E*,4′*E*)-*cyclo*-[(*N*Me-L-Ala)-(*N*Me-L-Val)-(N_αMe-(6*R*,7*R*)-6,7-dimethoxyocta-2,4-dienoyl-L-Orn)].

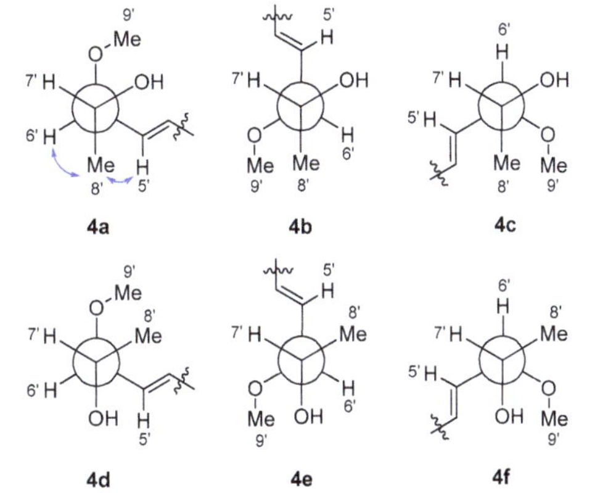

Figure 4. Newman projections for C-6′/C-7′ of **4**. All possible relative conformations are shown: 6′*R**,7′*R** (**4a–4c**) and 6′*S**, 7′*R** (**4d–4f**). Observed ROESY correlations are presented as arrowed line.

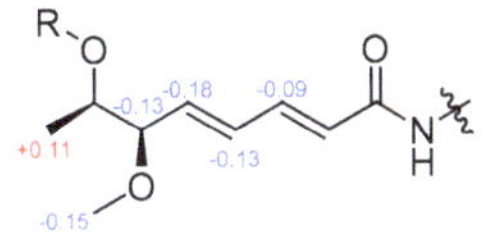

4g: R = *R*-MPA
4h: R = *S*-MPA

Figure 5. $\Delta\delta^{RS}$ values of **4g** and **4h**.

It is well-known that some artificial compounds are formed from natural compounds due to oxidation when exposed to air [10]. In order to verify the origin of compounds **1–6**, the fermentation broth of *A. insulicola* HDN151418 was dried under a freeze dryer and extracted by MeCN and further analyzed by LC-MS. Only compounds **1**, **2**, **5**, and **6** were detected. After compound **6** was dissolved in solvent mixture MeOH-H$_2$O and exposed to air for two weeks, compounds **4** and **5** were also detected from the product, indicating that compounds **4** and **5** could be formed from **6** during fermentation or isolation steps. The methoxy groups in **3** and **4** could come from methanol during isolation steps.

Compounds **1–6** were tested for their cytotoxic activities on 16 cancer cell lines (K562, BEL-7402, HCT-116, A549, Hela, L-02, GES-1, U87, ASPC-1, SH-SY5Y, PC-3, MGC-803, HO8910, MCF-7, MDA-MB-231, and NCI-H446); none of them showed activity (IC$_{50}$ > 30 μM). The antimicrobial activities were tested against eight pathogenic strains, including *Bacillus cereus*, *Proteus species*, *Mycobacterium phlei*, *Bacillus subtilis*, *Vibrio parahemolyticus*, *Edwardsiella tarda*, MRCNS, and MRSA. Compounds **1** and **2** showed broad inhibition against a panel of strains with MIC values ranging from 1.56 to 25.0 μM, while compound **3–6** were less active (Table 3), which indicated that the carboxyl group or its methyl ester display an important role for antibacterial activities. Notably, **1** and **2** showed potent activity against *M. phlei*, which provide potential candidates for antitubercular drug development. Additionally, no cytotoxicities further expands their pharmacological potential.

Table 3. Antimicrobial assays of compounds **1–5** (MIC μM).

No.	B. cereus	P. species	M. phlei	E. tarda	B. subtilis	MRCNS	MRSA	V. parahemolyticus
1	3.13	3.13	3.13	1.56	6.25	12.5	25.0	3.13
2	6.25	6.25	12.5	1.56	12.5	25.0	25.0	6.25
3	>50.0	>50.0	>50.0	25.0	>50.0	>50.0	>50.0	25.0
4	25.0	25.0	>50.0	25.0	>50.0	>50.0	>50.0	25.0
5	25.0	25.0	>50.0	25.0	>50.0	>50.0	>50.0	25.0
6	>50.0	>50.0	>50.0	>50.0	>50.0	>50.0	>50.0	>50.0
CIP [a]	0.780	0.195	0.780	0.0125	0.195	25.0	25.0	0.390

[a] Ciprofloxacin was used as positive drug.

3. Materials and Methods

3.1. General Experimental Procedures

By means of a JASCO P-1020 digital polarimeter developed by JASCO Corporation, Tokyo, Japan, optical rotations for all new compounds were calculated in methanol. Nuclear magnetic resonance data were obtained on a Bruker AVANCE NEO 400 MHz spectrometer made by Bruker Corporation, Karlsruhe, Germany, and an Agilent 500 MHz DD2 spectrometer by Agilent Technologies Inc., Santa Clara, CA, USA and a JEOL JNM-ECP600 spectrometer by JEOL, Tokyo, Japan using TMS as an internal standard. The ECD spectrum was measured on a JASCO J-815 spectropolarimeter made by JASCO Corporation, Tokyo, Japan. By using KBr discs in the Bruker Tensor-27 spectrophotometer made by Bruker Corporation, Karlsruhe, Germany, IR data were collected. In addition, HRESIMS data were recorded on a LTQ Orbitrap XL mass spectrometer made by Thermo Fisher Scientific, Waltham, MA, USA. UV spectra were carried out on Waters 2487 developed by Waters Corporation, Milford, MA, USA. Column chromatography was performed using the following chromatographic substrates: silica gel (300–400 mesh; Qingdao Marine Chemical Industrials, Qingdao, China), Sephadex LH-20 (developed by Amersham Biosciences, San Francisco, CA, USA). The compounds were purified by HPLC made by the Waters company equipped with a 2998 PDA detector and a C18 column (YMC-Pack ODS-A, 10 × 250 mm, 5 μm, 3 mL/min). LC-MS was recorded in ESI mode on an Acquity UPLC H-Class connected to a SQ Detector 2 mass spectrometer using a BEH C18 column (1.7 μm, 2.1 × 50 mm, 1 mLperminute) constructed by Waters Corporation, Milford, CT, USA.

3.2. Fungal Material and Fermentation

Aspergillus insulicola HDN151418 was isolated from an unidentified sponge sample collected 410 m deep from Prydz Bay, Antarctica at a latitude and longitude of E 68.7°, S 67.2° while identified as *Aspergillus insulicola* based on internal transcribed spacer DNA sequencing. The sequence is available with the accession number MT898544 at Genbank and has been submitted to the Key Laboratory of Marine Drugs working under the Ministry of Education of China, School of Medicine and Pharmacy, Ocean University of China.

To prepare the seed culture, the strain was cultured on potato dextrose agar (PDA) at 28 °C for 7 days and then was transferred to 30 mL potato dextrose broth (PDB) medium in a 100 mL flask. After fermentation for 3 days on a rotary shaker at 180 rpm at 28 °C, 1 mL aliquot of the liquid culture was transferred to 300 mL of PDB medium in a 1000 mL flask for scale-up. The culture was incubated in static condition for 30 days before extraction.

3.3. Isolation and Purification of the Compounds

The total fermentation broth (30 L) was harvested and the supernatant was separated from the mycelia by using a filter cloth. The solvent-associated extraction was performed, the supernatant was extracted with EtOAc (3 × 30 L), and the mycelia was crushed into small pieces by using an electric cutter and macerated with MeOH (3 × 15 L). Based on the corresponding HPLC and TLC profiles, both extracts were combined, and the subsequent removal of solvent afforded 32.3 g of reddish-brown crude extract. Moreover, the extract was fractioned by using vacuum chromatography on silica gel followed by stepped gradient elution via DCM-MeOH (10:0 to 0:10) solvent combination to obtain ten subfractions (Fr.1 to Fr.10). Then, Fr.3 was separated by an ODS column by using MeOH and H_2O in the form of a stepped gradient, 20:80 to 50:50 to obtain six subfractions (Fr.3-1 to Fr.3-6). Fr.3-3 was further subjected to a Sephadex LH-20 column and eluted with MeOH to provide four subfractions (Fr.3-3-1 to Fr.3-3-4). Fr.3-3-2 was purified by HPLC eluted with MeOH-H_2O (30:70) to obtain compounds **1** (10.2 mg, t_R = 15 min) and **5** (20.2 mg, t_R = 18 min). Likewise, Fr.5 was divided into three subfractions (Fr5-1 to Fr5-3) by MPLC using a stepped gradient elution of MeOH-H_2O (40:60 to 60:40). Fr.5-2 was separated by HPLC eluted with MeCN-H_2O (30:70) to obtain compounds **2** (5.7 mg, t_R = 23 min) and **4** (6.5 mg, t_R = 25 min). With the same procedure as used for fraction Fr.5, Fr.8 was purified by HPLC eluted with MeCN-H_2O (70:30) to yield compounds **3** (6.5 mg, t_R = 31 min) and **6** (12.0 mg, t_R = 33 min), respectively.

Sclerotiotide M (1): pale yellow, amorphous powder; $[\alpha]_D^{20}$ −66 (*c* 0.5, MeOH); UV (MeOH) λ_{max} (log ε) 206 (4.00), 270 (4.40) nm; IR ν_{max} 3394, 2937, 1681, 1527, 1205, 1135, 838, 584 cm^{-1}; ^{1}H and ^{13}C NMR data, Tables 1 and 2; HRESIMS *m/z* 435.2246 [M − H]$^-$ (calcd for $C_{21}H_{31}N_4O_6$ 435.2249).

Sclerotiotide N (2): pale yellow, amorphous powder; $[\alpha]_D^{20}$ −72 (*c* 0.2, MeOH); UV (MeOH) λ_{max} (log ε) 206 (4.05), 272 (4.35) nm; IR ν_{max} 3384, 2939, 1673, 1636, 1540, 1456, 1205, 1137, 1027, 839, 580 cm^{-1}; ^{1}H and ^{13}C NMR data, Tables 1 and 2; HRESIMS *m/z* 451.2556 [M + H]$^+$ (calcd for $C_{22}H_{35}N_4O_6$ 451.2551).

Sclerotiotide O (3): pale yellow, amorphous powder; $[\alpha]_D^{20}$ −59 (*c* 0.3, MeOH); UV (MeOH) λ_{max} (log ε) 206 (3.90), 258 (4.50) nm; IR ν_{max} 3336, 2931, 1679, 1534, 1453, 1206, 1138, 1027, 840, 801, 722 cm^{-1}; ^{1}H and ^{13}C NMR data, Tables 1 and 2; HRESIMS *m/z* 523.3484 [M + H]$^+$ (calcd for $C_{27}H_{47}N_4O_6$ 523.3490).

Sclerotiotide L (4): pale yellow, amorphous powder; $[\alpha]_D^{20}$ −52 (*c* 0.1, MeOH); UV (MeOH) λ_{max} (log ε) 206 (3.90), 258 (4.50) nm; IR ν_{max} 3384, 2972, 2265, 1684, 1450, 1204, 1139, 1029, 833, 800, 721 cm^{-1}; ^{1}H and ^{13}C NMR data, Tables 1 and 2; HRESIMS *m/z* 481.3017 [M + H]$^+$ (calcd for $C_{24}H_{41}N_4O_6$ 481.3021).

3.4. Absolute Configuration Assignments of Sclerotiotides M–N (1–2)

Compounds **1** and **2** (1.5 mg each) were reacted with 6 N HCl (1.5 mL) at 110 °C for 15 h. The solution was dried and separately dissolved in H_2O (100 µL). Then, 0.50 µM of FDAA (1-fluoro-2-4-dinitrophenyl-5-L-alanine amide) was added to 100 µL of acetone, and 1 N NaHCO$_3$ (50 µL) to form a mixture. The mixtures were heated for 2 h at 43 °C and then quenched by the addition of 2 N HCl (100 µL). Amino acid standards were derivatized with FDAA in a similar way. The resulting FDAA derivatives of compound **1**, compound **2**, L- and D-NMe-Ala, L- and D-NMe-Val, and L- and D-Orn were analyzed by HPLC eluted with a linear gradient of MeCN (A) and 0.10% aqueous TFA (B) from 50% to 100% in an over 30 min with UV detection at 320 nm. The measured retention times of amino acid standards are as follows (in min): 21.44 for D-NMe-Ala-FDAA, 19.36 for L-NMe-Ala-FDAA, 26.88 for D-NMe-Val-FDAA, 25.17 for L-NMe-Val-FDAA, 13.87 for D-Orn-FDAA, and 11.89 for L-Orn-FDAA. Retention times for the FDAA derivatives of **1** and **2** are as follows (in min): 25.17, 19.36, and 11.89, indicating all the amino acid residues in **1** and **2** are an *S* configuration [18,19].

3.5. Assay of Cytotoxicity Inhibitory Activity

Cytotoxicity of compounds **1–6** were screened against the 16 human cancer cell lines as previously reported in which Adriamycin was used as the positive control [9,21]. HeLa (human epithelial carcinoma cell line), HCT-116 (human colon cancer cell line), MCF-7 (human breast adenocarcinoma cell line), A549 (human lung adenocarcinoma cell line), SH-SY5Y (human neuroblastoma cell line), MDA-MB-231 (human breast cancer cell line), HO8910 (human ovarian carcinoma cell line), and MGC-803 (human stomach carcinoma cell line) were ordered from Shanghai Institute of Biochemistry and Cell Biology, Chinese Academy of Sciences (Shanghai, China). K562 (human leukaemia cell line), BEL-7402 (human hepatic carcinoma cell line), L-02 (human normal hepatic cell line), GES-1 (human gastric epithelial cell line), U87 (human primary glioblastoma cell line), ASPC-1 (human pancreatic cancer cell line), PC-3 (human prostate carcinoma cell line), and NCI-H446 (human small cell lung cancer cell line) were ordered from the American Type Culture Collection (ATCC, Gaithersburg, MD, USA).

3.6. Assay of Antimicrobial Activity

The antimicrobial activities of **1–6** against *Bacillus cereus*, *Proteus* species, *Mycobacterium phlei*, *Bacillus subtilis*, *Vibrio parahemolyticus*, *Edwardsiella tarda*, MRCNS, and MRSA were evaluated as previously reported by using the agar dilution method [22,23]. All experiments were performed in triplicates, and ciprofloxacin was used as a positive control. All strains were donated by the Qingdao municipal hospital.

3.7. Preparation of MPA Esters Derived from **4** (**4g** and **4h**)

The sample of compound **4** (0.5 mg each) was reacted with (*R*)- or (*S*)-MPA (5.0 mg) with *N*, *N*′-dicyclohexylcarbodiimide (DCC, 1.0 mg) and 4-dimethylaminopyridine (DMAP, 0.1 mg) in dry $CDCl_3$ (0.5 mL). After stirring for 2.0 h at 0 °C, the residue was evaporated under vacuum pressure and purified by HPLC eluted with 60% acetonitrile/H_2O to give the (*R*)-MPA ester (**4g**) and (*S*)-MPA ester (**4h**).

(*R*)-MPA Ester (**4g**): pale yellow powder; ^{1}H NMR (400 MHz, DMSO-d_6) δ_H 8.23 (d, *J* = 7.7 Hz, 1H), 7.48 (t, *J* = 6.7 Hz, 1H), 7.28-7.38 (m, 5H), 6.91 (dd, *J* = 15.1, 11.1 Hz, 1H), 6.24 (dd, *J* = 15.3, 11.2 Hz, 1H), 6.18 (d, *J* = 14.9 Hz, 1H), 5.71 (dd, *J* = 15.3, 7.5 Hz, 1H), 4.98 (d, *J* = 10.1 Hz, 1H), 4.88 (s, 1H), 4.74 (m, 1H), 4.51 (m, 1H), 4.11 (m, 1H), 3.65 (m, 1H), 3.31 (s, 3H), 3.06 (m, 1H), 3.03 (m, 1H), 2.88 (m, 1H), 2.85 (s, 3H), 2.85 (s, 3H), 2.18 (m, 1H), 1.98 (m, 2H), 1.63 (m, 2H), 1.49 (m, 1H), 1.39 (d, *J* = 7.1 Hz, 3H), 1.13 (d, *J* = 6.5 Hz, 3H), 0.81 (d, *J* = 6.4 Hz, 3H), 0.65 (d, *J* = 6.7 Hz, 3H). HRESIMS *m/z* 627.3387 [M − H]$^-$ (calcd for $C_{33}H_{49}N_4O_8$, 627.3399).

(*S*)-MPA Ester (**4h**): pale yellow powder; ^{1}H NMR (400 MHz, DMSO-d_6) δ_H 8.23 (d, *J* = 7.7 Hz, 1H), 7.50 (t, *J* = 6.7 Hz, 1H), 7.30-7.40 (m, 5H), 7.00 (dd, *J* = 14.9, 11.2 Hz, 1H), 6.37 (dd, *J* = 15.3, 11.2 Hz, 1H), 6.24 (d, *J* = 14.9 Hz, 1H), 5.89 (dd, *J* = 15.3, 7.5 Hz, 1H), 4.99 (d, *J* = 10.1 Hz, 1H), 4.86 (s, 1H), 4.74 (m, 1H), 4.51 (m, 1H), 4.11 (m, 1H), 3.78 (m, 1H), 3.29 (s, 3H), 3.21 (s, 3H), 3.03 (m, 1H), 2.88 (m, 1H), 2.85 (s, 3H), 2.85 (s, 3H), 2.21 (m, 1H), 1.98 (m, 2H), 1.63 (m, 2H), 1.48 (m, 1H), 1.39 (d, *J* = 7.1 Hz, 3H), 1.02 (d, *J* = 6.5 Hz, 3H), 0.81 (d, *J* = 6.4 Hz, 3H), 0.65 (d, *J* = 6.7 Hz, 3H); HRESIMS *m/z* 627.3398 [M − H]$^-$ (calcd for $C_{33}H_{49}N_4O_8$, 627.3399).

3.8. Chemical Transformation of **4**

MeI (100 µL) was mixed to 1 mL of THF solution of compound **4** (2.0 mg). Then, NaH (0.5 mg) was added, and the mixture was stirred at room temperature for 2 h. After that, the reaction was quenched with aqueous HCl. The mixture was treated with EtOAc for three times. The EtOAc part was dried under vacuum pressure and subjected to HPLC having an ODS column with 60% acetonitrile/H_2O to afford compound **4A** (2.8 mg, t_R = 13 min).

4. Conclusions

In summary, chemical investigation of the Antarctica sponge-derived fungus *Aspergillus insulicola* HDN151418 led to the isolation of three new aspochracin-type cyclic tripeptides, sclerotiotides M–O (**1–3**), along with the biogenetically related analogues, sclerotiotide L (**4**), sclerotiotide F (**5**), and sclerotiotide B (**6**). Among which, sclerotiotides M (**1**) and sclerotiotides N (**2**) represent the first example of aspochracin-type cyclic tripeptide, which was substituted by hexa-2,4-dienedioic acid/methyl ester moieties. Chemical derivatization indicated that compounds **4** and **5** could form from **6** during the fermentation or isolation steps. The antimicrobial activities of all the isolates were evaluated, and its structure–activity relationship (SAR) was also preliminary discussed. Our research results further expanded the members of the aspochracin-type cyclic tripeptide family, which again demonstrated that sponge-derived fungi are important producers of structurally diverse bioactive compounds.

Supplementary Materials: The following are available online at http://www.mdpi.com/1660-3397/18/11/532/s1: Figure S1: Structures of aspochracin-type cyclic tripeptides, Figure S2: HPLC analysis of the crude of *Aspergillus insulicola* HDN151418; Figure S3: The 18S rRNA sequences data of *Aspergillus insulicola* HDN151418; Figures S4–S38: 1D and 2D NMR spectra, HRESIMS spectra, IR spectra of compounds 1–4; Table S1. ^{1}H NMR (400 MHz) spectroscopic data for compound 4. Figures S39–S40: ^{1}H NMR spectra of *R*- and *S*- MPA esters of 4. Figure S41. ECD spectra of 3, 4, 4A. Figure S42. HSQMBC spectrum of 3. Table S1: ^{1}H NMR parameters of 1–4. Table S2: ^{13}C NMR parameters of 1–4.

Author Contributions: The contributions of the respective authors are as follows: C.S. drafted the work and performed isolation and structural elucidation of the extract. Z.Z., Z.R., L.Y., H.Z., and M.S. performed isolation and scale-up fermentation of the strain. Biological evaluations was performed by Y.H., G.Z., Q.C., T.Z. and D.L. checked the whole procedures of this work. D.L. and T.Z. designed the project and contributed to the critical reading of the manuscript. All authors have read and agreed to the published version of the manuscript.

Funding: This work was funded by the National Natural Science Foundation of China (41876216), National Key R&D Program of China (grants 2018YFC1406705), the National Science and Technology Major Project for Significant New Drugs Development (2018ZX09735004), Major national science and technology projects of the Ministry of science and technology (81991522), Qingdao Pilot National Laboratory for Marine Science and Technology (2018SDKJ0401-2), the Fundamental Research Funds for the Central Universities (201941001), the Taishan Scholar Youth Expert Program in Shandong Province (tsqn201812021) and the Youth Innovation Plan of Shandong province (2019KJM004).

Conflicts of Interest: The authors declare no conflict of interest.

References

1. Carroll, A.R.; Copp, B.R.; Davis, R.A.; Keyzers, R.A.; Prinsep, M.R. Marine natural products. *Nat. Prod. Rep.* **2020**, *37*, 175–223. [CrossRef] [PubMed]

2. Carroll, A.R.; Copp, B.R.; Davis, R.A.; Keyzers, R.A.; Prinsep, M.R. Marine natural products. *Nat. Prod. Rep.* **2019**, *36*, 122–173. [CrossRef] [PubMed]

3. Skropeta, D.; Wei, L. Recent advances in deep-sea natural products. *Nat. Prod. Rep.* **2014**, *31*, 999–1025. [CrossRef] [PubMed]

4. Rateb, M.E.; Ebel, R. Secondary metabolites of fungi from marine habitats. *Nat. Prod. Rep.* **2011**, *28*, 290–344. [CrossRef] [PubMed]

5. Velle, A.; Cebollada, A.; Macias, R.; Iglesias, M.; Gil-Moles, M.; Sanz Miguel, P.J. From imidazole toward imidazolium salts and *N*-heterocyclic carbene ligands: Electronic and geometrical redistribution. *ACS Omega* **2017**, *2*, 1392–1399. [CrossRef] [PubMed]

6. Wu, Y.; Liao, H.; Liu, L.Y.; Sun, F.; Chen, H.F.; Jiao, W.H.; Zhu, H.R.; Yang, F.; Huang, G.; Zeng, D.Q.; et al. Phakefustatins A-C: Kynurenine-bearing cycloheptapeptides as RXRalpha modulators from the marine sponge *Phakellia Fusca*. *Org. Lett.* **2020**, *22*, 6703–6708. [CrossRef]

7. Xu, W.J.; Liao, X.J.; Xu, S.H.; Diao, J.Z.; Pan, S.S. Isolation, structure determination, and synthesis of galaxamide, a rare cytotoxic cyclic pentapeptide from a marine algae *Galaxaura filamentosa*. *Org. Lett.* **2010**, *40*, 4569–4572. [CrossRef]

8. Teta, R.; Marteinsson, V.T.; Longeon, A.; Klonowski, A.M.; Groben, R.; Bourguet-Kondracki, M.L.; Costantino, V.; Mangoni, A. Thermoactinoamide A, an antibiotic lipophilic cyclopeptide from the icelandic thermophilic bacterium *Thermoactinomyces vulgaris*. *J. Nat. Prod.* **2017**, *80*, 2530–2535. [CrossRef]

9. Skehan, P.; Storeng, R.; Scudiero, D.; Monks, A.; McMahon, J.; Vistica, D.; Warren, J.T.; Bokesch, H.; Kenney, S.; Boyd, M.R. New colorimetric cytotoxicity assay for anticancer-drug screening. *J. Natl. Cancer Inst.* **1990**, *82*, 1107–1112. [CrossRef]

10. Zheng, J.; Xu, Z.; Wang, Y.; Hong, K.; Liu, P.; Zhu, W. Cyclic tripeptides from the halotolerant fungus *Aspergillus sclerotiorum* PT06-1. *J. Nat. Prod.* **2010**, *73*, 1133–1137. [CrossRef]

11. Motohashi, K.; Inaba, S.; Takagi, M.; Shin-ya, K. JBIR-15, a new aspochracin derivative, isolated from a sponge-derived fungus, *Aspergillus sclerotiorum Huber* Sp080903f04. *Biosci. Biotechnol. Biochem.* **2009**, *73*, 1898–1900. [CrossRef]

12. Liu, J.; Gu, B.; Yang, L.; Yang, F.; Lin, H. New Anti-inflammatory cyclopeptides from a sponge-derived fungus *Aspergillus violaceofuscus*. *Front. Chem.* **2018**, *6*, 226–233. [CrossRef] [PubMed]

13. Myokei, R.; Sakurai, A.; Chang, C.F.; Kodaira, Y.; Tamura, S. Structure of aspochracin, an insecticidal metabolite of Aspergillus ochraceus. *Tetrahedron Lett.* **1969**, *10*, 695–698. [CrossRef]

14. Zhou, H.; Li, L.; Wu, C.; Kurtan, T.; Mandi, A.; Liu, Y.; Gu, Q.; Zhu, T.; Guo, P.; Li, D. Penipyridones A-F, pyridone alkaloids from *Penicillium funiculosum*. *J. Nat. Prod.* **2016**, *79*, 1783–1790. [CrossRef]

15. Zhou, H.; Li, L.; Wang, W.; Che, Q.; Li, D.; Gu, Q.; Zhu, T. Chrodrimanins I and J from the antarctic moss-derived fungus *Penicillium funiculosum* GWT2-24. *J. Nat. Prod.* **2015**, *78*, 1442–1445. [CrossRef]

16. Shah, M.; Sun, C.; Sun, Z.; Zhang, G.; Che, Q.; Gu, Q.; Zhu, T.; Li, D. Antibacterial polyketides from antarctica sponge-derived fungus *Penicillium* sp. HDN151272. *Mar. Drugs* **2020**, *18*, 71. [CrossRef] [PubMed]

17. Wu, G.; Lin, A.; Gu, Q.; Zhu, T.; Li, D. Four new chloro-eremophilane sesquiterpenes from an Antarctic deep-sea derived fungus, *Penicillium* sp. PR19N-1. *Mar. Drugs* **2013**, *11*, 1399–1408. [CrossRef]

18. Marfey, P. Determination of D-amino acids. II. Use of a bifunctional reagent, 1,5-difluoro-2,4-dinitrobenzene. *Carlsberg Res. Commun.* **1984**, *49*, 591. [CrossRef]

19. Sun, C.; Ge, X.; Mudassir, S.; Zhou, L.; Yu, G.; Che, Q.; Zhang, G.; Peng, J.; Gu, Q.; Zhu, T.; et al. New glutamine-containing azaphilone alkaloids from deep-sea-derived fungus *Chaetomium globosum* HDN151398. *Mar. Drugs* **2019**, *17*, 253. [CrossRef]

20. Latypov, S.K.; Seco, J.M.; Quinoa, E.; Riguera, R. MTPA vs MPA in the determination of the absolute configuration of chiral alcohols by ^{1}H NMR. *J. Org. Chem.* **1996**, *61*, 8569–8577. [CrossRef]

21. Mosmann, T. Rapid colorimetric assay for cellular growth and survival: Application to proliferation and cytotoxicity assays. *J. Immunol. Methods* **1983**, *65*, 55–63. [CrossRef]

22. Andrews, J.M. Determination of minimum inhibitory concentrations. *J. Antimicrob. Chemother.* **2001**, *48*, 5–16. [CrossRef]

23. Yu, G.; Sun, Z.; Peng, J.; Zhu, M.; Che, Q.; Zhang, G.; Zhu, T.; Gu, Q.; Li, D. Secondary metabolites produced by combined culture of *Penicillium crustosum* and a *Xylaria* sp. *J. Nat. Prod.* **2019**, *82*, 2013–2017. [CrossRef]

Publisher's Note: MDPI stays neutral with regard to jurisdictional claims in published maps and institutional affiliations.

 *marine drugs*

Article

The Structure of the Lipid A of Gram-Negative Cold-Adapted Bacteria Isolated from Antarctic Environments

Flaviana Di Lorenzo [1,*], Francesca Crisafi [2], Violetta La Cono [2], Michail M. Yakimov [2], Antonio Molinaro [1] and Alba Silipo [1,*]

[1] Department of Chemical Sciences, University of Napoli Federico II, Complesso Universitario Monte S. Angelo, Via Cintia 4, I-80126 Napoli, Italy; molinaro@unina.it

[2] Marine Molecular Microbiology & Biotechnology Institute for Biological Resources and Marine Biotechnologies, CNR-IRBIM Sede di Messina, Spianata San Raineri 86, 98122 Messina, Italy; francesca.crisafi@irbim.cnr.it (F.C.); violetta.lacono@irbim.cnr.it (V.L.C.); mikhail.iakimov@cnr.it (M.M.Y.)

* Correspondence: flaviana.dilorenzo@unina.it (F.D.L.); silipo@unina.it (A.S.)

Received: 19 October 2020; Accepted: 25 November 2020; Published: 26 November 2020

Abstract: Gram-negative Antarctic bacteria adopt survival strategies to live and proliferate in an extremely cold environment. Unusual chemical modifications of the lipopolysaccharide (LPS) and the main component of their outer membrane are among the tricks adopted to allow the maintenance of an optimum membrane fluidity even at particularly low temperatures. In particular, the LPS' glycolipid moiety, the lipid A, typically undergoes several structural modifications comprising desaturation of the acyl chains, reduction in their length and increase in their branching. The investigation of the structure of the lipid A from cold-adapted bacteria is, therefore, crucial to understand the mechanisms underlying the cold adaptation phenomenon. Here we describe the structural elucidation of the highly heterogenous lipid A from three psychrophiles isolated from Terra Nova Bay, Antarctica. All the lipid A structures have been determined by merging data that was attained from the compositional analysis with information from a matrix-assisted laser desorption ionization (MALDI) time of flight (TOF) mass spectrometry (MS) and MS^2 investigation. As lipid A is also involved in a structure-dependent elicitation of innate immune response in mammals, the structural characterization of lipid A from such extremophile bacteria is also of great interest from the perspective of drug synthesis and development inspired by natural sources.

Keywords: psychrophiles; Antarctic bacteria; Lipopolysaccharide (LPS); lipid A; structural characterization; MALDI-TOF mass spectrometry

1. Introduction

Psychrophiles (or cold-adapted bacteria) are microorganisms able to thrive in permanently cold environments; for others, referred to as psychrotolerants, a larger range of growth temperature is tolerated [1]. In this context, Antarctica, the coldest and most inaccessible continent on the Earth, beside the low temperatures, is characterized by several other extremes including scarcity of nutrients, high or low pH, desiccation and osmotic stress, high levels of UVB radiation and a remarkably variable photoperiod, i.e., from no light to continuous light for 24 h a day [2]. Despite the hostile conditions, prokaryotes are the predominant biomass component in most Antarctic ecosystems including lakes, rivers, ponds, streams, rocks, and soils [3]. This necessarily implies that the ability of psychrophilic prokaryotes to survive and proliferate in Antarctica relies on a number of adaptive strategies aimed at maintaining vital cellular functions even at such prohibitive conditions.

Proteobacteria, a major phylum of Gram-negative bacteria, have been frequently found in Antarctic environments, and represent, with the Actinobacteria, the most abundant phylum isolated from

Antarctic soils [4]. Lipopolysaccharides (LPSs), exposed on the external leaflet of the Gram-negative outer membrane, are amphiphilic macromolecules indispensable for viability and survival, as they provide structural stabilization and protection to the whole bacterial envelope, in a dynamic interplay with the external environment [5]. Indeed, under adverse conditions, bacteria can colonize a hostile habitat by modifying their LPS primary structure, in order to reinforce the cell envelope to provide further protection and facilitate adaptation [6]. Consequently, several uncommon structural features have been observed in the LPS of bacteria inhabiting extreme environments, as in the case of cold-adapted bacteria.

LPSs display a tripartite structural architecture comprising: (i) a highly variable polysaccharide (the O-chain or O-antigen) covalently linked to (ii) an oligosaccharide moiety (the core OS), in turn linked to (iii) the glycolipid part (the lipid A) embedded in the outer leaflet of the outer membrane [5]. With the above three moieties, the LPS is defined as smooth-type LPS (or S-LPS), when lacking the O-chain is designated as rough-type LPS (or R-LPS) [5]. Importantly, LPS is widely known to interact with the mammalian innate immune system, with the lipid A moiety specifically recognized by the host innate immunity receptorial complex TLR4/MD-2 [7]. As a consequence of this interaction and depending on the lipid A fine structure, LPS differently activates the production of host pro-inflammatory cytokines, some causing excessive activation of the TLR4/MD-2 signaling, others exhibiting a weak or no immunopotency. This event can be beneficial to the host, enhancing resistance to infecting microbes, however massive and uncontrolled pro-inflammatory cytokines release can eventually lead to septic shock and multi-organ failure [7]. In this scenario, the search for novel lipid A structures which might possess modulatory activity towards TLR4/MD-2 dependent signaling cascade is considered of high relevance. Herein, LPSs that express uncommon structural features, as the case of psychrophilic bacteria, might act as potential immunomodulators of the TLR4/MD-2 complex.

From these significant structural and functional perspectives, we here report about the structural characterization of the lipid A from three different psychrophilic bacteria isolated from Terra Nova Bay, Antartica: *Pseudoalteromonas tetraodonis* strain SY174; *Psychromonas arctica* strain SY204b, and *Psychrobacter cryohalolentis* strain SY185. *Pseudoalteromonas tetraodonis* strain SY174 was isolated from platelet ice; *Psychromonas arctica* strain SY204b and *Psychrobacter cryohalolentis* strain SY185 were isolated from marine invertebrates belonged to the genus Holothuria and to the class Hydrozoa, respectively. The three strains were selected because of their abilities to reproduce and rapidly grow at subzero temperatures. Specifically, a generation time of less than 30 days was registered for all these strains at −2 °C, i.e., a temperature comparable to that of Antarctic seawater (field emission scanning electron microscopy micrographs of *P. tetraodonis* SY74 and *P. cryohalolentis* SY185 cultures growing at 0.5 °C are reported in Figure S1).

All the lipid A structures have been determined by merging information that was attained from the compositional analysis executed on pure LPS with information from a matrix-assisted laser desorption ionization (MALDI) time of flight (TOF) mass spectrometry (MS) investigation executed on the isolated lipid A fractions and directly on bacterial pellets. Finally, an in-depth MS^2 analysis has been conducted in order to detailly establish the location of the acyl chains with respect to the glucosamine disaccharide backbone of each isolated lipid A fraction.

2. Results

2.1. Isolation of the LPS and Compositional Analysis of the Lipid A from Cold-Adapted Bacteria

LPS material was extracted from dried bacterial cells and checked via SDS-PAGE after silver nitrate gel staining. This analysis highlighted the smooth-type nature of the LPS from *P. tetraodonis* as proven by the ladder-like pattern in the upper part of the gel, which is diagnostic for the occurrence of high molecular weight species; while a run to the bottom of the gel, typical of a low molecular mass R-type LPS, i.e., an LPS devoid of the O-chain moiety, was observed for *P. cryohalolentis* and *P. arctica*. After their purification, a detailed compositional analysis was performed to establish the fatty acid content (results are summarized in Table 1). This information was key in supporting the following elucidation of the lipid A structures by MALDI-TOF MS and MS^2 approaches.

Table 1. Fatty acid content of the lipopolysaccharide (LPS) isolated from the three cold-adapted bacteria examined in the current study. "+" and "-" indicate the presence and absence of the fatty acid in the lipid A, respectively. All the strains are characterized by a disaccharide of D-glucosamine as the lipid A sugar backbone. For the unsaturated acyl chains the position of the double bond or the stereochemistry remain to be defined.

Fatty Acid Component	*P. arctica* Strain SY204b	*P. cryohalolentis* Strain SY185	*P. tetraodonis* Strain SY174
3-hydroxylated fatty acids			
10:0 (3-OH)	-	-	+
11:0 (3-OH)	-	+	+
12:0 (3-OH)	-	+	+
13:0 (3-OH)	-	+	+
14:0 (3-OH)	+	+	-
non-hydroxylated fatty acids			
10:0	-	+	+
11:0	-	-	+
12:0	+	+	+
13:0	-	+	+
14:0	+	+	-
15:0	-	+	-
non-hydroxylated unsaturated fatty acids			
12:1	+	+	+
13:1	-	-	+
14:1	+	-	-

In order to investigate the structure of the lipid A portion, an aliquot of each LPS underwent a mild acid hydrolysis, typically performed to selectively cleave the acid labile glycosidic linkage between the core OS and the lipid A moiety. Once the lipid A of each strain was obtained, an aliquot was analyzed to define the nature of the lipid A sugar backbone (i.e., a D-glucosamine disaccharide for all the strains), whereas another aliquot underwent a detailed MALDI-TOF MS and MS^2 investigation to finally establish the fine structure of the lipid As. Moreover, an aliquot of a lyophilized cell pellet of each psychrophilic strain underwent a direct MALDI-TOF MS analysis (Figures S2–S4). This approach was essential to confirm the structures deduced by analyzing the isolated lipid A fractions, and to exclude any loss of structural data possibly occurring as a consequence of the chemical treatment used to isolated the lipid A (i.e., the mild acid hydrolysis of the LPS).

2.2. MALDI-TOF MS and MS^2 Analysis on the Isolated Lipid A from P. arctica Strain SY204b

The negative-ion reflectron MALDI-TOF MS spectrum of the lipid A from *P. arctica* is reported in Figure 1. The main ion peaks and the proposed interpretations of the fatty acids composing the lipid A are reported in Table 2. The spectrum three clusters of signals corresponding to mono- and bis-phosphorylated tetra- to hexa-acylated lipid A species showed in the range of *m/z* 1303.8–1822.1, highlighting a heterogenous lipid A blend. Indeed, besides the occurrence of minor peaks differing for 28 amu (i.e., a -CH$_2$CH$_2$- unit) representing variation in the acyl chains length, peaks differing for 2 amu were also identified, likely indicating the presence of lipid A species also bearing unsaturated acyl moieties, as also shown by compositional analysis (Table 1). Briefly, the main peak at *m/z* 1794.1 matched with a bis-phosphorylated lipid A species carrying 14:0 (3-OH) as primary fatty acids, and 12:0 and 14:1 as secondary acyl substituents, the corresponding mono-phosphorylated lipid A species was at *m/z* 1714.1. Similarly, mono- and bis-phosphorylated penta-acylated lipid A species lacking one primary 14:0 (3-OH) matched with peaks at *m/z* 1488.0 and 1568.0, respectively (Figure 1, Table 2), whereas a penta-acylated form devoid of the 12:0 acyl chain was detected at *m/z* 1612.0. The bis-phosphorylated tetra-acylated lipid A species, devoid of one primary 14:0 (3-OH) and the secondary 12:0, was assigned to peak at *m/z* 1385.9; the related mono-phosphorylated form was assigned to peak at *m/z* 1305.9. Interestingly, the negative-ion MALDI-TOF MS spectrum recorded on the intact *P. arctica* strain SY204b cell pellet (Figure S2) was similar to the above spectrum recorded on the mild acid hydrolysis product (Figure 1); this further confirmed our structural hypothesis and ruled out any loss of structural information.

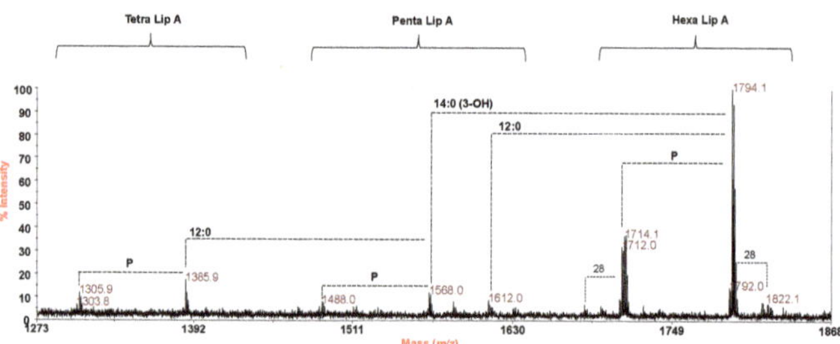

Figure 1. Reflectron MALDI-TOF mass spectrum, recorded in negative polarity, of lipid A from *P. arctica* strain SY204b obtained after acetate buffer treatment. The lipid A species are labelled as Tetra, Penta and Hexa Lip A indicating the degree of acylation. "**P**" indicates the phosphate group.

A detailed negative-ion MALDI-TOF MS^2 investigation on various peaks has been performed to define the location of the secondary acyl substituents. The MS^2 spectrum of the precursor ion at *m/z* 1794.1 (Figure 2a), corresponding to a bis-phosphorylated lipid A species carrying four 14:0 (3-OH), one 12:0 and one 14:1, showed an intense peak at *m/z* 1550.01 which was attributed to an ion originating from the loss of a 14:0 (3-OH) unit. Less intense peaks were detected at *m/z* 1594.01 and *m/z* 1696.01 and matched with ions originating from the loss of the 12:0 acyl chain (*m/z* 1594.01) and one phosphate unit (*m/z* 1696.01), respectively. An ion matching with the sequential loss of one primary 14:0 (3-OH) and the secondary 12:0 unit was also detected at *m/z* 1350.00, with the related fragment devoid also of one phosphate unit ascribed to peak at *m/z* 1251.91. Importantly, the peak at *m/z* 1025.79, matching with a fragment originated from the loss of one phosphate, one 14:0 (3-OH) and a whole unit of a 14:0 (3-OH) fatty acid bearing the secondary acyl substituent 12:0, suggested that the 12:0 was linked to a primary ester-bound acyl moiety. However, to strengthen the structural hypothesis on the exact location and nature of the secondary fatty acids, further MALDI-TOF MS^2 investigation was performed. The negative-ion MS^2 spectrum of the precursor ion at *m/z* 1712.0, chosen as a reference of a mono-phosphorylated hexa-acylated lipid A species, is reported in Figure 2b. The MS^2 spectrum, besides furnishing crucial information about the position of the secondary acyl substituents, also provided additional and unreported structural information about *P. arctica* lipid A. In detail, three intense peaks were detected at *m/z* 1467.94, 1485.90 and 1513.90 and were assigned as follows: the peak at *m/z* 1467.94 was attributed to an ion derived from the loss of a primary 14:0 (3-OH); the ion at *m/z* 1485.90 matched with a lipid A fragment originated from the loss of the secondary 14:1 fatty acid, whereas the peak at *m/z* 1513.90 was ascribed to an ion derived from the loss of a secondary 12:1 acyl moiety. The occurrence of two unsaturated fatty acids was in agreement with the mass difference of 2 amu with the respect to the species occurring at *m/z* 1714.1, i.e., the mono-phosphorylated form of the main lipid A species detected at *m/z* 1794.1. Importantly, the occurrence of both the Y$_1$ ion (*m/z* 710.19) [8], originated from the cleavage of the glycosydic linkage of the disaccharide backbone, and the ion at *m/z* 1287.80, which matched with the loss of a whole unit of 14:0 (3-OH) carrying the 12:1 acyl substituent, proved that the secondary fatty acids only decorate the primary acyl chains of the non-reducing glucosamine residue; moreover, these fragmentations suggested that the 14:1 moiety is present as a secondary fatty acid in an acyloxyacyl amide moiety. Therefore, *P. arctica* exhibited a more heterogeneous lipid A than previously reported [6], being composed of species also concomitantly decorated by two unsaturated secondary acyl chains, distributed in a 4 + 2 symmetry with the respect to the glucosamine disaccharide backbone.

Table 2. The main ion peaks observed in the MALDI-TOF MS spectra reported in Figures 1, 3 and 5, the predicted mass and the proposed interpretation of the substituting fatty acids and phosphates on the lipid A backbone. The observed masses reported in the table are compared to the calculated molecular weight (predicted mass, Da) of each ion based on the proposed lipid A structures.

	P. arctica Strain SY204b		
Predicted Mass (Da)	**Observed Ion Peaks (m/z)**	**Acyl Substitution**	**Proposed Fatty Acid/Phosphate Composition**
1794.20	1794.10	Hexa-acyl	$HexN_2P_2[14:0(3\text{-}OH)]^4$ (12:0) (14:1)
1792.18	1792.09	Hexa-acyl	$HexN_2P_2[14:0(3\text{-}OH)]^4$ (12:1) (14:1)
1714.23	1714.11	Hexa-acyl	$HexN_2P[14:0(3\text{-}OH)]^4$ (12:0) (14:1)
1712.21	1712.08	Hexa-acyl	$HexN_2P[14:0(3\text{-}OH)]^4$ (12:1) (14:1)
1612.03	1612.05	Penta-acyl	$HexN_2P_2[14:0(3\text{-}OH)]^4$ (14:1)
1568.00	1568.04	Penta-acyl	$HexN_2P_2[14:0(3\text{-}OH)]^3$ (12:0) (14:1)
1488.04	1488.04	Penta-acyl	$HexN_2P[14:0(3\text{-}OH)]^3$ (12:0) (14:1)
1385.84	1385.91	Tetra-acyl	$HexN_2P_2[14:0(3\text{-}OH)]^3$ (14:1)
1305.87	1305.91	Tetra-acyl	$HexN_2P[14:0(3\text{-}OH)]^3$ (14:1)

	P. cryohalolentis Strain SY185		
Predicted Mass (Da)	**Observed Ion Peaks (m/z)**	**Acyl Substitution**	**Proposed Fatty Acid/Phosphate Composition**
1782.16	1781.78	Hepta-acyl	$HexN_2P_2[12:0\ (3\text{-}OH)]^4$ (12:0) $(10:0)^2$
1628.02	1627.68	Hexa-acyl	$HexN_2P_2[12:0\ (3\text{-}OH)]^4$ (12:0) (10:0)
1642.04	1641.70	Hexa-acyl	$HexN_2P_2[12:0\ (3\text{-}OH)]^3$ [13:0 (3-OH)] (12:0) (10:0)
1548.06	1547.74	Hexa-acyl	$HexN_2P[12:0\ (3\text{-}OH)]^4$ (12:0) (10:0)
1429.86	1429.57	Penta-acyl	$HexN_2P_2[12:0\ (3\text{-}OH)]^3$ (12:0) (10:0)
1275.73	1275.46	Tetra-acyl	$HexN_2P_2[12:0\ (3\text{-}OH)]^3$ (12:0)

	P. tetraodonis Strain SY174		
Predicted Mass (Da)	**Observed Ion Peaks (m/z)**	**Acyl Substitution**	**Proposed Fatty Acid/Phosphate Composition**
1445.86	1445.59	Penta-acyl	$HexN_2P_2[12:0\ (3\text{-}OH)]^2$ $[10:0\ (3\text{-}OH)]^2$ (14:0)
1473.89	1473.61	Penta-acyl	$HexN_2P_2[12:0\ (3\text{-}OH)]^4$ (12:0)
1365.89	1365.64	Penta-acyl	$HexN_2P[12:0\ (3\text{-}OH)]^2$ $[10:0\ (3\text{-}OH)]^2$ (14:0)
1275.73	1275.49	Tetra-acyl	$HexN_2P_2[12:0\ (3\text{-}OH)]^2$ [10:0 (3-OH)] (14:0)
1195.76	1195.54	Tetra-acyl	$HexN_2P[12:0\ (3\text{-}OH)]^2$ [10:0 (3-OH)] (14:0)

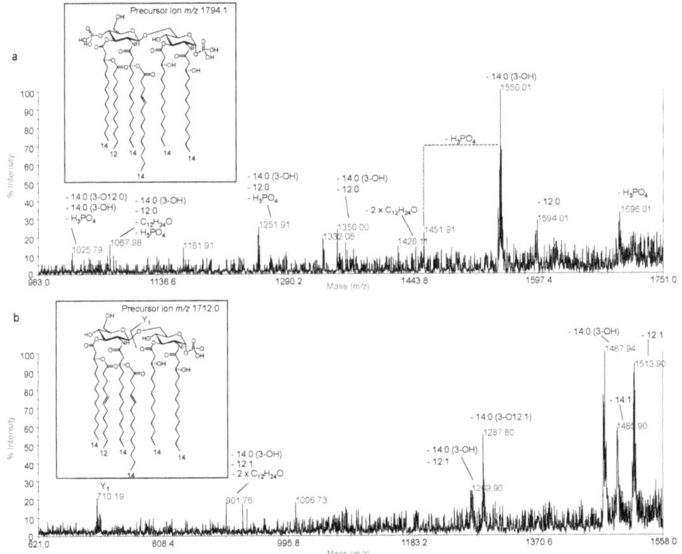

Figure 2. MALDI-TOF MS² analysis of hexa-acylated lipid A species from *P. arctica* strain SY204b. (**a**) Negative-ion MALDI MS² spectrum of precursor ion at m/z 1794.1, a representative ion peak of the cluster ascribed to hexa-acylated lipid A species decorated by two phosphates. (**b**) Negative-ion MALDI MS² spectrum of precursor ion at m/z 1712.0, a representative ion peak of the cluster ascribed to hexa-acylated lipid A species decorated by one phosphate. The assignment of main fragments is reported in both spectra. The proposed structure for the lipid A species is reported in each inset. The loss of $C_{12}H_{24}O$ (184 mass units) is a rearrangement typically occurring on primary 14:0 (3-OH) acyl chains only when their 3-OH group is free, thus contributing to the establishment of the location of the secondary acyl substitution.

2.3. MALDI-TOF MS and MS² Analysis on the Isolated Lipid A from P. cryohalolentis Strain SY185

The MALDI-TOF MS spectrum, recorded in negative polarity, of the lipid A isolated from *P. cryohalolentis* LPS is reported in Figure 3, whereas the spectrum recorded on intact bacteria is reported in Figure S3. The high heterogeneity of the lipid A was clearly visible in both spectra, which were similar and showed in the *m/z* range 1571.7–1697.9 a main, complex pattern of peaks relative to hexa-acylated lipid A species which differed in the nature of the fatty acids. This was proven by the mass difference of 14 amu occurring between the peaks of the cluster, in agreement with the heterogeneity observed in the lipid compositional analysis (Table 1). This family of peaks, with the main representative at *m/z* 1627.7, matched with a bis-phosphorylated lipid A species carrying four primary 12:0 (3-OH), and one 12:0 and one 10:0 as secondary acyl moieties (Table 2). Penta-acylated and tetra-acylated species devoid of one 12:0 (3-OH), or one 12:0 (3-OH) and one 10:0 were identified at *m/z* 1429.6 and 1275.5, respectively (Table 2). A less intense cluster of peaks at around *m/z* 1781.8 was ascribed to a hepta-acylated lipid A species carrying, in the case of the species at *m/z* 1781.8, an additional 10:0 unit (Figure 3 and Figure S3, Table 2).

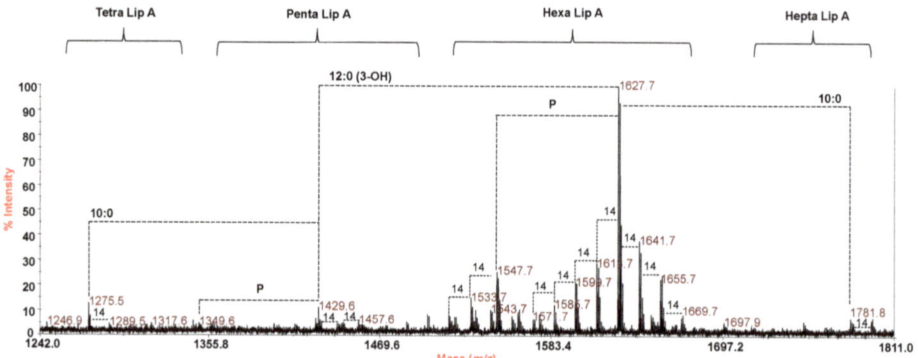

Figure 3. Reflectron MALDI-TOF mass spectrum, recorded in negative polarity, of lipid A from *P. cryohalolentis* strain SY185 obtained after acetate buffer treatment. The lipid A species are labelled as Tetra, Penta, Hexa and Hepta Lip A indicating the degree of acylation. Differences of 14 amu, i.e., a methylene group, are also indicated in the spectrum. "**P**" indicates the phosphate group.

The negative-ion MS² spectrum of precursor ion at *m/z* 1547.7 (Figure 4), corresponding to a mono-phosphorylated hexa-acylated lipid A species, confirmed the location of the secondary acyl chains with respect to the glucosamine backbone, that is only on the non-reducing glucosamine unit, as previously reported [9]. This structural hypothesis was corroborated by the observation in the MS² spectrum of an important ion derived from the sugar ring fragmentation ^{0,4}A$_2$ [8] which clearly indicates the location of the 10:0 and the 12:0 moieties on the non-reducing glucosamine. In parallel, the occurrence of the peak at *m/z* 1177.64, ascribable to an ion derived from the loss of a primary ester-bound 12:0 (3-OH) moiety decorated by the secondary 10:0, suggested the location of the latter fatty acid, and likewise of the other secondary acyl chain (12:0) as a substituent of the amide-bound primary 12:0 (3-OH). Similarly, the peak at *m/z* 961.66 was attributed to an ion originated from the loss of a 12:0 (3-OH) and a whole unit of 12:0 (3-OH) acylated by the secondary 10:0 moiety. However, the observation of an intense peak at *m/z* 1347.70 matching with an ion originating from the loss of the 12:0 unit, suggested that a minor lipid A species might exist with the inverted location of the secondary acyl chains (i.e., 12:0 in an acyloxyacyl ester moiety and 10:0 in an acyloxyacyl amide moiety).

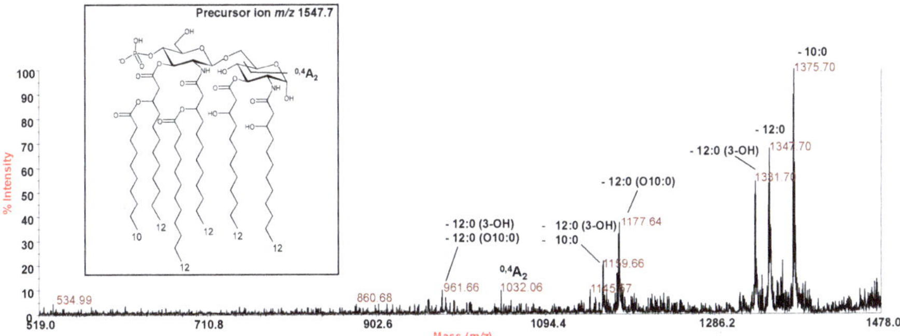

Figure 4. Negative-ion MALDI-TOF MS^2 spectrum of precursor ion at *m/z* 1547.7, main ion peak of the cluster ascribed to hexa-acylated lipid A species decorated by one phosphate. The assignment of the main fragments is reported in the spectrum. The proposed structure for the lipid A species is reported in the inset.

Therefore, as reported [9], *P. cryohalolentis* strain SY185 mainly expresses a highly heterogenous bis-phosphorylated hexa-acylated lipid A species whose primary fatty acids range from 11 to 14 carbon atoms in length, whereas as the secondary acyl chains 10:0, 12:0, 12:1, 14:0, and 15:0 have been found (Table 1).

2.4. MALDI-TOF MS and MS^2 Analysis on the Isolated Lipid A from P. tetraodonis Strain SY174

The negative-ion MALDI-TOF mass spectrum executed on *P. tetraodonis* lipid A is shown in Figure 5. The well-known heterogeneity of the lipid A was immediately evident, as previously reported by our group for the lipid A structure from other *Pseudoaltermonas* species [10,11], and in full accordance with compositional analysis. The MS spectrum showed two distinct clusters of peaks in the mass range *m/z* 1149.5–1487.6, each characterized by the occurrence of mass differences of 14 amu and/or 28 amu. As observed for *P. arctica*, peaks differing by 2 amu were also observed and were attributed to the presence of lipid A, also decorated by unsaturated acyl chains, in agreement with compositional analysis data and with previous reported data [10,11]. Mono- and bis-phosphorylated penta-acylated lipid A species were detected in the mass range *m/z* 1337.6 –1501.6, with the main species at *m/z* 1445.6 that matched with a bis-phosphorylated penta-acylated lipid A carrying two 12:0 (3-OH) and two 10:0 (3-OH) as primary fatty acids, and one 14:0 as secondary acyl substituent (Figure 5, Table 2). The related mono-phosphorylated species was detected at *m/z* 1365.6, whereas a bis-phosphorylated tetra-acylated form devoid of one primary 10:0 (3-OH) was ascribed to the peak at *m/z* 1275.5. As observed for the other bacteria analyzed in the present work, the negative-ion MALDI-TOF MS spectrum recorded on the *P. tetraodonis* strain SY174 cell pellet was similar to the one recorded on the isolated lipid A, and is reported in Figure S4.

The negative-ion MS^2 investigation of precursor ion at *m/z* 1445.6 (Figure 6) showed the occurrence of, among others, important ions originating from the loss of two primary O-linked 10:0 (3-OH) at *m/z* 1069.66 which suggested the location of the secondary 14:0 in an acyloxyacyl amide moiety. Moreover, the observation of the two ions at *m/z* 626.14 and 836.27 derived from the cleavage of the glycosidic linkage (Y$_1$ and C$_2$) [8] concurred to locate the secondary acyl substituent on the non-reducing glucosamine unit. However, in order to unequivocally establish the location of the secondary acyl substitution, an aliquot of lipid A, underwent a treatment with NH$_4$OH, which selectively removes the acyl and acyloxyacyl esters, leaving the acyl and acyloxyacyl amides unaffected [12]. The negative-ion MALDI-TOF MS spectrum, reported in Figure 7, showed a main peak at *m/z* 1105.4 matching with a bis-phosphorylated lipid A carrying the solely primary N-linked 12:0 (3-OH) chains with the secondary 14:0 moiety linked to the primary 12:0 (3-OH) of the non-reducing glucosamine unit (Figure 7), thus definitively confirming the location of the secondary acyl substitution in *P. tetraodonis* lipid A.

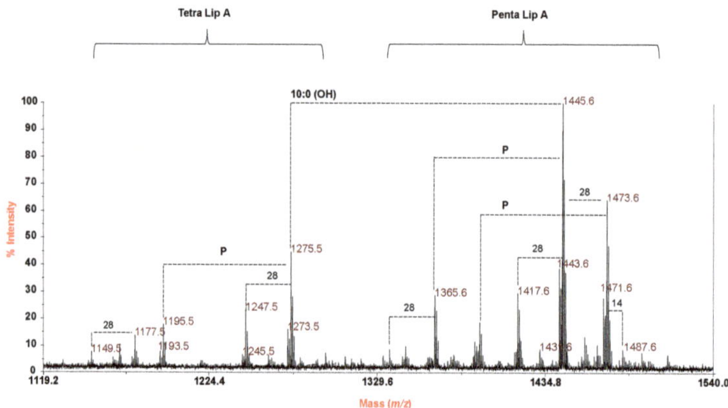

Figure 5. Reflectron MALDI-TOF mass spectrum, recorded in negative polarity, of lipid A from *P. tetraodonis* strain SY174 obtained after acetate buffer treatment. The lipid A species are labelled as Tetra and Penta Lip A indicating the degree of acylation. Differences of 14 and 28 amu are also reported in the spectrum. "**P**" indicates the phosphate group.

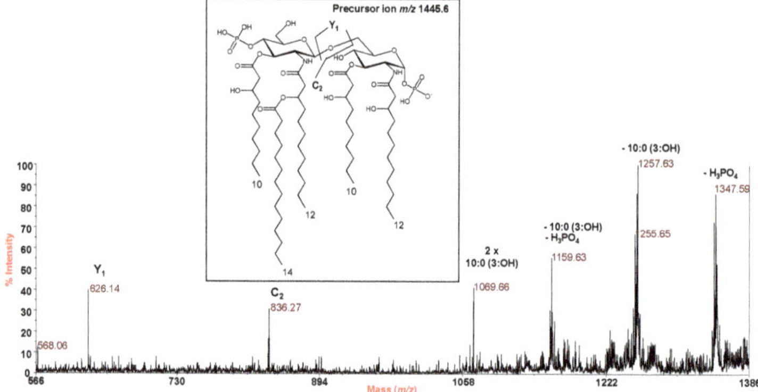

Figure 6. Negative-ion MALDI-TOF MS^2 spectrum of precursor ion at *m/z* 1445.6, main ion peak of the cluster ascribed to penta-acylated lipid A species decorated by two phosphates from *P. tetraodonis* strain SY174. The assignment of main fragments is reported in the spectrum. The proposed structure for the lipid A species is reported in the inset.

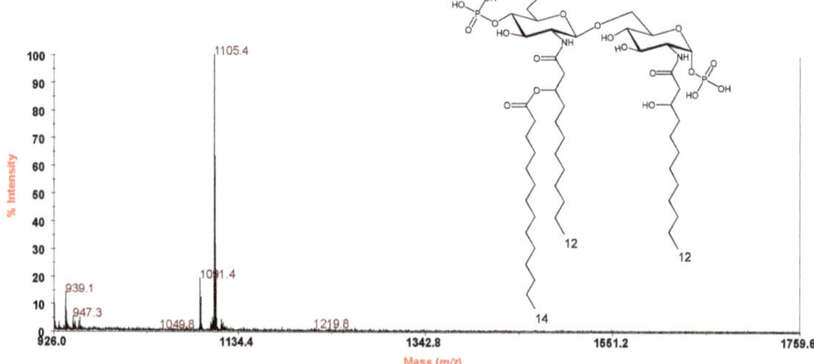

Figure 7. Reflectron MALDI-TOF mass spectrum, recorded in negative polarity, of lipid A from *P. tetraodonis* strain SY174 after treatment with NH$_4$OH. The structure of the main ion species is reported in the inset.

3. Discussion

The numerous survival strategies adopted by psychrophilic bacteria are considered as an excellent model system to appreciate the mechanisms underlying low-temperature adaptation. Indeed, the capability of some psychrophiles to experience temperatures near to or below the freezing point of water implies the production of cryotolerance-conferring compounds that have attracted the interest of researchers and manufacturers in various industries as potential molecules exploitable in several biotechnological applications [6,13]. In parallel, life in cold habitats also requires an array of adaptive modifications at nearly all levels of the cell envelope architecture and function [14]. Indeed, to endure the cold temperatures, bacterial cell membranes undergo a phase change from liquid to gel, which would dramatically increment the rigidity of the membranes themselves. Therefore, most bacteria modify their membrane lipids to maintain a certain degree of membrane fluidity [14]. Psychrophilic Gram-negative bacteria in this sense greatly modify the structure of the main component of their outer membrane, the LPS molecule, to increment the outer membrane fluidity, thus allowing protein movement, and hence function [6]. Here we reported on the lipid A structures of the LPS from three psychrophilic bacteria isolated from Terra Nova Bay, Antarctica; our analyses provided additional structural information about such an important structural component of the Gram-negative outer membrane component.

As reviewed elsewhere [6], the lipid A from a *P. arctica* strain isolated in the Svalbard islands in the Arctic Ocean was consistent with a blend of species ranging from penta- to hepta-acylated, with a predominant hexa-acylated form composed of the typical bis-phosphorylated diglucosamine skeleton bearing four 14:0 (3-OH), one 14:0 and one cyclopropyl-tetradecanoic acid unit. Here we also found a lipid A structure containing 14:0 (3-OH), as the primary acyl chains, whereas some differences were observed in the nature of the secondary acyl moieties. Indeed, in this study we have highlighted the occurrence of 14:1 acyl residue as well as the concomitant presence of 12:1 as a secondary acyl substitution not previously reported in *P. arctica*. Additionally, neither cyclopropyl-tetradecanoic acids nor hepta-acylated lipid A species [6,9] have been detected in the current study. The discovery of a higher degree of unsaturation found for the lipid A of this *P. arctica* strain isolated from the Antarctica environment is an important structural characteristic which confirms the tendency of cold-adapted bacteria to increment the outer membrane fluidity by modifying and desaturating their membrane lipids, decreasing the lipid packing and thus increasing the membrane fluidity.

As for *P. cryohalolentis*, previous studies on *P. cryohalolentis* K5 (ATCC BAA-1226) strain showed that this bacterium was able to synthesize at least seven major lipid A forms based on a high degree of acyl chains flexibility with a dominant hexa-acylated cluster, and minor clusters of the penta- and hepta-acylated forms [9]. Accordingly, we found a major hexa-acylated lipid A species with the main representative species, detected at *m/z* 1627.7 (Figure 3), corresponding to a bis-phosphorylated lipid A carrying four 12:0 (3-OH) and one 12:0 and one 10:0 secondary fatty acids in a 4 + 2 symmetry with the respect to the disaccharide backbone. Minor tetra-, penta- and hepta-acylated species were also here identified. Notably, the *P. cryohalolentis* K5 (ATCC BAA-1226) lipid A structure with a decreasing temperature showed a near-elimination of odd-numbered acyl chains and a shift toward shorter acyl moieties, a structural trend to favor bacterial membrane fluidity [15]. Here the observation in the spectrum (Figure 3) of the occurrence of lesser but significant MALDI-TOF peaks that differ by a single-methylene group flanking the most abundant peak at *m/z* 1627.7 demonstrated that this *P. cryohalolentis* strain tends to maintain a certain degree of odd-numbered acyl chains. On the other hand, a shortening of the acyl chains' moieties with respect to the most abundant lipid A species was observed, as proven by the occurrence of the peaks in the *m/z* range 1571.6–1613.7. However, lipid species with longer acyl chains were also detected (*m/z* 1641.7–1697.7) in this study. Finally, it is worth underlining that, in accordance with the tendency of cold-adapted bacteria to increase the desaturation of their membrane lipids, here we found that this Antarctic strain of *P. cryohalolentis* produces a lipid A species also carrying an unsaturated secondary fatty acid chain (12:1, Table 1) [9,16].

By contrast, no studies have been reported so far about *P. tetraodonis* lipid A. However, several papers described the structure of the lipid A from other *Pseudoalteromonas* species [10,11] which confirmed and strengthened our results on a heterogenous bis-phosphorylated penta-acylated lipid A species as the main form. Here, the main lipid A species was detected at *m/z* 1445.6 that matched with a bis-phosphorylated lipid A species bearing two O-linked 10:0 (3-OH) and two N-linked 12:0 (3-OH) as the primary fatty acids, whereas 14:0 was identified as the secondary acyl substituent. This structural assessment further proved the high heterogeneity of the lipid A species that characterizes *Pseudoalteromonas* sp. LPS because the same peak (at about *m/z* 1445.6) was previously observed for other *Pseudoaltermonas* sp. but a different structure was deduced [11]. Importantly, given the controversial data reported in the literature about the exact location of the secondary acyl moiety in the lipid A from *Pseudoalteromonas* species, here we showed that *P. tetraodonis* lipid A is characterized by the occurrence of the secondary fatty acid exclusively on the N-linked primary acyl chain of the non-reducing glucosamine.

It is worth underlining that all of the lipid A structural peculiarities discussed above, especially the occurrence of unsaturated and short acyl chains, are surely of great scientific interest from an evolutionary point of view, improving the comprehension of the molecular mechanisms allowing low temperature adaptation phenomena. Moreover, the discovery of such unusual structural features further proved the never-ending and still undiscovered chemical possibilities arising from microorganisms living in cold environments, thus also becoming incredibly attractive from a chemical point of view. This scenario becomes even more fascinating under the biological/immunological point of view considering that the TLR4/MD-2-mediated immunoactivity of an LPS is strictly related to the lipid A structure. Actually, hypoacylated lipid As (namely lipid A species expressing less than six acyl chains) usually only scantly activate the TLR4-MD-2-mediated immune response [6,7]. Moreover, a 4 + 2 symmetry of the acyl chains with respect to the glucosamine backbone, as in the case of *E. coli* LPS, has been correlated to a more potent immunostimulatory activity, whereas a 3 + 3 symmetry has been related to a reduction in the immunopotency of the whole LPS molecule [6,7]. In this context, previous studies have demonstrated that, despite expressing a bis-phosphorylated hexa-acylated lipid A with a 4 + 2 symmetry, *P. cryohalolentis* LPS exhibits a lower immunostimulatory capacity than *E. coli* LPS [16]; this was attributed to the shorter primary and secondary acyl chains (10–12 carbon atoms) compared with *E. coli* lipid A (12–14), which might not provide an equally efficient LPS activation of the TLR4/MD-2 mediated signaling. Similarly, several LPS from *Pseudoalteromonas* strains have been investigated for their immunological properties, revealing a very weak immunostimulant activity in accordance with the hypo-acylated nature of their lipid A; however, this LPS exhibited an interesting inhibitory capacity towards the toxic effects of the *E. coli* LPS [17]. The observation that, depending on the structure of the lipid A, LPSs isolated from psychrophilic bacteria are able to modulate the host immune response, in addition to the unlimited sources from which it is possible to isolate such microorganisms, represents an incentive for an in-depth investigation of cold-adapted Gram-negatives as a potential never-ending basin of natural immunomodulatory molecules. Therefore, a detailed study of the immunological properties of the particular LPS/lipid A here defined would be essential to improve the current knowledge about the role of the exact location and nature of the fatty acids in the LPS structure-immunoactivity relationship.

4. Materials and Methods

4.1. Bacterial Strains Isolation and Growth

Sampling was done during the XXXIII Antarctic Expedition in October/December 2017 in Terra Nova Bay, Antarctica. A sample of platelet ice was obtained from the bottom of 1.5 m thick and 2 m high ice core collected from Tethys Bay (74°40′01.3″ S and 164°11′58.6″ E); invertebrates were collected by Scuba-diving (74°41′24.5″ S and 164°06′15.4″ E). For cultivation, 1 mL of melted platelet ice was inoculated in 9 mL of sea water based medium (SWBM) (g/L^{-1}: 0.27 NH$_4$Cl, 0.089 Na$_2$HPO$_4$, 1.30 TAPSO, 0.002 FeCl$_2$ × 4H$_2$O, 0.002 Fe NH$_4$ citrate) supplemented with 300 mg (yeast extract, peptone, casamino acids, sucrose). Cultures were incubated at a 4 °C temperature for one week. 100 μL of positive enrichment was spread onto agar plates of the same medium. Invertebrates were rinsed in sterile sea water and one cm^3 of each sample was homogenized in 10 mL of SWBM medium in a sterile mortar. Then a ten-fold serial dilution was carried out to obtain dilutions varying from 10^{-1} to 10^{-6}. 100 μL of each dilution was spread onto plates of SWBM medium supplemented with 1.5 g (Yeast extract, peptone, casamino acids, sucrose). Cultivation was performed at 4 °C and pure cultures were obtained on plates. The colony transfer was repeated at least twice before the cultures were considered pure. Isolates were checked microscopically and by 16S rRNA sequencing. To obtain a sufficient amount of biomass for LPS isolation and characterization, growth of each strain was performed at 4 °C in ten 6 L flasks. The cultures were incubated for 5 days with periodic shaking.

4.2. LPS Isolation and Purification

The LPS from each bacterial strain was extracted by a modified enzyme-phenol-water protocol [18]. Each LPS was obtained as a precipitate after an ultracentrifugation step (200,000× *g*, 4 °C, 16 h). Then, an enzymatic digestion was executed to remove possible cell contaminants by using DNase (DN25-Sigma Aldrich®, St. Louis, MO, USA), RNase (R5503-Sigma Aldrich®), and protease (P4630-Sigma Aldrich®). The digested materials were then again extensively dialyzed (Spectra/Por®, Fisher Sci. Leicestershire, UK, cut-off 12–14 kDa) against distilled water. In order to remove any phospholipids that were possibly present, all the LPS underwent several washes with a mixture of $CHCl_3/CH_3OH$ (1:2, *v/v*) and $CHCl_3/CH_3OH/H_2O$ (3:2:0.25, *v/v*). After the removal of organic solvents and repeated cycles of lyophilization, an SDS-PAGE followed by gel staining with silver nitrate [19] was performed to establish the nature and the degree of purity of the extracted materials.

4.3. Chemical Analyses

The total fatty acid content was established by treating each LPS with 4 M HCl (100 °C, 4 h), followed by a treatment with 5 M NaOH (100 °C, 30 min). After the adjustment of the pH, an extraction in chloroform led to the collection of the fatty acids, which were then methylated with diazomethane and analyzed by a Gas Chromatography Mass Spectrometry (GC-MS). The ester-bound fatty acids, analyzed by GC-MS, were obtained after treatment with aqueous 0.5 M NaOH in CH_3OH (1:1, *v/v*, 85 °C, 2 h), followed by acidification of the products, extraction in chloroform and methylation with diazomethane. Furthermore, an aliquot of each LPS fraction was also methanolized with 1.25 M HCl/CH_3OH (80 °C, 16 h). The mixture was extracted three times with hexane. The hexane layer, containing the fatty acids as methyl esters derivatives, was then analyzed by GC-MS. The absolute configuration of the fatty acids was established as previously reported. Briefly, the 3-hydroxy fatty acids were released after treatment with 4 M NaOH (100 °C, 5 h), converted into the 3-methoxy acid L-phenylethylamides, and then analyzed by GC-MS [20]. The comparison of the retention times of authentic L-phenylethylamides of various standard fatty acids with those derived from the examined LPSs allowed the assignment of the (R) configuration to all of the fatty acids composing the lipid A from the three psychrophilic bacteria. The analyses were all executed on an Agilent Technologies gas chromatograph 6850A equipped with a mass selective detector 5973N and a Zebron ZB-5 capillary column (Phenomenex, Torrance, USA, 30 m × 0.25 mm internal diameter, flow rate 1 mL min^{-1}, He as carrier gas). The following temperature program was employed for the lipid analysis: 140 °C for 3 min, 140 °C → 280 °C at 10 °C min^{-1}.

4.4. Isolation of the Lipid A Fractions

An aliquot of each purified LPS was treated with acetate buffer (pH 4.4, 2 h, 100 °C) in order to separate the lipid A portion from the saccharide part of the LPS. A mixture of chloroform and methanol was added to the hydrolysis product to obtain a $CHCl_3/CH_3OH$/hydrolysate 2:2:1.8 (*v/v/v*) ratio. The mixture was then shaken and centrifuged. The chloroform phase, containing the lipid A, was collected and washed with the water phase of a freshly prepared Bligh/Dyer mixture ($CHCl_3/CH_3OH/H_2O$, 2:2:1.8) [21].

The organic phases were pooled, dried, and analyzed by MALDI-TOF MS (SCIEX, Concord, ON, Canada). In order to establish the nature of the sugar backbone of each lipid A, an aliquot of each lipid A fraction also underwent a methanolysis (1.25 M HCl/CH_3OH, 80 °C, 16 h) followed by acetylation (80 °C, 20 min) and GC-MS analysis [22].

4.5. MALDI-TOF Mass Spectrometry

All the MS and the MS^2 experiments were performed both in linear and reflectron mode, negative ion polarity on an ABSCIEX TOF/TOF 5800 Applied Biosystems (Foster City, CA, USA) mass spectrometer equipped with an Nd:YAG laser (λ = 349 nm), with a 3 ns pulse width and a repetition rate of up to 1000 Hz, and also equipped with delayed extraction technology. Lipid A fractions were dissolved in $CHCl_3$/CH_3OH (50:50, *v/v*). The matrix solution was 2,4,6-trihydroxyacetophenone in CH_3OH/0.1 % trifluoroacetic acid/CH_3CN (7:2:1, *v/v/v*) at a concentration of 75 mg/mL [23,24]. Bacterial pellet for MALDI preparation was treated as previously described [25,26] and the matrix solution was prepared by dissolving 2,5-dihydroxybenzoic acid (DHB) at a final concentration of 10 mg ml^{-1} in $CHCl_3$/CH_3OH (9:1, *v/v*) [25,26]. 0.5 µL of the sample and 0.5 µL of the matrix solution were deposited onto a stainless steel plate and left to dry at room temperature. Each spectrum in the MS experiments was a result of the accumulation of 1500 laser shots, whereas 5000–7000 shots were summed for the MS^2 spectra. Each experiment was performed in triplicate.

5. Conclusions

In the present paper we reported on the structural characterization of the LPS lipid A from three taxonomically different psycrophilic Gram-negative bacteria isolated from Terra Nova Bay, Antarctica. All the three cold-adapted bacteria, *P. tetraodonis* strain SY174, *P. arctica* strain SY204b, and *P. cryohalolentis* strain SY185 displayed structural features in their lipid As, likely supporting and favoring their ability to live and proliferate in Antarctica, the coldest continent on Earth. Among others, a high level of heterogeneity, both in the acylation and in the phosphorylation degree, has been observed for all the analyzed lipid As. In addition, in accordance with the tendency of psychrophiles to increase the desaturation of their membrane lipids, unsaturated fatty acids were detected as structural constituents of all the three lipid As. In particular, *P. arctica* strain SY204b showed an even more heterogenous lipid A than previously reported [6], being characterized by the occurrence of species also concomitantly carrying two unsaturated acyl moieties. Likewise, a tendency to decrease the length of the acyl chains, in order to maintain membrane fluidity at cold temperatures, was also observed, as proven by the occurrence of fatty acids ranging from 10 (*P. cryohalolentis* strain SY185 and *P. tetraodonis* strain SY174) to maximum 13 (*P. tetraodonis* strain SY174), 14 (*P. arctica* strain SY204b), or 15 (*P. cryohalolentis* strain SY185) carbon atoms in length. Finally, the structure of the lipid A from *P. tetraodonis* strain SY174 has been here described for the first time, and confirmed the high heterogeneity of the tetra- and penta-acylated lipid A species found in other *Pseudoalteromonas* species [10,11]. This study further supports and strengthens data describing the existence of lipid A modification systems that are often regulated by environmental conditions [27], and underlines the capability of extremophile bacteria to modify their outer membrane lipid components to deal with the detrimental stressors of the surrounding environment.

Supplementary Materials: The following are available online at http://www.mdpi.com/1660-3397/18/12/592/s1, Figure S1: Field emission scanning electron microscopy (FESEM) micrographs of *P. tetraodonis* SY74 (A) and *P. cryohalolentis* SY185 (B) cultures growing at 0.5 °C.; Figure S2. Reflectron MALDI-TOF mass spectrum, recorded in negative polarity, of the *P. arctica* strain SY204b cell pellet. Figure S3. Reflectron MALDI-TOF mass spectrum, recorded in negative polarity, of the cell pellet of *P. cryohalolentis* strain SY185. "P" indicates the phosphate group; Figure S4. Reflectron MALDI-TOF mass spectrum, recorded in negative polarity, of the cell pellet of *P. tetraodonis* strain SY174.

Author Contributions: Conceptualization, A.S. and F.D.L. and V.L.C. Data analysis, all authors. All authors have read and agreed to the published version of the manuscript.

Funding: This work was supported by the National Antarctic Research Program (PNRA), Italian Ministry of Education and Research (Research project PNRA16_00089).

Conflicts of Interest: The authors declare no conflict of interest.

Abbreviations

core OS	Core oligosaccharide
GC-MS	Gas Chromatography-Mass Spectrometry
Hepta Lip A	Hepta-acylated Lipid A
Hexa Lip A	Hexa-acylated Lip A
LPS	Lipopolysaccharide
MALDI-TOF MS	Matrix Assisted Laser Desorption Ionization-Time of Flight Mass Spectrometry
Penta Lip A	Penta-acylated Lipid A
R-LPS	Rough-type
SWBM	Sea water based medium
S-LPS	Smooth-type LPS
SDS-PAGE	Sodium Dodecyl Sulphate-Polyacrylamide Gel Electrophoresis
Tetra Lip A	Tetra-acylated Lipid A
TLR4/MD-2	Toll-Like Receptor 4/Myeloid Differentiation factor-2

References

1. Feller, G. Cryosphere and Psychrophiles: Insights into a Cold Origin of Life? *Life* **2017**, *7*, 25. [CrossRef]
2. Pearce, D.A. Extremophiles in Antarctica: Life at low temperatures. In *Adaptation of Microbial Life to Environmental Extremes: Novel Research Results and Application*; Stan-Lotter, H., Fendrihan, S., Eds.; Springer: Wien, Austria, 2012; pp. 99–131.
3. Núñez-Montero, K.; Barrientos, L. Advances in Antarctic Research for Antimicrobial Discovery: A Comprehensive Narrative Review of Bacteria from Antarctic Environments as Potential Sources of Novel Antibiotic Compounds against Human Pathogens and Microorganisms of Industrial Importance. *Antibiotics (Basel)* **2018**, *7*, 90. [CrossRef]
4. Aislabie, J.M.; Jordan, S.; Barker, G.M. Relation between soil classification and bacterial diversity in soils of the Ross Sea region, Antarctica. *Geoderma* **2008**, *144*, 9–20. [CrossRef]
5. Di Lorenzo, F.; De Castro, C.; Lanzetta, R.; Parrilli, M.; Silipo, A.; Molinaro, A. Lipopolysaccharides as Microbe-associated Molecular Patterns: A Structural Perspective. In *Carbohydrates in Drug Design and Discovery*; RSC Drug Discovery Series; Jimenez-Barbero, J., Javier Canada, F., Martın-Santamarıa, S., Eds.; Royal Society of Chemistry: London, UK, 2015; pp. 38–63.
6. Di Lorenzo, F.; Billod, J.-M.; Martín-Santamaría, S.; Silipo, A.; Molinaro, A. Gram-Negative Extremophile Lipopolysaccharides: Promising Source of Inspiration for a New Generation of Endotoxin Antagonists. *Eur. J. Org. Chem.* **2017**, *2017*, 4055–4073. [CrossRef]
7. Molinaro, A.; Holst, O.; Di Lorenzo, F.; Callaghan, M.; Nurisso, A.; D'Errico, G.; Zamyatina, A.; Peri, F.; Berisio, R.; Jerala, R.; et al. Chemistry of lipid A:at the heart of innate immunity. *Chem. Eur. J.* **2015**, *21*, 500–519. [CrossRef] [PubMed]
8. Domon, B.; Costello, C.E. A systematic nomenclature for carbohydrate fragmentations in FAB-MS/MS spectra of glycoconjugates. *Glycoconjug. J.* **1988**, *5*, 397–409. [CrossRef]
9. Sweet, C.R.; Alpuche, G.M.; Landis, C.A.; Sandma, B.C. Endotoxin Structures in the Psychrophiles *Psychromonas marina* and *Psychrobacter cryohalolentis* Contain Distinctive Acyl Features. *Mar. Drugs* **2014**, *12*, 4126–4147. [CrossRef]
10. Silipo, A.; Leone, S.; Lanzetta, R.; Parrilli, M.; Sturiale, L.; Garozzo, D.; Nazarenko, E.L.; Gorshkova, R.P.; Ivanova, E.P.; Gorshkova, N.M.; et al. The complete structure of the lipooligosaccharide from the halophilic bacterium *Pseudoalteromonas issachenkonii* KMM3549^T. *Carbohydr. Res.* **2004**, *339*, 1985–1993. [CrossRef] [PubMed]
11. Di Lorenzo, F. The lipopolysaccharide lipid A structure from the marine sponge-associated bacterium *Pseudoalteromonas* sp. 2A. *Anton. Leeuwenhoek. J. Microbiol.* **2017**, *110*, 1401–1412. [CrossRef] [PubMed]

12. Silipo, A.; Lanzetta, R.; Amoresano, A.; Parrilli, M.; Molinaro, A. Ammonium hydroxide hydrolysis: A valuablesupport in the MALDI-TOF mass spectrometry analysis of Lipid A fatty acid distribution. *J. Lipid Res.* **2002**, *43*, 2188–2195. [CrossRef] [PubMed]

13. Cavicchioli, R. Cold-adapted archaea. *Nat. Rev. Microbiol.* **2006**, *4*, 331–343. [CrossRef]

14. Bakermans, C.; Emili, L.A. Terrestrial systems of the Arctic as a model for growth and survival at low temperatures. In *Model Ecosystems in Extreme Environments*; Seckbach, J., Rampelotto, P., Eds.; Academic Press: Cambridge, MA, USA, 2019; pp. 1–21.

15. Sweet, C.R.; Watson, R.E.; Landis, C.A.; Smith, J.P. Temperature-Dependence of Lipid A Acyl Structure in *Psychrobacter cryohalolentis* and Arctic Isolates of *Colwellia hornerae* and *Colwellia piezophile*. *Mar. Drugs* **2015**, *13*, 4701–4720. [CrossRef]

16. Korneev, K.V.; Kondakova, A.N.; Arbatsky, N.P.; Arbatsky, N.P.; Novototskaya-Vlasova, K.A.; Rivkina, E.M.; Anisimov, A.P.; Kruglov, A.A.; Kuprash, D.V.; Nedospasov, S.A.; et al. Distinct biological activity of lipopolysaccharides with different lipid A acylation status from mutant strains of Yersinia pestis and some members of genus *Psychrobacter*. *Biochemistry* **2014**, *79*, 1333–1338. [CrossRef]

17. Maaetoft-Udsen, K.; Vynne, N.; Heegaard, P.M.; Gram, L.; Frokiaer, H. *Pseudoalteromonas* strains are potent immunomodulators owing to low-stimulatory LPS. *Innate Immun.* **2013**, *19*, 160–173. [CrossRef]

18. Apicella, M.A.; Griffiss, J.M.; Schneider, H. Isolation and Characterization of Lipopolysaccharides, Lipooligosaccharides, and Lipid A. *Methods Enzymol.* **1994**, *235*, 242–252.

19. Kittelberger, R.; Hilbink, F. Sensitive silver-staining detection of bacterial lipopolysaccharides in polyacrylamide gels. *J. Biochem. Biophys. Methods* **1993**, *26*, 81–86. [CrossRef]

20. Rietschel, E.T. Absolute configuration of 3-hydroxy fatty acids present in lipopolysaccharides from various bacterial groups. *Eur. J. Biochem.* **1976**, *64*, 423–428. [CrossRef]

21. Bligh, E.G.; Dyer, W.J. A rapid method of total lipid extraction and purification. *Can. J. Biochem. Physiol.* **1959**, *37*, 911–917. [CrossRef]

22. Di Lorenzo, F.; Sturiale, L.; Palmigiano, A.; Fazio, L.L.; Paciello, I.; Coutinho, C.P.; Sá-Correia, I.; Bernardini, M.L.; Lanzetta, R.; Garozzo, D.; et al. Chemistry and biology of the potent endotoxin from a *Burkholderia dolosa* clinical isolate from a cystic fibrosis patient. *ChemBioChem* **2013**, *14*, 1105–1115. [CrossRef]

23. Barrau, C.; Di Lorenzo, F.; Menes, R.J.; Lanzetta, R.; Molinaro, A.; Silipo, A. The Structure of the Lipid A from the Halophilic Bacterium *Spiribacter salinus* M19-40^{T}. *Mar. Drugs* **2018**, *16*, 124. [CrossRef]

24. Pallach, M.; Di Lorenzo, F.; Duda, K.A.; Le Pennec, G.; Molinaro, A.; Silipo, A. The Lipid A Structure from the Marine Sponge Symbiont *Endozoicomonas* Sp. HEX 311. *Chembiochem* **2019**, *20*, 230–236. [CrossRef]

25. Larrouy-Maumus, G.; Clements, A.; Filloux, A.; McCarthy, R.R.; Mostowy, S. Direct detection of lipid A on intact Gram-negative bacteria by MALDI-TOF mass spectrometry. *J. Microbiol. Methods* **2016**, *120*, 68–71. [CrossRef]

26. Pither, M.D.; McClean, S.; Silipo, A.; Molinaro, A.; Di Lorenzo, F. A chronic strain of the cystic fibrosis pathogen *Pandoraea pulmonicola* expresses a heterogenous hypo-acylated lipid A. *Glycoconj. J.* **2020**. [CrossRef]

27. Raetz, C.R.; Reynolds, C.M.; Trent, M.S.; Bishop, R.E. Lipid A modification systems in Gram-negative bacteria. *Annu. Rev. Biochem.* **2007**, *76*, 295–329. [CrossRef]

Publisher's Note: MDPI stays neutral with regard to jurisdictional claims in published maps and institutional affiliations.

marine drugs

Article

Biogenic Synthesis of Copper Nanoparticles Using Bacterial Strains Isolated from an Antarctic Consortium Associated to a Psychrophilic Marine Ciliate: Characterization and Potential Application as Antimicrobial Agents

Maria Sindhura John [1], Joseph Amruthraj Nagoth [1], Marco Zannotti [2,*], Rita Giovannetti [2], Alessio Mancini [1], Kesava Priyan Ramasamy [1], Cristina Miceli [1] and Sandra Pucciarelli [1,*]

[1] School of Biosciences and Veterinary Medicine, Biosciences and Biotechnology Division, University of Camerino, 62032 Camerino, Italy; mariasindhura.john@unicam.it (M.S.J.); josephamruthraj.nagoth@unicam.it (J.A.N.); alessio.mancini@unicam.it (A.M.); kesava.ramasamy@unicam.it (K.P.R.); cristina.miceli@unicam.it (C.M.)

[2] School of Sciences and Technology, Chemistry Division, University of Camerino, 62032 Camerino, Italy; rita.giovannetti@unicam.it

[*] Correspondence: marco.zannotti@unicam.it (M.Z.); sandra.pucciarelli@unicam.it (S.P.)

Citation: John, M.S.; Nagoth, J.A.; Zannotti, M.; Giovannetti, R.; Mancini, A.; Ramasamy, K.P.; Miceli, C.; Pucciarelli, S. Biogenic Synthesis of Copper Nanoparticles Using Bacterial Strains Isolated from an Antarctic Consortium Associated to a Psychrophilic Marine Ciliate: Characterization and Potential Application as Antimicrobial Agents. *Mar. Drugs* **2021**, *19*, 263. https://doi.org/10.3390/md19050263

Academic Editor: Daniela Giordano

Received: 31 March 2021
Accepted: 4 May 2021
Published: 8 May 2021

Abstract: In the last decade, metal nanoparticles (NPs) have gained significant interest in the field of biotechnology due to their unique physiochemical properties and potential uses in a wide range of applications. Metal NP synthesis using microorganisms has emerged as an eco-friendly, clean, and viable strategy alternative to chemical and physical approaches. Herein, an original and efficient route for the microbial synthesis of copper NPs using bacterial strains newly isolated from an Antarctic consortium is described. UV-visible spectra of the NPs showed a maximum absorbance in the range of 380–385 nm. Transmission electron microscopy analysis showed that these NPs are all monodispersed, spherical in nature, and well segregated without any agglomeration and with an average size of 30 nm. X-ray powder diffraction showed a polycrystalline nature and face centered cubic lattice and revealed characteristic diffraction peaks indicating the formation of CuONPs. Fourier-transform infrared spectra confirmed the presence of capping proteins on the NP surface that act as stabilizers. All CuONPs manifested antimicrobial activity against various types of Gram-negative; Gram-positive bacteria; and fungi pathogen microorganisms including *Escherichia coli*, *Staphylococcus aureus*, and *Candida albicans*. The cost-effective and eco-friendly biosynthesis of these CuONPs make them particularly attractive in several application from nanotechnology to biomedical science.

Keywords: green synthesis; biomaterials; metal; antibiotics; nanotechnology

1. Introduction

With the beginning of the 21st century, nanobiotechnology entered the scientific spotlight as a discipline for innovative materials and applications. Nanoparticles (hereafter called NPs) are becoming the fundamental building blocks of nanotechnology. Their small dimensions and high surface area to volume enable them to exhibit novel chemical and physical properties. Consequently, NPs can be used in applications that are different from those of their bulk materials, including but not limited to electrical resistivity and conductivity, chemical reactivity, and diverse and versatile biological processes [1,2]. Most research work in this field focused on silver and gold NPs. However, special interest has been taken in other metal NPs because these particles are being widely used as industrial catalysts, in chemical sensing devices, in medical applications, in cosmetics, and as microelectronics [1–4]. Among these, interest is growing for NPs from copper due to the attractive physical and chemical properties. Specifically, there are different types of copper NPs that can be synthesized via multiple methods: CuNPs, cupric oxide nanoparticles (CuONPs), and cuprous oxide nanoparticles

(Cu$_2$ONPs) [5–8]. In particular, CuONPs are p-type oxide semiconductors that, for its catalytic, optical, magnetic, and electrical properties, are largely used as catalysts; as sensors; in opto-electronics; as photocatalyst; and in antibacterial, antifungal, antiviral, anticancer, antioxidant, or drug delivery applications, among other [9,10]. In addition, CuONPs are significantly less toxic, with values L50 values reported between 64 and 840 mg/L vs. between 0.22 and 24 mg/L for CuNPs [11]. For the synthesis of CuNPs, many methods have been employed such as thermal evaporation, chemical synthesis, electrochemical synthesis, solvothermal route, and vapor–liquid–solid growth [5–10]. However, physical methods have some disadvantages such as the requirement of expensive and complicated vacuum techniques. Chemical synthesis limits NP applications in clinical fields due to the usage of toxic chemicals that may lead to environmental and biological hazards. By contrast, CuNP green synthesis is eco-friendly and cost effective and does not use toxic chemicals, which make CuNPs attractive in biomedical applications.

Cu is an essential micronutrient for living cells because it is a constituent of many metalloenzymes including cytochrome-c oxidase and superoxide dismutase (SOD) [12]. However, Cu can also generate reactive oxygen species [13,14], it can be poisonous at high concentrations, and can reduce microorganism growth. It is well established that, when microbes are kept in a toxic metal environment, they evolve mechanisms to survive in harsh conditions by transforming toxic metal ions into their corresponding nontoxic forms such as metal sulfide/oxides. For example, a new *Alcanivorax* sp. isolated from a shallow hydrothermal vent was resistant to copper toxicity [15]. A large group of biological resources such as bacteria, yeasts, fungi, algae, and plants can be used for the synthesis of NPs from metal ions; however, the detailed mechanisms involved in nanoscale transformation are not well established. Among all biological systems used until now, bacteria have acquired significant attention as they are easy to culture, are able to produce extracellular NPs with easy downstream processing (as purification steps), and have short generation times for NP synthesis. Furthermore, a large group of biological resources such as bacterial biomolecules acts as a reducing and stabilizing agent for NP synthesis, limiting particle growth, and prohibiting agglomeration, resulting in the formation of desired NPs. In this regard, copper oxide nanoparticles have different applications, such as antibacterial, antifungal, antiviral, anticancer, antioxidant, and drug delivery applications [10].

In the present study, we report the biosynthesis of CuONPs at low temperatures from bacterial strains isolated from a consortium associated with the Antarctic ciliate *Euplotes focardii* [16]. This microorganism is a free-swimming ciliate, endemic of the oligotrophic coastal sediments of the Terra Nova Bay [17]. All bacterial strains are identified as *Marinomonas* (MM) [18], *Rhodococcus* (RH), *Pseudomonas* (PM) [19–21], *Brevundimonas* (BM), and *Bacillus* (BC). All were named with the "ef1" suffix (ef stands for *Euplotes focardii* and 1 indicates that is the first strain of the genus isolated from this organism).

Although Antarctica is regarded as the last uncontaminated continent, it is not completely free from pollution [22]. Despite isolation of the continent by natural barriers such as circumpolar atmospheric and oceanic currents [23], contaminants such as heavy metals, pesticides, and other persistent organic pollutants (POPs) could reach Antarctica via long range atmospheric transport (LRAT) from other continents in the southern hemisphere and even beyond [24]. Previous work reported evidence that bacteria from Antarctica developed resistance to heavy metals [25,26]. Our results support this evidence since these bacterial strains shows resistance to up to 5 mM of CuSO$_4$ and produced CuONPs. Our study highlights an efficient strategy in obtaining bionanomaterials that can be used as antibiotics against a large number of drug-resistant pathogens bacteria, which has created serious concern across the globe due to the limited choices in antibiotic treatment [27].

2. Results and Discussion

2.1. Copper Tolerance and Growth Assessment for the Bacterial Strains

As a first step in this work, we assessed the Cu tolerance of all bacterial strains under study. Tables 1 and 2 report the bacterial growth assessments at different CuSO$_4$ concentrations and the maximum tolerated concentrations (MTCs) of heavy metals, respectively. Our results indicate that all strains tolerate $CuSO_4$ up to 3.5–4 mM.

Table 1. Growth assessment of bacteria with various concentrations of $CuSO_4$. High growth: +++, medium growth: ++, low growth: +, and no growth: −.

Organisms	$CuSO_4$ Concentration (mM)												
	0.0	0.5	1	1.5	2	2.5	3	3.5	4	4.5	5	5.5	6
Marinomonas ef1	+++	+++	+++	+++	+++	+++	++	++	++	++	+	−	−
Rhodococcus ef1	+++	+++	+++	+++	+++	+++	+++	++	+	+	−	−	−
Pseudomonas ef1	+++	+++	+++	+++	++	++	+	+	−	−	−	−	−
Brevundimonas ef1	+++	+++	+++	+++	+++	++	++	++	++	+	+	−	−
Bacillus ef1	+++	+++	+++	+++	+++	++	++	+	+	−	−	−	−

Table 2. Maximum tolerated $CuSO_4$ concentrations (MTCs).

Organisms	$CuSO_4$ (mM)
Marinomonas ef1	5
Rhodococcus ef1	4.5
Pseudomonas ef1	3.5
Brevundimonas ef1	5
Bacillus ef1	4

We also monitored bacterial growth in the presence of increasing $CuSO_4$ concentrations for each strain (Figure S1): increasing copper concentrations decreased the growth rate of all bacteria tested. The highest growth inhibition effect is visible at concentrations above 3.5 mM, in particular, for Pseudomonas ef1 and Bacillus ef1 (Figure S1).

2.2. Biosynthesis of CuNPs

All of the tested bacteria showed resistance to Cu up to 3.5 mM; thus, we reasoned that these strains may be able to synthesize copper NPs. With the addition of 1 mM $CuSO_4$ (final concentration) in the reaction medium, we observed a gradual change in the solution color from cyan to brown over a 48 h period of time (Figure S2). A similar change in color has been reported after the addition of 5 mM $CuSO_4$ to a flask containing *Morganella* sp. [28], or three different species of *Penicillium* and the white-rot fungus *Stereum hirsutum* [29]. Therefore, our result suggests the formation of copper NPs from $CuSO_4$ through microbial metabolisms.

2.3. Ultraviolet–Visible Absorption Spectroscopy (UV–Vis), Dynamic Light Scattering (DLS) Analysis, and Zeta Potential Measurements of CuNPs

UV–vis spectral analysis is the most important analysis method to detect the surface plasmon resonance (SPR) property of biosynthesized CuNPs. We applied UV–vis spectroscopy to all of the samples obtained from the different strains: a sharp peak with maximum absorption in the range of 381–383 nm was recorded in each sample and can be attributed to the formation of CuNPs [30] (Figure S3).

Our results are in agreement with previous reports on bacterial synthesized CuNPs. Indeed, broad absorption spectra peaks were observed at around 365 nm for CuNPs synthesized from *Escherichia coli* [31], at 310 nm for *Eichhornia crassipes* [32], and at 360 nm for CuNPs synthesized from Ixora coccinea leaf extracts [30].

Dynamic light scattering (DLS) was also performed to determine the size distribution and zeta potential for biosynthesized CuNPs. All of the average diameters are reported in Figure 1 for all of the different bacterial strains.

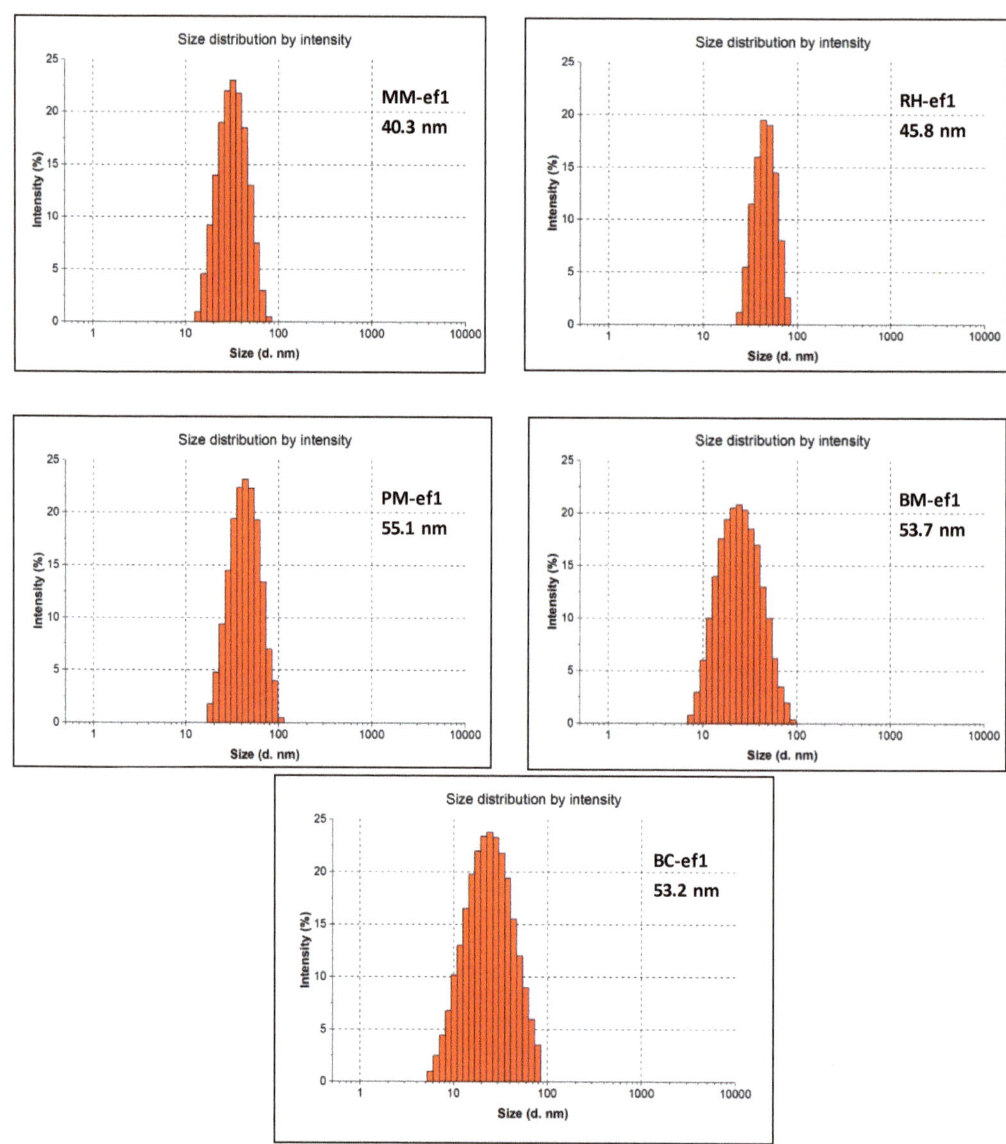

Figure 1. DLS of bio-CuNPs synthesized from the different bacterial strains: *Marinomonas* (MM), *Rhodococcus* (RH), *Pseudomonas* (PM), *Brevundimonas* (BM), and *Bacillus* (BC).

Similarly, the zeta potential values of CuNPs are in the range from -23.2 mV to -33.8 mV (Table 3). The zeta potential data demonstrated that biosynthesized NPs were stable in a liquid medium, and therefore, the tested bacterial strains represent a good source for production of narrow-sized CuNPs. The stability may be due to NP capping by biomolecules produced as byproducts. The higher negative zeta potential values obtained, thus, confirmed the repulsion between particles to thwart agglomeration.

Table 3. Zeta potential of CuNPs obtained from different strains.

Organisms	Zeta Potential (mV)
Marinomonas ef1	−23.2
Rhodococcus ef1	−33.8
Pseudomonas ef1	−33.1
Brevundimonas ef1	−33.6
Bacillus ef1	−25.1

2.4. XRD and FTIR Analyses

The powered XRD analysis of synthesized CuNPs with different strains was applied in order to investigate the crystalline phase of Cu nanostructures. In Figure 2, all XRD profiles of CuNPs are reported in comparison with simulation data (black line defined as theoretical). The obtained spectra reveal characteristic diffraction peaks indicating the formation of CuO in monoclinic and crystalline phase (the specific explanations are reported in Appendix A). In the CuO monoclinic structure, each Cu atom is situated at the center of four oxygen atoms positioned at the vertices of a rectangle with oxygen atoms at the center of a tetrahedron of Cu atoms [31]. The slight differences in peak positions with respect to the simulation data indicate that the different coatings of the obtained nanoparticles promote defects in the crystals, giving different crystalline orientations with respect to the theoretical ones. In addition, the XRD pattern demonstrate the absence of impurity and of peaks related to $Cu(OH)_2$ or Cu_2O phases (see Appendix A for more details).

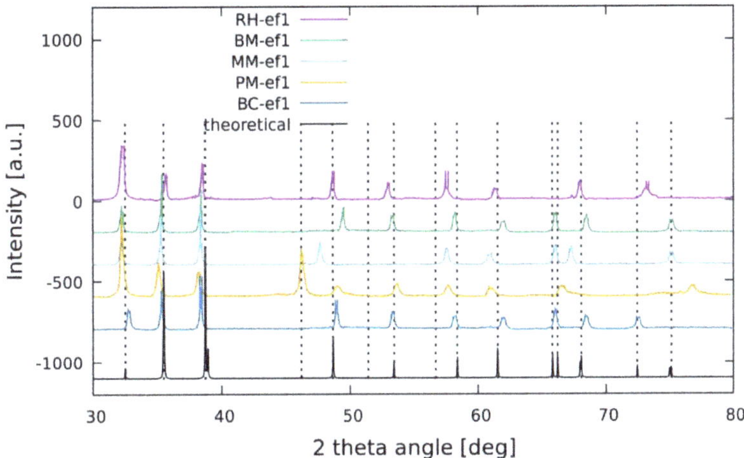

Figure 2. XRD profile of biosynthesized CuONPs. The obtained spectra reveal characteristic diffraction peaks indicating the formation of CuO in all bacterial NPs in comparison with simulation peaks (black line defined as the theoretical peaks).

FTIR spectroscopy was carried out to verify the possible involvement of functional groups or biomolecules in CuONP formation and stabilization. This technique is a powerful tool used to identify the chemical bonds in a molecule by producing an IR spectrum that is similar to a molecular fingerprint. The FTIR spectra of CuONPs biosynthesized by MM-ef1, RH-ef1, PS-ef1, BM-ef1, and BC-ef1 were obtained in the range between 400 and 4000 cm^{-1} (Figures S4–S8).

The IR spectra suggested that the biomolecules interacted with the biosynthesized CuONPs. The distinct bands analyzed indicated the presence of −OH, −NH, and $−CH_2$ scissor vibrations of aliphatic compounds and C=C bonds inside the biomolecules. From the analysis of the peaks, a carbonyl group (C=O) of the amide functional group was also detected. Therefore, the presence of carbonyl and NH groups is important for stabilization of the nanoparticles. In fact, the doublet of electrons present on both groups can be useful for the electrostatic stabilization of CuONP nanoparticles and thus to function as a capping agent [32].

2.5. Transmission Electron Microscopy (TEM)

Controlling NP size distribution is important for many applications, in particular for antimicrobial activity. We applied TEM analysis (Figure 3) to obtain additional insights in the morphology and size of the CuONPs from all of the strains.

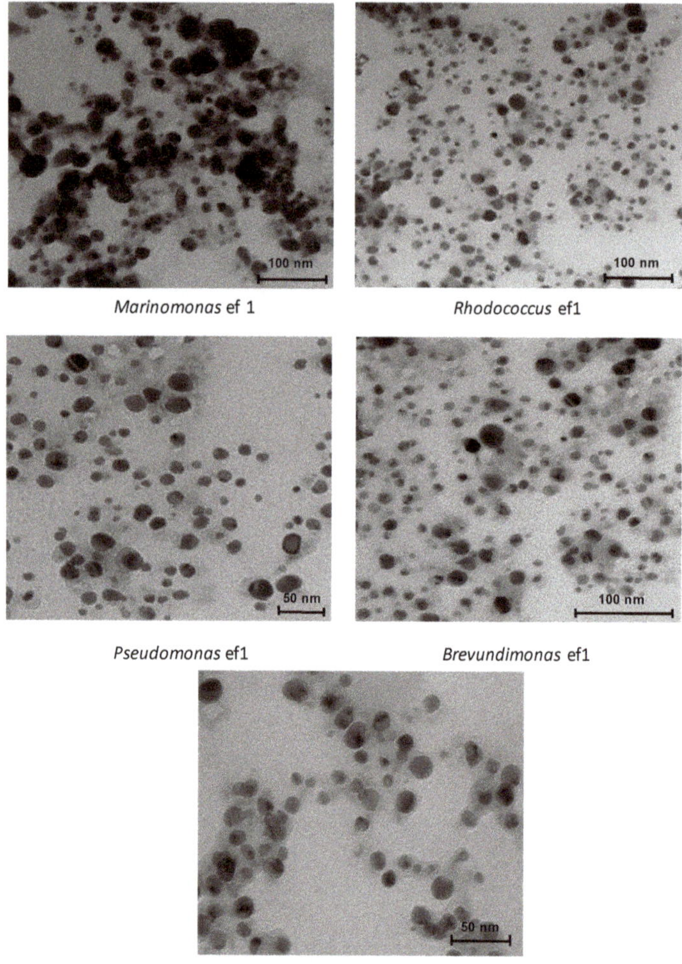

Figure 3. TEM images of biosynthesized CuONPs from the five bacterial strains: *Marinomonas* ef1, *Rhodococcus* ef1, *Pseudomonas* ef1, *Brevundimonas* ef1, and *Bacillus* ef1.

TEM micrographs of all bacterial CuONPs showed that the particles are mono-dispersed, spherical/ovoidal shapes. The micrographs suggested particle sizes around 10 nm to 70 nm, with the average size being 40 nm. The irregular shape and change in dimensions observed using different bacteria during the synthesis depends by the different bacterial metabolisms and of the secondary compounds produced that surround the nanoparticles. These compounds act as capping and dispersing agents influencing the shape and dimension of the nanostructures. All of the nanoparticles are well segregated except for the nanoparticles produced by *Marinomonas* ef1, where larger aggregates are observed (Figure 3).

2.6. Antimicrobial Activity of the Bacterial CuONPs

It has been reported that CuNPs of about 30 nm are more active against pathogens than those of smaller size [33]. CuONPs have received much attention in a wide range of applications, including their use as antimicrobial agents [9,34–37].

To test the potential antimicrobial activity of our biosynthesized CuONPs, we performed the disk diffusion test against *Staphylococcus aureus*, *Escherichia coli* (EC), *Klebsiella pneumoniae* (KP), *Pseudomonas aeruginosa* (PA), *Proteus mirabilis* (PrM), *Citrobacter koseri* (CK), *Acinetobacter baumanii* (AB), *Serratia marcescens* (SM), *Candida albicans* (CA), and *Candida parapsilosis* (CP).

The results of the disk diffusion test are reported in Figure 4, in which the presence of a clear zones around the CuONP disk is clearly visible, suggesting that all nanoparticles were able to inhibit the growth of bacterial and fungal pathogens.

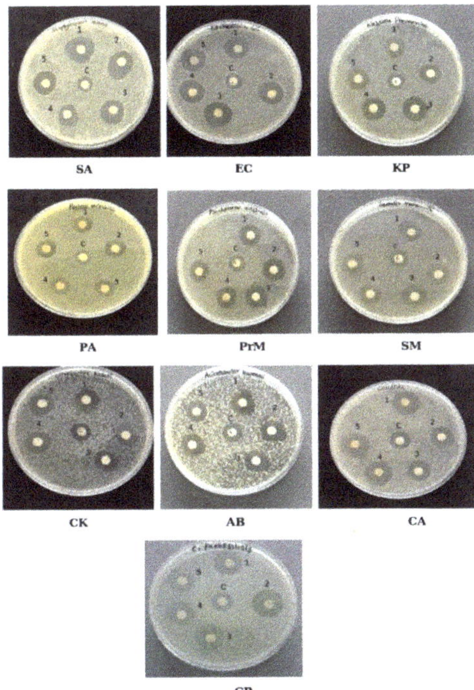

Figure 4. Antibacterial activity of bio-CuONPs against pathogenic bacteria: Staphylococcus aureus (SA), Escherichia coli (EC), Klebsiella pneumoniae (KP), Pseudomonas aeruginosa (PA), Proteus mirabilis (PrM), Serratia marcescens (SM), Citrobacter koseri (CK), Acinetobacter baumanii (AB), Candida albicans (CA), and Candida parapsilosis (CP). Kirby–Bauer disk diffusion test. 1. MM-ef1, 2. RH-ef1, 3. BM-ef1, 4. PM-ef1, 5. BC-ef1, and C—control (1 mM of $CuSO_4$).

All of the results for inhibition zone variation are summarized in Table 4 as a comparison with negative controls.

Table 4. Inhibition zone variation due to antibacterial activity of bio-CuONPs.

		MM-ef1	RH-ef1	BM-ef1	PM-ef1	BC-ef1	CuSO$_4$
		Gram-Positive Bacteria					
	SA	16 ± 0.2 *	16 ± 0.3 *	16 ± 0.1 *	13 ± 0.4 *	15 ± 0.2 *	9 ± 0.3 °
		Gram-Negative Bacteria					
	EC	15 ± 0.2 *	16 ± 0.4 *	17 ± 0.2 *	16 ± 0.3 *	16 ± 0.2 *	10 ± 0.1 °
	KP	16 ± 0.1 *	15 ± 0.1 *	17 ± 0.3 *	16 ± 0.4 *	16 ± 0.3 *	10 ± 0.3 °
	PM	16 ± 0.3 *	17 ± 0.4 *	16 ± 0.4 *	15 ± 0.4 *	15 ± 0.2 *	11 ± 0.2 °
	PrM	15 ± 0.1 *	15 ± 0.2 *	14 ± 0.2 *	15 ± 0.2 *	15 ± 0.4 *	10 ± 0.2 °
	CK	16 ± 0.3 *	15 ± 0.3 *	15 ± 0.2 *	16 ± 0.2 *	16 ± 0.3 *	10 ± 0.1 °
	AB	16 ± 0.2 *	17 ± 0.4 *	15 ± 0.1 *	16 ± 0.2 *	15 ± 0.2 *	11 ± 0.2 °
	SM	15 ± 0.2 *	15 ± 0.2 *	15 ± 0.2 *	14 ± 01 *	14 ± 0.2 *	10 ± 0.3 °
		Fungi					
	CA	18 ± 0.4 *	16 ± 0.4 *	16 ± 0.4 *	17 ± 0.4 *	18 ± 0.4 *	11 ± 0.2 °
	CP	17 ± 0.4 *	19 ± 0.2 *	17 ± 0.4 *	16 ± 0.4 *	16 ± 0.4 *	11 ± 0.2 °

Inhibition zone (Ø, in mm) with * CuONPs and with ° CuSO$_4$ (negative control).

We further evaluated the CuONP antibacterial activity by estimating the minimum inhibitory concentration (MIC) and minimum bactericidal concentration (MBC) values using the plate microdilution method [38]. The MIC is defined as the lowest concentration of antibacterial agent needed that inhibits the growth of bacteria. As showed in Table 5 and Figures S9–S13, the CuONPs synthesized by all bacteria showed an MIC against the Gram-negative bacteria from 3.12 to 25 µg/mL, against Gram-positive bacteria from 12.5 to 25 µg/mL, and against fungi from 12.5 to 25 µg/mL. MBC is defined as the lowest concentration of an antibacterial agent that kills the bacteria (no growth was observed on the agar plate). In the study, MBC values ranged from 12.5 to 25 µg/mL. These analyses showed that these CuONPs possess strong antimicrobial activity. The different MIC and MBC values could be attributed to the different shapes and dimensions of the nanoparticles determining their antibacterial activity.

Table 5. MIC and MBC values (µg/mL) of the CuONPs synthesized by MM-ef1, RH-ef1, PM-228 ef1, BM-ef1, and BC-ef1 against pathogenic bacteria.

	MM-ef1		RH-ef1		BM-ef1		PM-ef1		BC-ef1	
	MIC	MBC	MIC	MBC	MIC	MBC	MIC	MBC	MIC	MBC
	Gram Positive Bacteria									
SA	12.5 ± 0.2	25 ± 0.4	12.5 ± 0.2	25 ± 0.2	25 ± 0.2	25 ± 0.4	12.5 ± 0.4	25 ± 0.2	12.5 ± 0.4	12.5 ± 0.1
	Gram Negative Bacteria									
EC	12.5 ± 0.2	25 ± 0.3	12.5 ± 0.4	25 ± 0.3	25 ± 0.4	25 ± 0.3	25 ± 0.3	25 ± 0.2	12.5 ± 0.2	12.5 ± 0.2
KP	12.5 ± 0.1	25 ± 0.5	6.25 ± 0.3	12.5 ± 0.2	12.5 ± 0.2	25 ± 0.4	12.5 ± 0.2	25 ± 0.4	6.25 ± 0.2	6.25 ± 0.2
PM-sp	6.25 ± 0.2	12.5 ± 0.2	12.5 ± 0.2	25 ± 0.4	12.5 ± 0.2	25 ± 0.3	12.5 ± 0.1	12.5 ± 0.4	12.5 ± 0.3	12.5 ± 0.4
PM	3.12 ± 0.1	6.25 ± 0.2	6.25 ± 0.2	12.5 ± 0.1	6.25 ± 0.1	12.5 ± 0.2	6.25 ± 0.2	12.5 ± 0.2	6.25 ± 0.1	12.5 ± 0.1
CK	6.25 ± 0.2	6.25 ± 0.1	12.5 ± 0.3	12.5 ± 0.2	12.5 ± 0.2	25 ± 0.4	6.25 ± 0.1	12.5 ± 0.4	12.5 ± 0.4	12.5 ± 0.1
AB	12.5 ± 0.2	12.5 ± 0.2	12.5 ± 0.2	12.5 ± 0.2	12.5 ± 0.2	25 ± 0.4	12.5 ± 0.2	12.5 ± 0.2	12.5 ± 0.2	12.5 ± 0.4
SM	6.25 ± 0.1	12.5 ± 0.4	3.12 ± 0.1	12.5 ± 0.2	6.25 ± 0.1	12.5 ± 0.2	6.25 ± 0.1	12.5 ± 0.2	6.25 ± 0.2	12.5 ± 0.2
	Fungi									
CA	25 ± 0.4	25 ± 0.5	12.5 ± 0.2	25 ± 0.4	12.5 ± 0.2	25 ± 0.4	25 ± 0.2	25 ± 0.4	12.5 ± 0.2	25 ± 0.3
CP	12.5 ± 0.4	25 ± 0.3	25 ± 0.2	25 ± 0.2	12.5 ± 0.1	25 ± 0.2	12.5 ± 0.3	25 ± 0.1	6.25 ± 0.2	12.5 ± 0.2

It was observed that the bactericidal property of CuONPs is due to the generation of a large number of small molecules containing reactive oxygen species (ROS) by the nanoparticles attached to the bacterial cells. Although cells can detoxify ROS species within certain limits by promoting antioxidant activity such as specific enzymes or small molecules (e.g., ascorbic acid), the formation and accumulation of ROS species largely increased when the cell was constitutively exposed to intracellular oxidative stress [36].

3. Materials and Methods

3.1. Culture and Chemicals

All bacterial strains used in this work were isolated from a consortium associated with the Antarctic ciliate *E. focardii* and identified as *Marinomonas, Rhodococcus, Pseudomonas, Brevundimonas*, and *Bacillus* [17,18,20]. All strains were grown at 22 °C on agarized or liquid Luria–Bertani (LB) medium (tryptone 10g/L, yeast extract 5 g/L, and NaCl 5 g/L). All media were purchased from Liofilchem. Analytical grade copper (II) sulfate pentahydrate salt, $CuSO_4 \cdot 5H_2O$, and the other chemicals were purchased from Sigma Aldrich (Sigma Aldrich, St. Louis, MO, USA).

3.2. Determination of Copper Maximum Tolerated Concentrations (MTCs)

Copper (Cu) maximum tolerated concentrations (MTCs) were determined for each isolate on agarized LB medium in the presence of increasing concentrations (from 0 to 6 mM) of $CuSO_4$. With a sterile inoculating loop, each bacterial isolate was streaked on the Cu incorporated agarized LB medium. The plates were incubated at 22 °C and inspected at intervals up to 72 h. The MTCs were noted when the isolate failed to show growth on the plates after three days of incubation. All experimental setups were prepared in triplicate.

3.3. Estimation of Bacterial Growth Inhibition by Copper

Bacterial growth Cu inhibition was determined by monitoring the optical density at regular time intervals at 600 nm using a Shimadzu UV 1800 spectrophotometer. In total, 0.5 mL of an overnight active culture adjusted to OD600 = 0.1 was incorporated into 50 mL of LB medium supplemented with increasing concentrations of $CuSO_4$ from 0 up to the maximum concentration tolerated by each bacterium. Inoculated flasks were incubated on a rotary shaker at 22 °C. All experimental setups were prepared in triplicate.

3.4. Biosynthesis of CuNPs

Each strain was inoculated in LB medium (100 mL) and incubated at 22 °C on a rotatory shaker (200 rpm). After 24 h, $CuSO_4 \cdot 5H_2O$ was added to the microbial cell culture to a final concentration of 1mM. The reaction mixture was incubated for 24 to 48 h on a rotatory shaker at 150 rpm at 22 °C. LB medium with 1 mM $CuSO_4$ without the organism or heat-killed bacterial cultures were maintained as a control. During the incubation period, the solution color change from blue to dark green (which is indicative of the reduction of $CuSO_4$ to CuNPs) was monitored. The formation of CuNPs was also recorded by absorption spectra in the wavelength range of 200–800 nm at room temperature (23 °C) using a Shimadzu UV 1800 spectrophotometer.

3.5. Purification of CuNPs

After incubation, the culture was centrifuged at 5000 rpm at 4 °C for 20 min in a Beckman J2-21 with swinging rotor to separate the cell pellet from the cell-free supernatant. Nanoparticles were purified from both the supernatant and pellet. To recover CuNPs present in the cell-free supernatant, the solution was centrifuged at $17,000 \times g$ for 15 min in the same centrifuge with a fixed rotor. The CuNPs containing the pellet was then resuspended in double-distilled water (ddH$_2$O) and washed twice by repeated centrifugation steps.

To recover CuNPs from the cell pellet, ultrasonic wave shocks of short durations (15 s) were given to the ddH$_2$O-suspended pellet to rupture the microbial cell wall. After sonication, the sample was centrifuged 5000 rpm for 20 min (Beckman J2–21, Fullerton, California) and the NPs were recovered from the supernatant. This step was repeated three times to completely remove the cell debris from the supernatant. To recover CuNPs present in the cell-free supernatant, the solution was centrifuged at $17,000 \times g$ for 15 min in the same centrifuge with a fixed rotor. After being washed twice with deionized water and dried at 80 °C in an oven, the CuNPs were used for further characterization and experiments.

3.6. Dynamic Light Scattering, Zeta Potential Measurement, Transmission Electron Microscopy (TEM), X-ray Diffraction Analysis (XRD), and Fourier-Transform Infrared Spectroscopy (FTIR) Analyses

Zetasizer Nano ZS, (Malvern Instruments Ltd., Malvern, UK) was used to determine the size distribution of particles by measuring dynamic fluctuations of light scattering intensity caused by the Brownian motion of the particles. All measurements were carried out in triplicate with a temperature equilibration time of 2 min at 25 °C. Additionally, NP surface charge was measured using the zeta potential. The morphologies of the biosynthesized CuNPs were observed on a JEOL transmission electron microscope (TEM) system operating at 200 kV (JEM-2100, Hitachi Limited, Tokyo, Japan), with the acquirement of the particle size distribution ascertained from TEM micrographs based on professional software (Nano Measurer 1.2.5). The crystal structure of biosynthesized CuNPs was analyzed by powder X-ray diffraction (XRD) measurements performed using a Rigaku-D/MAX-PC 2500 X-ray diffractometer (Wilmington, MA, USA) with a Cu Kα (λ = 1.5405 Å) in the 2θ range from 20 to 80° at a scan rate of 0.03° S-1. FTIR spectra (Kyoto, Japan) were recorded (Shimadzu IR Affinity-1) to identify the possible interactions between CuNPs and the biomolecule. Analysis was carried out in the range of 400–4000 cm^{-1} at the resolution of 4 cm^{-1}.

3.7. Kirby–Bauer Disk Diffusion Susceptibility Test, Minimum Inhibitory Concentration (MIC), and Minimum Bactericidal Concentration (MBC) Evaluation

The antibacterial activity of biosynthesized CuNPs was tested against Staphylococcus aureus, Escherichia coli, Klebsiella pneumoniae, Pseudomonas aeruginosa, Proteus mirabilis, Citrobacter koseri, Acinetobacter baumanii, Serratia marcescens, Candida albicans, and Candida parapsilosis. All the strains were cultured in Mueller Hinton broth (MHB) (Merck, Darmstadt, Germany) at 37 °C. The antibacterial activity of CuNPs against the selected bacterial strains was assessed using the Kirby–Bauer disk diffusion susceptibility test method. Using a sterile cotton swab, the bacteria strains were spread on the Mueller–Hinton agar (MHA). The disks were loaded with 25 µL (25 µg) of 1 mg/mL CuNP solution and $CuSO_4$ solution (1 mM) and dried. The disks were then placed on the agar plate and incubated at 37 °C. The inhibition zone was observed after 24 h of incubation.

The MIC and MBC estimations of the CuNPs were performed using the method described in the guideline of CLSI 2012 [38]. The MIC test was performed on a 96-well round bottom microtiter plate using standard broth microdilution methods, while the MBC test was performed on MHA plates. The bacterial inoculums were adjusted to the concentration of 0.5 McFarland units. For the MIC test, CuNP stock solution was prepared by ultrasonication in sterilized deionized water to reach 200 µg/mL. A volume of 100 µL of stock solution was serially diluted twofold in 100 µL of MHB in the first row, and finally 100 µL was discarded such that the first well in the row of the microtiter plate contained the highest concentration of CuNPs while the last well of the row contained the lowest concentration. Similarly, CuNPs were prepared in all of the rows. The positive control contained the medium and bacterial inoculums (K^+), and the negative control contained only the medium (K^-). The microtiter plate was then incubated at 37 °C for 24 h. The MIC value was defined as the lowest concentration of antibacterial agents that inhibits the growth of bacteria. The MBC was taken as the lowest concentration of antibacterial agents that completely kills the bacteria. To check MBC, the suspension from each well of the microtiter plates was plated onto the MHA plate and were incubated at 37 °C for 24 h. The MBC value was taken as the lowest concentration with no visible growth on the MHA plate.

4. Conclusions

In this study, we reported the production of monodisperse, small, and highly pure bio-CuONPs using the green reduction of $CuSO_4$ at low temperatures, such as 22 °C, by using five Antarctic bacterial strains. The ability of these bacteria to synthesize CuONPs may represent a defense mechanism against this heavy metal. The results support the evidence that these bacterial strains are resistant to up to 5 mM of $CuSO_4$. All of the nanoparticles were fully characterized and tested for their antibacterial activity.

This study confirms that the tested Antarctic bacterial strains can be exploited in bioremediation to remove copper contamination from the environment and in the production of antibiotics against various types of pathogenic Gram-negative and Gram-positive bacteria, and fungi including *Escherichia coli*, *Staphylococcus aureus*, and *Candida albicans*. These results showed that the cost-effective and eco-friendly biosynthesis of these CuONPs make them particularly attractive in several applications including biomedical science.

5. Patents

The results of this paper are related to patent numbers 102019000014121 and 10201900-0024493. *Marinomonas* sp. ef1 and *Rhodococcus* sp. ef1 have been deposited at the Istituto Zooprofilattico Sperimentale della Lombardia e dell'Emilia Romagna "Bruno Ubertini"—IZSLER according to the Budapest treaty under access No. DPS RE RSCIC 17 and DPS RE RSCIC 4. *Brevundimonas* sp ef1 and *Bacilus* sp ef1 have be deposited under access No. DPS RE RSCIC 23 and DPS RE RSCIC 24.

Supplementary Materials: The following are available online at https://www.mdpi.com/article/10.3390/md19050263/s1. Figure S1, Bacterial growth in the presence of increasing $CuSO_4$ concentrations from 0.5 to 4 mM. Figure S2, Biosynthesis of Cu NPs from *Marinomonas* ef1, *Rhododccoccus* ef1, *Pseudomonas* ef1, *Brevundimonas* ef1, and *Bacillus* ef1. A. Control with a heat-killed bacterial culture using 1 mM of $CuSO_4$, B. control with the LB medium using 1 mM of $CuSO_4$, and C. biosynthesized CuNPs. Figure S3, UV−vis absorbance spectra of bio-CuONPs synthesized from (A) *Rhodococcus* ef1, (B) *Pseudomonas* ef1, (C) *Brevundimonas* ef1, (D) *Bacillus* ef1, and (E) *Marinomonas* ef1. Figures S4, FTIR spectrum of biosynthesized CuO NPs from *Marinomonas* ef1 and IR assignments. Figures S5, FTIR spectrum of biosynthesized CuO NPs from *Rhodococcus* ef1 and IR assignments. Figures S6, FTIR spectrum of biosynthesized CuO NPs from *Pseudomonas* ef1 and IR assignments. Figures S7, FTIR spectrum of biosynthesized CuO NPs from *Brevundimonas* ef1 and IR assignments. Figures S8, FTIR spectrum of biosynthesized CuO NPs from *Bacillus* ef1 and IR assignments. Figure S9, MIC values of *Marinomonas* CuO NPs. Figure S10, MIC values of *Rhodococcus* ef1 CuO NPs. Figure S11, MIC values of *Brevundimonas* ef1 CuO NPs. Figure S12, MIC values of *Pseudomonas* ef1 CuO NPs. Figure S13, MIC values of *Bacillus* ef1 CuO NPs.

Author Contributions: Study conception and design: M.S.J., J.A.N., S.P.; Acquisition of data: M.S.J. and J.A.N.; Analysis and interpretation of data: K.P.R., M.Z., R.G., A.M. and S.P.; Drafting of manuscript: M.S.J., J.A.N.; Critical revision: C.M., S.P. All authors have read and agreed to the published version of the manuscript.

Funding: This research work was supported by the European Commission Marie Sklodowska-Curie Actions H2020 RISE Metable—645693.

Acknowledgments: We are grateful to Ilidio Correia from the Faculty of Health Sciences, Universidade da Beira Interior, Covilhã (Portugal).

Conflicts of Interest: The authors declare no conflict of interest.

Appendix A. Comments on the X-ray Diffraction (XRD) Structural Analysis

The XRD analysis of *Marinomonas* ef1 synthesized CuO NPs show characteristic diffraction peaks at 2θ of 32.32°, 35.18°, 38.45°, 48.10°, 58.05°, 61.21°, 66.11°, 67.56° and 75.17° which were assigned to (110), (111), (111), (202), (202), (113), (022), (220) and (004) planes respectively. All the peaks correspond to the face centered cubic (fcc) structure and were in consistence with the standard JCPDS (No.04-0836) data. The strong intensity diffraction peaks clearly indicate that the CuO NPs are highly crystalline and in monoclinic phase.

The XRD pattern of CuO NPs synthesized from *Rhodococcus* ef1 demonstrate that the CuO NPs is crystalline in nature. Also the spectrum is similar to that of pure CuO, indicating the formation of single-phase CuO with monoclinic structure (JCPDS. 05-0661). In the present work, the diffraction patterns were observed to be at 2θ value of 32.37°, 35.39°, 38.66°, 48.61°, 53.30°, 58.21° and 61.35° and they were assigned to the corresponding (110), (111), (111), (202), (020), (202), (113), (113) and (113) planes, respectively. The reflection lines indicate the formation of monoclinic CuO NPs.

According to JCPDS data (80-0076), *Pseudomonas* ef1 exhibited diffraction peaks at 2θ of 32.49°, 35.50°, 38.70°, 48.72°, 58.28°, 61.54°,66.24°,68.03° and 75.16° which were assigned to (110), (111), (111), (202), (202), (113), (311), (220) and (004) planes respectively corresponds to different planes of monoclinic phase of CuONPs and correspond to the face centered cubic (FCC) structure. Diffraction peaks clearly indicate that the CuONPs are highly crystalline.

The phase purity and structural characteristics of the synthesized CuO NPs from *Brevundimonas* ef1 showed the XRD Diffraction patterns appeared at 2θ of 32.42°, 35.47°, 38.59°, 48.66°, 53.54°, 58.69°, 61.70°, 66.01°, 68.18° and 75.19° values and they were assigned to the corresponding (110), (002), (111), (112), (020), (202), (113), (310), (220) and (004) planes, respectively. The obtained XRD patterns indicated that the synthesized CuO NPs are highly crystalline with monoclinic structure of CuO, which was confirmed by the Joint Committee on Powder Diffraction Standards (JCPDS) (Card No.: 89-5895).

The XRD pattern of *Bacillus* ef1 revealed the orientation and crystalline nature of CuO NPs. The peak position with 2θ values of 32.41°, 35.39°, 38.86°, 48.67°, 53.40°, 58.28°, 61.43°, 65.69°, 68.18°, 72.37° and 75.10° are indexed as (110), (111), (111), (202), (020), (202), (113), (311), (220), (220), and (004) planes, which are in good agreement with those of powder CuO obtained from the International Center of Diffraction Data card (JCPDS-80-1916) confirming the formation of a crystalline monoclinic structure. No extra diffraction peaks of other phases are detected, indicating the phase purity of CuO NPs.

References

1. Grigore, M.E.; Biscu, E.R.; Holban, A.M.; Gestal, M.C.; Grumezescu, A.M. Methods of synthesis, properties and biomedical applications of CuO Nanoparticles. *Pharmaceuticals* **2016**, *9*, 75. [CrossRef]
2. Katwal, R.; Kaur, H.; Sharma, G.; Naushad, M.; Pathania, D. Electrochemical synthesized copper oxide Nanoparticles for enhanced photocatalytic and antimicrobial activity. *J. Ind. Eng. Chem.* **2015**, *31*, 173–184. [CrossRef]
3. Sastry, M.; Ahmad, A.; Khan, M.I.; Kumar, R. Biosynthesis of metal Nanoparticles using fungi and actinomycetes. *Curr. Sci.* **2003**, *85*, 162–170.
4. Mandal, D.; Bolander, M.E.; Mukhopadhyay, D.; Sarkar, G.; Mukherjee, P. The use of microorganisms for the formation of metal nanoparticles and their application. *Appl. Microbiol. Biotechnol.* **2006**, *69*, 485–492. [CrossRef]
5. Rahmatolahzadeh, R.; Aliabadi, M.; Motevalli, K. Cu and CuO nanostructures: Facile hydrothermal synthesis, characterization and photocatalytic activity using new starting reagents. *J. Mater. Sci. Mater. Electron.* **2017**, *28*, 148–156. [CrossRef]
6. Sagadevan, S.; Pal, K.; Chowdhury, Z.Z. Fabrication of CuO nanoparticles for structural, optical and dielectric analysis using chemical precipitation method. *J. Mater. Sci. Mater. Electron.* **2017**, *28*, 12591–12597. [CrossRef]
7. Zayyoun, N.; Bahmad, L.; Laânab, L.; Jaber, B. The effect of pH on the synthesis of stable Cu_2O/CuO Nanoparticles by sol–gel method in a glycolic medium. *Appl. Phys. A* **2016**, *122*, 488. [CrossRef]
8. Shi, G.; Bao, Y.; Chen, B.; Xu, J. Phenol hydroxylation over cubic/monoclinic mixed phase CuO Nanoparticles prepared by chemical vapor deposition. *React. Kinet. Mech. Cat.* **2017**, *122*, 289–303. [CrossRef]
9. Zhang, Q.; Zhang, K.; Xu, D.; Yang, G.; Huang, H.; Nie, F.; Liu, C.; Yang, S. CuO nanostructures: Synthesis, characterization, growth mechanisms, fundamental properties, and applications. *Progress Mater. Sci.* **2014**, *60*, 208–337. [CrossRef]
10. Waris, A.; Din, M.; Ali, A.; Ali, M.; Afridi, S.; Baset, A.; Khan, A.U. A comprehensive review of green synthesis of copper oxide nanoparticles and their diverse biomedical applications. *Inorgan. Chem. Commun.* **2021**, *123*, 1387–7003. [CrossRef]
11. Wu, F.; Harper, B.J.; Crandon, L.E.; Harper, S.L. Assessment of Cu and CuO nanoparticle ecological responses using laboratory small-scale microcosms *Environ. Sci. Nano* **2020**, *7*, 105–115. [CrossRef] [PubMed]
12. Lontie, R. *Copper Proteins and Copper Enzymes*; CRC Press: Boca Raton, FL, USA, 1984.
13. Kimura, T.; Nishioka, H. Intracellular generation of superoxide by copper sulphate in *Escherichia coli*. *Mutat. Res.* **1997**, *389*, 237–242. [CrossRef]
14. Nies, D.H. Microbial heavy-metal resistance. *Appl. Microbiol. Biotechnol.* **1999**, *51*, 730–750. [CrossRef]

15. Ramasamy, K.P.; Rajasabapathy, R.; Lips, I.; Mohandass, C.; James, R.A. Genomic features and copper biosorption potential of a new Alcanivorax sp. VBW004 isolated from the shallow hydrothermal vent (Azores, Portugal). *Genomics* **2020**, *112*, 3268–3273. [CrossRef] [PubMed]
16. Pucciarelli, S.; Devaraj, R.R.; Mancini, A.; Ballarini, P.; Castelli, M.; Schrallhammer, M.; Petroni, G.; Miceli, C. Microbial consortium associated with the Antarctic marine ciliate *Euplotes focardii*: An investigation from genomic sequences. *Microb. Ecol.* **2015**, *70*, 484–497. [CrossRef] [PubMed]
17. Pucciarelli, S.; La Terza, A.; Ballarini, P.; Barchetta, S.; Yu, T.; Marziale, F.; Passini, V.; Methé, B.; Detrich, H.W., 3rd; Miceli, C. Molecular cold-adaptation of protein function and gene regulation: The case for comparative genomic analyses in marine ciliated protozoa. *Mar. Genomics* **2009**, *2*, 57–66. [CrossRef]
18. John, M.S.; Nagoth, J.A.; Ramasamy, K.P.; Ballarini, P.; Mozzicafreddo, M.; Mancini, A.; Telatin, A.; Liò, P.; Giuli, G.; Natalello, A.; et al. Horizontal gene transfer and silver nanoparticles production in a new Marinomonas strain isolated from the Antarctic psychrophilic ciliate *Euplotes focardii*. *Sci. Rep.* **2020**, *10*, 1–14. [CrossRef]
19. John, M.S.; Nagoth, J.A.; Ramasamy, K.P.; Mancini, A.; Giuli, G.; Natalello, A.; Ballarini, P.; Miceli, C.; Pucciarelli, S. Synthesis of bioactive silver nanoparticles by a pseudomonas strain associated with the Antarctic psychrophilic protozoon *Euplotes focardii*. *Mar. Drugs* **2020**, *18*, 38. [CrossRef]
20. Orlando, M.; Pucciarelli, S.; Lotti, M. Endolysins from Antarctic *Pseudomonas* Display lysozyme activity at low temperature. *Mar. Drugs.* **2020**, *18*, 579. [CrossRef]
21. Ramasamy, K.P.; Telatin, A.; Mozzicafreddo, M.; Miceli, C.; Pucciarelli, S. Draft genome sequence of a new pseudomonas sp. Strain, ef1, Associated with the psychrophilic Antarctic ciliate *Euplotes focardii*. *Microbiol. Resour. Announ.* **2019**, *8*, e00867-19. [CrossRef]
22. Chu, W.L.; Dang, N.L.; Kok, Y.Y.; Yap, K.S.I.; Phang, S.M.; Convey, P. Heavy metal pollution in Antarctica and its potential impacts on algae. *Polar Sci.* **2019**, *20*, 75–83. [CrossRef]
23. Barker, P.F.; Thomas, E. Origin, signature and palaeoclimatic influence of the antarctic circumpolar current. *Earth-Sci. Rev.* **2004**, *66*, 143–162. [CrossRef]
24. Bargagli, R. Environmental contamination in Antarctic ecosystems. *Sci. Total Environ.* **2008**, *400*, 212–226. [CrossRef]
25. De Souza, M.J.; Nair, S.; Loka Bharathi, P.A.; Chandramohan, D. Metal and antibiotic-resistance in psychrotrophic bacteria from Antarctic Marine waters. *Ecotoxicology* **2006**, *15*, 379–384. [CrossRef]
26. Lo Giudice, A.; Casella, P.; Bruni, V.; Michaud, L. Response of bacterial isolates from Antarctic shallow sediments towards heavy metals, antibiotics and polychlorinated biphenyls. *Ecotoxicology* **2013**, *22*, 240–250. [CrossRef]
27. Mancini, A.; Pucciarelli, S. Antibiotic activity of the antioxidant drink effective Microorganism-X (EM-X) extracts against common nosocomial pathogens: An in vitro study. *Nat. Prod. Res.* **2018**, *4*, 1–6. [CrossRef]
28. Ramanathan, R.; Field, M.R.; O'Mullane, A.P.; Smooker, P.M.; Bhargava, S.K.; Bansal, V. Aqueous phase synthesis of copper nanoparticles: A link between heavy metal resistance and nanoparticle synthesis ability in bacterial systems. *Nanoscale* **2013**, *5*, 2300–2306. [CrossRef] [PubMed]
29. Honary, S.; Barabadi, H.; Ghara-El Fathabad, B.; Naguib, F. Green synthesis of copper oxide Nanoparticles using *Penicillium aurantiogriseum, Penicillium citrinum* and *Penicillium waksmanii*. *Digest. J. Nanomat. Bios.* **2012**, *7*, 999–1005.
30. Yedurkar, S.M.; Maurya, C.B.; Mahanwar, P.A. A biological approach for the synthesis of copper oxide nanoparticles by *Ixora Coccinea* leaf extract. *J. Mat. Environ. Sci.* **2017**, *8*, 1173–1178.
31. Dhineshbabu, N.R.; Rajendran, V.; Nithyavathy, N.; Vetumperumal, R. Study of structural and optical properties of cupric oxide nanoparticles. *Appl. Nanosci.* **2016**, *6*, 933–939. [CrossRef]
32. Jovanović, Ž.; Radosavljević, A.; Šiljegović, M.; Bibić, N.; Mišković-Stanković, V.; Kačarević-Popović, Z. Structural and optical characteristics of silver/poly(N-vinyl-2-pyrrolidone) nanosystems synthesized by γ-irradiation. *Radiat. Phys. Chem.* **2012**, *81*, 1720–1728. [CrossRef]
33. Datta Majumdar, T.; Singh, M.; Thapa, M.; Dutta, M.; Mukherjee, A.; Ghosh, C.K. Size-dependent antibacterial activity of copper nanoparticles against *Xanthomonas oryzae* pv. oryzae–A synthetic and mechanistic approach. *Colloid Interface Sci. Commun.* **2019**, *32*, 100190. [CrossRef]
34. Dadi, R.; Azouani, R.; Traore, M.; Mielcarek, C.; Kanaev, A. Antibacterial activity of ZnO and CuO nanoparticles against gram positive and gram negative strains. *Mater. Sci. Eng. C* **2019**, *104*, 109968. [CrossRef]
35. Ramzan, M.; Obodo, R.M.; Mukhtar, S.; Ilyas, S.Z.; Aziz, F.; Thovhogi, N. Green synthesis of copper oxide nanoparticles using Cedrus deodara aqueous extract for antibacterial activity. *Mater. Today Proc.* **2021**, *36*, 576–581. [CrossRef]
36. Applerot, G.; Lellouche, J.; Lipovsky, A.; Nitzan, Y.; Lubart, R.; Gedanken, A.; Banin, E. Understanding the antibacterial mechanism of CuO nanoparticles: Revealing the route of induced oxidative stress. *Small* **2012**, *8*, 3326–3337. [CrossRef] [PubMed]
37. Shuai, C.; Guo, W.; Gao, C.; Yang, Y.; Wu, P.; Feng, P. An nMgO containing scaffold: Antibacterial activity, degradation properties and cell responses. *Int. J. Bioprint.* **2018**, *4*, 120. [CrossRef] [PubMed]
38. *CLSI Document EP100-S23*; Clinical and Laboratory Standards Institute: Wayne, PA, USA, 2019.

marine drugs

Article

Svalbamides A and B, Pyrrolidinone-Bearing Lipodipeptides from Arctic *Paenibacillus* sp.

Young Eun Du [1], Eun Seo Bae [1], Yeonjung Lim [2], Jang-Cheon Cho [2], Sang-Jip Nam [3], Jongheon Shin [1], Sang Kook Lee [1], Seung-Il Nam [4] and Dong-Chan Oh [1,*]

1   Natural Products Research Institute, College of Pharmacy, Seoul National University, Seoul 08826, Korea; dye0302@snu.ac.kr (Y.E.D.); ddol1289@snu.ac.kr (E.S.B.); shinj@snu.ac.kr (J.S.); sklee61@snu.ac.kr (S.K.L.)
2   Department of Biological Sciences, Inha University, Incheon 22212, Korea; yj_lim@inha.edu (Y.L.); chojc@inha.ac.kr (J.-C.C.)
3   Department of Chemistry and Nanoscience, Ewha Womans University, Seoul 03760, Korea; sjnam@ewha.ac.kr
4   Korea Polar Research Institute, Incheon 21990, Korea; sinam@kopri.re.kr
*   Correspondence: dongchanoh@snu.ac.kr; Tel.: +82-880-2491; Fax: +82-762-8322

Abstract: Two new secondary metabolites, svalbamides A (**1**) and B (**2**), were isolated from a culture extract of *Paenibacillus* sp. SVB7 that was isolated from surface sediment from a core (HH17-1085) taken in the Svalbard archipelago in the Arctic Ocean. The combinational analysis of HR-MS and NMR spectroscopic data revealed the structures of **1** and **2** as being lipopeptides bearing 3-amino-2-pyrrolidinone, D-valine, and 3-hydroxy-8-methyldecanoic acid. The absolute configurations of the amino acid residues in svalbamides A and B were determined using the advanced Marfey's method, in which the hydrolysates of **1** and **2** were derivatized with L- and D- forms of 1-fluoro-2,4-dinitrophenyl-5-alanine amide (FDAA). The absolute configurations of **1** and **2** were completely assigned by deducing the stereochemistry of 3-hydroxy-8-methyldecanoic acid based on DP4 calculations. Svalbamides A and B induced quinone reductase activity in Hepa1c1c7 murine hepatoma cells, indicating that they represent chemotypes with a potential for functioning as chemopreventive agents.

Keywords: *Paenibacillus*; Arctic; Svalbard; Marfey's method; DP4 calculation; quinone reductase; lipopeptide; 3-amino-2-pyrrolidinone

Citation: Du, Y.E.; Bae, E.S.; Lim, Y.; Cho, J.-C.; Nam, S.-J.; Shin, J.; Lee, S.K.; Nam, S.-I.; Oh, D.-C. Svalbamides A and B, Pyrrolidinone-Bearing Lipodipeptides from Arctic *Paenibacillus* sp. *Mar. Drugs* **2021**, *19*, 229. https://doi.org/10.3390/md19040229

Academic Editor: Daniela Giordano

Received: 30 March 2021
Accepted: 15 April 2021
Published: 17 April 2021

Publisher's Note: MDPI stays neutral with regard to jurisdictional claims in published maps and institutional affiliations.

1. Introduction

Marine habitats were generally recognized as extreme environments exposing organisms to conditions of high salt, high pressure, and hypoxia, forcing them to develop unique physiologies in comparison to their terrestrial counterparts. Among the marine organisms, bacteria have always contributed significantly as a source for the discovery of new marine natural products, with 232 new compounds found in 2019 alone [1]. However, most (62.5%) of the new marine-bacterial molecules were derived from the single genus *Streptomyces*. We also reported the dimeric benz[*a*]anthracene thioethers donghaesulfins A and B, and rearranged angucyclinones donghaecyclinones A–C from marine-derived *Streptomyces* sp. SUD119 in 2019 and 2020, respectively [2,3]. Even though *Streptomyces* is chemically prolific and still provides numerous new bioactive compounds, the desire for compounds with greater structural diversity has brought about chemical investigations of a wider diversity of bacteria that extend beyond conventional phylogenetically biased chemical studies. The chemical examination of bacteria inhabiting the Arctic Ocean—a more extreme habitat than tropical or subtropical oceans that remains poorly investigated—represents a promising strategy for the discovery of new bioactive molecules.

In our continuing efforts to search for new bioactive microbial compounds from extreme marine environments, we explored the chemistry of bacterial strains from the Arctic Ocean. Our initial chemical profiling of Arctic strains led to the discovery of articoside and

C-1027 chromophore-V—two new benzoxazine-bearing compounds that inhibit *Candida albicans* isocitrate lyase—from *Streptomyces* sp. ART5 collected from the East Siberian continental margin [4]. In this study, we diversified the phylogeny of bacteria for chemical analysis and focused on non-*Streptomyces* bacterial strains inhabiting the Arctic Ocean. The *Paenibacillus* sp. SVB7 strain was isolated from sediment collected at the continental shelf (depth = 322 m) off Wijdefjorden, Svalbard, during a marine-geoscientific cruise to North Spitsbergen in 2017. Cultivation in liquid media and the LC/MS-based chemical examination of the strain *Paenibacillus* sp. SVB7 identified the production of previously unreported molecules with the molecular ions at m/z 384. Scaling-up of the culture enabled us to purify two new compounds, svalbamides A and B, and subsequently elucidate their structures by spectroscopic analysis, chemical derivatization, and quantum mechanics-based calculation. Here, we report the structural determination of svalbamides A and B (**1**, **2**; Figure 1) along with their biological activity.

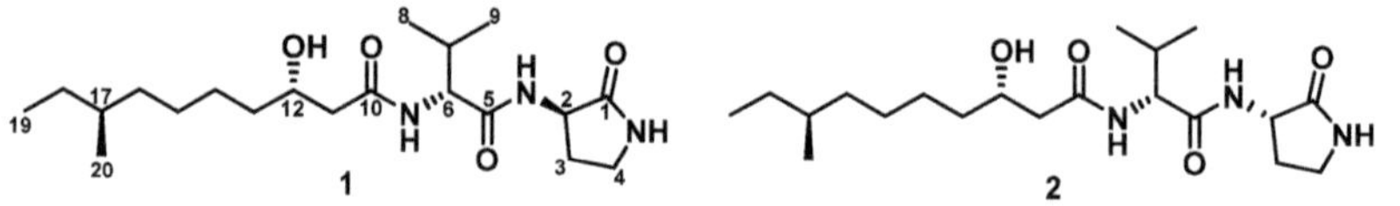

Figure 1. The structures of svalbamides A (**1**) and B (**2**).

2. Results and Discussion

2.1. Phylogenetic Analysis

Sequence comparison using the almost complete 16S rRNA gene sequence of strain SVB7 (1440 bp) in BLASTn and EzBioCloud searches revealed that strain SVB7 belongs to the genus *Paenibacillus* of the Paenibacillaceae family. According to 16S rRNA gene sequence similarities, strain SVB7 was most closely related to *P. maysiensis* SX-49^T (99.30% similarity), followed by *P. terrae* AM141^T (98.26%), and *P. peoriae* DSM 8320^T (97.42%). In all phylogenetic trees inferred by maximum likelihood, neighbor-joining, and minimum-evolution methods, strain SVB7 was located within the *Paenibacillus* clade and formed a robust clade with *P. maysiensis* and *P. terrae*, providing clear support for its genus being classified as *Paenibacillus* (Figure 2). Based on the formation of the robust clade with *P. maysiensis* SX-49^T and > 98.7% 16S rRNA gene sequence similarity, it is likely that strain SVB7 is a member of *Paenibacillus maysiensis*. However, inclusion of strain SVB7 must be confirmed using whole-genome sequencing analysis.

2.2. Structural Elucidation

Svalbamide A (**1**) was isolated as a white powder. The molecular formula of **1** was assigned as $C_{20}H_{37}N_3O_4$, which has an unsaturation number of 4 based on high-resolution electrospray ionization (HR-ESI) mass spectrometry ([M + H]$^+$ at *m/z* 384.2851, calculated as 384.2857) along with ^{1}H and ^{13}C NMR spectra. The ^{13}C NMR spectra of **1** showed three carbonyl carbon (δ_C 174.2, 171.0, and 170.8), one oxygenated methine carbon (δ_C 67.5), and two α-amino methine carbon (δ_C 57.2 and 49.4) signals. Further analysis of these spectra revealed the existence of eight methylene carbon resonances (δ_C 43.4–25.2), two more methine carbons (δ_C 33.7 and 30.7), and four methyl carbons (δ_C 19.3, 19.1, 18.0, and 11.2) in the aliphatic region. The ^{1}H and HSQC NMR spectra of **1** identified four exchangeable protons (δ_H 8.10, 7.83, 7.78, and 4.65), one carbinol proton (δ_H 3.78), two α-amino protons (δ_H 4.30 and 4.18), two more methine protons (δ_H 1.96 and 1.28), eight methylene protons (δ_H 1.05–3.16), and twelve methyl protons (δ_C 0.88, 0.84, 0.82, and 0.81). Based on the NMR spectroscopic features of the amide carbonyl carbons, α-amino groups, and many aliphatic signals, the structure of svalbamide A (**1**) was deduced as a peptide bearing an aliphatic chain.

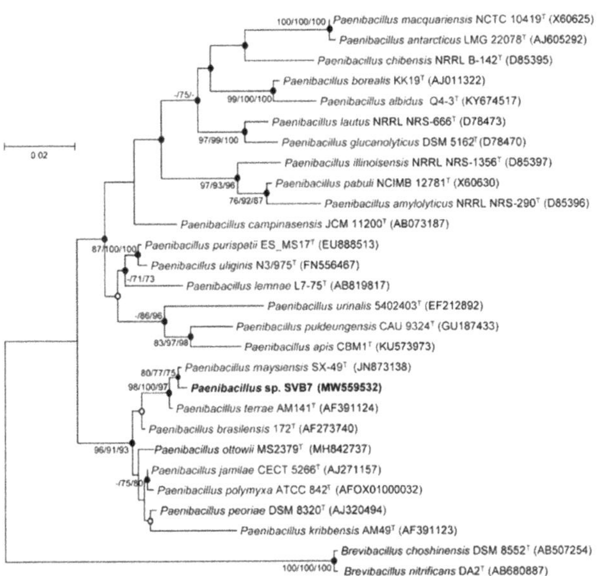

Figure 2. Maximum likelihood phylogenetic tree showing the position of *Paenibacillus* sp. SVB7. Bootstrap values (expressed as percentages of 1000 replications) over 70% are shown to the left of the node, and represent maximum likelihood, neighbor-joining, and minimum evolution (reading from left to right). Filled and open circles at each node indicate nodes recovered by all three treeing methods or by two treeing methods, respectively. Two 16S rRNA gene sequences of the genus *Brevibacillus* were used as outgroups. Bar, 0.02 substitutions per nucleotide position.

The interpretation of COSY, TOCSY, and HMBC NMR spectra enabled us to determine the partial structures of **1**. First, the 2-NH (δ_H 8.10)/H-2 (δ_H 4.30) COSY correlation connected 2-NH to the C-2 α-carbon (δ_C 49.4). The TOCSY and COSY correlations among H-2, H-3a and H-3b (δ_H 1.82 and 2.26), H_2-4 (δ_H 3.16), and 4-NH (δ_H 7.78) secured the spin system from 2-NH to 4-NH. The HMBC correlations from 4-NH to C-1 (δ_C 174.2), C-2 (δ_C 49.4), C-3 (δ_C 28.0), and C-4 (δ_C 38.0), and from 2-NH to C-1 (δ_C 174.2), led to elucidation of the substructure as a 3-amino-2-pyrrolidinone. The structure of valine was assigned based on the ^{1}H–^{1}H couplings of 6-NH (δ_H 7.83), H-6 (δ_H 4.18), H-7 (δ_H 1.96), H_3-8 (δ_H 0.84), and H_3-9 (δ_H 0.88) in the COSY and TOCSY NMR spectra along with the H-6/C-5 (δ_C 171.0) HMBC correlation. The remaining part of the molecule was composed of a lipophilic acyl chain. The protons in the substructure from C-11 to C-20 belong to a single spin system as revealed by COSY/TOCSY correlations of the protons. These protons showed correlation peaks with an exchangeable proton at δ_H 4.65 in the TOCSY spectrum, indicating that the exchangeable proton is also included in the spin system (Figure 3).

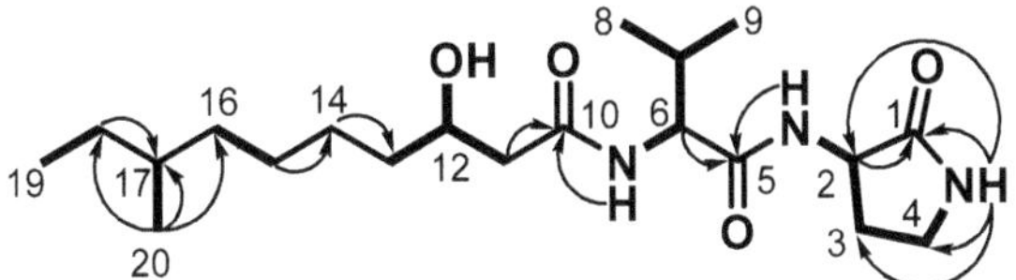

Figure 3. Key HMBC and COSY correlations of svalbamides A (**1**) and B (**2**).

The COSY correlation of H-11a and H-11b (δ_H 2.23 and 2.29)/H-12 (δ_H 3.78), H-12/H_2-13 (δ_H 1.33), and H-12/12-OH (δ_H 4.65) showed connectivity from C-11 to C-13, including 12-OH. The HMBC correlations of 12-OH (δ_H 4.65) to C-11 (δ_C 43.4), C-12 (δ_C

67.5), and C-13 (δ_C 36.7) confirmed the partial structure. C-11 was attached to the C-10 carbonyl carbon as inferred by the H-11a and H-11b/C-10 HMBC correlation. Due to the overlapping aliphatic signals from H_2-13 to H-18a and H-18b, HMBC correlations played a pivotal role in identifying the planar structure of this linear section. The HMBC correlations from H-14a and H-14b (δ_H 1.24 and 1.34) to C-13 (δ_C 36.7), from H_2-15 (δ_H 1.23) to C-14 (δ_C 25.2), and from H-16a and H-16b (δ_H 1.05 and 1.25) to C-15 (δ_C 26.5) revealed the connectivity from C-13 to C-16. In addition, the COSY ^{1}H–^{1}H couplings of H-17 (δ_H 1.28)/H_3-20 (δ_H 0.81) and H-18a and H-18b (δ_H 1.09 and 1.28)/H_3-19 (δ_H 0.82), along with the HMBC signals of H_3-20 (δ_H 0.81) to C-16 (δ_C 36.0), C-17 (δ_C 33.7), and C-18 (δ_C 28.9), were finally assigned to 3-hydroxy-8-methyldecanoic acid (Figure 3). Consequently, the four unsaturation equivalents were fully explained by one pyrrolidinone ring containing one carbonyl group and two more carbonyl functional groups. Thus, svalbamide A (**1**) must not possess an additional ring and comprises a combination of the three substructures as a linear molecule.

Once the partial structures of 3-amino-2-pyrrolidinone, valine, and 3-hydroxy-8-methyldecanoic acid were identified, they were assembled according to the HMBC correlations: 2-NH (δ_H 8.10) of pyrrolidinone was correlated with the amide carbonyl carbon C-5 (δ_C 171.0) belonging to the valine residue, connecting 3-amino-2-pyrrolidinone to valine. The HMBC correlation from 6-NH (δ_H 7.83) of valine to the carbonyl carbon C-10 (δ_C 170.8) of 3-hydroxy-8-methyldecanoic acid established the sequence from valine to 3-hydroxy-8-methyldecanoic acid. Therefore, the planar structure of svalbamide A (**1**) was finally elucidated as a previously unreported lipodipeptide (Figure 3).

Svalbamide B (**2**) was isolated as a white powder, and its molecular formula was determined to be $C_{20}H_{37}N_3O_4$, which contains four double bond equivalents, using high-resolution electrospray ionization (HR-ESI) mass spectrometry ([M + H]$^+$ at *m/z* 384.2845, calculated as 384.2857). This molecular formula was identical to that of svalbamide A (**1**). The ^{1}H and ^{13}C NMR data of **2** in DMSO-d_6 were extremely similar to those of **1** (Table 1), but distinct differences in chemical shifts were found mainly in the 3-amino-2-pyrrolidinone unit. Specifically, H-2 in **1** (δ_H 4.30) was shifted upfield in **2** (δ_H 4.27), while signals for H-3a and H-3b in **1**, at δ_H 1.82 and 2.26, were detected at δ_H 1.76 and 2.29 in **2**. C-3 (δ_C 28.0) also shifted slightly to the deshielded region by 0.3 ppm in svalbamide B (**2**). Comprehensive analysis of 1D and 2D NMR data indicated the planar structure of **2** to be the same as **1** (Figure 3). Based on the observation that the distinct chemical shift differences were found in 3-amino-2-pyrrolidone unit, the structure of svalbamide B (**2**) was expected to have stereochemical modification in this residue.

The absolute configurations at the α-carbons of the two amino acid units were determined by applying the advanced Marfey's method for derivatization with the L- and D-forms of 1-fluoro-2,4-dinitrophenyl-5-alanine amide (FDAA). LC/MS analysis of the FDAA derivatives of hydrolysates of **1** and **2** (Table S1) showed that they commonly possess D-valine. Because 3-amino-2-pyrrolidinone is converted into 2,4-diaminobutanoic acid during acid hydrolysis, an authentic sample of 2*S*,4-diaminobutanoic acid was derivatized with L- and D-FDAA to allow comparison. By comparing the retention times of the FDAA adducts of authentic 2*S*,4-diaminobutanoic acid, svalbamide A (**1**) was revealed to bear 3*R*-3-amino-2-pyrrolidinone, whereas svalbamide B (**2**) incorporates 3*S*-3-amino-2-pyrrolidinone (Figure 1).

The 3-hydroxy-8-methyldecanoic acid moiety contained stereogenic centers at C-12 and C-17. Initially, the modified Mosher's method was applied for the oxygen-bearing chiral center at C-12. However, multiple esterifying attempts at the hydroxy group by *S*- and *R*-MTPA-Cl were not successful. Therefore, DP4 calculation was used to determine the absolute configurations. Four possible diastereomers of 3-hydroxy-8-methyldecanoic acid of svalbamide A (**1**), namely **1a** (12*R* and 17*R*), **1b** (12*S* and 17*R*), **1c** (12*R* and 17*S*), and **1d** (12*S* and 17*S*), were constructed with the established 2*R* and 6*R* configurations (Figure 4). Following this, the ^{1}H and ^{13}C chemical shifts of 158 conformers were calculated and averaged with their Boltzmann populations. Our DP4 calculations, based on statistical comparisons of the calculated and experimental chemical shifts, indicated that the diastereomer **1d** (12*S* and 17*S*) was suitable for svalbamide A (**1**) with 96.0% probability (Figure 4). The absolute configuration of svalbamide B (**2**) was subsequently proposed as 2*S*, 6*R*, 12*S*, and 17*S*.

Table 1. ^{1}H and ^{13}C NMR data for svalbamides A (**1**) and B (**2**) in DMSO-d_6.

Position		Svalbamide A (1) [a]		Svalbamide B (2) [a]	
		δ_C, Type	δ_H, Mult (*J* in Hz)	δ_C, Type	δ_H, Mult (*J* in Hz)
3-amino-2-pyrrolidinone	1	174.2, C		174.2, C	
	2	49.4, CH	4.30, m	49.5, CH	4.27, m
	3a	28.0, CH$_2$	1.82, m	28.3, CH$_2$	1.76, m
	3b		2.26, m		2.29, m
	4	38.0, CH$_2$	3.16, m	38.0, CH$_2$	3.16, m
	4-NH		7.78, br s		7.81, br s
	2-NH		8.10, d (8.5)		8.21, d (8.5)
D-Val	5	171.0, C		171.0, C	
	6	57.2, CH	4.18, dd (9.0, 6.5)	57.2, CH	4.20, dd (9.0, 6.5)
	7	30.7, CH	1.96, m	30.6, CH	1.94, m
	8	18.0, CH$_3$	0.84, d (7.0)	18.0, CH$_3$	0.83, d (7.0)
	9	19.3, CH$_3$	0.88, d (7.0)	19.1, CH$_3$	0.84, d (7.0)
	6-NH		7.83, d (9.0)		7.85, d (9.0)
3-hydroxy-8-methyldecanoic acid	10	170.8, C		170.8, C	
	11a	43.4, CH$_2$	2.23, dd (14.0, 7.0)	43.4, CH$_2$	2.25, dd (14.0, 7.0)
	11b		2.29, dd (14.0, 5.0)		2.28 dd (14.0, 5.0)
	12	67.5, CH	3.78, m	67.6, CH	3.78, m
	13	36.7, CH$_2$	1.33, m [b]	36.7, CH$_2$	1.33, m [b]
	14a	25.2, CH$_2$	1.24, m [b]	25.2, CH$_2$	1.24, m [b]
	14b		1.34, m [b]		1.34, m [b]
	15	26.5, CH$_2$	1.23, m [b]	26.5, CH$_2$	1.23, m [b]
	16a	36.0, CH$_2$	1.05, m	36.0, CH$_2$	1.05, m
	16b		1.25, m [b]		1.25, m [b]
	17	33.7, CH	1.28, m [b]	33.7, CH	1.28, m [b]
	18a	28.9, CH$_2$	1.09, m	28.9, CH$_2$	1.09, m
	18b		1.28, m [b]		1.28, m [b]
	19	11.2, CH$_3$	0.82, t (7.0)	11.2, CH$_3$	0.83, t (7.0)
	20	19.1, CH$_3$	0.81, d (6.5)	19.1, CH$_3$	0.81, d (6.5)
	12-OH		4.65, d (5.0)		4.67, d (5.0)

[a] ^{1}H and ^{13}C NMR data were recorded at 800 and 200 MHz, respectively. [b] Overlapping signals.

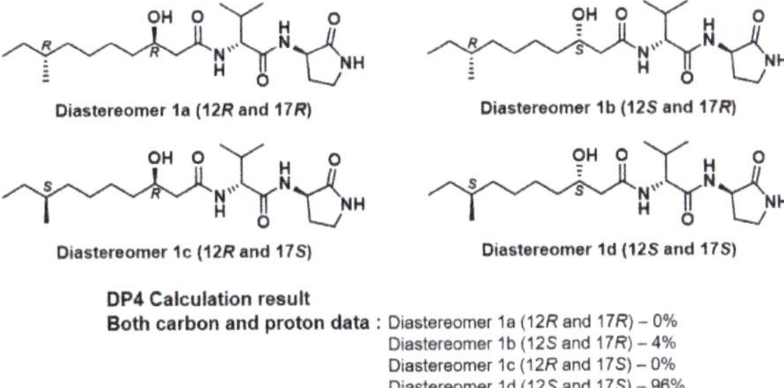

Figure 4. The simulated models of the four possible diastereomers (**a–d**) of svalbamide A (**1**) and the results of DP4 calculations.

2.3. Biological Evaluation

The biological activities of svalbamides A (**1**) and B (**2**) were evaluated in several ways. First, we measured cytotoxicity against various cancer cell lines [5], including HCT116 (human colorectal cancer cells), MDA-MB-231 (human breast cancer cells), A549 (human lung cancer cells), SK-HEP-1 (human liver cancer cells), and SNU-638 (human gastric cancer cells), but **1** and **2** showed no significant cytotoxicity against the tested cell lines even at 50 μM. Therefore, we evaluated the detoxification ability by measuring quinone reductase (QR) activity. QR is a major phase II detoxification enzyme, and the induction of QR activity is considered as a strategy to increase the chemoprevention effect. Svalbamide A (**1**) enhanced QR activity by 1.45-, 1.98-, and 2.54-fold at 10, 20, and 40 μM, respectively, in a concentration-dependent manner. In addition, svalbamide B (**2**) effectively induced QR activity by 1.93-, 2.64-, and 2.98-fold at 10, 20, and 40 μM, respectively (Figure 5). At a concentration of 40 μM, it exhibited a comparable level of QR activity induction in the positive control of 1 μM β-naphthoflavone (β-NF). These results suggest that **1** and **2** are potential chemotypes with chemopreventive activity.

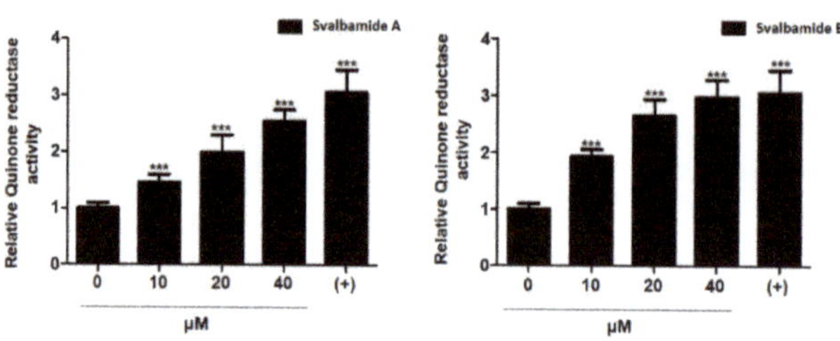

Figure 5. Induction of quinone reductase activity by svalbamide A (**1**) and B (**2**). Svalbamide A (**1**) and B (**2**) showed QR induction activity, with 2.54- and 2.98-fold increases, respectively, at 40 μM. All data represent the mean $\pm$ SD ($n = 3$). *** $p < 0.001$ compared to the control.

3. Materials and Methods

3.1. General Experimental Procedures

Optical rotations were measured using a JASCO P-2000 polarimeter (sodium light source, JASCO, Easton, PA, USA) with a 1 cm cell. IR spectra were obtained using a Thermo NICOLET iS10 spectrometer (Thermo, Madison, CT, USA). ^{1}H, ^{13}C, and 2D NMR spectra were recorded on a Bruker Avance 800 MHz spectrometer (Bruker, Billerica, MA, USA) at the Research Institute of Pharmaceutical Sciences, Seoul National University. ESI low-resolution LC/MS data were recorded using an Agilent Technologies 6130 Quadrupole mass spectrometer (Agilent Technologies, Santa Clara, CA, USA) coupled with an Agilent Technologies 1200 series high-performance liquid chromatography (HPLC) instrument using a reversed-phase $C_{18}(2)$ column (Phenomenex Luna, 100×4.6 mm). HR-ESI mass spectra were acquired on a high-resolution LC/MS–MS spectrometer (Q-TOF 5600) at the National Instrumentation Center for Environmental Management (NICEM) in the College of Agriculture and Life Sciences at Seoul National University.

3.2. Isolation, Cultivation, Phylogenetic Analysis, and Extraction of Bacteria

During the Korea–Norway Joint marine-geoscientific cruise with RV Helmer Hanssen to North Spitsbergen in 2017, a sediment core (HH17-1085) was taken at a water depth of 322 m with a giant box corer from the continental shelf off Wijdefjorden (80°16.469′ N, 016°12.625′ E) in Svalbard (Figure 6). Surface sediment corresponding to 1 cm depth was collected from core HH17-1085. A portion of the sample (2 g) was diluted in 20 mL sterilized water and vortexed. The mixture was spread on YEME isolation solid medium (500 mL of sterilized water, 9 g agar, 100 mg cycloheximide, 2 g yeast, 5 g malt, and 2 g glucose) for two weeks. Strain SVB7 was isolated in YEME medium after one week of incubation for further study.

The 16S rRNA gene sequence of strain SVB7 was obtained by Sanger sequencing using PCR products amplified with the universal primers 27F and 1492R. The resultant 16S rRNA gene sequence (1440 bp) was queried in a BLASTn search implemented at GenBank and was also identified by the "16S-based ID service" in the EzbioCloud database [6]. For phylogenetic analysis, sequences of strain SVB7 and its close relatives retrieved from the

EzBioCloud database were aligned with the SINA online aligner [7]. Using the aligned sequences, phylogenetic trees were inferred by maximum likelihood, neighbor-joining, and minimum-evolution algorithms implemented in MEGA software version 7.0 [8].

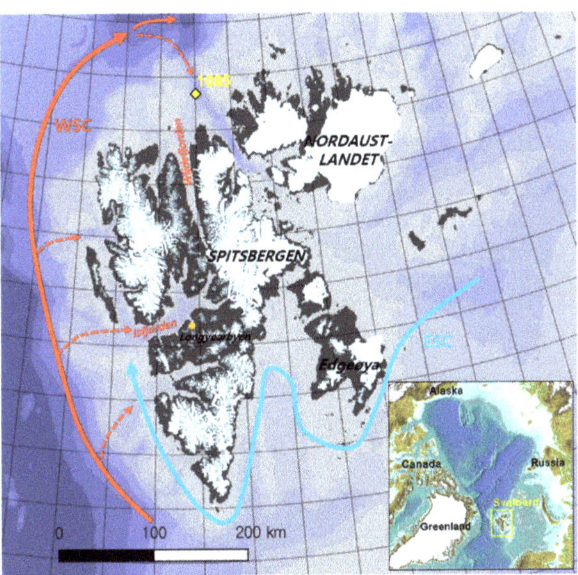

Figure 6. Map of the Svalbard archipelago with the core site and the main currents influencing Svalbard highlighted. The red and blue arrows indicate the West Spitsbergen Current (WSC) and East Spitsbergen Current (ESC), respectively, and the yellow rectangle indicates the coring site (HH17-1085-GC). The shaded white color represents the present glacier-covered areas on the archipelago.

The SVB7 strain was cultured in 50 mL modified K medium (4 g yeast extract, 5 g malt extract, 5 g soytone, 5 g soluble starch, 5 g mannitol, 2 g glucose, and 6 g glycerol in 1 L deionized water) in a 125 mL Erlenmeyer flask. After cultivation for 2 days on a rotary shaker at 200 rpm and 30 °C, 5 mL of the culture medium was inoculated in 200 mL of modified K medium in a 500 mL Erlenmeyer flask. After cultivation for 2 days under the same incubation conditions, 10 mL of the culture medium was inoculated in 1 L of modified K medium in 2.8 L Fernbach flasks (200 ea × 1 L) at 170 rpm and 30 °C for 6 days. The whole culture of SVB7 was extracted with 300 L of EtOAc. The EtOAc and water layers were separated, and the remaining water in the EtOAc layer was removed by adding anhydrous sodium sulfate. The extract was concentrated using a rotary evaporator, yielding 50 g of dry material.

3.3. Isolation of Svalbamides A and B

The crude extract was divided into ten equal parts and fractioned over a C_{18} reversed-phase open column (ϕ 6.5 × 10 cm) with 500 mL of 20%, 40%, 60%, 80%, and 100% MeOH–H_2O. The 80% MeOH fraction was subjected to a reversed-phase HPLC (Kromasil C_{18}, 5 µm, 250 × 10 mm, flow rate = 2 mL/min) using a gradient solvent system from 35% to 75% CH_3CN–H_2O over 40 min with 210 nm UV detection. Svalbamides A and B were collected as one broad peak at 25 min. The recorded ^1H NMR spectra of the peak indicated a diastereomeric mixture, prompting further purification on a chiral HPLC column (CHIRALPAK IB, 5 µm, 250 × 4.6 mm, flow rate = 0.6 mL/min) using a step gradient 45% CH_3CN isocratic solvent system over 10 min, followed by a 60% CH_3CN–H_2O isocratic solvent system from 10 to 50 min. Svalbamides A (10.0 mg) and B (7.8 mg) were isolated at 30 and 31.5 min, respectively.

Svalbamide A (1): white powder; [α]20D +19.5 (c 0.1, MeOH); IR (neat) ν_{max} 3289, 2926,1632 cm^{-1}; ^{1}H and ^{13}C NMR (800 MHz, DMSO-d_6) (Table 1); HR-ESI–MS m/z: [M + H]$^+$ Calcd for C$_{20}$H$_{38}$N$_3$O$_4$, 384.2857, found 384.2851.

Svalbamide B (2): white powder; [α]20D +25.2 (c 0.1, MeOH); IR (neat) ν_{max} 3276, 2928, 1612 cm^{-1}; ^{1}H and ^{13}C NMR (800 MHz, DMSO-d_6) (Table 1); HR-ESI–MS m/z: [M + H]$^+$ Calcd for C$_{20}$H$_{38}$N$_3$O$_4$, 384.2857, found 384.2845.

3.4. Conformational Search and DP4 Analysis

A conformational search was carried out using a mixed sampling method of torsional/low-mode using MacroModel (version 9.9, Schrödinger LLC) in the Maestro suite (version 9.9, Schrödinger LLC). A total of 158 conformers of the diastereomers were identified with relative potential energies below 10 kJ/mol using the MMFF force field. The shielding tensor values of the optimized conformers were calculated based on the equation below, where δ^x_{calc} is the calculated NMR chemical shift for nucleus x, and σ^o is the shielding tensor for the proton and carbon nuclei calculated at the B3LYP/6-31++ level. These values were averaged via the Boltzmann population with the associated Gibbs free energy and utilized for the DP4 analysis, which was facilitated using an Excel spreadsheet provided by the authors of [9] and as described in their publication.

$$\delta^x_{calc} = \frac{\sigma^o - \sigma^x}{1 - \sigma^o/10^6}$$

3.5. Quinone Reductase Assay

Hepa1c1c7 murine hepatoma cells (American Type Culture Collection, Manassas, VA, USA) were used to investigate QR induction activity. The test cells were seeded (3×10^4 cells/mL) and incubated at 37 °C for 24 h with 5% CO$_2$ containing humidified atmosphere. The plates were then exposed to svalbamides A and B (1, 2), including a positive control compound, β-naphthoflavone (β-NF). After 24 h, the media were decanted from the wells, and the cells in each well were lysed by incubation at 37 °C with 250 μL of a mixed solution consisting of 10 mM Tris-HCl pH 8.0, 140 mM NaCl, 15 mM MgCl$_2$, and 0.5% NP-40 (IGEPAL CA-630, Sigma, St. Louis, MO, USA) for 10 min. A 1 mL aliquot of the complete reaction mixture (12.5 mM Tris-HCl pH 7.4, 0.67 mg/mL bovine serum albumin (BSA), 0.01% Tween-20, 50 μM flavin adenine dinucleotide (FAD), 1 mM glucose-6-phosphate, 2 U/mL glucose-6-phosphate dehydrogenase, 30 μM NADP, 50 μg/mL 3-(4,5-dimethylthiazo-2-yl)-2,5-diphenyltetrazolium bromide (MTT), and 50 μM menadione) was added to each of the wells, and the plates were incubated at 25 °C for the colorimetric reaction. The rate of NADPH-dependent menadiol-mediated reduction of MTT in this reaction was measured at 610 nm, and cytotoxicity was determined by crystal violet staining of an identical set of the test plates. The quinone reductase activity was calculated from the following equation: absorbance change for MTT per min/absorbance of crystal violet × 3345 nmol/mg. The value of 3345 nmol/mg represents the ratio of the extinction coefficient of MTT and the proportionality constant of crystal violet. The relative QR activity was normalized using controls [10].

4. Conclusions

Our chemical study of the Arctic sediment-derived *Paenibacillus* sp. SVB7 led to the discovery and structural elucidation of two new pyrrolidinone-bearing lipodipeptides, svalbamides A (1) and B (2), in which QR activity could be induced. Based on spectroscopic analysis, advanced Marfey's analysis, and DP4 calculation, these two compounds were identified to have a diastereomeric relationship with alternative absolute configurations at the 3-amino-2-pyrrolidione unit. Svalbamides A and B are structurally unique as they contain 3-amino-2- pyrrolidinone amino acid. This amino acid unit was rarely reported in natural products, with the only example being of actinoramide E, an antimalarial peptide from a marine-derived *Streptomyces* strain [11]. 3-Hydroxy-8-methyldecanoic acid is another interesting component. This saturated fatty acid was occasionally found in natural products from *Paenibacillus* and related bacteria. For example, tridecaptins A–C were first isolated from *Bacillus polymyxa* in 1978 and were studied in various fields, with reports of new derivatives and biosynthesis undertaken to date [12]. A series of new tridecaptin compounds containing 3-hydroxy-8-methyldecanoic acid were discovered in a *Paenibacillus* strain collected in the deep oligotrophic Krubera-Voronja cave. However, the absolute configuration of this fatty acid was not determined [13–15]. Octapeptin and cerexin from the *Bacillus* sp. bear the same fatty acid, but no experiments have yet been conducted to reveal the absolute stereochemistry [16,17]. Therefore, svalbamides A and B are the first metabolites for which the stereochemistry of 3-hydroxy-8-methyldecanoic acid was addressed. Discovering these new bioactive secondary metabolites from *Paenibacillus* from the polar region indicates that chemical studies of underinvestigated bacterial taxa in marine extreme habitats, such as the Arctic Ocean, could lead to the discovery of significant natural chemical diversity with pharmaceutical potential in terms of drug discovery.

Supplementary Materials: The following are available online at https://www.mdpi.com/article/10.3390/md19040229/s1, Figure S1. ^{1}H NMR spectrum (800 MHz) of svalbamide A (**1**) in DMSO-d_6., Table S1: LC/MS analysis of D- and L-FDAA derivatives of the amino acid-derived units in svalbamide A (**1**), svalbamide B (**2**), L-2,4-diamino butanoic acid (**3**), D-valine (**4**) and L-valine (**5**) authentic samples. Retention times (min) are notified., Figure S2. ^{13}C NMR spectrum (200 MHz) of svalbamide A (**1**) in DMSO-d_6, Figure S3. COSY NMR spectrum (800 MHz) of svalbamide A (**1**) in DMSO-d_6, Figure S4. HSQC NMR spectrum (800 MHz) of svalbamide A (**1**) in DMSO-d_6, Figure S5. HMBC NMR spectrum (800 MHz) of svalbamide A (**1**) in DMSO-d_6, Figure S6. TOCSY NMR spectrum (800 MHz) of svalbamide A (**1**) in DMSO-d_6, Figure S7. ^{1}H NMR spectrum (800 MHz) of svalbamide B (**2**) in DMSO-d_6, Figure S8. ^{13}C NMR spectrum (200 MHz) of svalbamide B (**2**) in DMSO-d_6, Figure S9. COSY NMR spectrum (800 MHz) of svalbamide B (**2**) in DMSO-d_6, Figure S10. HSQC NMR spectrum (800 MHz) of svalbamide B (**2**) in DMSO-d_6, Figure S11. HMBC NMR spectrum (800 MHz) of svalbamide B (**2**) in DMSO-d_6, Figure S12. TOCSY NMR spectrum (800 MHz) of svalbamide B (**2**) in DMSO-d_6, Figure S13. The simulated models of four possible diastereomers (**a–d**) of svalbamide A (**1**) and the result of DP4 calculation, Table S1. LC/MS analysis of D- and L-FDAA derivatives of the amino acid-derived units in svalbamide A (**1**), svalbamide B (**2**) and L-2,4-diamino butanoic acid authentic sample (**3**), D-valine (**4**) and L-valine (**5**). Retention times (min) are notified, Table S2. The major conformers of diastereomers (**a–d**) of svalbamide A (**1**), identified by conformational searches in MMFF94 force field using MacroModel, Table S3. Experimental (Exp.) and calculated (Cal.) chemical shift values (CS, δ) of diastereomers (**a–d**) of **1** and svalbamide A (**1**).

Author Contributions: Conceptualization, D.-C.O. and Y.E.D.; methodology, D.-C.O., J.S., J.-C.C. and S.K.L.; software, Y.L., Y.E.D. and E.S.B.; validation Y.E.D. and E.S.B.; formal analysis, D.-C.O., Y.E.D., E.S.B., Y.L. and J.-C.C.; investigation, Y.E.D., D.-C.O., S.-I.N., E.S.B., Y.L., J.-C.C., S.-J.N. and S.K.L.; resources, S.-I.N. and D.-C.O.; data curation, Y.E.D., E.S.B., Y.L., J.-C.C., S.K.L. and D.-C.O.; writing—original draft preparation, D.-C.O., Y.E.D., E.S.B., Y.L., J.-C.C., S.K.L. and S.-I.N.; writing—review and editing, D.-C.O., Y.E.D., S.-J.N., S.-I.N. and J.S.; visualization, D.-C.O., Y.E.D., E.S.B., Y.L., S.K.L. and J.-C.C.; supervision, D.-C.O.; project administration, D.-C.O., Y.E.D.; funding acquisition, D.-C.O., S.-J.N. and S.-I.N. All authors have read and agreed to the published version of the manuscript.

Funding: This work was supported by the Collaborative Genome Program of the Korea Institute of Marine Science and Technology Promotion (KIMST) funded by the Ministry of Oceans and Fisheries (MOF) (No. 20180430) and the National Research Foundation of Korea (NRF) grants funded by the Ministry of Science and ICT (MSIT) (2021R1A4A2001251) and partly by the Basic Core Technology Development Program for the Oceans and the Polar Regions (NRF-2015M1A5A1037243).

Institutional Review Board Statement: Not applicable.

Data Availability Statement: All data is contained within this article and Supplementary Materials.

Acknowledgments: The captains and crews of *R/V Helmer Hanssen*, Matthias Forwick (UiT, Norway), and the cruise participants all supported the core sampling during the cruise in 2017.

Conflicts of Interest: The authors declare no conflict of interest.

References

1. Carroll, A.R.; Copp, B.R.; Davis, R.A.; Keyzers, R.A.; Prinsep, M.R. Marine natural products. *Nat. Prod. Rep.* **2021**, *38*, 362–413. [CrossRef] [PubMed]
2. Bae, M.; An, J.S.; Bae, E.S.; Oh, J.; Park, S.H.; Lim, Y.; Ban, Y.H.; Kwon, Y.; Cho, J.-C.; Yoon, Y.J.; et al. Donghaesulfins A and B, dimeric benz[*a*]anthracene thioethers from volcanic island derived *Streptomyces* sp. *Org. Lett.* **2019**, *21*, 3635–3639. [CrossRef] [PubMed]
3. Bae, M.; An, J.S.; Hong, S.-H.; Bae, E.S.; Chung, B.; Kwon, Y.; Hong, S.; Oh, K.-B.; Shin, J.; Lee, S.K.; et al. Donghaecyclinones A–C: New cytotoxic rearranged angucyclinones from a volcanic island-derived marine *Streptomyces* sp. *Mar. Drugs* **2020**, *18*, 121. [CrossRef] [PubMed]
4. Moon, K.; Ahn, C.-H.; Shin, Y.; Won, T.H.; Ko, K.; Lee, S.K.; Oh, K.-B.; Shin, J.; Nam, S.-I.; Oh, D.-C. New benzoxazine secondary metabolites from an arctic actinomycete. *Mar. Drugs* **2014**, *12*, 2526–2538. [CrossRef] [PubMed]
5. Kim, W.K.; Bach, D.-H.; Ryu, H.W.; Oh, J.; Park, H.J.; Hong, J.-Y.; Song, H.-H.; Eum, S.; Bach, T.T.; Lee, S.K. Cytotoxic activities of *Telectadium dongnaiense* and its constituents by inhibition of the Wnt/β-catenin signaling pathway. *Phytomedicine* **2017**, *34*, 136–142. [CrossRef] [PubMed]
6. Yoon, S.-H.; Ha, S.-M.; Kwon, S.; Lim, J.; Kim, Y.; Seo, H.; Chun, J. Introducing EzBioCloud: A taxonomically united database of 16S rRNA gene sequences and whole-genome assemblies. *Int. J. Syst. Evol. Microbiol.* **2017**, *67*, 1613–1617. [CrossRef] [PubMed]
7. Pruesse, E.; Peplies, J.; Glöckner, F.O. SINA: Accurate high-throughput multiple sequence alignment of ribosomal RNA genes. *Bioinformatics* **2012**, *28*, 1823–1829. [CrossRef] [PubMed]
8. Kumar, S.; Stecher, G.; Tamura, K. MEGA7: Molecular evolutionary genetics analysis version 7.0 for bigger datasets. *Mol. Biol. Evol.* **2016**, *33*, 1870–1874. [CrossRef] [PubMed]
9. An, J.S.; Lee, J.Y.; Kim, E.; Ahn, H.; Jang, Y.-J.; Shin, B.; Hwang, S.; Shin, J.; Yoon, Y.J.; Lee, S.K.; et al. Formicolides A and B, antioxidative and antiangiogenic 20-membered macrolides from a wood ant gut bacterium. *J. Nat. Prod.* **2020**, *83*, 2776–2784. [CrossRef] [PubMed]
10. Cuendet, M.; Oteham, C.P.; Moon, R.C.; Pezzuto, J.M. Quinone reductase induction as a biomarker for cancer chemoprevention. *J. Nat. Prod.* **2006**, *69*, 460–463. [CrossRef] [PubMed]
11. Cheng, K.C.-C.; Cao, S.; Raveh, A.; MacArthur, R.; Dranchak, P.; Chlipala, G.; Okoneski, M.T.; Guha, R.; Eastman, R.T.; Yuan, J.; et al. Actinoramide A identified as a potent antimalarial from titration-based screening of marine natural product extracts. *J. Nat. Prod.* **2015**, *78*, 2411–2422. [CrossRef] [PubMed]
12. Shoji, J.I.; Hinoo, H.; Sakazaki, R.; Kato, T.; Wakisaka, Y.; Mayama, M.; Matsuura, S.; Miwa, H. Isolation of tridecaptins A, B and C studies on antibiotics from the genus *Bacillus*. XXIII. *J. Antibiot.* **1978**, *31*, 646–651. [CrossRef] [PubMed]
13. Lebedeva, J.; Jukneviciute, G.; Čepaitė, R.; Vickackaite, V.; Pranckutė, R.; Kuisiene, N. Genome mining and characterization of biosynthetic gene clusters in two cave strains of *Paenibacillus* sp. *Front. Microbiol.* **2021**, *11*, 3433. [CrossRef] [PubMed]
14. Bann, S.J.; Ballantine, R.D.; Cochrane, S.A. The tridecaptins: Non-ribosomal peptides that selectively target Gram-negative bacteria. *RSC Med. Chem.* **2021**. [CrossRef]
15. Cochrane, S.A.; Lohans, C.T.; van Belkum, M.J.; Bels, M.A.; Vederas, J.C. Studies on tridecaptin B1, a lipopeptide with activity against multidrug resistant gram-negative bacteria. *Org. Biomol. Chem.* **2015**, *13*, 6073–6081. [CrossRef] [PubMed]
16. Kato, T.; Shoji, J.I. The structure of octapeptin D studies on antibiotics from the genus *Bacillus*. XXVIII. *J. Antibiot.* **1980**, *33*, 186–191. [CrossRef] [PubMed]
17. Shoji, J.I.; Kato, T.; Terabe, S.; Konaka, R. Resolution of peptide antibiotics, cerexins and tridecaptins, by high performance liquid chromatography studies on antibiotics from the genus *Bacillus*. XXVI. *J. Antibiot.* **1979**, *32*, 313–319. [CrossRef] [PubMed]

 *marine drugs*

Article

Dermacozine N, the First Natural Linear Pentacyclic Oxazinophenazine with UV–Vis Absorption Maxima in the Near Infrared Region, along with Dermacozines O and P Isolated from the Mariana Trench Sediment Strain *Dermacoccus abyssi* MT 1.1^T

Bertalan Juhasz [1], Dawrin Pech-Puch [2], Jioji N. Tabudravu [3], Bastien Cautain [4], Fernando Reyes [4], Carlos Jiménez [5], Kwaku Kyeremeh [6] and Marcel Jaspars [1,*]

[1] Marine Biodiscovery Centre, Department of Chemistry, University of Aberdeen, Old Aberdeen AB24 3UE, UK; r01bj16@abdn.ac.uk

[2] Departamento de Biología Marina, Universidad Autónoma de Yucatán, Km. 15.5, Carretera Mérida-Xmatkuil, A.P. 4-116 Itzimná, Mérida 97100, Yucatán, Mexico; dawrin.j.pech@udc.es

[3] School of Natural Sciences, Faculty of Science and Technology, University of Central Lancashire, Preston PR1 2HE, UK; JTabudravu@uclan.ac.uk

[4] Fundación MEDINA, Centro de Excelencia en Investigación de Medicamentos Innovadores en Andalucía, Avda. del Conocimiento 34, Edificio Centro de Desarrollo Farmacéutico y Alimentario, Parque Tecnológico de Ciencias de la Salud, 18016 Granada, Spain; cautainbastien@gmail.com (B.C.); fernando.reyes@medinaandalucia.es (F.R.)

[5] Centro de Investigacións Científicas Avanzadas (CICA) e Departmento de Química, Facultade de Ciencias, AE CICA-INIBIC, Universidad da Coruña, 15071 A Coruña, Spain; carlos.jimenez@udc.es

[6] Marine and Plant Research Laboratory of Ghana, Department of Chemistry, School of Physical and Mathematical Sciences, University of Ghana, Legon-Accra P.O. Box LG 56, Ghana; kkyeremeh@ug.edu.gh

* Correspondence: m.jaspars@abdn.ac.uk; Tel.: +44-1224-272-895

Citation: Juhasz, B.; Pech-Puch, D.; Tabudravu, J.N.; Cautain, B.; Reyes, F.; Jiménez, C.; Kyeremeh, K.; Jaspars, M. Dermacozine N, the First Natural Linear Pentacyclic Oxazinophenazine with UV–Vis Absorption Maxima in the Near Infrared Region, along with Dermacozines O and P Isolated from the Mariana Trench Sediment Strain *Dermacoccus abyssi* MT 1.1^T. *Mar. Drugs* **2021**, *19*, 325. https://doi.org/10.3390/md19060325

Academic Editor: Daniela Giordano

Received: 1 May 2021
Accepted: 30 May 2021
Published: 3 June 2021

Publisher's Note: MDPI stays neutral with regard to jurisdictional claims in published maps and institutional affiliations.

Abstract: Three dermacozines, dermacozines N–P (**1–3**), were isolated from the piezotolerant Actinomycete strain *Dermacoccus abyssi* MT 1.1^T, which was isolated from a Mariana Trench sediment in 2006. Herein, we report the elucidation of their structures using a combination of 1D/2D NMR, LC-HRESI-MS^n, UV–Visible, and IR spectroscopy. Further confirmation of the structures was achieved through the analysis of data from density functional theory (DFT)–UV–Visible spectral calculations and statistical analysis such as two tailed *t*-test, linear regression-, and multiple linear regression analysis applied to either solely experimental or to experimental and calculated ^{13}C-NMR chemical shift data. Dermacozine N (**1**) bears a novel linear pentacyclic phenoxazine framework that has never been reported as a natural product. Dermacozine O (**2**) is a constitutional isomer of the known dermacozine F while dermacozine P (**3**) is 8-benzoyl-6-carbamoylphenazine-1-carboxylic acid. Dermacozine N (**1**) is unique among phenoxazines due to its near infrared (NIR) absorption maxima, which would make this compound an excellent candidate for research in biosensing chemistry, photodynamic therapy (PDT), opto-electronic applications, and metabolic mapping at the cellular level. Furthermore, dermacozine N (**1**) possesses weak cytotoxic activity against melanoma (A2058) and hepatocellular carcinoma cells (HepG2) with IC$_{50}$ values of 51 and 38 μM, respectively.

Keywords: deep sea natural products; Mariana Trench; *Dermacoccus abyssi* MT 1.1^T; ^{13}C-NMR chemical shift linear and multiple regression; (DFT)-UV-Vis spectral calculation; phenoxazine; dermacozine; absorption maxima in the near infrared region

1. Introduction

Deep sea habitats have been shown to be an invaluable source of novel bacterial species [1]. Extreme environments (e.g., hyper-arid deserts, bathyal and hadal zones, hot volcanic lakes etc.) are capable of genetically segregating organisms due to their physical

Mar. Drugs **2021**, *19*, 325. https://doi.org/10.3390/md19060325

properties. Evolutionary adaptations to these extreme environments have generated novel biosynthetic pathways in extremophiles, giving novel structures that may find use in treating diseases [2–4]. *Dermacoccus abyssi* MT 1.1^T (Figure 1) is a piezotolerant Actinomycete isolated in 2006 from a Mariana Trench sediment, collected at a 10,898 m depth from the Challenger Deep by the remotely operated submersible *Kaiko* in 1998 [5].

Figure 1. Scanning electron micrograph of the strain *Dermacoccus abyssi* MT 1.1^T (Zeiss Gemini SEM 300).

Seven novel phenazines, dermacozines A–G (**4–10**), as novel phenazines originating from *Dermacoccus abyssi* strains MT 1.1^T and MT 1.2 were reported by our group in 2010. Subsequently, another four new derivatives: dermacozines H–J (**11–13**) and dermacozine M (**14**) were isolated and reported with the contribution of our group in 2014 and 2020, respectively (Figure 2) [6–8].

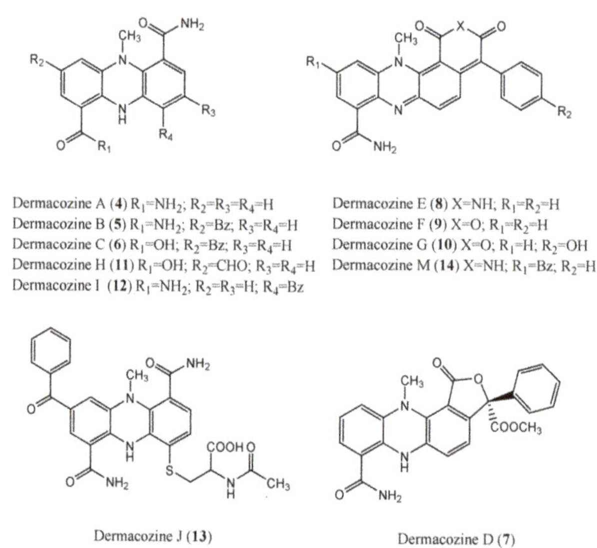

Dermacozine A (**4**) R$_1$=NH$_2$; R$_2$=R$_3$=R$_4$=H
Dermacozine B (**5**) R$_1$=NH$_2$; R$_2$=Bz; R$_3$=R$_4$=H
Dermacozine C (**6**) R$_1$=OH; R$_2$=Bz; R$_3$=R$_4$=H
Dermacozine H (**11**) R$_1$=OH; R$_2$=CHO; R$_3$=R$_4$=H
Dermacozine I (**12**) R$_1$=NH$_2$; R$_2$=R$_3$=H; R$_4$=Bz

Dermacozine E (**8**) X=NH; R$_1$=R$_2$=H
Dermacozine F (**9**) X=O; R$_1$=R$_2$=H
Dermacozine G (**10**) X=O; R$_1$=H; R$_2$=OH
Dermacozine M (**14**) X=NH; R$_1$=Bz; R$_2$=H

Dermacozine J (**13**)

Dermacozine D (**7**)

Figure 2. Chemical structures of known dermacozines (**4–14**) previously isolated from *Dermacoccus abyssi* MT 1.1^T and MT 1.2.

These highly pigmented dibenzo annulated pyrazines has kept our interest piqued toward finding further unknown derivatives. Properties of the previously discovered dermacozines include their radical scavenging ability, cytostatic activity against the K562 leukemia cell line, and theoretically calculated non-linear optical behavior [6–9]. Their IC_{50} cytotoxic activities were reported to be in the range from 7 to 220 µM against the K562 chronic myelogenous leukemia cell line; structure–activity relationship studies revealed a strong connection between the phenazine core linked to a cyclic carboxylic anhydride with their increased activity, but no positive correlation was found in relation to the carboxamide, lactone, or benzoyl moieties [10]. Recently, a synthetic study revealed that the modulation of dermacozine-1-carboxamides, especially with electron releasing substituents (e.g., chloro- and methoxy groups) increases the in vitro anti-tubulin activity of dermacozine-1-carboxamide derivatives comparable to or even superior to the nocodazole control. This evidence gave us reason to further investigate this strain for additional bioactive dermacozine derivatives [11].

2. Results and Discussion

Dermacoccus abyssi MT 1.1^T was phenotypically characterized following the initial isolation from the Mariana Trench sediment; the strain is capable of growing in the presence of 7.5% NaCl, which makes the strain halotolerant [5]. The production of new secondary metabolites has been reported by several authors when salt was added to the culture medium of the organisms capable of living in those conditions [12]. Xie et al. reported the isolation of a new sesquiterpene, ascotrichic acid, when the marine-derived fungus *Ascotricha* sp. ZJ-M-5 was cultivated in 33 g/L ocean salt containing medium [13]. In our recently published article on the full genome sequence of *Dermacoccus abyssi* MT 1.1^T and the isolation of dermacozine M (**14**), the strain was cultivated in a GYE seed culture medium initially, then subsequently large-scaled in 35 g/L ocean salt containing ISP2 medium [8]. Altering the preculture conditions has been shown to change the productivity of the strain when dermacozines H–J (**11–13**) were isolated [7]. Herein, we report on the isolation of further three dermacozines (**1–3**) produced by the strain *Dermacoccus abyssi* MT 1.1^T when a seed culture is grown in ISP2 medium containing 20 g/L NaCl, followed by a 35 g/L ocean salt supplemented ISP2 large-scale culture to approximate the deep-sea salinity of 34.7‰ [14].

2.1. Structure Determination of Dermacozine N (*1*)

Dermacozine N (**1**) was isolated as a pink amorphous powder. The molecular formula of **1** was established as $C_{21}H_{15}O_3N_5$ from LC-HRESI-MS^n of the [M + H]$^+$ ion at *m/z* 386.1247 (calculated *m/z* 386.1248, $\Delta = -0.3$ ppm) and its ^{13}C-NMR spectral data, corresponding to 17 degrees of unsaturation.

The analysis and the comparison of the molecular formula and calculated DBE with the previous metabolites reported from the same species along with its intense color suggested that **1** must be a dermacozine-like molecule. The presence of the characteristic phenazine substructure (**ABC** rings) in **1**—bearing two carboxamide groups in this particular case—present in most dermacozine structures, was deduced from the NMR data analysis and comparison to those of the reported dermacozines.

Thus, the ^{1}H-NMR spectrum of **1** showed the presence of an N-methyl group at δ_H 3.68 (3H, s, H-23), which has been a canonical part of the dermacozine structures thus far. The ^{1}H-^{13}C HMBC correlations from the H-23 methyl hydrogens at δ_H 3.68 to δ_C 134.4 (C-4a) and δ_C 135.5 (C-5a), confirmed the position of the N-methyl group in the pyrazine ring (**B** ring) of the phenazine core.

In accordance with other dermacozine structures bearing the phenazine substructure, the ^{1}H-^{1}H COSY of **1** revealed two characteristic aromatic signals corresponding to the aromatic **A** and **C** rings. The spin system with resonances at δ_H 7.88 (1H, dd, *J* = 7.6 and 1.3 Hz, H-2), 7.47 (1H, td, *J* = 8.3, and 7.6 Hz, H-3), and 7.55 (1H, dd, *J* = 8.3 and 1.3 Hz, H-4) were indicative of a trisubstituted **A** aromatic ring. The aromatic proton signals of the **A** and **C** rings were correlated to their corresponding carbon resonances at δ_C 125.9 (C-2), 128.4 (C-3), 118.0 (C-4), and 105.9 (C-9), respectively, through a ^{1}H-^{13}C HSQC experiment.

^{1}H-^{13}C HMBC correlations from H-2 at δ_H 7.88 to δ_C 166.3 (C-11), δ_C 118.0 (C-4), and δ_C 135.1 (C-10a); from H-3 at δ_H 7.47 to δ_C 128.9 (C-1) and δ_C 134.4 (C-4a); from H-4 at δ_H 7.55 to δ_C 125.9 (C-2) and δ_C 135.1 (C-10a) allowed us to fix the positions of the methines at C-2, C-3, and C-4 in relation to the non-protonated carbons at C-1, C-4a, and C-10a confirming the structure of the **A** ring (see ring system in Figure 3).

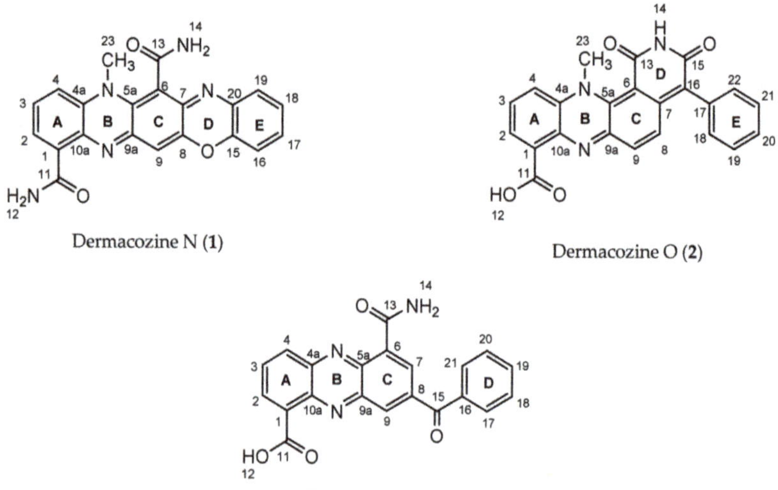

Dermacozine N (**1**)

Dermacozine O (**2**)

Dermacozine P (**3**)

Figure 3. Atom numbering for dermacozine N (**1**), dermacozine O (**2**), and dermacozine P (**3**).

On the other hand, a sharp singlet at δ_H 6.79 (1H, s, H-9), correlating to its corresponding carbon resonance at δ_C 105.9 (C-9), by a ^{1}H-^{13}C HSQC experiment, suggested the presence of a penta-substituted aromatic ring assigned to the **C** ring. The long range ^{1}H-^{13}C HMBC correlations from H-9 hydrogen at δ_H 6.79 to δ_C 151.6 (C-9a), δ_C 135.5 (C-5a), and δ_C 149.8 (C-8) were pivotal for the structure elucidation of the **C** aromatic ring as they were consistent with the linear annulation of the phenoxazine structure (Figure 4).

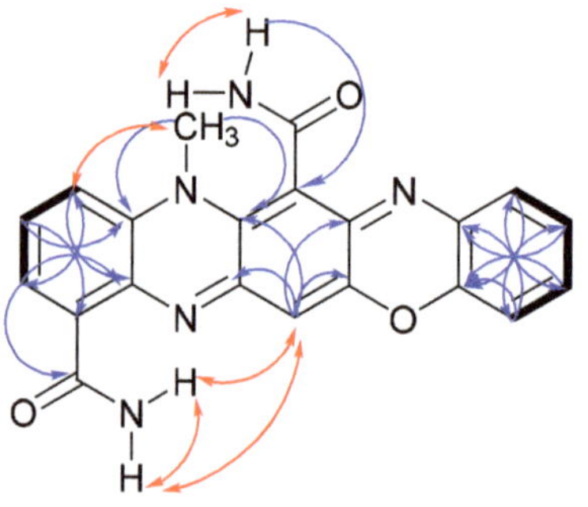

Figure 4. Key 2D NMR COSY (▬▬), NOESY (↔) and HMBC (H ⌒ C) correlations of dermacozine N (**1**).

The lack of proton resonances attached to C-7 (present in structures **4**, **5**, **6**, **11**, **12**, **13**) and C-8 (present in structures **4**, **5**, **6**, **7**, **8**, **9**, **10**, **11**, **12**, **13**, **14**) in the ^{1}H-NMR of **1** suggested that positions C-7 at δ_C 148.1 and C-8 at δ_C 149.8 of the **C** ring must be substituted at these positions.

The presence of the two carboxamide groups in **1**, as in most dermacozine structures, was deduced from a ^{1}H-^{15}N-NMR HMBC experiment showing two amide nitrogens at δ_N 112.6 and 118.7 that correlate to their corresponding ^{1}H-NMR signals at δ_H 7.70 (brs, NH-12a)/9.31 (brs, NH-12b) and δ_H 7.65 (brs, NH-14a)/7.98 (brs, NH-14b). However, only the C-11 amide carbonyl carbon at δ_C 166.3 could be detected in the ^{13}C-NMR spectrum of **1**. The ^{1}H-^{1}H COSY correlations between the two NH$_2$ hydrogens corresponding to each carboxamide group and the characteristic N–H stretch signal at 3439 cm^{-1} detected in the IR spectrum of **1** confirmed the presence of the two carboxamide groups in **1**. The two carboxamide groups were linked to C-6 and C-1 positions in **1** shown by the ^{1}H-^{13}C HMBC correlations from the NH-14b hydrogen at δ_H 7.98 (brs) to C-6 at δ_C 109.8 and from the H-2 hydrogen at δ_H 7.88 (dd) to the carbonyl of the carboxamide C-11 at δ_C 166.3, along with the ^{1}H-^{1}H NOESY correlation from NH$_2$-12 to H-9.

Once the phenazine substructure of **1** was confirmed, further NMR analysis allowed us to establish the remaining part of the molecule.

Eight aromatic carbons were found to be connected to heteroatoms. Four aromatic carbons at δ_C 134.4 (C-4a), 135.5 (C-5a), 151.6 (C-9a), and 135.1 (C-10a), linked to nitrogen atoms, were already located in the **B** ring and the two aromatic carbons—at δ_C 149.8 (C-8) linked to an oxygen atom, and 148.1 (C-7) linked to a nitrogen atom—were placed in the **C** ring. The locations of the two remaining aromatic carbons connected to heteroatoms with δ_C values of 143.3 (C-15) and δ_C 134.8 (C-20), attached to oxygen and nitrogen atoms, respectively, were determined as follows.

The ^{1}H-NMR spectrum of **1** also displayed the presence of an *ortho*-substituted benzene ring (**E** ring) by the aromatic proton resonances at δ_H 7.30 (1H, dd, *J* = 7.6 and 1.6 Hz, H-19), 7.15 (1H, ddd, *J* = 7.6, 7.4, and 1.5 Hz, H-18), 7.19 (1H, ddd, *J* = 7.6, 7.4, and 1.6 Hz, H-17), 7.12 (1H, dd, *J* = 7.6 and 1.5 Hz, H-16), which were correlated to their corresponding carbons at δ_C 127.2 (C-19), 125.3 (C-18), 127.9 (C-17), and, 115.0 (C-16), respectively, demonstrated by the ^{1}H-^{13}C HSQC spectrum of **1**. The non-protonated carbons at δ_C 143.3 and δ_C 134.8, assigned to the C-15 and C-20 carbons, respectively, completed the *ortho*-substituted benzene substructure of the **E** ring. Theoretical chemical shift increment values are consistent with an sp^3 oxygen atom substitution at C-15, whereas the C-20 chemical shift supports an sp^2 nitrogen atom substituent [15]. The carbons C-16 and C-18 are *ortho* and *para* position to the nearby oxygen atom, respectively, which possesses two lone pairs of electrons and participate in the positive mesomeric effect, resulting in the shielded carbon (and hydrogen) atoms at these positions. Key ^{1}H-^{13}C HMBC correlations from δ_H 7.12 (H-16) to C-15 (δ_C 143.3), C-20 (δ_C 134.8), and C-18 (δ_C 125.3); from δ_H 7.19 (H-17) to C-15 and C-19 (δ_C 127.2); from δ_H 7.15 (H-18) to C-16 (δ_C 115.0) and C-20; and from δ_H 7.30 (H-19) to C-15 and C-17 (δ_C 127.9) confirmed the *ortho*-substituted benzene ring.

The 106 unit mass loss observed in the LC-HRESI-MS/MS of **1**, which matched with a loss of a (-ON(C$_6$H$_4$)-) fragment agrees with the presence of this *ortho*-substituted aromatic **E** ring in **1** (Figures S2 and S3A). The presence of an electron rich π-electron configuration in the vicinity could explain why the C-6 carbon in the structure of **1** at δ_C 109.8 is shielded in relation to the other dermacozines lacking that ring.

According to the literature in polycyclic aromatic hydrocarbons when five benzene rings are annulated to each other in a linear fashion like in pentacene, the UV-Visible absorption maxima in the visible electromagnetic spectrum compared to the angularly fused benzo[*a*]anthracene shows bathochromic shift [16,17]. The UV–Vis spectra of dermacozine E (**8**), F (**9**), G (**10**), and M (**14**)—exhibiting angular annulation of four rings as seen in benzo[*a*]anthracene—displayed absorption maxima at 576, 566, 580, and 590 nm in the visible electromagnetic spectrum, respectively [6–8]. Consequently, the bathochromic shift displayed by the UV–Vis absorption maxima of **1** at 729 and 660 nm supports the annulation of its five aromatic rings. Thus, dermacozine N (**1**) shows a higher extended conjugation than that of dermacozine E (**8**), dermacozine F (**9**), dermacozine G (**10**), and dermacozine M (**14**), which were the dermacozines with the longest visible absorption maxima in their UV–Vis spectra observed to date. Based on this UV–Vis spectral comparison, the core phenazine structure of **1** must be annulated with the (-ON(C_6H_4)-) substructure. The correlations observed in the ^{1}H-^{1}H NOESY spectrum of **1** were crucial confirming this result. The calculated ca. 2.8 Å distance between H-9 and NH$_2$-12a and NH$_2$-12b hydrogen atoms in the molecular model enabled the linear fusion of the **A/B/C/D/E** rings (Theme S1).

In order to support the proposed structure for **1**, we carried out a linear regression analysis and (DFT)–UV–Vis spectral simulation as follows.

Twenty-three possible structures of **1** (**A–W**) could be drawn in ACD Labs (Figures S34–S38), satisfying the molecular formula of $C_{21}H_{15}O_3N_5$, the ^{1}H-NMR splitting pattern, 2D NMR, ^{15}N-NMR, and the presence of the (-ON(C_6H_4)-) substructure connected to the core phenazine in an angular or linear fashion. The phenazine biosynthesis occurs through the Shikimic acid pathway and with this as the suggested route of dermacozine biosynthesis, ten structures containing more heteroatoms in the phenazine core than N-5 and N-10 corresponding to **B, C, F, L, M, O, R, S, T** and **U** were excluded as possible solutions to the current structure. The remaining thirteen possible structures (**A, D, E, G, H, I, J, K, N, P, Q, V, W**) were subjected to statistical analysis. Their ^{13}C-NMR chemical shifts were calculated using the ACD Labs software simulation with the Neural Network Algorithm and then the obtained values were subjected to linear regression with the experimental ^{13}C-NMR data of **1**. Linear regression between experimental and ACD Labs software predicted ^{13}C-NMR chemical shifts has been shown to be an effective method in predicting the correct structures of natural products [18–20]. The strongest correlation was observed for structure **W** (12-methyl-12*H*-quinoxalino[2,3-*b*]phenoxazine-8,13-dicarboxamide) with $R^2 = 0.99$ (Figure 5), which matched the proposed structure of dermacozine N (**1**), as shown in Figure 4.

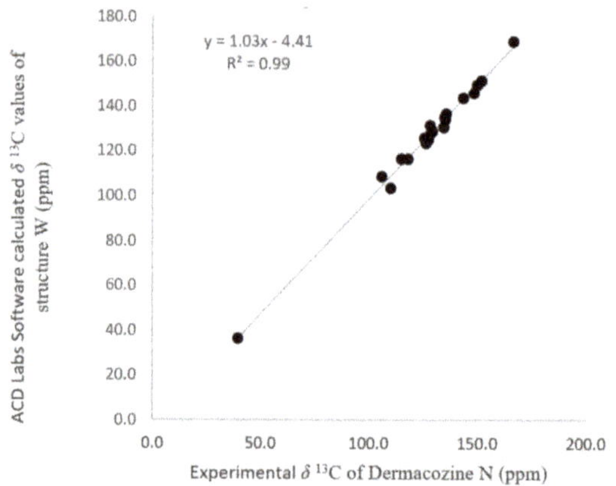

Figure 5. Correlation between experimental ^{13}C-NMR chemical shift values (ppm) of dermacozine N (**1**) and those of structure **W** (ACD Labs software (ACD/Structure Elucidator, version 2019.2.0, Neural Network Algorithm, DMSO-d_6).

Additionally, we carried out (DFT)–UV–Vis spectral calculations with the proposed structure of **1** and its second and third most likely isomer structures based on the aforementioned linear regression model (structures **K** and **Q**, $R^2 = 0.92$ and 0.91, respectively). Time-dependent DFT approach (TDDFT) calculations were used for generating the UV spectra. First, a conformational search of structures **K** and **Q** was performed in the Macromodel module implemented in Maestro Quantum mechanical software (Schrödinger). Using a 4.0 kcal/mol energy threshold from global minimum, ten and eight conformers were found, respectively. The geometry of all these conformers was optimized and their corresponding frequencies were calculated by using a density functional theory (DFT) method at the

HSEH1PBE/cc-pVDZ level (see Section 3.4). The resulting UV spectra were combined by Boltzmann weighting to give the composite spectra displayed in Figure 6 [21–23].

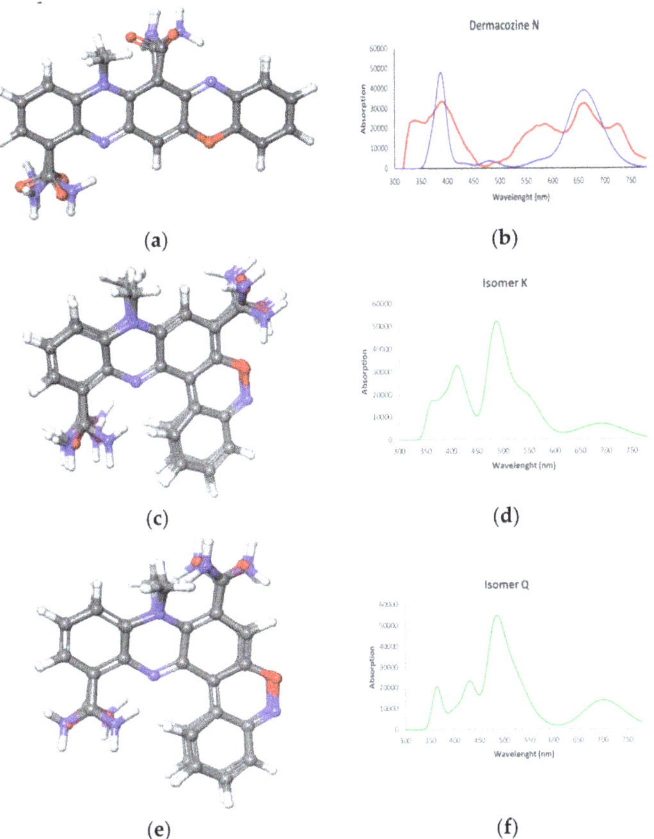

Figure 6. Conformers obtained from the conformational search for (**a**) structure **W** and structures (**c**) **K** and (**e**) **Q**; DFT calculated UV–Vis spectra of structures (**b**) **W** (————) (**d**) **K** and (**f**) **Q** (————) and experimental UV–Vis spectrum (————) of (**b**) dermacozine N (**1**).

The TD-DFT calculated UV–Vis absorption maxima of structure **W** at λ_{max} 358 and 659 nm were in excellent agreement with that of the experimental data and these independent calculations support the proposed structure for dermacozine N (**1**). Whereas the absorption maxima at λ_{max} 412 and 488 nm for structure **K** and λ_{max} at 364, 431, and 486 nm for structure **Q** were clearly different from the experimental curve of **1**. Therefore, on the basis of the combined data including 1D/2D NMR, LC-HRESI-MS^n, MS/MS, UV–Vis spectroscopy data, TD-DFT calculated UV–Vis spectrum, and linear regression made between the calculated and experimental ^{13}C-NMR data of **1**, we can conclude that the structure of dermacozine N (**1**) is 12-methyl-12*H*-quinoxalino[2,3-*b*]phenoxazine-8,13-dicarboxamide.

To the best of our knowledge, the framework of dermacozine N (**1**) is unprecedented in natural product chemistry. A synthetic compound bearing the same oxazinophenazine skeleton present in dermacozine N (**1**) was reported by Fischer and Hepp in 1895. The skeleton of this compound, referred to as *"triphenazinoxazine"* in the 19th century was synthesized through nucleophilic substitution of 2-amino-3-phenoxazinone with *ortho*-phenylenediamine [24]. The solvatochromic behavior of the oxazinophenazine and its derivatives in different solvents and the spectacular bathochromic color change upon addition of strong acids and a subsequent hypsochromic shift when this solution was mixed with glacial acetic acid has been described by various authors in the 19th century [24–28]. Fischer and Hepp observed dark red fluorescence of the "*triphenazinoxazine*" in "alcohol solution" and the aforementioned color change from red to violet upon making the solution acidic with mineral acids. In those times, induline derivatives with fluorescent properties were listed under the collective name of "*Fluorindine*"-s [24–28]. In 1978, G.B. Afanas'eva et al. reported the synthesis of the same oxazinophenazine skeleton when they carried out nucleophilic substitution of 2-ethoxy-3-phenoxazinone with *ortho*-phenylendiamine [29]. Given the compound absorption maxima of **1** in the near infrared spectrum (NIR), its fluorescence spectroscopic investigation would be intriguing in the future.

Fluorophores from nature have gained particular interest among scientists in visualizing physiological processes at the cellular level as the example of one of the most notable discoveries in this field—the isolation of the green fluorescent protein (GFP)—shows [30]. Phenoxazines were found to generate second harmonics (SHG, i.e., frequency doubling) whilst being investigated for non-linear optical properties [31]. Utilizing the optoelectronic features of phenoxazines is widespread in the literature. Benzo[*a*]phenoxazines have been reported to possess useful NIR absorption and emission spectral features as well as solvatochromic effects, thus making them useful in biosensing chemistry (e.g., in pH sensing, glucose sensing, organic biomolecule labelling, and in vivo cellular metabolic mapping [32–39]). Utilizing the long wavelength absorption and emission maxima in visualizing biological processes at the cellular level with fluorogenic probes has gained extreme importance in recent years [37–39]. The absorption maxima of the current benzo[*a*]phenoxazine probes is just about or under 700 nm whereas dermacozine N (**1**) at 729 nm possesses the longest absorption maximum in the visible electromagnetic spectrum among the phenoxazines currently being utilized for this purpose [36]. Biosensor molecules exhibiting red or NIR wavelength absorption and emission maxima, as in the case of dermacozine N (**1**), are required for the study of body fluids such as blood, serum, and urine where the matrix may possess components with long wavelength absorption maxima that can interfere with spectroscopic measurements [33].

The cytotoxic activity of dermacozine N (**1**) was investigated against a panel of five human tumor cell lines: human melanoma (A2058) and hepatocellular human carcinoma cell lines (HepG2) exhibiting weak activity, with IC_{50} values of 51 and 38 μM, respectively. Certain phenoxazine derivatives (e.g., Nile Blue analogues were shown to be useful in vitro in photodynamic therapy (PDT) against human bladder carcinoma cells (MGH-U1) acting via the singlet oxygen ($^1O_2$) pathway [40]. In addition to its cytotoxicity, the NIR absorption maxima of dermacozine N (**1**), which would interfere less with the tissue absorption maxima as detailed above, suggests that it may be a suitable compound for further research in the field of photodynamic therapy.

2.2. Structure Determination of Dermacozine O (2)

Dermacozine O (**2**) was isolated as an ink-blue amorphous powder. The LC-HRESI-MS^n measurement gave an *m/z* of 398.1125 for [M + H]$^+$, consistent with a molecular formula of $C_{23}H_{15}O_4N_3$ (calculated *m/z* 398.1135, Δ = −2.5 ppm), giving 18 degrees of unsaturation. Interestingly, this molecular formula matches that of previously reported dermacozine F (**9**).

Comparison of the NMR spectra of dermacozine O (**2**) to those of dermacozines E and F (**8** and **9**) showed the same **A**, **B**, **C**, and **E** rings for the three compounds. Indeed, the spin systems identified in the ^1H-^1H COSY spectra of **8** and **9** had high similarity to the one in **2**.

The first spin system belonging to the **A** ring (see ring system in Figure 3) was assigned to the H-2, H-3, H-4 hydrogens, which were correlated to their corresponding carbons according to the ^1H-^{13}C HSQC experiment with δ_H/ δ_C resonances at 7.87 (1H, dd, *J* = 7.5 Hz, 1.1 Hz)/126.6 (C-2); 7.78 (1H, td, *J* = 8.5 Hz, 7.5 Hz)/ 131.3 (C-3); 7.97 (1H, dd, *J* = 8.5 Hz, 1.1 Hz)/120.3 (C-4). The structure of the **A** ring was confirmed by the following ^1H-^{13}C HMBC correlations: from H-2 at δ_H 7.87 to C-11 at δ_C 167.2, C-4 at δ_C 120.3, and

C-10a at δ_C 135.3; from H-3 at δ_H 7.78 to C-1 at δ_C 129.6 and C-4a at δ_C 134.0; and from H-4 at δ_H 7.97 to C-2 at δ_C 126.6 and C-10a at δ_C 135.3. Furthermore, these correlations clearly confirmed the C-1, C-2, C-3, and C-4a, C-10a positions of the **A** ring in relation to the neighboring pyrazine moiety (**B** ring).

A methyl group that resonates as a sharp singlet at δ_H 3.67 (3H, s, H-23) in the ^{1}H-NMR spectrum of **2**, correlated by ^{1}H-^{13}C HSQC to δ_C 45.9 (C-23), which was assigned to the characteristic N-methyl attached to the pyrazine ring (**B** ring). The position of this N-methyl group relative to the pyrazine ring was confirmed by the ^{1}H-^{13}C HMBC correlations from H-23 at δ_H 3.67 to C-4a at δ_C 134.0 and C-5a δ_C 139.5.

The second spin system located in the **C** ring was found to be composed of the H-8 and H-9 hydrogens at δ_H 7.21 (1H, d, *J* = 9.7 Hz) and 7.24 (1H, d, *J* = 9.7 Hz), which were correlated to δ_C 134.5 (C-8) and δ_C 129.7 (C-9), respectively, by the ^{1}H-^{13}C HSQC experiment. The positions of the H-8/H-9 spin system relative to the C-5a, C-9a, C-6, C-7 in the **C** ring were confirmed by key ^{1}H-^{13}C HMBC correlations from H-9 at δ_H 7.24 to C-5a at 139.5 ppm, from H-8 at δ_H 7.21 to C-9a at δ_C 150.6, C-6 at δ_C 100.4, C-7 at δ_C 139.6, and C-16 at δ_C 122.6.

A monosubstituted benzene ring was assigned to the third spin system comprising the five hydrogens corresponding to the **E** ring: H-18/22 at δ_H 7.30 (2H, dd, *J* = 7.4, 1.3 Hz)/δ_C 131.3 (C-18/22); H-19/21 at δ_H 7.47 (2H, td, *J* = 7.4, 1.3 Hz)/ δ_C 128.2 (C-19/21); and H-20 δ_H 7.41 (1H, td, *J* = 7.4, 1.3 Hz)/δ_C 127.9 (C-20). The key ^{1}H-^{13}C HMBC correlation from H-18/22 at δ_H 7.30 to C-16 at δ_C 122.6 ppm allowed us to confirm the connection of the monosubstituted benzene **E** ring to C-16 of the probable **D** ring consisting of a cyclic carboximide (Figure 7).

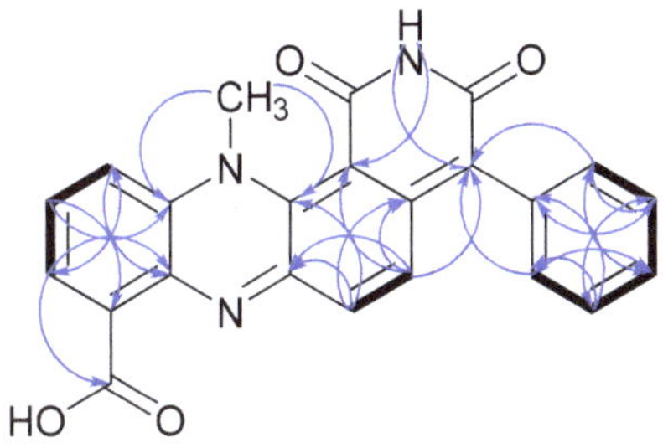

Figure 7. Key 2D NMR COSY (▬▬▬) and HMBC (H ⌒ C) correlations of dermacozine O (**2**).

Since the ^{1}H-^{1}H COSY spectrum of **8**, **9**, and **10** showed considerable similarity to the ^{1}H-^{1}H COSY of **2**, statistical comparison was made between the corresponding carbon chemical shifts of these dermacozines. We found that based on the two-tailed *t*-test and multiple regression analysis of the corresponding experimental δ_C values, the most similar structure to dermacozine O (**2**) is dermacozine E (**8**) ($p = 2.19 \times 10^{-8}$). Figure 8 shows the linearly correlated experimental ^{13}C-NMR resonances of **2** and **8** (see multiple regression in Figure S43).

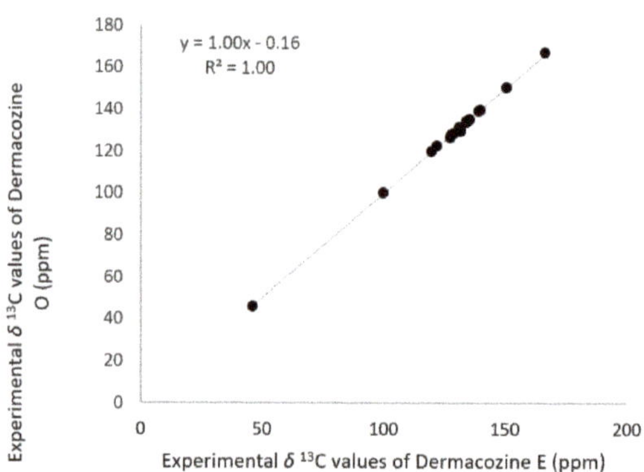

Figure 8. Correlation between the experimental ^{13}C-NMR chemical shift values (ppm) of dermacozine O (**2**) and the experimental ^{13}C-NMR chemical shift values (ppm) of dermacozine E (**8**).

The most important differences between the NMR spectral data of dermacozine O (**2**) to those of dermacozines E and F (**8** and **9**) were found related to the carbonyl functionalities at position C-11.

The ^{1}H-NMR spectrum of **2** showed the presence of a broad singlet at δ_{H} 11.27 (1H, brs, NH-14), which matched the cyclic carboximide group present in dermacozine E (**8**). The δ_{N} 170.2 resonance observed in the ^{1}H-^{15}N-NMR of **2** is in agreement with the presence of that functionality and this chemical shift is in the range of ^{15}N-NMR chemical shifts of the cyclic carboximide nitrogens present in the structures of dermacozines E (**8**) and M (**14**) [6,8]. Unfortunately, the ^{13}C-NMR signals of the C-13 and C-15 carbonyl carbons of **2** were not observed. However, upon comparison of the ^{13}C-NMR chemical shifts at position C-6 and C-16, one C–C bond distance from those carbonyl carbons, it was observed that they were closely correlated to the ^{13}C-NMR chemical shifts in similar dermacozine structures (**8–10** and **14**), providing indirect evidence that C-13 and C-15 must be carbonyl carbons [6,8]. The key ^{1}H-^{13}C HMBC correlations from H-14 at δ_{H} 11.27 of the cyclic carboximide group to C-6 at δ_{C} 100.4 and C-16 at δ_{C} 122.6 allowed us to confirm the structure of the **D** ring.

On the other hand, even though H-2 at δ_{H} 7.87 showed a ^{1}H-^{13}C HMBC correlation to C-11 at δ_{C} 167.2, the lack of signals in the ^{1}H-NMR spectrum of **2**, corresponding to the NH_2 group observed in dermacozines E and F (**8** and **9**), was indicative of the absence of the C-11 carboxamide group in **2**. This information, along with the fact that dermacozine F (**9**) has the same molecular formula as compound **2**, is indicative that the C-11 carboxamide group present in **9** was substituted by a C-11 carboxylic acid in **2**.

Upon assembling the **A–E** rings, the structure of dermacozine O (**2**) was determined to be a constitutional isomer of dermacozine F (**9**) and the biosynthetically more closely related C-1 carboxylated derivative of dermacozine E (**8**) with the IUPAC name of 12-methyl-1,3-dioxo-4-phenyl-1,2,3,12-tetrahydropyrido[3,4-*a*]phenazine-8-carboxylic acid. Dermacozine O (**2**) is the first isolated compound bearing a carboxylic acid in its structure among the dermacozine E (**8**), F (**9**), G (**10**), and M (**14**) type of structures.

Due to the absence of some experimental ^{13}C-NMR signals, we carried out an ACD Labs simulation using the Neural Network Algorithm, which provided the chemical shift carbon values for the carbonyl resonances of the cyclic carboximide moiety of **2** at δ_{C} 163.6 (C-13) and δ_{C} 163.1 (C-15), in perfect agreement with the C-13 and C-15 carbon shifts of the aforementioned resembling dermacozine structures.

Dermacozine O (**2**) displayed no cytotoxic activity when tested against A549 (lung carcinoma), A2058 (metastatic melanoma), MCF7 (breast adenocarcinoma), MIA PaCa-2 (pancreatic carcinoma), and HepG2 (hepatocyte carcinoma) cell lines.

The compound as discussed above resembles the structures of **8**, **9**, and **10**, which have been found to exhibit the strongest non-linear optical properties in computational chemistry studies among the previously isolated dermacozines [9].

Following the initial isolation of the dermacozines from the producing *Dermacoccus abyssi* MT 1.1^T and MT 1.2, their function in nature has not yet been determined [6–8]. Since the survival capability of bacteria that produce the phenazine type of compounds is greater, these compounds are proposed to have defensive functions that protect the producing organism [6,41]. By looking at the structures of dermacozine N (**1**) and dermacozine O (**2**) alongside the previously published dermacozines E (**8**), F (**9**), G (**10**), and M (**14**), the quinonoid **C** rings resemble the quinones playing important roles in electron shuttling in the respiratory process of the cell. Quinone derivatives are found in almost every living organism's lipid membrane with rare exceptions [42]. Certain phenazine derivatives have been reported to be able to mediate electron transfer from NADPH to molecular O_2 [43]. The function of dermacozines in nature is still unknown, but based on this observation, their participation in redox reactions helping the strain to survive (e.g., in microaerobic conditions) is potentially possible. This seems to be supported by the fact that the genome sequence of the bacterium revealed the existence of cytochrome *d* oxidase, which exhibits high affinity to O_2 and operates under low oxygen concentrations and multiple copies of succinate dehydrogenases, which can be used as electron donating systems in oxygen deprived conditions [8].

2.3. Structure Determination of Dermacozine P (3)

Dermacozine P (**3**) was isolated as a purplish amorphous powder. The LC-HRESI-MS^n of **3** showed a protonated $[M + H]^+$ ion at *m/z* 372.0990 $[M + H]^+$, consistent with a molecular formula of $C_{21}H_{13}O_4N_3$ (calculated *m/z* 372.0979, Δ= 3.0 ppm), possessing 17 degrees of unsaturation.

The ^{1}H-^{1}H COSY spectrum displayed the presence of three spin systems. The first spin system assigned to the **A** ring comprises the three aromatic hydrogens at δ_H 8.74 (1H, dd, *J* = 7.0 and 1.3 Hz, H-2), 8.21 (1H, td, *J* = 7.0 and 8.6 Hz, H-3), and 8.55 (1H, dd, *J* = 8.6 and 1.3 Hz, H-4). Key long range correlations in the ^{1}H-^{13}C HMBC spectrum of **3** from H-2 at δ_H 8.74 to C-10a at δ_C 141.2; from H-3 at δ_H 8.21 to C-1 at δ_C 131.5 and C-4a at δ_C 142.4; and from H-4 at δ_H 8.55 to C-10a at δ_C 141.2 and C-2 at δ_C 134.9, defined the H-2/H-3/H-4 spin system relative to quaternary C-4a and C-10a carbon atoms, completing the substructure of the **A** ring (see ring system in Figure 3).

A second spin system corresponded to the aromatic hydrogens H-7 and H-9 with resonances at δ_H 8.65 (1H, d, *J* = 1.9 Hz) and 8.95 (1H, d, *J* = 1.9 Hz). The $^4J_{H7/H9}$ coupling constant of 1.9 Hz, indicative of a *meta* relationship between these hydrogen atoms, along with ^{1}H-^{13}C HMBC correlations from H-7 to C-5a at δ_C 140.7, C-9 at δ_C 135.1, C-13 at δ_C 166.6 and from H-9 to C-15 at δ_C 194.4, C-7 at δ_C 131.3, and C-5a at δ_C 140.7 allowed us to determine the substructure of the **C** ring (Figure 9).

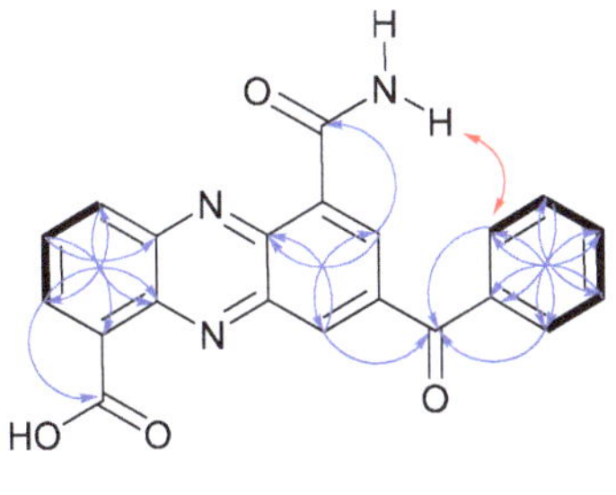

Figure 9. Key 2D NMR COSY (▬▬▬), NOESY (⟵) and HMBC (H ⟶ C) correlations of dermacozine P (**3**).

The aromatic hydrogens H-17/21, H-18/20, and H-19 at δ_H 7.95 (2H, dd, *J* = 7.6 Hz and 1.3 Hz); 7.66 (2H, td, *J* = 7.6 Hz and 1.3 Hz), and 7.78 (1H, td, *J* = 7.6 Hz and 1.3 Hz) of the third spin system were assigned to a monosubstituted benzene corresponding to the **D** ring. Those hydrogens were correlated to their corresponding carbons at δ_C 130.0

(C-17/21), δ_C 128.7 (C-18/20), and δ_C 133.5 (C-19) by a ^{1}H-^{13}C HSQC experiment of **3**. Finally, C-16—the connection point of the monosubstituted benzene at δ_C 136.3—was identified by the ^{1}H-^{13}C HMBC spectrum.

In contrast to all the dermacozine structures (**4–14**) reported thus far, the ^{1}H-NMR spectrum of **3** showed no evidence of the characteristic N-methyl group attached to the pyrazine ring (**B** ring). Three out of the four carbon atoms of the **B** ring were assigned next to nitrogen: δ_C 142.4 (C-4a), δc 140.7 (C-5a), and δ_C 141.2 (C-10a).

When the ^{1}H-^{13}C HSQC and ^{1}H-^{13}C HMBC spectra of **3** were overlaid, the presence of three carbonyl carbons at δ_C 165.5 (C-11), δ_C 166.6 (C-13), and at δ_C 194.4 (C-15) was revealed. One of them, C-13, was assigned to a carboxamide substituent from the characteristic primary amide protons that resonate as two broad singlets at δ_H 8.03 (1H, brs, H-14a) and δ_H 9.47 (1H, brs, H-14b), both correlated in the ^{1}H-^{1}H COSY spectrum. The IR band at 3440 cm^{-1} corresponding to a NH stretching confirmed the existence of that substituent. The ^{1}H-^{13}C HMBC correlation from H-7 at δ_H 8.65 to C-13 at δ_C 166.6, along with enhancement of the H-17/21 aromatic protons at δ_H 7.95 by selective NOE irradiation at the NH-14a at δ_H 8.03, allowed us to link the carboxamide substituent to position C-6 of the **C** ring.

The **C** ring of the phenazine moiety was connected at C-8 to a monosubstituted benzene **D** ring at C-16 through the C-15 ketone carbonyl group at δ_C 194.4, confirmed by the ^{1}H-^{13}C HMBC correlations from the H-9 and H-17/21 protons at δ_H 8.95 and 7.95, respectively, to C-15. The already mentioned *meta*-coupling (^4J = 1.9 Hz) between H-7 and H-9 of the **C** ring also supports the attachment of C-15 to C-8 (Figure 9).

The ^{13}C-NMR chemical shifts of C-6, C-8, and C-9a in **3** were not detected, so these missing values were calculated with ACD Labs software, with the Neural Network Algorithm as δ_C 129.8, 135.6, and 144.0 ppm, respectively. Upon assembling the structure of **3**, we had one more carbonyl group left with a δ_C of 165.5 ppm (C-11). Taking into consideration the molecular formula, it was assigned to a carboxylic acid functionality. The connection of this –COOH group was deduced from the ^{1}H-^{13}C HMBC correlation from H-2 at δ_H 8.74 to C-11 at δ_C 165.5. The carboxylic OH-12 was not observed in the ^{1}H-NMR spectrum of **3** in DMSO-d_6, therefore we applied a similar approach as in the case of **1** and **2** to confirm the proposed structure for **3**.

We modeled dermacozine P (**3**) with the ACD Labs software (Neural Network Algorithm, DMSO-d_6) and the linear regression curve was obtained between the calculated and experimentally observed ^{13}C-NMR chemical shifts. The statistical analysis provided an R^2 = 0.98 (Figure 10).

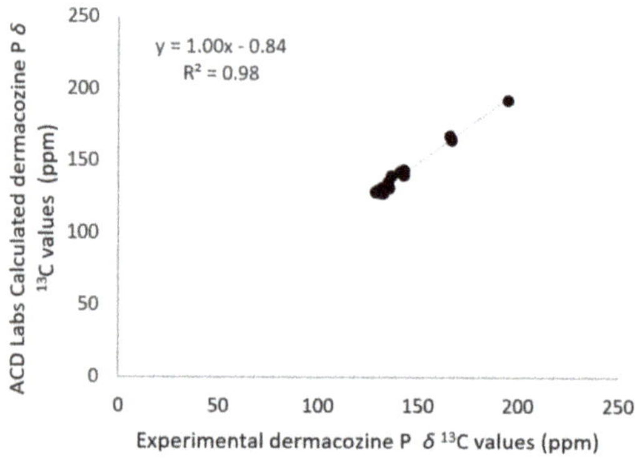

Figure 10. Correlation between the ACD Labs calculated (version 2019.2.0, Neural Network Algorithm, in DMSO-d_6) and experimental ^{13}C-NMR chemical shift values (ppm) of dermacozine P (**3**).

Upon comparison made between the experimental ^{13}C-NMR shifts of **3** versus those chemical shifts observed in the case of **5** and **6**, we could observe a negative mesomeric effect due to the carboxylic group at *ortho-* and *para* positions relative to that functionality in the case of **3** (Figure 11).

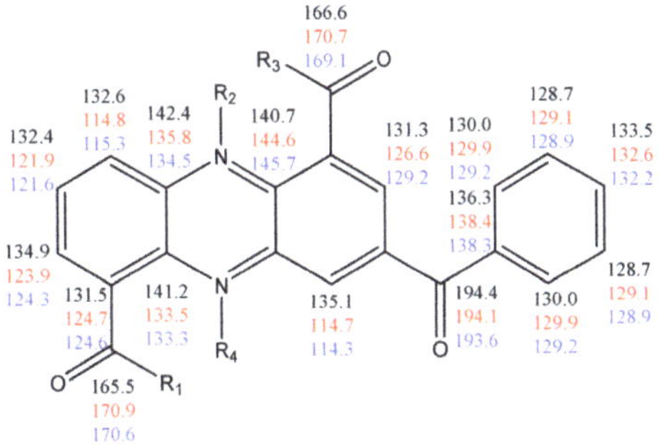

Dermacozine P R$_1$=OH, R$_2$, R$_4$=-; R$_3$=NH$_2$
Dermacozine B R$_1$, R$_3$=NH$_2$; R$_2$=CH$_3$; R$_4$=H
Dermacozine C R$_1$=NH$_2$; R$_2$=CH$_3$; R$_3$=OH; R$_4$=H

Figure 11. Experimental ^{13}C-NMR values (ppm) of dermacozine P (**3**), dermacozine B (**5**), and dermacozine C (**6**).

This trend of the ^{13}C-NMR chemical shifts provides additional evidence that the substituent at C-1 is more electron withdrawing compared to the one in the case of **5** and **6** and keeping with the carboxylic acid substitution at position C-1. Due to the oxidized phenazine substructure of **3**, the carbon atoms at C-7 and C-9 are more deshielded because of the nitrogen atoms conjugated to them three (N-5) and two bonds (N-10) away, respectively, as opposed to the more shielded aromatic carbons in the case of the more reduced dermacozine B (**5**) and C (**6**) structures (Figure 11). Comparison was made between the UV–Vis spectra of **3**, **5**, and **6** in order to obtain further evidence that the structure of **3** was correct. We found that the electronic structures of **3** and **6** possessed more similarity with the measured $\Delta\lambda_{max}$ = 5 nm in the visible electromagnetic spectrum, as opposed to the electronic structures of **3** and **5** based on the $\Delta\lambda_{max}$ = 46 nm measured difference between their visible absorption maxima [6]. This indicates that the 3-benzoyl-phenazine substructure of **3** must be substituted in similar fashion with electron withdrawing functional groups at C-1 and C-6, just like in the structure of **6**. Nevertheless, the hypsochromic shift of **6** of 5 nm compared to the one observed in **3** is in keeping with the more electron deficient phenazine core and the consequently higher degree of conjugation of **3**. To satisfy the observed ^{13}C-NMR chemical shifts of **3**, the fit of the linear regression analysis between the experimental and the ACD Labs software calculated (Neural Network Algorithm, DMSO-d_6) ^{13}C-NMR shifts of **3** as detailed above, the biosynthetic route considerations (the carboxamide and carboxylic acid functionalities occur at positions C-1 and C-6 in the shikimic acid biosynthetic pathway of dermacozines) as well as the comparison of the UV–Vis absorption maxima in the visible electromagnetic spectrum of **3**, **5**, and **6**, the carboxylic and carboxamide groups are needed to be positioned in the opposite way between the C-1 and C-6 carbon atoms as opposed to the positioning in dermacozine C (**6**). This assignment agrees with the result of the selective NOE experiment, thus placing the –CONH$_2$ group at position C-6.

The MS/MS fragmentation data of **3** agreed with our proposed structure, as shown in Figure 9. We were able to identify, among other molecular fragment ions of **3**, the $C_{20}H_{11}N_2O_2{}^+$ fragment (measured m/z of [M]$^+$ as 309) that agrees with the loss of an –OH and a –CONH$_2$ fragment keeping with the proposed –COOH group at position C-1. The benzoyl substituted core phenazine fragment molecular ion with a molecular formula of

$C_{19}N_2O^+$ (measured m/z of $[M]^+$ as 283) following the loss of the –COOH and –CONH$_2$ fragments was also apparent in the spectrum (Figure S25).

Therefore, dermacozine P (**3**) is the 5N-demethylated and 6-carboxylated, oxidized analogue of dermacozine B (**5**) and as such, it was chemically identified as 8-benzoyl-6-carbamoylphenazine-1-carboxylic acid.

Dermacozine B (**5**) and C (**6**) showed antioxidant activity in the DPPH assay as well as activity against the K562 leukemia cell line as previously described following their isolation [6]. Given the structural similarity of dermacozine P (**3**) to these substances, further investigation of **3** in anti-tumor and radical scavenging studies would be interesting, especially taking into consideration that the synthetic modulation of the related dermacozine-1-carboxamide derivatives led to increased biological activity in anti-tumor assays as mentioned earlier (see Introduction).

3. Materials and Methods

3.1. Microorganisms

Pure colonies of *Dermacoccus abyssi* MT 1.1^T were provided on an ISP2 agar plate by the School of Biology, University of Newcastle.

3.2. Fermentation and Initial Partitioning

A seed culture of *Dermacoccus abyssi* MT 1.1^T was prepared as follows: 25 mL of ISP2 medium (yeast extract 4 g, D-glucose 10 g, malt extract 10 g, MilliQ water 1 L, pH 7.0) was supplemented with 20 g/L NaCl. The seed culture incubation was carried out in a 50 mL Falcon tube at 28 °C and 150 rpm for five days. The large-scale fermentation was done in six 2 L Erlenmeyer flasks, each of them containing 1000 mL ISP2 medium (yeast extract 4 g, D-glucose 10 g, malt extract 10 g, MilliQ water 1 L, pH 7.0) supplemented with 35 g/L ocean salt (H2Ocean + Pro Formula with trace elements, The Aquarium Solution Ltd., Hainault Industrial Estate, Ilford Essex, UK). Each flask was inoculated with 1500 µL of the first stage seed culture incubated at 28 °C with agitation at 150 rpm for 14 days in an incubator with a transparent glass cover.

The fermentation broth (6 L) was harvested by the addition of 50 g/L Diaion HP20 resin (>250 µm, Alfa Aesar by Thermo Fisher Scientific, Heysham, Lancashire, U tation for 24 h in a shaker at 28 °C and 150 rpm. The HP20 resin was eluted with methanol (3 × 500 mL) and then with dichloromethane (3 × 500 mL). The successive methanol and dichloromethane extracts were combined and concentrated under reduced pressure yielding the crude material (10,256 mg). The crude material was subjected to liquid–liquid partitioning with the Kupchan method previously used during the isolation of dermacozine M (**14**) [8]. The crude material was suspended in 500 mL H$_2$O and extracted three times with an equal volume of dichloromethane in a 1 L separation funnel. The resultant H$_2$O layer was then extracted with 3 × 500 mL 2-butanol providing the water fraction (WF; 3498 mg) and the 2-butanol fraction (WB; 2295 mg). The dichloromethane layer was dried under reduced pressure. The dried material was dissolved in 300 mL 9:1 methanol:H$_2$O solution and extracted with an equal volume of n-hexane. The n-hexane fraction was dried under reduced pressure resulting in the n-hexane fraction (FH; 65 mg). The remainder 300 mL 9:1 methanol:H$_2$O solution after separating it from the n-hexane layer was adjusted to 1:1 methanol:H$_2$O solution with 120 mL 100% MilliQ water then extracted with 420 mL dichloromethane three times, which gave the dichloromethane fraction. The dichloromethane layer was separated and dried under reduced pressure (FD; 379 mg). The remainder of the methanol:H$_2$O layer was dried providing the methanol fraction (FM; 4019 mg). The dichloromethane fraction (FD; 379 mg) was further separated with silica gel chromatography (Fluorochem, 60A 40-63U MW = 60.083) with a 9:1 dichloromethane:methanol mobile phase as follows: a Quickfit XA42 29/32, 19/26, 19 mm × 500 mm (Fisher Scientific, Loughborough, UK) glass column was used, packed with 200 g silica (~550× of the mass of the FD fraction), with a flow rate of ~0.8 mL/min, collected into 10 mL vials. Bands were collected based on their visible colors but they also showed fluorescence under UV light (360 nm). In the silica chromatography of the FD fraction, 35 fractions were obtained (S1–S35). These fractions were combined based on TLC (thin layer chromatography) experiments and purified further, as detailed in the Section 3.5 (please see the TLC plate of the FD fraction with dichloromethane:methanol 9:1 mobile phase and the colorful bands in Theme S2).

3.3. Instrumentation

High performance liquid chromatography: Semi-preparative Gradient Agilent HPLC apparatus (1100 series, Agilent—Santa Clara, Cequipped with binary pump, diode array detector (G1315B, Agilent—Santa Clara, CSunfire C_{18} reversed phase column (5 µm, 10 × 250 mm), and ACE HL C_{18} reversed phase column (5 µm, 10 × 250 mm).

Nuclear magnetic resonance spectrometers: For the structure determination of **1** and **2**, a Bruker 800 MHz NMR spectrometer (Bruker Biospin—Billerica, Merating with a 5 mm TCI He Cryoprobe was used. 1D ^{1}H-NMR and magnitude 2 D NMR experiments were run apart from the ^{1}H-^{13}C HMBC of **1**, where an additional phase sensitive experiment was conducted. In the case of compound **3**, 1 D ^{1}H-NMR and magnitude mode 2 D NMR experiments were run including a selective NOESY experiment with a Bruker 600 MHz NMR spectrometer (Bruker Biospin—Billerica, MCE III HD operating with a N_2 Cryoprobe. HSQC experiments were obtained using 2D H-1/X correlation via the double inept transfer phase sensitive using Echo/Antiecho-TPPI gradient selection; with decoupling during acquisition; using trim pulses in inept transfer; with multiplicity editing during selection step, shaped pulses for inversion on f2—channel for matched sweep adiabatic pulses; for HMBC, experiments were obtained with 2D H-1/X correlation via heteronuclear zero and double quantum coherence, phase sensitive and magnitude mode in the case of **1**, and magnitude mode for **2** and **3**, using Echo/Antiecho gradient selection with two-fold low-pass J-filter to suppress one-bond correlations with no decoupling during acquisition [44–49].

Liquid chromatography-mass spectrometry: **1** and **2** were analyzed with a Bruker MAXIS II qTOF LC-MS instrument. LC-qTOF utilizes a Phenomenex Kinetex XB-C18 column (2.6 µm; 100 × 2.1 mm) with a mobile phase of 5% acetonitrile + 0.1% formic acid to 100% acetonitrile + 0.1% formic acid with a run time of 11 min. **3** was analyzed with a Thermo Instruments MS system (LTQ-XL Orbitrap Discovery) coupled to a 1290 Infinity Agilent UHPLC system, utilizing a Phenomenex Kinetex XB-C18 column (2.6 µm; 100 × 2.1 mm) with a mobile phase of 5% acetonitrile + 0.1% formic acid to 100% acetonitrile + 0.1% formic acid in 25 min. The analytes were diluted to 0.01 mg/mL concentration with methanol.

Infrared spectra were recorded in ethanol with a Perkin Elmer Spectrum Two FTIR spectrometer Perkin Elmer, MVis Spectroscopic data were recorded in ethanol using a Thermo Scientific Evolution 201 UV–Visible Spectrophotometer Thermo, M. Computational Calculations

A Macromodel module implemented in the Maestro Quantum mechanical software (Maestro Schrödinger LLC, New York, N was used for conformational searches [21]. The calculations were carried out by using an OPLS 2005 force field with ethanol as the solvent. A torsional enhanced sampling with 1000 or 10,000 steps was fixed using an energy window of 4.0 kcal/mol. Molecular geometry optimizations were performed at the DFT theoretical level using the Gaussian 09W package with a HSEH1PBE/cc-pVDZ auto for energy and frequency calculations. After removing redundant conformers and those with imaginary frequencies, theoretical Boltzmann energy population-weighted UV–Vis spectra were calculated by using PBEPBE/6-311++g(3d,2p) with 24 states [22]. The open software SpecDis V.1.71 (Berlin, Germany, 2017, https:/specdis-software.jimdo.com, (accessed on the 7th y 2020 and on the 14 April 2021)) was used to obtain the graphical theoretical UV–Vis curves [23].

3.4. Isolation of Compounds

Dermacozine N (**1**) was obtained with further purification of the initial silica fractions detailed in Section 3.2 as follows: the first fraction (S1) of the initial FD fractionation with silica chromatography gave dermacozine M (**14**) as reported by Abdel-Mageed et al. [8]. The S2–S3 deep purple colored fractions of the initial silica chromatography as described in Section 3.2 were combined based on TLC chromatography, dried in a N_2 drier (57.3 mg), and further partitioned with silica glass column chromatography. A Quickfit CR 20/30 19/26 glass column was used, packed with ~20 g silica (~350× of the mass of the initial mass of the combined S2–S3 silica chromatography FD fraction), and the mobile phase was 9:1 dichloromethane:methanol also in this case. In this experiment, fifteen (SA1–SA15) fractions were collected, each into 7 mL vials with a flow rate of ~0.4 mL/min. The SA9–SA11 pink colored fractions of this experiment were combined based on TLC. Keeping this fraction in a cold room at 4 °C, a pink precipitate appeared, which was centrifuged at 10,000 rpm for 15 min. Upon removal of the supernatant and drying the substance in N_2 drier, the substance appeared to be a pure compound (1.6 mg) according to the ^{1}H-NMR and the LC-HRESI-MS^n measurements (Figures S1 and S6).

Dermacozine O (**2**) was obtained with further purification of the initial silica fractions detailed in Section 3.2 as follows: the S5–S8 red colored silica fractions (28.2 mg) were combined based on TLC chromatography, dried in a N_2 drier, and subjected to HPLC purification. High performance liquid chromatography purification of dermacozine O (2) was carried out isocratically with a methanol/H_2O 60%/40% + 0.1% trifluoroacetic acid solvent system over 40 min with 1.8 mL/min flowrate, and with Sunfire C_{18} reversed phase column (5 µm; 110 × 250 mm). Peak detection and UV–Vis trace analysis was carried out at the wavelengths of 254 ± 5 nm, 280 ± 5 nm, 330 ± 5 nm, 350 ± 5 nm, and 530 ± 25 nm. Material collection was automated based on UV–Vis peak appearance versus time. The HPLC fraction with RT = 31.9 min yielded (3.1 mg) the pure compound (see Figures S13, S16 and Theme S3).

Dermacozine P (**3**) was obtained with further purification of the initial silica fractions detailed in Section 3.2 as follows: the initial S4 (7 mg) blue colored silica fraction of the dichloromethane Kupchan fraction (FD) was dried in N_2 drier and subsequently repurified with High performance liquid chromatography. The HPLC purification of dermacozine P (**3**) was carried out with a gradient from the initial solvent ratio of 0% methanol and 95% H_2O + 5% methanol + 0.05% trifluoro-acetic acid to 100% methanol and 0% H_2O over 40 min with 2 mL/min flowrate, and with an ACE C_{18} HL reversed phase column (5 µm; 110 × 250 mm). Peak detection and the UV–Vis trace analysis was carried out at the wavelengths of 250 ± 5 nm, 280 ± 5 nm, 500 ± 25 nm, 550 ± 25 nm, and 600 ± 25 nm. Material collection was automated based on UV–Vis peak appearance versus time. The HPLC fraction with RT = 15.9 min yielded (3.4 mg) the pure compound (see Figures S24, S26 and Theme S4).

3.5. Cytotoxic Activity of Dermacozines against Human Tumor Cell Lines

The cytotoxic activity of dermacozines N (**1**) and O (**2**) against five different human cancer cell lines, namely A549 (lung carcinoma), A2058 (metastatic melanoma), MCF7 (breast adenocarcinoma), MIA PaCa-2 (pancreatic carcinoma), and HepG2 (hepatocyte carcinoma) (obtained from ATCC, Manasas, V [50] was studied based on the MTT (3-(4,5-dimethylthiazol-2-yl)-2,5-diphenyltetrazolium bromide) assay [51]. The compounds were tested as a 10-point dose-response curve ($\frac{1}{2}$ serial dilutions) starting at a concentration of 20 µg/mL in triplicate. IC_{50} values were determined as previously described [50]. Dermacozine P (**3**) was not tested against cancer cell lines.

3.6. Chemical Characterization of Compounds

Dermacozine N (**1**): pink amorphous powder, 1.6 mg; LC-HRESI-MS^n (qTOF, Bruker), *m/z* measured 386.1247 [M + H]$^+$, 408.1068 [M + Na]+, 772.2448 [2M + H]$^+$, 793.2225 [2M + Na]$^+$, Δ = −0.3 ppm for calculated *m/z* of [M + H]$^+$ 386.1248; DBE = 17; MF = $C_{21}H_{15}O_3N_5$. UV λ_{max}^{EtOH} = 729, 660, 577, 364, 309 nm. IR^{EtOH} = 3439, 2988, 1662, 1066, 1039, 544, 521, 471, 463 cm^{-1}. ^{1}H- and ^{13}C-NMR data (DMSO-d_6), see Table 1.

Dermacozine O (**2**): ink blue amorphous powder, 3.1 mg; LC-HRESI-MS^n (qTOF, Bruker) *m/z* 398.1125 [M + H]+, Δ = −2.5 ppm for calculated *m/z* of [M + H]$^+$ 398.1135, DBE = 18; MF = $C_{23}H_{15}O_4N_3$. UV λ_{max}^{EtOH} = 644, 563, 463, 396(sh), 308 nm. IR^{EtOH} = 2971, 1989, 1949, 1091, 1048, 882, 524, 501, 481, 453 cm^{-1}. ^{1}H- and ^{13}C-NMR data (DMSO-d_6), see Table 1.

Dermacozine P (**3**): purplish amorphous powder, 3.4 mg; LC-HRESI-MS^n (Orbitrap, Xcalibur) *m/z* 372.0990 [M + H]$^+$, Δ = 3.0 ppm for calculated *m/z* of [M + H]$^+$ 372.0979; DBE = 17; MF = $C_{21}H_{13}O_4N_3$. UV λ_{max}^{EtOH} = 465, 415(sh), 363 nm. IR^{EtOH} = 3440, 3178, 1661, 872, 757, 603, 575, 556, 534, 523, 465, 437, 410 cm^{-1}. ^{1}H- and ^{13}C-NMR data (DMSO-d_6), see Table 1.

Table 1. NMR spectroscopic data for dermacozine N (**1**) (800 MHz, DMSO-d_6), dermacozine O (**2**) (800 MHz, DMSO-d_6), and dermacozine P (**3**) (600 MHz, *DMSO-d_6*).[†]: Calculated with ACD Labs software, Nural Nework Algrithm, DMSO-d_6 (Chemical shift is not observed).

N.	Dermacozine N (1)		Dermacozine O (2)		Dermacozine P (3)	
	δ_C, mult	δ_H, mult (*J* in Hz)	δ_C, mult	δ_H, mult (*J* in Hz)	δ_C, mult	δ_H, mult (*J* in Hz)
1	128.9, C		129.6, C		131.5, C	
2	125.9, CH	7.88 (dd, 7.6, 1.3)	126.6, CH	7.87 (dd, 7.5, 1.1)	134.9, CH	8.74 (dd, 7.0, 1.3)
3	128.4, CH	7.47 (td, 8.3, 7.6)	131.3, CH	7.78 (td, 8.5, 7.5)	132.4, CH	8.21 (td, 7.0, 8.6)
4	118.0, CH	7.55 (dd, 8.3, 1.3)	120.3, CH	7.97 (dd, 8.5, 1.1)	132.6, CH	8.55 (dd, 8.6, 1.3)
4a	134.4, C		134.0, C		142.4, C	
5a	135.5, C		139.5, C		140.7, C	
6	109.8, C		100.4, C		129.8, C [†]	
7	148.1, C		139.6, C		131.3, CH	8.65 (d, 1.9)
8	149.8, C		134.5, CH	7.21 (d, 9.7)	135.6, C [†]	
9	105.9, CH	6.79, s	129.7, CH	7.24 (d, 9.7)	135.1, CH	8.95 (d, 1.9)
9a	151.6, C		150.6, C		144.0, C [†]	
10a	135.1, C		135.3, C		141.2, C	
11	166.3, C		167.2, C		165.5, C	
12		A 7.70, brs B 9.31, brs		COOH, not observed		COOH, not observed
13	168.3, C [†]		163.6, C [†]		166.6, C	
14		A 7.65, brs B 7.98, brs		11.27, brs		A 8.03, brs B 9.47, brs
15	143.3, C		163.1, C [†]		194.4, C	
16	115.0, CH	7.12 (dd, 7.6, 1.5)	122.6, C		136.3, C	
17	127.9, CH	7.19 (ddd, 7.6, 7.4, 1.6)	134.1, C		130.0, CH	7.95 (dd, 7.6, 1.3);
18	125.3, CH	7.15 (ddd, 7.6, 7.4, 1.5)	131.3, CH	7.30 (dd, 7.4, 1.3)	128.7, CH	7.66 (td, 7.6, 1.3)
19	127.2, CH	7.30 (dd, 7.6, 1.6)	128.2, CH	7.47 (td, 7.4, 1.3)	133.5, CH	7.78 (td, 7.6, 1.3)
20	134.8, C		127.9, CH	7.41 (td, 7.4, 1.3)	128.7, CH	7.66 (td, 7.6, 1.3)
21			128.2, CH	7.47 (td, 7.4, 1.3)	130.0, CH	7.95 (dd, 7.6, 1.3);
22			131.3, CH	7.30 (dd, 7.4, 1.3)		
23	39.5, CH$_3$	3.68, s	45.9, CH$_3$	3.67, s	-	-

3.7. Regression Models Used in the Structure Elucidation

We used a linear regression model in the case of dermacozine N (**1**). The experimental* ^{13}C-NMR chemical shifts of **1** were correlated to the ACD Labs software calculated ones (ACD/Structure Elucidator, version 2019.2.0, Neural Network Algorithm, solvent used for simulation: DMSO-d_6) in their theoretically possible structures (**A–W**). The simple linear regression model showed clear difference between the possible structures keeping with the results of other modalities in the structure elucidation process. The correlation coefficient obtained was R^2 = 0.99 (See Section 2.1).

In the case of dermacozine O (**2**), we compared its experimental ^{13}C-NMR chemical shift values to the ones reported in structures **8**, **9**, and **10** [6] due to their structural similarity. The experimentally not observed ^{13}C-NMR shifts (e.g., C-13 and C-15 in **2**) were excluded from the regression analysis, and since in dermacozine M (**14**) C-3 was calculated with ACD Labs software and contains an additional benzoyl ring at C-3, we excluded this structure from the calculations. A multiple regression test was carried out that showed significant similarity between dermacozine O (**2**) and dermacozine E (**8**) based on the *p*-value and *t*-test (see Section 2.2).

In the case of dermacozines P (**3**), linear regression was done between the ACD Labs software calculated (ACD/Structure Elucidator, version 2019.2.0, Neural Network Algorithm, solvent used for simulation: DMSO-d_6) and its experimentally observed ^{13}C-NMR chemical shift values. The R^2 value obtained was $R^2 = 0.98$ (See Section 2.3).

*The experimental ^{13}C-NMR values for the statistical calculations were measured in DMSO-d_6 in the case of structures **1**, **2**, and **3**.

4. Conclusions

Dermacoccus abyssi MT 1.1^T is a piezotolerant, halotolerant Actinomycete from the deepest part of the Earth, the Mariana Trench, Challenger Deep. Since the discovery of the strain in 2006, more than a dozen novel, highly colored dermacozines have been isolated from it. This work provides evidence that the bacterium ability of producing new compounds is still not exhausted yet.

Herein, we report on the isolation of three new dermacozines; dermacozine N (**1**), dermacozine O (**2**), and dermacozine P (**3**) when the strain was seed cultured in 20 g/L containing NaCl medium, followed by the cultivation in ISP2 medium, approximating the salinity of its usual habitat of 34.7‰. The difficulty of structure elucidation arising from the polycyclic nature of dermacozines was overcome by the combined approach of utilizing linear/multiple linear regression of experimental and ACD Labs software calculated ^{13}C-NMR chemical shifts, (TD-DFT) UV–Vis spectral simulation, 1D/2D NMR, LC-HRESI-MS^n, biosynthetic route considerations, which step by step offered complementary evidence for our structural assignments.

In dermacozine N (**1**)—whose phenoxazine skeleton is unprecedented in nature to the best of our knowledge—the [*b*]benzopyrazine substitution of the core phenoxazine resulted in UV–Vis absorption maxima in the near infrared (NIR) region. Phenoxazines have been reported to exhibit non-linear optical properties; nevertheless the structure of dermacozine O (**2**) is highly similar to those of dermacozines E (**8**), F (**9**), G (**10**), which were suggested to also exhibit non-linear optic properties in computational chemistry studies. This gives a perspective of investigating **1** and **2** in opto-electronics and second harmonic generation. Given the NIR absorption spectrum of **1**, the reported solvatochromic behavior, and bathochromic shift observed in acidic conditions of related phenoxazine skeletons, **1** would be an outstanding candidate to conduct research on in biosensing chemistry, biomolecular labelling, and *in vivo* metabolic mapping at the cellular level. Compound **1** showed weak activity against melanoma (A2058) and hepatocellular human carcinoma cell lines (HepG2) with IC$_{50}$ values of 51 and 38 µM, respectively. Dermacozine P (**3**) is the oxidized, 5N-demethylated, and C-6 carboxylated derivative of dermacozine B (**5**). Based on the literature data and cytostatic activities of the previously reported dermacozine derivatives and that synthetic modulation of dermacozine-1-carboxamides increased the compounds' anti-tubulin activity, further anti-tumor studies involving synthetic modulation of the compounds of dermacozine N (**1**), O (**2**), and P (**3**) would be promising.

Fascinating research is being conducted in various fields of chemistry to be able to utilize the novelty of these vividly colored compounds including opto-electronics, computational chemistry, biosensing chemistry, and medicinal chemistry exhibiting exciting results. Whilst the potential use of the colorful dermacozines in chemistry keeps stimulating the interest of researchers, their biological role in the mesmerizing depth of the hadal zone remains unknown.

Supplementary Materials: The following are available online at https://www.mdpi.com/article/10.3390/md19060325/s1, Figure S1: LC-MS analysis of dermacozine N (**1**) (qTOF, Bruker), Figure S2: MS/MS of dermacozine N (**1**) (qTOF, Bruker), Figure S3A: Proposed MS fragmentation pathway of dermacozine N (**1**) showing loss of a (-ON(C_6H_4)-) fragment, Figure S3B: Mass error between the calculated (Chemdraw modeled) and experimental (qTOF) *m/z* ratio of dermacozine N (**1**), Figure S4: UV–Vis spectrum of dermacozine N (**1**) in EtOH, Figure S6: Dermacozine N (**1**) ^{1}H-NMR spectrum (800 MHz, DMSO-d_6), Figure S7: ^{1}H-^{13}C HSQC spectrum of dermacozine N (**1**) (800 MHz, DMSO-d_6), Figure S8: ^{1}H-^{13}C HMBC (magnitude mode) spectrum of dermacozine N (**1**) (800 MHz, DMSO-d_6), Figure S9: ^{1}H-^{13}C HMBC spectrum of dermacozine N (**1**) with *J* = 2 Hz (800 MHz, DMSO-d_6), Figure S10: ^{1}H-^{1}H COSY spectrum of dermacozine N (**1**) (800 MHz, DMSO-d_6), Figure S11: ^{1}H-^{1}H NOESY spectrum of dermacozine N (**1**) (800 MHz, DMSO-d_6), Figure S12: ^{1}H-^{15}N HMBC spectrum of dermacozine N (**1**) (800 MHz, DMSO-d_6), Figure S13: LC-MS analysis of dermacozine O (**2**) (qTOF, Bruker), Figure S14: UV-Vis spectrum of dermacozine O (**2**) in EtOH, Figure S15: IR spectrum of dermacozine O (**2**) in EtOH, Figure S16: ^{1}H-NMR spectrum of dermacozine O (**2**) (800 MHz, DMSO-d_6), Figure S17: ^{1}H-^{13}C HSQC spectrum of dermacozine O (**2**) (800 MHz, DMSO-d_6), Figure S18: ^{1}H-^{13}C HMBC spectrum of dermacozine O (**2**) (800 MHz, DMSO-d_6), Figure S19: Long Range ^{1}H-^{13}C HMBC correlations of dermacozine O (**2**) with *J* = 2 Hz (800 MHz, DMSO-d_6), Figure S20: ^{1}H-^{1}H COSY spectrum of dermacozine O (**2**) (800 MHz, DMSO-d_6), Figure S21: ^{1}H-^{15}N HMBC spectrum of dermacozine O (**2**) (800 MHz, DMSO-d_6), Figure S22: UV-Vis spectrum of dermacozine P (**3**) in EtOH, Figure S23: IR spectrum of dermacozine P (**3**) in EtOH, Figure S24: LC-MS analysis of dermacozine P (**3**) (Orbitrap, Xcalibur), Figure S25: MS/MS analysis of dermacozine P (**3**) (Orbitrap, Xcalibur), Figure S26: ^{1}H-NMR spectrum of dermacozine P (**3**) (600 MHz, DMSO-d_6), Figure S27: ^{1}H-^{13}C HSQC spectrum of dermacozine P (**3**) (600 MHz, DMSO-d_6), Figure S28: ^{1}H-^{13}C HMBC spectrum of dermacozine P (**3**) (600 MHz, DMSO-d_6), Figure S29: ^{1}H-^{1}H COSY spectrum of dermacozine P (**3**) (600 MHz, DMSO-d_6), Figure S30: 1D-NOESY spectrum of dermacozine P (**3**) (600 MHz, DMSO-d_6) from irradiation of the signal at 8.03 ppm, Figures S31-S33: ^{13}C-NMR δ calculated values of dermacozine A–P (**1–14**) using the ACD Labs software with Neural Network Algorithm, solvent DMSO-d_6, Figures S34-S38: ^{13}C-NMR δ calculated values of possible structures (**A–W**) of dermacozine N (**1**), using the ACD Labs software with Neural Network Algorithm, solvent DMSO-d_6, Table S1: Comparison between calculated vs. experimental ^{13}C-NMR δ values of dermacozines A–J (**4–13**), Figure S39: Linear regression graphics between ACD Labs (Neural Network Algorithm, DMSO-d_6) calculated vs. experimental ^{13}C-NMR δ values of dermacozine A-H (**4–11**), Figure S40: Linear regression graphics between ACD Labs (Neural Network Algorithm, DMSO-d_6) calculated vs. experimental ^{13}C-NMR δ values of dermacozine I & J (**12–13**), Table S2: Comparison between the experimental ^{13}C-NMR δ values of dermacozine N (**1**) vs. the calculated ones of possible structures **A, D, E, G, H, I, J, K, N, P, Q, V, W** by ACD Labs (Neural Network Algorithm, DMSO-d_6), Figure S41: Linear regression graphics between ^{13}C-NMR δ experimental values of dermacozine N (**1**) vs. the ACD Labs calculated ones (Neural Network Algorithm, DMSO-d_6) (**4–13**), Figure S42: Linear regression graphics between ^{13}C-NMR δ experimental values of dermacozine N (**1**) vs. the ACD Labs (Neural Network Algorithm, DMSO-d_6) calculated ones for possible structures of **N, P, Q, V, W**, Table S3: Absolute error and the standard deviation of the error from the mean between the calculated (Structure **W**) and the experimental dermacozine N (**1**) ^{13}C-NMR chemical shifts, Table S4: Experimental values of ^{13}C-NMR δ chemical shifts of dermacozines E (**8**), F (**9**), G (**9**), and O (**2**) for multiple regression and *t*-test, Figure S43: Multiple regression analysis between the experimental ^{13}C-NMR δ values of dermacozines E (**8**), F (**9**), G (**9**) as independent variables and those of dermacozine O (**2**) as dependent variables, Table S5: Residuals between the observed and predicted values of ^{13}C-NMR δ of dermacozines E (**8**), F (**9**), G (**9**) as independent variables and those of dermacozine O (**2**) as dependent variables, Table S6: Comparison between experimental and the ACD Labs calculated ^{13}C-NMR δ chemical shift values of dermacozines P (**3**) (Neural Network Algorithm, DMSO-d_6), Figure S44: Linear regression graphics between the experimental ^{13}C-NMR δ chemical shifts of dermacozine P (**3**) and the ACD Labs calculated ^{13}C-NMR δ values (Neural Network Algorithm, DMSO-d_6), Table S7: Absolute error and the standard deviation of the error from the mean between the calculated and the experimental dermacozine P (**3**) ^{13}C-NMR chemical shifts, Figure S45: Evaluation of the cytotoxic activity of dermacozine N (**1**) against human (A) Melanoma (A2058) and (B) Hepatocellular carcinoma (HepG2) cell lines (IC$_{50}$ graphs), Table S8: Experimental NMR spectroscopic data for dermacozine N (**1**) with HMBC, COSY, and NOESY correlations (800 MHz, DMSO-d_6), Table S9: Experimental NMR spectroscopic data for dermacozine O (**2**) with

HMBC and COSY correlations (800 MHz, DMSO-d_6), Table S10: Experimental NMR spectroscopic data for dermacozine P (**3**) with HMBC and COSY correlations (600 MHz, DMSO-d_6), Figure S46: Workflow of dermacozine N (**1**) structure determination, Figure S47: Workflow of dermacozine O (**2**) structure determination, Figure S48: Workflow of dermacozine P (**3**) structure determination, Theme S1: Dermacozine N (**1**) H-9 and NH$_2$-12 distance (modeled with Chemdraw, Chem 3D), Theme S2: TLC plate of the initial FD fraction following Kupchan liquid–liquid partitioning, showing the colorful bands where the dermacozines were isolated from (left) and the same TLC plate under UV 360 nm light (right), Theme S3: HPLC chromatogram for the isolation of dermacozine O (**2**), Theme S4: HPLC chromatogram for the isolation of dermacozine P (**3**).

Author Contributions: Conceptualization, B.J., M.J., D.P.-P., C.J.; methodology, B.J., M.J., D.P.-P., C.J., J.N.T., F.R., B.C., K.K.; software, B.J., M.J., D.P.-P., C.J., J.N.T., F.R., B.C.; validation, B.J., M.J., D.P.-P., C.J., F.R., B.C.; formal analysis, B.J., M.J., D.P.-P., C.J., J.N.T., F.R., B.C., K.K.; investigation, B.J., M.J., D.P.-P., C.J., J.N.T., F.R., B.C., K.K.; resources B.J., M.J., D.P.-P., C.J., J.N.T., F.R.; data curation, B.J., M.J., D.P.-P., C.J., J.N.T., F.R., B.C., K.K.; writing—original draft preparation, B.J., M.J., D.P.-P., C.J.; writing—review and editing, B.J., M.J., D.P.-P., C.J., J.N.T., F.R., B.C., K.K.; visualization, B.J., M.J., D.P.-P., C.J., J.N.T., F.R., B.C.; supervision, M.J., C.J., F.R.; project administration, B.J, M.J., D.P.-P., C.J.; funding acquisition, B.J., M.J., D.P.-P., C.J. All authors have read and agreed to the published version of the manuscript.

Funding: Dawrin Pech-Puch received his postdoctoral fellowship from the National Council of Science and Technology (CONACYT) of Mexico.

Data Availability Statement: Not applicable.

Acknowledgments: The authors wish to express their thanks to Michael Goodfellow, School of Natural and Environmental Sciences, Newcastle University for providing the pure colony of *Dermacoccus abyssi* MT 1.1$^\mathrm{T}$, to Wasu Pathom-Aree (Chiang Mai University, Thailand), who isolated the strain furthermore to the *Kaiko* operation team and the crew of M.S. Yokosuka, JAMSTEC, Yokosuka, Japan, for the sediment collection. The authors thank and acknowledge the significant contribution of John W. Still—SEM Specialist, ACEMAC Facility, The University of Aberdeen; Lucinda Wight—Microscope and Histology Specialist, School of Medicine, Medical Sciences and Nutrition, The University of Aberdeen for preparing the strain *Dermacoccus abyssi* MT 1.1$^\mathrm{T}$ for EM imaging; Russel Gray for NMR measurements—The University of Aberdeen, Department of Chemistry, Marine Biodiscovery Centre, NMR Laboratory; Bella Juraj for NMR measurements—The University of Edinburgh, Joseph Black Building, School of Chemistry, NMR Laboratory. Marcel Jaspars and Bertalan Juhasz acknowledge the help of the University of Edinburgh through the EPSR Grant: EP/R030065/1 for the 800 MHz NMR measurements with the 5 mm TCI He Cryoprobe of compounds **1** and **2**. Carlos Jimenéz and Dawrin Pech-Puch acknowledge the help of CESGA for the computational support. Bertalan Juhasz wishes to express his sincere gratitude to his supervisor, Marcel Jaspars, for his invaluable trust and support.

Conflicts of Interest: The authors declare no conflict of interest.

References

1. Kamjam, M.; Sivalingam, P.; Deng, Z.; Hong, K. Deep Sea Actinomycetes and Their Secondary Metabolites. *Front. Microbiol.* **2017**, *8*, 760. [CrossRef] [PubMed]
2. Sayed, A.; Hassan, M.; Alhadrami, H.; Hassan, H.; Goodfellow, M.; Rateb, M. Extreme environments: Microbiology leading to specialized metabolites. *J. Appl. Microbiol.* **2020**, *128*, 630–657. [CrossRef] [PubMed]
3. Wilson, Z.E.; Brimble, M.A. Molecules derived from the extremes of life. *Nat. Prod. Rep.* **2008**, *26*, 44–71. [CrossRef] [PubMed]
4. Wilson, Z.E.; Brimble, M.A. Molecules derived from the extremes of life: A decade later. *Nat. Prod. Rep.* **2021**, *38*, 24–82. [CrossRef] [PubMed]
5. Pathom-Aree, W.; Nogi, Y.; Sutcliffe, I.C.; Ward, A.C.; Horikoshi, K.; Bull, A.T.; Goodfellow, M. *Dermacoccus abyssi* sp. nov., a piezotolerant actinomycete isolated from the Mariana Trench. *Int. J. Syst. Evol. Microbiol.* **2006**, *56*, 1233–1237. [CrossRef] [PubMed]
6. Abdel-Mageed, W.M.; Milne, B.F.; Wagner, M.; Schumacher, M.; Sandor, P.; Pathom-Aree, W.; Goodfellow, M.; Bull, A.T.; Horikoshi, K.; Ebel, R.; et al. Dermacozines, a new phenazine family from deep-sea dermacocci isolated from a Mariana Trench sediment. *Org. Biomol. Chem.* **2010**, *8*, 2352–2362. [CrossRef] [PubMed]
7. Wagner, M.; Abdel-Mageed, W.M.; Ebel, R.; Bull, A.T.; Goodfellow, M.; Fiedler, H.-P.; Jaspars, M. Dermacozines H–J Isolated from a Deep-Sea Strain of *Dermacoccus abyssi* from Mariana Trench Sediments. *J. Nat. Prod.* **2014**, *77*, 416–420. [CrossRef] [PubMed]

8. Abdel-Mageed, W.M.; Juhasz, B.; Lehri, B.; Alqahtani, A.S.; Nouioui, I.; Pech-Puch, D.; Tabudravu, J.N.; Goodfellow, M.; Rodríguez, J.; Jaspars, M.; et al. Whole Genome Sequence of *Dermacoccus abyssi* MT1.1 Isolated from the Challenger Deep of the Mariana Trench Reveals Phenazine Biosynthesis Locus and Environmental Adaptation Factors. *Mar. Drugs* **2020**, *18*, 131. [CrossRef] [PubMed]

9. Milne, B.; Norman, P.; Nogueira, F.; Cardoso, C. Marine natural products from the deep Pacific as potential non-linear optical chromophores. *Phys. Chem. Chem. Phys.* **2013**, *15*, 14814. [CrossRef] [PubMed]

10. Schumacher, M.; Kelkel, M.; Dicato, M.; Diederich, M. A Survey of Marine Natural Compounds and Their Derivatives with Anti-Cancer Activity Reported in 2010. *Molecules* **2011**, *16*, 5629–5646. [CrossRef] [PubMed]

11. Ghanta, V.R.; Madala, N.; Pasula, A.; Pindiprolu, S.K.S.S.; Battula, K.S.; Krishnamurthy, P.T.; Raman, B. Novel dermacozine-1-carboxamides as promising anticancer agents with tubulin polymerization inhibitory activity. *RSC Adv.* **2019**, *9*, 18670–18677. [CrossRef]

12. Pan, R.; Bai, X.; Chen, J.; Zhang, H.; Wang, H. Exploring Structural Diversity of Microbe Secondary Metabolites Using OSMAC Strategy: A Literature Review. *Front. Microbiol.* **2019**, *10*, 294. [CrossRef]

13. Xie, L.-R.; Li, D.-Y.; Li, Z.-L.; Hua, H.-M.; Wang, P.-L.; Wu, X. A new cyclonerol derivative from a marine-derived fungus *Ascotricha* sp. ZJ-M-5. *Nat. Prod. Res.* **2013**, *27*, 847–850. [CrossRef]

14. Favali, P.; Beranzoli, L.; De Santis, A. *Seafloor Observatories*; Springer: Berlin, Germany, 2015; p. 449.

15. Pavia, D.L.; Lampman, G.M.; Kriz, G.S.; Vyvyan, J.R. *Introduction to Spectroscopy*, 4th ed.; Brooks/Cole: Pacific Grove, CA, USA, 2009.

16. Jones, R.N. The ultraviolet absorption spectra of some derivatives of 1,2-benzanthracene. *J. Am. Chem. Soc.* **1940**, *62*, 148–152. [CrossRef]

17. Hellner, C.; Lindqvist, L.; Roberge, P.C. Absorption spectrum and decay kinetics of triplet pentacene in solution, studied by flash photolysis. *J. Chem. Soc. Faraday Trans.* **1972**, *68*, 1928–1937. [CrossRef]

18. Elyashberg, M.; Williams, A.J.; Blinov, K. Structural revisions of natural products by Computer-Assisted Structure Elucidation (CASE) systems. *Nat. Prod. Rep.* **2010**, *27*, 1296–1328. [CrossRef]

19. Rateb, M.E.; Tabudravu, J.; Ebel, R. NMR characterisation of natural products derived from under-explored microorganisms. *Nucl. Magn. Reson.* **2016**, *45*, 240–268. [CrossRef]

20. Tabudravu, J.N.; Pellissier, L.; Smith, A.J.; Subko, K.; Autréau, C.; Feussner, K.; Hardy, D.; Butler, D.; Kidd, R.; Milton, E.J.; et al. LC-HRMS-Database Screening Metrics for Rapid Prioritization of Samples to Accelerate the Discovery of Structurally New Natural Products. *J. Nat. Prod.* **2019**, *82*, 211–220. [CrossRef]

21. Schrödinger suite, V. 2018-4. Schrödinger, LCC. Available online: https://www.schrodinger.com (accessed on 3 May 2021).

22. Jacquemin, D.; Laurent, A.D.; Perpète, E.A.; André, J.-M. An ab initio simulation of the UV/visible spectra of N-benzylideneaniline dyes. *Int. J. Quantum Chem.* **2009**, *109*, 3506–3515. [CrossRef]

23. Bruhn, T.; Schaumlöffel, A.; Hemberger, Y.; Bringmann, G. SpecDis: Quantifying the comparison of calculated and experimental electronic circular dichroism spectra. *Chirality* **2013**, *25*, 243–249. [CrossRef]

24. Fischer, O.; Hepp, E. Ueber die Fluorindine II. *Eur. J. Inorg. Chem.* **1895**, *28*, 293–301. [CrossRef]

25. Fischer, O.; Jonas, O. Beitrag zur Oxydation der aromatischen Orthodiamine und Orthoamidophenole. *Eur. J. Inorg. Chem.* **1894**, *27*, 2782–2785. [CrossRef]

26. Fischer, O.; Giesen, C. Ueber die Einwirkung von Basen auf Aposafranin. *Eur. J. Inorg. Chem.* **1897**, *30*, 2489–2494. [CrossRef]

27. Diepolder, E. Ueber Methyl-o-anisidin, Methyl-o-aminophenol und dessen Oxydationsproduct (*N*-Methylphenoxazin-*o*-chinon). *Eur. J. Inorg. Chem.* **1899**, *32*, 3514–3528. [CrossRef]

28. Diepolder, E. Ueber Oxydationsproducte des *o*-Aminophenols. *Eur. J. Inorg. Chem.* **1902**, *35*, 2816–2822. [CrossRef]

29. Afanas'Eva, G.B.; Postovskii, I.Y.; Viktorova, T.S. Research in the chemistry of phenoxazines. *Chem. Heterocycl. Compd.* **1978**, *14*, 966–968. [CrossRef]

30. Shimomura, O. Discovery of Green Fluorescent Protein (GFP) (Nobel Lecture). *Angew. Chem. Int. Ed.* **2009**, *48*, 5590–5602. [CrossRef]

31. Janjua, M.R.S.A. Non-linear Optical response of Phenoxazine-based Dyes: Molecular Engineering of Thiadiazole Derivatives as π-spacers. *J. Mex. Chem. Soc.* **2017**, *61*, 260–265.

32. Traven, V.F.; Cheptsov, D.A. Sensory effects of fluorescent organic dyes. *Russ. Chem. Rev.* **2020**, *89*, 713–749. [CrossRef]

33. Steiner, M.-S.; Duerkop, A.; Wolfbeis, O.S. Optical methods for sensing glucose. *Chem. Soc. Rev.* **2011**, *40*, 4805–4839. [CrossRef]

34. Sherman, D.B.; Pitner, J.B.; Ambroise, A.; Thomas, K.J.; Pitner, B. Synthesis of Thiol-Reactive, Long-Wavelength Fluorescent Phenoxazine Derivatives for Biosensor Applications. *Bioconjugate Chem.* **2006**, *17*, 387–392. [CrossRef] [PubMed]

35. Liu, W.; Sun, R.; Ge, J.-F.; Xu, Y.-J.; Xu, Y.; Lu, J.-M.; Itoh, I.; Ihara, M. Reversible Near-Infrared pH Probes Based on Benzo[a]phenoxazine. *Anal. Chem.* **2013**, *85*, 7419–7425. [CrossRef] [PubMed]

36. Jose, J.; Burgess, K. Benzophenoxazine-based fluorescent dyes for labeling biomolecules. *Tetrahedron* **2006**, *62*, 11021–11037. [CrossRef]

37. Knorr, G.; Kozma, E.; Herner, A.; Lemke, E.A.; Kele, P. New Red-Emitting Tetrazine-Phenoxazine Fluorogenic Labels for Live-Cell Intracellular Bioorthogonal Labeling Schemes. *Chem. A Eur. J.* **2016**, *22*, 8972–8979. [CrossRef] [PubMed]

38. Boonacker, E.; Van Noorden, C.J. Enzyme Cytochemical Techniques for Metabolic Mapping in Living Cells, with Special Reference to Proteolysis. *J. Histochem. Cytochem.* **2001**, *49*, 1473–1486. [CrossRef]

39. Yee, D.J.; Balsanek, V.; Bauman, D.R.; Penning, T.; Sames, D. Fluorogenic metabolic probes for direct activity readout of redox enzymes: Selective measurement of human AKR1C2 in living cells. *Proc. Natl. Acad. Sci. USA* **2006**, *103*, 13304–13309. [CrossRef]

40. Lin, C.W.; Shulok, J.R.; Wong, Y.K.; Schanbacher, C.F.; Cincotta, L.; Foley, J.W. Photosensitization, uptake, and retention of phenoxazine Nile blue derivatives in human bladder carcinoma cells. *Cancer Res.* **1991**, *51*, 1109–1116.

41. Laursen, J.B.; Nielsen, J. Phenazine Natural Products: Biosynthesis, Synthetic Analogues, and Biological Activity. *Chem. Rev.* **2004**, *104*, 1663–1686. [CrossRef] [PubMed]

42. Sarewicz, M.; Osyczka, A. Electronic Connection Between the Quinone and Cytochrome c Redox Pools and Its Role in Regulation of Mitochondrial Electron Transport and Redox Signaling. *Physiol. Rev.* **2015**, *95*, 219–243. [CrossRef] [PubMed]

43. Sinha, S.; Shen, X.; Gallazzi, F.; Li, Q.; Zmijewski, J.W.; Lancaster, J.R.; Gates, K.S. Generation of Reactive Oxygen Species Mediated by 1-Hydroxyphenazine, a Virulence Factor of *Pseudomonas aeruginosa*. *Chem. Res. Toxicol.* **2015**, *28*, 175–181. [CrossRef]

44. Willker, W.; Leibfritz, D.; Kerssebaum, R.; Bermel, W. Gradient selection in inverse heteronuclear correlation spectroscopy. *Magn. Reson. Chem.* **1993**, *31*, 287–292. [CrossRef]

45. Zwahlen, C.; Legault, P.; Vincent, S.J.F.; Greenblatt, J.; Konrat, R.; Kay, L.E. Methods for Measurement of Intermolecular NOEs by Multinuclear NMR Spectroscopy: Application to a Bacteriophage λ N-Peptide/boxB RNA Complex. *J. Am. Chem. Soc.* **1997**, *119*, 6711–6721.

46. Boyer, R.D.; Johnson, R.; Krishnamurthy, K. Compensation of refocusing inefficiency with synchronized inversion sweep (CRISIS) in multiplicity-edited HSQC. *J. Magn. Reson.* **2003**, *165*, 253–259. [CrossRef]

47. Cicero, D.O.; Barbato, G.; Bazzo, R. Sensitivity Enhancement of a Two-Dimensional Experiment for the Measurement of Heteronuclear Long-Range Coupling Constants, by a New Scheme of Coherence Selection by Gradients. *J. Magn. Reson.* **2001**, *148*, 209–213.

48. Shaw, A.A.; Salaun, C.; Dauphin, J.-F.; Ancian, B. Artifact-Free PFG-Enhanced Double-Quantum-Filtered COSY Experiments. *J. Magn. Reason.* **1996**, *120*, 110–115. [CrossRef]

49. Ancian, B.; Bourgeois, I.; Dauphin, J.-F.; Shaw, A.A. Artifact-Free Pure Absorption PFG-Enhanced DQF-COSY Spectra Including a Gradient Pulse in the Evolution Period. *J. Magn. Reson.* **1997**, *125*, 348–354. [CrossRef]

50. Koagne, R.R.; Annang, F.; Cautain, B.; Martín, J.; Pérez-Moreno, G.; Thierry, M.; Bitchagno, G.; González-Pacanowska, D.; Vicente, F.; Simo, I.K.; et al. Cytotoxicity and antiplasmodial activity of phenolic derivatives from *Albizia zygia* (DC.) J.F. Macbr. (*Mimosae*). *BMC Complement. Med. Ther.* **2020**, *20*, 8. [CrossRef]

51. Präbst, K.; Engelhardt, H.; Ringgeler, S.; Hübner, H. Methods in Molecular Biology. In *Basic Colorimetric Proliferation Assays: MTT, WST, and Resazurin*; Gilbert, D.F., Friedrich, O., Eds.; Springer Nature: New York, NY, USA, 2017; pp. 1–17.

Article

A Thermotolerant Marine *Bacillus amyloliquefaciens* S185 Producing Iturin A5 for Antifungal Activity against *Fusarium oxysporum* f. sp. *cubense*

Pratiksha Singh [1,†], Jin Xie [1,†], Yanhua Qi [1], Qijian Qin [1], Cheng Jin [1,2,3], Bin Wang [1,2,*] and Wenxia Fang [1,2,*]

1   State Key Laboratory of Non-Food Biomass and Enzyme Technology, Guangxi Academy of Sciences, Nanning 530007, China; pratiksha23@gxas.cn (P.S.); xiejin@gxas.cn (J.X.); qiyanhua@gxas.cn (Y.Q.); qqij@gxas.cn (Q.Q.); jinc@im.ac.cn (C.J.)
2   National Engineering Research Center for Non-Food Biorefinery, Guangxi Academy of Sciences, Nanning 530007, China
3   State Key Laboratory of Mycology, Institute of Microbiology, Chinese Academy of Sciences, Beijing 100101, China
*   Correspondence: bwang@gxas.cn (B.W.); wfang@gxas.cn (W.F.)
†   These authors have contributed equally to this work.

Abstract: Fusarium wilt of banana (also known as Panama disease), is a severe fungal disease caused by soil-borne *Fusarium oxysporum* f. sp. *cubense* (*Foc*). In recent years, biocontrol strategies using antifungal microorganisms from various niches and their related bioactive compounds have been used to prevent and control Panama disease. Here, a thermotolerant marine strain S185 was identified as *Bacillus amyloliquefaciens*, displaying strong antifungal activity against *Foc*. The strain S185 possesses multiple plant growth-promoting (PGP) and biocontrol utility properties, such as producing indole acetic acid (IAA) and ammonia, assimilating various carbon sources, tolerating pH of 4 to 9, temperature of 20 to 50 °C, and salt stress of 1 to 5%. Inoculation of S185 colonized the banana plants effectively and was mainly located in leaf and root tissues. To further investigate the antifungal components, compounds were extracted, fractionated, and purified. One compound, inhibiting *Foc* with minimum inhibitory concentrations (MICs) of 25 µg/disk, was identified as iturin A5 by high-resolution electrospray ionization mass spectrometry (HR-ESI-MS) and nuclear magnetic resonance (NMR). The isolated iturin, A5, resulted in severe morphological changes during spore germination and hyphae growth of *Foc*. These results specify that *B. amyloliquefaciens* S185 plays a key role in preventing the *Foc* pathogen by producing the antifungal compound iturin A5, and possesses potential as a cost-effective and sustainable biocontrol strain for Panama disease in the future. This is the first report of isolation of the antifungal compound iturin A5 from thermotolerant marine *B. amyloliquefaciens* S185.

Keywords: antifungal activity; *Bacillus amyloliquefaciens*; Panama disease; *Fusarium oxysporum* f. sp. *cubense*; bioactive compound; iturin A5

Citation: Singh, P.; Xie, J.; Qi, Y.; Qin, Q.; Jin, C.; Wang, B.; Fang, W. A Thermotolerant Marine *Bacillus amyloliquefaciens* S185 Producing Iturin A5 for Antifungal Activity against *Fusarium oxysporum* f. sp. *cubense*. *Mar. Drugs* **2021**, *19*, 516. https://doi.org/10.3390/md19090516

Academic Editors: Daniela Giordano and Bill J. Baker

Received: 17 August 2021
Accepted: 8 September 2021
Published: 11 September 2021

Publisher's Note: MDPI stays neutral with regard to jurisdictional claims in published maps and institutional affiliations.

1. Introduction

Crop wastage is primarily affected by plant diseases, which are mostly caused by soil-borne fungi leading to substantial yield reduction and income losses [1]. Banana (*Musa* spp.) is a major cash crop in the tropics and subtropics and is one of the top ten staple foods in the world [2,3]. The majority of vegetatively propagated bananas are susceptible to many pests and diseases, notably Fusarium wilt of banana (FWB). This disease is caused by the soil-borne fungus *Fusarium oxysporum* f. sp. *cubense* (*Foc*) and is recognized as the most damaging and extensively disseminated disease in banana-producing regions worldwide [4,5]. FWB has spread to the world's leading banana-producing areas since the 1990s [6]. In China, FWB has expanded over much in recent years, placing the banana industry's health and long-term sustainability under risk [7]. As the pathogen of FWB,

Foc is resistant to stress and capable of living in soil for up to 30 years [8]. *Foc* infects the xylem through the roots, resulting in widespread necrosis, plant mortality, and causing widespread orchard damage [4,6].

Currently, no suitable cultivar has been selected or bred to reduce the occurrence of FWB [9]. This is a worldwide issue with consequences that include losses in a wide variety of banana production methods and plant mortality in certain cases. Crop rotation, the selection of resistant varieties, and chemical or biological approaches are now the most common strategies for managing FWB [10,11]. However, due to the persistent spread of FWB throughout continents, countries, and regions, there is no effective control system [12]. Chemical fungicides are (relatively) the most effective tool for controlling this soil-borne disease [13]. However, scientists have currently found that long-term usage of fungicides might result in negative consequences such as environmental pollution, plant disease outbreaks, escalating production costs due to chemical overuse, and even human toxicity [14]. So far, there has not been a feasible way to manage the *Foc* pathogen in the field. Soil sterilization with fungicides has only been utilized on a small scale in intensive agriculture such as in greenhouses [6,15]. Therefore, ecologically and environmentally friendly cultural practices are urgently required [16]. In present agriculture, the biocontrol approach has been being investigated for several years, and many helpful bacteria have been proposed for controlling crop diseases. In addition, biocontrol is the only alternative disease-management strategy that explores plant growth-promoting (PGP) antagonistic bacteria or fungi that are regarded to be safe, ecologically friendly, and cost-effective [11,17], whereas many biocontrol agents are selective to host species, pathogen type, environmental circumstances, soil types, seasons, etc. [18]. Hence, a comprehensive study is necessary to find novel biocontrol agents that can respond to a wide range of environments [19].

The marine ecosystem is composed of marine invertebrates, plants, and their related microbial communities, providing a potential source for new bioactive compounds [20]. The marine microorganisms in the marine ecosystem produce the majority of secondary metabolites which have a wide range of functions, including antiviral, antifungal, immunesuppressive, anti-inflammatory, antitumor, and many other biotechnological and pharmaceutical applications [21,22]. The trend of discovering novel compounds from marine microbes has been increasing during the past decade [23]. Several strains of *Bacillus* spp., which have a suppressive effect against plant diseases caused by soil-borne phytopathogens, have expanded rapidly, and interested readers now have vast quantities of information on biocontrol strategies, as well as their applications and efficacies under a wide range of situations [24]. *Bacillus* spp. are chosen in agricultural systems due to their capacity to produce endospores that can withstand heat and desiccation, in addition to their capacity to be formed into stable dry powders with long shelf lives [25]. Furthermore, many *Bacillus* species have fast growth rates, common inhabitation of plant root microflora, and the ability to produce a large number of secondary metabolites, which are important in antibiosis against a variety of harmful pathogens [26]. Biocontrol methods using *Bacillus* spp. are divided into two categories: direct antagonism and indirect antagonism. Nutrition and geographical location competition, production of secondary metabolites with antibiosis activity, secretion of hydrolases, and the emission of volatile organic molecules are examples of direct antagonism, whereas indirect antagonism involves changing the variety of the soil or plant microbial population to induce plant tolerance and prevent pathogens [10]. *Bacillus* produces important antagonistic substances, such as surfactin [27], iturins [28], and fengycin [29], which are usually generated by non-ribosomal peptide synthetase (NRPS) and inducing plant eliciting system against infections [30].

In this study, we isolated a *B. amyloliquefaciens* strain S185 from offshore of the South Sea, China. This strain exhibits antagonism activity against *Foc*, possesses PGP traits and has other properties suitable for biocontrol utilization. Further analysis showed that iturin A5 is the main bioactive compound contributing to antifungal activity. To the best of our knowledge, this is the first report on the inhibition of *Foc* growth by compound iturin A5 extracted from a marine *B. amyloliquefaciens* strain, which is a potential biocontrol strain for FWB.

2. Results

2.1. A Thermotolerant Marine Bacterium S185 Possesses Antagonistic Activity against Foc

From sediment samples collected from offshore of South Sea, China, we isolated a thermotolerant bacterium S185. This strain is capable of growing at 28 °C, 37 °C, 42 °C, and 50 °C, respectively (Figure 1A). We then examined the antagonistic activity of S185 against *Foc*, the phytopathogen of FWB. As shown in Figure 1B, strain S185 displayed strong antifungal activity against *Foc* with an inhibition percentage of 78% on a dual culture plate (Table S1).

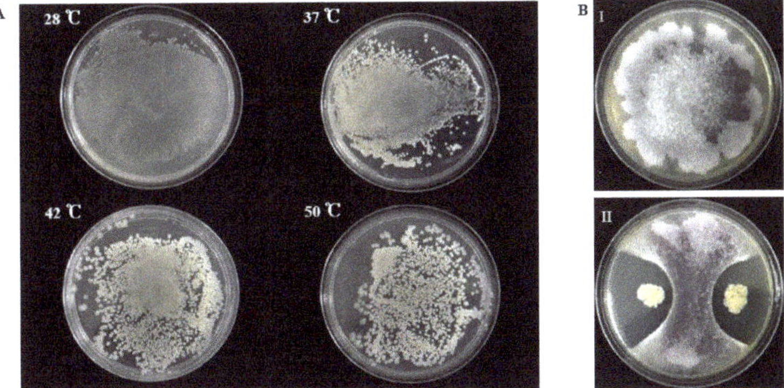

Figure 1. Thermotolerant property of S185 and its antagonistic activity towards *Foc*. (**A**). Strain S185 was cultured at 28 °C, 37 °C, 42 °C, and 50 °C, respectively. After 24 h cultivation plates were taken out for photographing. (**B**). Dual culture plate assay for the screening of strain S185 against *Foc*, (**I**). *Fusarium oxysporum* f. sp. *cubense* (*Foc*) control plate, (**II**). Growth inhibition of *Foc* pathogen by S185.

Phylogenetic analysis was conducted to identify the strain S185. After PCR of the 16S *rRNA* product and sequencing analysis, the BlastN program was used to compare the 16S *rRNA* sequence of strain S185 to nucleotide sequences in the NCBI GenBank database. The strain S185 displayed over 97% similarity with the aligned species. A phylogenetic tree was constructed using representative strains from BlastN (Figure 2). The phylogenetic tree indicated that the strain was *Bacillus amyloliquefaciens*.

2.2. S185 Displays PGP Traits and Tolerance to Abiotic Stresses

Bacterial strains releasing secondary metabolites such as ammonia are helpful in the prevention of fungal infections in plants [31]. The S185 strain indeed showed strong ammonia production in the medium, but was negative for phosphorus (P) solubilization, hydrogen cyanide (HCN), and siderophore production (Table S1). Indole acetic acid (IAA) production is a key property of plant growth-promoting bacteria (PGPB). Strain S185 produces 93.96 ± 2.28 and 57.13 ± 1.32 µg·ml^{-1} IAA in the presence and absence of tryptophan, respectively (Table S1).

In addition, strain S185 exhibited tolerance to abiotic stresses such as pH of 4–9, temperature of 20–50 °C, and salt concentration of 1–5%. The optimum growth of S185 was observed at pH 6, temperature 45 °C, and 5% of NaCl (Figure 3). These results confirm that S185 displays PGP traits and tolerance to abiotic stresses.

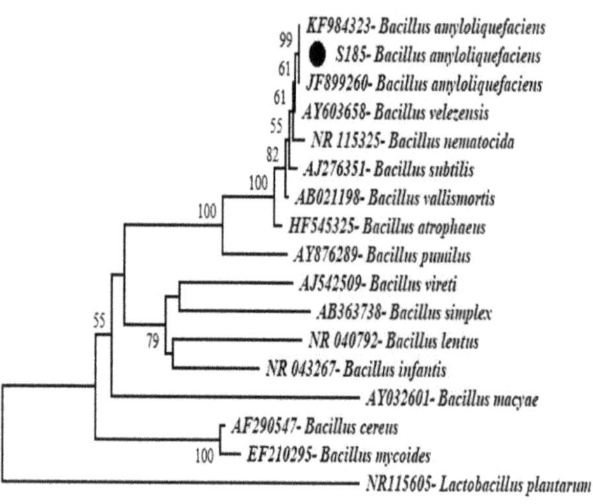

Figure 2. Phylogenetic tree representing the position of strain S185 compared to other strains inside the genus *Bacillus* with *Lactobacillus plantarum* as an outgroup. The tree was created via neighbor-joining with a scale bar of 0.01 substitutions per nucleotide.

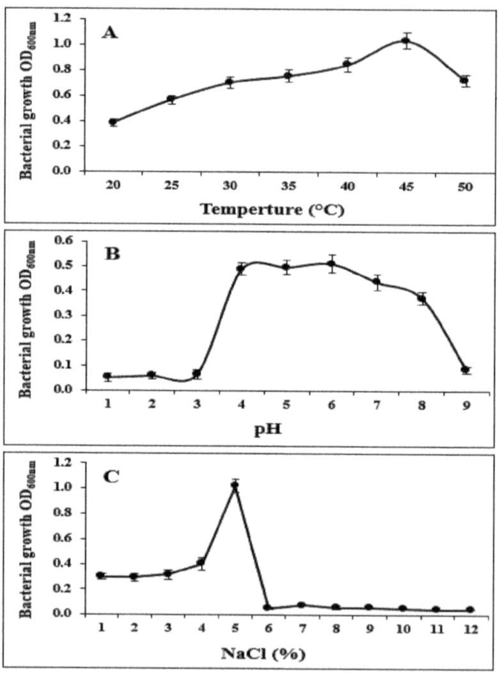

Figure 3. Tolerance of strain *B. amyloliquefaciens* S185 towards abiotic stresses. (**A**). Temperature (20–50 °C), (**B**). pH (1–9), (**C**). Salt (1–12%).

2.3. S185 Utilizes Broad Carbon Substrates

Strains having a broad metabolite tolerance are more appropriate for plant nodulation [32]. Carbon substrates utilization pattern of strain S185 was tested against GNIII Biolog plate. Strain S185 was positive in the assimilation of dextrin, D-turanose, D-salicin, *N*-acetyl-D-glucosamine, *N*-acetyl-β-D-mannosamine, *N*-acetyl-D-galactosamine, *N*-acetyl neuraminic acid, D-fructose, D-galactose, 3-methyl glucose, D-fucose, L-fucose, inosine, D-sorbitol, D-mannitol, D-arabitol, myo-inositol, glycerol, D-glucose-6-PO4, D-fructose-6-PO4, D-aspartic acid, D-serine, glycyl-L-proline, L-alanine, L-arginine, L-aspartic acid, L-glutamic acid, L-histidine, L-serine, lincomycin, niaproof 4, pectin, D-gluconic acid, D-glucuronic acid, glucuronamide, mucic acid, quinic acid, tetrazolium violet, tetrazolium blue, and, L-lactic acid on Biolog GN III plate (Table S2). These findings indicate that the S185 strain is capable of using a wide range of carbon substrates.

2.4. S185 Colonizes in All Tissues of Banana Plants

The capacity of bacteria to colonize plant tissues is important for disease control and plant growth development. Therefore, colonization of green fluorescent protein (GFP)-tagged S185 was assessed in tissues of banana plants by confocal laser scanning microscopy (CLSM). In control plants, without GFP-tagged S185 there was no GFP signal in leaf, stem, and root tissues of plants (Figure 4A,C,E), whereas the GFP-tagged S185 cells were observed as green spots in all plant tissues after 96 h of inoculation (Figure 4B,D,F). The strain S185 exhibited maximum colonization in roots followed by leaves and stems (Figure 4).

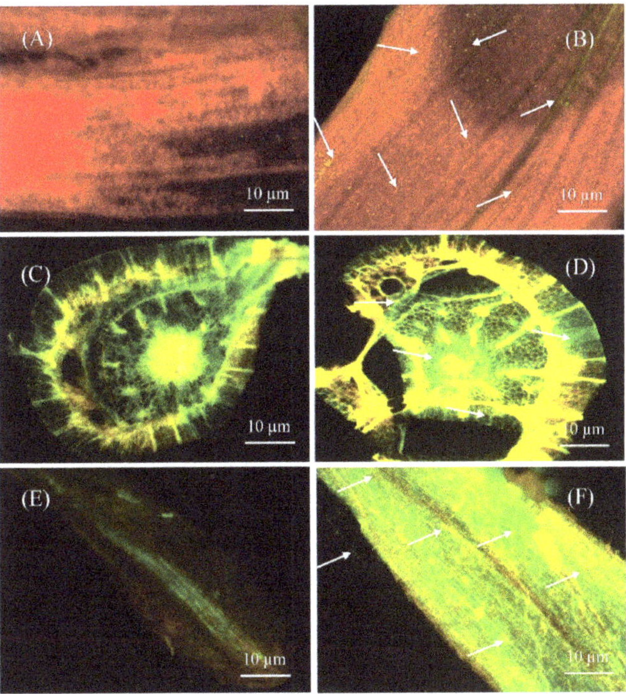

Figure 4. Confocal laser scanning microscopy (CLSM) showing colonization of *B. amyloliquefaciens* S185 in tissues of banana plants. (**A,C,E**) are sectioned leaf, stem, and root tissues of banana plants without S185 inoculation, (**B,D,F**) are sectioned leaf, stem, and root tissues of banana plants after 96 h of inoculation with S185. White arrow marks indicate colonization of strain S185 as small green dots in banana plants.

2.5. Antifungal Activity of Main Active Compound **1** Extracted from S185

Despite the strong antagonistic activity of S185 against *Foc*, we next conducted crude extraction of strain S185 to identify the active components. Guided by activity tracing, butanol extracted components were further fractionated by silica gel column chromatography and gel column chromatography; finally, the main active antifungal compound **1** was isolated from S185 with the yield of 20 mg/L (Figure 5A). Compound **1** displayed the best antifungal activity as shown by the inhibition zone (Figure 5B). At the concentration of 25 µg/disk, compound **1** started to inhibit the growth of *Foc* pathogen with a clear inhibition zone of 7.67 mm (Figure 5B). The diameter of the inhibition zone increased to 9.67 mm, 12.33 mm, 13.33 mm, and 14.33 mm at the 50, 100, 200, and 400 ug/disk of compound **1**, respectively (Table S3; Figure 5B). These results confirm that compound **1** produced by strain S185 has strong antifungal activity against *Foc*.

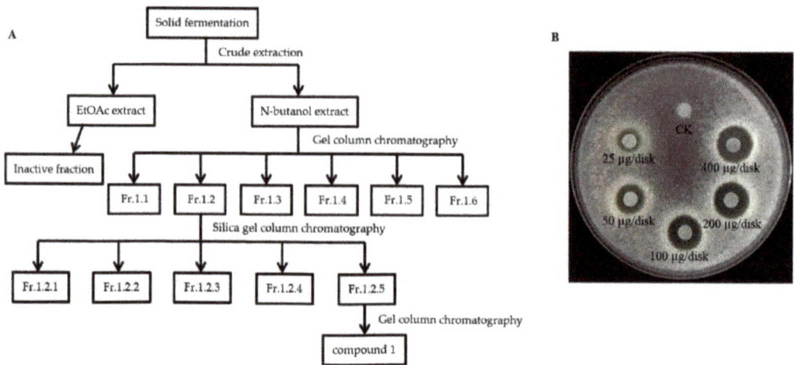

Figure 5. Isolation and antifungal activity of main active compound **1** of strain S185. (**A**) Flow chart showing extraction and separation procedure of compound **1** from strain S185 (**B**). Inhibitory effect of compound **1** on the growth of *Foc* at a different level of concentrations. CK; methanol control.

2.6. Identification of Compound **1** as Iturin A5

The extracted compound **1** from S185 was shown as a white powder and directly identified using high-resolution electrospray ionization mass spectrometry (HR-ESI-MS): m/z [M + H]$^+$, 1057.5671 (Figure 6). Based on this information, the molecular formula of this compound was predicted as $C_{49}H_{76}N_{12}O_{14}$ (Cacl. for $C_{49}H_{77}N_{12}O_{14}{}^+$: 1057.5676). ^{1}H nuclear magnetic resonance (NMR) reveals signals of the β-amino acid moiety (800 MHz, DMSO-d_6, δ) (Figure S1): 7.11 (ovl, 1H, β-amino acid NH), 3.98 (m, 1H, β-amino acid C_3H), 2.32 (d, 2H, β-amino acid C_2H_2), 1.41 (m, 2H, β-amino acid C_4H_2), 1.29–1.07 (m, β-amino acid aliphatic CH_2), 0.83 (t, *J* = 6.4 Hz, 3H, β-amino acid $C_{15}H_3$). ^{13}C NMR shows signals of (DMSO-d_6, δ) (Figure S2): 45.8 (β-amino acid, C_3), 42.2 (β-amino acid, C_2), 35.0 (β-amino acid, C_4), 31.8 (β-amino acid, C_{13}), 29.6 (β-amino acid), 29.5 (β-aminoacid), 29.2 (β-amino acid), 29.1 (β-amino acid), 25.8 (β-amino acid, C_5), 22.6 (β-amino acid, C_{14}), 14.4 (β-amino acid, C_{15}). All these data were compared with published NMR data and were consistent with the data for iturin A5 [33]. Thus, compound **1** was identified as iturin A5 (Figure 7, Table S4).

2.7. Iturin A5 Inhibited Spore Germination of Foc

Despite the antifungal activity from iturin series is known, we assessed the effect of iturin A5 on the growth morphology of *Foc* under an inverted microscope. As shown in Figure 8, compared to the 100% germination rate of *Foc* in the absence of iturin A5, the germination rate of *Foc* decreased to 66% and 37% in the presence of 62.5 µg/mL and 125 µg/mL iturin A5, respectively (Table S5). Moreover, the germ tubes displayed significantly distorted morphology such as ballooned tips and loss of polarity (Figure 8).

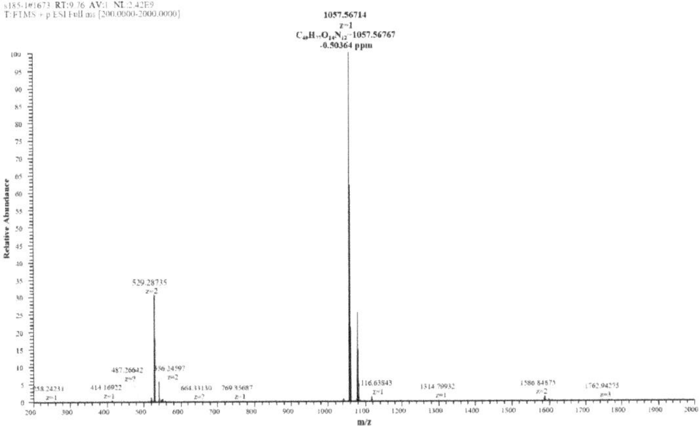

Figure 6. HR-ESI-MS of antifungal compound **1** isolated from *B. amyloliquefaciens* S185.

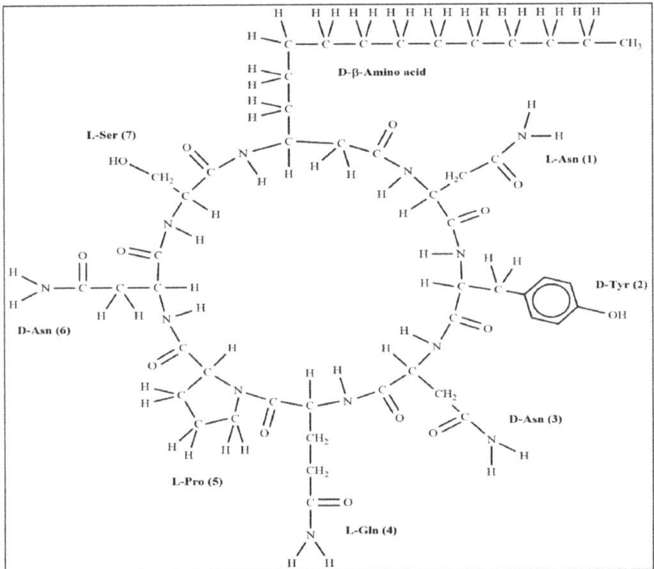

Figure 7. Structure of compound iturin A5 isolated from *B. amyloliquefaciens* S185.

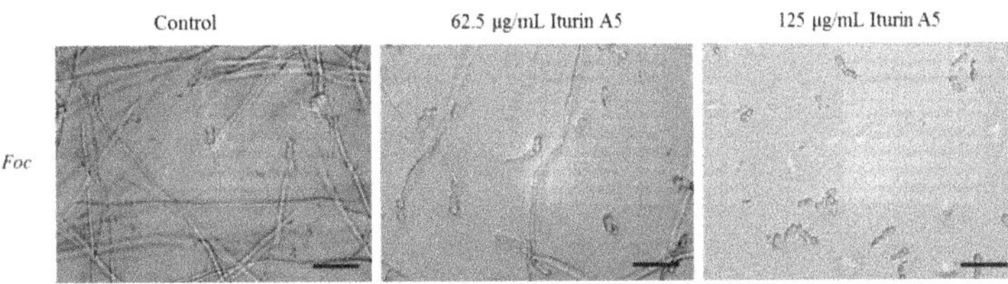

Figure 8. Inhibitory effect of antifungal compound iturin A5 on spore germination of *Fusarium oxysporum* f. sp. *cubense* (*Foc*). We incubated 2×10^5 spores/mL of *Foc* with 0, 62.5 and 125 µg/mL of iturin A5 at 28 °C. After 24 h incubation, images were taken under an inverted microscope. Control is DMSO. Scale bar, 40 µm.

3. Discussion

An essential goal of this study is focusing on the isolation and identification of the most effective biocontrol bacteria and antifungal compounds against *Foc*, the devastating pathogen for FWB. Marine microorganisms have been identified as promising natural sources for the development of biocontrol strains and agents [20]. In marine ecology, unusual circumstances such as extreme temperature and salinity provide a high success rate for the discovery of new and innovative microorganisms as well as their secondary metabolites. More than 20,000 distinct bioactive chemicals have been found in marine fauna and flora to date [34,35]. Previously, different species of *Bacillus* with confirmed biocontrol activity against *Foc* have been reported, including *B. amyloliquefaciens* GKT04 [36,37], W19 [38], NIN-6 [39], and NJN-6 [40], *B. velezensis* HN03 [41], and (iii) *B. subtilis* strains B26 [42], B04, B05 and B10 [43], and N11 [44]. However, an extensive screening of the biocontrol strains from marine bacteria to control FWB pathogen is limited.

Bacillus species are useful biocontrol candidates as they possess advantages of forming spores to survive for a long time in extreme environments, producing a wide range of physiologically active secondary metabolites that usually impede the growth of plant pathogens [45], and supporting the development and production of more stable commercial formulations over time from an agro-biotechnological perspective. Combining several *Bacillus* spp. strains is an attractive way to increase biocontrol efficiency under various cropping situations and environmental circumstances, given their adaptability and diverse biocontrol mechanisms. Because of the spore-forming abilities, the *Bacillus* group outperforms non-*Bacillus* species in terms of biocontrol efficiency. Spores can live in a variety of adverse conditions and remain stable during commercial production and maintain tolerance to fungicides [10,46]. *Bacillus*-mediated plant growth promotion is linked to the bacteria's ability to produce phytohormones such as gibberellic acid and IAA, which improve host nutrient absorption and increase plant defense responses to biotic and abiotic stresses [47]. IAA produced by *B. amyloliquefaciens* FZB42 promotes root growth and development of lateral roots, resulting in increased nutrient intake from the rhizosphere [48]. Consistent with previous reports, in this study, we observed IAA production by strain S185 in the presence and absence of tryptophan in the medium. Several bacterial strains produce secondary metabolites such as HCN and ammonia, which are helpful in the management of fungal diseases in a variety of plants [31,49–51]. Similarly, the strain S185 showed strong ammonia production as well as potent antifungal activities against *Foc*, demonstrating the dual role of S185 in controlling FWB.

Being a potential biocontrol strain, one key essential is the colonization capacity of the strain to banana plants. Previously, CLSM of GFP-tagged strains was utilized to illustrate the colonization pattern of *B. megaterium* in rice [52], *B. megaterium* and *B. mycoides* in sugarcane [50]. Similarly, when using this strategy, we observed that the GFP-labelled S185 strain was successfully colonized in all banana plant tissues including leaves, stems, and roots (Figure 4). Previous studies have also suggested that strains with a wider metabolite tolerance are preferable for plant nodulation and growth [32,53]. The BIOLOG metabolic profiling study is a useful technique for identifying microbial diversity using different types of substrates. In this study, carbon utilization pattern of strain S185 was observed on the GNIII Biolog plate and the resulting phenotypes indicated that S185 can use a variety of carbon substrates (Table S2).

Variation in ambient temperatures between crop seasons influenced disease outcomes by affecting the biocontrol agent's ability to kill pathogens [54]. In China, bananas are mostly grown in Guangdong, Guangxi, Hainan, Fujian, Yunnan, and Taiwan. However, diseases and adverse weather conditions continue to pose major obstacles to Chinese banana production. In this study, we found the strain S185 was able to grow between 20–50 °C, suggesting that the strain can operate consistently in temporally variable environments, thus making it a suitable biocontrol agent against *Foc* for naturally changing field circumstances. Wei et al. have reported that bacterial strains maintaining their activity in a variety of environments are promising candidates for consistent biocontrol applications [54].

Producing a wide range of secondary metabolites as active components is one of the main properties of *B. amyloliquefaciens*. Therefore, we performed the crude extract from S185 and examined antifungal activity against *Foc*. Not surprisingly, the extract displayed strong antifungal activity, suggesting antimicrobials in the S185 secondary metabolites. Subsequently, a compound **1** purified from the crude extract was identified using HR-ESI-MS (Figure 6). In combination with ^{13}C- and ^{1}H- NMR spectra the compound **1** was identified as iturin A5 (Figure 7).

Iturins are well-known antifungal metabolites generated by *Bacillus* strains [45,55,56], which comprise an amphiphilic peptide ring that is made up of seven chiral amino acids [57]. Iturin A, a member of the iturin family, has a high antibiotic action as well as a broad antifungal spectrum, making it a promising biological control agent for decreasing chemical pesticide use in agriculture [57]. Iturin A has also been found to be a useful drug in human and animal clinical studies because of its broad antifungal range, low toxicity, and low allergic impact [58]. Despite having so many benefits over chemical agents, iturin A has had very few practical uses too far, owing to poor strain productivity and very expensive manufacturing costs. To date, many iturin A producing strains have been identified such as *B. amyloliquefaciens*, *B. licheniformis*, *B. methyltrophicus*, *B. subtilis*, and *B. thuringiensis* [59–63]. However, this is the first report of iturin A5 lipopeptides produced from a thermotolerant marine *B. amyloliquefaciens* S185, which exhibited strong antifungal activity against *Foc*.

Iturin A disrupts the fungal cytoplasmic membrane by interacting mostly with ergosterol molecules, resulting in the formation of transmembrane channels that allow essential ions such as K^+ to be released. For example, iturin A has been shown to change the permeability of membranes and the lipid content of *S. cerevisiae* cells [64]. Similarly, Hiradate et al. reported iturin A5 secreted by *B. amyloliquefaciens* RC-2 inhibited the development of mulberry anthracnose caused by the fungus, *Colletotrichum dematium* [33]. We also studied the antifungal activity of purified iturin A5 on the growth of *Foc* pathogen. Different concentrations of iturin A5 produced by S185 was given to test the growth of *Foc*, revealing that MIC of iturin A5 was 25 µg/disk for *Foc*. Earlier, MICs of iturin A for other pathogens were also measured, such as iturin A produced by *B. amyloliquefaciens* PPCB004 inhibited *F. aromaticum* and *Botryosphaeria* sp. with MICs of 1.0 and 1.5 mg/mL [65]. Aberrant conidial and spore germination in fungi was observed when treated by iturins produced by *Bacillus* strains [30,65,66]. Optical and fluorescent microscopy revealed morphological alterations in conidia and significant deformities in *F. graminearum* hyphae treated with iturin A [67]. Wang et al. observed *Phytophthora infestans* mycelia was damaged and had a rough and swollen shape after iturin A treatment [68]. Similarly to previous findings, we also observed the major changes and deformations of *Foc* hyphae after treatment with iturin-A5 produced by S185 (Figure 8).

4. Materials and Methods

4.1. Collection of Bacterial Strains

The sediment samples were collected from offshore of South Sea, China. The isolation procedure was described below. Ten grams of sediment from each sample were suspended in 90 mL of autoclaved seawater in a flask and shaken at 100 rpm for 1 h at 25 °C. Then, the mixture was serially diluted 10^{-4} times, and 100 µL of each dilution were distributed on 2216E agar plates. The colony morphology was evaluated by incubating all the plates at 25 °C for 2–4 days. Antagonistic activity was tested for all isolated strains against *Foc*. *B. amyloliquefaciens* S185 was chosen for identification. 16s *rRNA* sequence of S185 was submitted to the NCBI GenBank database under accession number MZ333473. All collected strains were stored in 25% glycerol solution at −80 °C.

4.2. In Vitro Screening for Antifungal, Plant Growth-Promoting (PGP) Traits, and Abiotic Stress Tolerance

A dual culture plate method was used to test the antifungal efficacy of strain S185 against *Foc*, following the procedure of Singh et al. [69]. Bacterial strain S185 was spotted 3 cm away from 5 mm diameter of actively growing fungal pathogens spotted in the center point of the yeast extract medium (YAG) (yeast extract-5 g, dextrose-20 g, 1000× trace element-1 mL, agar-18 g, and ultrapure water-1000 mL; 1000× trace element: $ZnSO_4 \cdot 7H_2O$-2.2 g, H_3BO_3-1.1 g, $MnCl_2 \cdot 4H_2O$-0.5 g, $FeSO_4 \cdot 7H_2O$-0.5 g, $CoCl_2 \cdot 5H_2O$-0.16 g, $CuSO_4$-0.16 g, $(NH_4)_6Mo_7O_{24} \cdot 4H_2O$-0.11 g, EDTA-5.0 g, and ultrapure water-100 mL). Plate containing only fungal disk served as the control and was incubated at 28 °C for 5–7 days. The percentage of inhibition was measured using the formula; $[(R1-R2)/R1] \times 100$, where R1 is radial growth of the fungal pathogen in control plate and R2 is radial growth of fungal pathogen in the presence of test strain.

PGP traits including P-solubilization, production of IAA, siderophore, ammonia, and HCN by strain S185 were estimated by performing standard protocols of Brick et al. [70], Glickmann and Dessaux [71], Schwyn and Neilands [72], Dey et al. [73] and Lorck [74]. The ability of S185 to solubilize P was tested qualitatively using Pikovskaya medium added with tricalcium phosphate. Strain S185 was inoculated in the plate and observed for the clear hallow zone formation surrounding the isolate after 5–7 days incubation at 30 °C. IAA production was evaluated with colorimetric technique in the presence (0.5%) and absence of tryptophan in the medium. Siderophore production of strain S185 was detected on chrome azurol S (CAS) medium and the formation of a halo zone on the medium confirmed siderophore production. For ammonia production, strain S185 was cultured for 72 h in 10% sterile peptone water at 30 °C and changes in yellow color by adding 0.5 mL of Nessler's reagent established the ammonia production. HCN production ability of strain S185 was tested on a nutrient broth (NB) medium containing 4.4 g/L glycine to produce hydrocyanic acid. A filter paper soaked with 0.5% picric acid and 2% Na_2CO_3 was placed on a cover plate, then sealed with Parafilm and incubated at 30 °C, and alteration of filter paper color displayed HCN production. All assays were repeated three times with five replications.

Strain S185 was further tested in NB for its capacity to survive under different abiotic stress conditions, such as pH, temperature, and NaCl, and the medium without inoculation was employed as a blank. Temperature tolerance was determined by incubating 0.1 mL fresh bacterial solution of strain S185 in NB medium (5 mL) for 36 h at 20, 25, 30, 35, 40, 45, and 50 °C in a shaker incubator at 120 rpm, and O.D. was measured at 600 nm. For pH tolerance, pH of the NB medium was adjusted to 1, 2, 3, 4, 5, 6, 7, 8, and 9 with sterile buffers. Fresh culture of strain S185 was placed into 5 mL of LB broth medium with varied pH levels and incubated at 37 °C in an incubator shaker at 120 rpm for 36 h, with growth monitored at 600 nm. In addition, the salt tolerance property of strain S185 was tested in 5 mL of NB medium supplemented with 1, 2, 3, 4, 5, 6, 7, 8, 9, 10, 11, and 12% of NaCl concentration. 0.1 mL bacterial suspension (0.1 mL) was transferred in NB tubes and kept at 37 °C at 120 rpm in a shaker incubator and growth was measured after 36 h at 600 nm.

4.3. Phylogenetic Study of Strain S185

Phylogenetic analysis of strain S185 based on 16S *rRNA* sequences was performed with NCBI GenBank reference sequences. Sequence alignment was performed using the ClustalW and BlastN search algorithms, and closely related sequences were retrieved from NCBI. The Neighbor-Joining approach [75] was used to prepare phylogenetic analysis based on 16S *rRNA* sequence using MEGAX version X [76]. The number of differences method [77] was used to compute the genetic evolutionary distances, and the bootstrap test (1000 replicates) was performed as stated by Felsenstein [78].

4.4. Carbon Utilization Pattern of Strain S185

Carbon substrates utilization pattern of strain S185 was studied using GENIII BIOLOG[R] Phenotype Micro-ArrayTM plate. Strain S185 was cultivated at 30 °C on Luria-Bertani

(LB) agar medium, then washed and suspended in an inoculation fluid (IF) to achieve a transmittance of 90–98 percent as per the protocol. The phenotypic fingerprint was formed by incubating a 100 µL of cell suspension in the 96 wells of the GNIII Micro-Plate at 30 °C for 48 h. During incubation, the wells' respiration rate increases, allowing the cells to use a variety of carbon sources while growing. Increased respiration causes the tetrazolium dye to be reduced, resulting in purple color. Readings were recorded using an automated BIOLOG(R) Micro-Station Reader following the manufacturer guidelines after incubation.

4.5. Tagging of S185 with GFP-pPROBE-pTetr-TT

The pPROBE-pTetr -TT plasmid expressing green fluorescent protein (GFP) was provided by the Agriculture College, Guangxi University, Nanning, China. Tissue culture banana plantlets obtained from Guangxi Academy of Agricultural Sciences, Nanning, China were used for this experiment. Freshly grown cultures of S185 and GFP-pPROBE-pTetr-TT in LB medium were mixed (1:2) and then incubated for 36–48 h at 32 °C with a 160 rpm orbital shaker. Following the incubation, 100 µL of bacterial broth was disseminated on LB agar plate and kept overnight to assess the purity of the tagged strain, as well as validate the tagging using confocal laser scanning microscopy (CLSM). Fluorescent strain S185, when exposed to UV light, was chosen for further investigation.

Colonization of S185 in Banana Plantlets

Before bacterial inoculation, banana tissue culture plantlets were rinsed with autoclaved distilled water. Plantlets were placed in an autoclaved glass container comprising 50 mL of MS liquid media (mixture of basal salt and sucrose) at 30 °C in a growth chamber. After 3–4 days, plantlets were carefully moved to a new bottle comprising tagged bacterial suspension ($\sim$2.0 $\times$ 10^5/mL cell count). The uninoculated banana plantlets (control) were placed in a growth chamber set at 30 °C with a 14 h photoperiod and 60 μmol·m^{-2}·s^{-1} photon flux density. Plantlets were removed after 72–96 h of growth, rinsed with distilled water, and CLSM was used to check for bacterial colonization. Banana leaf, stem, and root (inoculated and uninoculated) samples were sliced into 1 cm lengths and put on the bridge slide with 10% (v/v) of glycerol before being detected with CLSM at varying emission rates depending on the intensity of autofluorescence UV light (Leica DMI 6000, Mannheim, Germany) [79].

4.6. Cultivation and Extraction of Antifungal Compounds

Strain S185 was cultured in 12 L of YAG medium at 25 °C for 9 days. The solid fermentation products were cut into small pieces and carefully extracted with ethyl acetate (EtOAc)/methanol (MeOH)/acetic acid (HAc) solution (80:15:5, $v/v/v$). The extraction was performed three times for the preparation of crude extracts. Crude extracts were dissolved in water and first extracted three times with EtOAc and then extracted three times with n-butanol. We obtained 7.981 g of EtOAc extract and 4.674 g of n-butanol extract after condensation and evaporation with a rotary evaporator. The n-butanol extract residue (4.674 g) was purified by Sephadex LH-20 (Amersham Pharmacia) (MeOH, 2 $\times$ 100 cm, 1 mL·min^{-1}) to produce 6 fractions (Fr.1.1-Fr.1.6). Fr.1.2 (592.7 mg) was applied to a silica gel column (Qingdao Marine Chemical Factory) (GF254, 3 $\times$ 50 cm, 70 g) using a chloroform: methanol (90:10, 80:20, 70:30) solvent gradient system with ammonia water (0.3%) to generate fractions Fr.1.2.1-Fr.1.2.5. Finally, Fr.1.2.5 (326 mg) was purified by Sephadex LH-20 (MeOH, 2 $\times$ 100 cm, 1 mL·min^{-1}) to obtain compound **1** (240.8 mg).

4.6.1. Antifungal Assay of Extracted Compound 1

Antifungal activity of selected compound **1** was assessed by disk diffusion method against pathogenic fungi *Foc* [80]. The extracted compound was dissolved in methanol and added in 6 mm diameter of paper disks at different levels of concentrations, i.e., 25, 50, 100, 200, and 400 µg/disk, respectively. A disc containing only methanol was used as the control. The dried paper discs were applied onto the surface of YAG plates dispersed with *Foc* at 28 °C for 24 h to 48 h. After incubation, the diameter of the inhibition zone was determined. Each treatment was repeated three times.

4.6.2. Structure Identification of Compound 1

The Nuclear Magnetic Resonance (NMR) spectra were generated using an Agilent NMR system 800 MHz NMR spectrometer (Agilent Technologies Inc., Colorado Springs, CO, USA) to determine the mass of the antifungal compound **1**. Electrospray ionization mass spectrometry (ESI-MS) and high-resolution electrospray ionization mass spectrometry (HR-ESI-MS) were performed using a Thermo Scientific Orbitrap Elite MS spectrometer (Thermo Fisher Scientific, Waltham, MA, USA) to identify the structure of compound **1** isolated from S185.

4.7. Effect of Purified Iturin A5 on Spore Germination of Foc

The inhibitory effect of purified iturin A5 on spore germination of *Foc* was performed following the standard EUCAST MIC determination method [81] and observed under an inverted microscope (Leica Microsystems CMS GmbH, Ernst-Leitz-Str. 17–37, D-35578, Wetzlar). *Foc* was cultured on a YAG medium at 28 °C for 7 days and then mixed with 10 mL of 0.2% Tween-80 solution. Subsequently, we gently scraped the surface of the fungal colony with a sterile loop, and spores were collected. Finally, the spore concentration was diluted to 2×10^5 CFU/mL. 10 mg iturin A5 was dissolved in 1 mL DMSO to obtain the concentration of 10 µg/µL stock solution. $2 \times$ RPMI1640 medium with 2% glucose was used for spore germination, which was carried out in 96-well plates. Firstly, 300 µL liquid spores were added to the Eppendorf tube and mixed with 270 µL of $2 \times$ RPMI1640 medium with 2% glucose. For control, 30 µL DMSO was added to the Eppendorf tube. For treatment, different concentration of iturin A5 (62.5, 125, 250, and 500 µg/mL) was used. 200 µL of prepared different spore suspensions were added into 96-well plates. After 24 h of incubation at 28 °C, spore germination was observed under an inverted microscope.

5. Conclusions

Bacillus strains are potential candidates for controlling FWB because of their wide range of secondary metabolites to inhibit the growth of pathogens as well as colonization and PGP properties. Searching for cost-effective and environmentally friendly *Bacillus* to manage banana pathogenic outbreaks is urgent and practical. In this study, a marine bacterium, *B. amyloliquefaciens* S185, exhibited thermotolerance, IAA, and ammonia production, diverse carbon utilization patterns, and good colonizing capacity in banana plants. In parallel, the S185 strain displayed strong antagonistic activity against *Foc*. The compound iturin A5 from S185 is the main bioactive component for antifungal activity by delaying spore germination and disrupting the polarity establishment of *Foc*. In summary, strain S185 and its iturin A5 lipopeptide is a promising biocontrol strain and agent for FWB. However, field testing of this strain is required to evaluate its efficacy in controlling Foc and promoting banana growth in the future.

Supplementary Materials: The following are available online at https://www.mdpi.com/article/10.3390/md19090516/s1, Figure S1: The ^{1}H spectra of compound **1** in DMSO-d$_6$. Figure S2: The ^{13}C spectra of compound **1** in DMSO-d$_6$. Table S1: Antifungal and plant growth promoting features of thermotolerant marine *Bacillus amyloliquefaciens* S185. Table S2: List of carbon sources utilized by S185 in GNIII Biolog plate. Table S3: Inhibitory effect of compound **1** extracted from S185 on the growth of *Fusarium oxysporum* f. sp. *cubense*. Table S4: NMR spectroscopic data for compound **1**, iturin A2 and iturin A5 in DMSO-d$_6$. Table S5: Effect of iturin A5 on spore germination rate of *Fusarium oxysporum* f. sp. *cubense*.

Author Contributions: Conceptualization, B.W. and W.F.; methodology, P.S. and J.X.; software, J.X.; validation, P.S. and J.X.; formal analysis, P.S. and J.X.; investigation, B.W., C.J. and W.F.; resources, Y.Q. and Q.Q.; data curation, P.S. and J.X.; writing—original draft preparation, P.S. and J.X.; writing—review and editing, P.S., J.X., B.W. and W.F.; visualization, W.F.; supervision, W.F.; project administration, B.W. and W.F.; funding acquisition, B.W., C.J. and W.F. All authors have read and agreed to the published version of the manuscript.

Funding: This work was supported by National Natural Science Foundation of China (31960032, 32071279), Guangxi Natural Science Foundation (2020GXNSFDA238008) to W.F., Research Start-up Funding of Guangxi Academy of Sciences (2017YJJ026) to B.W., Bagui Scholar Program Fund (2016A24) of Guangxi Zhuang Autonomous Region to C.J.

Data Availability Statement: Data sharing does not apply to this article as no new data were created or analyzed in this study.

Acknowledgments: We are very grateful to Sugarcane Research Center, Guangxi Academy of Agricultural Sciences, Nanning, China for a colonization study using confocal laser scanning microscopy (CLSM).

Conflicts of Interest: All the authors report no conflict of interest relevant to this editorial.

References

1. Passera, A.; Venturini, G.; Battelli, G.; Casati, P.; Penaca, F.; Quaglino, F.; Bianco, P.A. Competition assays revealed *Paenibacillus pasadenensis* strain R16 as a novel antifungal agent. *Microbiol. Res.* **2017**, *198*, 16–26. [CrossRef]
2. Butler, D. Fungus threatens top banana. *Nature* **2013**, *504*, 195–196. [CrossRef]
3. FAOSTAT. FAOSTAT Database. 2020. Available online: http://www.fao.org/faostat/en/#data/QC (accessed on 5 July 2021).
4. Dita, M.; Barquero, M.; Heck, D.; Mizubuti, E.S.G.; Staver, C.P. Fusarium wilt of banana: Current knowledge on epidemiology and research needs toward sustainable disease management. *Front. Plant Sci.* **2018**, *9*, 1468. [CrossRef] [PubMed]
5. Pegg, K.G.; Coates, L.M.; O'Neill, W.T.; Turner, D.W. The epidemiology of *Fusarium* wilt of banana. *Front. Plant Sci.* **2019**, *10*, 1395. [CrossRef]
6. Shen, Z.Z.; Ruan, Y.Z.; Chao, X.; Zhang, J.; Li, R.; Shen, Q.R. Rhizosphere microbial community manipulated by 2 years of consecutive biofertilizer application associated with banana *Fusarium wilt* disease suppression. *Biol. Fertil. Soils.* **2015**, *51*, 553–562. [CrossRef]
7. Guo, L.; Yang, L.; Liang, C.; Wang, J.; Liu, L.; Huang, J. The G-protein subunits FGA2 and FGB1 play distinct roles in development and pathogenicity in the banana fungal pathogen *Fusarium oxysporum* f. sp. *cubense*. *Physiol. Mol. Plant Pathol.* **2016**, *93*, 29–38. [CrossRef]
8. Duan, Y.; Chen, J.; He, W.; Chen, J.; Pang, Z.; Hu, H.; Xie, J. Fermentation optimization and disease suppression ability of a *Streptomyces* ma. FS-4 from banana rhizosphere soil. *BMC Microbiol.* **2020**, *20*, 24. [CrossRef] [PubMed]
9. Wei, Y.X.; Liu, W.; Hu, W.; Liu, G.Y.; Wu, C.J.; Liu, W.; Zeng, H.Q.; He, C.Z.; Shi, H.T. Genome-wide analysis of autophagy-related genes in banana highlights *MaATG8s* in cell death and autophagy in immune response to *Fusarium* wilt. *Plant Cell Rep.* **2017**, *36*, 1237–1250. [CrossRef]
10. Bubici, G.; Kaushal, M.; Prigigallo, M.I.; Gómez-Lama Cabanás, C.; Mercado-Blanco, J. Biological control agents against *Fusarium* wilt of banana. *Front. Microbiol.* **2019**, *10*, 616. [CrossRef]
11. Damodaran, T.; Rajan, S.; Muthukumar, M.; Ram, G.; Yadav, K.; Kumar, S.; Ahmad, I.; Kumari, N.; Mishra, V.K.; Jha, S.K. Biological management of banana *Fusarium* wilt caused by *Fusarium oxysporum* f. sp. *cubense* tropical race 4 using antagonistic fungal isolate CSR-T-3 (*Trichoderma reesei*). *Front. Microbiol.* **2020**, *11*, 595845. [CrossRef]
12. Wang, B.; Yuan, J.; Zhang, J.; Shen, Z.; Zhang, M.; Li, R.; Ruan, Y.; Shen, Q.R. Effects of novel bioorganic fertilizer produced by *Bacillus amyloliquefaciens* W19 on antagonism of Fusarium wilt of banana. *Biol. Fertil. Soils.* **2013**, *49*, 435–446. [CrossRef]
13. Shen, Z.Z.; Zhong, S.T.; Wang, Y.G.; Wang, B.B.; Mei, X.L.; Li, R.; Ruan, Y.Z.; Shen, Q.R. Induced soil microbial suppression of banana fusarium wilt disease using compost and biofertilizers to improve yield and quality. *Eur. J. Soil Biol.* **2013**, *57*, 1–8. [CrossRef]
14. Dadrasnia, A.; Usman, M.M.; Omar, R.; Ismail, S.; Abdullah, R. Potential use of *Bacillus* genus to control of bananas diseases: Approaches toward high yield production and sustainable management. *J. King Saud Univ.-Sci.* **2020**, *32*, 2336–2342. [CrossRef]
15. Shen, Z.Z.; Xue, C.; Taylor, P.W.J.; Ou, Y.N.; Wang, B.B.; Zhao, Y.; Ruan, Y.Z.; Li, R.; Shen, Q.R. Soil pre-fumigation could effectively improve the disease suppressiveness of biofertilizer to banana *Fusarium wilt* disease by reshaping the soil microbiome. *Biol. Fertil. Soils* **2018**, *54*, 793–806. [CrossRef]
16. Xiong, W.; Li, R.; Ren, Y.; Liu, C.; Zhao, Q.; Wu, H.; Jousset, A.; Shen, Q. Distinct roles for soil fungal and bacterial communities associated with the suppression of vanilla *Fusarium wilt* disease. *Soil Biol. Biochem.* **2017**, *107*, 198–207. [CrossRef]
17. Van Bruggen, A.H.C.; Sharma, K.; Kaku, E.; Karfopoulos, S.; Zelenev, V.V.; Blok, W.J. Soil health indicators and Fusarium wilt suppression in organically and conventionally managed greenhouse soils. *Appl. Soil Ecol.* **2015**, *86*, 192–201. [CrossRef]
18. Prasad, R.; Chandra, H.; Sinha, B.K.; Srivastava, J. Antagonistic effect of *Pseudomonas fluorescens* isolated from soil of doon valley (Dehradun–India) on certain phyto-pathogenic fungi. *Octa J. Biosci.* **2015**, *3*, 92–95.
19. Chandra, H.; Kumari, P.; Bisht, R.; Prasad, R.; Yadav, S. Plant growth promoting *Pseudomonas aeruginosa* from *Valeriana wallichii* displays antagonistic potential against three phytopathogenic fungi. *Mol. Biol. Rep.* **2020**, *47*, 6015–6026. [CrossRef]

20. Adnan, M.; Alshammari, E.; Patel, M.; Ashraf, S.A.; Khan, S.; Hadi, S. Significance and potential of marine microbial natural bioactive compounds against biofilms/biofouling: Necessity for green chemistry. *PeerJ* **2018**, *6*, 5049. [CrossRef]

21. Imhoff, J.F.; Labes, A.; Wiese, J. Bio-mining the microbial treasures of the ocean: New natural products. *Biotechnol. Adv.* **2011**, *29*, 468–482. [CrossRef] [PubMed]

22. Taylor, M.W.; Radax, R.; Steger, D.; Wagner, M. Sponge-associated microorganisms: Evolution, ecology, and biotechnological potential. *Microbiol. Mol. Biol. Rev.* **2007**, *71*, 295–347. [CrossRef]

23. Carroll, A.R.; Copp, B.R.; Davis, R.A.; Keyzers, R.A.; Prinsep, M.R. Marine natural products. *Nat. Prod. Rep.* **2020**, *37*, 175–223. [CrossRef]

24. Aloo, B.N.; Makumba, B.A.; Mbega, E.R. The potential of Bacilli rhizobacteria for sustainable crop production and environmental sustainability. *Microbiol. Res.* **2019**, *219*, 26–39. [CrossRef]

25. Chowdhury, S.P.; Dietel, K.; Rändler, M.; Schmid, M.; Junge, H.; Borriss, R.; Hartmann, A.; Grosch, R. Effects of *Bacillus amyloliquefaciens* FZB42 on Lettuce growth and health under pathogen pressure and its impact on the rhizosphere bacterial community. *PLoS ONE* **2013**, *8*, e68818. [CrossRef]

26. Wu, L.; Wu, H.J.; Qiao, J.; Gao, X.; Borriss, R. Novel routes for improving biocontrol activity of *Bacillus* based bioinoculants. *Front. Microbiol.* **2015**, *6*, 1395. [CrossRef]

27. Ongena, M.; Jourdan, E.; Adam, A.; Paquot, M.; Brans, A.; Joris, B.; Arpigny, J.-L.; Thonart, P. Surfactin and fengycin lipopeptides of *Bacillus subtilis* as elicitors of induced systemic resistance in plants. *Env. Microbiol.* **2007**, *9*, 1084–1090. [CrossRef]

28. Hsieh, F.C.; Lin, T.C.; Meng, M.; Kao, S.S. Comparing methods for identifying *Bacillus* strains capable of producing the antifungal lipopeptide iturin A. *Curr. Microbiol.* **2008**, *56*, 1–5. [CrossRef]

29. Steller, S.; Vollenbroich, D.; Leenders, F.; Stein, T.; Conrad, B.; Hofemeister, J.; Vater, J. Structural and functional organization of the fengycin synthetase multienzyme system from *Bacillus subtilis* b213 and A1/3. *Chem. Biol.* **1999**, *6*, 31–41. [CrossRef]

30. Kumar, A.; Saini, S.; Wray, V.; Nimtz, M.; Prakash, A.; Johri, B.N. Characterization of an antifungal compound produced by *Bacillus sp. strain A*(5)F that inhibits *Sclerotinia sclerotiorum*. *J. Basic Microbiol.* **2012**, *52*, 670–678. [CrossRef]

31. Olanrewaju, O.S.; Glick, B.R.; Babalola, O.O. Mechanisms of action of plant growth promoting bacteria. *World J. Microb. Biot.* **2017**, *33*, 197. [CrossRef]

32. Wielbo, J.; Marek-Kozaczuk, M.; Kubik-Komar, A.; Skorupska, A. Increased metabolic potential of *Rhizobium* spp. is associated with bacterial competitiveness. *Can. J. Microbiol.* **2007**, *53*, 957–967. [CrossRef]

33. Hiradate, S.; Yoshida, S.; Sugie, H.; Yada, H.; Fujii, Y. Mulberry anthracnose antagonists (iturins) produced by *Bacillus amyloliquefaciens* RC-2. *Phytochemistry* **2002**, *61*, 693–698. [CrossRef]

34. Bibi, F.; Yasir, M.; Al-Sofyani, A.; Naseer, M.I.; Azhar, E.I. Antimicrobial activity of bacteria from marine sponge *Subereamollis* and bioactive metabolites of *Vibrio* sp. EA348. *Saudi J. Biol. Sci.* **2020**, *27*, 1139–1147. [CrossRef]

35. Choudhary, A.; Naughton, L.; Montánchez, I.; Dobson, A.; Rai, D. Current status and future prospects of marine natural products (MNPs) as antimicrobials. *Mar. Drugs* **2017**, *15*, 272. [CrossRef]

36. Tian, D.D.; Zhou, W.; Li, C.S.; Wei, D.; Qin, L.Y.; Huang, S.M.; Wei, L.P.; Long, S.F.; He, Z.F.; Wei, S.L. Isolation and identification of lipopetide antibiotic produced by *Bacillus amyloliquefaciens* GKT04 antagonistic to banana *Fusarium* wilt. *J. South China Agric. Univ.* **2020**, *51*, 1122–1127.

37. Tian, D.; Song, X.; Li, C.; Zhou, W.; Qin, L.; Wei, L.; Di, W.; Huang, S.; Li, B.; Huang, Q.; et al. Antifungal mechanism of *Bacillus amyloliquefaciens* strain GKT04 against *Fusarium* wilt revealed using genomic and transcriptomic analyses. *Microbiol. Open* **2021**, *10*, e1192. [CrossRef]

38. Wang, B.; Shen, Z.; Zhang, F.; Raza, W.; Yuan, J.; Huang, R.; Ruan, Y.Z.; Li, R.; Shen, Q.R. *Bacillus amyloliquefaciens* strain W19 can promote growth and yield and suppress *Fusarium wilt* in banana under greenhouse and field conditions. *Pedosh* **2016**, *26*, 733–744. [CrossRef]

39. Xue, C.; Ryan Penton, C.; Shen, Z.; Zhang, R.; Huang, Q.; Li, R.; Ruan, Y.; Shen, Q. Manipulating the banana rhizosphere microbiome for biological control of Panama disease. *Sci. Rep.* **2015**, *5*, 11124. [CrossRef]

40. Yuan, J.; Li, B.; Zhang, N.; Waseem, R.; Shen, Q.; Huang, Q. Production of bacillomycin-and macrolactin-type antibiotics by *Bacillus amyloliquefaciens* NJN-6 for suppressing soilborne plant pathogens. *J. Agric. Food Chem.* **2012**, *60*, 2976–2981. [CrossRef]

41. Wu, L.; Wu, H.; Chen, L.; Yu, X.; Borriss, R.; Gao, X. Difficidin and bacilysin from *Bacillus amyloliquefaciens* FZB42 have antibacterial activity against *Xanthomonas oryzae* rice pathogens. *Sci. Rep.* **2015**, *5*, 12975. [CrossRef]

42. Aminian, H.; Irandegani, J. Biological control of *Fusarium verticillioides* agent of *Fusarium* wilt of banana by *Pseudomonas fluorescens* and *Bacillus subtilis* isolates. In *Biological Control in Agriculture and Natural Resources*; Elsevier: Amsterdam, The Netherlands, 2013.

43. Hadiwiyono, W.S. Endophytic *Bacillus*: The potentiality of antagonism to wilt pathogen and promoting growth to micro-plantlet of banana in vitro. *Biomirror* **2012**, *3*, 1–4.

44. Zhang, N.; Wu, K.; He, X.; Li, S.Q.; Zhang, Z.H.; Shen, B.; Yang, X.M.; Zhang, R.F.; Huang, Q.W.; Shen, Q.R. A new bioorganic fertilizer can effectively control banana wilt by strong colonization with *Bacillus subtilis* N11. *Plant Soil* **2011**, *344*, 87–97. [CrossRef]

45. Ongena, M.; Jacques, P. *Bacillus* lipopeptides: Versatile weapons for plant disease biocontrol. *Trends Microbiol.* **2008**, *16*, 115–125. [CrossRef]

46. Nicholson, W.L.; Munakata, N.; Horneck, G.; Melosh, H.J.; Setlow, P. Resistance of *Bacillus* endospores to extreme terrestrial and extra-terrestrial environments. *Microbiol. Mol. Biol. Rev.* **2000**, *64*, 548–572. [CrossRef]

47. Harman, G.E. Multifunctional fungal plant symbionts: New tools to enhance plant growth and productivity. *New Phytol.* **2011**, *189*, 647–649. [CrossRef]

48. Idris, S.E.; Iglesias, D.J.; Talon, M.; Borriss, R. Tryptophan-dependent production of indole-3-acetic acid (IAA) affects level of plant growth promotion by *Bacillus amyloliquefaciens* FZB42. *Mol. Plant-Microbe Interact.* **2007**, *20*, 619–626. [CrossRef]

49. Singh, R.K.; Singh, P.; Li, H.B.; Guo, D.J.; Song, Q.Q.; Yang, L.T.; Malviya, M.K.; Song, X.P.; Li, Y.R. Plant-PGPR interaction study of plant growth-promoting diazotrophs *Kosakonia radicincitans* BA1 and *Stenotrophomonas maltophilia* COA2 to enhance growth and stress-related gene expression in *Saccharum* spp. *J. Plant Interact.* **2020**, *15*, 427–445. [CrossRef]

50. Singh, R.K.; Singh, P.; Li, H.B.; Song, Q.Q.; Guo, D.J.; Solanki, M.K.; Verma, K.K.; Malviya, M.K.; Song, X.P.; Lakshmanan, P.; et al. Diversity of nitrogen-fixing rhizobacteria associated with sugarcane, a comprehensive study of plant-microbe interactions for growth enhancement in *Saccharum* spp. *BMC Plant Biol.* **2020**, *20*, 220. [CrossRef]

51. Singh, P.; Singh, R.K.; Guo, D.J.; Sharma, A.; Singh, R.N.; Li, D.P.; Malviya, M.K.; Song, X.P.; Lakshmanan, P.; Yang, L.T.; et al. Whole genome analysis of sugarcane root-associated endophyte *Pseudomonas aeruginosa* B18- a plant growth-promoting bacterium with antagonistic potential against *Sporisorium scitamineum*. *Front. Microbiol.* **2021**, *12*, 628376. [CrossRef] [PubMed]

52. Liu, X.; Zhao, H.; Chen, S. Colonization of maize and rice plants by strain *Bacillus megaterium* C4. *Curr. Microbiol.* **2006**, *52*, 186–190. [CrossRef]

53. Mazur, A.; Stasiak, G.; Wielbo, J.; Koper, P.; Kubik-Komar, A.; Skorupska, A. Phenotype profiling of *Rhizobium leguminosarum* bv. *trifolii* clover nodule isolates reveal their both versatile and specialized metabolic capabilities. *Arch. Microbiol.* **2013**, *195*, 255–267. [CrossRef] [PubMed]

54. Wei, Z.; Huang, J.; Yang, T.; Jousset, A.; Xu, Y.; Shen, Q.; Friman, V.P. Seasonal variation in the biocontrol efficiency of bacterial wilt is driven by temperature-mediated changes in bacterial competitive interactions. *J. Appl. Ecol.* **2017**, *54*, 1440–1448. [CrossRef]

55. Guardado-Valdivia, L.; Tovar-Pérez, E.; Chacón-López, A.; López-García, U.; Gutiérrez-Martínez, P.; Stoll, A.; Aguilera, S. Identification and characterization of a new *Bacillus atrophaeus* strain B5 as biocontrol agent of postharvest anthracnose disease in soursop (*Annona muricata*) and avocado (*Persea americana*). *Microbiol. Res.* **2018**, *210*, 26–32. [CrossRef]

56. Ruiz-Sánchez, E.; Mejía-Bautista, M.Á.; Serrato-Díaz, A.; Reyes-Ramírez, A.; Estrada Girón, Y.; Valencia-Botín, A.J. Antifungal activity and molecular identification of native strains of *Bacillus subtilis*. *Agrociencia* **2016**, *50*, 133–148.

57. Maget-Dana, R.; Peypoux, F. Iturins, a special class of pore-forming lipopeptides. Biological and physicochemical properties. *Toxicology* **1994**, *87*, 151–174. [CrossRef]

58. Yao, S.; Gao, X.; Fuchsbauer, N.; Hillen, W.; Vater, J.; Wang, J. Cloning, sequencing, and characterization of the genetic region relevant to biosynthesis of the lipopeptides iturin A and surfactin in *Bacillus subtilis*. *Curr. Microbiol.* **2003**, *47*, 272–277. [CrossRef]

59. Bernat, P.; Paraszkiewicz, K.; Siewiera, P.; Moryl, M.; Płaza, G.; Chojniak, J. Lipid composition in a strain of *Bacillus subtilis*, a producer of iturin A lipopeptides that are active against uropathogenic bacteria. *World J. Microbiol. Biotechnol.* **2016**, *32*, 157. [CrossRef]

60. Kalai-Grami, L.; Karkouch, I.; Naili, O.; Slimene, I.B.; Elkahoui, S.; Zekri, R.B.; Touati, I.; Mnari-Hattab, M.; Hajlaoui, M.R.; Limam, F. Production and identification of iturin A lipopeptide from *Bacillus methyltrophicus* TEB1 for control of Phomatracheiphila. *J. Basic Microbiol.* **2016**, *56*, 864–871. [CrossRef]

61. Kim, P.I.; Ryu, J.; Kim, Y.H.; Chi, Y.T. Production of biosurfactant lipopeptides Iturin A.; fengycin and surfactin A from *Bacillus subtilis* CMB32 for control of *Colletotrichum gloeosporioides*. *J. Microbiol. Biotechnol.* **2010**, *20*, 138–145. [CrossRef] [PubMed]

62. Perez, K.J.; Viana, J.D.; Lopes, F.C.; Pereira, J.Q.; Santos, D.M.; Oliveira, J.S.; Velho, R.V.; Crispim, S.M.; Nicoli, J.R.; Brandelli, A.; et al. *Bacillus* spp. isolated from puba as a source of biosurfactants and antimicrobial lipopeptides. *Front. Microbiol.* **2017**, *8*, 61. [CrossRef] [PubMed]

63. Zhang, Q.X.; Zhang, Y.; Shan, H.H.; Tong, Y.H.; Chen, X.J.; Liu, F.Q. Isolation and identifcation of antifungal peptides from *Bacillus amyloliquefaciens* W10. *Environ. Sci. Pollut. Res. Int.* **2017**, *24*, 25000–25009. [CrossRef]

64. Latoud, C.; Peypoux, F.; Michel, G. Action of iturin A, an antifungal antibiotic from *Bacillus subtilis* on the yeast *Saccharomyces cerevisiae*. Modifications of membrane permeability and lipid composition. *J. Antibiot.* **1987**, *40*, 1588–1595. [CrossRef]

65. Arrebola, E.; Jacobs, R.; Korsten, L. Iturin A is the principal inhibitor in the biocontrol activity of *Bacillus amyloliquefaciens* PPCB004 against postharvest fungal pathogens. *J. Appl. Microbiol.* **2010**, *108*, 386–395. [CrossRef]

66. Gu, Q.; Yang, Y.; Yuan, Q.; Shi, G.; Wu, L.; Lou, Z.; Huo, R.; Wu, H.; Borriss, R.; Gao, X. Bacillomycin D produced by *Bacillus amyloliquefaciens* is involved in the antagonistic interaction with the plant-pathogenic fungus *Fusarium graminearum*. *Appl. Environ. Microbiol.* **2017**, *83*, e01075-17. [CrossRef] [PubMed]

67. Gong, A.D.; Li, H.P.; Yuan, Q.S.; Song, X.S.; Yao, W.; He, W.J.; Zhang, J.B.; Liao, Y.C. Antagonistic mechanism of iturin A and plipastatin A from *Bacillus amyloliquefaciens* S76-3 from wheat spikes against *Fusarium graminearum*. *PLoS ONE* **2015**, *10*, e0116871. [CrossRef]

68. Wang, T.; Wang, X.; Zhu, X.; He, Q.; Guo, L. A proper PiCAT2 level is critical for sporulation, sporangium function, and pathogenicity of *Phytophthora infestans*. *Mol. Plant Pathol.* **2020**, *21*, 460–474. [CrossRef] [PubMed]

69. Singh, R.K.; Kumar, D.P.; Solanki, M.K.; Singh, P.; Srivastava, S.; Kashyap, P.L.; Kumar, S.; Srivastva, A.K.; Singhal, P.K.; Arora, D. KMultifarious plant growth promoting characteristics of chickpea rhizosphere associated *Bacilli* help to suppress soil-borne pathogens. *Plant Growth Regul.* **2014**, *73*, 91–101. [CrossRef]

70. Brick, J.M.; Bostock, R.M.; Silverstone, S.E. Rapid in situ assay for indole acetic acid production by bacteria immobilized on nitrocellulose membrane. *Appl. Environ. Microbiol.* **1991**, *57*, 535–538. [CrossRef] [PubMed]

71. Glickmann, E.; Dessaux, Y. A critical examination of the specificity of the Salkowski reagent for indolic compounds produced by phytopathogenic bacteria. *Appl. Environ. Microbiol.* **1995**, *61*, 793–796. [CrossRef]
72. Schwyn, B.; Neilands, J.B. Universal chemical assay for the detection and determination of siderophores. *Anal. Biochem.* **1987**, *160*, 47–56. [CrossRef]
73. Dey, R.; Pal, K.K.; Bhatt, D.M.; Chauhan, S.M. Growth promotion and yield enhancement of peanut (*Arachis hypogaea* L.) by application of plant growth promoting rhizobacteria. *Microbiol. Res.* **2004**, *159*, 371–394. [CrossRef]
74. Lorck, H. Production of hydrocyanic acid by bacteria. *Physiol. Plant.* **1948**, *1*, 142–146. [CrossRef]
75. Saitou, N.; Nei, M. The neighbor-joining method, a new method for reconstructing phylogenetic trees. *Mol. Biol. Evol.* **1987**, *4*, 406–425. [PubMed]
76. Kumar, S.; Stecher, G.; Li, M.; Knyaz, C.; Tamura, K. MEGA X, molecular evolutionary genetics analysis across computing platforms. *Mol. Biol. Evol.* **2018**, *35*, 1547–1549. [CrossRef] [PubMed]
77. Nei, M.; Kumar, S. *Molecular Evolution and Phylogenetics*; Oxford University Press: New York, NY, USA, 2000.
78. Felsenstein, J. Confidence limits on phylogenies, an approach using the bootstrap. *Evolution* **1985**, *39*, 783–791. [CrossRef] [PubMed]
79. Singh, P.; Singh, R.K.; Li, H.B.; Guo, D.J.; Sharma, A.; Lakshmanan, P.; Malviya, M.K.; Song, X.P.; Solanki, M.K.; Verma, K.K.; et al. Diazotrophic bacteria *Pantoea dispersa* and *Enterobacter asburiae* promote sugarcane growth by inducing nitrogen uptake and defense-related gene expression. *Front. Microbiol.* **2021**, *11*, 600417. [CrossRef] [PubMed]
80. Espinel-Ingroff, A.; Pfaller, M.; Messer, S.A.; Knapp, C.C.; Killian, S.; Norris, H.A.; Ghannoum, M.A. Multicenter comparison of the sensititre Yeast One Colorimetric Antifungal Panel with the National Committee for Clinical Laboratory standards M27-A reference method for testing clinical isolates of common and emerging *Candida* spp., *Cryptococcus* spp., and other yeasts and yeast-like organisms. *J. Clin. Microbiol.* **1999**, *37*, 591–595.
81. EUCAST 2008. EUCAST technical note on the method for the determination of broth dilution minimum inhibitory concentrations of antifungal agents for conidia-forming moulds. *Clin. Microbiol. Infect.* **2008**, *14*, 982–984. [CrossRef]

 *marine drugs*

Article

Anti-Food Allergic Compounds from *Penicillium griseofulvum* MCCC 3A00225, a Deep-Sea-Derived Fungus

Cui-Ping Xing [1], Dan Chen [2], Chun-Lan Xie [1], Qingmei Liu [3], Tian-Hua Zhong [1], Zongze Shao [1], Guangming Liu [3], Lian-Zhong Luo [2,*] and Xian-Wen Yang [1,*]

[1] Key Laboratory of Marine Biogenetic Resources, Third Institute of Oceanography, Ministry of Natural Resources,184 Daxue Road, Xiamen 361005, China; xingcuiping123@126.com (C.-P.X.); xiechunlanxx@163.com (C.-L.X.); zhongtianhua@tio.org.cn (T.-H.Z.); shaozongze@tio.org.cn (Z.S.)

[2] Fujian Universities and Colleges Engineering Research Center of Marine Biopharmaceutical Resources, Xiamen Medical College, 1999 Guankouzhong Road, Xiamen 361023, China; cd@xmmc.edu.cn

[3] College of Food and Biological Engineering, Jimei University, 43 Yindou Road, Xiamen 361021, China; liuqingmei1229@163.com (Q.L.); gmliu@jmu.edu.cn (G.L.)

* Correspondence: llz@xmmc.edu.cn (L.-Z.L.); yangxianwen@tio.org.cn (X.-W.Y.); Tel.: +86-592-636-5150 (L.-Z.L.); +86-592-219-5319 (X.-W.Y.)

Abstract: Ten new (**1–10**) and 26 known (**11–36**) compounds were isolated from *Penicillium griseofulvum* MCCC 3A00225, a deep sea-derived fungus. The structures of the new compounds were determined by detailed analysis of the NMR and HRESIMS spectroscopic data. The absolute configurations were established by X-ray crystallography, Marfey's method, and the ICD method. All isolates were tested for in vitro anti-food allergic bioactivities in immunoglobulin (Ig) E-mediated rat basophilic leukemia (RBL)-2H3 cells. Compound **13** significantly decreased the degranulation release with an IC_{50} value of 60.3 µM, compared to that of 91.6 µM of the positive control, loratadine.

Keywords: deep-sea microorganism; fungus; *Penicillium griseofulvum*; anti-food allergy; fungal metabolites; marine natural products

Citation: Xing, C.-P.; Chen, D.; Xie, C.-L.; Liu, Q.; Zhong, T.-H.; Shao, Z.; Liu, G.; Luo, L.-Z.; Yang, X.-W. Anti-Food Allergic Compounds from *Penicillium griseofulvum* MCCC 3A00225, a Deep-Sea-Derived Fungus. *Mar. Drugs* **2021**, *19*, 224. https://doi.org/10.3390/md19040224

Academic Editor: Daniela Giordano

Received: 25 March 2021
Accepted: 12 April 2021
Published: 16 April 2021

Publisher's Note: MDPI stays neutral with regard to jurisdictional claims in published maps and institutional affiliations.

1. Introduction

For the past decade, the trend to discover new compounds from marine microorganisms continues to rise [1], especially from marine fungi [2,3], which accounted for 68% of the reported new marine natural products in 2019 [4]. Of particular importance is the *Penicillium* species, which are recognized as the richest source for the discovery of biologically important and structurally unique secondary metabolites [5–8].

As our ongoing research for novel and bioactive secondary metabolites from the deep sea-derived microorganisms [8–11], the fungal strain *Penicillium griseofulvum* isolated from the Indian Ocean sediment was selected for a systematic chemical examination. As a result, five carotanes, four naphthalenes, and three viridicatol derivates were obtained [12,13]. A continuous study, however, led to the isolation of 10 new (Figure 1) and 26 known compounds. Herein, we report the isolation, structure elucidation, and biological activity of these compounds.

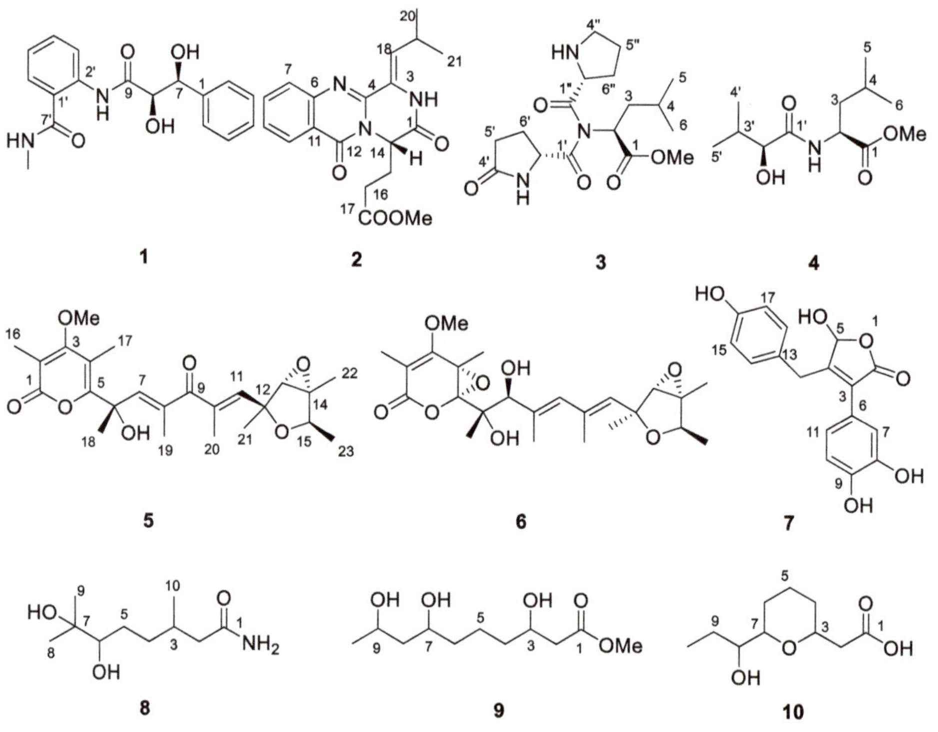

Figure 1. Compounds **1–10** from *Penicillium griseofulvum* MCCC 3A00225.

2. Results and Discussion

Compound **1** was isolated as a white powder. Its molecular formula was established as $C_{17}H_{18}N_2O_4$ according to the protonated molecule peak at m/z 337.1176 [M + Na]$^+$ in its (+)−HRESIMS (High Resolution Electrospray Ionization Mass Spectroscopy) spectrum, requiring ten degrees of unsaturation. The ^{1}H and ^{13}C NMR spectroscopic data (Figures S1 and S2, Table 1) displayed 17 carbons, characteristics of one mono-substituted aromatic unit [δ_H 7.24 (1H, br t, J = 7.4 Hz, H-4), 7.33 (2H, dd, J = 7.8, 7.3 Hz, H-3, 5), 7.46 (2H, d, J = 7.8 Hz, H-2, 6); δ_C 127.5 (d × 2, C-2/C-6), 128.3 (d, C-4), 129.1 (d × 2, C-3/C-5), 143.2 (s, C-1)], one ortho-disubstituted benzene moiety [δ_H 7.16 (1H, td, J = 7.6, 1.0 Hz, H-5′), 7.47 (1H, td, J = 7.8, 1.5 Hz, H-4′), 7.60 (1H, dd, J = 7.8, 1.4 Hz, H-6′), 8.51 (1H, d, J = 8.1 Hz, H-3′); δ_C 122.4 (d, C-3′), 124.2 (s, C-1′), 124.7 (d, C-5′), 128.8 (d, C-6′), 132.7 (d, C-4′), 138.8 (s, C-2′)], one methyl [δ_H 2.89 (3H, s, 7′-NMe); δ_C 26.8 (q, 7′-NMe)], two oxygenated methines [δ_H 4.25 (1H, d, J = 2.3 Hz, H-8); 5.16 (1H, d, J = 2.0 Hz, H-7) δ_C 75.6 (d, C-7), 77.8 (d, C-8)], and two carbonyls [δ_C 171.3 (s, C-7′), 174.0 (s, C-9)]. In the ^{1}H–^{1}H COSY (Correlation Spectroscopy) spectrum, correlations of H-2 (H-6)/H-3 (H-5)/H-4, H-3′/H-4′/H-5′/H-6′, and H-7 (δ_H 5.16, d, J = 2.0 Hz)/H-8 (δ_H 4.25, d, J = 2.3 Hz) confirmed the two benzene units and deduced another fragment of C-7/C-8. By the HMBC (Heteronuclear Multiple-bond Correlation) correlations of H-7 (δ_H 5.16) to C-1/C-2/C-6/C-9 and H-6′ (δ_H 7.60)/7′-NMe (δ_H 2.89) to C-7′, **1** was then assigned a phenylpropionyl moiety and a benzamide groups (Figure 2). However, the limited HMBC correlations hindered the connection of these two fragments. Fortunately, crystals of 1 were obtained. By the single X-ray crystallography (Figure 3), the absolute configuration of **1** was then unambiguously assigned as 2-(2*R*,3*S*-dihydroxy-3-phenyl-propionylamino)-*N*-methyl-benzamide, and named penigrisamide.

Table 1. ^{1}H (400 MHz) and ^{13}C (100 MHz) NMR spectroscopic data of **1**, **3**, **4**, **8**, and **9** in CD$_3$OD.

No.	1 δ_C	1 δ_H	3 δ_C	3 δ_H	4 δ_C	4 δ_H	8 δ_C	8 δ_H	9 δ_C	9 δ_H
1	143.2 C		174.7 C		174.5 C		178.8 C		173.7 C	
2	127.5 CH	7.46 (d, 7.8)	52.3 CH	4.41 (dd, 8.9, 6.2)	51.6 CH	4.52 (dd, 9.8, 4.8)	44.0 CH$_2$	2.24 (dd, 12.8, 5.2)1.98 m	42.2 CH$_2$	2.49 (dd, 15.1, 4.3)2.42 (dd, 15.1, 8.9)
3	129.1 CH	7.33 (dd, 7.8, 7.3)	41.4 CH$_2$	1.60 m	41.6 CH$_2$	1.67 m	32.1 CH	1.94 m	75.9 CH	3.79 (tdd, 8.9, 4.4, 2.0)
4	128.3 CH	7.24 (br t, 7.4)	25.9 CH	1.74 m	26.0 CH	1.68 m	35.4 CH$_2$	1.64 m; 1.22 m	32.1 CH$_2$	1.62 m; 1.22 m
5	129.1 CH	7.33 (dd, 7.8, 7.3)	23.3 CH$_3$	0.95 (d, 6.6)	23.3 CH$_3$	0.95 (d, 6.2)	29.6 CH$_2$	1.66 m; 1.23 m	24.3 CH$_2$	1.82 m; 1.58 m
6	127.5 CH	7.46 (d, 7.8)	21.9 CH$_3$	0.91 (d, 6.6)	21.7 CH$_3$	0.92 (d, 6.2)	79.8 CH	3.21 (d, 9.5)	32.5 CH$_2$	1.57 m; 1.21 m
7	75.6 CH	5.16 (d, 2.0)					73.8 C		78.5 CH	3.54 (tdd, 10.2, 3.7, 1.7)
8	77.8 CH	4.25 (d, 2.3)					25.8 CH$_3$	1.16 s	45.9 CH$_2$	1.61 m; 1.48 (dt, 14.0, 4.4)
9	174.0 C						24.8 CH$_3$	1.12 s	67.4 CH	3.94 m
10							20.2 CH$_3$	0.96 (d, 6.2)	23.1 CH$_3$	1.14 (d, 6.2)
1'	124.2 C		172.9 C		176.7 C					
2'	138.8 C		56.2 CH	4.55 (dd, 8.8, 4.0)	77.0 CH	3.86 (d, 3.7)				
3'	122.4 CH	8.51 (d, 8.1)			33.0 CH	2.07 m				
4'	132.7 CH	7.47 (td, 7.8, 1.5)	181.6 C		19.5 CH$_3$	1.00 (d, 7.0)				
5'	124.7 CH	7.16 (td, 7.6, 1.0)	30.3 CH$_2$	2.36 m; 2.30 m	16.3 CH$_3$	0.84 (d, 6.8)				
6'	128.8 CH	7.60 (dd, 7.8, 1.4)	25.5 CH$_2$	2.47 m; 2.16 m						
7' (1'')	171.3 C		174.4 C							
2''			61.3 CH	4.47 (dd, 8.4, 2.8)						
4''			48.1 CH$_2$	3.63 m						
5''			25.9 CH$_2$	2.02 m						
6''			30.3 CH$_2$	2.18 m; 2.00 m						
NMe/ OMe	26.8 CH$_3$	2.89 s	52.6 CH$_3$	3.69 s	52.7 CH$_3$	3.70 s			52.1 CH$_3$	3.65 s

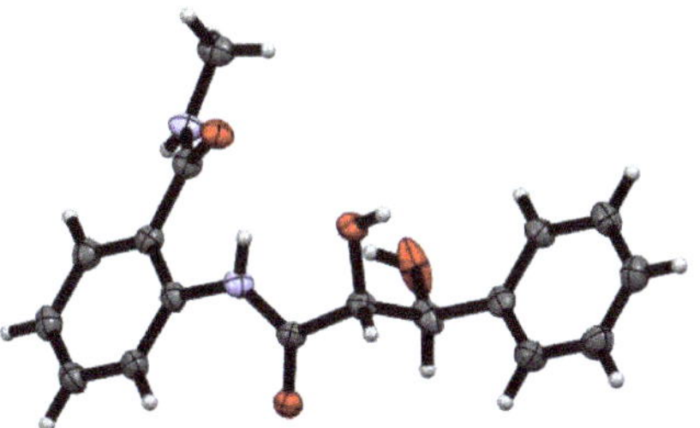

Figure 2. The key ^{1}H–^{1}H COSY (bold) and HMBC (arrow) correlations of **1**.

Figure 3. The X-ray crystallography of **1**.

Compound **2** was afforded as a colorless oil. The molecular formula C$_{19}$H$_{21}$N$_3$O$_4$ was deduced from (+)-HRESIMS data (*m/z* 378.1418 for [M + Na]$^+$), indicative of eleven degrees of unsaturation. The ^{1}H and ^{13}C NMR spectroscopic data (Figures S7 and S8 from the Supplementary Materials, Table 2) exhibited 19 carbons, including three methyl singlets (one oxygenated), two methylenes, seven methines (five olefinic), and seven non-protonated carbons (one carbonyl and two ketone groups). These signals were closely similar to those of aurantiomide C (11) [14], except that the terminal amino group in **11** was replaced by the methoxy unit (δ_C 52.2) in **2**. The assumption was confirmed by the

HMBC correlation of 17-OMe (δ_H 3.46) to C-17 (δ_C 173.9). Accordingly, the structure of **2** was determined as 17-deamino-17-methoxylaurantiomide C, and named aurantiomoate C.

Table 2. ^{1}H (400 MHz) and ^{13}C (100 MHz) NMR spectroscopic data of **2**, **5**, **6**, **7**, and **10**.

No.	2 [a] δ_C	2 [a] δ_H	5 [a] δ_C	5 [a] δ_H	6 [a] δ_C	6 [a] δ_H	7 [b] δ_C	7 [b] δ_H	10 [a] δ_C	10 [a] δ_H
1	167.7 C		167.1 C		167.9 C				175.9 C	
2			111.1 C		109.1 C		170.6 C		42.8 CH$_2$	2.41 (d, 6.5)
3	126.3 C		171.2 C		168.3 C		120.3 C		75.8 CH	3.76, m
4	147.0 C		113.4 C		80.4 C		157.7 C		32.2 CH$_2$	1.64, m; 1.21, m
5			160.8 C		109.1 C		96.5 CH	5.77 (d, 6.8)	24.6 CH$_2$	1.84, m; 1.59, m
6	121.1 C		75.2 C		82.1 C		127.8 C		32.8 CH$_2$	1.52, m; 1.21, m
7	128.4 CH	7.64 (d, 8.1)	147.7 CH	6.38 (d, 1.4)	90.5 CH	4.09 s	116.1 CH	6.92 (d, 1.8)	76.2 CH	3.57, m
8	136.0 CH	7.77 (td, 8.4, 1.4)	137.3 C		134.1 C		145.1 C		46.5 CH$_2$	1.48, m
9	128.0 CH	7.46 (t, 7.6)	202.8 C		133.3 CH	5.84 s	146.1 C		65.3 CH	3.93, m
10	127.6 CH	8.14 (dd, 8, 1.1)	139.3 C		136.6 C		115.6 CH	6.79 (d, 8.2)	23.9 CH$_3$	1.13 (d, 6.3)
11	148.7 C		143.7 CH	6.28 (d, 1.4)	133.2 CH	5.53 s	120.2 CH	6.75 (dd, 8.2, 1.8)		
12	162.1 C		81.3 C		81.6 C		31.0 CH$_2$	3.85 (d, 15.1); 3.57 (d, 15.1)		
13			68.0 CH	3.64 s	68.7 CH	3.55 s	126.5 C			
14	56.2 CH	5.34 (t, 6.6)	68.7 C		68.7 C		129.6 CH	6.99 (d, 8.4)		
15	28.7 CH$_2$	2.65 m; 2.15 m	78.5 CH	4.08 (dt, 6.8, 6.8)	78.3 CH	4.05 (d, 6.8)	115.5 CH	6.69 (d, 8.4)		
16	30.6 CH$_2$	2.44 m	14.6 CH$_3$	2.03 s	10.2 CH$_3$	1.84 s	156.1 C			
17	173.9 C		11.1 CH$_3$	2.09 s	20.8 CH$_3$	1.61 s	115.5 CH	6.69 (d, 8.4)		
18	129.6 CH	6.33 (d, 10.4)	27.0 CH$_3$	1.68 (d, 0.8)	19.7 CH$_3$	1.28 s	129.6 CH	6.99 (d, 8.4)		
19	27.1 CH	2.87 m	13.4 CH$_3$	1.67 s	15.8 CH$_3$	1.86 (d, 0.9)				
20	22.5 CH$_3$	1.13 (d, 6.6)	10.3 CH$_3$	2.02 (d, 1.4)	19.1 CH$_3$	1.93 s				
21	22.6 CH$_3$	1.16 (d, 6.6)	21.1 CH$_3$	1.38 s	22.1 CH$_3$	1.37 s				
22			13.7 CH$_3$	1.45 s	13.8 CH$_3$	1.45 s				
23			19.3 CH$_3$	1.17 (d, 6.8)	19.2 CH$_3$	1.20 (d, 6.8)				
OMe	52.2 CH$_3$	3.46 s	61.3 CH$_3$	3.87 s	61.1 CH$_3$	3.92 s				

[a] CD$_3$OD. [b] DMSO-d_6.

Compound **3** was obtained as a colorless oil. Its molecular formula was established as $C_{17}H_{27}N_3O_5$ on the basis of the protonated molecule peak at m/z 376.1841 [M + Na]$^+$ in its (+)-HRESIMS spectrum, requiring six degrees of unsaturation. Diagnostic NMR data for **3** suggested the presence of a pyroglutamylleucinmethylester (**20**) [15]. Moreover, the ^{1}H–^{1}H COSY correlation of H$_2$-4″ (δ_H 3.63 m)/H$_2$-5″ (δ_H 2.02 m) and H$_2$-6″ (δ_H 2.18 m, 2.00 m)/H-2″ (δ_H 4.47, dd, J = 8.4, 2.8 Hz), with HMBC correlations from H-2″ (δ_H 4.47, dd, J = 8.4, 2.8 Hz) to C-4″/C-5″, and H-6″ (δ_H 2.18 m, 2.00 m) to C-1″/C-4″, allowed for the presence of another pyroglutamyl moiety. The absolute configuration of **3** was determined by the hydrolysis and derivation using Marfey's reagent, and Nα-(2,4-dinitro-5-fluorophenyl)-l-alaninamide (FDDA) derivatives were compared with the retention times of standard FDDA-amino acids (Figure 4). On the basis of the above evidences, **3** was then assigned as *N,N*-pyroglutamylleucinmethylester.

Compound **4** was obtained as a colorless oil. Its molecular formula was established as $C_{12}H_{23}NO_4$ based on the sodium adduct ionic peak at m/z 268.1526 [M + Na]$^+$ in its positive HRESIMS spectrum, requiring two degrees of unsaturation. Its ^{1}H and ^{13}C NMR spectra were very similar to those of pyroglutamylleucinmethylester (**20**) [15], except for a 2-hydroxy-3-methylbutanoyl unit instead of a pyroglutamyl moiety in **4**. This was confirmed by the ^{1}H–^{1}H COSY correlations of H$_3$-4′ (δ_H 1.00, d, J = 7.0 Hz) and H$_3$-5′ (δ_H 0.84, d, J = 6.8 Hz) via H-3′ (δ_H 2.07 m) to H-2′ (δ_H 3.86, d, J = 3.7 Hz). Via detailed analysis of the HMBC spectroscopic data and using Marfey's method (Figure 5), the absolute configuration of **4** was then assigned as methyl-2S-hydroxy-3-methylbutanoyl-L-leucinate.

The molecular formula of **5** was established as $C_{24}H_{32}O_7$ by the ion peak at m/z 455.2040 [M + Na]$^+$ in its positive HRESIMS. The ^{1}H and ^{13}C NMR spectra exhibited 24 carbons, including three doublets and five singlet methyls, one methoxyl, four methines (two oxygenated and two olefinic), and eleven quaternary carbons (six olefinic and two carbonyl carbons). These signals were closely similar to those of penicyrone A [16] except that the hydroxy (δ_C 82.6) at the C-9 position in penicyrone A was replaced by the carbonyl (δ_C 202.8) in **5**. This was confirmed by the HMBC correlations from H-7 (δ_H 6.38, d, J = 1.4 Hz)/H-11(δ_H 6.28, d, J = 1.4 Hz)/H$_3$-19 (1.67, s)/H$_3$-20 (2.02, d, J = 1.4 Hz) to δ_C

202.8. Accordingly, **5** was established to be 9-dehydroxy-9-oxopenicyrone A, and named verrucosidinol A.

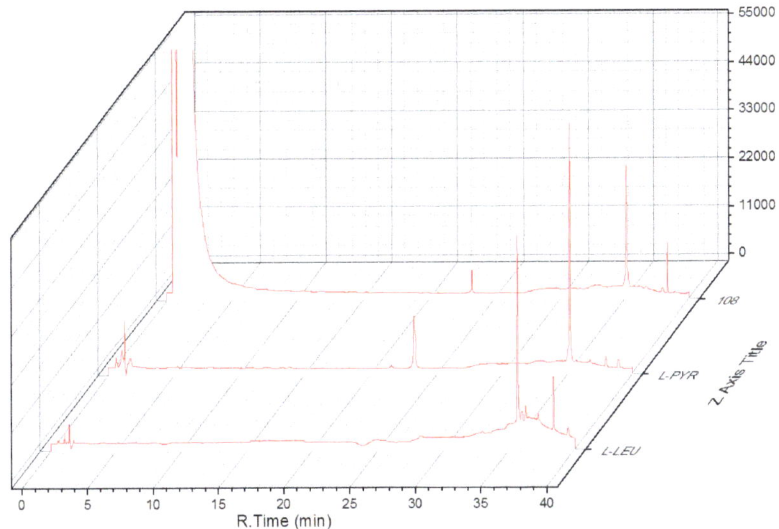

Figure 4. FDDA derivatives of **3** compared with the retention times of standard FDDA-amino acids.

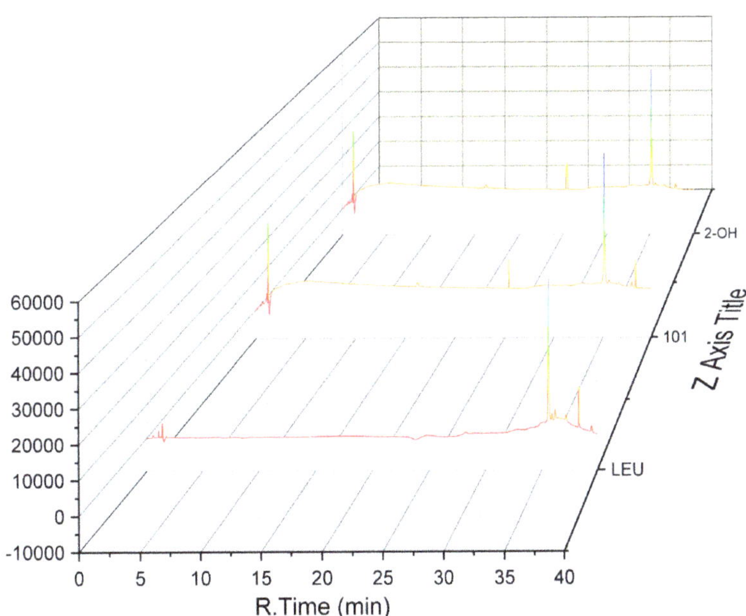

Figure 5. FDDA derivatives of **4** compared with the retention times of standard FDDA-amino acids.

Compound **6** presented a molecular formula of $C_{24}H_{34}O_8$ by positive HRESIMS at *m/z* 473.2140 [M + Na]$^+$. Comparison of the ^{1}H and ^{13}C NMR spectra of **6** with those of verrucosidinol (**25**) [17] showed they were very similar except that two olefinic carbons at C-4 and C-5 in **25** were replaced by an epoxy group in **6**. This was evidenced by the HMBC correlations from H$_3$-16 (δ_H 1.84) to C-1/C-2/C-3, H$_3$-17 (δ_H 1.61) to C-3/C-4/C-5, and H$_3$-18 (δ_H 1.28) to C-5/C-6/C-7. Therefore, **6** was established as 4,5-dihydro-4,5-epoxyverrucosidinol, and named verrucosidinol B.

Compound **7** had a molecular formula C$_{17}$H$_{14}$O$_6$ as assigned by its positive HRESIMS at *m/z* 337.0690 [M + Na]$^+$. Its ^{1}H and ^{13}C NMR spectroscopic data greatly resembled those of helvafuranone [18] except for an additional hydroxy substituent at the C-8 position. By detailed analysis of its 1D and 2D NMR spectroscopic data, **7** was then established as 8-hydroxyhelvafuranone.

Compound **8** gave a molecular formula C$_{10}$H$_{21}$NO$_3$ as deduced by the protonated molecule peak at *m/z* 202.1504 [M − H]$^-$ in its negative HRESIMS spectrum. The ^{1}H NMR spectrum exhibited one methyl doublet at δ_H 0.96 (3H, d, *J* = 6.2 Hz, H-10), and two methyl singlets at δ_H 1.12 (3H, s, H-9) and δ_H 1.16 (3H, s, H-8). The ^{13}C and DEPT spectra revealed the presence of 10 carbons, including three methyls, two methylenes, three methines, and one oxygenated and one carbonyl non-protonated carbon. In the ^{1}H–^{1}H COSY spectrum, correlations were found of H-6 via H-5 to H-4/H-3 and of H-3 to H$_3$-10/H-2. By the HMBC correlations of H$_2$-2 (δ_H 2.24, dd, *J* = 12.8, 5.2 Hz; 1.98, m) to C-1/C-4/C-10, H-6 (δ_H 3.21, d, *J* = 9.5 Hz) to C-4/C-7, and H$_3$-8 (δ_H 1.16, s)/H$_3$-9 (δ_H 1.12, s) to C-6/C-7, the planar structure of **8** was then established. To determine the absolute configuration of C-6, a dimolybdenum tetraacetate [Mo$_2$(OAc)$_4$]-induced circular dichroism (ICD) experiment was employed. The ICD spectrum exhibited a positive Cotton effect at 310 nm (Figure 6). The sign of the diagnostic band at about 310 nm was correlated to the absolute configuration of the chiral centers in the 1,2-diol moiety. According to the rule proposed by Snatzke, the positive sign suggested a positive torsional angle for the O-C-C-O moiety. It was ascertained that the 6*R*-form could maintain the favored conformation in which the bulkyl moiety and O-C-C-O center stayed away from each other. Based on the above evidence, the structure of **8** was then designated as 6*R*,7-dihydroxy-3,7-dimethyloctanamide.

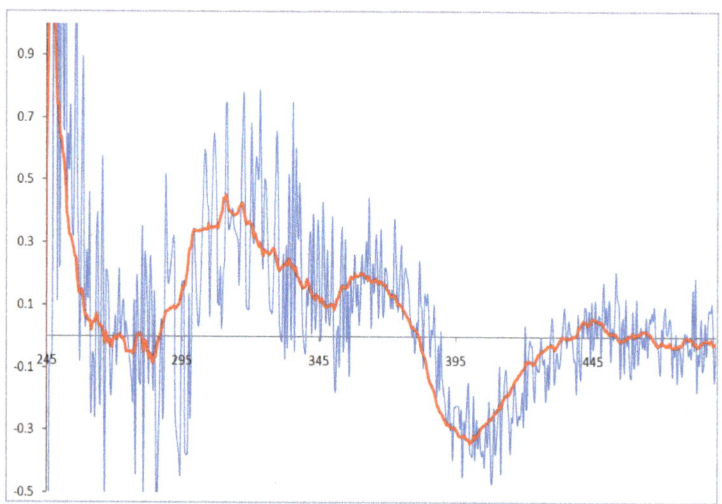

Figure 6. The induced CD spectrum of **8** in DMSO solution of Mo$_2$(OAc)$_4$.

Compound **9** was obtained as a white powder. The molecular formula C$_{11}$H$_{22}$O$_5$ was deduced from (+)-HRESIMS data at *m/z* 257.1237 ([M + Na]$^+$), indicative of one degree of unsaturation. The ^{1}H NMR spectrum showed a methyl at δ_H 1.14 (d, *J* = 6.2 Hz, H-10) and a methoxyl at δ_H 3.65 (s, H-11). The ^{13}C NMR and DEPT (Distortionless Enhancement by Polarization Transfer) data displayed 11 carbons, including one methyl, one methoxyl, five methylenes, three methines, and one carbonyl. In the ^{1}H–^{1}H COSY spectrum, two isolated spin systems were observed as H$_2$-2 (δ_H 2.49, 2.42)/H-3 (δ_H 3.79)/H$_2$-4 (δ_H 1.22)/H$_2$-5 (δ_H 1.82) and H$_2$-6 (δ_H 1.21)/H-7 (δ_H 3.54)/H$_2$-8 (δ_H 1.61, 1.48)/H-9 (δ_H 3.94)/H$_3$-10 (δ_H

1.14). These two fragments could be connected by the HMBC correlations of H_2-2 (δ_H 2.49, 2.42) and H-11(δ_H 3.65) to C-1 (δ_C 173.7). Therefore, **9** was established as methyl-3,7,9-trihydroxydecanate.

Compound **10** was obtained as a colorless oil. Its molecular formula was established as $C_{10}H_{18}O_4$ on the basis of the protonated molecule peak at m/z 225.1109 $[M + Na]^+$ in its positive HRESIMS spectrum, requiring two degrees of unsaturation. The ^{13}C NMR spectrum in association with the DEPT spectrum indicated 10 carbon signals ascribed to one methyl doublet (δ_C 23.9, C-10), five sp^3 methylenes (δ_C 42.8, C-2; 32.2, C-4; 24.6, C-5; 32.8, C-6; 46.5, C-8), three sp^3 methines (δ_C 75.8, C-3; 76.2, C-7; 65.3, C-9), and one carbonyl (δ_C 175.9, C-1). In the ^1H–^1H COSY spectrum, a long chain of C-2/C-3/C-4/C-5/C-6/C-7/C-8/C-9/C-10 could be deduced by correlations of H_2-2 (δ_H 2.41)/H-3 (δ_H 3.76)/H_2-4 (δ_H 1.64, 1.21)/H_2-5 (δ_H 1.84, 1.59)/H_2-6 (δ_H 1.52, 1.21)/H-7 (δ_H 3.57)/H_2-8 (δ_H 1.48)/H-9 (δ_H 3.93)/H_3-10 (δ_H 1.13). In the HMBC spectrum, H-3 (δ_H 3.76) was correlated to C-7 and C-1, which constructed a hexacyclic ring via an ether bond between C-1 and C-7. Accordingly, **10** was established as 9-hydroxy-3,7-epoxydecanoic acid.

By comparison of the NMR and MS data with those published in the literatures, 26 known compounds were determined to be aurantiomide C (**11**) [14], cyclopenin (**12**) [19], (−)-cyclopenol (**13**) [20], (3*S*)-1,4-benzodiazepine-2,5-diones (**14**) [21], 3-benzylidene-3,4-dihydro-4-methyl-lH-l,4-benzodiazepine-2,5-dione (**15**) [22], 3-methyl-3,4-dihydroquinazoline-4-one (**16**) [23], 1,2-dihydro-2,3-dimethyl4(3H)quinazolinone (**17**) [24], *N,N'*-1,2-phenylenebis-acetamide (**18**) [25], aconicarpyrazine B (**19**) [26], pyroglutamylleucinmethylester (**20**) [15], cyclo-(L-Trp-L-Phe) (**21**) [27], fructigenine A (**22**) [28], fructigenine B (**23**) [28], brevicompanine B (**24**) [29], verrucosidinol (**25**) [17], (*S*)-penipratynolene (**26**) [30], (*S*)-4-(2-hydroxybutynoxy)benzoic acid (**27**) [31], (*S*)-4-(2-hydroxybutoxy)benzoic acid (**28**) (CAS:1357392-03-0), (*S*)-2,4-dihydroxy-1-butyl(4- hydroxy)benzoate (**29**) [32], methyl *p*-hydroxybenzeneacetate (**30**) [33], 2-hydroxy phenyl acetic acid (**31**) [34], methyl homogentisate (**32**) [35], 5-hydroxymethyl-furaldehyde (**33**) [36], leptosphaerone A (**34**) [37], 3-methyl-2-penten-5-olide (**35**) [38], and (*R*)-mevalonolactone (**36**) [39].

All isolated compounds (**1−36**) were evaluated for their antifood allergic activities in RBL-2H3 cells. Compound **13** showed potent degranulation-inhibitory activity with an IC_{50} value of 60.3 μM, which was stronger than the commercially available antifood allergy medicine, loratadine (IC_{50} = 91.6 μM), while **14** and **29** showed weak effects with IC_{50} values of 167.0 and 134.0 μM, respectively (Table 3).

Table 3. Inhibition effects of compounds **1–36** on RBL-2H3 cell degranulation (*n* = 3).

Compound	IC_{50} (μM)
13	60.3
14	167.0
29	134.0
Others [a]	≥ 200
Loratadine [b]	91.6

[a] Other compounds, including **1–12**, **15–28**, and **30–36**. [b] Loratadine was a commercially available anti-food allergic medicine.

3. Materials and Methods

3.1. General Experimental Procedures and Fungal Fermentation

Penicillium griseofulvum, isolated from a sediment sample of the Indian Ocean at a depth of 1420 m, was deposited at the Marine Culture Collection of China (MCCC) with the accession number MCCC 3A00225. It was cultivated on corn medium in 100 × 1 L Erlenmeyer flasks for 62 days. The detailed general experimental procedures, fungal fermentation, and extraction were reported previously [12].

3.2. Isolation and Purification

The defatted extract (55.4 g) was separated by column chromatography (CC) over silica gel (500 g) using a CH_2Cl_2-MeOH gradient (0→100%, 49 mm × 460 mm) to give six fractions (Fr.1−Fr.6). Fr.2 (1.9 g) was subjected to ODS (octadecylsilyl) (H_2O-MeOH, 5→100%, 15 × 460 mm, 0.5 L for each fraction) to attain five subfractions (sfrs) (sfrs.2.1–sfrs.2.5). Sfr.2.3 (155.0 mg) was purified by column chromatography on Sephadex LH-20 (100 g) (MeOH, 2.0 × 120 cm, 300 mL) to afford **26** (12.7 mg). Fr.3 (2.1 g) was subjected to column chromatography (CC) on ODS (70 g) (H_2O-MeOH, 5→100%, 15 × 460 mm, 0.5 L for each fraction) to attain eleven subfractions (sfrs) (sfrs.3.1–sfrs.3.11). Sfr.3.3 (111.6 mg) was subjected to CC over Sephadex LH-20 (70 g) (MeOH, 2.0 × 120 cm, 300 mL) and silica gel (PE-EtOAc, 2:1, 17 × 305 mm) to yield **35** (25.7 mg). Sfr.3.5 (131.2 mg) was chromatographed on Sephadex LH-20 (100 g) (MeOH, 2.0 cm × 180 cm, 500 mL) resulting in two sub-subfractions (ssfrs) (ssfrs.3.5.1− ssfrs.3.5.2). Ssfr.3.5.1 (3.8 mg) was further purified by HPLC using gradient MeOH-H_2O (20→70%, 10 × 250 mm, 4 mL/min) to provide **17** (2.4 mg). Ssfr.3.5.2 (41.0 mg) was purified using preparative TLC (CH_2Cl_2-Me2CO, 20:1) to give **16** (16.6 mg). Compound **15** (30.7 mg) was isolated from Sfr.3.6 (66.9 mg) by CC over Sephadex LH-20 (70 g) (MeOH, 2.0 × 120 cm, 300 mL). Sfr.3.8 (52.4 mg) was chromatographed on a Sephadex LH-20 (70 g) (MeOH, 2.0 × 120 cm, 300 mL) to give two sub-subfractions (ssfrs) (ssfrs.3.8.1− ssfrs.3.8.2), ssfrs.3.8.1 and ssfrs.3.8.2 were purified by preparative TLC on silica gel (CH_2Cl_2-MeOH, 20:1) to provide **24** (4.9 mg) and **9** (1.7 mg), respectively. Fr.4 (4.9 g) was subjected to ODS (130 g) (H_2O-MeOH, 10→100%, 26 × 310 mm, 1.5 L for each fraction) to obtain twelve subfractions (sfrs) (sfrs.4.1− sfrs.4.12). Compound **12** (216.4 mg) was isolated from sfr.4.1 (304.0 mg) by CC over Sephadex LH-20 (100 g) (MeOH, 2.0 × 180 cm, 500 mL). Sfr.4.2 (906.0 mg) was chromatographed on a Sephadex LH-20 (225 g) column (MeOH, 3.5 × 180 cm, 800 mL) and silica gel (PE-EtOAc, 2:1, 46 × 457 mm) to yield **36** (152.9 mg). Sfr.4.3 (644.0 mg) was fractionated by CC over Sephadex LH-20 (225 g) (MeOH, 3.5 × 180 cm, 800 mL) to attain three sub-subfractions (ssfrs) (ssfrs.4.3.1–ssfrs.4.3.3), ssfr.4.3.3 (78.2 mg) was purified by Sephadex LH-20 (70 g) (MeOH, 2.0 × 120 cm, 200 mL), followed by preparative TLC (CH_2Cl_2- Me2CO, 10:1) to provide **33** (10.0 mg) and **34** (10.5 mg). Sfr.4.4 (270.9 mg) was subjected to CC over Sephadex LH-20 (100 g) (MeOH, 2.0 × 180 cm, 500 mL), further purified using preparative TLC (PE-EtOAc, 1:2) to obtain **14** (48.6 mg). Sfr.4.5 (33.3 mg) was purified by Sephadex LH-20 (70 g) (MeOH, 2.0 cm × 120 cm, 300 mL) to yield **18** (8.6 mg). Sfr.4.6 (270.9 mg) and sfr.4.7 (342.5 mg) were subjected to CC over Sephadex LH-20 (225 g) (MeOH, 3.5 × 180 cm, 800 mL) to attain **32** (4.0 mg) and **31** (2.9 mg), respectively. Sfr.4.9 and sfr.4.10 (376.6 mg) were fractionated by CC on Sephadex LH-20 (225 g) (MeOH, 3.5 × 180 cm, 800 mL) to obtain four sub-subfractions (ssfrs) (ssfrs.4.10.1–ssfrs.4.10.4). Ssfr.4.10.1 (191.0 mg) was subjected to Sephadex LH-20 (100 g) (MeOH, 2.0 × 180 cm, 500 mL) to attain **22** (117.2 mg), while **28** (3.5 mg) was isolated from ssfr.4.10.3 (10.3 mg) by preparative TLC (CH_2Cl_2-MeOH, 5:1). Sfr.4.12 (239.5 mg) was chromatographed on Sephadex LH-20 (100 g) (MeOH, 2.0 cm × 180 cm, 500 mL), further purified using preparative TLC (PE-EtOAc, 2:1) to yield **23** (46.2 mg). Fr.5 (40.0 g) separated by column chromatography (CC) over ODS (650 g) (H_2O-MeOH, 5→80%, 49 × 460 mm, 3 L for each fraction) to obtain fifteen subfractions (sfrs.5.1–sfrs.5.15). Sfr.5.2 (1.7 g) was separated by CC over Sephadex LH-20 (225 g) (CH_2Cl_2-MeOH, 1:1, 3.5 × 180 cm, 1000 mL) to give three sub-subfractions (ssfrs) (ssfrs.5.2.1− ssfrs.5.2.3), ssfr.5.2.2 (126.0 mg) was subjected to Sephadex LH-20 (100 g) (MeOH, 2.0 × 180 cm, 500 mL), followed by preparative TLC (CH_2Cl_2-MeOH, 20:1) to provide **19** (2.9 mg). Ssfr.5.2.3 (103.0 mg) was purified by preparative TLC (CH_2Cl_2-MeOH, 10:1) to attain **29** (8.6 mg). Sfr.5.3 (625.0 mg) was subjected to CC over Sephadex LH-20 (225 g) (MeOH, 3.5 × 180 cm, 800 mL) to furnish five sub-subfractions (ssfrs) (ssfrs.5.3.1–ssfrs.5.3.5), ssfr.5.3.1 (228.0 mg) was separated by silica gel (CH_2Cl_2-MeOH 50:1→10:1, 46 mm × 305 mm), then subjected to HPLC (MeOH-H_2O, 55→65%, 10 × 250 mm, 5 mL/min) to yield **20** (22.8 mg). Compounds **27** (9.3 mg) and **30** (5.1 mg) were isolated from ssfr.5.3.3 (54.0 mg) and ssfr.5.3.5 (29.9 mg) by preparative TLC (CH_2Cl_2-MeOH, 20:1), respectively, while **10** (3.7 mg) was isolated from ssfr.5.3.4

(29.5 mg) by preparative TLC (EtOAc -MeOH, 50:1), and further purified by preparative TLC (CH$_2$Cl$_2$-MeOH, 20:1). Sfr.5.4 (3.3 g) was fractionated by CC over Sephadex LH-20 (225 g) (3.5 × 180 cm, CH$_2$Cl$_2$-MeOH 1:1, 1200 mL) to attain five sub-subfractions (ssfrs) (ssfrs.5.4.1−ssfrs.5.4.5), ssfr.5.4.2 (73.0 mg) was purified by preparative TLC (CH$_2$Cl$_2$-MeOH, 20:1) to provide **3** (11.0 mg). Ssfr.5.4.3 (1.6 g) was subjected to CC over Sephadex LH-20 (225 g) (3.5 × 180 cm, MeOH, 1200 mL) and preparative TLC (CH$_2$Cl$_2$-MeOH, 20:1) to yield **11** (29.4 mg). Sfr.5.5 (484.0 mg) was subjected to HPLC (MeOH-H$_2$O, 20→40%, 10 × 250 mm, 5 mL/min), followed by preparative TLC on silica gel (CH$_2$Cl$_2$-MeOH, 10:1) to attain **7** (4.5 mg), **13** (34.1 mg), and **8** (4.2 mg). Sfr.5.7 (180 mg) was chromatographed on a Sephadex LH-20 (100 g) (MeOH, 2.0 × 180 cm, 500 mL), and then subjected to preparative TLC (CH$_2$Cl$_2$-MeOH, 20:1) to obtain **1** (1.5 mg). Sfr.5.11 (753.0 mg) was purified by CC over repeated Sephadex LH-20 (225 g) (MeOH, 3.5 × 180 cm, 800 mL) to obtain four sub-subfractions (ssfrs) (ssfrs.5.11.1- ssfrs.5.11.4), **21** (30.9 mg) was isolated from ssfr.5.11.2 (127.5 mg) by preparative TLC on silica gel using CH$_2$Cl$_2$-MeOH (10:1), while **5** (6.1 mg) was isolated from ssfr.5.11.3 (235.7 mg) by preparative TLC on silica gel (PE-EtOAc, 1:1). Sfr.5.12 (3.5 g) was separated by CC over SephadexLH-20 (CH$_2$Cl$_2$-MeOH, 1:1, 3.5 × 180 cm, 1200 mL) to attain three sub-subfractions (ssfrs) (ssfrs.5.12.1–ssfrs.5.12.3). Ssfr.5.12.1 (489.0 mg) was purified by Sephadex LH-20 (225 g) (MeOH, 3.5 × 180 cm, 800 mL) and silica gel (PE-EtOAc, 5:1→1:1, 46 × 305 mm), finally, by preparative TLC (CH$_2$Cl$_2$-MeOH, 10:1) to provide **6** (6.9 mg) and **25** (22.9 mg). Ssfr.5.12.2 (1.6 g) was purified by CC over repeated Sephadex LH-20 (225 g) (MeOH, 3.5 × 180 cm, 1000 mL) and preparative TLC (CH$_2$Cl$_2$-MeOH, 20:1) to yield **4** (3.1 mg) and **2** (24.6 mg).

Penigrisamide (**1**): Colorless needles; $[\alpha]_D^{25}$ +34.5 (c 0.20, MeOH); UV (MeOH) λ_{max} (log ε) 212 (3.03), 252 (2.77) nm; ECD (ACN) $\Delta\varepsilon_{195}$ +3.67, $\Delta\varepsilon_{203}$ +1.78, $\Delta\varepsilon_{203}$ +1.78, $\Delta\varepsilon_{213}$ +4.40, $\Delta\varepsilon_{225}$ −0.62, $\Delta\varepsilon_{250}$ +1.98; ^{1}H and ^{13}C NMR data, see Table 1; (+)-HRESIMS m/z 337.1176 [M + Na]$^+$ (calculated for C$_{17}$H$_{18}$N$_2$O$_4$Na, 337.1164).

Aurantiomoate C (**2**): Colorless oil; $[\alpha]_D^{25}$ −20.8 (c 1.20, MeOH), $[\alpha]_D^{25}$+19.4 (c 1.20, CHCl$_3$); UV (MeOH) λ_{max} (log ε) 211 (4.40), 305 (3.94) nm; ECD (ACN) $\Delta\varepsilon_{191}$ −20.6, $\Delta\varepsilon_{228}$ +10.7, $\Delta\varepsilon249$ −7.66, $\Delta\varepsilon_{272}$ −1.60, $\Delta\varepsilon_{294}$ −2.81, $\Delta\varepsilon_{330}$ +2.17; ^{1}H and ^{13}C NMR data, see Table 2; (+)-HRESIMS m/z 378.1418 [M + Na]$^+$ (calculated for C$_{19}$H$_{21}$N$_3$O$_4$Na, 378.1430).

5-Deoxypyroglutamyl-pyroglutamylleucinmethylester (**3**): colorless oil; $[\alpha]_D^{25}$−85.6 (c 0.27, MeOH); UV (MeOH) λ_{max} (log ε) 205 (3.77) nm; ECD (ACN) $\Delta\varepsilon_{217}$ +1.96, $\Delta\varepsilon_{235}$ −0.39, $\Delta\varepsilon_{249}$ +0.16; ^{1}H and ^{13}C NMR data, see Table 1; (+)-HRESIMS m/z 376.1841 [M + Na]$^+$ (calculated for C$_{17}$H$_{27}$N$_3$O$_5$Na, 376.1848).

Methyl-2-hydroxy-3-methylbutanoyl-L-leucinate (**4**): colorless oil; $[\alpha]_D^{25}$ −42.9 (c 0.27, MeOH); UV (MeOH) λ_{max} (log ε) 203 (3.31) nm; ECD (ACN) $\Delta\varepsilon_{210}$ +0.98, $\Delta\varepsilon_{234}$ −0.11; ^{1}H and ^{13}C NMR data, see Table 1; (+)-HRESIMS m/z 268.1526 [M + Na]$^+$ (calculated for C$_{12}$H$_{23}$NO$_4$Na, 268.1525).

Verrucosidinol A (**5**): Colorless oil; $[\alpha]_D^{20}$ +86.8 (c 0.22, MeOH), $[\alpha]_D^{25}$ +82.7 (c 0.22, MeOH); UV (MeOH) λ_{max} (log ε) 205 (4.13), 231 (4.00), 298 (3.67) nm; ECD (ACN) $\Delta\varepsilon_{187}$ +1.57, $\Delta\varepsilon_{205}$ −7.27, $\Delta\varepsilon_{296}$ +7.91; ^{1}H and ^{13}C NMR data, see Table 2; (+)-HRESIMS m/z 455.2040 [M + Na]$^+$ (calculated for C$_{24}$H$_{32}$O$_7$Na, 455.2046).

Verrucosidinol B (**6**): Colorless oil; $[\alpha]_D^{20}$ + 32.3 (c 0.35, MeOH), $[\alpha]_D^{25}$ +34.6 (c 0.35, MeOH); UV (MeOH) λ_{max} (log ε) 240 (3.98) nm; ECD (ACN) $\Delta\varepsilon_{195}$ +0.66, $\Delta\varepsilon_{214}$ −0.89, $\Delta\varepsilon_{254}$ +4.30; ^{1}H and ^{13}C NMR data, see Table 2; (+)-HRESIMS m/z 473.2140 [M + Na]$^+$ (calculated for C$_{24}$H$_{34}$O$_8$Na, 473.2151).

8-Hydroxyhelvafuranone (**7**): Colorless oil; $[\alpha]_D^{25}$ − 16.7 (c 0.03, MeOH); UV (MeOH) λ_{max} (log ε) 204 (4.28) nm; ECD (MeOH) $\Delta\varepsilon_{193}$ +2.23; ^{1}H and ^{13}C NMR data, see Table 2; (+)-HRESIMS m/z 337.0690 [M + Na]$^+$ (calculated for C$_{17}$H$_{14}$O$_6$Na, 337.0688).

6,7-Dihydroxy-3,7-dimethyloctanamide (**8**): Colorless oil; $[\alpha]_D^{25}$ −7.3 (c 0.15, MeOH); UV (MeOH) λ_{max} (log ε) 203 (3.09) nm; ECD (MeOH) $\Delta\varepsilon_{225}$ +0.02; ^{1}H and ^{13}C NMR data, see Table 1; (−)-HRESIMS m/z 202.1504 [M − H]$^-$ (calculated for C$_{10}$H$_{20}$NO$_3$, 202.1443).

Methyl-3,7,9-trihydroxydecanate (**9**): White powder; $[\alpha]_D^{20}$ −6.8 (c 0.19, MeOH), $[\alpha]_D^{20}$ −8.9 (c 0.19, CHCl$_3$); UV (MeOH) λ_{max} (log ε) 205 (2.21) nm; ECD (MeOH) $\Delta\varepsilon_{210}$ +0.11; ^{1}H and ^{13}C NMR data, see Table 1; (+)-HRESIMS*m/z* 257.1237 [M + Na]$^+$.

9-Hydroxy-3,7-epoxydecanoic acid (**10**): Colorless oil; $[\alpha]_D^{25}$ +15.7 (c 0.21, MeOH); UV (MeOH) λ_{max} (log ε) 205 (3.10) nm; ECD (MeOH) $\Delta\varepsilon_{211}$ +0.18; ^{1}H and ^{13}C NMR data, see Table 2; (+)-HRESIMS *m/z* 225.1109[M + Na]$^+$ (calculated for C$_{10}$H$_{18}$O$_4$Na, 225.1103).

3.3. X-ray Crystallography of 1

Compound **1** was obtained as colorless needles from MeOH. Its crystallographic data were measured by an Xcalibur and Gemini single-crystal diffractometer with Cu Kα radiation (λ = 1.54184 Å). Space group P2$_1$2$_1$2$_1$, a = 4.7555(2) Å, b = 14.7379(7) Å, c = 22.971(1) Å, α = β = γ = 90°, V = 1609.95(12) Å^3, Z = 4, D$_{calcd}$ = 1.371 mg/cm^3; μ = 0.847 mm^{-1}, F (000) = 704. The final R indicates R = 0.0484 (2682), wR_2 = 0.1337 (3174). Crystallographic data of 1 have been deposited in the Cambridge Crystallographic Data Center, with deposition number 2072655. Copies of the data can be obtained, free of charge, on application to CCDC, 12 Union Road, Cambridge CB21EZ, U.K. (fax +44(0)-1233-336033; email: deposit@ccdc.cam.ac.uk).

3.4. Maryer's Method

As reported [40], compounds **3** and **4** (each for 1 mg) were separately dissolved in HCl (1 mL) and incubated for 24 h. The hydrolysate was dried and dissolved in acetone. Then NaHCO$_3$ and FDAA were added to incubate for 1 h. After being cooled, the mixture was dissolved in 50% aqueous CH$_3$CN to yield FDDA derivatives. The corresponding standard amino acids were treated with the same procedures. The FDAA derivates were analyzed by HPLC at 254 and 340 nm by comparing the retention times with those of standards.

3.5. Induced CD (ICD) Experiment

Compound **8** and dimolybdenum tetracetate [Mo$_2$(OAc)$_4$] were resolved in dried DMSO. Their CD spectra were recorded immediately. Then the ICD spectra were measured every 3 min until they were stationary. The inherent CD data of compound **8** was subtracted to provide its induced CD spectrum as described previously [41,42].

3.6. Anti-Food Allergic Experiment

The in vitro anti-food allergic experiment was conducted according to the reported method [43]. Briefly, IgE-sensitized RBL-2H3 cells were treated with tested compounds for 1 h. Then cells were stimulated with dinitrophenyl-bovine serum albumin. The bioactivities were quantified by measuring the fluorescence intensity of the hydrolyzed substrate in an Infinite M200PRO fluorometer (Tecan, Zurich, Switzerland). Phosphate-buffered saline (PBS) buffer and loratadine were used as negative and positive controls, respectively.

4. Conclusions

From the deep sea-derived fungus *Penicillium griseofulvum* MCCC 3A00225, 10 new and 26 known compounds were obtained. The structures of the new compounds were determined by extensive analysis of their NMR and HRESIMS spectra, the absolute configurations were confirmed by different methods including the single X-ray crystallography, Marfey's method, and ICD experiment etc. (−)-Cyclopenol (**13**) showed the strongest in vitro anti-food allergic activity with an IC$_{50}$ value of 60.3 μM in IgE-mediated RBL-2H3 cells.

Supplementary Materials: The following are available online at https://www.mdpi.com/article/10.3390/md19040224/s1, Figures S1–S60: 1D and 2D NMR spectra of **1**−**10**.

Author Contributions: X.-W.Y. designed the project; C.-P.X. isolated all compounds. Q.L. and G.L. performed the bioactive experiments. Z.S. provided the fungus. C.-L.X. conducted fermentation. D.C. and L.-Z.L. performed the ICD and Marfey's methods. T.-H.Z. obtained NMR data. C.-P.X.,

L.-Z.L., and X.-W.Y. wrote the paper, while critical revision of the publication was performed by all authors. All authors have read and agreed to the published version of the manuscript.

Funding: The work was supported by grants from the National Natural Science Foundation of China (21877022), the COMRA program (DY135-B2-08), and the Xiamen Southern Oceanographic Center (17GYY002NF02).

Conflicts of Interest: The authors declare no conflict of interest.

References

1. Carroll, A.R.; Copp, B.R.; Davis, R.A.; Keyzers, R.A.; Prinsep, M.R. Marine natural products. *Nat. Prod. Rep.* **2020**, *37*, 175–223. [CrossRef]
2. Tang, Y.; Liu, Y.; Ruan, Q.; Zhao, M.; Zhao, Z.; Cui, H. Aspermeroterpenes A–C: Three meroterpenoids from the marine-derived fungus *Aspergillus terreus* GZU-31-1. *Org. Lett.* **2020**, *22*, 1336–1339. [CrossRef] [PubMed]
3. Jiao, W.H.; Xu, Q.H.; Ge, G.B.; Shang, R.Y.; Zhu, H.R.; Liu, H.Y.; Cui, J.; Sun, F.; Lin, H.W. Flavipesides A–C, PKS-NRPS hybrids as pancreatic lipase inhibitors from a marine sponge symbiotic fungus *Aspergillus flavipes* 164013. *Org. Lett.* **2020**, *22*, 1825–1829. [CrossRef]
4. Carroll, A.R.; Copp, B.R.; Davis, R.A.; Keyzers, R.A.; Prinsep, M.R. Marine natural products. *Nat. Prod. Rep.* **2021**, *38*, 362–413. [CrossRef]
5. Xie, C.L.; Zhang, D.; Lin, T.; He, Z.H.; Yan, Q.X.; Cai, Q.; Zhang, X.K.; Yang, X.W.; Chen, H.F. Antiproliferative sorbicillinoids from the deep-sea-derived *Penicillium allii-sativi*. *Front. Microbiol.* **2021**, *11*, 636948. [CrossRef] [PubMed]
6. Kong, F.D.; Fan, P.; Zhou, L.M.; Ma, Q.Y.; Xie, Q.Y.; Zheng, H.Z.; Zheng, Z.H.; Zhang, R.S.; Yuan, J.Z.; Dai, H.F.; et al. Penerpenes A–D, four indole terpenoids with potent protein tyrosine phosphatase inhibitory activity from the marine-derived fungus *Penicillium* sp. KFD28. *Org. Lett.* **2019**, *21*, 4864–4867. [CrossRef]
7. Frank, M.; Hartmann, R.; Plenker, M.; Mándi, A.; Kurtán, T.; Özkaya, F.C.; Müller, W.E.G.; Kassack, M.U.; Hamacher, A.; Lin, W.; et al. Brominated azaphilones from the sponge-associated fungus *Penicillium canescens* strain 4.14.6a. *J. Nat. Prod.* **2019**, *82*, 2159–2166. [CrossRef] [PubMed]
8. Niu, S.; Fan, Z.; Xie, C.L.; Liu, Q.; Luo, Z.H.; Liu, G.; Yang, X.W. Spirograterpene A, a tetracyclic spiro-diterpene with a fused 5/5/5/5 ring system from the deep-sea-derived fungus *Penicillium granulatum* MCCC 3A00475. *J. Nat. Prod.* **2017**, *80*, 2174–2177. [CrossRef]
9. Zou, Z.B.; Zhang, G.; Li, S.M.; He, Z.H.; Yan, Q.X.; Lin, Y.K.; Xie, C.L.; Xia, J.M.; Luo, Z.H.; Luo, L.Z.; et al. Asperochratides A–J, Ten new polyketides from the deep-sea-derived *Aspergillus ochraceus*. *Bioorg. Chem.* **2020**, *105*, 104349. [CrossRef]
10. Niu, S.; Xie, C.L.; Xia, J.M.; Liu, Q.M.; Peng, G.; Liu, G.M.; Yang, X.W. Botryotins A–H, Tetracyclic diterpenoids representing three carbon skeletons from a deep-sea-derived *Botryotinia fuckeliana*. *Org. Lett.* **2019**, *22*, 580–583. [CrossRef]
11. Xie, C.L.; Liu, Q.; He, Z.H.; Gai, Y.B.; Zou, Z.B.; Shao, Z.Z.; Liu, G.M.; Chen, H.F.; Yang, X.W. Discovery of andrastones from the deep-sea-derived *Penicillium allii-sativi* MCCC 3A00580 by OSMAC strategy. *Bioorg. Chem.* **2021**, *108*, 104671. [CrossRef]
12. Xing, C.P.; Xie, C.L.; Xia, J.M.; Liu, Q.M.; Lin, W.X.; Ye, D.Z.; Liu, G.M.; Yang, X.W. Penigrisacids A–D, four new sesquiterpenes from the deep-sea-derived *Penicillium griseofulvum*. *Mar. Drugs* **2019**, *17*, 507. [CrossRef] [PubMed]
13. Shu, Z.; Liu, Q.; Xing, C.; Zhang, Y.; Zhou, Y.; Zhang, J.; Liu, H.; Cao, M.; Yang, X.; Liu, G. Viridicatol isolated from deep-sea *Penicillium griseofulvum* alleviates anaphylaxis and repairs the intestinal barrier in mice by suppressing mast cell activation. *Mar. Drugs* **2020**, *18*, 517. [CrossRef] [PubMed]
14. Xin, Z.H.; Fang, Y.C.; Du, L.; Zhu, T.J.; Duan, L.; Chen, J.; Gu, Q.Q.; Zhu, W.M. Aurantiomides A-C, quinazoline alkaloids from the sponge-derived fungus *Penicillium aurantiogriseum* SP0-19. *J. Nat. Prod.* **2007**, *70*, 853–855. [CrossRef] [PubMed]
15. Orlowska, A.; Witkowska, E.; Izdebski, J. Sequence dependence in the formation of pyroglutamyl peptides in solid phase peptide synthesis. *Int. J. Pept. Protein Res.* **2009**, *30*, 141–144. [CrossRef]
16. Bu, Y.Y.; Yamazaki, H.; Takahashi, O.; Kirikoshi, R.; Ukai, K.; Namikoshi, M. Penicyrones A and B, an epimeric pair of α-pyrone-type polyketides produced by the marine-derived *Penicillium* sp. *J. Antibiot.* **2015**, *69*, 57–61. [CrossRef]
17. Yu, K.; Ren, B.; Wei, J.; Chen, C.; Sun, J.; Song, F.; Dai, H.; Zhang, L. Verrucisidinol and verrucosidinol acetate, two pyrone-type polyketides isolated from a marine derived fungus, *Penicillium aurantiogriseum*. *Mar. Drugs* **2010**, *8*, 2744–2754. [CrossRef]
18. Furukawa, T.; Fukuda, T.; Nagai, K.; Uchida, R.; Tomoda, H. Helvafuranone produced by the fungus *Aspergillus nidulans* BF0142 isolated from hot spring-derived soil. *Nat. Prod. Commun.* **2016**, *11*, 1001–1003. [CrossRef]
19. Hodge, R.P.; Harris, C.M.; Harris, T.M. Verrucofortine, a major metabolite of *Penicillium verrucosum* var. cyclopium, the fungus that produces the mycotoxin verrucosidin. *J. Nat. Prod.* **1988**, *51*, 66–73. [CrossRef]
20. Fremlin, L.J.; Piggott, A.M.; Lacey, E.; Capon, R.J. Cottoquinazoline A and cotteslosins A and B, metabolites from an Australian marine-derived strain of *Aspergillus versicolor*. *J. Nat. Prod.* **2009**, *72*, 666–670. [CrossRef]
21. Sugimori, T.; Okawa, T.; Eguchi, S.; Kakehi, A.; Yashima, E.; Okamoto, Y. The first total synthesis of (-)-benzomalvin A and benzomalvin B via the intramolecular aza-wittig reactions. *Tetrahedron* **1998**, *54*, 7997–8008. [CrossRef]
22. Sun, W.N.; Chen, X.T.; Tong, Q.Y.; Zhu, H.C.; He, Y.; Lei, L.; Xue, Y.B.; Yao, G.M.; Luo, Z.W.; Wang, J.P.; et al. Novel small molecule 11 β-HSD1 inhibitor from the endophytic fungus *Penicillium commune*. *Sci. Rep.* **2016**, *6*, 1–10. [CrossRef]

23. Spulak, M.; Pourova, J.; Voprsalova, M.; Mikusek, J.; Kunes, J.; Vacek, J.; Ghavre, M.; Gathergood, N.; Pour, M. Novel bronchodilatory quinazolines and quinoxalines: Synthesis and biological evaluation. *Eur. J. Med. Chem.* **2014**, *74*, 65–72. [CrossRef]

24. Moehrle, H.; Seidel, C.M. ChemInform Abstract: Dehydrogenation of *N*-secondary cyclic aminals and acylaminals. *Chem. Inf.* **1976**, *7*, 471–479. [CrossRef]

25. Park, J.; Lee, J.; Chang, S. Iterative C−H functionalization leading to multiple amidations of anilides. *Angew. Chem. Int. Ed.* **2017**, *56*, 4256–4260. [CrossRef] [PubMed]

26. Guo, L.; Peng, C.; Dai, O.; Geng, Z.; Guo, Y.P.; Xie, X.F.; He, C.J.; Li, X.H. Two new pyrazines from the parent roots of *Aconitum carmichaelii*. *Biochem. Syst. Ecol.* **2013**, *48*, 92–95. [CrossRef]

27. Kimura, Y.; Tani, K.; Kojima, A.; Sotoma, G.; Okada, K.; Shimada, A. Cyclo-(l-tryptophyl-l-phenylalanyl), a plant growth regulator produced by the fungus *Penicillium* sp. *Phytochemistry* **1996**, *41*, 665–669. [CrossRef]

28. Arai, K.; Kimura, K.; Mushiroda, T.; Yamamoto, Y. Structures of fructigenines A and B, new alkaloids isolated from *Penicillium fructigenum* takeuchi. *Chem. Pharm. Bull.* **1989**, *37*, 2937–2939. [CrossRef]

29. Kusano, M.; Sotoma, G.; Koshino, H.; Uzawa, J.; Chijimatsu, M.; Fujioka, S.; Kawano, T.; Kimura, Y. Brevicompanines A and B: New plant growth regulators produced by the fungus, *Penicillium brevicompactum*. *J. Chem. Soc. Perkin Trans. 1* **1998**, *1*, 2823–2826. [CrossRef]

30. Jian, Y.J.; Wu, Y. On the structure of penipratynolene and WA. *Tetrahedron* **2010**, *66*, 637–640. [CrossRef]

31. Oh, H.; Swenson, D.C.; Gloer, J.B.; Shearer, C.A. New bioactive rosigenin analogues and aromatic polyketide metabolites from the freshwater aquatic fungus *Massarina tunicata*. *J. Nat. Prod.* **2003**, *66*, 73–79. [CrossRef]

32. Xin, Z.H.; Zhu, W.M.; Gu, Q.Q.; Fang, Y.C.; Duan, L.; Cui, C.B. A new cytotoxic compound from *Penicillium auratiogriseum*, symbiotic or epiphytic fungus of sponge *Mycale plumose*. *Chin. Chem. Lett.* **2005**, *16*, 1227–1229.

33. Mandava, N.; Finegold, H. ^{1}H and ^{13}C Nuclear magnetic resonance spectra of phenylacetic acid derivatives. *Spectrosc. Lett.* **1980**, *13*, 59–68. [CrossRef]

34. Gutierrez-Lugo, M.T.; Woldemichael, G.M.; Singh, M.P.; Suarez, P.A.; Maiese, W.M.; Montenegro, G.; Timmermann, B.N. Isolation of three new naturally occurring compounds from the culture of *Micromonospora* sp. P1068. *Nat. Prod. Res.* **2005**, *19*, 645–652. [CrossRef] [PubMed]

35. Dai, J.; Kardono, L.B.; Tsauri, S.; Padmawinata, K.; Pezzuto, J.M.; Kinghorn, A.D. Phenylacetic acid derivatives and a thioamide glycoside from *Entada phaseoloides*. *Phytochemistry* **1991**, *30*, 3749–3752. [CrossRef]

36. Zhang, Z.; Wang, N.; Zhao, Y.; Gao, H.; Hu, Y.H.; Hu, J.F. Fructose-derived carbohydrates from *Alisma orientalis*. *Nat. Prod. Res.* **2009**, *23*, 1013–1020. [CrossRef] [PubMed]

37. Davison, J.; Al Fahad, A.; Cai, M.; Song, Z.; Yehia, S.Y.; Lazarus, C.M.; Bailey, A.M.; Simpson, T.J.; Cox, R.J. Genetic, molecular, and biochemical basis of fungal tropolone biosynthesis. *Proc. Natl. Acad. Sci. USA* **2012**, *109*, 7642–7647. [CrossRef]

38. Shimomura, H.; Sashida, Y.; Mimaki, Y.; Adachi, T.; Yoshinari, K. A new mevalonolactone glucoside derivative from the bark of Prunus buergeriana. *Chem. Pharm. Bull.* **1989**, *37*, 829–830. [CrossRef]

39. Kishida, M.; Yamauchi, N.; Sawada, K.; Ohashi, Y.; Eguchi, T.; Kakinuma, K. Diacetone-glucose architecture as a chirality template. Part 9.1 Enantioselective synthesis of (*R*)-mevalonolactone and (*R*)-[2H9]mevalonolactone on carbohydrate template. *J. Chem. Soc. Perkin Trans. 1* **1997**, *1*, 891–896. [CrossRef]

40. Krasnoff, S.B.; Keresztes, I.; Gillilan, R.E.; Szebenyi, D.M.E.; Donzelli, B.G.G.; Churchill, A.C.L.; Gibson, N.M. Serinocyclins A and B, cyclic heptapeptides from *Metarhizium anisopliae*. *J. Nat. Prod.* **2007**, *70*, 1919–1924. [CrossRef]

41. Xia, M.W.; Cui, C.B.; Li, C.W.; Wu, C.J. Three new and eleven known unusual C25 steroids: Activated production of silent metabolites in a marine-derived fungus by chemical mutagenesis strategy using diethyl sulphate. *Mar. Drugs* **2014**, *12*, 1545–1568. [CrossRef] [PubMed]

42. Niu, S.; Peng, G.; Xia, J.M.; Xie, C.L.; Li, Z.; Yang, X.W. A new pimarane diterpenoid from the *Botryotinia fuckeliana* fungus isolated from deep-sea water. *Chem. Biodivers.* **2019**, *16*, e1900519. [CrossRef] [PubMed]

43. Xie, C.L.; Liu, Q.; Xia, J.M.; Gao, Y.; Yang, Q.; Shao, Z.Z.; Liu, G.; Yang, X.W. Anti-allergic compounds from the deep-sea-derived actinomycete *Nesterenkonia flava* MCCC 1K00610. *Mar. Drugs* **2017**, *15*, 71. [CrossRef] [PubMed]

Article

Effect of Drying Methods on Lutein Content and Recovery by Supercritical Extraction from the Microalga *Muriellopsis* sp. (MCH35) Cultivated in the Arid North of Chile

Mari Carmen Ruiz-Domínguez [1,*], Paola Marticorena [2], Claudia Sepúlveda [2], Francisca Salinas [1], Pedro Cerezal [1] and Carlos Riquelme [2]

[1] Laboratorio de Microencapsulación de Compuestos Bioactivos (LAMICBA), Departamento de Ciencias de los Alimentos y Nutrición, Facultad de Ciencias de la Salud, Universidad de Antofagasta, Antofagasta 1240000, Chile; francisca.salinas@uantof.cl (F.S.); pedro.cerezal@uantof.cl (P.C.)

[2] Centro de Bioinnovación, Facultad de Ciencias del Mar y Recursos Biológicos, Universidad de Antofagasta, Antofagasta 1240000, Chile; marticorena.paola@gmail.com (P.M.); claudia.sepulveda@uantof.cl (C.S.); carlos.riquelme@uantof.cl (C.R.)

* Correspondence: maria.ruiz@uantof.cl; Tel.: +56-552-633-660

Received: 30 September 2020; Accepted: 14 October 2020; Published: 26 October 2020

Abstract: In this study, we determined the effect of drying on extraction kinetics, yield, and lutein content and recovery of the microalga *Muriellopsis* sp. (MCH35) using the supercritical fluid extraction (SFE) process. The strain was cultivated in an open-raceways reactor in the presence of seawater culture media and arid outdoor conditions in the north of Chile. Spray-drying (SD) and freeze-drying (FD) techniques were used for dehydrating the microalgal biomass. Extraction experiments were performed by using Box-Behnken designs, and the parameters were studied: pressure (30–50 MPa), temperature (40–70 °C), and co-solvent (0–30% ethanol), with a CO_2 flow rate of 3.62 g/min for 60 min. Spline linear model was applied in the central point of the experimental design to obtain an overall extraction curve and to reveal extraction kinetics involved in the SFE process. A significant increase in all variables was observed when the level of ethanol (15–30% *v/v*) was increased. However, temperature and pressure were non-significant parameters in the SFE process. The FD method showed an increase in lutein content and recovery by 0.3–2.5-fold more than the SD method. Overall, *Muriellopsis* sp. (MCH35) is a potential candidate for cost-effective lutein production, especially in desert areas and for different biotechnological applications.

Keywords: microalgae; *Muriellopsis*; spray drying; freeze-drying; lutein; supercritical fluid extraction

1. Introduction

In the past few decades, demand for bioprospection of microorganisms isolated from harsh environmental conditions has been increased because of their diverse biotechnological applications. Among microorganisms, microalgae are the most diversified photosynthetic organisms with high adaptability to different environmental conditions [1]. They are mainly classified as *Cyanophyta* (cyanobacteria), *Rhodophyta* (red algae), *Chlorophyta* (green algae), and *Chromophyta* (brown algae) [2,3].

Microalgae are characterized as natural sources of bioactive molecules such as phycobiliproteins, polysaccharides, carotenoids, lipids, fatty acids, polyphenols, and vitamins [4,5]. These compounds exhibit health benefits such as antibacterial, antifungal, antioxidant, and anticancer activities that are essential in pharmacological, nutraceutical, food, and biotechnological development [5–7]. Carotenoids belong to the class of terpenoids and are derived from the 40-carbon polyene chain and xanthophylls as their oxygenated derivatives in the presence of –OH groups (e.g., lutein), oxi-groups (e.g., canthaxanthin), or both (e.g., astaxanthin) [8–10]. Pure lutein is an orange-yellow, crystalline, and lipophilic solid whose chemical name is β, ε-carotene-3,3′-diol ($C_{40}H_{56}O_2$). It is beneficial to human health due to its potential to ameliorate cardiovascular diseases [11], various types of cancer [12], and age-related macular degeneration [13] because of its antioxidant potential. The species known to accumulate carotenoids efficiently are *Chlorella* sp., *Chlamydomonas* sp., *Dunaliella* sp., *Muriellopsis* sp., and *Haematococcus* sp.

In this study, we used *Muriellopsis* sp. (MCH35), having the potential of producing carotenoids, especially lutein, as the main oxygenated carotenoid. This strain was isolated from an arid region of the north of Chile, as described previously by Marticorena et al. [14]. This area is located in Antofagasta, a well-known region for its high solar radiation, an environmental factor that favors microalgal growth [15]. Several studies have reported strategies to enhance microalgal carotenoid production using unfavorable environmental conditions such as nutrient deficiency, intense irradiation, and salinity to photo-bioreactors design [16–18]. These factors affect not only photosynthesis and productivity of cell biomass, but also pathways and cellular metabolism, and thus alter the cell composition [16]. The isolation and selection of microalgae is a prerequisite for the successful industrial production of biomass and beneficial compounds. Two factors are important in their successful industrial production: lutein content and biomass productivity. Other factors, such as the presence of cell wall or content of other carotenoids, can also be considered. A high lutein content in microalgae is necessary to achieve cost-effective and adequate amounts of extraction [19].

After the harvesting process, microalgal biomass pretreatment and extraction of bioactive compounds are most important because of their effect on bioactive molecule recovery. A drying process is the most common pretreatment method that increases the shelf life of naturally occurring bioactive biomass [20,21]. A dehydration process represents an essential step for reducing microbial growth, avoiding oxidative reactions, and improving bioactive compound extraction, although they are expensive [22–24]. However, these methods can alter the stability of labile bioactive compounds because of thermal breakdown [25]. Therefore, techniques such as spray-drying (SD) or freeze-drying (FD) that can adjust drying temperature have been recommended by researchers [23]. For the extractions of bioactive compounds, such as carotenoid from microalgae, many conventional methods, including maceration or soxhlet extraction, are used [19]. However, these methods are extensive and require a relatively huge amount of solvents, and thus are expensive and less eco-friendly [26]. Therefore, the use of green extraction technologies is increasing for the extraction of bioactive compounds. Green extraction technologies enhance extraction time, recovery, selectivity, and mass transfer with decreasing consumption of solvents. Supercritical fluid extraction (SFE), pressurized liquid extraction, microwave-assisted extraction, ultrasound-assisted extraction, and high-pressure homogenization are some examples of green extraction technologies [26].

The aim of this work is to study the effect of two different methods of biomass drying on lutein content and recovery, extraction kinetics, and yield of *Muriellopsis* sp. extracted by using the SFE process. The microalga was cultured in an open-raceways reactor under adapted outdoor conditions in the arid north of Chile. After harvesting of cells, SD and FD methods were employed for removing water content from the microalgal biomass and for knowing an individual carotenoid benchmark profile by using conventional methods. In the case of SFE, the parameters such as temperature (40–70 °C), pressure (30–50 MPa), and percentage of co-solvent (0–30% ethanol) were studied on the basis of Box-Behnken designs. In a similar way, the Spline linear model was used to control parameters for optimal lutein recovery from *Muriellopsis* sp. (MCH35). Drying pretreatment and optimal lutein extraction parameters for the microalgal biomass were studied with a focus on using the strain for future biotechnological applications and minimizing production cost when grown on a pilot-scale in a harsh environment.

2. Results and Discussion

2.1. Growth Parameters and Carotenoid Profile of Muriellopsis sp. (MCH35)

Photoautotrophic cultivation is a growing condition wherein light is an energy source, and carbon dioxide is an inorganic carbon source used to form chemical energy by using photosynthesis [27]. To evaluate the growth performance and parameters of *Muriellopsis* sp. (MCH35), batch tests were performed by using UMA5 culture medium [28]. The specific growth rate (μ) was found to be 0.085 d^{-1}, as calculated by using Equation (1),

$$\mu = Ln\ (C)/(Ci)/t \tag{1}$$

and is in accordance with previously reported data on the green microalga *Nannochloropsis gaditana* with similar environmental conditions, such as sufficient light and nutrient availability [29]. Therefore, the UMA5 culture medium is suitable for the cultivation of *Muriellopsis* sp. (MCH35) as it contains an adequate amount of nutrients required for the microalga. Nitrogen concentration and nitrogen:phosphorus ratio are widely recognized as determining factors for the growth and composition of microalgae [30,31]. An exponential increase was observed from day 0 to day 12 in biomass concentration starting with 0.44 g/L and reaching to the final concentration of 1.34 g/L (Figure 1A).

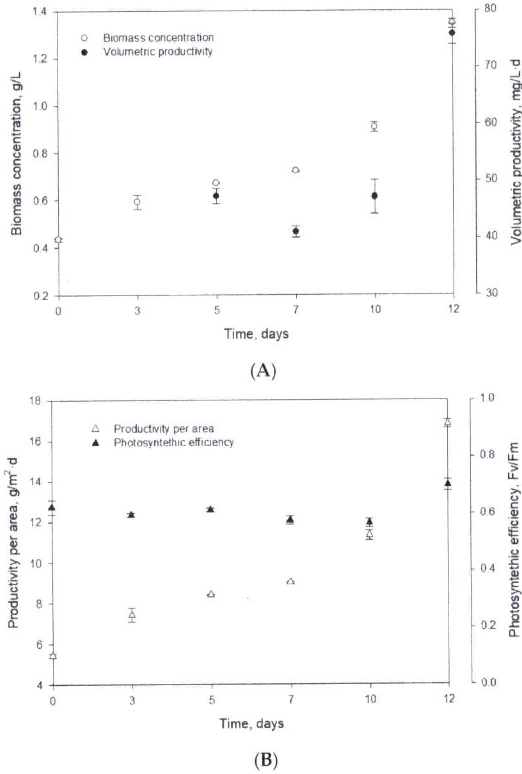

Figure 1. Evolution of (**A**) biomass concentration and volumetric productivity, and (**B**) productivity per area and photosynthetic efficiency of the strain *Muriellopsis* sp. (MCH35) in batch mode culture adapted under outdoor conditions in the arid north of Chile.

Chew et al. [32] reported the *Dunaliella* sp. biomass concentration of 1.5 g/L (in a volume of 3.4 L) at the laboratory scale. For *Phaeodactylum* sp., the biomass concentration of 1.38 g/L in 5 L medium was observed, while for *Chlorella vulgaris*, it was in the range of 0.6 to 1.08 g/L·d (in a volume of 1.5 L). Similar values were obtained for the first two strains, and the difference in biomass productivity of *Muriellopsis* sp. and *Chlorella vulgaris* can be due to the optimized laboratory conditions and differences in sizes of the two strains. The volumetric productivity of *Muriellopsis* sp. culture reached 75.73 mg/L·d on day 12, and the productivity per area was 16.81 g/m^2·d (Figure 1A,B). High productivity was reported in the *Nannochloropis gaditana* cultures with 400 mg/L·d and incident irradiance of 1.100 µE/m^2·s [33]. Our experiments on *Muriellopsis* sp. were carried out with an incident irradiance of 1.400 µE/m^2·s, under outdoor conditions. The difference in productivities can be due to the different cultivation modes used, as a batch mode can affect productivity compared with semi-continuous mode. During the cultivation, it is essential to determine the variation in chlorophyll fluorescence (Fv/Fm) by focusing on the suitability of culture conditions. This parameter represents a measure of the quantum yield of PSII (photosystem II) and identifies any damage to the protein complex caused by photo-inhibition [34]. Figure 1B shows the values of photosynthetic efficiency ranging from 0.58 to 0.7 during the cultivation. An acceptable level of physiological acclimatization exhibited by *Muriellopsis* sp. was confirmed in our study (Figures 1B and 2). In optimal cultivation conditions, the productivity of *Muriellopsis* sp. was ~0.60 at a pH of 7.9–8.2 and temperature of 16.3–19.6 °C, when cultivated in a semi-continuous mode by using Arnon culture medium [35]. Del Campo et al. [36] determined the limiting growth conditions such as pH of 6–9 and the temperature of 33 °C that stimulated carotenogenesis in *Muriellopsis* sp.

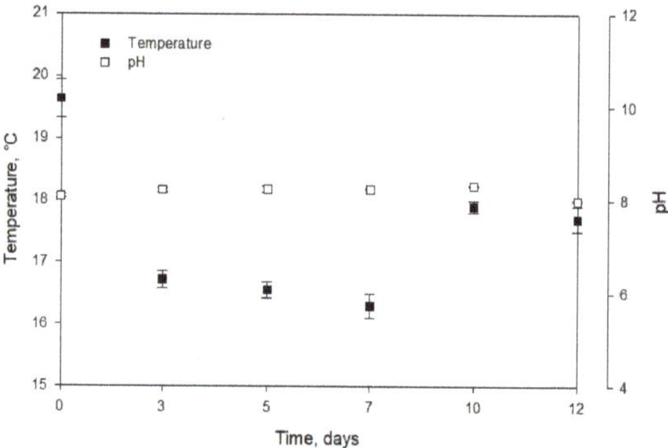

Figure 2. Temperature and pH and evolution of *Muriellopsis* sp. (MCH35) in a batch mode culture adapted under outdoor conditions in the arid north of Chile.

The main carotenoids present in the SD and FD biomass are shown in Table 1. These data are supported by Figure 3, showing an HPLC chromatogram with a diode array detector from the FD microalga.

Table 1. Individual carotenoid profiles present in *Muriellopsis* sp. (MCH35) from two modes of dry biomass by the conventional method of extraction.

Biomass	Carotenoids Content (mg/g Biomass) *				
	Lutein	Zeaxanthin	Violaxanthin	Astaxanthin	β-carotene
Spray-dried (SD)	3.45 ± 0.20 [b]	0.60 ± 0.04 [b]	0.15 ± 0.10 [a]	0.45 ± 0.15 [a]	0.45 ± 0.04 [b]
Freeze-dried (FD)	4.20 ± 0.30 [a]	0.75 ± 0.06 [a]	0.30 ± 0.20 [a]	0.30 ± 0.02 [b]	0.60 ± 0.05 [a]

* Benchmark extraction. Mean values in the same column followed by different letters (a-b) are significantly different ($p < 0.05$).

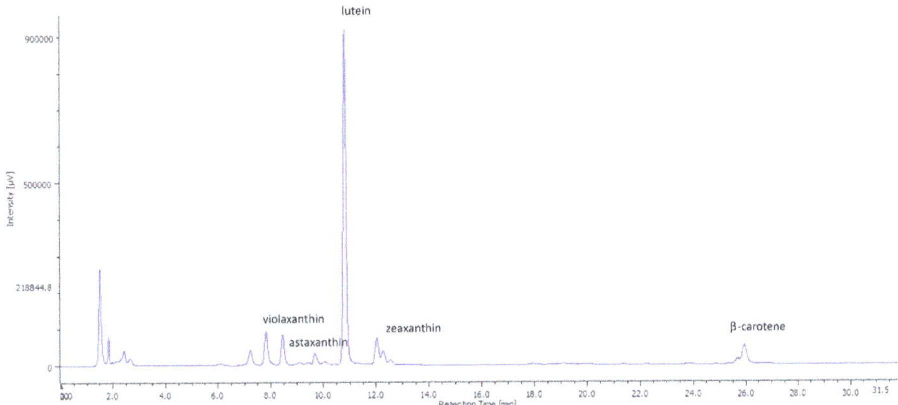

Figure 3. HPLC (High Performance Liquid Chromatography) chromatogram of individual carotenoids present in *Muriellopsis* sp. (MCH35), extracted under conventional extraction (freeze-dried (FD) biomass) and measured at 450 nm by a diode array detector.

The high individual carotenoid content was observed in FD biomass, except for astaxanthin that was 1.5 times higher for the SD method than the FD method. A similar drying effect was reported by Ryckebosch et al. [37] in which they evaluated the effect of the SD and FD methods on fresh biomass and the storage stability of lipids and carotenoids in the diatom *Phaeodactylum tricornutum*. Their study showed better results on carotenoid contents in the fresh and FD biomass of algae than that in SD biomass at 48 h (denominated as short-term storage). Lutein is the main carotenoid profile that *Muriellopsis* sp. shows [19,36,38]. Our results on lutein content (particularly lyophilized cells) were in a similar range with that reported previously (in range of 4.0 to 6.0 mg/g dry weight) [39,40]. Better lutein content was obtained by using both the drying methods than that reported by Molino et al. [41], who studied lutein production in different microalgae in a comparative manner. Our results were in a similar range as that reported by Del Campo et al. [8] and had an advantage that our cultures were produced at a large scale using seawater. This is relevant when developing massive cultures in areas with scarce water resources, such as desert areas (Antofagasta Region).

We found other carotenoids such as zeaxanthin, violaxanthin, astaxanthin, and β-carotene (can be seen in Figure 3) in low levels in *Muriellopsis* sp., similar to those reported by Del Campo et al. [36], with a total concentration of 5.10 ± 0.53 and 6.15 ± 0.63 mg/g for the SD and FD processes, respectively. Despite their low levels in *Muriellopsis* sp., their presence can contribute to pharmacological, nutraceutical, food, and biotechnological applications [7,42]. Lutein and zeaxanthin reduce age-related macular degeneration [43], while β-carotene prevents cataracts, skin diseases, and other illnesses like cancer [42,44]. Moreover, violaxanthin is a potential anti-photoaging agent acting against ultraviolet-B radiation (UV-B; λ 280–315 nm) [45], and astaxanthin is considered super vitamin E [46] for its stronger antioxidant activity (500 times more effective than α-tocopherol), preventing arteriosclerosis, coronary heart disease, and ischemic brain development [42,46]. Therefore, although it would be beneficial to optimize the extraction technology of lutein as suggested by Di Caprio et al. [47], our results provide preliminary information on the carotenoid profile of *Muriellopsis* sp. (MCH35).

2.2. Effects of Drying Processes on the Extraction Yield of Muriellopsis sp. (MCH35) by SFE

The experimental extraction conditions and results of the Box-Behnken designs from two-mode dry biomass by the SFE process are given in Table 2. The range of extraction yield (Y) was 0.11–7.84% (*w/w*) for the SD biomass and 0.76–6.05% (*w/w*) for the FD biomass. Table 2 shows the effect of co-solvent on extraction yield at all the conditions. The low levels of extracts were obtained without the addition of the modifier, such as 0.11–0.44% for the SD and 0.76–2.07% for the FD biomass.

Table 2. Extraction yields (Y) and lutein content and recovery by SFE from spray- and freeze-dried *Muriellopsis* sp. (MCH35) using Box-Behnken experimental design. The general parameters were biomass loading = 2.0 g, CO_2 flow rate = 3.62 g/min, and extraction time = 60 min.

Biomass	Run	T (°C)	P (MPa)	Ethanol (%, v/v)	Yield (%, w/w)	Lutein Content (mg/g Extract)	Lutein Recovery (%, w/w)
	1	40	30	15	1.52 ± 0.08	45.47 ± 1.45	20.05 ± 0.48
	2	70	30	15	1.84 ± 0.09	43.95 ± 0.10	23.38 ± 0.13
	3	40	50	15	1.63 ± 0.08	34.27 ± 0.36	16.23 ± 0.04
	4	70	50	15	2.17 ± 0.02	34.70 ± 0.21	21.79 ± 0.92
	5	40	40	0	0.11 ± 0.01	2.60 ± 0.03	0.09 ± 0.3 × 10^{-3}
	6	70	40	0	0.42 ± 0.02	6.28 ± 0.06	0.76 ± 1.2 × 10^{-3}
Spray-dried	7	40	40	30	1.87 ± 0.09	27.69 ± 0.31	15.04 ± 0.05
	8	70	40	30	2.31 ± 0.08	23.69 ± 0.34	15.86 ± 0.14
	9	55	30	0	0.32 ± 0.10	1.27 ± 0.02	0.12 ± 0.03
	10	55	50	0	0.44 ± 0.02	7.80 ± 10^{-3}	0.99 ± 0.01
	11	55	30	30	7.84 ± 0.27	13.30 ± 0.11	30.23 ± 0.45
	12	55	50	30	1.72 ± 0.06	43.76 ± 0.41	21.76 ± 0.30
	13	55	40	15	1.67 ± 0.03	31.21 ± 0.12	15.12 ± 0.52
	14	55	40	15	1.70 ± 0.09	47.14 ± 0.75	23.25 ± 0.18
	15	55	40	15	1.57 ± 0.03	38.68 ± 0.10	17.65 ± 0.63

Biomass	Run	T (°C)	P (MPa)	Ethanol (%, v/v)	Yield (%, w/w)	Lutein Content (mg/g Extract)	Lutein Recovery (%, w/w)
	16	40	30	15	4.79 ± 0.24	25.96 ± 1.18	29.62 ± 0.71
	17	70	30	15	4.55 ± 0.16	46.00 ± 0.26	49.84 ± 1.54
	18	40	50	15	1.62 ± 0.08	60.69 ± 2.06	23.39 ± 0.29
	19	70	50	15	2.32 ± 0.05	53.69 ± 2.41	29.64 ± 0.20
	20	40	40	0	2.07 ± 0.10	13.42 ± 0.31	6.62 ± 0.01
	21	70	40	0	1.15 ± 0.05	8.03 ± 0.39	2.20 ± 0.04
Freeze-dried	22	40	40	30	2.69 ± 0.13	39.82 ± 1.29	25.52 ± 0.28
	23	70	40	30	3.77 ± 0.13	42.25 ± 2.55	37.90 ± 0.91
	24	55	30	0	0.76 ± 0.03	4.72 ± 0.30	0.86 ± 3.0 × 10^{-3}
	25	55	50	0	1.18 ± 0.04	7.91 ± 1.00	2.22 ± 0.20
	26	55	30	30	6.05 ± 0.30	40.98 ± 4.49	59.04 ± 2.21
	27	55	50	30	5.74 ± 0.29	54.48 ± 0.83	74.52 ± 0.47
	28	55	40	15	3.91 ± 0.16	38.14 ± 1.82	35.55 ± 0.58
	29	55	40	15	3.33 ± 0.07	41.33 ± 0.33	32.72 ± 1.43
	30	55	40	15	3.07 ± 0.12	37.58 ± 1.09	27.47 ± 0.07

Acronyms: Temperature (T), Pressure (P), and Ethanol (co-solvent). Standard deviation was less than 5% in all operating conditions (SD ≤ 5%, n = 3).

Supplementary Table S1 shows the statistical data of both drying processes with a p-value in terms of the goodness of fit of the model. It can be seen that ethanol, as a co-solvent, was a significant variable in the extraction process for two drying methods of biomass. These data are reinforced with the mathematical equations obtained in the statistical section and are summarized for the significant factors and their interactions (ethanol). Equations (2) and (3) were developed for an approximate mathematical model to maximize the extraction yield from *Muriellopsis* sp. dehydrated using the SD (Equation (2)) and FD (Equation (3)) methods:

$$\text{Yield (\%, w/w)} = 0.88 + 0.49 \cdot Ethanol + 0.0001 \cdot T \cdot Ethanol - 0.001 \cdot P \cdot Ethanol + 0.0007 \cdot Ethanol^2 \quad (2)$$

$$\text{Yield (\%, w/w)} = 7.85 + 0.10 \cdot Ethanol + 0.002 \cdot T \cdot Ethanol + 0.0001 \cdot P \cdot Ethanol - 0.002 \cdot Ethanol^2 \quad (3)$$

where Yield is extraction yield in %, w/w, Ethanol is co-solvent in % v/v, T is temperature in °C, and P is pressure in MPa. These regression equations mathematically approach a model to maximize the yield from the two biomass drying methods based on the experimental results obtained in this work. Figure 4 also shows the results of extraction yields obtained by using the Pareto chart and RSM (Response Surface Methodology) using conditions such as 40–70 °C temperature and 0–30% v/v ethanol (extractant) with the optimal pressure of 30 MPa (original ranging from 30–50 MPa) using the two modes of dry biomass by SFE. Figure 4A,B represent the extraction yield of *Muriellopsis* sp. SD biomass, for which the optimum value was 6.40% *w/w* at 57.14 °C and 30 MPa pressure with 30% *v/v* ethanol as a co-solvent. These data are similar to the experimental conditions shown in Table 2 (run 11) with a value of 7.84% (55 °C, 30 MPa, and 30% ethanol). Figure 4C,D shows the significant factors obtained by the Pareto charts and the extraction yield obtained by RSM from the FD biomass. Its optimum value was similar to that of SD biomass with 5.96% at 59.67 °C temperature and 30 MPa pressure with 30% *v/v* ethanol as a co-solvent. In this case, run 26 (see Table 2) was the experimental condition closest to the optimum value, with 6.05% *w/w*. In the SFE process, the highest extraction yield and recovery of bioactive compounds can be achieved by optimizing some critical parameters. Indeed, many studies have shown an ability to modulate CO_2 polarity by using co-solvents such as ethanol, and thereby increasing the extraction yields [48,49]. Our results showed an increase in the extraction yields using maximum extractant volume (30% *v/v*), similar to other studies. However, no significant interaction was found between temperature and pressure for the extraction yield in both the drying methods. Conversely, other reports accentuate the effect of factors including pressure or temperature on the extraction yield. Mehariya et al. [50] evaluated the pressure factor for measuring lutein extraction by the SFE process from the green microalga *Scenedesmus almeriensis*. They reported an increase in its yield with an increase in pressure from 25 to 55 MPa at 50 °C and above 65 °C (CO_2 flow rate of 7.24 g/min).

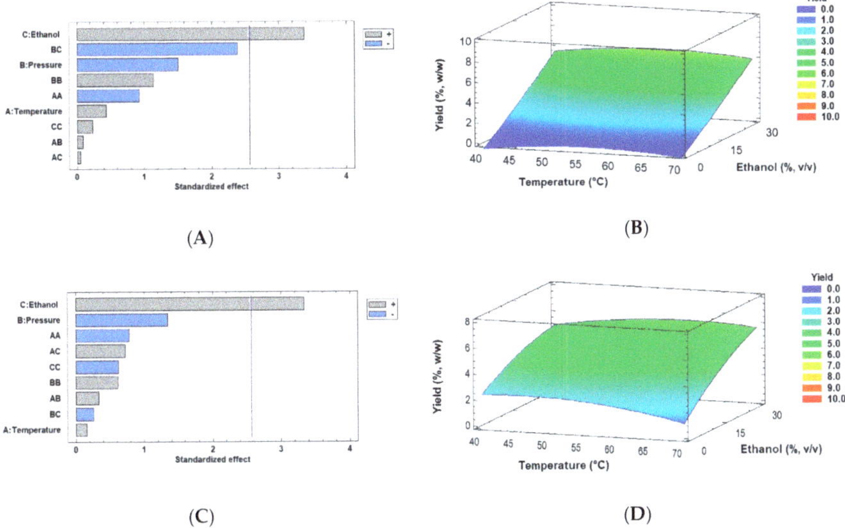

Figure 4. Pareto Charts and response surface curves of the combined effects of temperature (40–70 °C), pressure (30–50 MPa), and ethanol as co-solvent (0–30% *v/v*) on extraction yield from (**A,B**) spray- and (**C,D**) freeze-dried biomass of *Muriellopsis* sp. (MCH35), respectively. The ± signs are interpreted in the Pareto graph according to the area of significance of factor or interaction in the experimental design and Response surface curves were drawn at 30 MPa as optimal pressure.

Generally, an improving trend in the yield was observed in the case of FD biomass. In spite of various alternatives for drying methods available, both SD and FD are the most commonly used drying methods for high-value products [51]. In particular, FD is the best method for water removal and for the conservation of biochemical composition, although it is the most expensive process [51,52]. However, the SD process is faster than other methods, but it can damage different thermolabile components [37]. Therefore, our results showed that the solid-state of water during FD could protect the primary structure of the microalgal biomass with minimal reduction in volume and improving its general extraction yield.

2.3. The Effect of Drying Methods on Lutein Recovery of Muriellopsis sp. (MCH35) by SFE

SFE experiments based on Box-Behnken design were also performed to evaluate lutein content and recovery from *Muriellopsis* sp. (MCH35) using both drying processes. The SFE study focused on lutein as the main carotenoid as other individual carotenoids were in low levels. The effects of drying methods on lutein recovery are shown in Table 2 and complemented with Supplementary Material (data shown in Table S1). The solvent factor was the most significant in both drying processes. Indeed, the interaction was stronger in the SD process as a quadratic factor. Both drying processes showed almost similar lutein content, with slightly more lutein content in FD biomass. FD biomass showed the highest lutein content (60.69 mg/g extract) during run 18 at 40 °C temperature, 50 MPa pressure, and 15% ethanol, followed by run 27 at 55 °C temperature, 50 MPa pressure, and 30% ethanol (54.48 mg/g extract). However, SD biomass showed lutein content of 47.14 mg/g extract (run 14) under 55 °C, 40 MPa pressure, and 15%, followed by run 1 (45.47 mg/g extract) at 40 °C temperature, 30 MPa pressure, and 15% ethanol. High pressure caused an increase in lutein content in FD biomass, contrary to SD biomass. In both processes, the temperature employed was between 40 and 55 °C and the amount of ethanol ranged from 15% to 30%, *v/v*. It is evident that the main difference in the lutein yield was due to the drying methods. Particularly, high pressure causes a major variation in solvent density so that it can have a good extraction yield for lutein. Moreover, the amount of free water present in *Muriellopsis* sp. (MCH35) can affect lutein extraction because of the difference in the degree of dehydration between SD and FD methods [52].

In accordance with the above results, lutein recovery obtained by using SD and FD methods is given in Table 2. Figure 5 shows the results obtained by Pareto charts and RSM curves for the combined effects of temperature (40–70 °C), pressure (30–50 MPa), and ethanol as a co-solvent (0–30% *v/v*) on lutein recovery from SD (Figure 5A,B) and FD biomass (Figure 5C,D). The presence of ethanol was significant in both methods. A similar trend was observed in the SD method because, again, the quadratic ethanol factor was relevant in lutein recovery (can be confirmed by Supplementary Material such as Table S1). The mathematical models are included in Equations (4) and (5) for the optimization of lutein recovery from SD and FD microalga, respectively. The equations summarize factors that were significant (ethanol as co-solvent), including their interactions.

$$\text{Lutein recovery (\%, } w/w) = 32.06 + 2.48 \cdot Ethanol + 0.0002 \cdot \text{T} \cdot Ethanol - 0.002 \cdot \text{P} \cdot Ethanol - 0.04 \cdot Ethanol^2 \qquad (4)$$

$$\text{Lutein recovery (\%, } w/w) = 8.06 + 0.43 \cdot Ethanol + 0.019 \cdot \text{T} \cdot Ethanol + 0.002 \cdot \text{P} \cdot Ethanol - 0.03 \cdot Ethanol^2 \qquad (5)$$

where Lutein recovery is measured as %, *w/w*, Ethanol is co-solvent in % *v/v*, T is temperature in °C, and P is pressure in MPa. These regression equations again approach a mathematical model obtained by the statistical software to maximize lutein recovery from both drying methods based on the experimental results. The highest lutein recovery obtained in this study was 49.84–74.52% *w/w* for the FD biomass in the presence of ethanol (15–30%, *v/v*). Conversely, the pressure and temperature factors were variable and hence were non-significant in the process. Moreover, this set of experimental factors also obtained the highest extraction yield. The optimal factors determined by the statistical analysis for the FD biomass were similar to those cited above. However, optimal lutein recovery was lower than the results mentioned above for the FD biomass, such as 60.47% *w/w* at 60 °C, 50 MPa, and 29.9% ethanol (*v/v*), approximately. On the other hand, the SD method achieved poor lutein recovery compared to that obtained by the lyophilization method. Thus, we obtained the best experimental result between 23.25% and 30.23% *w/w* in the presence of ethanol and at intermediate to high temperatures. A low-pressure trend was defined by the SD biomass in the presence of other optimal conditions (58.14 °C, 30 MPa, and 25.5% ethanol), with a value of 28.24% lutein recovery (*w/w*).

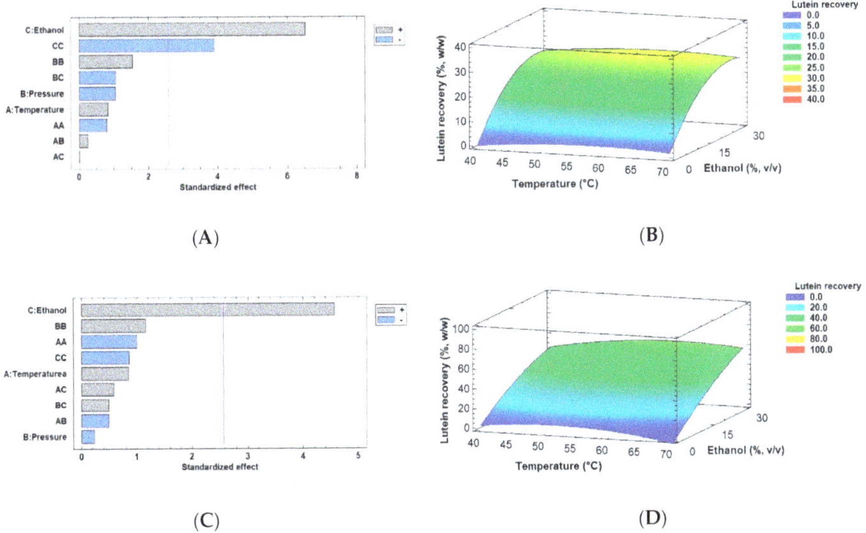

Figure 5. Pareto Charts and response surface curves of the combined effects of temperature (40–70 °C), pressure (30–50 MPa), and ethanol as a co-solvent (0–30% *v/v*) on lutein recovery from (**A,B**) spray- and (**C,D**) freeze-dried biomass of *Muriellopsis* sp. (MCH35), respectively. The ± signs are interpreted in the Pareto graph according to the area of significance of factor or interaction in the experimental design and Response surface curves were made at 30 MPa for spray-dried and 50 MPa for freeze-dried microalga as an optimal pressure.

Other groups of scientists have also studied the effect of supercritical fluid parameters such as pressure, temperature, solvent, and CO_2 flow rate on lutein recovery and purity from other sources. Mehariya et al. [50] investigated the effect of pressure (25–55 MPa), temperature (50 and 65 °C), and CO_2 flow rate (7.24 and 14.48 g/min) on the green microalga *Scenedesmus almeriensis*. Their results showed improvements in the lutein recovery (~98%) and purity (~34%) with an increase in temperature, pressure, and CO_2 flow rate. Yen et al. [53] reported 76.7% lutein recovery from *Scenedesmus* sp. in the presence of 70 °C, 40 MPa, and ethanol (30 mol%), which is in the similar range of our results. Wu et al. [54] extracted 87.0% of lutein from *Chlorella pyrenoidosa* by SFE in 4 h in the presence of 50 °C, 25 MPa, and modified CO_2 with 50% ethanol. The factors optimized for lutein recovery from microalgae by the SFE process in our study are similar to those reported in previous studies. The effect of water content on the *Muriellopsis* biomass for lutein recovery has been evidenced in our SFE results. Some studies have reported that the optimal parameters for SFE can be obtained from samples that exhibit 3% to 12% of water content [55–57]. The presence of moisture in samples can act as a barrier for the diffusion of supercritical CO_2 and extracted compounds [57]. However, other studies showed that the presence of water can enhance extraction kinetics and yields from plants and microalgae such as *Nannochloropsis oculata* [58,59], and water can play the role of co-solvent for polar compounds according to the type of matrix. We got the results emphasizing that optimal supercritical extraction was obtained in the samples with low moisture (lyophilized cells). Although FD is an expensive process for dehydration, an increase of lutein recuperation was significant. This can coalesce with growing the endogenous *Muriellopsis* sp. in seawater and under arid outdoor conditions to counteract the high costs of the dehydration process. Therefore, *Muriellopsis* sp. (MCH35) can be used as an efficient lutein producer for biotechnological applications, especially in desert areas.

2.4. Global Yield and Kinetic Curve of Muriellopsis sp. (MCH35)

Based on the yield results obtained for SFE by using Box-Behnken designs from the two modes of dry biomass of *Muriellopsis* sp., the condition selected for kinetic study was the central point of the experimental design, that is, 55 °C, 40 MPa, and CO_2 + ethanol (85:15 *v/v*) flow rate, since temperature and pressure were non-significant ($p < 0.05$).

Figure 6A,B show overall extraction curves (OEC) of SD and FD biomass. A recovery of 2.54% and 5.10% of extracts was obtained after 150 and 160 min of extraction for SD and FD biomass, respectively. Similar weights (~2.0 g) of dry biomass were used for extraction, indicating that a double amount of extract was obtained by FD compared with that of SD. This gives an advantage to the biomass obtained by FD over SD since the resulting final powder is better in quality and quantity. This is due to the fact that drying by sublimation, as in the case of the FD technique, is better.

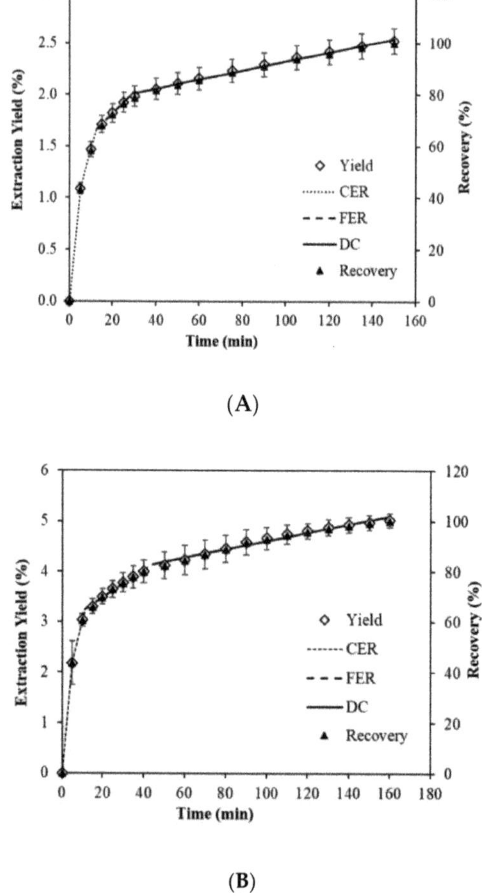

(A)

(B)

Figure 6. Overall extraction curves (OECs) of the experimental (◊) and predicted data for Spline (···· CER, — FER, — DC) models, and recovery (▲), at P = 40 MPa, T = 55 °C, CO_2 + ethanol (85:15 *v/v*) flow rate (3.305 g/min), for Spray-Drying biomass (**A**), and Freeze-Drying biomass (**B**). Abbreviations: constant extraction rate period (CER); falling extraction rate period (FER); diffusion-controlled rate period (DC); Pressure (P) and Temperature (T).

The OEC plotted for SD and FD biomass followed the SFE kinetics that were established by Meireles [60] and Jesus et al. [61]. The extraction process began with the CER period, characterized by the removal of easily extractable compounds by solvent and co-solvent, which was mainly controlled by the convective mass transfer in the fluid film around the powder particles. Following the CER period, the transition period began with a reduced extraction rate, wherein the extraction rate was controlled by mass transfer mechanisms through both convection and diffusion. This period is commonly called the FER period. When easily accessible solute became scarce in the microalgae matrix, intra-particle diffusion became the main mass transfer mechanism during SFE, and hence the OEC assumed a typical shape of diffusion curve with reduced extraction rate.

From the fitted data by the Spline linear model in Figure 6A,B, the OEC parameters were estimated as given in Table 3. The calculated t_{CER} were 12.54 and 11.24 min with the accumulated extracts of 1.66% and 3.25%, and the recovery of 65.85% and 64.72% for SD and FD biomass, respectively. This is in agreement with a previous study that reported the recovery between 50% and 90% in the CER period [62]. Although t_{CER} of SD differs by less than 1 min and about 1.13% of the recovery compared to FD, the accumulated extract in FD was double that in SD. The calculated t_{FER} was 29.01 and 44.11 min for the accumulated extract of 2.01% and 4.15%, and the total recovery of 79.63% and 82.70% for SD and FD biomass, respectively. At this stage, the difference between t_{FER} of SD and FD was ~15 min, and the accumulated extract of FD was more than double that for SD. In our research, a recovery > 75% was achieved in the FER period. The M_{CER} and M_{FER} values represent the extraction rate of the CER and FER periods respectively [61], with values of 0.0026 and 4.6×10^{-4} g/min for SD, and 0.0057 and 5.3×10^{-4} g/min for FD. Therefore, the extraction rate that produced CO_2 + Ethanol flow in the FD biomass was more than double that for SD. These values of the extraction rates were lower than the values reported for M_{CER} in peach almond oil (0.0084–0.0752 g/min) [63] and M_{CER} and M_{FER} of chañar almond oil (0.066–0.0124 g/min) [64].

Table 3. Adjusted parameters of the spline linear model to SFE from *Muriellopsis* sp. biomass at 55 °C, 40 MPa, and CO_2 + ethanol (85:15 *v/v*) flow rate.

| Parameters | Stages of the General Extraction Curve | | | | | |
| | Spray-Drying | | | Freeze-Drying | | |
	CER	FER	DC	CER	FER	DC
Time (min.)	12.54	29.01	150.0	11.24	44.11	160.0
Accumulated extract (%)	1.66	2.01	2.54	3.25	4.15	5.10
Recovery (%)	65.85	13.78	20.37	64.72	17.98	17.30
Total Recovery (%)	65.85	79.63	100.0	64.72	82.70	100.0
M (g/min)	0.0026	4.6×10^{-4}	9.1×10^{-5}	0.0057	5.3×10^{-4}	1.7×10^{-4}
Y (mg extract/g biomass)	15.91	3.79	5.50	30.76	9.45	9.53
Y (g extract/g (CO_2 85%+ ethanol 15%)	1.5×10^{-3}	2.4×10^{-4}	5.4×10^{-5}	3.5×10^{-3}	3.2×10^{-4}	1.0×10^{-4}
R^2	0.9273	1.0000	1.0000	0.9451	1.0000	1.0000

Drying Methods	b_o	a_1	a_2	a_3
Spray-Drying	0.7098	0.0758	−0.0547	−0.0167
Freeze-Drying	1.3251	0.1711	−0.1437	−0.0192

b_0: Linear coefficient of the first line (CER); a_1, a_2 and a_3 : Slopes of the lines 1, 2, and 3 corresponding to the periods CER, FER, and DC, respectively; t_{CER}, and t_{FER} : Times in the intercepts of the lines 1 and 2 and the lines 2 and 3, respectively; $m_{EXT}(t)$: mass of the extract at time t in each period; Y_t: Variable response of the sixth, and the seventh row, for the considered stage (CER, FER, and DC).

On the other hand, Y values (mg extract/g biomass) in the CER and FER periods of FD showed a similar trend as above, since they were 1.9 and 2.5 times higher than the SD values. Y value (g extract/g CO_2 + ethanol 85:15 *v/v*) represents the extract ratio in the supercritical phase at the bed outlet and were 2.3 and 1.3 times higher for the CER and FER periods respectively, in the FD biomass than in SD.

In the final stage of the process or DC period, the difference between the FD and SD biomass was observed. Table 3 shows the coefficients of determination (R^2) for all periods, such as CER, FER, and DC obtained by means of the Spline linear model and their corresponding coefficients, such as b_o, a_1,

a_2, and a_3. The CO_2 + ethanol (85:15 *v/v*) flow rate was 3.305 g/min and solvent to feed ratio (S/F) of 8.2 to 247.1 (5–150 min) and 8.2 to 260.5 (5–160 min) for SD and FD, respectively. In different studies performed by using the SFE-CO_2 technique, Sanzo et al. [65] reported the lutein recovery of ~47% from the *Haematococcus pluvialis* dry biomass with the CO_2 flow rate of 3.62 g/min, 50 °C temperature, and 40 MPa pressure in 120 min. Yen et al. [53] reported the lutein recovery of 76.65% from the *Scenedesmus* sp. dry biomass with the CO_2 flow rate of 1.45 g/min, 30% ethanol, 47.5 °C temperature, and 40 MPa pressure in 60 min.

The CO_2 + ethanol mixture was efficient for the solubilization of lutein from *Muriellopsis* sp. (MCH35) under the established conditions studied by us and also supported by findings in other studies on *Haematococcus pluvialis* [65], *Scenedesmus* sp. [53], and *Scenedesmus almeriensis* [50]. Meireles [60] recommended to extend the SFE process up to the end of CER period, where the extraction rate is the highest, and sometimes it is necessary to extend the extraction process more than the CER period to attain the lowest production cost depending on the characteristics of the product to be extracted [66]. In the present study, the process period was up to 60 min to ensure the complete recovery of the extract of more than 85%.

3. Material and Methods

3.1. Microalgal Strain and Chemicals

The microalga *Muriellopsis* sp. (MCH35) was selected for this research and was isolated from freshwater in the arid north of Chile (Antofagasta Region), as previously described by Marticorena et al. [14]. This strain was deposited in the Spanish algae bank with accession number BEA_IDA_0063B. The UMA5 culture medium was of analytical grade and compounds were purchased from Merck (Darmstadt, Germany). The chemicals used for SFE were carbon dioxide (99% purity), purchased from Indura Group Air Products (Santiago, Chile), and ethanol co-solvent (99.5%), from Merck (Darmstadt, Germany). Other chemicals such as ultrapure water, ethanol, methanol, hexane, and acetone were of chromatographic grade (Sigma-Aldrich, Santiago, Chile) for the HPLC (Jasco Inc, Tokyo, Japan) system. Individual carotenoid standards such as lutein, zeaxanthin, violaxanthin, astaxanthin, and β-carotene were also procured from Sigma-Aldrich (Santiago, Chile).

3.2. Microalgal Culture Conditions

Muriellopsis sp. (MCH35) was maintained under controlled conditions in 20 L bottles at 20 ± 2 °C, under constant illumination at 80 μE/m²·s provided by fluorescent lamps, with constant aeration of 0.1 *v/v*/min without CO_2 supply. The culture medium UMA5 was adapted to natural seawater conditions as described by Riveros et al. [28] and Marticorena et al. [14]. The inoculum was sub-cultured during the exponential growth phase on every 12th day by taking 10% of the old culture and 90% of fresh culture medium and were scaled up to open-raceway ponds of the surface of 36 m² and capacity of 5.4 m³. Subsequently, *Muriellopsis* sp. (MCH35) cells were adapted to outdoor conditions under natural illumination with incident irradiance being evaluated by the light availability present at the installations of the Universidad Antofagasta (Antofagasta, Chile), where the reactor was located. Subsequently, the culture was maintained as a batch mode for 12 days with controlled pH by using the automatic injection of CO_2. Finally, *Muriellopsis* sp. (MCH35) was harvested during its exponential growth phase by using a batch centrifuge (GEA separator, AS-1936076 model, Oede, Westphalia, Germany) at a flow rate of 2 m³/h and a maximum pressure of 0.3 MPa.

3.3. Growth Measurements

Dry biomass concentration (C_b) was measured using 50 mL of the culture sample. The samples were passed through fiberglass filters (Ø1.6 μ Munktell Filter, Falun, Sweden) and were washed with distilled water. They were then dried in an oven at 105 ± 2 °C for 2 h (in triplicate) until the weight was stabilized. Biomass concentration was determined gravimetrically. In addition, biomass productivity was calculated in volumetric terms (Pb). Batch mode Equation (6) is mentioned below:

$$Pb = (Cf - Ci)/(tf\text{-}ti), \tag{6}$$

where C is the biomass concentration in g/L and t is the time in days. The subscripts i and f denote initial and final measurements, respectively. On the other hand, an equation for calculating the specific growth rate (μ) of the culture was mentioned in the text as Equation (1).

In this equation, μ is the specific growth rate, C_i is the initial biomass concentration, and C is the biomass concentration at any time t during the exponential growth phase. The photosynthetic performance (Fv/Fm) was measured to determine cell viability. The maximum quantum yield of photosynthetic efficiency of photosystem II was achieved in the case of samples that were previously adapted to darkness for 15 min. The AquaPen-C fluorometer (Photon Systems Instruments, Drásov, Czech Republic) was used for this experiment, as described previously by Riveros et al. [28].

3.4. Drying Treatment on Microalgal Biomass

The biomass was divided into two groups: SD and FD. In the first group, the SD process was performed by using a LPG-5 Speed centrifuge spray-dryer (Jiangsu, China). The operating conditions for drying were as follows: air temperature of 185 ± 5 °C, outlet air temperature of 80 ± 5 °C, and a flow rate of 4 L/h. In the second group, the FD process was performed by using a Labconco FreeZone 18 L Benchtop Dry System (Labconco, Kansas City, MO, USA) at a temperature of −48 ± 5 °C and pressure of 2 Pa. Finally, all samples were packed in vacuum sealing plastic bags and stored at 4 ± 2 °C in the dark until use.

3.5. Detecting Individual Carotenoids

Individual carotenoids were extracted by using 5 mg of two-mode dry biomass for conventional extraction (benchmark extraction) or 50 mg from supercritical fluids (SF) extracts. Saponification of samples was then performed, and a tricomponent solution was added in all the samples as described by Cerón-García et al. [67]. This tricomponent solution was composed of ethanol:hexane:water in a proportion of 77:17:6 *v/v/v* and contained 0–60% *w/w* potassium hydroxide [68]. The supernatant was transferred into an amber vial for chromatographic analysis. Subsequently, individual carotenoids were separated and identified by using the HPLC system (Jasco Inc, Tokyo, Japan). It was equipped with a quaternary pump (PU-2089 s Plus), diode array detector, and RP-18 column (Lichrosphere, 5 µm × 150 mm) by using a method described by Cerón-García et al. [67]. In the mobile phase, solvent A was water/methanol (2:8, *v/v*), solvent B was acetone/methanol (1:1, *v/v*), and the detection wavelength was 450 nm at 25 °C of column temperature. External standards (Sigma-Aldrich) and their corresponding calibration curves were used to identify and quantify individual carotenoids such as lutein, zeaxanthin, violaxanthin, astaxanthin, and β-carotene. It was performed in triplicate ($n = 3$).

3.6. Recovery of Individual Carotenoids

The effect of operating conditions on the extraction of individual carotenoids was expressed in terms of recovery that was calculated on the basis of the initial mass of each compound, as per Equation (7) given below:

$$\text{Recovery (\%)} = (Wc/Wt) \times 100 \tag{7}$$

where W_C is the mass of the extracted compound in mg, and W_t is the theoretical mass of the compound extracted conventionally (mg). Total carotenoids were extracted by the conventional method described in Section 3.5 and defined as benchmark extraction in Table 1.

3.7. SFE

The extraction was also performed by using a Speed Helix supercritical extractor (Applied Separation, Allentown, PA, USA), which was designed by Ruiz-Domínguez et al. [69], and the extraction process is described in detail in Figure 7. For each extraction, 2 g of SD or FD biomass of *Muriellopsis* sp. (MCH35) was used. It was previously ground and sieved using a standard sieve of 35 mesh of the Tyler series (particle size ≤ 0.354 mm), along with polypropylene wool and glass beads (ϕ = 1 mm), which was then inserted into a 24 mL stainless-steel extraction cell. In all the cases, the CO_2 flow rate of 3.62 g/min was maintained, and each extraction was performed for 60 min. Extraction conditions for the microalga were selected on the basis of preliminary kinetic studies performed on *Muriellopsis* sp. (MCH35) and were set for 150–160 min to ensure the complete removal of bioactive compounds. The resulting extracts were collected in vials under dark conditions. The residual ethanol was evaporated under an N_2 gas stream avoiding the oxidation of biomolecules in the extracts by using Flexivap Work-Station (Model 109A YH-1, Glas-Col, Terre Haute, IN, USA) for calculating the extraction yield. Then, the dried extracts with N_2 atmosphere were stored at −20 ± 2 °C and in the dark until further analysis (at maximum 2 h).

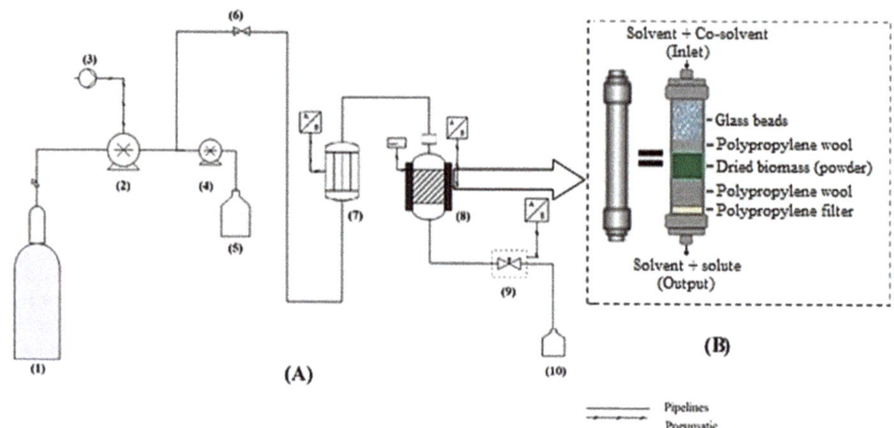

Figure 7. (**A**) Diagram of the SFE equipment (Applied Separations, Spe-ed, Allentown, PA). (1) CO_2 cylinder, (2) CO_2 pump, (3) Compressor, (4) Modifier pump, (5) Solvent tank, (6) Inlet valve, (7) Heater, (8) Extraction vessel and oven vessel, (9) Micrometric valve, (10) Sample collection. (**B**) Schematic diagram of the extraction vessel showing the way in which all elements are arranged.

3.7.1. Experimental Design

Two Box-Behnken designs were implemented in random run order, generating 15 experimental conditions for each biomass drying mode independently (30 runs in total, refer to Table 2). Considering both the designs, three factors were evaluated at 3 different experimental levels, such as temperature (40, 55, and 70 °C), pressure (30, 40, and 50 MPa), and percentage of ethanol as a co-solvent (0%, 15%, and 30% *v/v*). The effect of the factors on different response variables such as extraction yield (Y) and lutein content and recovery were determined in triplicate (n = 3). Other carotenoids were excluded from the study because of their low quantities.

3.7.2. Mathematical Modeling of Overall Extraction Curve (OEC) and Spline Linear Model

For kinetic analysis, an OEC between the optimal extraction time versus accumulated extract and lutein recovery was plotted. Extraction kinetics was performed at the central point of the experimental design (40 MPa, 55 °C, 15% ethanol *v/v*, and flow rate Q_T = 2 L/min = 3.62 g CO_2/min), as described by Gilbert-López et al. [70]. Each SD and FD biomass sample was collected at preselected intervals of 150 and 160 min, respectively. In this assay, the extraction yield (Y%) and lutein recovery (as the majority carotenoid in the profile) were calculated at each point of the curve from both the modes. This assay was performed in duplicate with 15–20 points per sample.

The OEC was fitted to a Spline linear model containing three straight lines Equation (8), as shown in Equations (9)–(11). An adjustment was performed by using PROC REG and PROC NLIN of SAS University Edition Software (https://www.sas.com/en_us/software/university-edition.html) Finally, the fitted data from Equation (7) were plotted by using a Microsoft Excel-2016 spreadsheet. Each fitted line represents the following extraction stages related to the mass transfer mechanism: constant extraction rate (CER) period, falling extraction rate (FER) period that represents the stage at which both convection and diffusion in the solid substratum control the process, and diffusion-controlled (DC) periods, as described by Meireles [60]. In the CER period, the mass transfer rate for the CER period (M_{CER}), as well as the time corresponding to the interception of the two lines (t_{CER}), was computed from the Spline linear model. A similar procedure was followed for the FER period, and finally, the DC stage was computed. The experimental data obtained from the OEC were fitted. The mass ratio of the solute in the supercritical phase at the equilibrium cell outlet (Y_{CER}) was obtained by dividing M_{CER} by the mean solvent flow rate for the CER period. A similar procedure was employed by Salinas et al. [64] in the mathematical calculations for obtaining almond oil from the chañar fruit (*Geoffroea decorticans*) by SFE.

$$y = m_{Ext} = \left(b_0 - \sum_{i=1}^{i=N} C_i\, a_{i+1}\right) + \sum_{i=1}^{i=N} a_i t \tag{8}$$

For one straight line:

$$y = m_{Ext} = b_0 + a_1 t \quad \text{for } t \leq t_{CER} \tag{9}$$

For two straight lines:

$$y = m_{Ext} = b_0 - t_{CER} a_2 + (a_1 + a_2)t \quad \text{for } t_{CER} < t \leq t_{FER} \tag{10}$$

For three straight lines:

$$y = m_{Ext} = b_0 - t_{CER} a_2 - t_{FER} a_3 + (a_1 + a_2 + a_3)t \quad \text{for } t_{FER} < t \tag{11}$$

where y = response variable = m_{Ext} is the mass of extract, a_i (i = 0, 1, 2, 3) = linear coefficients of lines, t = time (min), t_{CER} = CER time (min), and t_{FER} = FER time (min). C_i for i = 1, 2 are the intercepts of these lines (for instance, C_1 is the intercept of the first and second lines, and C_2 is the intercept of the second and third lines).

Using the adjusted parameters, y_{CER} and y_{FER} calculated from t_{CER} and t_{FER} were calculated. Recoveries were then calculated at each time according to Equation (12) given below:

$$Recovery\ (\%) = \frac{y_t}{y_{time\ final\ OEC}}\ (100) \tag{12}$$

3.8. Statistical Analysis

Experimental designs and data analysis were performed by response surface methodology (RSM) and using the Statgraphics Centurion XVI® (StatPoint Technologies, Inc., Warrenton, VA, USA) software. The effects of the factors on response variables in the separation process were assessed by using the pure error and considering a confidence interval of 95% for all the variables. The effect of each factor on response variables and its statistical significance were analyzed by using analysis of variance (ANOVA) (included in the Supplementary Materials) and the standardized Pareto chart.

The response surfaces of the respective mathematical models were also obtained, and a *p*-value of ≤ 0.05 was considered significant. All measurements were performed in triplicate ($n = 3$).

The mathematical relationship of the response with three factors, X_1, X_2, and X_3, involved in the design was approximated by using quadratic polynomial Equation (13) (second degree):

$$Z = \beta_0 + \beta_1 X_1 + \beta_2 X_2 + \beta_3 X_3 + \beta_{12} X_1 X_2 + \beta_{13} X_1 X_3 + \beta_{23} X_2 X_3 + \beta_{11} X_1^2 + \beta_{22} X_2^2 + \beta_{33} X_3^2 \quad (13)$$

where Z = estimate response, β_0 = constant, β_1, β_2, and β_3 = linear coefficients, β_{12}, β_{13}, and β_{23} = interaction coefficients between the three factors, and β_{11}, β_{22}, and β_{33} = quadratic coefficients. The multiple regression analysis was performed to obtain coefficients and equations that can be used to predict the responses.

4. Conclusions

The effect of drying methods as a pretreatment for *Muriellopsis* sp. (MCH35) biomass was studied for the optimization of lutein recovery extracted by the SF process. The production of microalga isolated from the arid north Chile was undertaken in seawater medium (UMA5) and arid outdoor conditions, with the focus on reduction of the operational costs. The production conditions were suitable especially for desert areas where solar radiation is high and fresh water is limited. The strain showed high biomass content and volumetric productivity values between 1.34 g/L and 75.73 mg/L·d respectively, after 12 days. Moreover, the *Muriellopsis* cells showed an average photosynthetic efficiency of 0.65, confirming the microalga was able to adapt to harsh environmental conditions. Moreover, the lutein content was in a similar range with that reported by other lutein-producing microalgae in outdoor conditions. The supercritical experimental outcomes showed that the modifier (ethanol) played a crucial role in the extraction in terms of yield, and lutein content and recovery. However, the parameters such as extraction temperature and pressure were non-significant in the extraction. The maximum optimal yield was similar under both the drying methods at the temperature of the medium to high, low pressure, and maximum extractant (30% *v/v* ethanol). The maximum lutein recovery of 74% (*w/w*) for the FD biomass was 2.5-fold higher than that for the SD biomass. Therefore, FD was the optimal pre-treatment to enhance the high-valuable extracts of *Muriellopsis* sp. (MCH35) as microalga adaptable to hostile growth environments for biotechnological applications.

Supplementary Materials: The following are available online at http://www.mdpi.com/1660-3397/18/11/528/s1, Table S1: Regression coefficients (values of variables are specified in their original units) extraction yields (Y), lutein content, and recovery from spray- (SD) and freeze-dried (FD) biomass, statistics for the fit obtained by multiple linear regression.

Author Contributions: Conceptualization, M.C.R.-D. and P.C.; methodology, F.S. and P.M.; software, M.C.R.-D. and P.C.; validation, M.C.R.-D., P.C. and C.R.; formal analysis, M.C.R.-D., P.M. and P.C.; resources, C.S. and C.R.; data curation, P.M., C.S. and F.S.; writing—original draft preparation, M.C.R.-D.; writing—review and editing, P.C.; supervision, P.C. and C.R.; funding acquisition, P.C. and C.R. All authors have read and agreed to the published version of the manuscript.

Funding: This research was funded by projects with public funds of Chile (ANID, National Agency for Research and Development of Chile) such as grants FONDEQUIP EQM-160073 and FONDEF ID18I10095.

Conflicts of Interest: The authors declare no conflict of interest.

References

1. Forján, E.; Navarro, F.; Cuaresma, M.; Vaquero, I.; Ruíz-Domínguez, M.C.; Gojkovic, Ž.; Vázquez, M.; Márquez, M.; Mogedas, B.; Bermejo, E. Microalgae: Fast-growth sustainable green factories. *Crit. Rev. Environ. Sci. Technol.* **2015**, *45*, 1705–1755. [CrossRef]

2. Tomaselli, L. The microalgal cell. In *Handbook of Microalgal Culture: Biotechnology and Applied Phycology*; Wiley Online Library: Hoboken, NY, USA, 2004; Volume 1, pp. 3–19. [CrossRef]

3. Singh, S.K.; Kaur, R.; Bansal, A.; Kapur, S.; Sundaram, S. Biotechnological exploitation of cyanobacteria and microalgae for bioactive compounds. In *Biotechnological Production of Bioactive Compounds*; Elsevier: Amsterdam, The Netherlands, 2020; pp. 221–259.

4. Liu, R.H. Potential synergy of phytochemicals in cancer prevention: Mechanism of action. *J. Nutr.* **2004**, *134*, 3479S–3485S. [CrossRef]

5. Michalak, I.; Chojnacka, K. Algae as production systems of bioactive compounds. *Eng. Life Sci.* **2015**, *15*, 160–176. [CrossRef]

6. Simopoulos, A.P. Nutrition tid-bites: Essential fatty acids in health and chronic disease. *Food Rev. Int.* **1997**, *13*, 623–631. [CrossRef]

7. Kini, S.; Divyashree, M.; Mani, M.K.; Mamatha, B.S. Algae and cyanobacteria as a source of novel bioactive compounds for biomedical applications. In *Advances in Cyanobacterial Biology*; Elsevier: Amsterdam, The Netherlands, 2020; pp. 173–194.

8. Del Campo, J.A.; García-González, M.; Guerrero, M.G. Outdoor cultivation of microalgae for carotenoid production: Current state and perspectives. *Appl. Microbiol. Biotechnol.* **2007**, *74*, 1163–1174. [CrossRef] [PubMed]

9. Gong, M.; Bassi, A. Carotenoids from microalgae: A review of recent developments. *Biotechnol. Adv.* **2016**, *34*, 1396–1412. [CrossRef]

10. Guedes, A.C.; Amaro, H.M.; Malcata, F.X. Microalgae as sources of carotenoids. *Mar. Drugs* **2011**, *9*, 625–644. [CrossRef]

11. Dwyer, J.H.; Navab, M.; Dwyer, K.M.; Hassan, K.; Sun, P.; Shircore, A.; Hama-Levy, S.; Hough, G.; Wang, X.; Drake, T. Oxygenated carotenoid lutein and progression of early atherosclerosis: The Los Angeles atherosclerosis study. *Circulation* **2001**, *103*, 2922–2927. [CrossRef]

12. Heber, D.; Lu, Q.-Y. Overview of mechanisms of action of lycopene. *Exp. Biol. Med.* **2002**, *227*, 920–923. [CrossRef]

13. Landrum, J.T.; Bone, R.A. Lutein, zeaxanthin, and the macular pigment. *Arch. Biochem. Biophys.* **2001**, *385*, 28–40. [CrossRef]

14. Marticorena, P.; Gonzalez, L.; Riquelme, C.; Silva Aciares, F. Effects of beneficial bacteria on biomass, photosynthetic parameters and cell composition of the microalga *Muriellopsis* sp. adapted to grow in seawater. *Aquac. Res.* **2020**. [CrossRef]

15. Rocha, S.; Candia, O.; Valdebenito, F.; Espinoza-Monje, J.F.; Azócar, L. Biomass quality index: Searching for suitable biomass as an energy source in Chile. *Fuel* **2020**, *264*, 116820. [CrossRef]

16. Hu, Q. *Environmental Effects on Cell Composition*; Wiley Online Library: Hoboken, NY, USA, 2004; Volume 1, pp. 83–93.

17. Solovchenko, A.; Khozin-Goldberg, I.; Recht, L.; Boussiba, S. Stress-induced changes in optical properties, pigment and fatty acid content of Nannochloropsis sp.: Implications for non-destructive assay of total fatty acids. *Mar. Biotechnol.* **2011**, *13*, 527–535. [CrossRef] [PubMed]

18. Minhas, A.K.; Hodgson, P.; Barrow, C.J.; Adholeya, A. A review on the assessment of stress conditions for simultaneous production of microalgal lipids and carotenoids. *Front. Microbiol.* **2016**, *7*, 546. [CrossRef] [PubMed]

19. Fernández-Sevilla, J.M.; Fernández, F.A.; Grima, E.M. Biotechnological production of lutein and its applications. *Appl. Microbiol. Biotechnol.* **2010**, *86*, 27–40. [CrossRef]

20. Orikasa, T.; Koide, S.; Okamoto, S.; Imaizumi, T.; Muramatsu, Y.; Takeda, J.-I.; Shiina, T.; Tagawa, A. Impacts of hot air and vacuum drying on the quality attributes of kiwifruit slices. *J. Food Eng.* **2014**, *125*, 51–58. [CrossRef]

21. Viswanathan, T.; Mani, S.; Das, K.; Chinnasamy, S.; Bhatnagar, A.; Singh, R.; Singh, M. Effect of cell rupturing methods on the drying characteristics and lipid compositions of microalgae. *Bioresour. Technol.* **2012**, *126*, 131–136. [CrossRef]

22. Oliveira, E.G.; Duarte, J.H.; Moraes, K.; Crexi, V.T.; Pinto, L.A. Optimisation of Spirulina platensis convective drying: Evaluation of phycocyanin loss and lipid oxidation. *Int. J. Food Sci. Technol.* **2010**, *45*, 1572–1578. [CrossRef]

23. Stramarkou, M.; Papadaki, S.; Kyriakopoulou, K.; Krokida, M. Effect of drying and extraction conditions on the recovery of bioactive compounds from Chlorella vulgaris. *J. Appl. Phycol.* **2017**, *29*, 2947–2960. [CrossRef]

24. Chen, C.-L.; Chang, J.-S.; Lee, D.-J. Dewatering and drying methods for microalgae. *Dry. Technol.* **2015**, *33*, 443–454. [CrossRef]

25. Grima, E.M.; Fernández, F.A.; Medina, A.R. 10 Downstream Processing of Cell-mass and Products. In *Handbook of Microalgal Culture: Biotechnology and Applied Phycology*; Wiley Online Library: Hoboken, NY, USA, 2004; Volume 215.

26. Poojary, M.M.; Barba, F.J.; Aliakbarian, B.; Donsì, F.; Pataro, G.; Dias, D.A.; Juliano, P. Innovative alternative technologies to extract carotenoids from microalgae and seaweeds. *Mar. Drugs* **2016**, *14*, 214. [CrossRef] [PubMed]

27. Huang, G.; Chen, F.; Wei, D.; Zhang, X.; Chen, G. Biodiesel production by microalgal biotechnology. *Appl. Energy* **2010**, *87*, 38–46. [CrossRef]

28. Riveros, K.; Sepulveda, C.; Bazaes, J.; Marticorena, P.; Riquelme, C.; Acién, G. Overall development of a bioprocess for the outdoor production of Nannochloropsis gaditana for aquaculture. *Aquac. Res.* **2018**, *49*, 165–176. [CrossRef]

29. San Pedro, A.; González-López, C.; Acién, F.; Molina-Grima, E. Outdoor pilot production of Nannochloropsis gaditana: Influence of culture parameters and lipid production rates in raceway ponds. *Algal Res.* **2015**, *8*, 205–213. [CrossRef]

30. González-Garcinuño, Á.; Tabernero, A.; Sánchez-Álvarez, J.M.; del Valle, E.M.M.; Galán, M.A. Effect of nitrogen source on growth and lipid accumulation in Scenedesmus abundans and Chlorella ellipsoidea. *Bioresour. Technol.* **2014**, *173*, 334–341. [CrossRef] [PubMed]

31. Sepúlveda, C.; Acien, F.G.; Gómez, C.; Jiménez-Ruiz, N.; Riquelme, C.; Molina-Grima, E. Utilization of centrate for the production of the marine microalgae Nannochloropsis gaditana. *Algal Res.* **2015**, *9*, 107–116. [CrossRef]

32. Chew, K.W.; Chia, S.R.; Show, P.L.; Yap, Y.J.; Ling, T.C.; Chang, J.-S. Effects of water culture medium, cultivation systems and growth modes for microalgae cultivation: A review. *J. Taiwan Inst. Chem. Eng.* **2018**, *91*, 332–344. [CrossRef]

33. Camacho-Rodríguez, J.; González-Céspedes, A.; Cerón-García, M.; Fernández-Sevilla, J.; Acién-Fernández, F.; Molina-Grima, E. A quantitative study of eicosapentaenoic acid (EPA) production by Nannochloropsis gaditana for aquaculture as a function of dilution rate, temperature and average irradiance. *Appl. Microbiol. Biotechnol.* **2014**, *98*, 2429–2440. [CrossRef]

34. Murata, N.; Takahashi, S.; Nishiyama, Y.; Allakhverdiev, S.I. Photoinhibition of photosystem II under environmental stress. *Biochim. Biophys. Acta* **2007**, *1767*, 414–421. [CrossRef]

35. Morales-Amaral, M.d.M.; Gómez-Serrano, C.; Acién, F.G.; Fernández-Sevilla, J.M.; Molina-Grima, E. Production of microalgae using centrate from anaerobic digestion as the nutrient source. *Algal Res.* **2015**, *9*, 297–305. [CrossRef]

36. Del Campo, J.A.; Moreno, J.; Rodríguez, H.; Vargas, M.A.; Rivas, J.; Guerrero, M.G. Carotenoid content of chlorophycean microalgae: Factors determining lutein accumulation in Muriellopsis sp.(Chlorophyta). *J. Biotechnol.* **2000**, *76*, 51–59. [CrossRef]

37. Ryckebosch, E.; Muylaert, K.; Eeckhout, M.; Ruyssen, T.; Foubert, I. Influence of drying and storage on lipid and carotenoid stability of the microalga Phaeodactylum tricornutum. *J. Agric. Food Chem.* **2011**, *59*, 11063–11069. [CrossRef]

38. Jin, E.; Polle, J.E.; Lee, H.K.; Hyun, S.M.; Chang, M. Xanthophylls in microalgae: From biosynthesis to biotechnological mass production and application. *J. Microbiol. Biotechnol.* **2003**, *13*, 165–174.

39. Blanco, A.M.; Moreno, J.; Del Campo, J.A.; Rivas, J.; Guerrero, M.G. Outdoor cultivation of lutein-rich cells of Muriellopsis sp. in open ponds. *Appl. Microbiol. Biotechnol.* **2007**, *73*, 1259–1266. [CrossRef] [PubMed]

40. Del Campo, J.A.; Rodríguez, H.; Moreno, J.; Vargas, M.Á.; Rivas, J.; Guerrero, M.G. Lutein production by Muriellopsis sp. in an outdoor tubular photobioreactor. *J. Biotechnol.* **2001**, *85*, 289–295. [CrossRef]

41. Molino, A.; Mehariya, S.; Iovine, A.; Casella, P.; Marino, T.; Karatza, D.; Chianese, S.; Musmarra, D. Enhancing biomass and lutein production from Scenedesmus almeriensis: Effect of carbon dioxide concentration and culture medium reuse. *Front. Plant Sci.* **2020**, *11*, 415. [CrossRef] [PubMed]

42. Begum, H.; Yusoff, F.M.; Banerjee, S.; Khatoon, H.; Shariff, M. Availability and utilization of pigments from microalgae. *Crit. Rev. Food Sci. Nutr.* **2016**, *56*, 2209–2222. [CrossRef]

43. Schalch, W.; Cohn, W.; Barker, F.M.; Köpcke, W.; Mellerio, J.; Bird, A.C.; Robson, A.G.; Fitzke, F.F.; van Kuijk, F.J. Xanthophyll accumulation in the human retina during supplementation with lutein or zeaxanthin–the LUXEA (LUtein Xanthophyll Eye Accumulation) study. *Arch. Biochem. Biophys.* **2007**, *458*, 128–135. [CrossRef]

44. Spolaore, P.; Joannis-Cassan, C.; Duran, E.; Isambert, A. Commercial applications of microalgae. *J. Biosci. Bioeng.* **2006**, *101*, 87–96. [CrossRef]

45. Kim, H.M.; Jung, J.H.; Kim, J.Y.; Heo, J.; Cho, D.H.; Kim, H.S.; An, S.; An, I.S.; Bae, S. The Protective Effect of Violaxanthin from Nannochloropsis oceanica against Ultraviolet B-Induced Damage in Normal Human Dermal Fibroblasts. *Photochem. Photobiol.* **2019**, *95*, 595–604. [CrossRef]

46. Jyonouchi, H.; Sun, S.; Gross, M. Effect of carotenoids on in vitro immunoglobulin production by human peripheral blood mononuclear cells: Astaxanthin, a carotenoid without vitamin a activity, enhances in vitro immunoglobulin production in response to at-dependent stimulant and antigen. *Nutr. Cancer* **1995**, *23*, 171–183. [CrossRef] [PubMed]

47. Di Caprio, F.; Altimari, P.; Pagnanelli, F. Sequential extraction of lutein and β-carotene from wet microalgae biomass. *J. Chem. Technol. Biotechnol.* **2020**, *95*, 3024–3033. [CrossRef]

48. Corzzini, S.C.; Barros, H.D.; Grimaldi, R.; Cabral, F.A. Extraction of edible avocado oil using supercritical CO_2 and a CO_2/ethanol mixture as solvents. *J. Food Eng.* **2017**, *194*, 40–45. [CrossRef]

49. Sosa-Hernández, J.E.; Escobedo-Avellaneda, Z.; Iqbal, H.; Welti-Chanes, J. State-of-the-art extraction methodologies for bioactive compounds from algal biome to meet bio-economy challenges and opportunities. *Molecules* **2018**, *23*, 2953. [CrossRef] [PubMed]

50. Mehariya, S.; Iovine, A.; Di Sanzo, G.; Larocca, V.; Martino, M.; Leone, G.P.; Casella, P.; Karatza, D.; Marino, T.; Musmarra, D. Supercritical fluid extraction of lutein from Scenedesmus almeriensis. *Molecules* **2019**, *24*, 1324. [CrossRef] [PubMed]

51. Ratti, C. Hot air and freeze-drying of high-value foods: A review. *J. Food Eng.* **2001**, *49*, 311–319. [CrossRef]

52. Irzyniec, Z.; Klimczak, J.; Michalowski, S. Freeze-drying of the black currant juice. *Dry. Technol.* **1995**, *13*, 417–424. [CrossRef]

53. Yen, H.-W.; Chiang, W.-C.; Sun, C.-H. Supercritical fluid extraction of lutein from Scenedesmus cultured in an autotrophical photobioreactor. *J. Taiwan Inst. Chem. Eng.* **2012**, *43*, 53–57. [CrossRef]

54. Wu, Z.; Wu, S.; Shi, X. Supercritical fluid extraction and determination of lutein in heterotrophically cultivated Chlorella pyrenoidosa. *J. Food Process Eng.* **2007**, *30*, 174–185. [CrossRef]

55. Smyth, T.J.; Zytner, R.; Stiver, W. Influence of water on the supercritical fluid extraction of naphthalene from soil. *J. Hazard. Mater.* **1999**, *67*, 183–196. [CrossRef]

56. Nagy, B.; Simándi, B. Effects of particle size distribution, moisture content, and initial oil content on the supercritical fluid extraction of paprika. *J. Supercrit. Fluids* **2008**, *46*, 293–298. [CrossRef]

57. Mouahid, A.; Crampon, C.; Toudji, S.-A.A.; Badens, E. Effects of high water content and drying pre-treatment on supercritical CO_2 extraction from Dunaliella salina microalgae: Experiments and modelling. *J. Supercrit. Fluids* **2016**, *116*, 271–280. [CrossRef]

58. Crampon, C.; Mouahid, A.; Toudji, S.-A.A.; Lépine, O.; Badens, E. Influence of pretreatment on supercritical CO_2 extraction from Nannochloropsis oculata. *J. Supercrit. Fluids* **2013**, *79*, 337–344. [CrossRef]

59. Ivanovic, J.; Ristic, M.; Skala, D. Supercritical CO_2 extraction of Helichrysum italicum: Influence of CO_2 density and moisture content of plant material. *J. Supercrit. Fluids* **2011**, *57*, 129–136. [CrossRef]

60. Meireles, M.A.A. Extraction of bioactive compounds from Latin American plants. In *Supercritical Fluid Extraction of Nutraceuticals and Bioactive Compounds*; CRC Press—Taylor & Francis Group: Boca Raton, FL, USA, 2008; pp. 243–274.

61. Jesus, S.P.; Calheiros, M.N.; Hense, H.; Meireles, M.A.A. A simplified model to describe the kinetic behavior of supercritical fluid extraction from a rice bran oil byproduct. *Food Public Health* **2013**, *3*, 215–222. [CrossRef]

62. Pereira, C.G.; Meireles, M.A.A. Supercritical fluid extraction of bioactive compounds: Fundamentals, applications and economic perspectives. *Food Bioprocess Technol.* **2010**, *3*, 340–372. [CrossRef]

63. Mezzomo, N.; Martínez, J.; Ferreira, S.R. Supercritical fluid extraction of peach (Prunus persica) almond oil: Kinetics, mathematical modeling and scale-up. *J. Supercrit. Fluids* **2009**, *51*, 10–16. [CrossRef]

64. Salinas, F.; Vardanega, R.; Espinosa-Álvarez, C.; Jimenéz, D.; Muñoz, W.B.; Ruiz-Domínguez, M.C.; Meireles, M.A.A.; Cerezal-Mezquita, P. Supercritical fluid extraction of chañar (Geoffroea decorticans) almond oil: Global yield, kinetics and oil characterization. *J. Supercrit. Fluids* **2020**, 104824. [CrossRef]

65. Sanzo, G.D.; Mehariya, S.; Martino, M.; Larocca, V.; Casella, P.; Chianese, S.; Musmarra, D.; Balducchi, R.; Molino, A. Supercritical carbon dioxide extraction of astaxanthin, lutein, and fatty acids from Haematococcus pluvialis microalgae. *Mar. Drugs* **2018**, *16*, 334. [CrossRef]

66. Del Valle, J.M. Extraction of natural compounds using supercritical CO2: Going from the laboratory to the industrial application. *J. Supercrit. Fluids* **2015**, *96*, 180–199. [CrossRef]

67. Cerón-García, M.C.; González-López, C.V.; Camacho-Rodríguez, J.; López-Rosales, L.; García-Camacho, F.; Molina-Grima, E. Maximizing carotenoid extraction from microalgae used as food additives and determined by liquid chromatography (HPLC). *Food Chem.* **2018**, *257*, 316–324. [CrossRef] [PubMed]

68. Fernández Sevilla, J.M.; Acién Fernández, F.G.; Molina Grima, E.; Cerón García, M.d.C. Extracción de carotenoides mediante el uso de mezclas ternarias. ES200703145A, 13 November 2016.

69. Ruiz-Domínguez, M.C.; Cerezal, P.; Salinas, F.; Medina, E.; Renato-Castro, G. Application of Box-Behnken Design and Desirability Function for Green Prospection of Bioactive Compounds from Isochrysis galbana. *Appl. Sci. Basel* **2020**, *10*, 2789. [CrossRef]

70. Gilbert-López, B.; Mendiola, J.A.; Fontecha, J.; van den Broek, L.A.; Sijtsma, L.; Cifuentes, A.; Herrero, M.; Ibáñez, E. Downstream processing of Isochrysis galbana: A step towards microalgal biorefinery. *Green Chem.* **2015**, *17*, 4599–4609. [CrossRef]

Publisher's Note: MDPI stays neutral with regard to jurisdictional claims in published maps and institutional affiliations.

Article

A Crustin from Hydrothermal Vent Shrimp: Antimicrobial Activity and Mechanism

Yujian Wang [1,2,3], Jian Zhang [1,4], Yuanyuan Sun [1,2] and Li Sun [1,2,3,*]

1   CAS and Shandong Province Key Laboratory of Experimental Marine Biology, Institute of Oceanology, Center for Ocean Mega-Science, Chinese Academy of Sciences, Qingdao 266071, China; wangyujian@qdio.ac.cn (Y.W.); zhangjian@ytu.edu.cn (J.Z.); sunyuanyuan@qdio.ac.cn (Y.S.)
2   Laboratory for Marine Biology and Biotechnology, Pilot National Laboratory for Marine Science and Technology (Qingdao), Qingdao 266237, China
3   College of Earth and Planetary Sciences, University of Chinese Academy of Sciences, Beijing 100049, China
4   School of Ocean, Yantai University, Yantai 264005, China
*   Correspondence: lsun@qdio.ac.cn; Tel.: +86-532-8289-8829

Abstract: Crustin is a type of antimicrobial peptide and plays an important role in the innate immunity of arthropods. We report here the identification and characterization of a crustin (named Crus1) from the shrimp *Rimicaris* sp. inhabiting the deep-sea hydrothermal vent in Manus Basin (Papua New Guinea). Crus1 shares the highest identity (51.76%) with a Type I crustin of *Penaeus vannamei* and possesses a whey acidic protein (WAP) domain, which contains eight cysteine residues that form the conserved 'four-disulfide core' structure. Recombinant Crus1 (rCrus1) bound to peptidoglycan and lipoteichoic acid, and effectively killed Gram-positive bacteria in a manner that was dependent on pH, temperature, and disulfide linkage. rCrus1 induced membrane leakage and structure damage in the target bacteria, but had no effect on bacterial protoplasts. Serine substitution of each of the 8 Cys residues in the WAP domain did not affect the bacterial binding capacity but completely abolished the bactericidal activity of rCrus1. These results provide new insights into the characteristic and mechanism of the antimicrobial activity of deep sea crustins.

Keywords: crustin; antimicrobial peptides; shrimp; deep-sea hydrothermal vent

Citation: Wang, Y.; Zhang, J.; Sun, Y.; Sun, L. A Crustin from Hydrothermal Vent Shrimp: Antimicrobial Activity and Mechanism. *Mar. Drugs* **2021**, *19*, 176. https://doi.org/10.3390/md19030176

Academic Editor: Daniela Giordano

Received: 17 February 2021
Accepted: 19 March 2021
Published: 23 March 2021

1. Introduction

Antimicrobial peptides (AMPs) are a class of evolutionarily conserved molecules that exist in almost all organisms as mediators of innate immunity. Invertebrates, which lack the adaptive immune system, rely particularly on AMPs and other innate immune factors to resist invading pathogens [1,2]. Functionally, AMPs can destroy invading microbial pathogens and directly kill bacteria, fungi, viruses, and parasites [3,4]. Unlike traditional antibiotics, which are well known to induce resistance in the target bacteria, AMPs are intrinsic components of organisms and target the inner and/or outer membranes of bacteria in a non-receptor-specific manner, with a rate of resistance several orders of magnitude lower than that of conventional antibiotics [5–7].

AMPs are highly diverse in structure and function, and usually have a low molecular mass (<10 kDa) [8]. AMPs possess biochemical features, such as amino acid composition, size, amphipathicity, and cationic charge, that allow them to have a high propensity for selective membrane-interaction [8]. Extensive studies indicate that AMP-mediated permeabilization/disruption of the microbial cytoplasmic membrane is the main mechanism of cell killing for most AMPs [9–11]. Some non-membrane permeable AMPs can inhibit or destroy the key processes of intracellular targets (DNA, RNA or protein), inactivate essential intracellular enzymes, or affect the formation of membrane compartments and cell wall synthesis [8,12–14].

Crustin is categorized as a type of AMP that plays a vital role in the immune defense of crustaceans. Crustin is generally a cationic peptide of 7–22 kDa and contains twelve conserved cysteine residues, eight of which comprise a typical whey acid protein (WAP) domain [15]. The WAP domain forms a four-disulfide bond core arrangement at the C-terminus and is potentially associated with multiple functions [16,17]. At present, a large number of crustins have been reported, which exhibit various antibacterial or protease inhibitory functions [18–20]. However, in many cases, the specific bactericidal mechanisms of the crustins remain to be investigated.

The deep sea is the largest ecosystem on earth, with many unique biological resources, including microorganisms and invertebrates [21–23]. Marine invertebrates, such as shrimp, have been considered as promising sources for the discovery of bioactive materials [24]. Shrimps of the family Alvinocarididae inhabit the deep waters in Atlantic, Pacific, and Indian Oceans, especially the hydrothermal vents and cold seeps, where they are often found to be the dominant fauna [25,26]. Recently, a novel anti-Gram-positive crustin, Re-crustin, was identified from the extremophile Pleocyemata shrimp, *Rimicaris exoculata*, collected from the hydrothermal vent site of the Mid-Atlantic Ridge (MAR) [27]. In a previous study, we identified several crustin-like genes from the transcriptome of the shrimp *Rimicaris* sp (Alvinocarididae family) from a hydrothermal vent in Desmos, manus basin [26]. In this study, we characterized one of these crustins (designated Crus1). We investigated the structural feature and antimicrobial effect of recombinant Crus1, and identified the key cysteine residues required for bactericidal activity. Our results add new knowledge to the antimicrobial mechanism of deep sea crustins.

2. Results

2.1. Sequence and Structure Characterization of Crus1

The deduced amino acid sequence of Crus1 contains 109 residues, with a calculated molecular weight of 12.05 kDa and a predicted pI of 7.82. Crus1 possesses a signal peptide in the N-terminus (residues 1 to 20) and a WAP domain in the C-terminus, in which a 'four-disulfide core' structure can be formed by C64–C93, C70–C97, C80–C92, and C86–C103. Protein BLAST showed that Crus1 shares the highest identity (51.76%) with PvCrus, a Type I crustin of *Penaeus vannamei* (GenBank accession No. MT375562). The sequence identity between Crus1 and Re-crustin, the Type II crustin identified in the shrimp from MAR [27], is 24.74%. The sequence alignment between Crus1 and representative Type I crustins indicated that the conserved WAP domain, in particular the 8 cysteines that form the four-disulfide core structure, was shared among the crustins (Figure 1A). In addition, four cysteines in the N-terminus (corresponding to C31, C35, C44, and C45 in Crus1) were also conserved among the Type I crustins (Figure 1A). Phylogenetic analysis showed that Crus1 was grouped into the clade of Type I crustin (Figure 1B). The predicted protein structure of Crus1 contains mostly random coils, with a very little amount of α-helix and β-pleated sheet (Figure 1C).

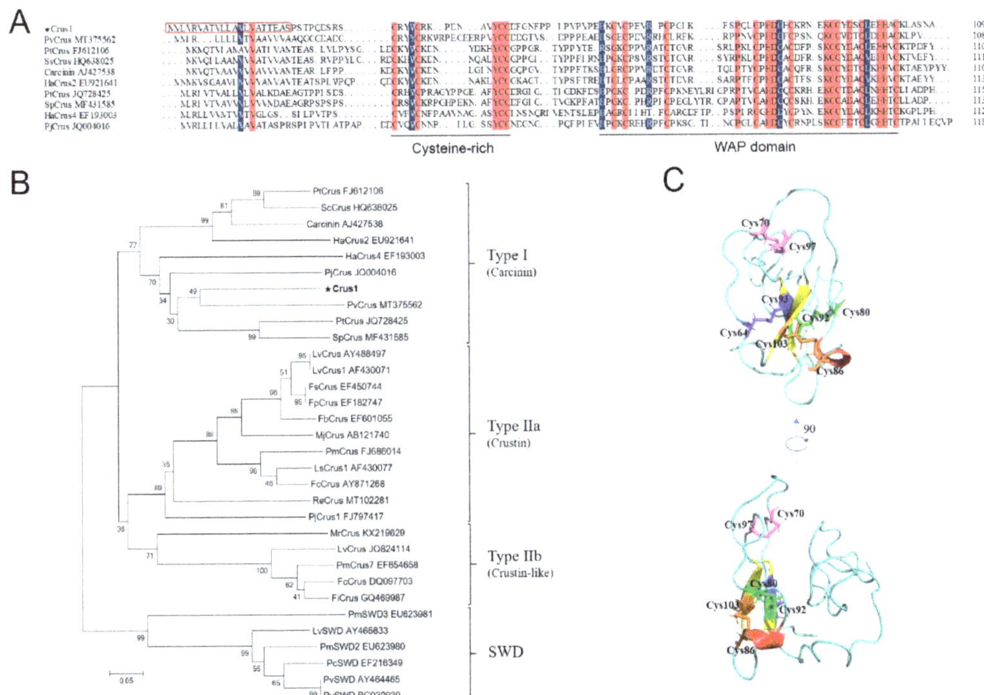

Figure 1. Sequence, phylogenetic, and structural analysis of Crus1. (**A**) Alignment of Crus1 with Type I crustins. Dots denote gaps introduced for maximum matching. The consensus residues are shaded red, the residues that are ≥75% identical among the aligned sequences are shaded blue. The signal peptide sequence of Crus1 is boxed with red lines. (**B**) Phylogenetic analysis of Crus1 homologues. The phylogenetic tree was constructed with MEGA 6.0 using the neighbor-joining method. Numbers beside the internal branches indicate bootstrap values based on 1000 replications. The GenBank accession numbers of the crustins used in (**A**,**B**) are indicated after the names of the crustins. (**C**) The predicted structure of Crus1 was built using I-TASSER. The disulfide bonds in the WAP domain are shown in blue (C64–C93), pink (C70–C97), green (C80–C92) and orange (C86–C103).

2.2. Antimicrobial Activity of rCrus1 and Its Dependence on Temperature, pH, and Disulfide Bonds

Recombinant Crus1 (rCrus1) was purified from *E. coli* as a His-tagged protein (Figure S1). The antibacterial activity of rCrus1 was tested against a variety of Gram-positive and Gram-negative bacteria, including those from deep sea environments, by measuring the minimal inhibitory concentration (MIC) and minimal bactericidal concentration (MBC) against each of the bacteria. As shown in Table 1, rCrus1 exhibited apparent inhibitory and killing activities against Gram-positive bacteria, but not against Gram-negative bacteria. The most potent activity was detected against *M. luteus*, with MIC and MBC values of 2.5 and 5 μM, respectively. Temperature dependence analysis showed that when rCrus1 was incubated with *M. luteus* at 4 °C, 16 °C, 37 °C, and 42 °C, the survival rates of the bacteria were similar (Figure 2A). pH dependence analysis showed that the bactericidal activity of rCrus1 against *M. luteus* was retained at pH 5 and 7, but completely lost at pH 9 and 11 (Figure 2B). In contrast, no apparent bactericidal activity of rCrus1 against the Gram-negative bacterium *Vibrio harveyi* was detected at either of these temperature/pH conditions (Figure S2). With *M. luteus* as the target bacterium, the killing effect of rCrus1 under the optimal condition (pH 7, 37 °C) was time-dependent (Figure S3). To examine whether the disulfide linkages were required for the bactericidal activity of rCrus1, the protein was treated with dithiothreitol (DTT), which reduces disulfide bond. The results showed that DTT treatment completely abolished the bactericidal effect of rCrus1 on *M. luteus* (Figure 2C).

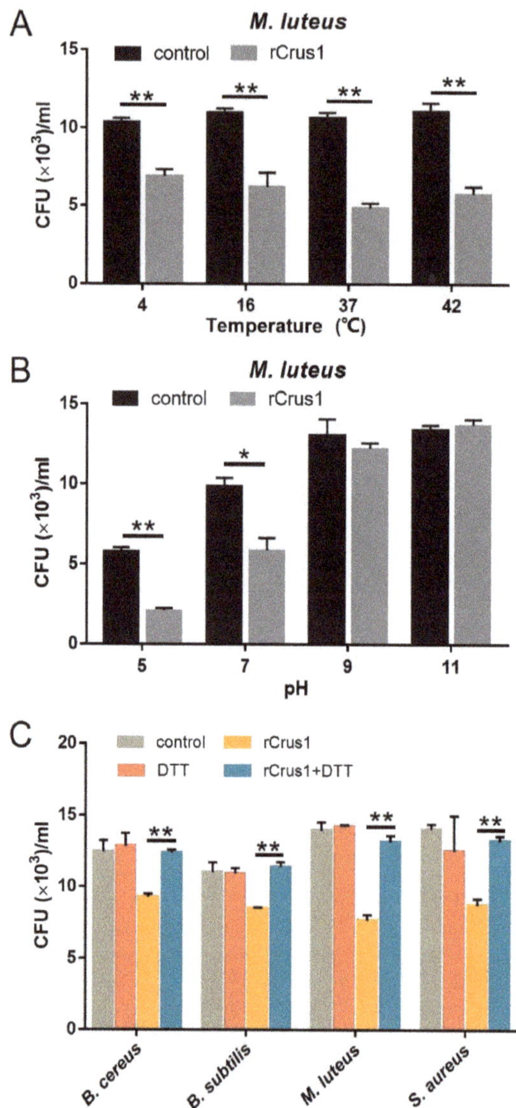

Figure 2. Effect of temperature, pH and disulfide bond on the antibacterial activity of rCrus1. (**A**) *Micrococcus luteus* was incubated with or without (control) rCrus1 (2.5 μM) at various temperatures for 2 h, and bacterial survival was determined by plate count. (**B**) *M. luteus* was incubated with or without (control) rCrus1 (2.5 μM) at various pH for 2 h, and bacterial survival was determined as above. (**C**) Bacteria were incubated with or without (control) rCrus1 (2.5 μM), DTT-treated rCrus1 (2.5 μM), or DTT for 2 h, and bacterial survival was determined as above. Values are shown as means ± SD (*N* = 3). N, the number of replicates. ** *p* < 0.01, * *p* < 0.05 (Student's *t* test).

Table 1. The minimal inhibitory concentration (MIC) and the minimal bactericidal concentration (MBC) of rCrus1 against Gram-positive and Gram-negative bacteria.

Bacteria	MIC (μM)	MBC (μM)
Gram-positive		
Bacillus subtilis WB800N	20	40
Bacillus subtilis G7	40	40
Bacillus wiedmannii SR52	20	40
Bacillus cereus MB1	20	40
Bacillus toyonensis P18	30	60
Bacillus sp	30	60
Micrococcus luteus	2.5	5
Staphylococcus aureus	10	20
Streptococcus iniae	15	30
Gram-negative		
Escherichia coli	—	—
Vibrio harveyi	—	—
Edwardsiella tarda	—	—
Vibrio anguillarum	—	—
Pseudoalteromonas sp	≥200	—
Pseudomonas fluorescens	—	—

—: No inhibitory or bactericidal activity was detected at the tested concentrations (1.25–60 μM).

2.3. Binding of rCrus1 to Bacterial Cell Wall Components and Its Effect on Bactericidal Activity

Enzyme-linked immunosorbent assay (ELISA) showed that rCrus1 bound well to Gram-positive bacteria, including *M. luteus*, *S. aureus*, *B. subtilis*, *B. cereus* and *S. iniae* (Figure 3A). rCrus1, at the same concentration, also bound to Gram-negative bacteria, but the binding was much weaker than that to Gram-positive bacteria (Figure S4). Consistent with the relative strong binding between rCrus1 and Gram-positive bacteria, rCrus1 exhibited apparent and comparable bindings to peptidoglycan (PGN) and lipoteichoic acid (LTA) (Figure 3B). The binding of rCrus1 to the mixture of PGN and LTA was similar to that of the binding to PGN or LTA alone (Figure 3B). Treatment of rCrus1 with DTT had no significant effect on the binding of rCrus1 to PGN, LTA, or bacteria (Figure 3C). In the presence of exogenously added LTA or PGN, especially the former, the bactericidal effect of rCrus1 was markedly decreased (Figure 3D).

2.4. Effects of rCrus1 on the Morphology and Membrane Integrity of Bacteria

Electron microscopy showed that treatment of *B. cereus* with rCrus1 caused rapid changes in cell morphology. As revealed by scanning electron microscope (SEM) and transmission electron microscope (TEM), after 2 h treatment, the cells exhibited shrunken surface and much reduced cytoplasmic density (Figure 4A,B). Propidium iodide (PI) staining showed that after incubation of rCrus1 with *B. cereus* and *M. luteus*, a large amount of PI was able to penetrate into the bacteria cells (Figure 5A), suggesting damage of the cellular membrane. The ability of rCrus1 to mediate membrane damage was further investigated by using the membrane potential sensitive probe DiSC3 (5), which can monitor depolarization of the cell plasma membrane [28,29]. When DiSC3(5)-tr-eated bacteria were incubated with rCrus1, DiSC3(5) was found to be released from the cells, although the amount of released DiSC3(5) was much less than that released from the bacteria incubated with valinomycin, a strong depolarizer of membrane potential (Figure 5B). These results indicated that rCrus1 could cause membrane depolarization in a manner similar to, though in a lesser degree, valinomycin. In contrast, rCrus1 treatment of the protoplasts of *B. cereus* and *M. luteus* caused no apparent damage (Figure S5A). Consistently, rCrus1 did not bind to the protoplasts (Figure S5B).

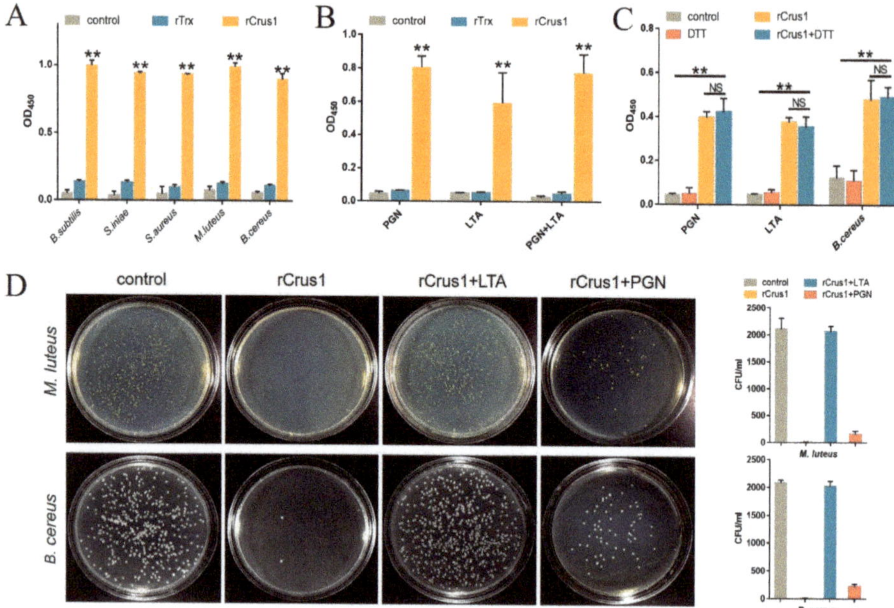

Figure 3. Binding of rCrus1 to bacteria and cell wall components. (**A**) Bacteria were incubated with rCrus1, recombinant Thioredoxin (rTrx), or phosphate buffer saline (PBS) (control), and the bound rCrus1 was detected by ELISA. (**B**) PGN and LTA were incubated with rCrus1, rTrx, or PBS, and the bound rCrus1 was detected as above. (**C**) PGN, LTA, and *Bacillus cereus* were incubated with or without (control) rCrus1, DTT-treated rCrus1, or DTT, and the bound rCrus1 was detected as above. Values are shown as means $\pm$ SD ($N = 3$). N, the number of replicates. ** $p < 0.01$, * $p < 0.05$, NS, not significant (Student's *t* test). (**D**) *Micrococcus luteus* and *B. cereus* were treated with or without (control) rCrus1, rCrus1 plus PGN, or rCrus1 plus LTA for 1 h. The bacteria were plated on LB plates and observed after 20–24 h incubation. The number of colony-forming units (CFU) was determined and shown on the right panels.

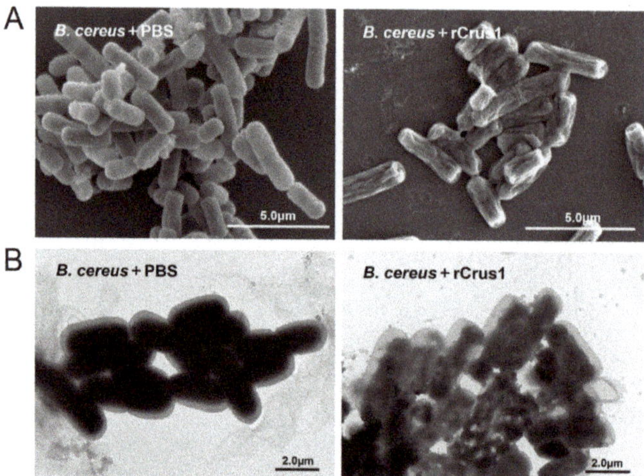

Figure 4. Morphological changes of the bacterial cells treated with rCrus1. *Bacillus cereus* was incubated with rCrus1 or PBS for 2 h, and the bacterial cells were inspected with a scanning (**A**) or transmission (**B**) electron microscope.

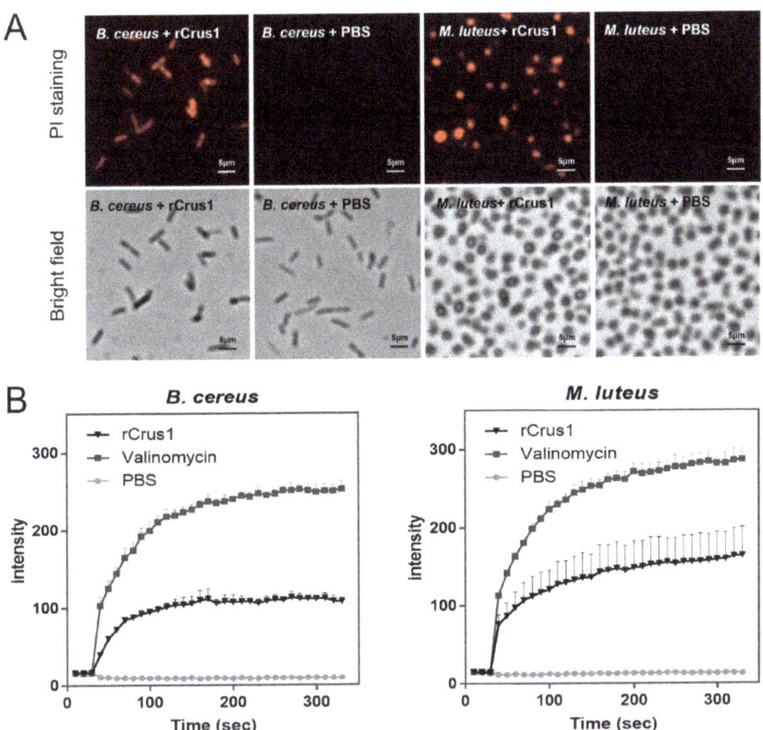

Figure 5. The effect of rCrus1 on bacterial cell membrane integrity. (**A**) *Bacillus cereus* and *Micrococcus luteus* were incubated with rCrus1 or PBS for 2 h. The cells were stained with PI and observed for PI uptake (upper panels) with a fluorescence microscope. The bright field image is shown in the lower panels. (**B**) *B. cereus* and *M. luteus* were pre-incubated with DiSC3(5) and then treated with rCrus1, valinomycin, or PBS, and the fluorescence of the cells was subsequently determined.

2.5. The Conserved Cysteine Residues in the WAP Domain Are Essential to the Antimicrobial Activity of rCrus1

To evaluate the functional importance of the conserved cysteine residues in the WAP domain, the eight Cys residues in this domain were mutated individually to Ser. The bactericidal activities of the resulting mutants, i.e., rCrus1-C64S, rCrus1-C70S, rCrus1-C80S, rCrus1-C86S, rCrus1-C92S, rCrus1-C93S, rCrus1-C97S, and rCrus1-C103S, were examined. None of the mutants exhibited apparent bactericidal activity at the MBC of rCrus1 (Table S1), or inhibited the growth of *M. luteus* even at the high concentration of 8 × MIC of rCrus1 (Figure 6A). However, all mutants were still able to bind to bacteria and bacterial cell wall components in a manner comparable to that of rCrus1 (Figure 6B–D). To examine whether the mutation changed the structure of rCrus1, the secondary structures of rCrus1 and rCrus1-C103S were subjected to circular dichroism (CD) analysis. Both rCrus1 and rCrus1-C103S showed a CD profile indicative of the formation of random coil structure; however, a fraction of the random coil differed slightly between rCrus1 and rCrus1-C103S (Figure 6E,F).

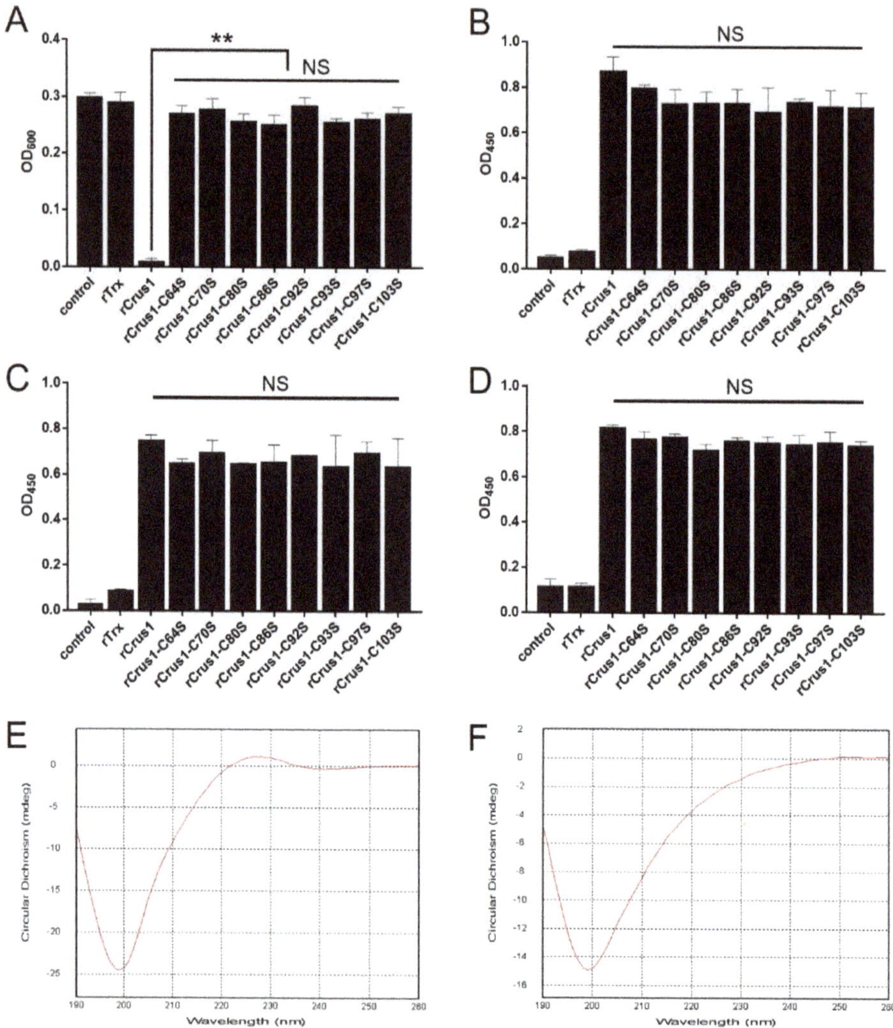

Figure 6. Microbial inhibitory and binding activity and structure characteristics of rCrus1 variants. (**A**) *Micrococcus luteus* was incubated with or without (control) 20 μM rCrus1, rCrus1 mutants, or rTrx for 18–20 h. Bacterial growth was then determined by measuring absorbance at OD_{600}. (**B–D**) PGN (**B**), LTA (**C**), or *Bacillus cereus* (**D**) were incubated with or without (control) rCrus1, rCrus1 mutants, or rTrx, and the bound proteins were detected by ELISA. (**E** and **F**) Circular dichroism (CD) spectra of rCrus1 (**E**) and rCrus1-C103S (**F**) in PB buffer. Values are shown as means ± SD ($N = 3$). N, the number of replicate. ** $p < 0.01$, NS, not significant (one-way ANOVA).

3. Discussion

Crustins are a large and diverse family of AMPs. In this study, we identified and analyzed a crustin, designated Crus1, from the shrimp of a deep-sea hydrothermal vent. Like typical crustins, Crus1 possesses a WAP domain, which contains eight conserved Cys capable of forming a four-disulfide core structure. It is interesting that, based on its cysteine-rich region [30], Crus1 was classified by phylogenetic analysis as a member of the Type I crustin, which has been mainly found in crabs and lobsters [31]. This observation of Crus1 is in contrast to that of the recently reported Re-crustin from the hydrothermal vent in MAR, which is a Type II crustin [27] and, as shown in our study, shares a low sequence identity with Crus1. These results suggest the possible existence of diverse forms of crustins in deep sea hydrothermal shrimp. Structural modeling showed that Crus1 formed mainly random coil, with very few α-helix and β-pleated sheet, which suggests a possibility that Crus1 may function via a unique mechanism.

The antimicrobial properties of crustins have been reported by many research groups [32,33]. Generally, crustins exhibit a broad-spectrum of antibacterial activities against Gram-positive and Gram-negative bacteria and fungi [34]. However, most members of the Type I crustins appear to have a spectrum of activity restricted to Gram-positive bacteria [19]. Similarly, in our study, we found that rCrus1 effectively killed Gram-positive bacteria from land, coastal waters, and deep sea, but had very limited killing effect on Gram-negative bacteria. The activity of rCrus1 was stable at acidic to neutral pH and over a wide range of temperatures, especially 4°C, which is close to the ambient temperature of the shrimp habitat. These results suggest that Crus1 likely functions as an active AMP under the native condition.

Binding to the target bacteria is a prerequisite for the antimicrobial activity of AMPs. In our study, rCrus1 bound strongly to Gram-positive bacteria as well as the major cell wall components of Gram-positive bacteria. We observed that the presence of added free PGN and LTA markedly reduced the bactericidal effect of rCrus1, suggesting that the binding between rCrus1 and the bacterial cell is likely mediated by PGN and LTA. Electron microscopy revealed that rCrus1-treated bacteria were shrunken and crinkled on the surface, resembling the formation of cracks on cells [35]. Consistently, PI staining indicated that rCrus1 induced membrane rupture in the bound bacterial cells, which was corroborated by the depolarization of membrane potential in rCrus1-treated bacteria. It is notable that, in contrast to the cell walled bacteria, the protoplasts of the bacteria were resistant to the binding and damage of rCrus1, which supported the above conclusion that it was the bacterial cell wall components, i.e., PGN and LTA, that rCrus1 interacted with directly. The importance of PGN and LTA is likely due to the reason that binding of AMPs to teichoic acids may initiate bacterial killing by facilitating the entry of the peptides toward the cytoplasmic membrane [36,37], and by building a poly anionic ladder, LTA and WTA may help poly cationic peptides, such as AMPs, to traverse from the outside to the cytoplasmic membrane [38]. The AMPs may further disrupt the cytoplasmic membrane by interfering with PGN biosynthesis [39]. It has been shown that the membrane bound PGN precursor lipid II could act as a docking moiety to attract cationic peptide to the bacterial membrane and promote peptide insertion into the membrane, leading eventually to membrane permeation [40,41].

For crustins, the WAP domain with its tetra-disulfide bond structure is thought to be vital to function [16,17]. In our study, we found that mutation of either of the eight cysteine residues in the WAP domain abolished the bactericidal activity of rCrus1, but neither of the mutations affected the ability of the protein to bind to bacteria or bacterial cell wall components. This finding indicates that bacterial binding and killing are via different mechanisms in rCrus1. Considering the importance of PGN and LTA in the binding of rCrus1 to bacteria, it is possible that rCrus1-bacteria interaction is mediated largely by ionic interaction, which is little affected by the Cys-to-Ser substitution, while bacterial killing is mediated by the interaction of the WAP domain with the bacterial membrane, which depends on the four-disulfide bonds. Circular dichroism showed that mutation of C103S caused a mild but distinct change in the secondary structure of rCrus1, which further supports the importance of the disulfide bonds of the WAP domain in the functioning of Crus1. It is possible that other residues besides these cysteines, such as those highly conserved in the WAP domain, may also paly vital roles in the structuring and functioning of Crus1.

In conclusion, our study demonstrates that, as an AMP, Crus1 binds bacteria probably via the bacterial cell wall components in a fashion that is independent of the WAP structure, but kills bacteria in a manner that requires the disulfide-based structural integrity of the WAP domain. These results add new insights into the immunological property and bactericidal mechanism of deep sea crustins.

4. Materials and Methods

4.1. Bacterial Strains and Culture Conditions

The bacteria used in this study are listed in Table 1. The Gram-positive bacteria (Bacillus subtilis WB800N, Bacillus subtilis G7, Bacillus wiedmannii SR52, Bacillus toyonensis P18, Staphylococcus aureus, Streptococcus iniae, and Micrococcus luteus) and the Gram-negative bacteria (*Escherichia coli*, *Edwardsiella tarda*, *Vibrio harveyi*, *Vibrio anguillarum*, and *Pseudomonas fluorescens*) have been reported previously [42–45]. Of these bacteria, *B. subtilis* G7, *B. wiedmannii* SR52, and *B. toyonensis* P18 are from deep sea hydrothermal vents. In addition, three other bacteria, i.e., *Pseudoalteromonas* sp., *Bacillus cereus* MB1, and *Bacillus* sp. are also from deep sea environments. *S. iniae* was cultured in TSB medium (Hopobio, Qingdao, China) at 28 °C. *E. tarda*, *B. subtilis* G7, *B. wiedmannii* SR52, *B. cereus* MB1, *Pseudoalteromonas* sp., and *Bacillus* sp. were cultured in marine 2216E medium (Hopobio, Qingdao, China) at 28 °C. All other bacterial strains were cultured in Luria-Bertani broth (LB) medium at 37 °C (for *E. coli*, *B. subtilis* WB800N, *M. luteus* and *S. aureus*) or 28 °C (for *P. fluorescens*, *V. anguillarum*, and *V. harveyi*). When used for determining the antibacterial activity of rCrus1, the bacteria were cultured in Mueller-Hinton broth (MHB) medium.

4.2. Bioinformatics Analysis and Structural Modeling of Crus1

The nucleotide sequence of Crus1 has been deposited to GenBank (accession number MW448473). The deduced amino acid sequence of Crus1 was analyzed with DNAMAN 6.0 (Lynnon Biosoft, San Ramon, CA, USA). Homology searches of deduced amino acid sequences were performed using the Protein BLAST algorithm of the NCBI. Signal peptide was identified using the SignalP program [46]. Multiple alignments of amino acid sequences were created with ClustalX 2.0 (SFI, Dublin, Ireland), and the output pattern was generated using DNAMAN 6.0 (Lynnon Biosoft, San Ramon, CA, USA). The neighbor-joining phylogenetic tree was constructed with MEGA 6.0 (Mega Limited, Auckland, New Zealand), and 1000 bootstraps were selected to assess reliability. The full-length atomic model of Crus1 was constructed with iterative template-based fragment assembly simulations using I-TASSER [47], and the spatial structure was edited with PyMOL 3.7 (Schrödinger, New York, NY, USA).

4.3. Protein Expression and Purification

To construct pETCrus1, the plasmid expressing rCrus1, the coding sequence of Crus1 without signal peptide and C-terminally tagged with six histidine residues was synthesized by BGI Technology (Beijing, China). The sequence was inserted into the expression plasmid pET28a (Sangon Biotech, Shanghai, China) at the NdeI/NotI sites, resulting in pETCrus1. pETCrus1 and pET32a (Novagen, Madison, WI, USA), which expresses rTrx, were separately introduced into *E. coli* BL21(DE3) (TransGen Biotech, Beijing, China) by transformation. The transformants were cultured in LB medium at 37 °C to OD600 0.6. Isopropyl-beta-D-thiogalactoside (IPTG) was added to the culture at a final concentration of 0.06 mM to induce protein expression. The culture was continued overnight at 16 °C with shaking (120 rpm), and the cells were then harvested by centrifugation. His-tagged rCrus1 and rTrx were purified as described previously [48]. Briefly, the cells were disrupted by sonication on ice, and the lysate was centrifuged to collect the supernatant. The His-tagged recombinant protein in the supernatant was purified under native conditions using nickel-nitrilotriacetic acid (Ni-NTA) columns (QIAGEN, Germantown, MD, USA) as recommended by the manufacturer. The protein was also treated with Triton X-114 to remove endotoxin as reported previously [49]. The purified protein was dialyzed against PBS for 36 h at 4 °C and concentrated with Amicon Ultra Centrifugal Filter (Millipore, Bedford, MA, USA). The purified protein was analyzed by sodium dodecyl sulfate-polyacrylamide gel electrophoresis (SDS-PAGE). The protein concentration was determined using the BCA Protein Assay Kit (Beyotime, Shanghai, China) according to the manufacturer's instruction.

For the preparation of Crus1 mutants, site-directed serine substitutions of Cys64 (C64S), Cys70 (C70S), Cys80 (C80S), Cys86 (C86S), Cys92 (C92S), Cys93 (C93S), Cys97 (C97S), and Cys103 (C103S) were performed using the Q5 Site-Directed Mutagenesis Kit (New England Biolabs, Beverly, MA) according to the manufacturer's instruction. The mutant proteins were expressed and purified as described above. The primers used in mutagenesis are shown in Table S2.

4.4. Antibacterial Activity Assay

The MIC and MBC were determined with the microdilution broth method [50]. In short, bacteria were cultured in MHB to an OD600 of 0.5–0.6 and diluted to 10^4 CFU/mL. The bacteria were mixed with a 2-fold dilution series of rCrus1 in a 96-well microtiter plate. The plate was incubated at 37 °C for 18 h. MIC is defined as the lowest concentration of rCrus1 that rendered no visible bacterial growth. To determine the MBC, 10 µL of bacteria-rCrus1 mixture was removed from the above wells corresponding to 1 × MIC, 2 × MIC and 4 × MIC, and plated on MH agar plates. MBC is defined as the concentration of rCrus1 that killed 99.9% of bacteria after 18 h incubation. The assay was performed at least three times.

To examine the effect of temperature on rCrus1 activity, rCrus1 (1 × MIC) or PBS was incubated with 1×10^4 CFU/mL *M. luteus* in MHB at 4 °C, 16 °C, 37 °C, or 42 °C for 2 h. Bacterial survival was then examined by plate count as described above. To examine the effect of pH on rCrus1 activity, rCrus1 or PBS was mixed with *M. luteus* as above and incubated in MHB (37 °C) at pH 5, 7, 9, or 11 for 2 h. Bacterial survival was determined as above. To examine the effect of disulfide bond elimination on rCrus1 activity, DTT (final concentration 50 mM) was added to rCrus1 in PBS and incubated for 30 min at 37 °C. The mixture was named DTT-treated rCrus1 and used immediately in the subsequent bactericidal assay. For the bactericidal assay, rCrus1 (2.5 µM), DTT-treated rCrus1 (2.5 µM rCrus1 plus 12.5 mM DTT), DTT (12.5 mM), or PBS (control) was incubated with *M. luteus* as above for 2 h. Bacterial survival was determined as above. To examine the effect of LTA and PGN on the bactericidal activity of rCrus1, LTA and PGN (Sigma, St Louis, MO, USA) (final concentration of 1 mg/mL) were each incubated with 1 × MBC rCrus1 for 1 h at room temperature, and then bacteria was added to the mixture. After incubation for 2 h, bacterial survival was determined as above by plate count. To examine the time-dependent bactericidal activity of rCrus1, *M. luteus* was cultured in MHB to an OD600 of 0.2 and diluted to 10^6 CFU/mL in fresh MHB. rCrus1 (final concentration of 10 µM) was added to the bacterial dilution, followed by incubation at 37 °C for 5 h. Every 30 min, the number of bacteria was determined by plate count. All experiments were performed in triplicate.

4.5. Protein Binding to Bacteria and Cell Wall Components

Bacteria were cultured to an OD600 of 0.8 and resuspended in coating buffer (0.159% Na_2CO_3 and 0.293% $NaHCO_3$, pH 9.6) to 10^8 CFU/mL. LTA and PGN (Sigma, USA) were dissolved in coating buffer to 200 µg/mL. Binding of rCrus1 (20 µM) to bacteria and cell wall components was determined with ELISA as reported previously [51,52] with modifications. Briefly, a 96-well microtiter plate containing each of the bacteria or cell wall components (100 µL/well) was incubated overnight at 4 °C. The plate was washed 3 times with PBST (PBS with 0.05% Tween 20), and blocked with 200 µL 5% skim milk powder (Solarbio, Beijing, China) in PBST at 37 °C for 2 h. The plate was washed three times as above, and 10 µM protein (rCrus1 or rTrx) or PBS (control) was added to the wells. The plate was incubated at 37 °C for 1 h and washed as above. HRP-conjugated mouse anti-His antibody (1/1000 dilution) (ABclonal, Hubei, China) was added to the wells. The plate was incubated at 37 °C for 1 h and washed 5 times with PBST. Color development was performed using TMB substrate solution (Tiangen, Beijing, China) and terminated by adding ELISA stop solution (Solarbio, Beijing, China). The absorbance at

450 nm was measured using a multifunctional microplate reader. To examine the effect of disulfide bond elimination on the ability of rCrus1 to interact with bacteria and bacterial components, the binding assay was performed as above using rCrus1 (10 μM), DTT-treated rCrus1 (10 μM rCrus1 plus 50 mM DTT), DTT (50 mM), or PBS (control).

4.6. Electron Microscopy and PI Staining Assay

Electron microscopy was performed based on previous methods [53]. For microscopy with SEM (S-3400N, Hitachi, Tokyo, Japan), *B. cereus* MB1 and *M. luteus* were cultured in LB medium to logarithmic phase, and the cells were washed and resuspended in PBS to 1×10^6 CFU/mL. The cells were pretreated with $1 \times$ MBC rCrus1 or PBS at 37 °C for 2 h. After treatment, the cells were fixed with 5% glutaraldehyde in PBS (pH 7.4) for 2 h and dehydrated in a series of increased concentration of ethanol (30, 50, 70, 80, 90 and 100%) for 10 min at 4 °C. The cells were treated with isoamyl acetate for 10 min, critical point-dried (Hitachi-HCP, Hitachi, Japan), sputter-coated with platinum (MC1000, Hitachi, Japan) and examined with a SEM. For microscopy with TEM (HT7700, Hitachi, Tokyo, Japan), *B. cereus* and *M. luteus* were pretreated with rCrus1 or PBS as above. TEM microscopy was performed as previously reported [54]. For the PI assay, *B. cereus* MB1 and *M. luteus* were pretreated with rCrus1 or PBS as above. The sample was stained with a PI staining kit (BestBio, Shanghai, China) following the manufacturer's instruction. The cells were observed with a fluorescence microscope (TiS/L100, Nikon, Tokyo, Japan).

4.7. Bacterial Cytoplasmic Membrane Depolarization

The cytoplasmic membrane depolarization activity of the rCrus1 was determined as reported previously d [29]. Briefly, *B. cereus* MB1 and *M. luteus* were grown at 37 °C to an OD600 of 0.6 and harvested by centrifugation. The cells were washed three times with HEPES buffer (5 mM HEPES with 20 mM glucose, pH 7.4), and resuspended in HEPES buffer containing 100 mM KCl to an OD600 of 0.05. DiSC3(5) (Macklin, Shanghai, China) was added to the bacterial suspension at a final concentration of 0.4 μM. The mixture was incubated in the dark for 30 min and then quenched at room temperature. rCrus1 ($2 \times$ MBC), valinomycin (a potassium ionophore), or PBS was added to the mixture. Membrane depolarization was monitored by observing change in the intensity of fluorescence (λex = 622 nm, λem = 670 nm).

4.8. Protoplast Preparation and Lysis Assay

Preparation of bacterial protoplast was performed as previously reported [55]. In short, *B. subtilis* G7 and *M. luteus* were cultured in LB broth to an OD600 of 0.9–1. The cells were pelleted by centrifugation at 12,000 rpm for 1 min. The cells were washed twice with pre-warmed (37 °C) steady buffer (20 mM sodium malate, 20 mM MgCl$_2$, and 500 mM sucrose, pH 6.5). The cells were resuspended in steady buffer containing lysozyme (2.0 mg/mL) to an OD600 of 0.5–0.8 and incubated at 37 °C for 2–2.5 h. After washing twice with steady buffer, the cells were incubated with rCrus1 ($1 \times$ MBC) for 1 h at 37 °C. Triton X-100 (1%) was used as a positive control for maximal cell lysis. The OD600 of the protoplasts was measured and statistically calculated.

4.9. Immunofluorescence Microscopy

B. subtilis G7 and *B. subtilis* G7 protoplasts were diluted to 10^8 CFU/mL with PBS and steady buffer, respectively. The cells were dropped onto adhesion microscope slides (CITOTEST, Jiangsu, China) at 4 °C and allowed to stand overnight. The slides were washed with PBS or steady buffer and incubated with rCrus1 (20 μM) at 37 °C for 1 h. The slides were washed as above, and anti-His-FITC antibody (Abcam, Cambridge, MA, USA) (1/200 dilution) was added to the slide. The slides were incubated at 37 °C for 2 h and washed as above. The slides were observed with a fluorescence microscope (Carl Zeiss LSM710, Jena, Germany).

4.10. Circular Dichroism (CD) Spectroscopy

CD spectroscopy was performed by Sangon Biotech Co., Ltd. (Shanghai, China). Briefly, the protein sample was diluted to 0.2 mg/mL with PB buffer (20 mM $Na_2HPO_4 \cdot 12H_2O$). The spectra were collected on a Chirascan Plus CD spectrometer (Applied Photophysics, Leatherhead, UK) using a 0.5 mm path cell. The data were obtained from 190 to 260 nm at an interval of 1.0 nm and a speed of 2 nm/s.

4.11. Statistical Analysis

Statistical analysis was carried out using GraphPad Prism 7.0 (GraphPad, San Diego, CA, USA). Statistical significance was determined with Student's *t* test for two groups or one-way analysis of variance (ANOVA) for more than two groups. All data are presented as mean $\pm$ SD. $p < 0.05$ was considered statistically significant.

Supplementary Materials: The following are available online at https://www.mdpi.com/1660-339 7/19/3/176/s1, Figure S1: SDS-PAGE analysis of rCrus1, Figure S2. Effect of temperature and pH on the antibacterial activity of rCrus1 against *Vibrio harveyi*, Figure S3. Time-dependent bactericidal activity of rCrus1 against *Micrococcus luteus*, Figure S4. Binding of rCrus1 to Gram-negative bacteria, Figure S5: The potential effect of rCrus1 on bacterial protoplasts, Table S1. Bactericidal activity of rCrus1 variants, Table S2. Primers used in point mutation.

Author Contributions: L.S. and Y.W. conceived and designed the research work. J.Z. provided some of the original materials. Y.W. and Y.S. conducted the experiments. Y.W. analyzed the data. Y.W. wrote the first draft of the manuscript. L.S. edited the manuscript. All authors have read and agreed to the published version of the manuscript.

Funding: This research was funded by the grants from the Strategic Priority Research Program of the Chinese Academy of Sciences (XDA22050403), Qingdao National Laboratory for Marine Science and Technology (QNLM2016ORP0309), and the Taishan Scholar Program of Shandong Province.

Institutional Review Board Statement: Not applicable.

Informed Consent Statement: Not applicable.

Data Availability Statement: The Crus 1 sequence data of this study are available from GenBank under the accession number MW448473.

Conflicts of Interest: The authors declare no conflict of interest.

References

1. Huang, Y.; Ren, Q. Research progress in innate immunity of freshwater crustaceans. *Dev. Comp. Immunol.* **2019**, *104*, 103569. [CrossRef]
2. Lemaitre, B.; Hoffmann, J. The host defense of Drosophila melanogaster. *Annu. Rev. Immunol.* **2007**, *25*, 697–743. [CrossRef]
3. Brown, K.L.; Hancock, R.E. Cationic host defense (antimicrobial) peptides. *Curr. Opin. Immunol.* **2006**, *18*, 24–30. [CrossRef]
4. Benincasa, M.; Runti, G.; Mardirossian, M.; Scocchi, M. Non-Membrane Permeabilizing Modes of Action of Antimicrobial Peptides on Bacteria. *Curr. Top. Med. Chem.* **2016**, *16*, 76–88.
5. Giuliani, A.; Pirri, G.; Nicoletto, S.F. Antimicrobial peptides: An overview of a promising class of therapeutics. *Cent. Eur. J. Biol.* **2007**, *2*, 1–33. [CrossRef]
6. Hancock, R.E.; Hancock, R.E.W. Cationic peptides: Effectors in innate immunity & novel antimicrobials. *Lancet Infect. Dis.* **2001**, *1*, 156–164. [PubMed]
7. Boman, H.G. Antibacterial peptides: Basic facts and emerging concepts. *J. Intern. Med.* **2010**, *254*, 197–215. [CrossRef] [PubMed]
8. Essig, A.; Hofmann, D.; Munch, D.; Gayathri, S.; Kunzler, M.; Kallio, P.T.; Sahl, H.G.; Wider, G.; Schneider, T.; Aebi, M. Copsin, a Novel Peptide-based Fungal Antibiotic Interfering with the Peptidoglycan Synthesis. *J. Biol. Chem.* **2014**, *289*, 34953–34964. [CrossRef]
9. Zhang, L.; Rozek, A.; Hancock, R.E.W. Interaction of Cationic Antimicrobial Peptides with Model Membranes. *J. Biol. Chem.* **2001**, *276*, 35714–35722. [CrossRef]
10. Arias, M.; Jensen, K.V.; Nguyen, L.T.; Storey, D.G.; Vogel, H.J. Hydroxy-tryptophan containing derivatives of tritrpticin: Modification of antimicrobial activity and membrane interactions. *Biochim. Biophys. Acta Biomembr.* **2015**, *1848*, 277–288. [CrossRef]
11. Novkovic, M. Selective antimicrobial activity and mode of action of adepantins, glycine-rich peptide antibiotics based on anuran antimicrobial peptide sequences. *Biochim. Biophys. Acta* **2013**, *1828*, 1004–1012.

12. Hilde, U.; Ørjan, S.; Haukland, H.H.; Manuela, K.; Vorland, L.H. Lactoferricin B inhibits bacterial macromolecular synthesis in *Escherichia coli* and *Bacillus subtilis*. *FEMS Microbiol. Lett.* **2004**, *237*, 377–384.

13. Haney, E.F.; Petersen, A.P.; Lau, C.K.; Jing, W.; Storey, D.G.; Vogel, H.J. Mechanism of action of puroindoline derived tryptophan-rich antimicrobial peptides. *Biochim. Biophys. Acta* **2013**, *1828*, 1802–1813. [CrossRef]

14. Schneider, T.; Kruse, T.; Wimmer, R.; Wiedemann, I.; Kristensen, H.-H. Plectasin, a Fungal Defensin, Targets the Bacterial Cell Wall Precursor Lipid II. *Science* **2010**, *328*, 1168–1172. [CrossRef] [PubMed]

15. Smith, V.J.; Fernandes, J.M.O.; Kemp, G.D.; Hauton, C. Crustins: Enigmatic WAP domain-containing antibacterial proteins from crustaceans. *Dev. Comp. Immunol.* **2008**, *32*, 758–772. [CrossRef] [PubMed]

16. Hauton, C.; Brockton, V.; Smith, V.J. Cloning of a crustin-like, single whey-acidic-domain, antibacterial peptide from the haemocytes of the *European lobster, Homarus gammarus*, and its response to infection with bacteria. *Mol. Immunol.* **2006**, *43*, 1490–1496. [CrossRef]

17. Sallenave, J.-M. The role of secretory leukocyte proteinase inhibitor and elafin (elastase-specific inhibitor/skin-derived antileuko-protease) as alarm antiproteinases in inflammatory lung disease. *Respir. Res.* **2000**, *1*, 5. [CrossRef]

18. Wang, H.; Zhang, J.X.; Wang, Y.; Fang, W.H.; Wang, Y.; Zhou, J.F.; Zhao, S.; Li, X.C. Newly identified type II crustin (SpCrus2) in Scylla paramamosain contains a distinct cysteine distribution pattern exhibiting broad antimicrobial activity. *Dev. Comp. Immunol.* **2018**, *84*, 1–13. [CrossRef]

19. Supungul, P.; Tang, S.; Maneeruttanarungroj, C.; Rimphanitchayakit, V.; Hirono, I.; Aoki, T.; Tassanakajon, A. Cloning, expression and antimicrobial activity of crustinPm1, a major isoform of crustin, from the black tiger shrimp *Penaeus monodon*. *Dev. Comp. Immunol.* **2008**, *32*, 61–70. [CrossRef]

20. Arockiaraj, J.; Gnanam, A.J.; Muthukrishnan, D.; Gudimella, R.; Milton, J.; Singh, A.; Muthupandian, S.; Kasi, M.; Bhassu, S. Crustin, a WAP domain containing antimicrobial peptide from freshwater prawn *Macrobrachium rosenbergii*: Immune characterization. *Fish Shellfish Immunol.* **2013**, *34*, 109–118. [CrossRef]

21. Jobstvogt, N.; Hanley, N.; Hynes, S.; Kenter, J.; Witte, U. Twenty thousand sterling under the sea: Estimating the value of protecting deep-sea biodiversity. *Ecol. Econ.* **2014**, *97*, 10–19. [CrossRef]

22. Folkersen, M.V.; Fleming, C.M.; Hasan, S. The economic value of the deep sea: A systematic review and meta-analysis. *Mar. Policy* **2018**, *94*, 71–80. [CrossRef]

23. Rodrigo, A.P.; Costa, P.M. The hidden biotechnological potential of marine invertebrates: The Polychaeta case study. *Environ. Res.* **2019**, *173*, 270–280. [CrossRef] [PubMed]

24. Romano, G.; Costantini, M.; Sansone, C.; Lauritano, C.; Ruocco, N.; Ianora, A. Marine microorganisms as a promising and sustainable source of bioactive molecules. *Mar. Environ. Res.* **2016**, *128*, 58–69. [CrossRef] [PubMed]

25. Iván, H.Á.; Marie-Anne, C.B.; Florence, P.; Sébastien, D. Morphology of First Zoeal Stage of Four Genera of Alvinocaridid Shrimps from Hydrothermal Vents and Cold Seeps: Implications for Ecology, Larval Biology and Phylogeny. *PLoS ONE* **2015**, *10*, e0144657.

26. Zhang, J.; Sun, Q.L.; Luan, Z.D.; Lian, C.; Sun, L. Comparative transcriptome analysis of Rimicaris sp. reveals novel molecular features associated with survival in deep-sea hydrothermal vent. *Sci. Rep.* **2017**, *7*, 2000. [CrossRef] [PubMed]

27. Bloa, S.L.; Boidin-Wichlacz, C.; Cueff-Gauchard, V.; Rosa, R.D.; Tasiemski, A. Antimicrobial Peptides and Ectosymbiotic Relationships: Involvement of a Novel Type IIa Crustin in the Life Cycle of a Deep-Sea Vent Shrimp. *Front. Immunol.* **2020**, *11*, 1511. [CrossRef]

28. Torrent, M.; Navarro, S.; Moussaoui, M.; Nogués, M.V.; Boix, E. Eosinophil Cationic Protein High-Affinity Binding to Bacteria-Wall Lipopolysaccharides and Peptidoglycans. *Biochemistry* **2008**, *47*, 3544–3555. [CrossRef]

29. Bellemare, A.; Vernoux, N.; Morin, S.; Gagné, S.M.; Bourbonnais, Y. Structural and antimicrobial properties of human pre-elafin/trappin-2 and derived peptides against *Pseudomonas aeruginosa*. *BMC Microbiol.* **2010**, *10*, 253. [CrossRef]

30. Zhao, X.F.; Wang, J.X. The antimicrobial peptides of the immune response of shrimp. *Invertebr. Surviv. J.* **2008**, *5*, 4.

31. Imjongjirak, C.; Amparyup, P.; Tassanakajon, A.; Sittipraneed, S. Molecular cloning and characterization of crustin from mud crab *Scylla paramamosain*. *Mol. Biol. Rep.* **2009**, *36*, 841–850. [CrossRef]

32. Bandeira, P.T.; Vernal, J.; Matos, G.M.; Farias, N.D.; Rosa, R.D. A Type IIa crustin from the pink shrimp *Farfantepenaeus paulensis* (crusFpau) is constitutively synthesized and stored by specific granule-containing hemocyte subpopulations. *Fish Shellfish Immunol.* **2019**, *97*, 294–299. [CrossRef]

33. Liu, N.; Zhang, R.R.; Fan, Z.X.; Zhao, X.F.; Wang, X.W. Characterization of a type-I crustin with broad-spectrum antimicrobial activity from red swamp crayfish *Procambarus clarkii*. *Dev. Arative Immunol.* **2016**, *61*, 145–153. [CrossRef]

34. Krusong, K.; Poolpipat, P.; Supungul, P.; Tassanakajon, A. A comparative study of antimicrobial properties of crustinPm1 and crustinPm7 from the black tiger shrimp *Penaeus monodon*. *Dev. Comp. Immunol.* **2011**, *36*, 208–215. [CrossRef] [PubMed]

35. Wang, K.; Dang, W.; Yan, J.; Chen, R.; Liu, X.; Yan, W.; Zhang, B.; Xie, J.; Zhang, J.; Wang, R. Membrane Perturbation Action Mode and Structure-Activity Relationships of Protonectin, a Novel Antimicrobial Peptide from the Venom of the Neotropical Social Wasp. *Chem. Funct. Proteins* **2013**, *57*, 4632–4639. [CrossRef]

36. Koprivnjak, T.; Weidenmaier, C.; Peschel, A.; Weiss, J.P. Wall Teichoic Acid Deficiency in Staphylococcus aureus Confers Selective Resistance to Mammalian Group IIA Phospholipase A2 and Human β-Defensin 3. *Infect. Immun.* **2008**, *76*, 2169. [CrossRef] [PubMed]

37. Nermina, M.; Karl, L. Antimicrobial Peptides Targeting Gram-Positive Bacteria. *Pharmaceuticals* **2016**, *9*, 59.

38. Brown, S.; John Maria, S.P.; Suzanne, W. Wall Teichoic Acids of Gram-Positive Bacteria. *Annu. Rev. Microbiol.* **2013**, *67*, 313–336. [CrossRef]
39. Oppedijk, S.F.; Martin, N.I.; Breukink, E. Hit'em where it hurts: The growing and structurally diverse family of peptides that target lipid-II. *Biochim. Biophys. Acta* **2016**, *1858*, 947–957. [CrossRef]
40. Kruijff, B.D.; Dam, V.V.; Breukink, E. Lipid II: A central component in bacterial cell wall synthesis and a target for antibiotics. *Prostaglandins Leukot. Essent. Fat. Acids* **2008**, *79*, 117–121. [CrossRef] [PubMed]
41. Martin, N.I.; Breukink, E. Expanding role of lipid II as a target for lantibiotics. *Future Microbiol.* **2007**, *2*, 513–525. [CrossRef]
42. Gu, H.J.; Sun, Q.L.; Jiang, S.; Zhang, J.; Sun, L. First characterization of an anti-lipopolysaccharide factor (ALF) from hydrothermal vent shrimp: Insights into the immune function of deep-sea crustacean ALF. *Dev. Comp. Immunol.* **2018**, *84*, 382–395. [CrossRef] [PubMed]
43. Zhao, Y.; Chen, C.; Gu, H.J.; Zhang, J.; Sun, L. Characterization of the Genome Feature and Toxic Capacity of a Bacillus wiedmannii Isolate From the Hydrothermal Field in Okinawa Trough. *Front. Cell. Infect. Microbiol.* **2019**, *9*, 370. [CrossRef]
44. Wang, Y.J.; Miao, Y.Q.; Hu, L.P.; Kai, W.; Zhu, R. Immunization of mice against alpha, beta, and epsilon toxins of Clostridium perfringens using recombinant rCpa-b-x expressed by *Bacillus subtilis*. *Mol. Immunol.* **2020**, *123*, 88–96. [CrossRef] [PubMed]
45. Luo, J.C.; Long, H.; Zhang, J.; Zhao, Y.; Sun, L. Characterization of a deepsea Bacillus toyonensisisolate: Genomicand pathogenicfeatures. *Front. Cell. Infect. Microbiol.* **2021**, *11*, 107. [CrossRef]
46. Nielsen, H.; Tsirigos, K.D.; Brunak, S.; von Heijne, G. A brief history of protein sorting prediction. *Protein J.* **2019**, *38*, 200–216. [CrossRef] [PubMed]
47. Roy, A.; Kucukural, A.; Zhang, Y. I-TASSER: A unified platform for automated protein structure and function prediction. *Nat Protoc.* **2010**, *5*, 725–738. [CrossRef] [PubMed]
48. Yu, C.; Zhang, P.; Zhang, T.; Sun, L. IL-34 regulates the inflammatory response and anti-bacterial immune defense of Japanese flounder Paralichthys olivaceus. *Fish Shellfish Immunol.* **2020**, *104*, 228–236. [CrossRef]
49. Zhang, B.C.; Sun, L. Tongue sole (*Cynoglossus semilaevis*) prothymosin alpha: Cytokine-like activities associated with the intact protein and the C-terminal region that lead to antiviral immunity via Myd88-dependent and -independent pathways respectively. *Dev. Comp. Immunol.* **2015**, *53*, 96–104. [CrossRef]
50. Rončević, T.; Čikeš-Čulić, V.; Maravić, A.; Capanni, F.; Gerdol, M.; Pacor, S.; Tossi, A.; Giulianini, P.; Pallavicini, A.; Manfrin, C. Identification and functional characterization of the astacidin family of proline-rich host defence peptides (PcAst) from the red swamp crayfish (*Procambarus clarkii*, Girard 1852). *Dev. Comp. Immunol.* **2020**, *105*, 103574. [CrossRef]
51. Li, W.R.; Guan, X.L.; Jiang, S.; Sun, L. The novel fish miRNA pol-miR-novel_171 and its target gene FAM49B play a critical role in apoptosis and bacterial infection. *Dev. Comp. Immunol.* **2020**, *106*, 103616. [CrossRef] [PubMed]
52. Amparyup, P.; Sutthangkul, J.; Charoensapsri, W.; Tassanakajon, A. Pattern recognition protein binds to lipopolysaccharide and β-1,3-glucan and activates shrimp prophenoloxidase system. *J. Biol. Chem.* **2016**, *291*, 10949. [CrossRef] [PubMed]
53. Zhang, T.; Jiang, S.; Sun, L. A Fish Galectin-8 Possesses Direct Bactericidal Activity. *Int. J. Mol. Sci.* **2020**, *22*, 376. [CrossRef] [PubMed]
54. Li, M.F.; Jia, B.B.; Sun, Y.Y.; Sun, L. The Translocation and Assembly Module (TAM) of Edwardsiella tarda Is Essential for Stress Resistance and Host Infection. *Front. Microbiol.* **2020**, *11*, 1743. [CrossRef] [PubMed]
55. Jiang, S.; Gu, H.; Zhao, Y.; Sun, L. Teleost Gasdermin E Is Cleaved by Caspase 1, 3, and 7 and Induces Pyroptosis. *J. Immunol.* **2019**, *203*, ji1900383. [CrossRef] [PubMed]

MDPI

St. Alban-Anlage 66

4052 Basel

Switzerland

Tel. +41 61 683 77 34

Fax +41 61 302 89 18

www.mdpi.com

Marine Drugs Editorial Office

E-mail: marinedrugs@mdpi.com

www.mdpi.com/journal/marinedrugs